Lecture Notes in Physics

90

Sixth International Conference on Numerical Methods in Fluid Dynamics

Proceedings of the Conference,
Held in Tbilisi (U.S.S.R.)
June 21–24, 1978

Edited by
H. Cabannes, M. Holt and V. Rusanov

Springer-Verlag
Berlin Heidelberg New York 1979

Editors
H. Cabannes
Université de Paris VI
Mécanique Theorique
Tour 66 – 4, Place Jussieu
F-75230 Paris, Cedex 05

M. Holt
College of Engineering and
Mechanical Engineering
University of California
Berkeley, CA 94720/USA

V. Rusanov
Academy of Sciences USSR
Leninskii Prospect, 14
Moscow
USSR

ISBN 3-540-09115-7 Springer-Verlag Berlin Heidelberg New York
ISBN 0-387-09115-7 Springer-Verlag New York Heidelberg Berlin

Printed in Germany

Printing and binding: Beltz Offsetdruck, Hemsbach/Bergstr.
2153/3140-543210

Editors' Preface

The present volume of Lecture Notes in Physics is devoted to the Sixth International Conference on Numerical Methods in Fluid Dynamics, held in Tbilisi, Georgian Soviet Socialist Republic, USSR, from June 20-25, 1978. The volume includes all papers presented by authors from Western Europe, United States, Canada, Australia and Japan (written in English or French). In addition, the majority of the authors from USSR, Poland, Bulgaria and Czechoslovakia prepared English versions of their Conference lectures and all such manuscripts are presented here. The original versions of the latter papers, together with all remaining manuscripts not translated into English or French, are being printed in separate Proceedings prepared by the Computing Center of the USSR Academy of Sciences.

The Conference was well attended. The majority of those registered were from USSR and other Socialist countries, but strong delegations attended from USA, France, Germany and many other Western countries. A list of those registered is given at the end of the volume.

The President of the Conference was O. M. Belotserkovskii. The Conference was efficiently organized by a Committee under Dr. Belotserkovskii's direction, including N. N. Yanenko, V. V. Rusanov, V. D. Kupradze, T. S. Vashakmadze and many others. Dr. Vashakmadze was in charge of local arrangements and the undersigned express their deep gratitude to him and his Georgian colleagues for the excellent arrangements and warm hospitality.

Travel support for most of the U.S. delegation was provided through the U.S. National Academy of Sciences with funds provided by the Office of Naval Research and the Air Force Office of Scientific Research. Travel support for some delegates from Western Europe was provided by the European Research Office of the U.S. Army and by a fund established after the Fifth Conference at Twente Universite by Professor van der Vooren and Professor Zandbergen. We are indebted to all these sources for their support.

The Proceedings begin with three invited papers by Roesner, Samarskii and Temam. The remaining contributions are presented in alphabetical order.

We are indebted to Dr. W. Beiglböck and the editorial staff of Springer Verlag for valuable assistance in preparing these Proceedings.

December 14, 1978

Henri Cabannes
Maurice Holt
Viktor V. Rusanov
(Editors)

Contents

Invited Lectures

Communications

NUMERICAL CALCULATION OF HYDRODYNAMIC STABILITY PROBLEMS WITH TIME-DEPENDENT BOUNDARY CONDITIONS

K. G. Roesner

Institut fuer Stroemungslehre und Stroemungsmaschinen
Universitaet Karlsruhe, Federal Republic of Germany

TABLE OF CONTENT

Introduction

In the past an increasing interest in rotating flows has led to a rapidly growing number of contributions in this special field, which is characterized by a large number of parameters describing the actual problem. Starting with the famous TAYLOR-vortices between concentric cylinders the geometrical configurations have changed resulting in a hugde amount of different problems which have one in common: the fact that some of the boundary surfaces rotate around a fixed axis.

In the present article it is intended to summarize and comment on work in the field of numerical calculation of rotating flows which can be thought of as a numerical experiment to simulate realistic flow conditions. The aim of this investigation is to show the progress which was made in the numerical simulation of solutions of initial-boundary value problems for the NAVIER-STOKES equations during the last four years and to collect experimental and theoretical results especially for rotating fluid flows inside of simple geometries. It will be shown in which cases the theoretical results are still unsatisfactory and which problems are not yet solved.

Generally speaking an experiment in the laboratory and a numerical experiment with the aid of the computer represent two different but powerful tools for solving fluid dynamic problems. Sometimes – as for rotating fluid flows – the experiments in the laboratory serve as an important guideline for the theory to focus the main interest on those phenomena which have a key-position in the understanding of the whole experiment. This is true e.g. for the simulation of the flow inside of cavities with simple boundaries. These flows have the remarkable feature that the boundary conditions can be stated very accurately. Some of the commonly treated geometrical shapes of rotating vessels are shown in **Fig. 1** and **Fig. 2**.

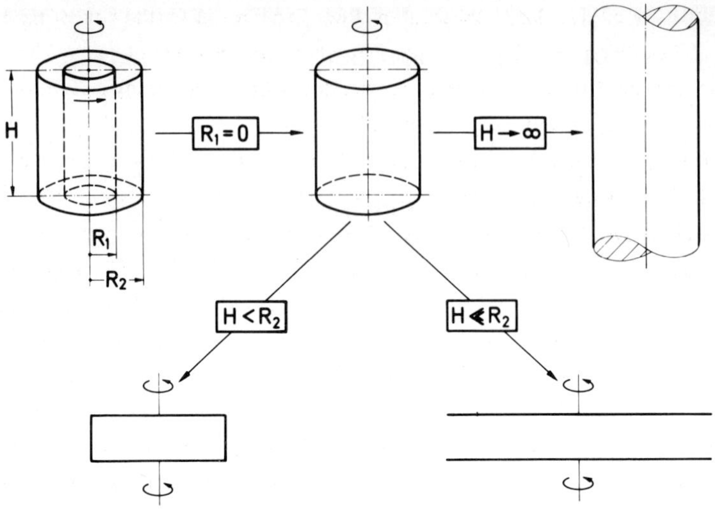

Fig. 1 Typical geometrical shapes encountered for rotating containers

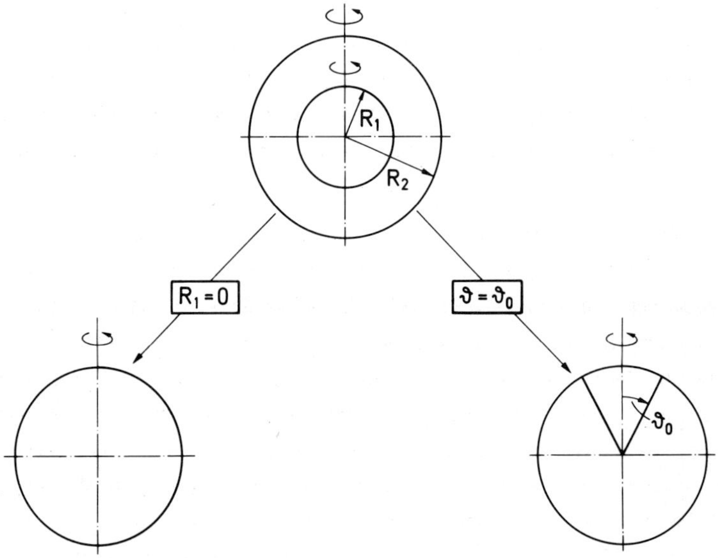

Fig. 2 Different spherical geometrical configurations taken into account by Israeli [38]

The main part of the present investigation will be devoted to spherical gap flow problems which are described accurately by a lot of reliable experiments:

Khlebutin [47] published experimental data for thick and thin fluid layers between two concentric spheres and gave an experimentally derived relation for the dependence of the critical REYNOLDS-number on the fluid layer thickness. Earlier experiments due to Sorokin, Khlebutin, and Shaidurov [90] were devoted to investigate boundary layer formation in spherical gap flow. Experiments for the ratio of the gap width related to the inner radius $(R_2 - R_1)/R_1 = 0.6$ and of smaller values were performed by Wimmer [103/104] showing for the first time that there is a strong relationship between the steady state of the rotating fluid layer in the supercritical region and the acceleration law prescribed for the angular velocity. These results have their origin in earlier experiments due to Sawatzki and Zierep [84] for a rotating inner sphere and a fixed outer sphere. For a wide range of the main parameters: ROSSBY-number and REYNOLDS-number the spherical gap flow was studied experimentally in great detail by Yavorskaya, Belaev, and Monachov [107]. The ratio $(R_2 - R_1)/R_1$ varies between 0.11 and 1.33. The REYNOLDS-number covers the interval from 0 to 10^6, and the ROSSBY-number can be chosen arbitrarily large. The spherical surfaces can rotate in the same or in opposite direction. A great variety of stationary secondary flows are described which exist at very high REYNOLDS-numbers and which show a strong periodicity with respect to the circumferential variable. Further work in the same problem was done by Astafeva, Vvedenskaya, and Yavorskaya [2] both theoretically and experimentally. The contributions due to Bonnet and Alziary de Roquefort [13] and to Israeli [38] both with respect to theory and experiment will be discussed later in more detail.

Experiments for the cylindrical geometry in the supercritical region can be found in the excellent publication by Burkhalter and Koschmieder [18].

From the theoretical point of view it is difficult to understand all the phenomena which are observed in experiments and even more difficult to predict them by a numerical simulation. It is well known that both the computer experiment and the experiment in the laboratory have their inherent limits. Sometimes it is possible to overcome some of the arising numerical difficulties and natural limitations of the computer. Therefore it seems from time to time to be worth while to demonstrate in which direction great strides were made.

The main problem for rotating fluid flows in cavities is to predict the steady secondary flow pattern – if it exists – when a set of physical parameters is given which describe a supercritical state of the system. For rotating fluid flows such parameters are e.g. the REYNOLDS-number or the EKMAN-number, the ROSSBY-number and other geometrical and kinematic parameters. In addition to these data the prescribed initial and boundary conditions will strongly influence the solution. The question arises whether these data determine uniquely the final steady state of a rotating fluid in a vessel or not. Normally the answer to this question is given within the framework of the hydrodynamic stability theory. But since TAYLOR's famous investigations new questions arose from similar experiments which can not be explained satisfactorily by the theory at the moment. For example the experimental and analytical results due to Sawatzki and Zierep [84], Ritter [81], and Wimmer [103/104] are hard to understand from the point of view of a linear stability analysis. The history of the inner rotating boundary, starting from the state of rest and leading to the same final angular velocity is responsible for the final flow pattern. Astafeva, Vvedenskaya, and Yavorskaya [2] and Yavorskaya, Belaev, and Monachov [107] demonstrate experimentally the nonuniqueness of states which exist in the supercritical region of the parameter space. A picture taken from the publication [107] shows the puzzling situation that at one point in the parameter space three different types of secondary flows can be realized.

The present investigation will be concerned with the question whether such types of flow can be simulated on the base of the NAVIER-STOKES equations and whether a computer experiment can lead to the same results as observed in the experiment.

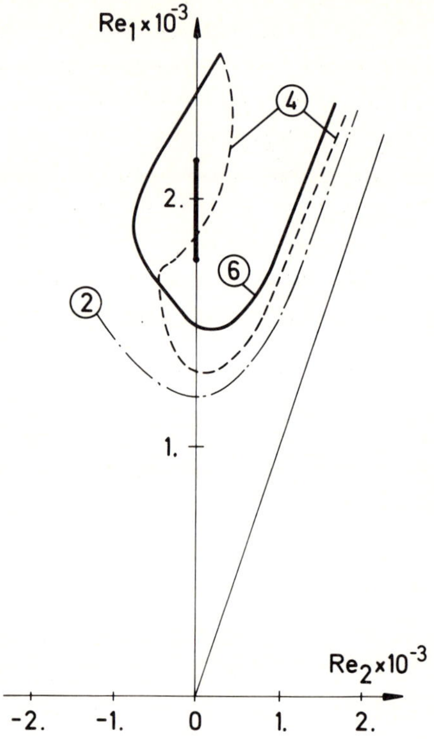

Fig. 3 Regions where different flow patterns exist.
Two, four or six vortices can be observed for the same REYNOLDS-numbers Re_1, Re_2 in the whole gap (both the spheres are allowed to rotate, $R_2/R_1 = 1.11$, see [107]).

Following this line some general questions arise in connection with the treatment of rotating fluid flows. Some questions are basic and can not be answered at the moment, others are apparent and can be answered by means of our present knowledge.

The first question is of practical interest: Which are the applications of the phenomena under consideration? Then it is necessary to ask which experimental results are known nowadays in this special field of research. Can these data be completely understood by the theory? Is the computer a useful tool for the simulation of the observed types of flow? Are the NAVIER-STOKES equations for NEWTONIAN fluids the right base for a theoretical interpretation of the phenomena? Some of the posed questions can be answered, others are so fundamental that they will be open.

Analytical methods have been used in many cases to analyse rotating flows: Aldridge [1] gets the eigenfrequencies for thick spherical layers from a variational principle. His results were later confirmed by Israeli [38] who could find numerically the resonances corresponding to experimental values. Bark, Meijer, and Cohen [4] apply a perturbation technique for small MACH-numbers to investigate the spin-up problem for a rapidly rotating gas in a cylinder. A combined analytical-numerical approach for the fluid motion induced by an impulsively changed angular velocity of a rotating disk is used by Chawla [20]. The flow induced by an impulsively started cylinder in a TAYLOR-flow apparatus was analysed by Chen and Kirchner [21] using the unsteady perturbation equations and the quasi-steady approach. A very early contribution to the analytical investigation of viscous flow problems is given by Helmholtz and Pitrowski [37] for small REYNOLDS-numbers. The onset of flow instability in the closed region of two concentric spheres was discussed by Bratukhin [15] using the method of small perturbations. Sorokin [89] considers the closed flows near critical REYNOLDS-numbers. Munson and Joseph [64] presented results on incompressible flow between concentric rotating spheres for small REYNOLDS-numbers using analytical methods. Stewartson and Roberts [92] discuss the flow in an almost spherical rigid body after an impulsive start by an analytical investigation for small REYNOLDS-numbers. St. Maurice and Veronis introduce the concept of multi-scaling for the analysis of the response of a rotating fluid to a sudden change in the angular velocity [93].

Besides these analytical methods applied to treat rotating flow problems the numerical methods can be divided into two distinct groups: One group includes those methods which allow the determination of a steady state solution based on the stationary NAVIER-STOKES equations. It is not intended to describe the iterative methods for the direct solution of the stationary NAVIER-STOKES equations as discussed by Roache [82] or by Greenspan [33].

The other group represents time-dependent methods. Especially to these methods the present investigation is devoted. In the references both the groups are taken into consideration.

The various phenomena encountered in rotating fluid flows are of basic interest for several branches of science where rotational effects play a dominant role. So the answer to the question concerning the field of applications can be given by listing authors and their contributions from different disciplines:

Astrophysics

A central problem is the investigation of the atmosphere of planets. Gierasch [30], Kalnay de Rivas [45], Turikov and Chalikov [94] present models for the VENUSIAN atmosphere. Yavorskaya and Simuni [106] give an interpretation of the equatorial jet phenomena on the JOVIAN surface. Weir [99] includes thermal effects, and his results are applicable to the understanding of convective flow in the core of a rotating star.

General Relativity

Butterworth and Ipser [19] have worked out a numerical method for the computation of the structure of rotating stars. Their contribution is of great importance for the understanding of the stellar evolution.

Geophysics

Processes in the earth's core can be interpreted by analysing the spin-up problem. The excellent review article written by Benton and Clark [11] is devoted to the spin-up problem for cylindrical flows. In [12] Benton discusses the behavior of an electrically conducting fluid between two disks with an axial homogeneous magnetic field. Israeli and Orszag [43] apply the spectral method to find trapped oscillations in rotating spherical layers. These phenomena can serve as a model for the geomagnetism or for geophysical fluid flow. Janjic and Wiin-Nielsen [44] analyse the cylindrical flow problem analytically to illustrate the wind fields and their time development. Stewartson's and Roberts' results [91] are strongly related to geophysical flow problems in the earth.

Centrifuge flow

Isotope separation plays an important role in nuclear power station supply. One possible way to separate isotopes is realised by a centrifugal set-up. From the hudge amount of publications in this special field only three examples shall be cited. Harada and Ozaki [36] take into account thermal effects and treat conducting and insulating sidewalls of a rapidly rotating cylinder. They use a rotating coordinate system and give a numerical solution for the meridional circulating flow. Lahargue and Soubbaramayer [54] present for compressible and viscous flows in a centrifuge a computer program of universal character on the base of the finifinite element analysis. Matsuda and Hashimoto [58] treat thermal effects using a linear approach and boundary-layer technique to describe the axisymmetric motion in a rapidly rotating cylinder.

Meteorology

For the atmospheric dynamics on the earth, the formation of jet flows and thermal wind generation a detailed description of rotating flows on spherical surfaces in a central task. Morel [62] discusses in his review article the simulation of atmospheric circulation in great detail. The stability of large-scale ROSSBY-waves is investigated by Sadournay [83] using a finite difference model.

Oceanography

Bottom topography effects are the subject of different investigations. Görtler [32] has investigated the influence of the viscosity on oceanic and atmospheric flows over barriers. His analytical solution is exact for the twodimensional case. Boyer [14] considered the same problem for different types of barriers on a rotating flat plate.

Fluid rotor gyroscopes

An important field of application of rotating flows is the study of motions inside gyroscopes. Pfeiffer [75] discusses an approximate solution for the set of basic equations transformed into a rotating coordinate system.

In the near future experiments with fluid filled containers are planned on SPACELAB I. Vreeburg [96] has proposed some experiments with partially filled vessels. Because of the free surfaces involved these problems are far from being easily simulated by numerical model calculations!

As the field of applications is widespread it is more than likely, that the literature collected in this paper is not complete. In their very useful review article Orszag and Israeli [69] summarize the literature up to 1973. This was done under the more general point of view of the simulation of any viscous incompressible flow. The present contribution will cut out only one narrow aspect of this broader field of numerical methods.

The given references are confined to the problem of flows inside cavities neglecting the other phenomena with occur when a body is accelerated in an infinitely large flow region. A typical example for this type of flow is given by Lugt and Haussling [57] for the acceleration of thin cylindrical bodies imbedded in a large outer flow region. Experimentally the phenomena outside of an impulsively started cylinder are studied by Chen and Christensen [22]. A theoretical analysis of this experiment is contained in [21].

I. Geometry of the Cavities

A variety of geometrical shapes of fluid-filled containers can be encountered in the present literature. **Fig.** 1 and **Fig.** 2 show schematically several commonly treated configurations. Effects of eccentricity as treated by Munson [63] will not be taken into account. A large number of publications is devoted to the TAYLOR-type configuration of two concentric vertical cylinders of finite length H. Bennets and Jackson [10] give ex-experimental and theoretical results for the rotating annulus. Quon [78] includes internally heating effects for the cylindrical gap flow applying a semi-implicit finite difference algorithm on variable grids which are condensed in the edges of the rectangular cross section. Quon solves the NAVIER-STOKES equations in a nonrotating coordinate system with respect to the primitive variables u, v, w, and p. The conduction dominated and the convection dominated case are treated and discussed. The differences of the solutions to the normally treated wall-heated annulus are significant. In [79] Quon discusses also a combined spectral and finite difference model for the circular COUETTE-flow with differentially heated side walls. Compared with earlier results this model saves computing time and storage requirements. Briley and Walls [17] present numerical results for time-dependent rotating flow in a cylindrical gap. Their method is based on Pearson's method [72] using 26×51 and 26×101 grid points for the spin-up and the spin-down experiment, respectively. In a later published paper Briley [16] presents a solution for internal flow problems at high REYNOLDS-numbers which can be extended also to determine the flow field in a curved duct. The flow in a cylindrical annulus is investigated also in [21] and [95] from a theoretical point of view. Burkhalter and Koschmieder [18] have investigated intensively the TAYLOR-type flow in the supercritical region performing very accurate experiments. Other cylindrical flow problems which are solved theoretically or experimentally can be found in the following contributions: Fasel and Booz [28] have applied an implicit finite difference method using three point backward differences and a relaxation procedure to analyse unsteady flows between rotating coaxial cylinders. Hall [35] takes into account a small periodic variation of the inner angular velocity. Kasagi and Hirata [46] consider the stability of the time-dependent flow between cylindrical surfaces on the base of a time-dependent linear stability analysis. Experimental data are also given by Kirchner and Chen [48] for the same problem. The behavior of two capillary liquids under complete weightlessness in a rotating cylinder is analysed by Kopachevskii and Radyakin [49]. Launder and Ying [55] include the eccentricity effect for two parallel cylinders. The dependence of the steady state on the initial conditions is analysed by Meyer [60/61], and counterrotating cylindrical walls are considered. Pustalov and Sparrow [76] have treated the problem of rotationally induced convection in a cylindrical container using the full NAVIER-STOKES equations. The spin-up and spin-down problem for a rotating cylinder is considered by Weidman [98], including the flat plate geometry for an abrupt start. Warn-Varnas, Fowlis, Piacsek, and Lee [97] give experimental and numerical data for the spin-up problem of a cylinder. Moderate

ROSSBY-numbers are included ($\leqslant 0.33$). The numerical treatment guarantees the correct resolution of the EKMAN-layer.

For the disk geometry the following contributions can be mentioned: Barrett [6] has proposed a numerical procedure which can be applied to large REYNOLDS-numbers and which solves singular perturbation problems of two-point-boundary value type. Impulsive change effects in the angular velocity are discussed analytically in Chawla's paper [20] when only one disk is assumed. An analytical treatment for the laminar flow between two rotating disks is given by Gorin and Shilyaev [31]. The same problem is attacked by Nguyen, Ribault, and Florent [67] with the assumption that one disk is porous. An asymptotic analysis for large ROSSBY-numbers is contained in [88]. Viscoelastic effects play an important role in the contribution due to Leonov, Lipkina, Paskhin, and Prokunin [56]. A numerical study for the flow between two disks with suction can be found in Wilson's and Schreyer's paper [102]. They treat the case where the angular velocity is gradually increased to reach an end state. Several other authors contribute also to the disk flow problem: De Socio, Sparrow, and Eckert [26], Kubicek, Holodniok, and Hlavacek [51], Pearson [72], Quemard, Cousteix [77], Raal [80], Siegmann [87], and Williams and Kurth [101].

II. Numerical Methods

II.1. Finite Element Methods

In the last years the finite element method has been applied with considerable success also in the field of fluid dynamics. Nowadays it is no longer true that finite element methods remain a dark horse as it was in the past. Only two fluid dynamic problems will be mentioned in detail as examples for rotating flows where the numerical simulation of the flow field with the aid of the computer is based on this young and powerful method. Bar-Yoseph, Blech, and Solan [9] calculate the flow inside a cylindrical container where the bottom and the container rotate at different angular velocities. The three-dimensional NAVIER-STOKES equations in cylindrical coordinates are solved for the variables u, v, w, and p. Spatial oscillations as they normally occur for large element REYNOLDS-numbers in connection with GALERKIN methods are suppressed by applying various upwind schemes. For $Re = 100$ ($Re := (\omega R^2)/\nu$) the case of a rotating disk in a stationary container was treated giving velocity profiles which agree very well with other calculations. For a stationary disk and a rotating container the $Re = 400$ was treated satisfactorily. In the latter case the element REYNOLDS-number was about $Re_h > 5.5$. Secondary flows which exist in the inside a cylinder with rotating bottom plate can be treated without difficulties for moderate REYNOLDS-numbers using the finite element technique.

Based on the finite element approach a powerful numerical method was developed by Lahargue and Soubbaramayer [54]. They include compressibility effects besides thermal effects and viscosity. The problem to be solved is the flow inside a gas centrifuge. By the same method also countercurrent flow in the container can be simulated and the stability of such flows can be investigated. With a relatively coarse mesh (20 meshes in radial and 31 meshes in axial direction) the flow can be represented inside a cylinder. According to the strong gradients of the flow field quantities near the side wall a smaller mesh size is used. By this stretching technique the resolution of steep gradients can be guaranteed in the vicinity of the top and bottom cap, too. This leads to an accurate treatment of the STEWARTSON and EKMAN-layers. The radial spacing is governed by the principle that every loop should contain equal mass. Examples are given for thermally and mechanically driven flows for two concentric cylinders and one cylindrical container only. This finite element method covers all real effects which occur in the case of a gas in strong rotation. If one wants to describe the stationary state of centrifugal flow in simple geometries this method seems to be superior to others, as far as the author is informed.

The algorithm due to Soubbaramayer et al. [91] can be used to make sure whether an asymptotic solution produced by a time-dependent calculation has reached its limit. When taken as an initial distribution the asymptotic solution must be reproduced by the algorithm based on the finite element method. Such a numerical test on the time-independency of a limit solution using two different methods has not yet been carried out.

Barrett [5] discusses the stream-function-vorticity formulation of the NAVIER-STOKES equations for time-dependent flows. On the base of the finite element approach the usual difficulties of finding an accurate finite difference approximation for the boundary conditions at a wall can be avoided. This is an important result because normally the main difficulties arise from an incorrect representation of the boundary values of the vorticity (see also **III. Concluding Remarks**).

II.2. Spectral (GALERKIN) Methods

Orszag and Israeli [71] have reviewed the whole subject of spectral methods with respect to the solution of planetary scale flows beginning with the year 1954. This article contains more than that. For spherical flow problems it contains a new spectral method on the base of the linearized motion of a homogeneous fluid in a rotating spherical annulus. The main problem which is discussed in this investigation is devoted to the so called resonant trapped modes of motion in the spherical layer. For $Re = 36000$ and $Ro = 0$ plots of contours of the stream function are shown in a relatively small gap $(R_2 - R_1)/R_1 = 1/8$. The time interval was chosen to be $\Delta t = 1/(4\Omega_0)$ and the radial mesh interval was assumed to be $R_2/240$. A number of 12 to 44 modes was taken into account. The results are limited to symmetric velocity flow fields with respect to the equator. As only the linearized equations written in a rotating coordinate system are used the influence of nonlinear terms on the asymptotic solution is eliminated. In [43] Israeli and Orszag discuss the same case of very small ROSSBY-number using the linearized equations about the state of uniform rotation. For $Re = 2250, \ldots, 576000$ the calculations were carried out with 30 modes. A maximal number of 65 points in radial direction was used together with a strong stretching of the mesh intervals. The used cut-off was found to be sufficient for all problems under consideration. The set of equations for the components of the spectral representation was solved using a discretization in radial direction by a second order centered formula and in time by a centered implicit (CRANK-NICOLSON) scheme. The number of iterations necessary for numerical convergence was found to be about 2 or 3 for the solution of the system of coupled linear equations.

A basic discussion of spectral methods for simple boundaries can be found in [70]. Orszag extends also the results to more complicated geometries using especially CHEBYCHEV expansions. The application of pseudospectral methods is considered for the NAVIER-STOKES equations in a rotating frame. The similarity of this method with the collocation method is mentioned.

Another method based on series expansion for the nonlinear NAVIER-STOKES equations is presented by Astafeva, Vvedenskaya, and Yavorskaya [2]. This method is based on the early work due to Babenko, Vvedenskaya, and Orlova [3]. In [2] especially for different gap sizes a numerical analysis for the axisymmetric viscous spherical COUETTE flow is carried out. Only the inner sphere is assumed to rotate with a constant angular velocity. The REYNOLDS-number covers an interval between 200 and 2200. For the cases discussed the plots for the streamlines, the lines of constant angular velocity, and the energy spectra are given. Also the torque needed for the rotation of the inner sphere at a given REYNOLDS-number is presented. Concerning the question of nonuniqueness of the steady solutions of the NAVIER-STOKES equations for the small gap size examples are given. For the wide gap experiments and calculations are performed and lead to an understanding of the stability behavior in the range of Re-numbers mentioned above. $R_2/R_1 = 1.11$ and 2.33 were taken into account. For the small gap width in **Fig. 4** a contour plot of the stream function and angular velocity is given. The calculations show in the supercritical region of the REYNOLDS-number that different stationary solutions can be found.

The same flow pattern is found in the experiments which are performed by Belaev and Monachov [107].

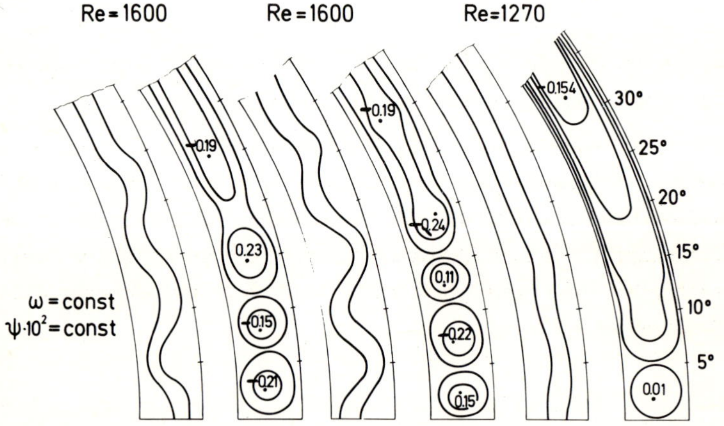

Fig. 4 Contour plots of the stream function and the angular velocity due to Astafeva, Vvedenskaya, Yavorskaya [2]

II.3. Finite Difference Approximations

II.3.1. Explicit Methods

On the base of an explicit finite difference method Krause and Bartels [50] compare numerical results for spherical gap flow at high REYNOLDS-numbers for two different gap widths ($R_1/R_2 = 0.15$, $Re = 1500$ and $R_1/R_2 = 0.08$, $Re = 2000$). These data correspond to the experiments due to Sawatzki and Zierep [84] and Wimmer [103/104]. In two earlier papers Bartels and Krause [8] and Bartels [7] presented results for the same type of flow gained by the explicit difference method. The agreement of the numerical date to the experimental results for $R_1/R_2 = 0.15$ is excellent. The set of the basic equations are formulated in the variables: stream function, vorticity, and angular momentum. An explicit method necessarily demands a control of the time step. This time step is smaller than derived from the von-NEUMANN-stability condition with frozen coefficients in the equations. In radial direction 20 meshes are chosen while in the ϑ-direction 98 intervals are assumed. For the acceleration of the inner sphere different functions of time for the angular velocity $\omega(t)$ were prescribed. Also sudden start experiments were simulated besides constant accelerations during a finite interval of time. Above the REYNOLDS-number 1150 two different stationary states of the gap flow were observed. For an acceleration rate of $\dot{\omega}_i = 0.105\ \omega_0^2$ two vortices with axis parallel to the equator were observed. When the acceleration was raised a slightly distorted flow similar to the basic flow appeared. $Re = 800$ was the smallest value for which the appearence of vortices could be found. For the interval $800 \leqslant Re \leqslant 1150$ only one type of secondary flow was found showing two vortices.

It is an open question up to now whether the secondary flow which consists of only one vortex near the equator in the northern hemisphere with a source-like character in the vicinity of the equator plane can be simulated by the assumption that the whole flow field is symmetric with respect to the equator plane. Krause and Bartels [50] argue that this case can be found only if the symmetry condition is violated. They can show that a single vortex flow near the equator exists if the symmetry condition is slightly changed passing the critical REYNOLDS-number. If the REYNOLDS-number was below 700 and the symmetry condition was strictly fulfilled at the equator plane the vortex did not exist. These results imply that perhaps it is necessary to treat this flow field as a three-dimensional one.

It is evident that the spatial resolution of the flow field calculations influences the appearence of computed vortices. A grid of 21×99 points leads for $Re = 1500$ and $R_1/R_2 = 0.85$ to two vortices in the northern hemisphere while the reduction of the number of grid points to 11×33 gives a slightly distorted basic flow. This was demonstrated by Bartels [7]. Modes with uneven numbers of vortices in the spherical gap near the equator could only be represented numerically when the values of the circumferential velocity had its minimum in the equatorial plane.

II.3.2. Implicit Methods

From all other numerical methods based on finite difference approximations the contribution due to Bonnet and Alziary de Roquefort [13] gives excellent results which agree very well to Wimmer's experimental data [103/104]. The numerical simulation of the flow of a fluid in a gap between rotating spherical shells uses the stream-function-vorticity formulation of the NAVIER-STOKES equations. The calculations are performed with respect to a nonrotating coordinate system. The stationary flow is found by an asymptotic limit solution of the time-dependent equations. Rotational symmetry of the flow field around the axis of revolution together with a symmetrical behavior of the solution with respect to the equatorial plane is assumed. The pressure is eliminated from the basic equations resulting in two parabolic equations and one elliptic equation for the stream function, the vorticity, and the angular momentum per unit mass.

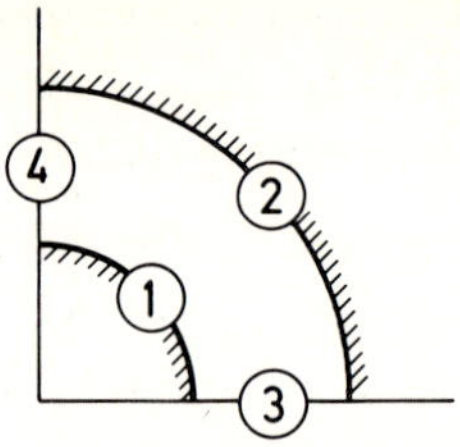

		ψ	$\frac{\partial\psi}{\partial r}$	$\frac{\partial\psi}{\partial\vartheta}$	$\frac{\partial^2\psi}{\partial\vartheta^2}$	Ω	$\frac{\partial\Omega}{\partial\vartheta}$
inner sphere	(1)	0	0			$R_1^2\sin^2\vartheta$ $\omega_1 R_1^2\sin^2\vartheta$	
outer sphere	(2)	0	0			0 $\sin^2\vartheta$	
equator	(3)	0			0		0
axis of rotation	(4)	0		0		0	

Fig. 5 Boundary conditions posed by Bonnet and Alziary de Roquefort [13] admitting only symmetrical solutions with respect to the equatorial plane.

The numerical iterative procedure is based on the discretization by central differences with fractional step approximation in time. The POISSON equation is solved using fast FOURIER transformation. For the vorticity the boundary values on the spherical surfaces are determined iteratively. The boundary condition $\partial\psi/\partial r = 0$ on the spheres is not fulfilled exactly so that a small slip of the fluid may exist. The whole set of boundary conditions is summarized in **Fig. 5**.

The calculations were performed under the following restrictions:

$$|u|\, Re \cdot \Delta r \leqslant 2 \quad , \qquad \left| v\, Re + \frac{\mathrm{cotg}\,\vartheta}{r} \right| r\, \Delta\vartheta \leqslant 2 \quad .$$

(u = velocity component in radial direction, v = velocity component in ϑ-direction). A net of 65×65 grid points was used in the meridional plane. Typical computing time was 10 m on the IBM 370/168 for Re = 1000.

Bonnet and Alziary de Roquefort reproduce numerically Wimmer's results in an excellent manner. The centrifugal instabilities are investigated for supercritical Re-numbers 700, 900 and 1500. Starting from rest the numerical results show for Re = 700 one vortex and for Re = 900 two vortices parallel to the equator. When the REYNOLDS-number is raised to 1500 the nonuniqueness of the end state is apparent. If the calculation is starting from the state of rest the same secondary flow appears as for the REYNOLDS-number 700 showing only one equatorial jet. If the initial state is chosen according to the flow pattern of the REYNOLDS-number 900 again a double vortex is reproduced.

In the calculations the intermediate values of the angular velocity are not prescribed according to a given acceleration law. Only the final end state is of interest and the 'history of the flow' is not followed exactly. This can be seen from the order of the errror bounds of the difference approximations in space and time.

Concerning the numerical description of the supercritical end states of the fluid flow in spherical gaps the results of Bonnet and Alziary de Roquefort show a close correspondence to the experiments due to Wimmer. The numerical method gives reliable results even for higher Re-numbers up to 8000 and for smaller and larger gap widths. The TAYLOR-number for an onset of instabilities is determined as 48.4 which is in good agreement to the otherwise derived theoretical and experimental values [47], [84], [108].

Following the ideas of numerical approximation which are developed by Israeli [38], [39], [40], [41], [42], a simulation of the acceleration process was tried for the case where only the inner sphere is started from rest leading to end states which are characterized by the same supercritical REYNOLDS-numbers. The geometry of the problem is shown in **Fig. 6**.

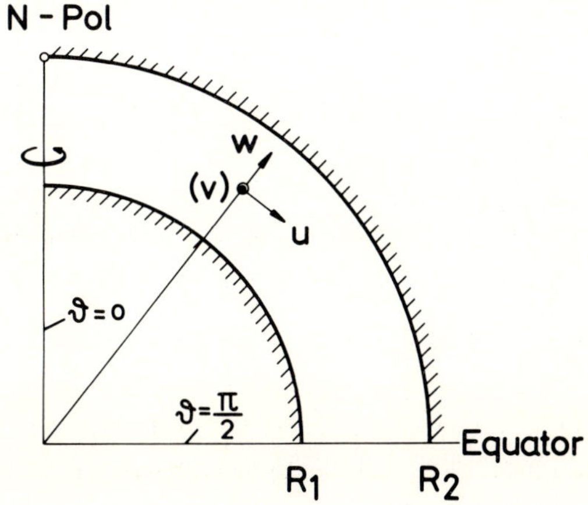

Fig. 6 Geometry of the problem

Israeli introduces also the stream-function-vorticity formulation using the formulas

$$\begin{cases} \dfrac{\partial \psi}{\partial \vartheta} := r^2 \cdot \sin\vartheta \cdot w \quad , \\[2ex] \dfrac{\partial \psi}{\partial r} := -r \cdot \sin\vartheta \cdot u \quad , \\[2ex] \Omega := r \cdot \sin\vartheta \cdot v \quad . \end{cases}$$

ψ is the stream function and Ω the angular momentum per unit mass.

The basic equations used for the numerical analysis of the time-dependent flow is given by the following set of relations:

$$\frac{\partial\zeta}{\partial t}-\frac{J(\psi,\zeta)}{r^2\sin\vartheta}+\frac{J(\Omega^2,r\sin\vartheta)+2\zeta\cdot J(\psi,r\sin\vartheta)}{r^3\sin^2\vartheta}=\nu D^2\zeta$$

$$\frac{\partial\Omega}{\partial t}-\frac{J(\psi,\Omega)}{r^2\sin\vartheta}=\nu D^2\Omega$$

$$\zeta := D^2\psi,$$

J = Determinant of JACOBIAN,

$$D^2 := \frac{\partial^2}{\partial\rho^2}+\frac{1}{\rho^2}\cdot\frac{\partial^2}{\partial\vartheta^2}-\frac{\operatorname{cotg}\vartheta}{\rho^2}\,\frac{\partial}{\partial\vartheta}$$

Basic equations

A spherical coordinate system is chosen and ζ represents the vorticity.

Introducing dimensionless variables by the following relations:

$$r = \bar{r}\cdot R_2$$

$$\psi = \bar{\psi}\cdot\omega_0 R_2^3$$

$$t = \bar{t}\cdot\frac{1}{\omega_0}$$

$$\Omega = \bar{\Omega}\cdot R_2^2\,\omega_0$$

and reformulating the set of equations in a rotating coordinate system the following coupled system of equations for the vorticity, stream function and angular momentum per unit mass is derived (the bar $^-$ is omitted again in the following formulas):

$$\frac{\partial\zeta}{\partial t}-\varepsilon\frac{J(\psi,\zeta)}{r^2\sin\vartheta}+2\frac{J(\Omega,r\sin\vartheta)}{r}+2\varepsilon\frac{\Omega\cdot J(\Omega,r\sin\vartheta)+\zeta\cdot J(\psi,r\sin\vartheta)}{r^3\sin^2\vartheta}=E\cdot D^2\zeta$$

$$\frac{\partial\Omega}{\partial t}-\varepsilon\frac{J(\psi,\Omega)}{r^2\sin\vartheta}-2\frac{J(\psi,r\sin\vartheta)}{r}=E\cdot D^2\Omega$$

ε = ROSSBY-number

Rotating coordinate system

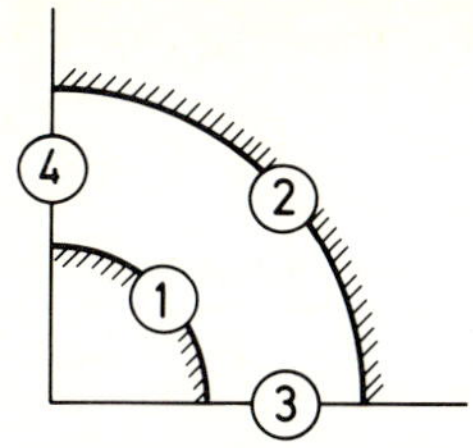

	ψ	$\frac{\partial \psi}{\partial r}$	$\frac{\partial \psi}{\partial \vartheta}$	ζ	Ω	$\frac{\partial \Omega}{\partial \vartheta}$
(1)	0	0	0		$\omega(t) \times r^2 \sin^2 \vartheta$	
(2)	0	0	0		0	
(3)	0			0		0
(4)	0			0	0	

Fig. 7 Boundary conditions posed by Israeli [38] for the spherical gap flow.

The boundary conditions for the case where the outer sphere is at rest are given in **Fig.** 7.

The EKMAN-number $E := \nu/(\omega_0 R_2^2)$ and the ROSSBY-number ϵ are the two dimensionless parameters which describe the physical and kinematic situation. In the numerical experiment the inverse EKMAN-number was chosen to 800 and 1270. The ROSSBY-number was chosen equal to 1. So all nonlinear terms were taken into account.

The numerical algorithm for the solution of the finite difference approximation of the equations is shown schematically in a flow chart in **Fig.** 8. The calculation starts with an extrapolation of ψ and ζ with respect to time. Ω is then assumed at the boundaries. Then Ω and ζ are determined by the fractional step method and ψ is found solving the POISSON equation. A special step is necessary to find the boundary values of ζ.

Using Israeli's program a typical acceleration law for the angular velocity of the inner sphere was simulated as shown in **Fig.** 9.

Different numerical experiments were performed using Israeli's program. The results can be summarized as follows: When the ROSSBY-number is zero, which means that a linearized set of equations is used then no secondary flow is found in the supercritical region (Re = 800, 1270). The nonlinear terms are responsible

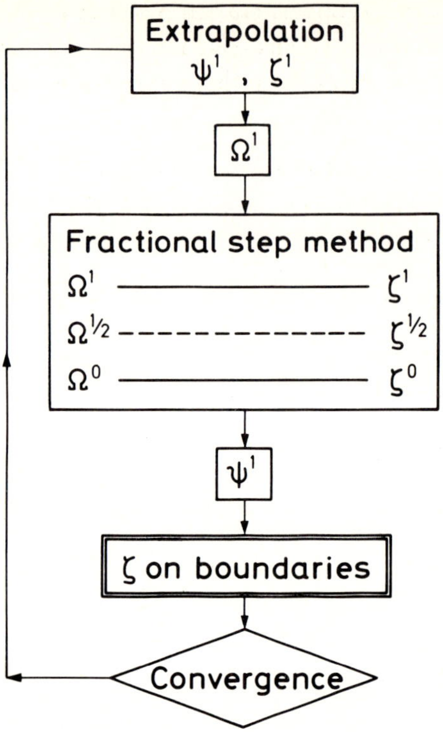

Fig. 8 Flow chart of the algorithm due to Israeli [38]

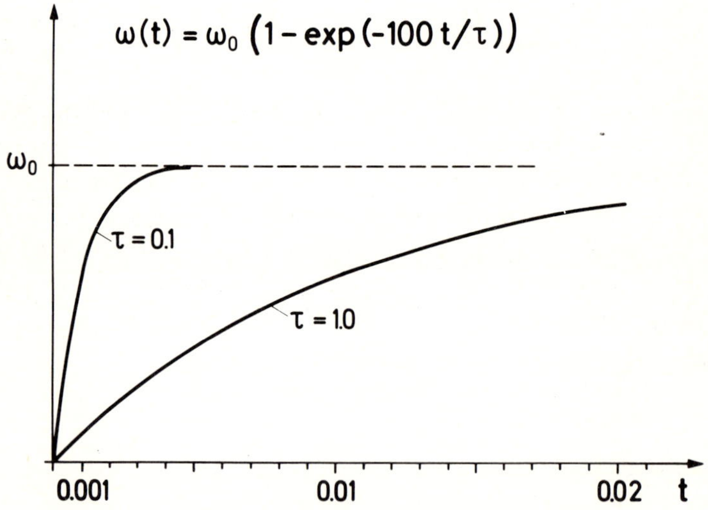

Fig. 9 Typical acceleration curves for the inner sphere

for the appearance of the secondary flows at moderate REYNOLDS-numbers. One typical example is given in **Fig. 10**. Starting from rest, in the vicinity of the equator two stationary vortices occur. At higher angles of the latitude additional vortices are created. These have not yet reached a fully stationary state when the calculation was terminated.

The number of grid points in radial direction was varying between 37 and 73. In ϑ-direction of coarse meshing of 15 grid points was found to be sufficient. The computing time for the picture shown in **Fig. 10** was about 40 minutes on the UNIVAC 1108 computer.

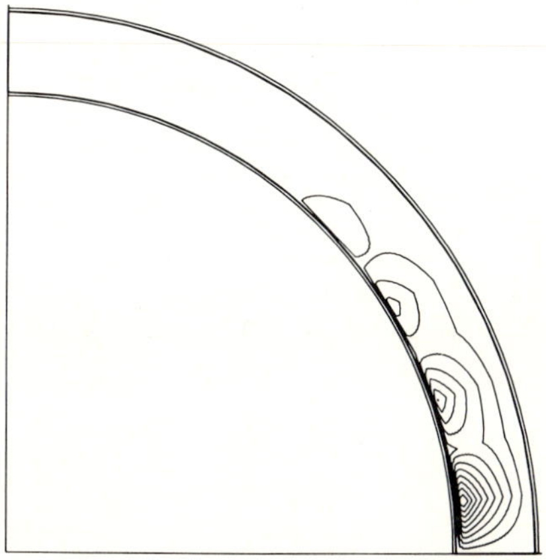

Fig. 10 Secondary flow pattern in a spherical fluid layer at REYNOLDS-number 1270, $(R_2 - R_1)/R_1 = 0.18$.

From the mathematical point of view the efficiency of a program for the numerical treatment of the NAVIER-STOKES equations depends strongly on the POISSON-solvers applied. Israeli suggests a block tri-diagonal method and finds that it is superior to SOR- and ADI-methods. For this essential part of the whole calculation there are nowadays a lot of other methods available. Schumann [85] shows comparisons of different methods applied from different authors to the same problem of solving a POISSON equation in a simple region. For three-dimensional problems Wilhelmson and Erickson [100] give comparisons of several methods: cyclic reduction, FOURIER analysis and a combination of both of them. Changing this central part of the procedure may lead to a faster computation of the flow field.

III. Concluding Remarks

Results on the spherical COUETTE-flow problem were discussed besides the others which are combined with different geometrical shapes of the container walls. Experiments and numerical investigations in the supercritical region exist which show the stability behavior of the spherical rotating fluid flow. In some cases the simulation of the viscous flow by an experiment on the computer shows the same behavior of the flow field as in the experiment in the laboratory. The dependence of the secondary flow on the history of the whole flow during the acceleration time starting from the state of rest can be established numerically and experimentally.

It is well known that the most troublesome point in the numerical simulation of the solution of the NAVIER-STOKES equations using the stream-function-vorticity formulation comes from the vorticity transport equation. The correct approximation of the boundary values of the vorticity plays an important role for the stability of the whole calculation. As Wu [105] has pointed out the violation of the integral law for the vorticity is a reason for instabilities in the numerical process. Wu derives a condition for the rate of change of the total vorticity in the fluid domain which is connected with the boundary condition of the vorticity by the relation:

$$\frac{\partial}{\partial t} \int_V \zeta \, dV = \nu \oint_S \frac{\partial \zeta}{\partial n} \, dS \quad .$$

V is the flow volume, ζ the vorticity, ν the kinematic viscosity, S the boundary surface of the flow field. Using this additional condition Wu's results for the RAYLEIGH-problem (impulsively started flat plate) where in excellent agreement with the exact solution, which is available in this special case. A similar check of the spherical gap flow formulas should be used to exclude wrong approximations of the vorticity at the boundaries.

Not all existing methods were discussed in great detail in this summarizing report. There are other different approaches to the numerical solution of the NAVIER-STOKES equations: Shestakov [86] e.g. gives a new hybrid method for two-dimensional problems with low viscosity. This method is based on the vortex method and the finite difference approximation in flow regions far away from the boundaries. It is possible to treat the square cavity flow problem where one side wall is moving in its own plane. The steady state solution was reached as an asymptotic limit of the time-dependent equations. REYNOLDS-numbers 400, 1000, 2000 were chosen and gave satisfactory results.

No results are known about the application of the "artificial" stabilization method as it was called by Dorodnicyn [27] for the calculation of rotating fluid flows. The special feature of this procedure is that a system of three time-dependent parabolic equations is treated for which a steady state for $t \to \infty$ is calculated numerically. The boundary conditions are the same as in the "real" stabilization method, where a system of two parabolic and one elliptic equation are solved. Chien [24] proposes a method which can be used when there is a local analytical solution available for the NAVIER-STOKES equations.

The comparison of implicit contra explicit methods for time-dependent problems is contained in the notable review article written by Cheng [23]. Cheng emphasizes the need for a careful and critical interpretation of numerical results because especially steady state solution on the computer may not represent the desired asymptotic flow field. As in the present investigation the rotating sphere experiments show very clearly the real steady state solution one can judge easily whe ther a numerical solution represents the physical situation in the spherical gap or not.

Again we come back to the questions at the beginning of this article:

The applications of rotating fluid flows can be found in many fields of science, and there is still a growing interest in the solution of special problems connected with the spherical geometry. The experimental results are numerous and raise a large number of questions which are not yet answered. As an open question one can consider the problem of the dependence of a reached supercritical steady state on the rate of change of the angular velocity of the inner sphere. The numerical simulation of these phenomena is only a very modest contribution. A deep understanding is only possible by using mathematical rigorous proofs for existence and uniqueness of the solutions of NAVIER-STOKES equations as performed by Ladyzhenskaya [52], [53]. When the "contaminations" introduced by the computational approach are less severe than the contaminations which are present in any laboratory experiment, then the computer can serve as an equivalent instrument in the investigation of hydrodynamic stability equations as formulated here.

A final remark concerning the references:

The following references mentioned in this article can also be found in the review article [69]: Greenspan's book [34] which is basic for any treatment of rotating fluids, the important experimental and theoretical contribution due to Munson and Joseph [64/65] devoted to spherical gap flow, Orszag's investigation on spectral representations [70] and the basic contributions due to Israeli [38/39/40/41] concerning the numerical calculation of viscous flows inside several geometrically different bodies as shown in **Fig.** 2. Pearson's early papers [72/73/74] are cited because they represent the starting point for a lot of later numerical investigations. Roache's well-known book [82] summarizes as a reference book the results on the stream-function-vorticity formulation of the NAVIER-STOKES equations and the results concerning the computation of boundary values of the vorticity. Most of the other references cover the time period of 1973 –1978. The author is sure that there are still gaps in the list of publications of such a rapidly growing field of applications. With special emphasis those publications were compiled which can be found in reports or in the proceedings of conferences which are not easily available to the majority of the interested readers.

The author is grateful to his colleagues D. Martin and R.E. Kelly for their help to get reference to a great number of publications which would have been undetected otherwise. If the reader will find that important papers are omitted the author must confess that he could acquire only a very limited knowledge of the literature devoted to the problem under consideration.

References

[1] Aldridge, K.D.: Axisymmetric inertial oscillations of a fluid in a rotating spherical shell. Mathematika, 19, 1972, pp. 163–168.

[2] Astafeva, N.M.; Vvedenskaya, N.D.; Yavorskaya, I.M.: Nonlinear axisymmetric fluid flows in spherical layers (Russian). Report Nr. 385, Institute of Cosmic Research of AN SSSR, 1977.

[3] Babenko, K.I.; Vvedenskaya, N.D.; Orlova, M.G.: Calculation of the stationary flow of a viscous fluid around a circular cylinder (Russian). Zh. vych. mat. mat. fiz., 15, 1975.

[4] Bark, F.H.; Meijer, P.S.; Cohen, H.I.: Spin up of a rapidly rotating gas. Phys. Fluids, 21, 1978, pp. 531–539.

[5] Barrett, K.E.: A Variational Principle for the Stream-Function-Vorticity Formulation of the Navier-Stokes-Equations Incorporating No-Slip Conditions. J. Comp. Phys., 26, 1978, pp. 153–161.

[6] Barrett, K.E.: Numerical Study of the Flow Between Rotating Coaxial Discs. ZAMP, 26, 1975, pp. 807–817.

[7] Bartels, F.: Flow Between Two Rotating Concentric Spheres. Proc. II. GAMM-Conf. on Num. Meth. Fluid Mech., DFVLR, Köln, 1977, pp. 7–13.

[8] Bartels, F.; Krause, E.: Unsteady Rotating Flows in Fluid-Filled Spheres. Proc. I. GAMM-Conf. on Num. Meth. Fluid Mech., DFVLR, Köln, 1975, pp. 15–20.

[9] Bar-Yoseph, P.; Blech, J.J.; Solan, A.: Upwind Schemes for the Finite Element Solution of the Navier-Stokes-Equations in Rotating Flow. Int. Conf. on Num. Meth. in Lam. and Turb. Flow, Swansea, 1978.

[10] Bennetts, D.A.; Jackson, W.D.N.: Source-Sink Flows in a Rotating Annulus – A Combined Laboratory and Numerical Study. J. Fluid Mech., 66, 1974, pp. 689–705.

[11] Benton, E.R.; Clark, Jr., A.: Spin-up. Ann. Rev. Fluid Mech., 6, 1974, pp. 257–280.

[12] Benton, E.R.: Nonlinear hydrodynamic and hydromagnetic spin-up driven by Ekman-Hartmann boundary layers. J. Fluid. Mech., 57, 1973, pp. 337–360.

[13] Bonnet, J.-P.; Alziary de Roquefort, T.: Ecoulement entre deux sphères concentriques en rotation. J. Méch., 15, 1976, pp. 373–397.

[14] Boyer, D.L.: Bottom Topography Effects on Rotating Flows. Proc. X. Symp. on Adv. Probl. and Meth. in Fluid Mech., Rynia, Poland, 1971, part II, pp. 27–51.

[15] Bratukhin, Yu.K.: On estimating the critical Reynolds number for fluid flow between two rotating spherical surfaces. PMM, 25, 1961, pp. 1286–1299.

[16] Briley, W.R.: The Computation of Three-Dimensional Viscous Internal Flows. Proc. III. Int. Conf. on Num. Meth. in Fluid Mech., Vol. II, 1973, pp. 33–38.

[17] Briley, W.R.; Walls, H.A.: A Numerical Study of Time-Dependent Rotating Flow in a Cylindrical Container at Low and Moderate Reynolds Numbers. Proc. II. Int. Conf. on Num. Meth. in Fluid Dyn., Berkeley, Calif., 1970. Springer-Verlag, 1971, pp. 377–384.

[18] Burkhalter, J.E.; Koschmieder, E.L.: Steady supercritical Taylor vortex flow. J. Fluid Mech., 58, 1973, pp. 547–560.

[19] Butterworth, E.M.; Ipser, J.R.: On the structure and stability of rapidly rotating fluid bodies in general relativity. I – The numerical method for computing structure and its application to uniformly rotating homogeneous bodies. Astrophys. J., 204, 1976, pp. 200–223.

[20] Chawla, S.S.: Spin-up from a rotating disk.
J. Fluid Mech., 78, 1976, pp. 609 – 619.

[21] Chen, C.F.; Kirchner, R.P.: Stability of time-dependent rotational Couette Flow – Part 2. Stability analysis.
J. Fluid Mech., 48, 1971, pp. 365 – 384.

[22] Chen, C.F.; Christensen, D.C.: Stability of flow induced by an impulsively started rotating cylinder.
Phys. Fluids, 10, 1967, pp. 1845 – 1846.

[23] Cheng, Sin-I: A Critical Review of Numerical Solution of Navier-Stokes Equations.
Progress in Num. Fluid Dyn., Springer-Verlag, 1975, pp. 78 – 225.

[24] Chien, J.C.: A General Finite-Difference Formulation With Application to Navier-Stokes-Equations.
Comp. and Fluids, 5, 1977, pp. 15 – 31.

[25] Dennis, S.C.R.: The Numerical Solution of the Vorticity Transport Equation.
Proc. III. Int. Conf. on Num. Meth. in Fluid Mech., Vol. II, Springer-Verlag, 1973, pp. 120 – 129.

[26] De Socio, L.M.; Sparrow, E.M.; Eckert, E.R.G.: The contrived transient-explicit method for solving steady-state flows – Application to a rotating, recirculating flow.
Comp. and Fluids, 1, 1973, pp. 273 – 287.

[27] Dorodnicyn, A.A.: Review of methods for solving Navier-Stokes-equations.
Proc. III. Int. Conf. on Num. Meth. in Fluid Mech., Springer-Verlag, 1973, pp. 1 – 11.

[28] Fasel, H.; Booz, O.: An Implicit Finite Difference Method for the Solution of the Navier-Stokes Equations for Unsteady Flows Between Rotating Coaxial Cylinders.
Proc. II. GAMM-Conf. on Num. Meth. in Fluid Mech., DFVLR, Köln, 1977, pp. 28 – 35.

[29] Fowlis, W.W.; Martin, P.-J.: A rotating laser Doppler velocimeter and some new results on the spin-up experiment.
Geophys. Fluid Dyn., 7, 1975, pp. 67 ...

[30] Gierasch, P.J.: Meridional Circulation and the Maintenance of the Venus Atmospheric Rotation.
J. Atmosph. Sci., 32, 1975, pp. 1038 – 1044.

[31] Gorin, A.V.; Shilyaev, M.I.: Laminar Flow Between Rotating Disks.
Fluid Dyn., 11, 1976, pp. 219 – 224.

[32] Görtler, H.: Einfluß der Bodentopographie auf Strömungen über der rotierenden Erde.
ZAMM, 21, 1941, pp. 279 – 303.

[33] Greenspan, D.: Numerical Studies of Steady, Viscous, Incompressible Flow Between Two Rotating Spheres.
Comp. and Fluids, 3, 1975, pp. 69 – 82.

[34] Greenspan, H.P.: The Theory of Rotating Fluids.
Cambridge University Press, 1968.

[35] Hall, P.: The stability of unsteady cylinder flow.
J. Fluid Mech., 67, 1975, pp. 29 – 63.

[36] Harada, I.; Ozaki, N.: A numerical study of the thermal spin-up of a stratified fluid in a rapidly rotating cylinder.
Proc. IV. Int. Conf. Num. Meth. Fluid Dyn., Colorado, 1974.
Springer-Verlag, 1975, pp. 197 – 203.

[37] Helmholtz, H.; Pitrowski, G.V.: Über Reibung tropfbarer Flüssigkeiten.
Wiener Berichte (2), 40, 1860, pp. 607 ff.

[38] Israeli, M.: Time Dependent Motions of Confined Rotating Fluids.
Ph. D. Thesis, Massachusetts Institute of Technology, 1971.

[39] Israeli, M.: A Fast Implicit Numerical Method for Time Dependent Viscous Flows.
Studies in Appl. Math., 49, 1970, pp. 327 – 349.

[40] Israeli, M.: On the Evaluation of Iteration Parameters for the Boundary Vorticity.
Studies in Appl. Math., 51, 1972, pp. 67 – 71.

[41] Israeli, M.: On Trapped Modes of Rotating Fluids in Spherical Shells.
Studies in Appl. Math., 51, 1972, pp. 219 – 237.

[42] Israeli, M.: Nonlinear motions of a confined rotating fluid.
Proc. IUTAM Quebec, 1971, pp. 1287 ...

[43] Israeli, M.; Orszag, S.A.: Numerical Investigation of Viscous Effects on Trapped Oscillations in a Rotating Fluid.
Proc. V. Int. Conf. on Num. Meth. in Fluid Dyn., Enschede, 1976.
Springer-Verlag, 1976, pp. 241–246.

[44] Janjic, Z.; Wiin-Nielsen, A.: On Geostrophic Adjustment and Numerical Procedures in a Rotating Fluid.
J. Atmos. Sc., 34, 1977, pp. 297–310.

[45] Kalnay, De Rivas, E.: Numerical Models of the Circulation of the Atmosphere of Venus.
J. Atm. Sc., 30, 1973, pp. 763–779.

[46] Kasagi, N.; Hirata, M.: Stability of Time-Dependent Flow around a Rotating Cylinder.
1975 Joint JSME-ASME Appl. Mech. West. Conf., Honolulu, Hawaii, 1975, pp. 431–438.

[47] Khlebutin, G.N.: Stability of Fluid Motion Between a Rotating and a Stationary Concentric Sphere.
Izv. AN SSSR Mekh. Zhidk. i Gaza, 3, 1968, pp. 53–56.

[48] Kirchner, R.P.; Chen, C.F.: Stability of time-dependent rotational Couette flow.
Part 1. Experimental investigation.
J. Fluid Mech., 40, 1970, pp. 39–48.

[49] Kopachevskii, N.D.; Radyakin, N.K.: Free vibrations of two capillary liquids rotating in a cylindrical vessel.
Izv. AN SSSR Mekh. Zhidk. i Gaza, 11, 1976, pp. 730–735.

[50] Krause, E.; Bartels, F.: Rotationssymmetrische Strömungen im Kugelspalt.
In: Recent Developments in Theoretical and Experimental Fluid Mechanics – Compressible and Incompressible Flows (ed.: Müller, U.; Roesner, K.G.; Schmidt, B.).
Springer-Verlag, 1978.

[51] Kubicek, M.; Holodniok, M.; Hlavacek, V.: Calculation of flow between two rotating coaxial disks by differentiation with respect to an actual parameter.
Comp. and Fluids, 4, 1976, pp. 59–64.

[52] Ladyzhenskaya, O.A.: Mathematical analysis of Navier-Stokes equations for incompressible liquids.
Ann. Rev. Fluid Mech., 1975, pp. 249–272.

[53] Ladyzhenskaya, O.A.: Some recent mathematical results concerning the Navier-Stokes equations.
Arch. Mech., Archivum Mech. Stos., 30, 1977, pp. 217–224.

[54] Lahargue, J.P.; Soubbaramayer: Application of Numerical Methods to the Investigation of the Flow Field in a Rapidly Rotating Gas Centrifuge.
Proc. II. GAMM Conf. on Num. Meth. in Fluid Mech., DFVLR, Köln, 1977, pp. 89–96.

[55] Launder, B.E.; Ying, W.M.: Numerical Solutions of Flow Between Rotating Cylinders.
(Convective Transport Terms Effect on Laminar Flowfield of Newtonian Fluid Between Rotating Cylinders, Using Adapted Finite Difference Solution Technique)
J. Mech. Eng. Sc., 14, 1972, pp. 400–403.

[56] Leonov, A.I.; Lipkina, E.Kh.; Paskhin, E.D.; Prokunin, A.N.: Theoretical and Experimental Investigation of the Flow of a Viscoelastic Fluid in the Gap Between two Rotating Disks.
Fluid Dyn., 11, 1976, pp. 196–201.

[57] Lugt, H.J.; Haussling, H.J.: The Acceleration of Thin Cylindrical Bodies in a Viscous Fluid.
J. Appl. Mech., 100, 1978, pp. 1–6.

[58] Matsuda, T.; Hashimoto, K.: Thermally, mechanically or externally driven flows in a gas centrifuge with insulated horizontal end plates.
J. Fluid Mech., 78, 1976, pp. 337–354.

[59] Menguturk, M.: Flow in a spherical annulus.
Ph. D. Dissertation, Duke University, Durham, N.C.

[60] Meyer, K.A.: Three-Dimensional Study of Flow Between Concentric Rotating Cylinders.
High-Speed Comp. in Fluid Dyn., Phys. Fluids Suppl. II, 1969, pp. 165–170.

[61] Meyer, K.A.: Time-Dependent Numerical Study of Taylor Vortex Flow.
Phys. Fluids, 10, 1967, pp. 1874–1879.

[62] Morel, P.: Atmospheric Dynamics and the Numerical Simulation of Atmospheric Circulation.
Proc. III. Int. Conf. on Num. Meth. in Fluid Mechanics, I, Springer-Verlag, 1973, pp. 12–71.

[63] Munson, B.: Viscous incompressible flow between eccentric coaxially rotating spheres.
Phys. Fluids, 17, 1974, pp. 528–531.

[64] Munson, B.R.; Joseph, D.D.: Viscous incompressible flow between concentric rotating spheres.
Part 1. Basic flow.
J. Fluid Mech., 49, 1971, pp. 289–303.

[65] Munson, B.R.; Joseph, D.D.: Viscous incompressible flow between concentric rotating spheres.
Part 2. Hydrodynamic stability.
J. Fluid Mech., 49, 1971, pp. 305–318.

[66] Munson, B.; Menguturk, M.: Viscous incompressible flow between concentric rotating spheres.
Part 3. Linear stability.
J. Fluid Mech., 69, 1975, pp. 705–719.

[67] Nguyen, N.D.; Ribault, J.P.; Florent, P.: Multiple Solutions for Flow Between Coaxial Disks.
J. Fluid Mech., 68, 1975, pp. 369–388.

[68] Nikolskii, A.A.: Hydrodynamik der rotierenden Flüssigkeiten.
Proc. Int. Council of the Aeron. Sc., 2nd Congress, Zürich, Pergamon Press, New York 1962, pp. 323–333.

[69] Orszag, S.A.; Israeli, M.: Numerical Simulation of Viscous Incompressible Flows.
Ann. Rev. Fluid Mech., 6, 1974, pp. 281–318.

[70] Orszag, S.A.: Numerical Simulation of Incompressible Flows Within Simple Boundaries.
I. Galerkin (Spectral) Representations.
Stud. in Appl. Math., 50, 1971, pp. 293–327.

[71] Orszag, S.A.; Israeli, M.: Numerical Flow Simulation by Spectral Methods.
Num. Models of Ocean Circulation, ISBN 0-309-02225-8, Nat. Acad. of Sc., 1975, pp. 284–300.

[72] Pearson, C.E.: Numerical solutions for the time-dependent viscous flow between two rotating coaxial disks.
J. Fluid Mech., 21, 1965, pp. 623–633.

[73] Pearson, C.E.: A Computational Method for Viscous Flow Problem.
J. Fluid Mech., 21, 1965, pp. 611–622.

[74] Pearson, C.E.: A numerical study of the time-dependent viscous flow between two rotating spheres.
J. Fluid Mech., 28, 1967, pp. 323–336.

[75] Pfeiffer, F.: An approximate procedure for liquid-filled gyros.
Ing. Arch., 43, 1974, pp. 306–316.

[76] Pustovalov, V.N.; Sparrow, E.M.: Natural Convection in a Closed Rotating Cylinder.
Int. J. Heat and Mass Trans., 17, 1974, pp. 1623–1626.

[77] Quemard, C.; Cousteix, J.: Calculation Methods of Three-Dimensional Boundary Layers With and Without Rotation of the Walls.
ONERA, TP No. 1135, 1972, 13 p.

[78] Quon, C.: Axisymmetric States of an Internally Heated Rotating Annulus – Atmospheric Dynamics.
Tellus, 29, 1977, pp. 83–96.

[79] Quon, C.: A mixed spectral and finite difference model to study baroclinic annulus waves.
J. Comp. Physics, 20, 1976, pp. 442–479.

[80] Raal, J.D.: Radial source flow between parallel disks.
J. Fluid Mech., 85, 1978, pp. 401–416.

[81] Ritter, C.-F.: Berechnung der zähen, inkompressiblen Strömung im Spalt zwischen zwei konzentrischen, rotierenden Kugelflächen.
Dissertation, Universität Karlsruhe, 1973.

[82] Roache, P.J.: Computational Fluid Dynamics.
Hermosa Publishers, Albuquerque, N. M. 87108, 1972.

[83] Sadournay, R.: Compressible model flows on the sphere – Potential enstrophy conservation effect on Rossby wave stability.
J. Atm. Sc., 32, 1975, pp. 2103–2110.

[84] Sawatzki, O.; Zierep, J.: Das Stromfeld im Spalt zwischen zwei konzentrischen Kugelfächen, von denen die innere rotiert.
Acta Mech., 9, 1970, pp. 13–35.

[85] Schumann, U. (ed.): Computers, fast elliptic solvers and applications.
Proc. GAMM-Workshop on Fast Solution Methods for the Discretized Poisson Equation, Karlsruhe, 1977,
Advance Publ. Ltd. London, 1978.

[86] Shestakov, A.I.: Numerical Solution of the Navier-Stokes equations at high Reynolds numbers.
Proc. III. Int. Conf. on Num. Meth. in Fluid Mechanics,
Lecture Notes in Physics, Springer-Verlag, 1976, pp. 404–409.

[87] Siegmann, W.L.: Evolution of unstable shear layers in a rotating fluid.
J. Fluid Mech., 64, 1974, pp. 289–305.

[88] Smith, S.H.: The non-linear split-disc problem – Shear layer with Ekman condition in rotating fluids.
Quart. J. Mech. Appl. Math., 29, 1976, pp. 399–414.

[89] Sorokin, V.S.: Nonlinear phenomena in closed flows near critical Reynolds numbers.
PMM, 25, 1961, pp. 248–258.

[90] Sorokin, M.P.; Khlebutin, G.N.; Shaidurov, G.F.: On the stability of fluid motion between two rotating spherical surfaces.
PMTF (J. Appl. Mech. Tech. Phys.), 6, 1966, pp. 73 ...

[91] Soubbaramayer; Lahargue, J.P.: CENTAURE, A numerical model for the computation of the flow and isotopic concentration fields in a gas centrifuge.
Proc. II. Workshop on Gases in Strong Rotation, Cadarache, France, 1977.

[92] Stewartson, K.; Roberts, P.H.: On the motion of a liquid in a spheroidal cavity of a precessing rigid body.
J. Fluid Mech., 17, 1963, pp. 1–20.

[93] St-Maurice, J.P.; Veronis, G.: A multiscaling analysis of the spin-up problem.
J. Fluid Mech., 68, 1975, pp. 417–445.

[94] Turikov, V.G.; Chalikov, D.V.: Computation of the general circulation of the Venusian atmosphere. (Numerical modeling of Venus atmospheric circulation, taking into account short wave radiation absorption, boundary layers, mesoscale convection and horizontal friction)
Izv. Fiz. Atm. i Okeana, 7, 1971, pp. 705–772.

[95] Urintsev, A.L.: Numerical analysis of the branching of COUETTE flow between concentric cylinders at different wave numbers.
Fluid Dyn., 11, 1976, pp. 455–457.

[96] Vreeburg, J.P.B.: Contained-Liquid Experiments on Spacelab I.
Proc. CNES-ESA Conf. on Attitude Control of Space Vehicles – Technological and dynamical problems associated with the presence of liquids, Toulouse, 1977, ESA SP-129, pp. 31–37.

[97] Warn-Varnas, A.; Fowlis, W.W.; Piacsek, St.; Lee, S.M.: Numerical solutions and laser-Doppler measurements of spin-up.
J. Fluid Mech., 85, 1978, pp. 609–639.

[98] Weidmann, P.D.: On the spin-up and spin-down of a rotating fluid.
Part 1. Extending the Wedemeyer model, Part 2. Measurements and Stability.
J. Fluid Mech., 77, 1976, pp. 685–735.

[99] Weir, A.D.: Axisymmetric convection in a rotating sphere.
J. Fluid Mech., 75, 1976, pp. 49–79.

[100] Wilhelmson, R.B.; Ericksen, J.H.: Direct Solution for Poisson's Equation in Three Dimensions.
J. Comp. Phys., 25, 1977, pp. 319–331.

[101] Williams, R.T.; Kurth, R.P.: Formation of Discontinuities in a Stratified Rotating Atmosphere.
J. Geophys. Res., 81, 1976, pp. 1133–1140.

[102] Wilson, L.O.; Schryer, N.L.: Flow between a stationary and a rotating disk with suction. J. Fluid Mech., 85, 1978, pp. 479–496.

[103] Wimmer, M.: Experiments on a viscous fluid flow between concentric rotating spheres. J. Fluid Mech., 78, 1976, pp. 317–335.

[104] Wimmer, M.: Experimentelle Untersuchungen der Strömung im Spalt zwischen zwei konzentrischen Kugeln, die beide um einen gemeinsamen Durchmesser rotieren. Dissertation, Universität Karlsruhe, 1974.

[105] Wu, J.C.: Numerical Boundary Conditions for Viscous Flow Problems. AIAA-Journ., 14, 1976, pp. 1042–1049.

[106] Yavorskaya, I.M.; Simuni, L.M.: A possible interpretation of the mechanism of equatorial jet formation on the Jovian surface. (Russian) Akad. Nauk SSSR, Dokl., 233, 1977, pp. 60–63.

[107] Yavorskaya, I.M.; Belaev, Yu,N.; Monachov, A.A.: Investigation of the stability and secondary flow in rotating spherical layers at arbitrary Rossby-numbers. Akad. Nauk SSSR, Dokl., 237, 1977, pp. 804–807.

[108] Yakushin, V.I.: Motion of a fluid between two rotating concentric spheres. Izv. AN SSSR, Mekh. Zhidk. i Gaza, 3, 1968, pp. 59–61.

ON NUMERICAL SIMULATION IN FLUID DYNAMICS

A.A. Samarskii
Institute of Applied Mathematics,
Academy of Sciences,
Moscow, USSR

This paper discusses some questions of numerical simulation of experiments in physics and fluid dynamics problems.

I. Recently the problems in magnetohydrodynamics, radiative fluid dynamics, plasma flow in strong electric fields, etc. have become of considerable interest along with the classical issues of continuum mechanics, i.e. the problems of the elasticity theory, gas dynamics or flow around a body.

The formulation of many such problems that demands intensive applying the numerical methods has originated with plasma physics. As a rule, the mathematical models in plasma physics are nonlinear. In general, two limiting cases most widely used can be emphasized:

a) a model of dense plasma - equations of radiative magneto-gas-dynamics (RMGD);

b) a model of collisionless plasma in the Vlasov equation approximation.

At present the problem of controlled thermonuclear fusion (CTR) should be considered as most urgent in plasma physics. A number of approaches to solving this problem has been proposed (stationary Tokamak-type installations, fast plasma heating and compressing in inertial pulse systems by a laser pulse or relativistic electron beam, etc.).

The CTR problem is closely associated with problems of utilizing the thermonuclear energy released, so the problem of constructing an effective reactor is in prospect.

2. Currently a numerical experiment (NE) is the basis for comprehensive theoretical investigations of problems in fluid mechanics [I] . The NE enables us not only to explain some known experimental facts or confirm the theoretical concepts but, in some cases, to predict new physical effects. A new physical phenomenon, the T-layer effect, discovered through the NE may be an example [2] . The main point is that under certain conditions a self-maintained,

high-temperature, electro-conductive gas layer forms and develops during the magneto-hydrodynamic plasma motion. The conditions of the T-layer formation predicted by theoretical investigation permitted to reveal the effect later in laboratory experiments. The NE may be considered as a sequence of a number of the following stages:

I) choice of a physical approach and mathematical formulation of a problem (choice of a mathematical model); as a rule the equations describing the mathematical model express the conservation laws (for mass, energy, momentum, charge, etc.) and are the partial differential equations of mathematical physics;

2) development of a computational algorithm;

3) algorithm programming;

4) computing;

5) analysis of computational results; comparison with experimental and theoretical results; revision and correction of the mathematical model; improvement of the calculation technique.

The typical features of NE are as follows:

I) Within the mathematical model chosen, a number of runs (not one) is computed in the required interval of the parameters involved.

2) The mathematical model may be repeatedly varied.

In fact, one may speak of a new approach to performing the theoretical investigations on the basis of NE, which provides a proper connection between the mathematical model and a physical experiment through computing.

3) The multi-variant nature of computations in NE imposes strict requirements on the computational algorithms and corresponding software as well. While the same problem is being solved, the form of non-linearity or equations type may vary, discontinuities may arise, interact and dissipate, domain geometry and topology may change, and so on.

3. The numerical methods should have a sufficient resolution, i.e. accuracy for an admissible amount of computations, to describe correctly the main characteristics of complex, nonlinear processes.

In discretizing the continuum problems, i.e. changing differential equations by the difference ones, the natural requirement is that the discrete model obtained should reflect the basic properties of continuum correctly. In particular, the conservation of mass, momentum and total energy is such a property (these laws hold on the grid for conservative difference schemes [I], [3]) as well as the balance equations for internal and kinetic energies, electromagnetic field energy, etc. The difference schemes with such properties we shall call the completely conservative schemes (CCS) [4] . CCS have proved to be highly effective and permitted to obtain the solutions sufficiently accurate for the MHD - and RMGD-problems in both cases of low and high temperature plasmas.

A heuristic approach to obtaining CCS, proposed by Popov and Samarskii (1969), allowed to construct CCS for other problems too, e.g. the Landau kinetic equation.

4. At present an integro-interpolation method, i.e. a balance method [3] , as well as projection and variational methods are used for obtaining CCS of the desired quality for the classical equations of mathematical physics.

In case of arbitrary dimension the variational method has proved to be effective for obtaining CCS for the MHD-problems [5] . The hydrodynamic equations result from a variational principle similar to that of least action in classical mechanics. For example, let us consider the liquid volume Ω of an ideally conductive adiabatic plasma, moving in the $x-y$ plane. In Lagrangian coordinates (α, β) the volume Ω corresponds to the domain $G(\alpha, \beta)$. The Lagrangian $L(t)$ of the volume Ω is taken as the integral in $G(\alpha, \beta)$ of the expression proportional to $(u^2+v^2)/2 - \varepsilon - |H|^2/8\pi\rho$,where ε is the internal energy, $(u^2+v^2)/2$ is the kinetic energy, $|H|^2/8\pi\rho$ is the magnetic field energy per unit mass, ρ is the density. The integral of action $S=\int_{t_0}^{t_1} L(t)\,dt$ is varied with taking into account the additional constraints, i.e. equations of continuity and freezing-in of magnetic field. By setting the first variation δS equal to zero we obtain the Euler equation. Note that the energy and momentum conservation results from the absence of an explicit dependence of the Lagrangian and constraint equations upon time and coordinates.

The CCS are obtained in a similar way. The Lagrangian and constraint equations are approximated on the grid $\omega_h\{\alpha_i, \beta_j\}$. Varying the integral of action with the constraint equations taken into account gives us a system of differential-difference equations. After replacing the derivatives in t by the difference relations and introducing an artificial dissipation we obtain the homogeneous CCS of the first- and second-order approximation in t (the schemes of run-through computations). The above approach permitted to obtain CCS for the MHD-equations in arbitrary cirvilinear coordinates with heat conductivity, magnetic field diffusion and other factors involved, and to apply them to solving a number of physical problems [6] ,

5. Taking into account the radiation transfer, i.e. the transformation of MHD-problems into the RMGD ones, requires additional efforts for constructing and implementing the CCS.

Let us consider a technique proposed by Goldin and Chetverushkin [8] for solving the non-stationary problems (including two-dimensional ones) of radiative gas dynamics. The radiation transfer is considered in the multi-group diffusion approximation.

Solving the set of corresponding RMGD-equations is performed in three steps: I) solving the proper gasdynamic equations, 2) solving the transport and diffusion equations, and 3) obtaining the temperature from the combined solution of the energy and diffusion equations.

We now dwell on difficulties arising when the radiation is involved, i.e. the second and third steps. Due to the complex structure of absorption coefficients the general number of diffision equations is rather great. Solving the non-linear difference diffusion equations on each step in time is quite a problem that requires applying the specific iterational methods. In a number of cases the most simple explicit scheme for solving the combined diffusion and energy equations requires a very small time step to provide the stability of computations. While the technique having been developed the difficulties were overcome. The diffusion approximation used in the problems of radiative gas dynamics describes correctly the radiative contribution in the energy equations. In the numerical implementation the direct use of transport equation for determining the radiation field may cause an undesirable effect, "the beam effect" (it arises in case when the zone generating, in general, the radiation is small as compared with the whole region under investigation).

Recently a technique has been developed to compute the radiation transfer equations with the use of Vladimirov's self-adjoint equations and the above effect eliminated.

6. The verified methods for solving the gasdynamics problems being available, the hydrodynamic approximation can be used to solve other physical problems [9] . In nonlinear optics and plasma physics, of great interest are the problems on wave field dynamics, energy localization in the vicinity of a certain point (for example, light self-focusing in a nonlinear medium, collapse of Langmuir waves). For an amplitude of electric field in plasma, the nonlinear Schrödinger-type equation

$$2i\frac{\partial E}{\partial t}+\frac{1}{r^\nu}\frac{\partial}{\partial r}\left(r^\nu\frac{\partial E}{\partial r}\right)+|E|^2E=0$$

$$0<r<R\ ,\quad t>0\ ,\quad \nu=0,1$$

is reduced to the "gasdynamic" equations

$$\frac{\partial\rho}{\partial t}+\frac{1}{r^\nu}\frac{\partial}{\partial r}(r^\nu\rho v)=0,\quad \frac{\partial v}{\partial t}+v\frac{\partial v}{\partial r}+\frac{1}{\rho}\frac{\partial P}{\partial r}+g=0$$

where $\rho=|E|^2$ has the sense of density, and the "state" equation has the form

$$P=-\frac{1}{4}\rho^2\left[\frac{1}{\rho r^\nu}\frac{\partial}{\partial r}\left(\frac{r^\nu}{\rho}\frac{\partial\rho}{\partial r}+1\right)\right],$$

$$E=\rho^{1/2}e^{i\psi},\quad v=\frac{\partial\psi}{\partial r},\quad g=g\left(\rho,\frac{\partial\rho}{\partial r}\right).$$

By introducing the energy Lagrangian coordinate $\ell=\int_0^r\rho r^\nu dr$ the set of equations obtained may be solved using the known schemes on the Lagrangian grid.

The implementation of the above approach made it possible to investigate the dynamics of self-focusing in a cubic medium as well as successfully construct the computational algorithms for Langmuir turbulence.

7. Let us discuss particularly such important questions as the "physical attributes" of a numerical experiment, the thermodynamic properties of a medium.

The RMGD-equations describe some macroscopic phenomena in a moving medium. In fact, primary are the microscopic motions of particles (electrons, ions, atoms and photons) in the field created by the external sources and other particles. Averaging the microscopic motions yields the coefficients of RMGD-equations, i.e. thermodynamic functions, electric conductivity and other properties of the matter. It is evident that the best gasdynamic calculation procedures do not give the correct solution to a physical problem if they contain incorrect properties of the matter. Finding the properties of the matter is a complex individual problem.

In solving the RMGD-problems one has to know the properties of various substances: hydrogen, inert gases, water and metal vapors, products of coal and oil combustion, etc. In the processes of interest for physicists, most various conditions develop, temperatures from the room to stellar ones and densities from those of gases to those of solids. Most of these conditions are so different from the room ones that the direct experimental determination of the matter properties seems to be impossible. On the other hand, the simplified models similar to that for an ideal gas, which are commonly treated in the theoretical physics, often cannot be used under those conditions. Therefore, we have to develop new physical approximations to determine the matter properties of interest.

To provide this part of our gasdynamic computations it was necessary to construct the mathematical description of continuum (in this connection we had to solve a number of most complex problems in quantum mechanics) and to develop the proper computational algorithms. At present quantum, exchange and oscillation (envelope) corrections to the Tomas-Fermi model have been obtained at arbitrary temperatures for chemical elements and arbitrary compounds. The computational algorithms have been developed and calculations of the above corrections performed. This work is carried out under the direction of N.N. Kalitkin.

To obtain the coefficients of electrical conductivity and electron heat conductivity a semi-classical model of electron transfer was constructed, involving the most essential quantum effects [IO] . Now the methods have been developed (under the direction of A.Ph. Nikifforov and V.V. Uvarov) to calculate the photon free paths in a high-temperature plasma of any complex composition with not only the inverse bremsstrahlung and photoeffect taken into account but the line absorption as well. One can conceive a complexity of this problem by noting that solving the Hartry-Fok equation for multi-electron atoms is a part of it.

The tables for the equations of states, conductivity, electron heat conductivity and photon (and other particles) free paths being available, numerical computations are in good quantitative agreement with physical experiments in many complex problems of plasma physics.

8. The CCS developed for the MHD- and RMGD-problems permitted to carry out NEs for investigating a number of problems in plasma physics and radiative gas dynamics. We restrict ourselves by listing some problems with brief comments supplied.

I) The investigation of pulsed radiating discharges in inert gases [II] .

The RMGD-model coupled with an electric circuit was chosen. The accurate treatment of the physical properties of substances involved and the proper description of the electric circuit with the photon transport taken into account allow to solve the $\mathfrak{Z}$ -pinch and inverse $\mathfrak{Z}$ -pinch problems with detailed quantitative agreement between the experimental and computed values of total current, Joule heat released, location of shock and boundaries of emitting regions in discharge etc. The spectral effects were found to be of major importance in heat transfer, and the luminosity boundary observed corresponds to the ionization front, and so on.

2) The T-layer stability. The T-layer was shown to be stable with respect to two-dimensional disturbances. The investigation was performed by means of CCS for the MHD-problems [I2] .

3) The study of compressing and heating a thermonuclear target by a laser pulse [I3] . The stability of compression was investigated with respect to the disturbances of initial target form and laser pulse [7] .

4) The investigation of the Rayleigh-Taylor instability.

5) The problems of laser "pumping" and searching for periodic solutions I4 .

Here we speak mainly of two-dimensional problems of gas dynamics and MHD. Among the people engaged in gas- and aerodynamics the complexity of a problem is usually defined by the number of dimensions.

However in the plasma physics problems, such as compressing a thermonuclear target by laser pulse or electron beam and burning in a target, the basic difficulties are in comprising various physical effects. Since the question of mathematical model choice is not clear it is necessary to consider different approaches to describing the same physical effect which implies comparing a few mathematical models. In conducting the numerical experiment we stick to the following principle:

a) taking into account all physical effects in "one-dimensional" approximation (e.g., under the assumption of spherical or cylindrical symmetry);

b) considering the influence of two- or three-dimension on the physical process under investigation.

In some cases it is expedient to carry out an one-dimensional physical experiment to compare it with an one-dimensional NE.

R E F E R E N C E S

I. Samarskii, A.A. The theory of finite-difference schemes, M., "Nauka", I977.

2. Tihonov, A.N., Samarskii, A.A., Zaklyazminskii, A.A., Volosevich, P.P., Degtyaryov, L.M., Kurdyumov, S.P., Popov, Yu.P., Sokolov, V.S., Favorskii, A.P. Dokl.Akad.Nauk SSSR, I967, I973, N 4.

3. Tihonov, A.N., Samarskii, A.A. Zh.Vychisl. Mat. i Mat.Fiz., I96I, I, N I, 6-64.

4. Samarskii, A.A., Popov, Yu.N. The finite-difference schemes of gas dynamics, M., "Nauka", I976.

5. Goloviznin, V.M., Samarskii, A.A., Favorskii, A.P. Dokl. Acad. Nauk SSSR, I977, 235, N 6, I285-I288.

6. Volkova, R.A., Goloviznin, V.M., Ulinich, F.R., Favorskii, A.P. Preprint IPM Akad. Nauk SSSR, I976, N III.

7. Volosevich, P.P., Gamalii, Ye.G., Gulin, A.V., Rusanov, V.B., Samarskii, A.A., Tyurina, N.N., Favorskii, A.P., Pis'ma v JETP, 24, Vyp. 5, 285-286.

8. Goldin, V.Ya., Chetverushkin, B.N. On numerical simulation of the problems of radiative gas dynamics, M., Preprint IPM Akad. Nauk SSSR, N I7 (I973).

9. Dyegtyarev, L.M., Krylov, V.V. Zh.Vychisl. Mat. i Mat.Fiz., I977, I7, N 4.

I0. Kalitkin, N.N., Kuzmina L.V., Rogov, V.S. Tables of thermodynamic functions and transport coefficients of plasma, IPM Akad. Nauk SSSR, I972.

II. Andreev, S.I., Goldin, V.Ya., Goldina, D.A., Zobov, E.A., Kalitkin, N.N., Samarskii, A.A., Sokolov, V.G., Chetverushkin, B.N. Pulse radiative discharges in inert gases, Dokl. Akad. Nauk SSSR, I976, v.226, N 6, I045-I048.

I2. Volkova, R.A., Gulin, A.V., Sosnin, N.V., Favorskii, A.P., Numerical investigation of the T-layer structure.

I3. Volosevitch P.P., Degtyarev, L.M., Levanov, E.I., Kurdjumov, S.P., Popov, Yu.P., Samarskii, A.A., Favorskii, A.P., Process of super-high compression of the matter and ignition of thermonuclear fusion by a powerful laser pulse, Fizika plasmy, I976, v.2, vyn. 2, 883-897.

I4. Velikhov, E.P., Mikhailova, N.V., et al., Dokl. Akad. Nauk SSSR, I975, v.228, I300.

SOME FINITE ELEMENT METHODS IN FLUID FLOW

R. TEMAM

Laboratoire d'Analyse Numérique

Université de Paris-Sud

91405 - ORSAY, France

INTRODUCTION

We describe in this paper some finite element methods which has been recently introduceed for the computation of viscous and inviscid incompressible flows. The reader is assumed to be somehow familiar with the general theory of finite element methods and is referred to the basic references on the subject for the fundamental aspects of the theory.

Section 1 is devoted to the description of some finite elements methods of the displacement type which apply to viscous flows, with emphasize on a linear non conforming finite element. We start by recalling the principle of the weak (or variational) formulation for the Stokes and Navier-Stokes equations, and their Galerkin-finite element discretization, and then we describe some appropriate finite elements which approximate the space of divergence free vector functions.

In the case of fluid flow problems, all the numerical difficulties are not automatically overcomed by introducing a finite element discretization, and in some cases some further procedures may be desirable. Precisely Section 2 is devoted to numerical procedures which are related to the effective treatment of the constraint div u = 0 , in relation with the solution of the discrete problems arising in Section 1. On this occasion we consider a penalty method, an extrapolation algorithm, some iteration algorithms based on optimization procedures, and the augmented lagrangian method which combines the different features. Combining the technics introduceed in the first two Sections, some problems corresponding to moderate Reynolds numbers has been treated in references which are indicated in the text.

In Section 3 we introduce an interesting finite element applying to the Euler equations, which somehow, takes into account the hyperbolic character of the equations, and allows tangential discontinuities.

Plan

1. Discretization by the Finite Element Method

1.1. Weak (or Variational) Formulation of the Stokes and Navier-Stokes Equations.

Let Ω be an open bounded domain in the plane R^2 and let us denote by Γ its boundary. Steady incompressible flows in Ω are governed by the Navier-Stokes equations which we write in the nondimensional form

(1.1) $$-\frac{1}{Re}\Delta u + (u.\nabla)u + \nabla p = f \quad \text{in} \quad \Omega ,$$

(1.2) $$\nabla u = 0 \quad \text{in} \quad \Omega ,$$

(1.3) $$u = g \quad \text{on} \quad \Gamma$$

where $u = \{u_1,u_2\}$ is the velocity vector and p is the pressure.

For the sake of simplicity we will sometime assume that $g = 0$ (which is never an essential simplification) or we will consider as a model the Stokes equations obtained by droping the nonlinear term $(u.\nabla)u$ in (1.1).

In order to get the weak formulation of this problem, we consider a vector test function v which vanishes on Γ and is divergence free ; we multiply (1.1) by v, integrate over Ω, integrate by part using Green's formula and we obtain

$$-\frac{1}{Re}\int_\Omega \Delta u.v\, dx = -\frac{1}{Re}\sum_{i=1}^{2}\int_\Gamma \frac{\partial u_i}{\partial \nu} v_i\, d\ell + \frac{1}{Re}\int_\Omega (\nabla u_1.\nabla v_1 + \nabla u_2.\nabla v_2)\, dx ,$$

$$\int_\Omega \nabla p.v\, dx = \int_\Gamma pv.\nu\, d\ell - \int_\Omega p(\nabla v)\, dx = 0$$

($dx = dx_1\, dx_2$) so that

$$\frac{1}{Re} a(u,v) + b(u,u,v) = (f,v) \tag{1.4}$$

where

$$a(u,v) = \int_\Omega (\nabla u_1.\nabla v_1 + \nabla u_2.\nabla v_2)\, dx , \tag{1.5}$$

$$b(u,v,w) = \int_\Omega [(u.\nabla)v]w\, dx , \tag{1.6}$$

$$(f,v) = \int_\Omega fv\, dx . \tag{1.7}$$

The relation (1.4) will hold for each v such that

$$\nabla.v = 0 \text{ in } \Omega , \quad v = 0 \text{ on } \Gamma . \tag{1.8}$$

Conversely it can be proved that if $\nabla u = 0$, $u = g$ on Γ and (1.4) holds for all such v's, then there exists p such that (1.1) is satisfied. The relation (1.4) which characterizes u, represents merely the virtual work principle, v standing for some arbitrary increment δu of u.

The problem of finding u which satisfies (1.2)-(1.3) and (1.4) for each v of type (1.8) is called <u>the weak (or variational) formulation of problem</u> (1.1)-(1.3).

<u>Remark 1.1</u>. The weak formulation of the equations only involves u. The pressure p will then be found a posteriori, once the function u, solution of (1.4),

is known.

Remark 1.2. In the nonlinear case, the relation (1.4) does not correspond to any minimization or any variational property of the function u . The terminology "variational formulation" is however admitted by analogy with similar situations in elasticity or the calculus of variations.

In the linear case (i.e. when $(u.\nabla)u$ in (1.1) and $b(u,u,v)$ in (1.4) are deleted), the relation (1.4) characterizes u as the minimizing function of the expression

$$E(w) = \frac{1}{2Re} a(w,w) - (f,w) \tag{1.9}$$

among all the functions w which satisfy (1.2)-(1.3). This property will be useful.

1.2. Galerkin and Finite Element Methods

Let us denote by $\mathbb{V}(g)$ the space of vector functions u which satisfy (1.2)-(1.3).

A Galerkin approximation of Problem (1.4) consists in choosing a subspace $\mathbb{V}_h(g)$ of finite dimension m , and looking for u_h in $\mathbb{V}_h(g)$ which satisfies (1.4) for each $v \in \mathbb{V}_h(g)$. The problem then reduces to a set of quadratic equations for the components of u_h in a basis of $\mathbb{V}_h(g)$.

The conforming finite element methods consist in particular choices of the subspaces $\mathbb{V}_h(g)$, for which these equations are simplified (sparse matrices for their linearized form). More generally in the non conforming finite element methods, we introduce a similar problem with a finite dimensional space $\mathbb{V}_h(g)$ which is not included in $\mathbb{V}(g)$ but only "close", at the limit, to such a space.

The domain Ω is assumed to be covered by a family $\mathcal{T}_h$ of triangles K ; the following conditions are required :

$$\Omega_h = \bigcup_{K \in \mathcal{T}_h} K \subset \Omega . \tag{1.10}$$

(1.11) If K and $K' \in \mathcal{T}_h$ then either $K = K'$, or $K \cap K'$ is empty or $K \cap K'$ is exactly a vertice or a complete edge of K and K' .

We will use the following notations

$$\rho_h = \sup_{K \in \mathcal{T}_h} \rho(K) , \qquad \rho'_h = \inf_{K \in \mathcal{T}_h} \rho'(K)$$

$$\theta_h = \inf_{K \in \mathcal{T}_h} \theta(K)$$

where $\rho(K)$ is the diameter of K , $\rho'(K)$ the diameter of the greatest ball included in K and $\theta(K)$ is the smallest angle in the triangle K .

As usual we will consider mesh refinments corresponding to $\rho_h \rightarrow 0$ with

$$\theta_h \geqslant \theta_0 > 0 .$$

1.3. Linear Non conforming Elements

For simplicity we set $g = 0$.

We introduce the set $\mathbb{W}_h(0)$ of functions v_h of the following type :

- v_h is linear on each triangle ;
- the nodal values of v_h are the values of v_h at the mid-edges of the triangles $K \in \mathcal{T}_h$;
- $v_h(M) = 0$ at the boundary nodes (i.e. the boundary mid-sides) and $v_h(M) =$ an arbitrary vector, for each interior node (= interior mid-side).

Note that v_h is generally discontinuous along each side.

In order to take into account the constraint $\nabla . v = 0$, we also introduce the following subspace $\mathbb{V}_h(0)$ of $\mathbb{W}_h(0)$: $v_h \in \mathbb{V}_h(0)$ if $v_h \in \mathbb{W}_h(0)$ and $\nabla . v_h$ (which is piecewise constant) $= 0$ on each triangle.

The appropriate version of (1.4) (with $g = 0$) is :

To find $u_h \in \mathbb{V}_h(0)$ such that

(1.12) $$\frac{1}{Re} a_h(u_h,v_h) + b_h(u_h,u_h,v_h) = (f,v_h) \ , \ \forall\, v_h \in \mathbb{V}_h(0) \ ,$$

where

$$a_h(u_h,v_h) = \sum_K \int_K (\nabla u_{1h}\, \nabla v_{1h} + \nabla u_{2h}\, \nabla v_{2h})\, dx$$

$$b_h(u_h,v_h,w_h) = \sum_K \int_K [(u_h.\nabla)v_h].w_h \, dx \ .$$

The problem (1.12) is not easy to solve because of the constraints $\nabla.u_h = \nabla.v_h = 0$ on each K . We will describe in Section 2 appropriate algorithms for the solution of these discrete equations.

Remark 1.3. i) These elements had been introduceed and studied by Crouzeix-Raviart [2]. It is shown in [13] that in the linear case the L^2-norm of the error between u and u_h is of the order of ρ_h^2 for the functions themselves and of the order of ρ_h for their derivatives.

ii) A basis of $\mathbb{V}_h(0)$ may be useful for the solution of the Stokes problem, or in the nonlinear case for the solution of similar problems arising during an iterative procedure. This basis was explicited by Crouzeix and used by Thomasset (cf. [15]).

iii) These non conforming linear elements are the simplest elements which take into account the constraint $\nabla.u = 0$.

One can think of using simpler elements : the conforming linear elements, i.e. elements which are continuous, linear on each triangle (the nodal values are the values at the vertices), and which satisfy $\nabla.v_h$ (which is constant) $= 0$ on each K . Unfortunately it was

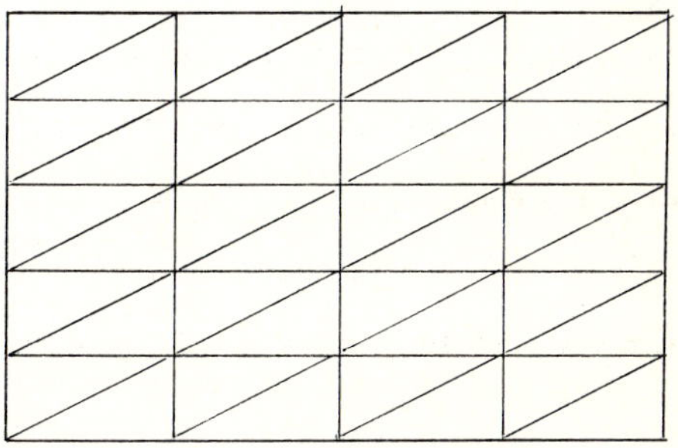

Fig.1

found by M. Fortin [4] that this impose too many linear constraints so that the space $\mathbb{V}_h(0)$ can be reduceed to $\{0\}$ (this is already the case for a rectangular domain with the triangulation shown in figure 1). To overcome this difficulty, a weaker form of the condition $\operatorname{div} v_h = 0$ has been proposed : average of $\operatorname{div} v_h = 0$, average taken on all the triangles around one vertice.

iv) Let us indicate an estimation of the dimension of the space $\mathbb{V}_h(0)$, when the number of triangles is sufficiently large.

Let T, E, Y denote the total number of triangles, edges and vertices in the triangulation. Then $T - E + Y$ is a constant depending only on Ω (= 1 if Ω is simply connected). If the number of triangles is sufficently large so that the number of boundary triangles is negligible, then $E \approx \frac{3T}{2}$ and $Y \approx \frac{T}{2}$. For linear non conforming elements, the dimension of $\mathbb{V}_h$ is the number of nodal values less the number of independent linear constraints : $\dim \mathbb{V}_h(0) \approx 2E - T \approx 2T \approx 4Y$.

1.4. Other Finite Elements

Let us mention some other elements which has been proposed.

The quadratic elements has been proposed by M. Fortin (cf. [4], [13]). The functions $v_h \in \mathbb{V}_h(0)$ are continuous and reduce one each triangle to a polynomial of degree 2 ; the nodal values are the values of v_h at the vertices and the mid edges (the nodal values are 0 on the boundary nodes). Furthermore $\operatorname{div} v_h = 0$ on each triangle.

In the linear case the error between a solution u of (1.4) and u_h of (1.12), is not optimal : the L^2 norm of the error of the derivatives is ρ_h and the L^2 norm of the function themselves is ρ_h^2 .

Crouzeix and Raviart [2] have proposed to add on each triangle a bulb function, which is a polynomial of degree 3 vanishing identically on the boundary of K ; in the linear case the error becomes optimal : ρ_h^2 for the L^2 norm of the error between the derivatives, ρ_h^3 for the L^2 norm error for the functions.

Finally M. Bercovier introduces in [1] a quadrangular finite element. The sets $K \in \mathcal{T}_h$, $K \subset \Omega$, are quadrangles and we assume for simplicity that they are rectangles.

On each rectangle v_h is a polynomial of degree ≤ 2 with respect to each variable x_1, x_2, and the nodal values are the $2 \times 8 = 16$ values of v_h at the vertices and mid-edges. The condition on the divergence is that $\mathrm{div}_h\, v_h = 0$ on each rectangle where

$$\int_K (\mathrm{div}_h\, v_h) r\, dx = \int_K (\mathrm{div}\, v_h) r\, dx$$

for each rectangle K and each polynomial function $r \in Q_1$ (polynomes of degree ≤ 1 with respect to each variable). It is shown (cf. [1]) that $\mathrm{div}_h\, v_h$ is defined by its value at the 4 points α_i, $1 \leq i \leq 4$, of Gauss integration on K and at these points we have precisely

$$\mathrm{div}_h\, v_h(\alpha_i) = \mathrm{div}\, v_h(\alpha_i)\ .$$

The conditions imposed on the v_h are then

$$\mathrm{div}\, v_h(\alpha_i) = 0\ , \quad i = 1,\ldots,4\ .$$

Also, when non divergence free vector functions have to be considered, we note that

$$\int_K (\mathrm{div}_h\, u)^2\, dx = \sum_{i=1}^{4} \omega_i\, (\mathrm{div}\, u(\alpha_i))^2$$

which means that the integral in the left hand side is exactly computed by a Gauss integration formula.

The L^2 norm error between the derivatives of u and u_h is of the order of ρ_h^2 (cf. [1]).

1.5. Approximation of the pressure

To make our notations systematic, we will denote by $\mathbb{W}_h(g)$ the spaces of finite elements previously introduceed (without any condition on $\mathrm{div}\, v_h$)

and by $\mathbb{V}_h(g)$ the corresponding space of functions satisfying in some sense $\text{div } v_h = 0$.

The approximate pressure appears as a Lagrange multiplier for the constraints "$\text{div } v_h = 0$" . By applying a Lagrange multiplier argument (cf. [13]) we find a function p_h belonging to some appropriate space X_h , and some kind of approximation $\text{div}_h\, v_h$ of $\text{div } v_h$, such that (compare to (1.12))

$$(1.13) \qquad \frac{1}{Re}\, a(u_h,v_h) + b_h(u_h,u_h,v_h) - (p_h,\text{div}_h\, v_h) = (f,v_h) \ , \forall v_h \in \mathbb{W}_h(0) \ .$$

For the linear non conforming elements, X_h is made of functions which are constant on each triangle K and $\text{div}_h\, v_h = \text{div } v_h$ on each K (= constant).

1.6. Time-dependent problems

We may also consider the evolution Navier-Stokes equations

$$(1.14) \qquad \frac{\partial u}{\partial t} - \frac{1}{Re}\,\Delta u + (u.\nabla)u + \nabla p = f \quad \text{in} \quad \Omega \times (0,T) \ ,$$

$$(1.15) \qquad \nabla u = 0 \quad \text{in} \quad \Omega \times (0,T) \ ,$$

$$(1.16) \qquad u = g \quad \text{on} \quad \Gamma \times (0,T) \ ,$$

$$(1.17) \qquad u(x,0) = u_0(x) \quad \text{in} \quad \Omega \ .$$

We consider a semi-discretization in time of the problem (for instance a Cranck-Nickolson or a fully implicit or a multistep scheme). At each step we obtain a problem similar to (1.1)-(1.4) and the same procedure is applicable.

2. Numerical Algorithms

2.1. The artificial time-dependent problems

It is classical to write that the solutions u, p, of (1.1)-(1.3) are the limit as $t \to \infty$ of the solutions of the evolution Navier-Stokes equations. Actually u, p, can be also considered as the limit as $t \to \infty$ of many other evolution systems which are more convenient for approximation than the real system (1.14)-(1.17).

For instance we may consider the artificial equations

$$(2.1)\qquad \begin{cases} -\frac{1}{Re}\Delta u(x,t) + (u(x,t).\nabla)\, u(x,t) + \nabla p(x,t) = f(x,t) \\[2ex] \frac{\partial p}{\partial t}(x,t) + \nabla u(x,t) = 0 \\[2ex] u = 0 \quad \text{on} \quad \Gamma \end{cases}$$

or

$$(2.2)\qquad \begin{cases} -\frac{\partial \Delta u}{\partial t} - \frac{1}{Re}\Delta u + (u.\nabla)u + \nabla p = f \\[2ex] \frac{\partial p}{\partial t} + \nabla u = 0 \\[2ex] u = 0 \quad \text{on} \quad \Gamma\,. \end{cases}$$

In each case it is clear that u(x,t) will not describe accurately the flow during the transient period but only at the limit $t \to \infty$. The above equations are easier to solve than the real evolution Navier-Stokes equations and, as far as stationary problems are considered, the procedure is legitimate.

Now we can take the discrete form of one of the previous equations either at the level of the continuous problem (1.4) or directly at the level of the finite dimensional problem (1.12)-(1.13). For instance we get

$$(2.3)\quad \left[\begin{array}{l} u_h^{m+1} \in W_h(0) \text{ satisfies} \\ \frac{1}{Re}\, a_h(u_h^{m+1}, v_h) + b_h(u_h^{m+1}, u_h^{m+1}, v_h) - (p_h^m, \mathrm{div}_h\, v_h) = (f, v_h) \;\forall\, v_h \in \mathbb{W}_h(0) \\ p_h^{m+1} \text{ satisfies} \\ p_h^{m+1} - p_h^m + \rho\, \mathrm{div}\, u_h^{m+1} = 0 \end{array}\right.$$

(Uzawa algorithm) or for (2.2)

$$(2.4)\quad \left[\begin{array}{l} u_h^{m+1} \in \mathbb{W}_h(0) \text{ satisfies} \\ a_h(u_h^{m+1} - u_h^m, v_h) + \rho\, \nu\, a_h(u_h^m, v_h) + \rho\, b_h(u_h^m, u_h^{m+1}, v_h) - \rho(p_h^m, \mathrm{div}_h\, v_h) = \\ \qquad = \rho(f, v_h)\;,\; \forall\; v_h \in \mathbb{W}_h(0)\;, \\ p_h^{m+1} \text{ satisfies} \\ \alpha(p_h^{m+1} - p_h^m) + \rho\, \mathrm{div}\, u_h^{m+1} = 0 \end{array}\right.$$

(Arrow-Hurwicz algorithm).

The parameter ρ in (2.3), the parameters ρ and ρ/α in (2.4) play the role of a discretization parameter (discretization in time) and have to be specified.

When the solution of (1.1)-(1.3) is unique, these algorithms converge under appropriate conditions on ρ and α (cf. [13]).

2.2. The penalty method and Application

Using the concept of penalization in the calculus of variation, we can consider $\nabla u = 0$ as a constraint and introduce the following equations :

$$(2.5)\qquad -\frac{1}{Re}\,\Delta u + (u.\nabla)u - \frac{1}{\varepsilon}\,\nabla(\nabla u) = f \quad \text{in } \Omega$$

(2.6) $$u_\varepsilon = 0 \quad \text{on} \quad \Gamma ,$$

where $\varepsilon > 0$ is a small parameter.

It is shown that, for $\varepsilon \to 0$, the solution u_ε of (2.5)-(2.6) converges to the solution u of (1.1)-(1.3) ; cf. [13]. An error estimate and a full asymptotic expansion is given in M.C. Pélissier [10]; cf. also M. Bercovier [1] and R.S. Falk [3].

Let us mention some other application.

Extrapolation procedure.

Using the principle of the penalty method, R.S. Falk proposes in [3] an extrapolation procedure which is obtained as follows (the linear and discrete version of (2.5)-(2.6) is considered $u_{h\varepsilon} \in \mathbb{W}_h(0)$) :

We set $\varepsilon = \gamma^{-1} \rho_h^\sigma$, $\gamma > 0$, $\sigma \geq 0$ and the solution of the linear discrete form of (2.5)-(2.6) is denoted $u_h(\gamma)$. Then some distinct parameters γ_0, ..., γ_k, are considered and the parameters a_k which satisfy

$$\sum_{i=o}^{k} a_i = 0 , \quad \sum_{i=o}^{k} a_i \gamma_i^{-j} = 0 , \quad 1 \leq j \leq k .$$

Then the extrapolation is defined by

$$u_h^{(k)}(\gamma) = \sum_{i=o}^{k} a_i \vec{u}_h(\gamma_i) .$$

If $\gamma_i = 2^i \gamma$, then

$$u_h^k(\gamma) = \frac{2^k u_h^{(k-1)}(2\gamma) - u_h^{(k-1)}(\gamma)}{2^k - 1} .$$

Augmented Lagrangian Method.

(cf. M. Fortin [5], [7], R.T. Rockafellar [11], F. Thomasset - R. Temam [14]).

Combining the idea of penalty and the algorithms considered in Section 2.1, we are led (cf. the above references) to consider the following algorithm which appears as very efficient (compare to (2.3) and (2.5)):

$$(2.7)\quad \left[\begin{array}{l} u_h^{m+1} \in \mathbb{W}_h(0) \text{ satisfies} \\ \frac{1}{Re} a_h(u_h^{m+1}, v_h) + b_h(u_h^{m+1}, u_h^{m+1}, v_h) + \frac{1}{\varepsilon}(\mathrm{div}_h\, u_h^{m+1}, \mathrm{div}_h\, v_h) \\ \qquad - (p_h^m, \mathrm{div}_h\, v_h) = (f, v_h) \quad \forall\, v_h \in \mathbb{W}_h(0) \\ p_h^{m+1} \text{ satisfies} \\ p_h^{m+1} - p_h^m + \rho\, \mathrm{div}_h\, u_h^{m+1} = 0 . \end{array}\right.$$

Cf. some computations made with this method in the Appendix of [13].

3. Mixed Finite Elements for the Euler Equations

We pass now to a mixed finite element for the Euler Equations proposed by Fortin in [6]. This scheme is an extension of a method of Lesaint [9] applying to linear hyperbolic systems.

We assume for simplicity that $\Omega \subset R^2$ is simply connected. Then the curl operator

$$\psi \longmapsto \{\frac{\partial\psi}{\partial x_2}, -\frac{\partial\psi}{\partial x_1}\}$$

gives a one to one mapping between the space $W(0)$ of function vanishing on Γ and the space $\mathbb{H}(0)$ of free divergence vector functions on Ω, such that $u.n = 0$ on Γ, n = the unit outward normal on Γ.

If $W_h(0)$ is an internal approximation of $W(0)$ (for instance conforming finite elements of Lagrange type), then the space

$$\mathbb{H}_h(0) = \{v_h, v_h = \text{curl}\ \psi_h, \psi_h \in W_h(0)\}$$

is an approximation of $\mathbb{H}(0)$ and $\mathbb{H}_h(0) \subset \mathbb{H}(0)$.

It is easily seen that if $u_h \in \mathbb{H}_h(0)$ then $u_h.n = 0$ on $\partial\Omega_h$ and at the interface of two adjacent triangles K and K' , $u_h.n$ is continuous while $u_h - (u_h.n)n = u_{h\tau}$ has no reason for being continuous. The discretization will take advantage of this discontinuity of the tangential components of u_h which is allowed in this space.

It is easily seen that the weak formulation of the Euler equations, analogous to (1.4) is :

$$(3.1) \qquad \left[\begin{array}{l} u : [0,T] \mapsto \mathbb{H}_h(0) \text{ satisfies} \\ \\ \dfrac{d}{dt}(u,v) + b(u,u,v) = (f,v) \ , \quad 0 < t < T \ , \ \forall v \in \mathbb{H}_h(0) \end{array}\right.$$

(we assumed $g = 0$).

Now let $\mathcal{T}_h$ be a triangulation of Ω and let K be a triangle, $K \in \mathcal{T}_h$, with boundary ∂K . We have :

$$\begin{aligned} b(u,u,v) &= \sum_{i,j=1}^{2} \int_\Omega u_i \frac{\partial u_j}{\partial x_i} v_j \, dx \\ &= \sum_{i,j=1}^{2} \int_\Omega \frac{\partial}{\partial x_i}(u_i u_j) v_j \, dx \\ &= \sum_{i,j=1} \sum_{K \in \mathcal{T}_h} \int_K \frac{\partial}{\partial x_i}(u_i u_j) v_j \, dx \ . \end{aligned}$$

After performing an integration by part on K , we find

$$(3.2) \qquad \int_K \frac{\partial}{\partial x_i}(u_i u_j) v_j \, dx = - \int_K (u_i u_j) \frac{\partial v_i}{\partial x_j} dx + \int_{\partial K} (u.n)(u.v) \, d\ell \ .$$

The idea used by Lesaint and Fortin is to distinguish on the boundary ∂K of K two parts ∂K_- and ∂K_+ where respectively $u.n < 0$ and $u.n > 0$, i.e.

the flux is coming in or is going out. On two adjacent triangles this gives the situation represented on the figure

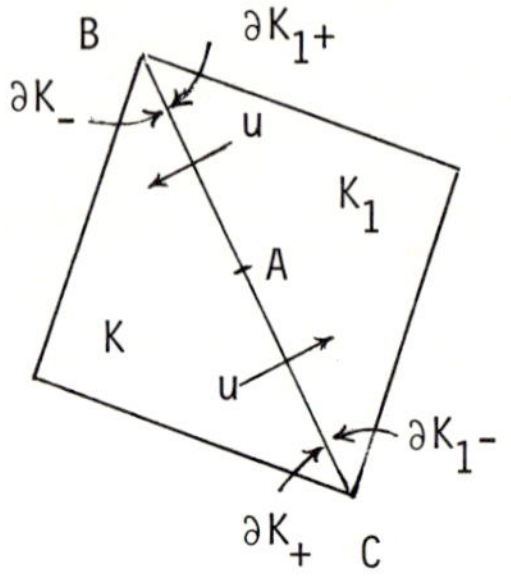

($u.n = 0$ at A , $u.n > 0$ for K_1 on AB and for K on AC).

Because of the hyperbolic character of the equations it is considered that the flow is influenced in K by the neighbouring triangles through ∂K_-. For this reason we set on $S = \partial K \cap \partial K_1$:

$$u^+(x) = \begin{cases} \text{value of } u(x) \text{ in } K \text{ if } x \in \partial K_+ \cap S \\ \text{value of } u(x) \text{ in } K_1 \text{ if } x \in \partial K_- \cap S . \end{cases}$$

and the expression (3.2) is replaced by

$$- \int_K (u_i u_j) \frac{\partial v_i}{\partial x_j} dx + \int_{\partial K} (u.n)(u^+.v) \, d\ell \quad .$$

This leads to the scheme

$$\text{(3.3)} \quad \frac{d}{dt}(u,v) + \sum_{i,j=1}^{2} \sum_K \{- \int_K (u_i u_j) \frac{\partial v_i}{\partial x_j} dx + \int_{\partial K} (u.n)(u^+ v) \, d\Gamma\} = (f,v) \; ,$$

$$\forall \; v \in \mathbb{H}_h(0) \; .$$

More generally, for $0 \leq \alpha \leq 1$ one can consider

$$\text{(3.4)} \quad \frac{d}{dt}(u,v) + \sum_{i,j=1}^{2} \sum_K \{- \int_K (u_i u_j) \frac{\partial v_i}{\partial x_j} + \int_{\partial K} (u.n)((1-\alpha)u^- + \alpha u^+) \, d\ell\} = (f,v) \; .$$

For $\alpha = \frac{1}{2}$ we obtain "centered" discrete differentiations ; for $\alpha < \frac{1}{2}$ the decentered discretization is against the flow, for $\alpha > 1/2$ the discretization is decentered in the direction of the flow.

It is shown in [6] that the scheme is dissipative for $\alpha > \frac{1}{2}$ conservative for $\alpha = \frac{1}{2}$, and of course unstable for $\alpha < \frac{1}{2}$.

Numerical experiments have been done by F. Thomasset. A version of this scheme applying to the Navier-Stokes equations is given in [6], and [15].

Remark 3.1. For other aspects of the application of finite elements to fluid mechanic, cf. the lecture of J.J. Chattot - J. Guiu and Glowinski - Pirronneau at this Conference.

This lecture is based on the material of an AGARD course on computational fluid mechanic given at the Von Karman Institute, Belgium and at Dayton, Ohio, U.S.A.

References

[1] M. Bercovier - Thèse, Université de Rouen 1976, and articles to appear.

[2] M. Crouzeix, P.A. Raviart - *Conforming and nonconforming finite element methods for solving the Stationary Stokes equations.* R.A.I.R.O., R.3, 1973, p.33-76.

[3] R.S. Falk - *An analysis of the penalty method and extrapolation for the stationary Stokes equations.* Advances in Computer Methods for P.D.E.'s, R. Vichnevelsky editor, Proceedings of A.I.C.A. Symposium.

[4] M. Fortin - *Calcul numérique des écoulements des fluides de Bingham et des fluides newtoniens incompressibles par la méthode des éléments finis.* Thèse, Université de Paris 1972.

[5] M. Fortin - *Minimization of some non differentiable Functionals by the augmented Lagrangian Method of Hestenes and Powell.* Applied Math. and Optimization, vol.2, N°3, 1975/76, p. 236-250.

[6] M. Fortin - *Résolution numérique des équations de Navier-Stokes par des éléments finis de type mixte.* IRIA - LABORIA, Rapport de Recherche n°184, Août 1976.

[7] M. Fortin - Cours de 3ème cycle, Université de Paris-Sud, Orsay 19/5/76.

[8] M. Fortin, R. Glowinski - To appear.

[9] P. Lesaint - *Sur la résolution des systèmes hyperboliques du premier ordre par des méthodes d'éléments finis.*

[10] M.C. Pélissier - *Résolution numérique de quelques problèmes raides en mécanique des milieux faiblement compressibles.* Calcolo, 12, 1975, p.275-314.

[11] R.T. Rockafellar - *Augmented Lagrange multiplier functions and duality in nonconvex programming.* SIAM J. Control.

[12] R. Temam - *On the Euler equations of incompressible perfect fluids.* J. Funct. Anal., vol.20, 1975, p.32-43.

[13] R. Temam - *Navier-Stokes equations, Theory and Numerical Analysis.* North Holland Publ. Comp., Amsterdam 1977. Russian translation, to appear.

[14] R. Temam, J.F. Thomasset - *Numerical solution of Navier-Stokes equations by a finite element Method.* Conference at Rapallo, Italy June 1976, on finite element Methods in Fluid Flows. To appear.

[15] F. Thomasset - *Numerical solution of the Navier-Stokes equations by Finite Element Methods.* Lecture at this Conference.

FIGURES

All the computations were made by F.Thomasset (cf [15])

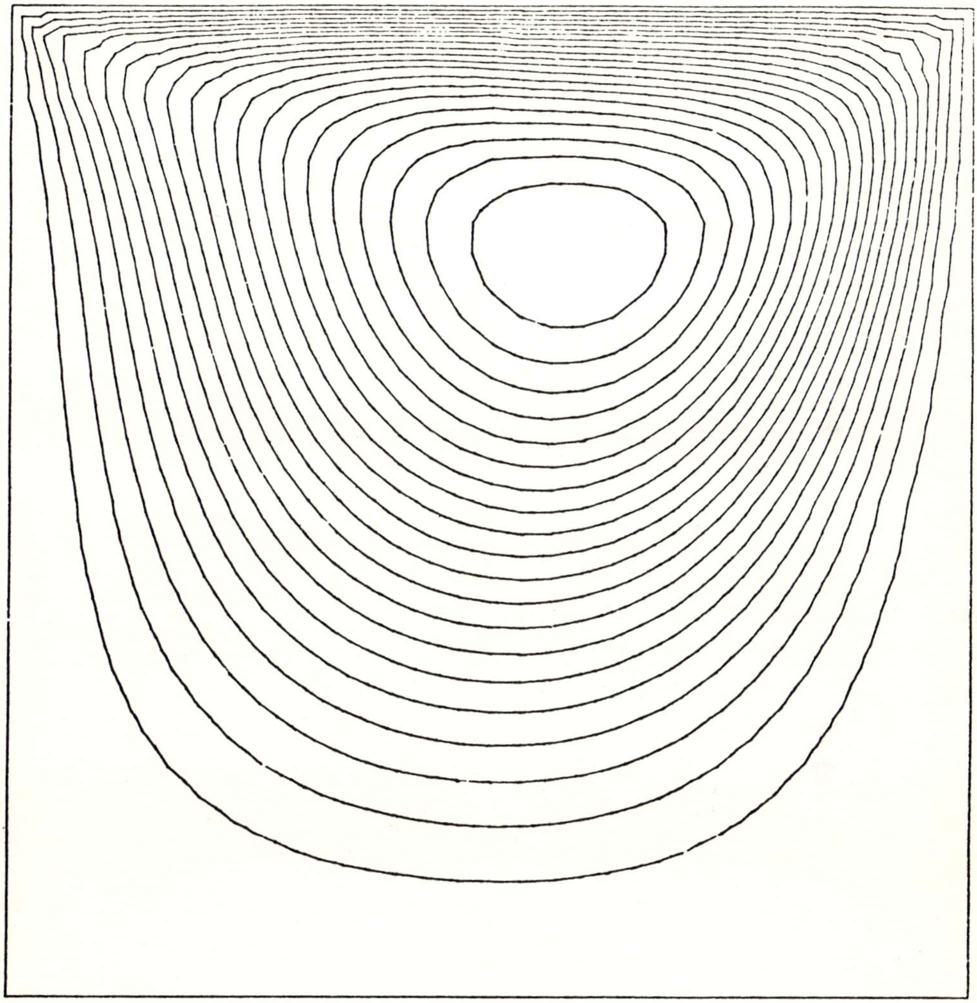

Fig.1 Streamlines for the flow in a square cavity, computed by the non conforming finite element method ($\nu = 10^{-2}$, length of the edge = 1, upper side moving with velocity 1)

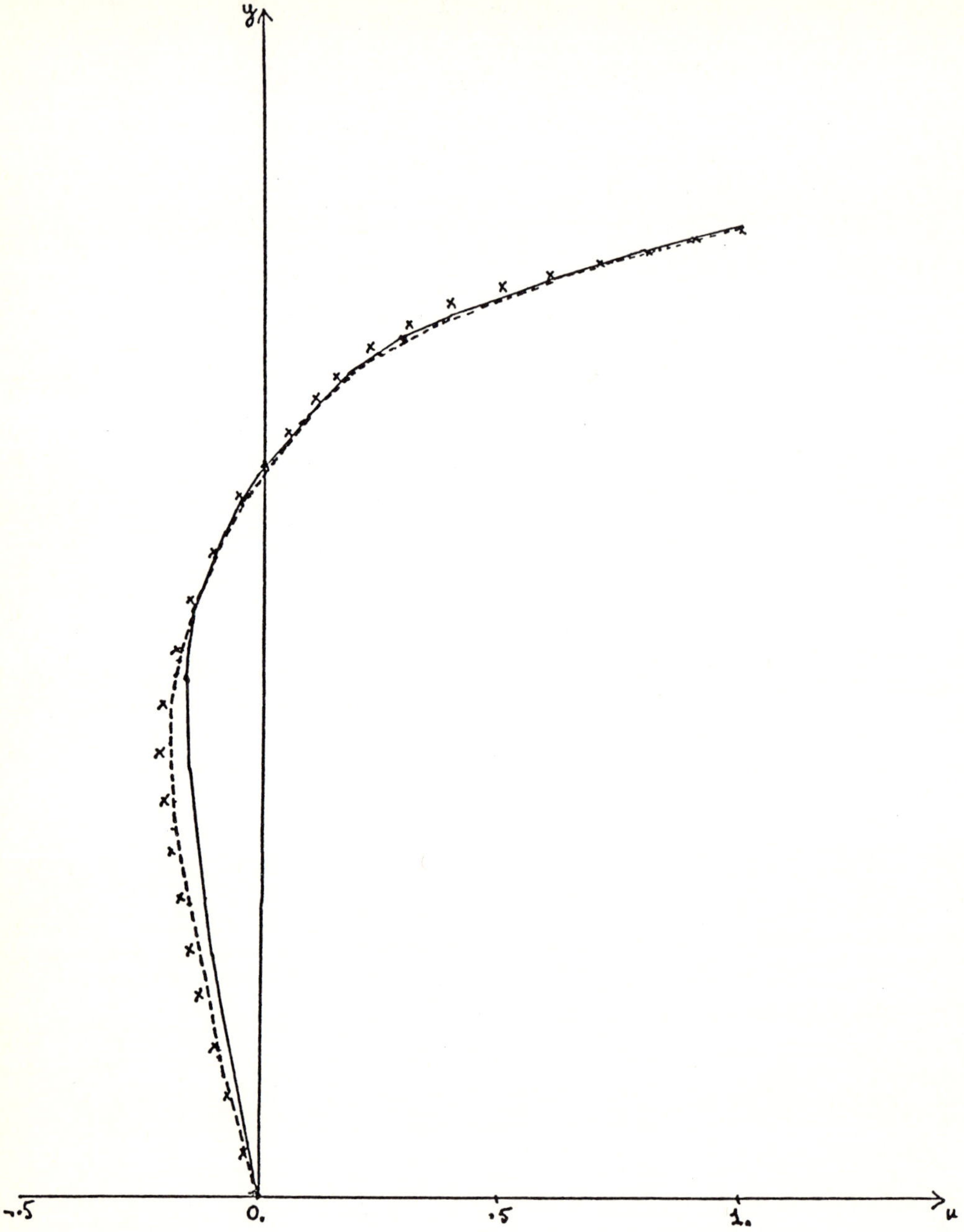

Fig.2 Flow in a square cavity (same conditions as in Figure 1): velocity u along the axis x = 0.5. Results with the non conforming finite element method, - - - - 2048 triangles, —— 512 triangles; x = results of O.R.Burggraf (J. Fluid Mech., Vol.24, Part I, pp.113-151, 1966)

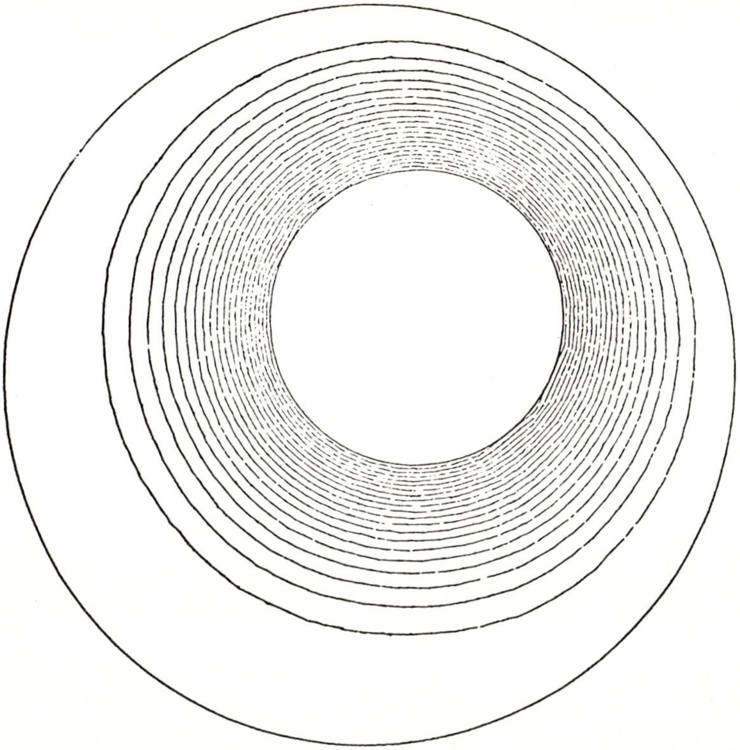

Fig.3 Streamlines for the flow between two rotating cylinders. Angular speed of interior circle = 1, outer circle at rest, $\nu = .5$, Re = 80. Non conforming finite elements

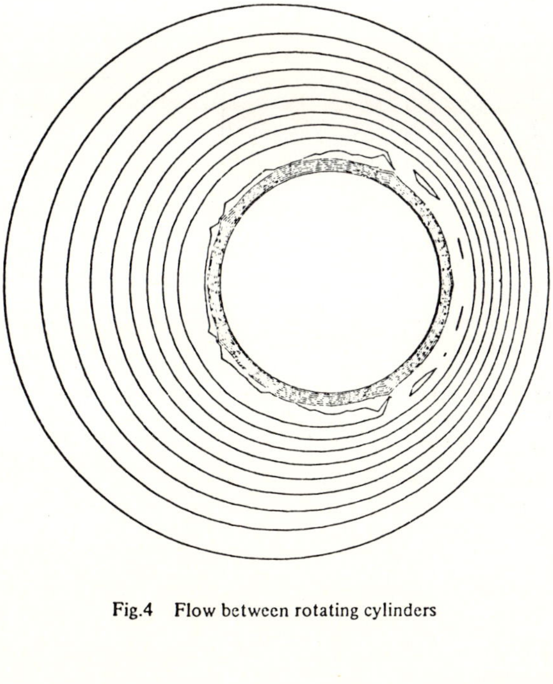

Fig.4 Flow between rotating cylinders

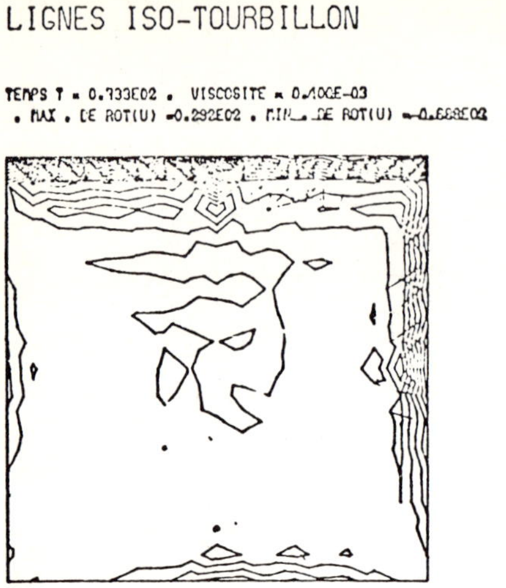

Figure 5

Figure 6

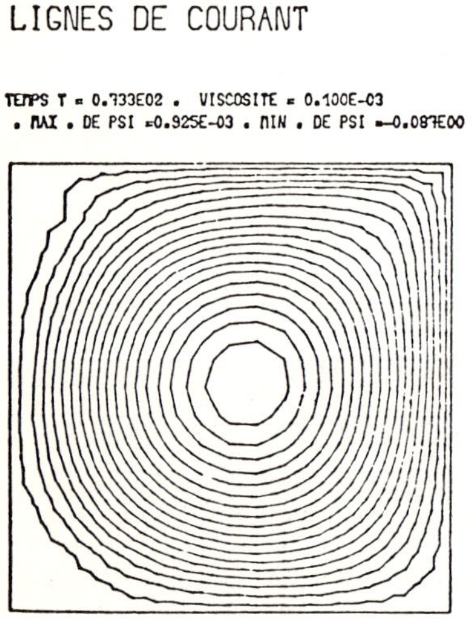

Figure 7

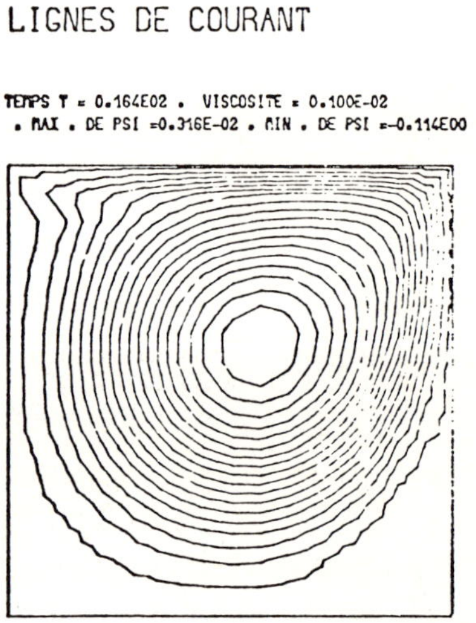

Figure 8

Figs.5, 6, 7, 8 Vortex lines and streamlines, flow in a cavity (same conditions as in Figure 1) mixed finite elements

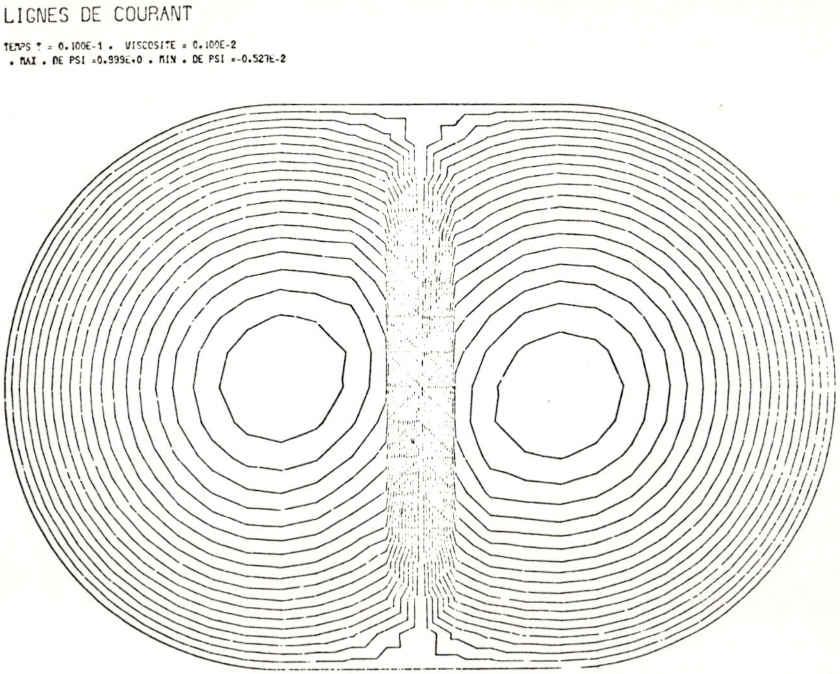

Figure 9

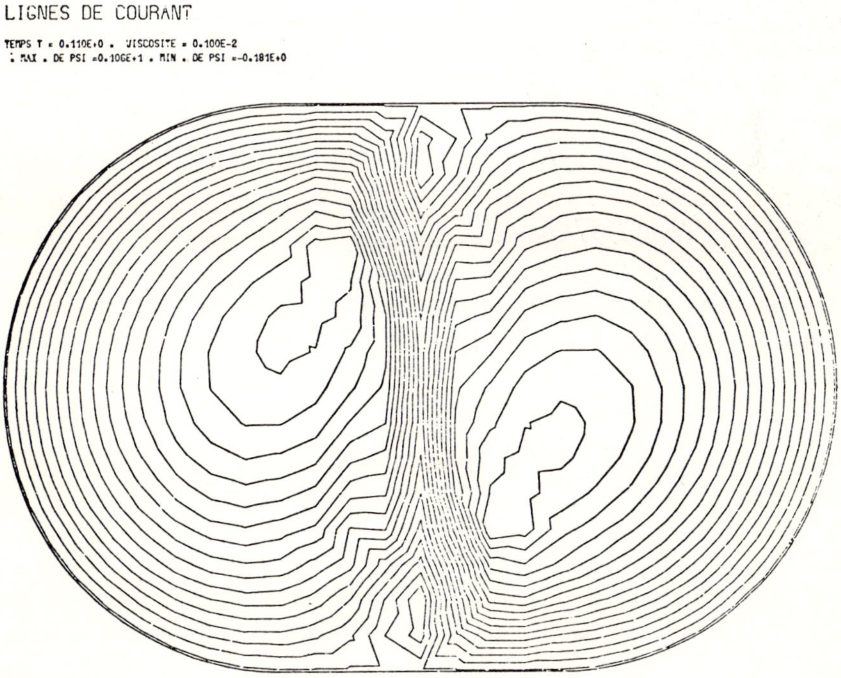

Figure 10

Figs.9 and 10 Collision of two vortices (streamlines). Mixed finite elements

NUMERICAL STUDY OF NONLINEAR AXISYMMETRIC FLOW OF FLUID BETWEEN TWO CONCENTRIC ROTATING SPHERES

N.M. Astaf'eva, N.D. Vvedenskaja, I.M. Yavorskaya

Institute of Space Research
Institute for Problems of Information Transmission
Moscow, USSR

The steady state flow of viscous incompressible fluid between two concentric spheres is studied using a numerical method. The spheres are rotating with different angular velocities. The solution of the stationary problem is found as a limit of the solution of the non-stationary one. The method uses the expansion of the solution into series in Legendre polynomials and associated functions, the coefficients of the series being functions of r and t (where r is the radius and t is time). The equations for these coefficients are solved by finite difference method. For some planes in the space of the main parameters of the problem stability curves are obtained. The non-uniqueness of the solution of the problem for the supercritical values of the parameters is shown. The domains of existence of different regimes of the flow are found. Numerical results are in fair agreement with the experimental data.

The steady state axisymmetric flow of viscous incompressible fluid is considered between two concentric spheres rotating with different angular velocities about the same axis.

This flow depends on three dimensionless parameters: the Reynolds number $Re=\omega_1 r_1^2/\nu$, the gap size $\delta=(r_2-r_1)/r_1$ and the Rossby number $\varepsilon=(\omega_2-\omega_1)/\omega_1$ (here r_1 and r_2 are the radii of the inner and outher spheres, ω_1 and ω_2 are their angular velocities and ν is the kinematic viscosity).

The solution of this nonlinear boundary-value problem is found by the method of stabilization, i.e. as the limit for $t\to\infty$ of the solution of the Cauchy problem for nonstationary Navier-Stokes equations

$$\frac{\partial \bar{u}}{\partial t} + (\bar{u}\nabla)\bar{u} = -\nabla p + \frac{1}{Re}\Delta \bar{u},$$

$$div\, \bar{u} = 0,$$

$$\bar{u}\Big|_{r=1} = \sin\theta \cdot \bar{\theta}, \quad \bar{u}\Big|_{r=1+\delta} = (1+\varepsilon)(1+\delta)\sin\theta \cdot \bar{\theta},$$

$$\bar{u}\Big|_{t=0} = \bar{u}_0 .$$

Here $\bar{u}(u_r, u_\theta, u_\varphi)$ denotes the velocity vector, p is the pressure, r, θ and φ are the spherical coordinates. The length, time, velocity and pressure scales are equal to r_1, ω_1^{-1}, $\omega_1 r_1$ and $\rho\omega_1^2 r_1^2$ respectively (here ρ is the fluid density).

The unknown functions are represented in series of Legendre polynomials and associated functions $P_l^0(x)$ and $P_l^1(x)$ $(x = \cos\theta)$, the coefficients being functions of r and t:

$$u_r(r,\theta,t) = \sum_{l=0}^{L} U_l(r,t)\, P_l^0(x), \qquad \text{etc.}$$

As can be seen from this formula, only a finite number of terms are taken. These series are considered as interpolation sums, namely it is supposed that the values of the sums coincide with those of the functions on the rays $\theta = \theta_j$, $j = 0,\dots,L$, $\theta_j = j\pi/L$ and that the equations and the boundary conditions hold on the same rays [1,2]. The nonlinear terms are also computed on these rays and then the sums of the same form are found which coincide with them on the rays:

$$F_r(r,\theta,t)\Big|_{\theta=\theta_j} = -\left\{ u_r \frac{\partial u_r}{\partial r} + \frac{u_\theta}{r}\frac{\partial u_r}{\partial \theta} - \frac{u_\theta^2 + u_\varphi^2}{r} \right\}\Big|_{\theta=\theta_j},$$

$$F_r(r,\theta,t) = \sum_{l=0}^{L} \Phi_{r,l}(r,t)\, P_l^0(x), \qquad \text{etc.}$$

The Cauchy problem for the coefficients is solved using the finite difference method. The viscous terms are computed on the (n + 1)-th time layer whereas the inertia terms are computed on the n-th time layer. This makes it possible to solve the linear difference equations in r using the Guassian elimination process (the maximum rank of the matrices to be inverted is equal to 3). The most time consuming part of the computation is thus the calculation of the nonlinear terms.

The computations were performed for the layer with $\delta = 0.11$ for different values of Re and $Re_2 = (1+\delta)^2(1+\varepsilon)Re$ and also for the layers with $0.049 \leq \delta \leq 1.33$. In the last case only the inner sphere rotated $(\varepsilon = -1,\ Re_2 = 0)$. For constant values of the angular velocities of the spheres the number $\mathcal{L}$ of therms to be taken increases as the gap δ decreases. For a fixed δ the number $\mathcal{L}$ increases together with Re. To get the real flow picture in the case when Re passes through the critical value Re_{cr} for which instability occurs the number of terms $\mathcal{L}$ in the series should be made several times greater than in the case when $Re < Re_{cr}$. This is due to the fact that the length scale of the secondary flow in a thin layer is several times as small as that of the main flow. To make computations more accurate the mash size of the difference scheme is made smaller near the boundaries where the flow gradient is usually larger. In the most cases the number of the nodes of the difference scheme is equal to 10 for thin layers and to 20 for thick layers. As a rule the initial values are chosen to be equal to the steady flow corresponding to a close value of Re.

The streamlines of the meridional flow are found as well as the lines of constant angular velocity, the spectral distribution of the kinetic energy and velocity components in r direction. The integral characteristics of the flow are also computed, namely the total kinetic energy

$$\mathcal{E} = \pi \int_0^{\pi} \int_1^{1+\delta} (u_r^2 + u_\theta^2 + u_\varphi^2)\, r^2 \sin^2\theta \, dr \, d\theta$$

the rotational moment of the inner sphere

$$\mathcal{M} = 2\pi r \nu \rho \int_0^{\pi} \left(\frac{\partial u_\varphi}{\partial r} - \frac{u_\varphi}{r} \right) \sin^2\theta \, d\theta \Big|_{r=r_1} .$$

The chart (Fig. 1) demonstrates the numerical experiment for the case $\delta = 0.11$, $\varepsilon = -1$. The values of Re for which the computations are made are marked with hatches ($1100 \leq Re \leq 2600$). For $Re > 2600$ the steady state solution is not obtained. In Fig. 1-3 the main flow characterized by one-vortex meridional circulation is marked by 0. The secondary flow having i pairs of torus-like axisymmetric vortices which are symmetric about the equator (Fig. 2) is marked by i.

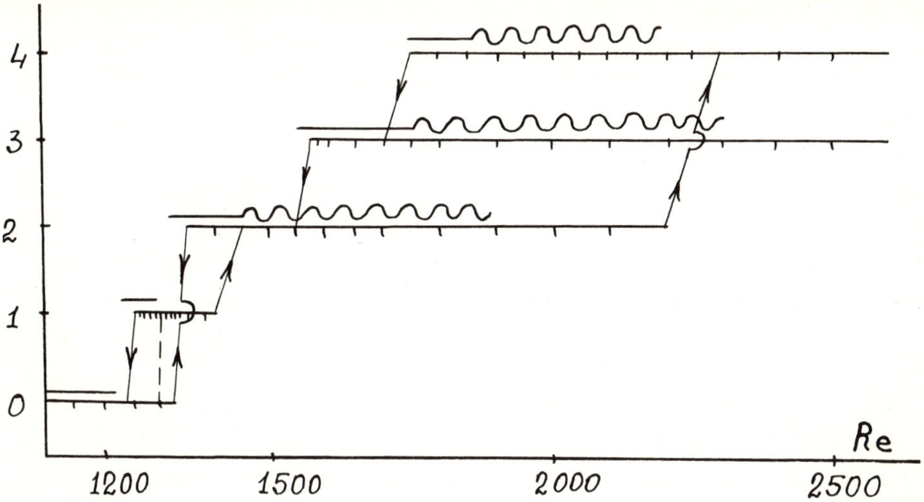

Figure I.

The arrows on Fig. 1 show the change of the flow-type which is due to the change of Re without any additional perturbation of the flow. The upper line indicates the values of Re for which the corresponding type of the flow is observed in physical experiments [3]. The wavy part of the upper line indicates the sinusoidal three-dimensional perturbations of the torus-like vortices observed in the experiments.

The Figs 2 and 3 represent the streamlines in the meridional plane and the spectral distribution of the total kinetic energy for different kinds of flow regimes. The results of the numerical and physical experiments are in fair qualitative agreement. The relative sizes of torus-like vortices are the same for both kinds of the experiments. Fig. 1 shows that in the physical experiment the main flow exists for $Re < 1230$ whence is seen that in the numerical experiment $Re^{(n)}_{cr}$ slightly exceeds $Re^{(p)}_{cr}$ in the physical experiment (for $\delta = 0.11$ and $\varepsilon = -1$ we have $Re^{(p)}_{cr} = 1225 \pm 15$ and $Re^{(n)}_{cr} = 1255 \pm 5$). The numerical regims 1,2,3 and 4 exist for $Re > Re^{(n)}_{cr}$. Both physical and numerical results show that there exist several different steady state flows for the same supercritical value of Re. This indicates the non-uniqueness of the steady solution of Navier-Stokes equations in these regions of the parameters of the problem. It should be noted

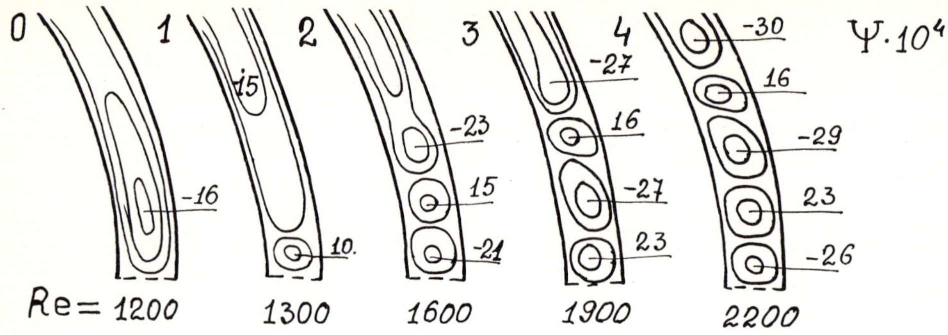

Figure 2.

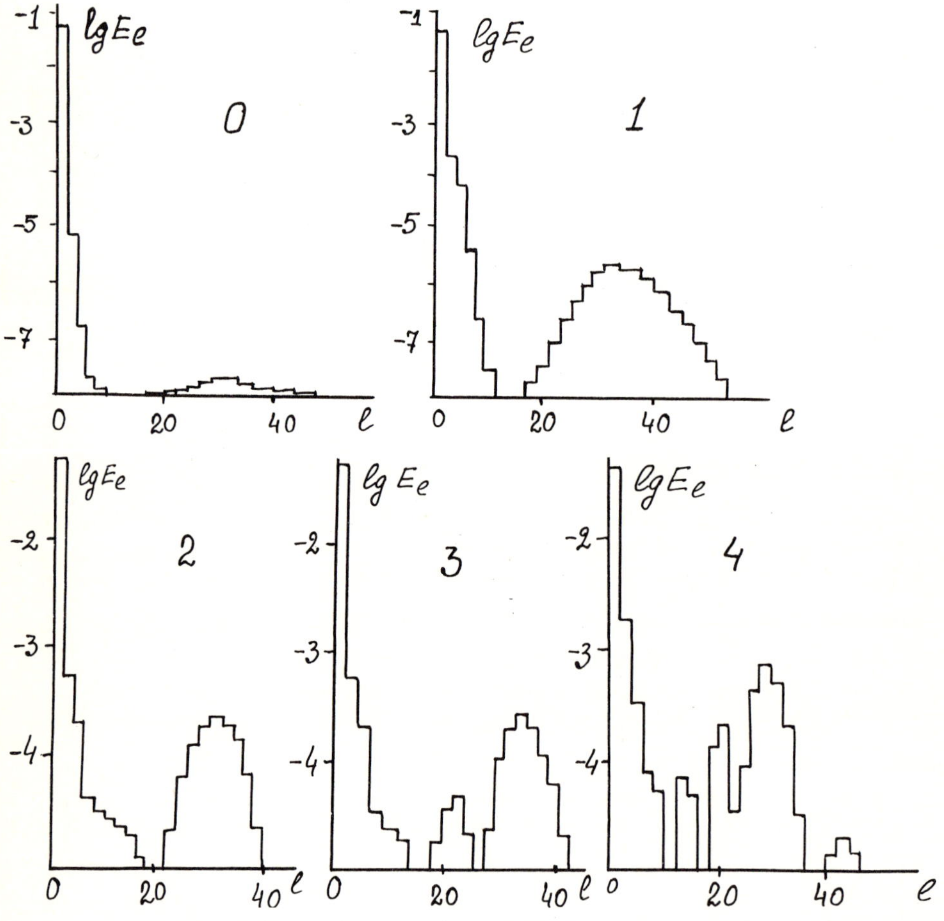

Figure 3.

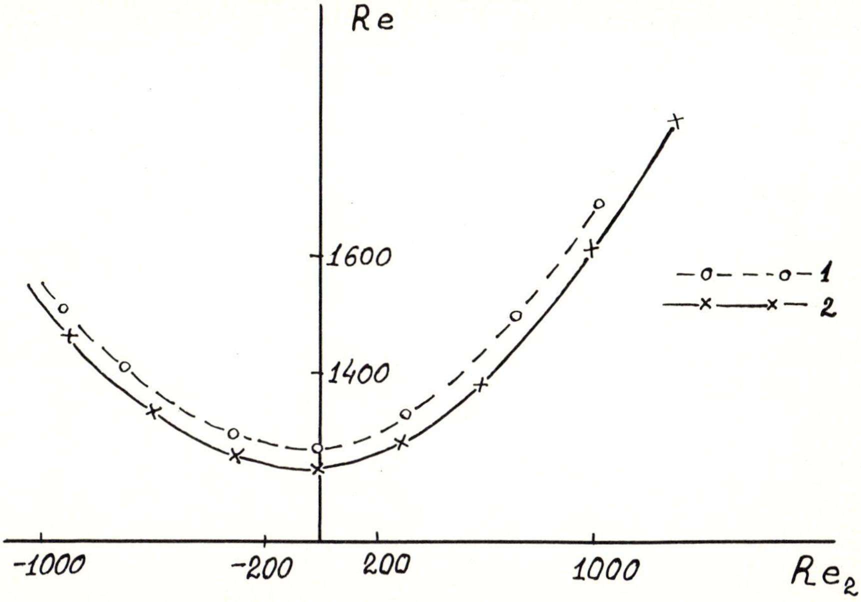

Figure 4, I-computations, 2-experiments.

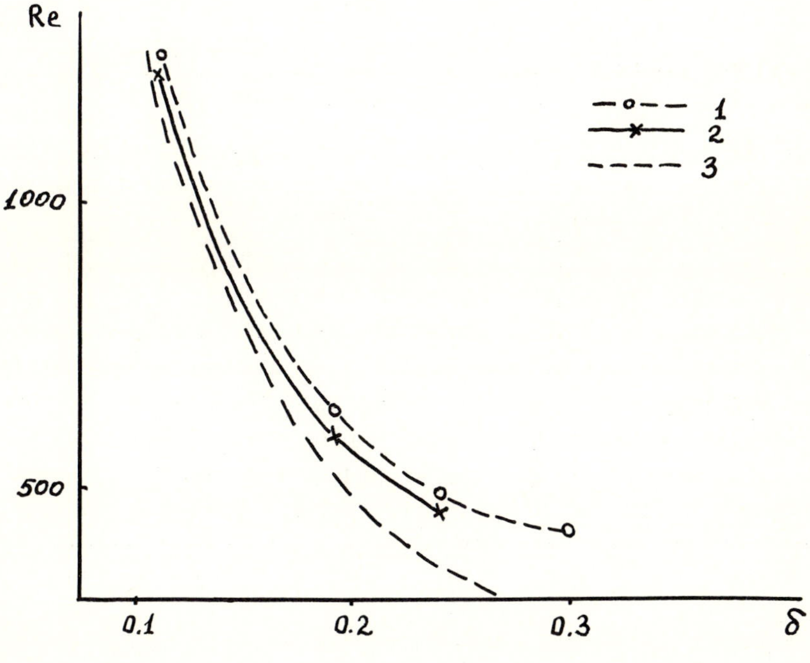

Figure 5, I-computations, 2-experiments, 3-cylinders.

that for $Re > Re_{cr}$ the existence of the flow with one pair of torus-like vortices (regime 1) is known from the physical experiment; this flow is caused by the instability of the main flow. To obtain such a flow in the numerical experiment it is necessary to perturb the main flow in a special way to make the initial conditions more or less close to flow-regime 1. In physical experiment the main flow is unstable under the perturbations that are either non-axisymmetric or asymmetric about the equator. The numerical computations show that regime 1 exists for 1260 Re 1400. The lower bound of this interval is identified with the critical value of the Reynolds number for the physical experiments show no hysteresis in regime 1.

For the layer with $\delta = 0.11$ the computations are performed for the cases in which both spheres rotate with different angular velocities. The stability curve in the plane Re, Re_2 is obtained (Fig. 4). These results differ from the physical ones by 2-3%. The computations for $\varepsilon = -1$ and for several values of δ provide the stability curve in the variables Re, δ. In Fig. 5 these results are compared with the physical ones. The critical values of Re (as function of δ) for the cylindrical Couette flow are given in the same Fig. 5. It should be stressed that in the numerical computations are observed the same kinds of flow as in the physical experiments. The advantage of the numerical experiments is that they make it possible to investigate the structure of the flow in detail and to get the spectra of kinetic energy.

A number of computations for the thick layer with $\delta = 1.33$ and in the case when only the inner sphere rotates are performed. The structure of the streamlines and the lines of equal angular velocities as well as the integral characteristics as functions of Re are obtained. These results and those of the physical experiments [4] lead to the conclusion that the main flow is stable for $Re \leq 400$.

References

1. K.I.Babenko, N.D.Vvedenskaja, M.G.Orlova. Numerical study of a stationary flow of viscous fluid past a circular cylinder. Zurnal vychislitelnoj matematiki i matematicheskoj fiziki, 15, № 1,, 1975.

2. I.M.Yavorskaya, N.M.Astaf'eva, N.D.Vvedenskaya. On stability and non-uniqueness of fluid flows in rotating spherical layers. Doklady AN SSSR, 241 , N 1 , 1978 .

3. I.M.Yavorskaya, Yu.N.Belyaev, A.A.Monakhov. Study of stability and secondary flows in rotating spherical gaps for arbitrary values of Rossby number. Doklady AN SSSR, 237 , N 4 , 1977 .

4. Yu.N.Belyaev, A.A.Monakhov, I.M.Yavorskaya . Stability of spherical Couette flow in thick layer when inner sphere rotates. Mekhanika zhidkhosti i gaza , N 2, 1978 .

INVESTIGATION OF HYDRODYNAMIC STABILITY BY MEANS OF COMPUTERS

Babenko K.I.
Institute of Applied Mathematics
Acad. Nauk USSR, Moscow

I. There is a number of problems in the theory of hydrodynamic stability which can be successfully soleved by means of computers. However, for that very reason one must dispose of perfect algorithms and a rational theory of constructing and investigating the efficiency of computational algorithms. We shall briefly discuss a subject of constructing computational algorithms, and thereupon we shall dwell on some specific problems of hydrodynamic stability. And finally as an illustration we shall analyze some numerical examples.

2. A great many computational problems can be schematized as follows. Let F and G be two metric spaces and let X, Y be some compacta in them. Assume that the mapping

$$A : X \rightarrow Y$$

is a mapping into Y. Now, given $x \in X$, the problem is to seek for an element $y \in Y$ such that $y = Ax$. Any computational algorithm for approximate calculation of the mapping A is based on some method of approximation of the elements of the compacta X and Y. The method of approximation which we choose defines the way of discretization of the mapping A. Thus the numerical method, its properties and efficiency are essentially determined by the approximate character of the compacta X and Y. The approximate character of compacta is determined by the asymptotic of their Alexandrov widths. Various methods of approximation are characterized by appropriate widths such as, for example, Kolmogorov widths, fore-table widths, etc. The efficiency of a method of approximation is high if the asymptotic of appropriate widths is like the asymptotic of Alexandrov widths. The method of approximation of the compacta X and Y we have chosen does not define uniquely the way of discretization of the mapping A. When making the mapping A discrete we want the discretization to have properties similar to those of the mapping A. When the problem is being solved on a computer the following step of the discretization is carried out: jump from finite dimentional compacta to sets consisting of a finite number of elements that is to tables. But the jump to a table is not quite a harmless moment and we shall try to explain the whole significance it has. The matter is that the jump to tables considerably simplifies quite a number of problems of the theory of computational methods and allows us to work out some of them. It is first of all the question of the lower bound of the error.

Let $H_\varepsilon(X)$ and $H_\varepsilon(Y)$ be ε - entropies of the compacta X and Y. We assume that $Y=A(X)$ and this is not limitation. The characteristics of the mapping A are defined by inter-relation between $H_\varepsilon(X)$ and $H_\varepsilon(Y)$. Thus, if $H_\varepsilon(X)/H_\varepsilon(Y)\to\infty$, $\varepsilon\to 0$, then the mapping A is called correct, otherwise it is called incorrect. Let us put

$$\varepsilon_*(N) = \inf\{\varepsilon : H_\varepsilon(Y) \leq N\}.$$

It is not difficult to prove that whatever algorithm for the computation of the mapping A we have, its accuracy ε always satisfies the inequality $\varepsilon \geq \varepsilon_*(N)$, if a table of length $\leq N$ is given as its input data.

Thus we know the utmost we can do, if the ε - entropies of the compacta X and Y are known. This fact is of particular importance in the case when we are studying problems of hydrodynamic stability or exploring weak turbulent regimes. In this particular cases X is a compactum of virtual initial states of a system, Y is a compactum of virtual states of a system at a given moment, and A is a trajectory translation operator. Since the asymptotic of the ε -entropy is defined by the smoothness of the elements of a compactum, then, if these characteristics are given, we can determine the extent of the error with which the fluid flow will be calculated and relate it to the Reynolds number.

3. Unfortunately we does not always have full information of the smoothness of functions calculated. Therefore it is very important to know how to construct such algorithms without saturation which can respond to any potential smoothness of the solution required. This is especially important in problems where there is a large parameter, and in eigen-value problems for partial differential equations as well. The problem of saturation of numerical algorithms is treated in the work by the author [I] where the general definition of saturation of numerical algorithm is given and causes which are at the bottom of saturation are pointed out as well. At the same work methods of construction of algorithms without saturation are considered and, in particular, some aspects of the stability of

plane Poiseuille flow and Rayleigh-Taylor instability are touched upon as well. The above ideas and the results obtained are stated in works by the author [2] , [3] , [4] and in works by the author et al [5] , [6] , [7] .

4. One of the most important and interesting problems of hydrodynamic stability is the problem of determination of a critical Reynolds' number and the loss of stability of flow past a smooth body. It is advisable to consider first the oversimplified version of the problem, namely the case of plane flow. The complicacy of the corresponding spectral problem is that one must find the eigen-value of a not self-adjoint system of differential equations in an unbounded domain. The problem next in complicacy is to find time-periodic solutions bifurcating from a stationary solution. Since the loss of stability of a flow past a body takes place at small Reynolds' numbers there is hope to advance in this problem by means of a computer.

5. Consider viscous non-compressible flow in a plane channel. We find the time-periodic flow, which bifurcates from Poisseulle flow. and is close to it. The mathematical formulation of this problem is as follows: inside the domain $\{x, y : -\infty < x < \infty, -1 \leq y \leq 1\}$ one must find the eigen-function ψ and eigen-value C of the following non-linear differential equation

$$R\left[(U-c)\frac{\partial \Delta\psi}{\partial x} - U''\frac{\partial \psi}{\partial x}\right] + \frac{\partial(\Delta\psi, \psi)}{\partial(x, y)} - \Delta^2\psi \equiv 0,$$

with the boundary conditions

$$\psi|_{y=\pm 1} = 0, \quad \frac{\partial \psi}{\partial y}\Big|_{y=\pm 1} = 0;$$

$$\psi(x+2\pi|\alpha, y) \equiv \psi(x, y), \; -\infty < x < \infty, -1 \leq y \leq 1.$$

Here $U = U(y)$ is the velocity of Poiseuille flow, Δ is Laplacian. We expend the solution ψ of the problem in Fourier series in the variable x and then expend the coefficients of the series in series in some parameter. As a result we obtain an infinite triangular system of ordinary differential equations from which the first one is Orr-Sommerfeld equation.

$$\ell_1^2\chi - i\alpha R[(U-c)\ell_1\chi - U''\chi] = 0, \quad \ell_1 = \frac{d^2}{dy^2} - \alpha^2, \quad y \in [-1, 1]. \qquad (1)$$

We have the following boundary conditions

$$\chi(\pm 1) = \chi'(\pm 1) = 0. \qquad (2)$$

The other equations are either non-homogeneous equations with left hand side as defined in (I) or non-homogeneous equations with left hand side following from (I) where α is replaced by $k\alpha$, k is a natural number or zero. Thus we have to deal with both the eigen--value problem and to solve a non-homogeneous equation when the corresponding homogeneous equation has a non-trivial solution. The essence of our method is as follows. We first apply the regularizing operator to the left hand side of (I) taking into account the boundary conditions. As a result we have the integral equation

$$\chi(y) + \int_{-1}^{1} K(y,\xi)\chi(\xi)d\xi - c_0 \int_{-1}^{1} L(y,\xi)\chi(\xi)d\xi = 0 \quad (3)$$

where the kernels K and L can be easily defined. Then we carry out the discretization of equation (3). Let $T_m(y)$ be a Chebyshev polynomial of the first kind, y_k, $1 \leq k \leq m$ be its zeroes, and $P_m(y,\zeta)$ be an interpolational polynomial such that $P_m(y_k,\zeta) = \zeta_k$, $1 \leq k \leq m$, $\zeta = (\zeta_1, \zeta_2, \ldots, \zeta_m)$.

We introduce the mapping $\mathcal{J}: C[-1,1] \to R^1$, where $C[-1,1]$ is a normed space of continuous functions defined in the interval $[-1, 1]$,

$$\mathcal{J} f = (f(y_1), \ldots, f(y_m)), \quad \forall f \in C[-1,1].$$

Let us put $\chi(y) = P_m(y, \mathcal{J}\chi) + \varrho_m(y,\chi)$ and denote the linear integral operator with the kernel $K(L)$ by $K\chi$ $(L\chi)$. Then from (3) we have

$$\zeta + \mathcal{J}KP_m(\cdot,\zeta) - c_0 \mathcal{J}LP_m(\cdot,\zeta) = -\mathcal{J}[K\varrho_m(\cdot,\chi) - c_0 L\varrho_m(\cdot\chi)],$$

where $\zeta = \mathcal{J}\chi$. Neglecting the error of the approximation in the right hand side we obtain the algebraic eigen-value problem which can be solved by means of one of the well known methods. Note that the error of our approximation is defined by the value $\max_y |\varrho_m(y,\chi)|$. By Lebesgue theorem

$$|\varrho_m(y,\chi)| \leq (1+\Lambda_m) E_m(\chi),$$

where $E_m(\chi)$ is the best approximation to χ by polynomials of degree $\leq m-1$, and Λ_m is the Lebesgue constant of the interpolation. It is known that $\Lambda_m = \frac{4}{\pi} \ln m + O(1)$. By Jackson's theorem $E_m(\chi) \leq A_\nu \max|\chi^{(\nu)}|/m^\nu$ for $\nu \leq m-1$,

where A_ν is a constant depending only on ν .
Therefore

$$E_m(\chi) \leq \min_{\nu \leq m-1} \{ A_\nu \max |\chi^{(\nu)}(y)| / m^\nu \}. \tag{4}$$

On the other hand by induction from (I) with $\alpha \leq 2$ we can deduce the inequality $\max |\chi^{(\nu)}(y)| \leq \beta_0 (\beta \sqrt{R})^\nu, \; \nu \leq \sqrt{R}$. It follows from inequality (4) that under the appropriate choice of m the rapid growth of the derivatives of the function χ will be neutralized. Explicit developments show that to obtain the estimate $E_m(\chi) \leq A \exp[-\gamma \sqrt{R} \ln 2]$, where $\gamma \simeq 0.5$, $A \leq 1.63 \cdot 2^8$, for $R \simeq 10^4$ it is sufficient to take $m = 200 \div 300$. Using these estimates and fulfilling some calculations on computer one can obtain the absolutely strict estimate of the divergence between the true and calculated values of c_0 and χ ,and, in particular, rigorously prove the existance of the critical Reynolds' number and estimate it with the accuracy$\leq$0.01%. Numerous calculations [5] demonstrate that already for $m = 60 \div 80$ we obtain the solution of the Orr-Sommerfeld equation with high accuracy.

Having calculated the critical Reynolds' number R_c and the corresponding eigen-function $\chi_0(y)$, normed by the condition $\chi_0(0) = 1$, we proceed with the calculation of the periodic solution ψ and the phase velocity c . We seek for the function ψ and the velocity c in the forms

$$\psi = \sum_{j=0}^{\infty} h^{j+1} \psi_j(x,y),$$

$$c = \frac{R_c}{R} c_0 + R^{-1} \sum_{j=1}^{\infty} c_{2j} h^{2j},$$

where $h = |R - R_c|^{1/2}$, and

$$\psi_j = \sum_{k \leq j+1}' \chi_{jk}(y) \exp(ik\alpha x), \quad j = 0, 1, \ldots ;$$

the primed sum means that the summation is carried out in $k \equiv (j+1)$ mod **2.** It is clear that χ_{01} satisfies both equality (I) and the boundary conditions (2) whence $\chi_{01}(y) = \gamma \chi_0(y)$, where the constant γ can be considered real. The function $\chi_{jk}(y)$ can be found as the solution of the equation

$$L_k \chi_{jk} = F_{jk}, \; L_k = \ell_k^2 - i\alpha k R_c [(U - c_0)\ell_k - U''], \; k = 0, 1, \ldots \tag{5}$$

with boundary conditions as defined in (2). Here F_{jk} is calculated with respect to the functions $\chi_{\ell k}$ with $\ell \le j-1$. Thus we have an infinite triangular system of equations. For $k = 1$ equation (5) with boundary conditions (2) is unsolvable since the homogeneous problem is solvable. Therefore the function F_{j1} must be orthogonal to the solution of the adjoint problem. These conditions permit us to calculate the constants γ, c_2, c_4, ... The discretization of problem (5), (2) is carried out in absolutely the same way as that of problem (I), (2) and the discretization of the condition of compatibility is carried out with regard to the assumed approximation. This method is explicitly described in reprint [8] which presents the results of a great many calculations for Poiseuille flow in plane channel as tables of the function χ_{jk}. We omit citing the above results here because of shortage of space.

6. Now let us consider the problem about the interface between two heavy fluids having different density. Let us assume that the interface originally coincides with horizontal plane and the acceleration of gravity is directed from above downwards. Assuming that the flow is plane, let us denote be $y = y(x,t)$ the equation of the intersurface $\Gamma(t)$. Assume that the motions considered are 2π - periodic in the variable x. Then we can find the velocity potential in the form

$$\varphi(z,t) = \operatorname{Re} F(z,t) = \operatorname{Re}\left[\frac{1}{\pi i}\int_0^{2\pi} \frac{\mu(\xi,t)\, d e^{iz(\xi,t)}}{e^{iz(\xi,t)} - e^{iz}}\right], \quad z = x+iy,$$

where $z(\xi,t) = \xi + iy(\xi,t) \in \Gamma(t)$ and $\mu(\xi,t)$ is a real function. Caushy-Lagrange integral together with the conditions on the intersurface yield for the functions μ and y a system of two integro-differential equations and Caushy problem for this system. Letting $\mu(x,t)$ and $y(x,t)$ be small in $C^2[0,2\pi]$ it is not difficult to linearize the system obtained. Omitting details we state the final result which is as follows

$$\begin{aligned} & y_t - H(\mu_\xi) = 0, \\ & \mu_t + \delta\left[y + \frac{1}{2\pi}\int_0^{2\pi} \mu_t(\xi,t)\, d\xi\right] = 0, \end{aligned} \tag{6}$$

where $\delta = (\chi-1)/(\chi+1)$, $\chi = \varrho_1/\varrho_2$, ϱ_1, ϱ_2 are the densities of the fluids above and under the intersurface, respectively; H is the Hilbert operator

$$H\cdot = \frac{1}{\pi}\int_0^{2\pi} \frac{\cdot\, d\xi}{2\,tg\,\frac{\xi - x}{2}} .$$

Assuming that $y = C_1 \exp(\lambda t + inx)$, $\mu = C_2 \exp(\lambda t + inx)$ we have the following relation $P(\lambda, n) = \lambda^2 - \delta|n|$, where $P(\lambda, n)$ is the symbol of system of pseudo-differential equations (6). So for $\delta > 0$, i.e. $\chi > 1$, the system is elliptic and the solution of the problem exists in $[0, T)$ if and only if the initial data $\mu(x, 0)$, $y(x, 0)$ satisfy the following condition: if

$$\mu(x, 0) = \sum_{-\infty}^{\infty} a_n e^{inx} , \quad y(x, 0) = \sum_{-\infty}^{\infty} b_n e^{inx}$$

then the inequality

$$\sum_{-\infty}^{\infty} |\sqrt{|n|}\, a_n - b_n| \exp(\sqrt{|n|\delta}\, T) < \infty \tag{7}$$

is valid. The local solution exists if the stronger condition

$$\sum_{-\infty}^{\infty} (\sqrt{|n|}\, |a_n| + |b_n|) \exp(|n|\delta T) < \infty \tag{8}$$

is satisfied. Thus the space of the 2π - periodic functions allowing analytical continuation into the strip $D_T = \{z : z = x + iy, |y| < T\}$ is the class of correctness. At the moment of time $t > 0$ the width of the strip of the analyticity of the functions y and μ is at least $2(T - t)$.

In this situation when carrying out the discretization of the problem the following principal sircumstance should be taken into account. When solving problems which are incorrectly posed in the class of functions of finite smoothness it is the form of the table used in the computational algorithm that is of primary importance. The right construction of the computational algorithm is possible only in that case when tables and their deciphering algorithms are chosen in a proper way. The optimal table for the functions f allowing analytical continuation into the strip D_ℓ and which are restricted in it by the constant M is the set of its Furier coefficients approximated in a proper way. The table of approximated values of the function at the nodes $x_k = 2\pi k/(2n+1)$, $k = 0, 1, \dots, 2n$ with the deciphering algorithm based on calculations of the Lagrange interpolational polynomial

$$P_n(x;\hat{f})=\frac{2}{2n+1}\sum_{k=0}^{2n}\hat{f}(x_k)D_n(x-x_k),\quad D_n(x)=\frac{1}{2}+\sum_{k=1}^{n}\cos kx, \quad (9)$$

where $\hat{f}(x_k)$ is the approximated value of the function $f(x)$ at the node x_k, is somewhat more extensive. It can be seen from the calculation that $\hat{f}$ should be taken with the error $\delta \leq \frac{8}{\pi}M \times \times \exp[-(n+1)\ell]$, so the representation of $\hat{f}(x_k)$ requires at most $(n+1)\ell\log_2 e + O(\ln M)$ bits. With such length of the word at the node we obtain our function with the error

$$\varepsilon \leq \left(1+\frac{4}{\pi}\ln n\right)\frac{8}{\pi}M\exp[-(n+1)\ell].$$

Expression (9) is just that very form in which the functions y and μ are sought for. Putting the polynomials $P_n(x;\hat{y})$ and $P_n(x;\hat{\mu})$ into the initial system we obtain a system of ordinary differential equations the solution of which can be found by use of one of high precision methods.

From the above discussion it is clear that to obtain the satisfactory numerical solution of the problem we must keep the proper relation between the dimension n of the problem and the length of the computer word used for storage of $\hat{f}(x_k)$. The limitation on the length of the word produces the bound of possible values of n.

There is a widely used method for working out an incorrectly posed problem according to which the problem should be regularized beforehand. In our case the regard for the forces of surface tension leads to the regularization of the problem and no other regularization is required, since any extra regularization will either misrepresent the picture of the phenomenon or be ineffective.

If ℓ is the scale of the problem and a is a capillary constant then the relation $n > \ell / a$ holds, and so, the constant a being given, we obtain for the computer we are working with the possible maximum values of ℓ. For the discussion of Rayleigh-Taylor problem see reprint [9] where a computational algorithm is analized and results of numerous calculations are adduced as well. For the reason mentioned above we cannot present any appropriate graphs and tables and so the reader is referred to reprint [9].

R E F E R E N C E S

I. Babenko K.I. Doklady Acad. Nauk of the USSR, vol. 24I, No 3, I978.

2. Babenko K.I. Reprint Inst.Prikl.Math., No 7, I974.

3. Babenko K.I. Computer Meth. Appl.Mech., 7 (I976), 47-73.

4. Babenko K.I. Reprint Inst. Prikl. Math. No 29, I977.

5. Babenko K.I., Stebunov V.A. Reprint Inst. Prikl.Math., No 93, I975.

6. Babenko K.I. Doklady Akad.Nauk of the USSR, vol.22I, No I, I975.

7. Babenko K.I., Algazin S.D. Reprint Inst.Prikl.Math., No 46, I978.

8. Babenko K.I., Orlova M.G., Stebunov V.A., Reprint Inst.Prikl. Math., No 55, I978.

9. Babenko K.I., Petrovich B.Yu. Reprint Inst.Prikl.Math., No 68, I978.

IMPLICIT FINITE-DIFFERENCE SIMULATIONS OF STEADY AND UNSTEADY TRANSONIC FLOWS

W. F. Ballhaus
Aeromechanics Laboratory
U.S. Army R&T Laboratories (AVRADCOM)
Ames Research Center, Moffett Field, California 94035, U.S.A.

and

T. L. Holst and J. L. Steger
Ames Research Center, Moffett Field, California 94035, U.S.A.

INTRODUCTION AND SCOPE

Implicit algorithms are often used for the efficient solution of physical problems when only the large-time-scale or the steady-state character of the solution is required. An example of such a problem, and one that is of interest to aeroelasticians and helicopter aerodynamicists, is the solution of the transonic flow about an airfoil oscillating at a low reduced frequency. This flow field can be computed efficiently by using an implicit finite-difference scheme with a relatively large time step of integration. Numerical stability considerations would restrict an explicit scheme to a much smaller time step; so, in principle, substantially more computer time would be expended with no noticeable increase in accuracy over the implicit scheme. For applications in which only a steady-state solution is sought, no time scales need be resolved. But, as such problems are often formulated in real or artificial-time fashion, properly constructed implicit schemes can frequently be used to achieve rapid and reliable convergence.

Implicit solution algorithms have recently been developed and applied to a hierarchy of mathematical formulations for flow-field simulation.[1-11] Listed in order of increasing complexity (and in order of decreasing approximation), these are: (1) transonic small disturbance, (2) full potential, (3) Euler, or the inviscid conservation law equations, and (4) Reynolds-averaged Navier-Stokes, the viscous conservation-law equations. The first two are velocity potential formulations; only a single dependent variable is involved and, hence, only scalar (rather than block) matrix operations are required.

Although there are substantial differences in these formulations, the basic steps involved in constructing implicit schemes for them are similar. These steps, which are illustrated here with examples, are: (1) selection of linearly stable, accurate, implicit difference operators, (2) time-linearization of nonlinear terms, and (3) approximate factorization[12-14] of the implicit operators into easily solved systems of equations.

For illustrative purposes, these steps are detailed first for the low-frequency, small disturbance equation, written in the form

$$\phi_{xt} = F_x + \phi_{yy} \tag{1}$$

where ϕ is the disturbance velocity potential and $F = -\phi_x^2$. In the first step, a set of stable difference operators are selected that have evolved from studies of model linear equations and from experience. If $\phi_x < 0$ (subsonic flow) both spatial derivatives can be approximated using conventional second-order-accurate central differences. At points where $\phi_x > 0$ (supersonic flow), F is approximated using upwind differences. The implicit algorithm (for $\phi_x < 0$) is written

$$\Delta t^{-1}\overleftarrow{\delta}_x(\phi^{n+1}_{i,j} - \phi^{n}_{i,j}) = \delta_x F^{n+1}_{i,j} + \overleftarrow{\delta}_y\overrightarrow{\delta}_y\phi^{n+1}_{i,j} \tag{2}$$

where i and j are gridpoint indices in the x and y directions; and $\delta_x F_{i,j} = (F_{i+1/2,j} - F_{i-1/2,j})\Delta x^{-1}$; $\overleftarrow{\delta}_x \phi_{i,j} = (\phi_{i,j} - \phi_{i-1,j})\Delta x^{-1}$; $\overleftarrow{\delta}_y \overrightarrow{\delta}_y \phi_{i,j} = \delta_{yy}\phi_{i,j} = (\phi_{i,j+1} - 2\phi_{i,j} + \phi_{i,j-1})\Delta y^{-2}$. The difference approximation to ϕ_{xt} is a first-order-accurate one that leads to a stable implicit algorithm. A second-order-accurate approximation can also be used without loss of stability,[1] but it has a form that is unduly complex for this illustration.

In implicitly advancing the solution one time level, large, but sparse, nonlinear systems of equations must be solved. This task is avoided, while maintaining first- (or second-) order time accuracy and conservative form, by use of a proper time linearization. Several approaches are possible, but the Taylor series approximation is usually the most reliable in terms of stability. Let

$$F^{n+1} = F^n + \left(\frac{dF}{d\phi_x}\right)^n (\phi_x^{n+1} - \phi_x^{n}) + O(\Delta t^2) \tag{3}$$

then

$$\delta_x F^{n+1}_{i,j} = -\overleftarrow{\delta}_x (2\phi_x^{\,n}\phi_x^{\,n+1})_{i+1/2,j} + \overleftarrow{\delta}_x (\phi_x^{\,n})^2_{i+1/2,j} + O(\Delta t^2) \tag{4}$$

The system of equations, though time linear, still requires a costly inversion process using direct elimination. This difficulty is avoided by the technique of approximate factorization. As in any factorization concept, one searches for common elements. Note that the first term in Eq. (4) can be written in the form $-2\,\Delta x^{-1}\left(\phi^n_{x_{i+1/2,j}} E_x^{+1} - \phi^n_{x_{i-1/2,j}}\right)\overleftarrow{\delta}_x \phi^{n+1}_{i,j}$, where $E_x^{+1}\phi^n_{i,j} \equiv \phi^n_{i+1,j}$. The operator $\overleftarrow{\delta}_x$ is common to the approximations for ϕ_{xt} and F_x. Hence, the finite difference equation, Eq. (2), can be approximately factored as

$$\left[1 + 2\Delta t\,\Delta x^{-1}\left(\phi^n_{x_{i+1/2,j}} E_x^{+1} - \phi^n_{x_{i-1/2,j}}\right)\right](\overleftarrow{\delta}_x - \Delta t\,\delta_{yy})\phi^{n+1}_{i,j}$$
$$= \overleftarrow{\delta}_x \phi^n_{i,j} + \left(\phi^{n^2}_{x_{i+1/2,j}} - \phi^{n^2}_{x_{i-1/2,j}}\right)\Delta x^{-1}\Delta t - 2\Delta t^2\,\Delta x^{-1}\left(\phi^n_{x_{i+1/2,j}} E_x^{+1} - \phi^n_{x_{i-1/2,j}}\right)\delta_{yy}\phi^n_{i,j} \tag{5}$$

Notice that the approximate factorization introduces an additional term

$$-2\Delta t^2\,\Delta x^{-1}\left(\phi^n_{x_{i+1/2,j}} E_x^{+1} - \phi^n_{x_{i-1/2,j}}\right)\delta_{yy}(\phi^{n+1}_{i,j} - \phi^n_{i,j}) = O\left[\Delta t^3\left(\phi_x\phi_{yyt}\right)_x\right] \tag{6}$$

which does not reduce the formal order of accuracy of a first- or second-order-accurate scheme. Furthermore, this term does not affect the steady-state solution.

As a result of the factorization, only two simple matrix equations need be solved to determine $\phi^{n+1}_{i,j}$. The difference equations are now in the form $\underline{\underline{U}}\,\underline{\underline{T}}\,\underline{\phi}^{n+1} = \underline{c}$, where $\underline{\underline{U}}$ and $\underline{\underline{T}}$ are upper bidiagonal and tridiagonal matrices, respectively. The solution is obtained using the following two-step procedure:

step 1: $\underline{\underline{U}}\,\underline{f} = \underline{c}$ is solved for $\underline{f}(\equiv \underline{\underline{T}}\,\underline{\phi}^{n+1})$

step 2: $\underline{\underline{T}}\,\underline{\phi}^{n+1} = \underline{f}$ is solved for $\underline{\phi}^{n+1}$

The same steps are followed in constructing implicit approximate-factorization schemes for the other formulations, although for the Euler and Navier-Stokes equations, vector rather than scalar quantities are involved. For example, for the Euler formulation, the governing equations can be written in the form

$$q_t + E_x + F_y = 0 \tag{7}$$

where q, E, and F are the flux vectors. Central spatial differences are chosen for stability reasons — the real eigenvalues associated with E and F can have mixed signs and are thus unstable if one-sided spatial operators are used. The implicit differencing is represented as

$$q^{n+1} - q^n + \Delta t(\delta_x E^{n+1} + \delta_y F^{n+1}) = 0 \tag{8}$$

where again for illustrative purposes, first-order time differencing is used.

To avoid solving nonlinear equations at each time level, the time linearizations

$$\left.\begin{aligned} E^{n+1} &= E^n + A^n(q^{n+1} - q^n) + O(\Delta t^2) \\ F^{n+1} &= F^n + B^n(q^{n+1} - q^n) + O(\Delta t^2) \end{aligned}\right\} \tag{9}$$

are introduced, where A and B are the 4×4 Jacobian matrices ($A = \partial E/\partial q$, $B = \partial F/\partial q$). Equation (8) then becomes

$$[I + \Delta t(\delta_x A^n + \delta_y B^n)](q^{n+1} - q^n) = -\Delta t(\delta_x E^n + \delta_y F^n) \tag{10}$$

Here, terms have been arranged in a correction (or delta) form.

Finally, the difference equations are approximately factored to avoid costly inversions (the common element in this case is the identity matrix, I)

$$(I + \Delta t\delta_x A^n)(I + \Delta t\delta_y B^n)(q^{n+1} - q^n) = -\Delta t(\delta_x E^n + \delta_y F^n) \tag{11}$$

Again, a second-order term $\Delta t^2\delta_x A^n\delta_y B^n(q^{n+1} - q^n) \approx O(\Delta t^3\delta_x A^n\delta_y B^n q_t)$ results from the approximate factorization. The factored equation is solved as a two-step procedure as before except that here more costly block (rather than scalar) tridiagonal matrix operations must be performed. Details for higher order spatial differencing, viscous terms, and arbitrary geometry can be found in Refs. 5-11.

Special consideration should be given to the effect of an approximate factorization on the enforcement of boundary conditions. For example, in the process of generating f in step 1 of Eq. (5), the quantity $(\overleftarrow{\delta}_x - \Delta t\delta_{yy})\phi^{n+1}$ must be specified on the maximum extent of the computational domain in the x-direction. If the boundary condition is one for which ϕ^{n+1} is specified (Dirichlet), then $-\Delta t\delta_{yy}\phi^{n+1}$ is known at the boundary. However, $\overleftarrow{\delta}_x\phi^{n+1}$ is unknown and must be estimated by extrapolation or evaluated explicitly. Stability limitations on the integration time step can sometimes result depending on how such terms are handled.[1]

ILLUSTRATIVE RESULTS

Solutions obtained from the implicit procedures for two representative airfoil problems are considered. The first is the computation of unsteady (periodic) pressures on the surface of an NACA 64A006 airfoil with an oscillating trailing-edge flap. Computed results from the small-disturbance, Euler, and Reynolds-averaged Navier-Stokes formulations are compared in Fig. 1 at two instances during the flap motion cycle. (No implicit time-accurate scheme has yet been developed for the full potential formulation.) At $\omega t = 50°$ the shock strength is maximum, while at $\omega t = 270°$ the shock has disappeared only to reappear at a subsequent time. The two inviscid solutions compare favorably, which is not surprising in this thin airfoil

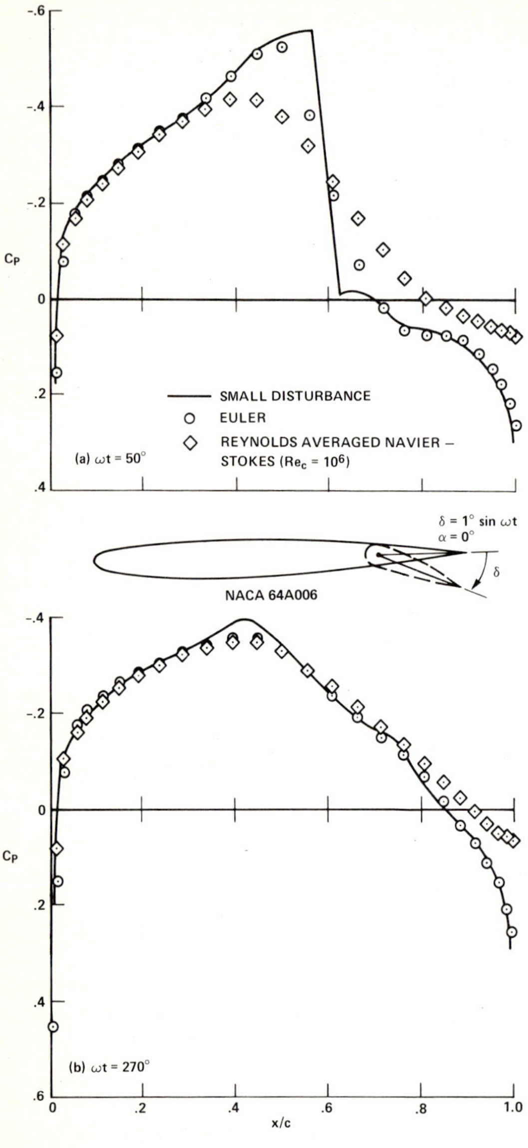

Fig. 1. Upper surface pressure coefficient, M_∞ = 0.854; k = 0.358 (reduced frequency, $\omega c/U_\infty$).

example. For both the Euler and Navier-Stokes cases, the initial finite-difference grids were computed numerically. They were then distorted, using a simple shearing transformation, to follow the motion of the flap. In the Navier-Stokes case, a two-layer eddy viscosity model was used. The influences of viscosity — the shock-boundary-layer interaction and the thickening of the boundary layer near the trailing edge — are apparent. The lift coefficient time history is shown in Fig. 2.

Integration time steps were on the order of from 1/2° (Navier-Stokes) to 3° (small disturbance) of flap motion. These time steps are substantially larger than stability restrictions would permit for explicit schemes with equivalent spatial grids.

As one considers successively thicker airfoils, a point is reached beyond which the transonic small disturbance theory is no longer a suitable approximation. This is illustrated in Fig. 3 for the second airfoil problem. Steady-state shock location (X_s) is plotted as a function of airfoil thickness-to-chord ratio for the circular-arc airfoil family. The error in small disturbance predicted shock wave location, as expected, increases with thickness.

Ultimately, a point is reached at which viscous phenomena dominate, and the full potential and Euler approximations are no longer valid. This is dramatically evident in the 18% thickness ratio case, because for this particular Mach number the viscous solution buffets. The extent of shock excursions is shown in Fig. 3; the viscous-induced unsteadiness of the flow field is apparent in Fig. 4. Lines of constant Mach number show intermittent trailing-edge and shock-induced separation. Figure 4(a) indicates boundary-layer separation on the upper surface at the shock; the lower surface boundary layer separates at some point downstream of the shock. At a subsequent time, Fig. 4(b), the shock waves move forward and aft on the upper and lower surfaces, respectively. Forward motion strengthens the shock, increasing the degree of separation; aft motion weakens the shock, decreasing its strength, so that the lower surface flow remains attached, for this particular point in time. Incidentally, the buffet predicted for this case computationally has been confirmed by experiment.

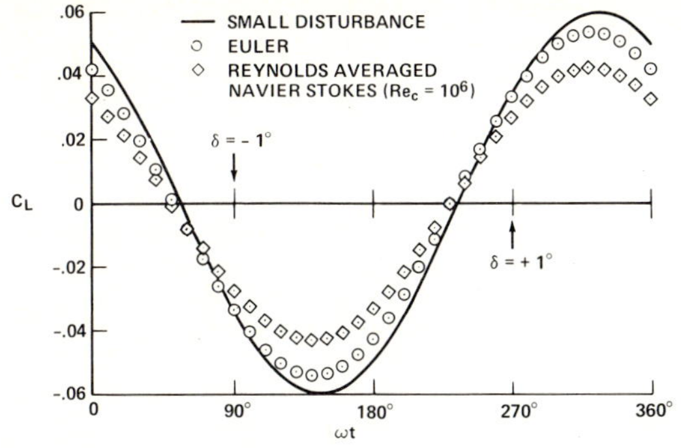

Fig. 2. Lift coefficient vs time ($M_\infty = 0.854$, k = 0.358).

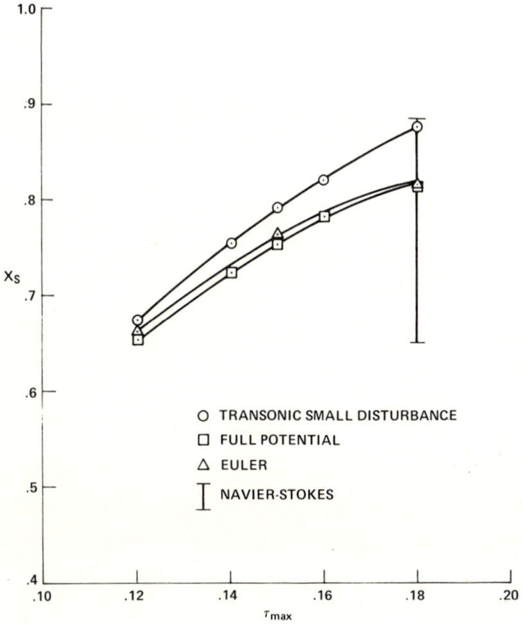

Fig. 3. Shock position vs airfoil thickness (circular arc airfoil, $M_\infty = 0.783$, $\alpha = 0°$).

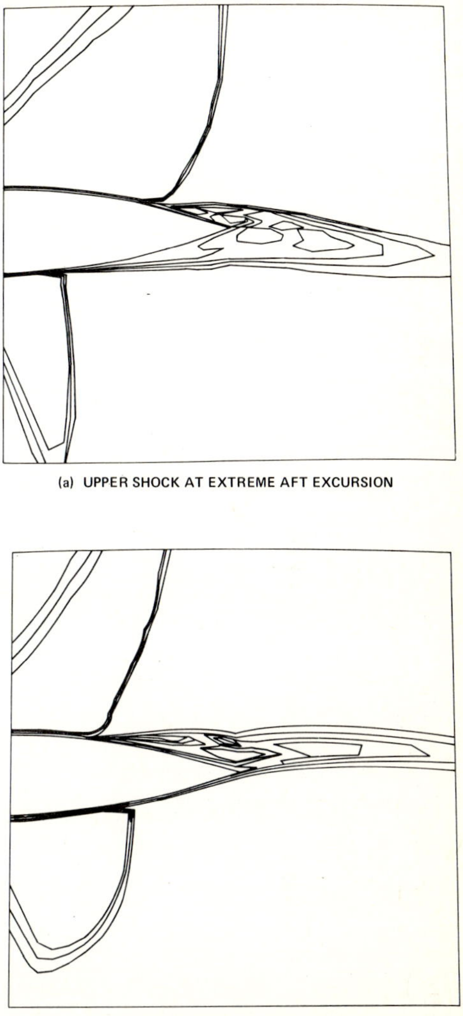

(a) UPPER SHOCK AT EXTREME AFT EXCURSION

(b) UPPER SHOCK MOVING FORWARD

Fig. 4. Instantaneous Mach contours during buffet cycle ($M_\infty = 0.783$, $\alpha = 0°$, 18% circular arc airfoil, $Re_c = 11 \times 10^6$).

Convergence histories of the three inviscid implicit methods for the 15% thick case are compared in Fig. 5. The comparison is in terms of (normalized) root-mean-squared error in airfoil surface pressure versus iteration. Integration time step sequences were used for the two potential formulations; a constant time step was used for the Euler solution. The full potential procedure[4] was formulated in terms of an artificial time. The number of grid points used for the small disturbance, full potential, and Euler formulations was 5525, 7236, and 2233, respectively. The convergence rates for the potential methods are many times greater than can be achieved with the standard (semi-implicit) successive line overrelaxation procedure. The performance of the implicit Euler solution procedure is superior to that of explicit schemes. However, it is retarded by (nonlinear) stability restrictions on the

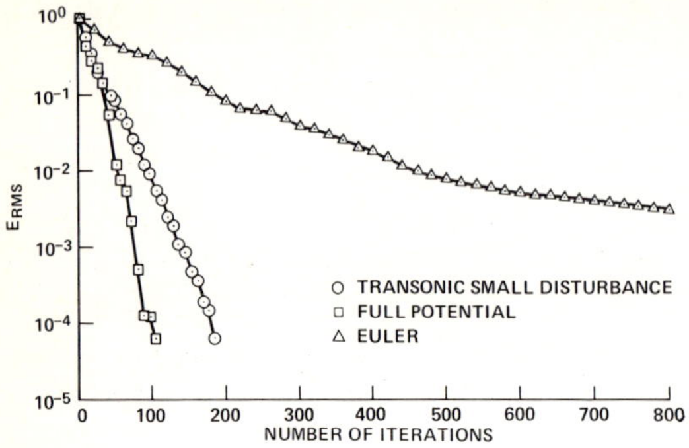

Fig. 5. Solution convergence rate.

integration time step, which preclude use of an effective sequence of time steps. Elimination of these restrictions should result in significant improvement.

To summarize, implicit methods for several fluid dynamic formulations have been developed and successfully applied to steady-state and low-frequency transonic flows. For the simpler formulations, these schemes have proved to be very efficient. Improvements can be expected for the Euler and Navier-Stokes formulations.

REFERENCES

1. Ballhaus, W. F. and Steger, J. L., NASA TM X-73,082, Nov. 1975.

2. Ballhaus, W. F., Jameson, A., and Albert, J., AIAA Paper 77-634, June 1977.

3. Ballhaus, W. F. and Goorjian, P. M., AIAA J., 16, 1978.

4. Holst, T. L. and Ballhaus, W. F., NASA TM 78469, Feb. 1978.

5. Beam, R. M. and Warming, R. F., J. Comp. Physics, 22, 1976.

6. Beam, R. M. and Warming, R. F., AIAA J., 16, 1978.

7. Briley, R. and MacDonald, H., United Aircraft Res. Lab. Rept., M911363-6, 1973.

8. Warming, R. F. and Beam, R. M., SIAM-AMS Proceedings, 11, Proceedings of the Symposium on Computational Fluid Mechanics, 1977.

9. Steger, J. L. and Kutler, P., AIAA J., 15, 1977.

10. Steger, J. L., AIAA Paper 77-665, 1977.

11. Pulliam, T. H. and Steger, J. L., AIAA Paper 78-10, 1978.

12. Yanenko, N. N., The Method of Fractional Steps, Springer-Verlag, 1971.

13. Peaceman, D. W. and Rachford, H. H., J. Soc. Indust. Appl. Math., 3, 1955.

14. Douglas, J., J. Soc. Indust. Appl. Math., 3, 1955.

NUMERICAL INVESTIGATION OF SOME GAS DYNAMICS PROBLEMS BY NET-CHARACTERISTIC METHOD

O. M. Belotserkovskii, A. S. Kholodov
(Moscow, USSR)

The problems facing computer scientists at the present time are manifold and are rapidly becoming more complicated. This situation requires the permanent improvement of numerical methods already developed and the creation of new ones. The properties a numerical method is to be endowed with from the viewpoint of modern developments are diverse and they are difficult to implement in any particular method. In view of this, complexes of numerical methods based upon a single approach should be available.

A large and important class of problems is described by multi-dimensional systems of equations of hyperbolic type. A fundamental concept underlying these systems is their characteristic properties which, in some form, are to be taken into account in the construction of numerical approximations for such systems. Similar equations in continuum mechanics are given in the form of some conservation laws so that a substantial circumstance here is to impart relevant features to the numerical methods of their solution as well.

Finally, in the numerical solution of the given problems it is not less significant to have a reasonable combination of homogeneous schemes (for "through" calculations arising in the unknown solutions of discontinuities) explicitly singling out the more important features and giving them adequate boundary conditions.

1. Among other difference schemes for equations of hyperbolic type schemes with positive approximation first introduced in [1] play an important role. Using a method of indefinite coefficients and the characteristic properties of equations of hyperbolic type one can write a rather large class of schemes of this kind and perform then a comparative analysis, in particular, with respect to the magnitude of "approximation viscosity" (see, e.g., [2]). These schemes possess a number of undoubted advantages, the main ones being the absence of oscillations of a difference origin on unsmooth solutions, the possibility of the construction of efficient algorithms for calculating boundary points and so on. However, they all have the first order accuracy which, in some cases, in the utilization of even the best of the schemes [3,4] (having a minimal "approximation viscosity") requires the choice of a great number of calculation points of a difference net and, as a consequence, voluminous calculations.

The same approach in which difference schemes are treated as components of linear space of indefinite coefficients can be applied for the construction of some schemes of a higher order accuracy on the solutions of initial equations.

The basic idea of the method is illustrated by the simplest equation of transfer

$$v_t + \lambda v_x = 0 \quad , \quad \lambda = \text{const.} \quad , \quad \lambda > 0 \tag{1}$$

and a pattern typical of explicit schemes (Fig. 1) comprising six points

$$(t^{n+1},x_m)(t^n,x_{m-2}),\dots(t^n,x_{m+2}) \quad . \tag{2}$$

Write linear difference schemes permitted by the pattern of the form

$$v_m^{n+1} = \sum_{\mu,\nu} \alpha_\mu^\nu v_{m+\mu}^{n+\nu} = \sum_{\mu=-2}^{2} \alpha_\mu v_{m+\mu}^n \tag{3}$$

and using approximation conditions

$$\sum_{\mu,\nu} (\mu-\nu\sigma)\alpha_\mu^\nu = -\sigma \quad , \quad \sum_{\mu,\nu} \alpha_\mu^\nu = 1 \quad , \quad \sigma = \lambda\tau/h \geqq 0 \quad , \tag{4}$$

exclude, e.g., coefficients α_{-1} and α_1.

In the space of the coefficients that remain indefinite $\alpha = \{\alpha_{-2}, \alpha_0, \alpha_2\}$ (Figs. 2a,b) any point gives rise to a difference scheme of the first order accuracy for Eq. (1), closed polyhedrons $A_1A_3,\ldots,A_6$ with $0 \leqq \sigma \leqq 1$ (Fig. 2a) and $A_2A_3A_4A_6$ with $1 < \sigma \leqq 2$ (Fig. 2b) incorporate difference schemes with positive approximation for which all coefficients

$$\alpha_\mu^\nu \geqq 0 \quad . \tag{5}$$

To points A_1 with $0 \leqq \sigma \leqq 1$ (scheme [3]) and A_2 with $1 < \sigma \leqq 2$ correspond difference schemes with the least "approximation viscosity" with the smallest value of the coefficient at v_{xx} in the first differential approximation (3) [2].

Plane $B_1,\ldots,B_9$ (Fig. 2)

$$\sum_{\mu,\nu} (\mu-\nu\sigma)^2 \alpha_\mu^\nu = \sigma^2 \quad \text{or} \quad \alpha_0 = 3(\alpha_{-2}+\alpha_2) + 1 - \sigma^2 \tag{6}$$

for the pattern (2) constitutes a two-parametric family of difference schemes of the second order accuracy on solutions (1).

In Figs. 2a,b the boundaries of the region of steady schemes with an approximation order higher than the first are shown by dashed lines on plane (6). It is seen that with $0 \leqq \sigma \leqq 2$ this region is not empty.

It is shown in [5] that there are no difference schemes of form (3) having the second order approximation (a higher order accuracy as well, see [2]) and satisfying the constraints in Eq. (5). (Plane (6) does not intersect a closed polyhedron (4), (5).)

A straight line $C_1,\ldots,C_5$ being an intersection line (6) and a plane

$$\sum_{\mu,\nu} (\mu-\nu\sigma)^3 \alpha_\mu^\nu = -\sigma^3 \tag{7}$$

incorporates a one-parametric family of schemes of the third order accuracy.

With $0 \leqq \sigma \leqq 1$ the segment C_1C_4 (Fig. 2a) contains steady schemes of the third order accuracy. Point C_1 on the pattern given is the only difference scheme of the fourth order accuracy. The familiar Lax-Wendroff scheme is indicated by point B_5 in Figs. 2a,b.

2. Evidently, the various kinds of oscillations on unsmooth solutions that are observed in familiar difference schemes with a higher approximation order than the first and that are not present in schemes with positive approximation are due to the fact that some of the coefficients α_μ^ν in difference expressions of type (3) for schemes of a higher order accuracy are negative.

It is natural to suppose that the behavior of a particular scheme on unsmooth solutions (the amplitude of oscillations, their character) is determined by the distance of a point (corresponding to this scheme) in the given space of indefinite coefficients from the region of difference schemes with positive approximation (from polyhedrons $A_1A_3,\ldots,A_6$ (Fig. 2a), $A_2A_3A_4A_6$ (Fig. 2b), etc.). That is, it is proposed that the "non-monotonicity" of difference schemes be characterized by the value

$$\gamma = |\alpha - \alpha_A| \quad , \tag{8}$$

where $\alpha\{\alpha_\mu^\nu\}$ is a set of coefficients in Eq. (3) corresponding to the difference scheme in question (in the case given to schemes with a higher approximation order than the first situated on plane (6), and α_1 is a set of coefficients in Eq. (3) corresponding to the vertex of a polyhedron in Eqs. (4), (5) giving difference schemes of the first order accuracy with positive approximation (point A_1 in Fig. 2a with point A_2 in Fig. 2b with $0 \leqq \sigma \leqq 1$, etc.). That is, schemes of the first order accuracy should be chosen as point A, they should be constructed on three-point patterns incorporating a calculated point (t^{n+1}, x_m) and two knots of the difference scheme nearest to characteristics $dx = \lambda\, dt$ lying on both sides of these characteristics.

For difference schemes of the second order accuracy on solutions (1) the coefficients α_μ^ν in such a scheme are found by a conventional geometric construction in space $\alpha = \{\alpha_{-2}, \alpha_0, \alpha_2\}$ of the point of intersection with plane (6) of the normal drawn from point A. With $0 \leqq \sigma \leqq 1$ ($\alpha_A = \{0, 1-\sigma, 0\}$) this procedure results in a difference scheme of the second order accuracy-point B_6 in Fig. 2a. With $1 \leqq \sigma \leqq 2$ to such a scheme corresponds point B_9 in Fig. 2b ($\alpha_A = \{\sigma_{-1}, 0, 0\}$).

For difference schemes of the third order accuracy a scheme with the smallest value γ which is steady at $0 \leqq \sigma \leqq 1$ is found by constructing the point of intersection with a straight line (6), (7) of the normal drawn from point $\alpha_A = \{0, 1-\sigma, 0\}$ (point C_3 in Fig. 2a).

The calculation of the simplest modelling problems reveals that the schemes built in this manner as compared to other schemes of the second and third order accuracy have the shortest amplitude of oscillations with fast damping.

Bearing in mind that Eq. (1) is, in essence, an ordinary differential equation along the characteristics $dx = \lambda\, dt$

$$\frac{dv}{dt} = 0 \quad , \quad \frac{d}{dt} = \frac{\partial}{\partial t} + \lambda \cdot \frac{\partial}{\partial x}$$

and, consequently, at point (t^{n+1}, x_m) the exact value is

$$v_m^{n+1} = v(t^m, \tilde{x}) \quad , \quad \tilde{x} = x_m - \lambda\tau \quad ,$$

the difference expressions in the right-hand part (3) are essentially interpolation formulas for calculating $v(t^n, \tilde{x})$. From this viewpoint the right-hand part in Eq. (3) for schemes with the smallest value γ may be treated as interpolation polynomials of the second and third order for filling in the function $v(t^n, \tilde{x})$; these polynomials diverge in the standard (8) from a piece-linear polynomial of the first order relevant to difference schemes of the first order accuracy with positive approximation. A more detailed description of the approach in hand including "implicit" patterns is supposed to be given in one of the later publications.

3. The construction of difference schemes in Sec. 1,2 is realized for a separate condition of compatibility along characteristics $dx = \lambda_i dt$ of a one-dimensional system of equations of hyperbolic type

$$v_t + Av_x = f \quad , \quad A = \Omega^{-1} \Lambda \Omega \quad , \tag{9}$$

which, in a conventional way, is reduced to a canonical form

$$w_i v_t + \lambda_i w_i v_x = w_i f \quad , \quad i = 1, \ldots, I \quad . \tag{10}$$

Here $v = \{v_1, \ldots, v_I\}$ is the vector of unknown functions; $f = \{f_1, \ldots, f_I\}$ is a vector-column of right-hand parts; $\Lambda = \{\lambda_i\}$ is a diagonal matrix from eigenvalues of matrix A; $\Omega = \{w_i\}$ is a nonsingular matrix whose lines are linear-independent eigenvectors w_i of matrix A.

If matrix A has fixed components and $f = 0$, difference schemes (3) are generalized, in an obvious manner, for the case of system (9)

$$v_m^{n+1} = \sum_{\mu,\nu} D_\mu^\nu v_{m+\mu}^{n+\nu} D_\mu^\nu = \Omega^{-1} A_\mu^\nu \Omega \quad ,$$

where $A_\mu^\nu = \{(\alpha_\mu^\nu)_i\}$ are diagonal matrices.

For a general quasilinear case of system (9) ($A = A(t,x,v)$, $f = f(t,x,v)$) it is required to develop a proper method of integration of "ordinary" differential equations (10) with account taken of the dependence λ_i, w_i, f upon the unknown solution v and independent variables t,x. Different approaches may be used here. One of them intended for the construction of explicit difference schemes on the pattern of type (2) is the Runge-Kutta method used in [7] and in a number of subsequent papers (see, e.g., [8]).

In [7] for a divergent system

$$v_t + F_x(t,x,v) = \phi \quad , \tag{11}$$

on the pattern (2) a one-parametric family of difference schemes of the third order accuracy is constructed

$$v^{n+\alpha}_{m+\frac{1}{2}} = (v^n_m + v^n_{m+1})/2 - \alpha\tau(F^n_m - F^n_m)/h + \alpha\tau(\phi^n_{m+1} + \phi^n_m)/2 \quad , \tag{12}$$

$$v^{n+\beta}_m = v^n_m - \beta\tau(F^{n+\alpha}_{m+\frac{1}{2}} - F^{n+1}_{m-\frac{1}{2}})/h + \beta\tau(\phi^{n+\alpha}_{m+\frac{1}{2}} + \phi^{n+\alpha}_{m-\frac{1}{2}})/2 \quad , \tag{13}$$

$$v^{n+1}_m = v^n_m + \frac{3\tau}{4}\left[-\frac{(F^{n+\beta}_{m+1} - F^{n+\beta}_{m-1})}{2h} + \phi^{n+\beta}_m\right] + \frac{\tau}{4}\left[-\frac{(F^n_{m+1} - F^n_{m-1})}{2h} + \phi^n_m\right]$$

$$+ \tau(F^n_{m-2} - 2F^n_{m+1} + 2F^n_{m-1} - F^n_{m-2})/12h + g(v^n_{m-2} - 4v^n + 6v^n_m - 4v^n_{m+2} + v^n_{m+2}) \quad ,$$

$$\alpha = 1/3 \quad , \quad \beta = 2/3 \quad . \tag{14}$$

It is also shown there that for providing the stability of these schemes a scalar parameter g is necessary to choose on the condition that

$$-1/8 \leqq g \leqq \sigma^2_*(\sigma^2_*-4)/24 \quad , \quad \sigma_* = \tau \max_{m,i}|\lambda_i|/n \quad . \tag{15}$$

According to the analysis carried out in Sec. 1,2 for linear equations it is preferable to choose a matrix

$$\Omega^{-1}\cdot G\Omega \quad , \quad G = \{g_i\} \quad , \quad \Omega = \{w_i\} \quad ,$$
$$g_i = |\sigma_i|\cdot(5|\sigma_i|-24)/152 \quad , \quad \sigma_i = \lambda_i\tau/n \quad , \quad i = 1,\dots,I \tag{16}$$

instead of a scalar factor in the last term (14) which is the same for all characteristics. Here and in Eq. (15) λ_i are eigenvalues of matrix $A = \partial f/\partial v$, Ω is a matrix derived above from eigenvectors, G is a diagonal matrix. The given modification of the scheme (12-14) in a linear case is a scheme discussed in Sec. 2 with the smallest value γ. We should also note that the last term in (14) modified in accordance with (16) is more reasonable to write in a divergent form

$$(\Omega^{-1}G\Omega)^n_{m+\frac{1}{2}}\cdot(-v^n_{m+1} + 3v^n_m - 3v^n_{m+1} + v^n_{m+2})$$
$$- (\Omega^{-1}G\Omega)^n_{m-\frac{1}{2}}\cdot(-v^n_{m-2} + 3v^n_{m-1} - 3v^n_m + v^n_{m+1}) \quad . \tag{17}$$

On the same pattern (2) there can be generalized for the case of quasilinear systems (11) difference schemes of the second order accuracy which are "closer" to schemes with positive approximation, for example, in the form of a two-step system with a predictor, the Lax scheme (2) with $\alpha = 0{,}5$ and a corrector

$$
\begin{aligned}
v_m^{n+1} &= (\Omega^{-1}A_2\Omega)^n_{m+\frac{1}{2}}(-v^n_{m-1} + 3v^n_m - 3v^n_{m+1} + v^n_{m+2}) \\
&\quad - (\Omega^{-1}A_2\Omega)^n_{m-\frac{1}{2}}(v^n_{m-2} + 3v^n_{m+1} - 3v^n_m + v^n_{m+1}) \\
&\quad + v^n_m - \tau(F^{n+\frac{1}{2}}_{m+\frac{1}{2}} - F^{n+\frac{1}{2}}_{m-\frac{1}{2}})/n + \tau(\phi^{n+\frac{1}{2}}_{m+\frac{1}{2}} + \phi^{n+\frac{1}{2}}_{m-\frac{1}{2}})/2 \quad ,
\end{aligned}
\tag{18}
$$

which is stable fulfilling a condition

$$
\sigma_* = \tau \max_{m,i}|\lambda_i|/n \leqq 2 \quad . \tag{19}
$$

Here $A = \{(\alpha_{-2})_i\}$, $A_2 = \{(\alpha_2)_i\}$ are diagonal matrices with elements

$$
\begin{aligned}
(\alpha_2)_i &= 3(1-\sigma_i)(2-\sigma_i)/19 \;, \quad (\alpha_{-2})_i = (\alpha_{-2})_i - (1-\sigma_i) \quad \text{for} \quad 1 < \sigma_i \leqq 2 \;, \\
(\alpha_{-2})_i &= 3(1+\sigma_i)(2+\sigma_i)/19 \;, \quad (\alpha_2)_i = (\alpha_{-2})_i - (1+\sigma_i) \quad \text{for} \; -2 \quad \sigma_i \quad -1 \;, \\
(\alpha_2)_i &= (\alpha_{-2})_i = -3|\sigma_i|(-1-|\sigma_i|)/19 \quad \text{for} \quad |\sigma_i| \leqq 1 \quad .
\end{aligned}
\tag{20}
$$

It is seen that satisfying a more strict condition $\sigma_* \leqq 1$ than (19) the scheme (12), (18), (20) is a divergent (conservative) one. The last three terms in Eq. (18) coincide with the familiar Lax-Wendroff scheme [6] and can be replaced by other modifications of analogous scheme.

As an example in Fig. 3a the results of the numerical solution $(t/\tau = 50)$ with the help of the schemes of the second order accuracy (12), (18), (20) (scheme II), Lax-Wendroff schemes [6], MacCormak schemes [9] are compared with the accurate solution (dashed lines) of a problem concerned with the destruction of a discontinuity for one-dimensional gas dynamics equations and initial conditions: $v(0,x) = 0$, $\rho(0,x) = p(0,x) = 2$ with $x \leqq 0$, $\rho(0,x) = p(0,x)$ with $x > 0$.

In Fig. 3b an analogous comparison is given for scheme (12)-(14) with $(\sigma_* = 1)$ choosing g_i from (16) (scheme III) and $g_i = \sigma_*^2(\sigma_*^2-4)/24 = -1/8$. In the same figure a scheme of the first order accuracy with positive approximation [10] (scheme I) is compared with an accurate solution and schemes of the third order accuracy.

Here, as well as in a linear case, one observes the improvement of "oscillation" properties in schemes II, III as compared to other schemes of the second and third order accuracy. It is also seen that scheme I as far as its accuracy is concerned is quite comparable with schemes II, III in the calculation of shock waves; however, it requires a very fine difference net for through calculations of contact discontinuities.

4. Without dwelling here upon the generalization of the difference schemes in questions for a multidimensional case it should only be noted that for schemes of the first order accuracy with positive approximation (3), (4) necessary construction procedures are performed rather formally (see, e.g., [3,10,11]). For difference schemes of a higher order accuracy there also exist various efficient approaches described in detail in the scientific publications.

In the past years systematic investigations on the aerogasdynamics of bodies of a complex geometric configuration, on the numerical modelling of some problems in plasma physics, on the numerical solution of dynamic problems in the theory of elasticity and so on have been conducted on the basis of some of the difference schemes discussed here. The priminent features in such problems are the complex structure of an unknown solution, the presence of regions with large gradients, discontinuous functions, etc.

Nondivergent schemes of the first order accuracy [11] prove to be sufficiently effective for the numerical solution of problems with a comparatively small number of surfaces of discontinuities in an unknown solution; they can be explicitly defined formulating on them proper boundary conditions.

As an example, in Fig. 4a, in the plane of symmetry of flow there are shown stationary patterns of a supersonic space inviscous nonthermoconducting gas flow (adiabatic index $x = 1{,}4$) around a cone with a half-angle $\theta_R = 10^o$ which is blunt along a sphere on one side and has segmental blunting on the other side ($\theta_c = 35^o$). The general length of the body is equal to 5.5 radii of spherical blunting. The Mach number of an incoming flow $M = 2$, an angle of attack varies from 0 to 180°.

Figure 4b presents resulting pressure distributions along a body referred to a double speed pressure. Other problems of a space supersonic flow around blunt bodies solved by means of a numerical method in [11] with the consideration of equilibrium physico-chemical processes, radiation, nonstationarity are described in detail in [12-16].

The introduction in a difference scheme from [11] of conservative elements (see, e.g., [2,10]) allows us with a desirable accuracy to obtain a solution in problems with singularities (discontinuities) within an integration region without singling them out explicitly.

A number of one- and two-dimensional problems of the interaction of laser light with matter is considered with the help of this scheme. As a typical example of calculations of this kind (taken from [10]) Fig. 5 presents at a time instant appropriate to the termination of an impulse $t = 10^{-10}$ sec isochores $\rho/\rho_o = \text{const.}$ (Fig. 5a) and isotherms in Kev (Te = const., solid curves, Ti = const. - dashed curves, Fig. 5b) in the interaction of a symmetrical impulse of laser light of energy E = 300j with a spherical envelope of variable thickness.

Here Te, Ti are electronic and ionic plasma temperatures, the initial disturbances of the envelope are equal to 50% its thickness; the following physical processes are taken into account: the absorption of outer laser radiation, nonlinear electronic thermal conduction, electron-ion collisional relaxation. The terms in energy equations for electronic and ion components related to thermal conduction and energy exchange between the components are approximated in an implicit way. The direct application for the solution of similar problems of the schemes in [11] results in a significant perturbation of an integral mass balance, impulse, energy and the solution in the vicinity of a nonstationary shock wave moving along a cold background is unsatisfactory.

As the last example of the application of difference schemes whose construction is based upon the utilization of characteristic features of equations of hyperbolic type and a fixed difference net (net-characteristic methods) in Fig. 6 in time $t = t_* = 0.029$ there is shown an instantaneous picture of the distribution of components $\sigma_{yy}, \sigma_{xy}, \sigma_{xx}$ of a stress tensor in an elastic layer of finite thickness. They appear at the nonstationary loading of part of the upper boundary of a layer on a perfectly rigid base (a flat deformed condition).

Boundary conditions in this problem are given as follows: On the portion AB′ of the upper boundary

$$v_y(t,x,1) = -\frac{1}{m_0}\int_0^t dt[P+2\int_A^B \sigma_{yy}(t,x,1)dx] \quad ,$$

$\sigma_{xy}(t,x,1) = 0$ on the other portion of the upper boundary $\sigma_{yy}(t,x,1) = \sigma_{xy}(t,x,1) = 0$, on the lower boundary $v_y(t,x,0.5) = \sigma_{xy}(t,x,0.5) = 0$, where AO is a plane of symmetry, $m,p = 1$ are constants.

In conclusion, the authors would like to express their sincere gratitude to V. V. Demchenko and I. B. Petrov who were helpful in conducting the numerical solution of some of the problems discussed here.

REFERENCES

[1] K. O. Fridrichs (1954) Comm. Pure and Appl. Math., Vol. 7, pp. 345-392.

[2] A. S. Kholodov (1978) J. Vych. Mat. i Mat. Fiz., Vol. 18, No. 5.

[3] R. Courant, E. Isacson, M. Rees (1952) Comm. Pure and Appl. Math., Vol. 5, pp. 243-254.

[4] H. B. Keller, B. Wendroff (1957) Comm. Pure and Appl. Math., Vol.10, pp. 567-582.

[5] S. K. Godunov (1959) Mat. sb., Vol. 3, pp. 271-306.

[6] P. D. Lax, B. Wendroff (1960) Comm. Pure and Appl. Math., Vol. 13, pp. 217-306.

[7] V. V. Rusanov (1968) Dokl. AN SSSR, Vol. 180, No. 6.

[8] V. B. Balakin (1970) J. Vych. Mat. i Mat. Fiz., Vol. 10, pp. 1512-1519.

[9] R. W. MacCormak (1969) AIAA Paper, 69-354.

[10] O. M. Belotserkovskii, V. V. Demchenko, V. I. Kosarev, A. S. Kholodov (1978) J. Vych. Mat. i Mat. Fiz., Vol. 18, pp. 420-444.

[11] K. M. Magomedov, A. S. Kholodov (1969) J. Vych. Mat. i Mat. Fiz., Vol. 9, pp. 373-386.

[12] O. M. Belotserkovskii (ed.) (1974) Numerical Investigation of Modern Gas Dynamics Problems., M., "Nauka."

[13] O. M. Belotserkovskii, S. D. Osetrova, V. N. Fomin, A. S. Kholodov (1974) J. Vych. Mat. i Mat. Fiz., Vol. 14, pp. 992-1003.

[14] V. S. Kostrykin, V. N. Fomin, A. S. Kholodov (1976) J. Vych. Mat. i Mat. Fiz., Vol. 16, pp. 451-459.

[15] A. V. Krasilnikov, A. N. Nikulin, A. S. Kholodov (1975) Izv. Ac. Nauk SSSR, MJG, No. 5, pp. 179-181.

[16] O. M. Belotserkovskii, L. I. Turchak, A. S. Kholodov (1975) Izv. Nauk SSSR, MJG, No. 5, pp. 110-115.

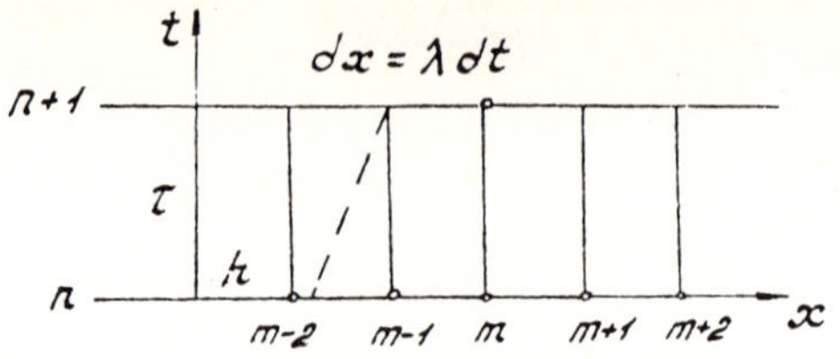

Fig. 1

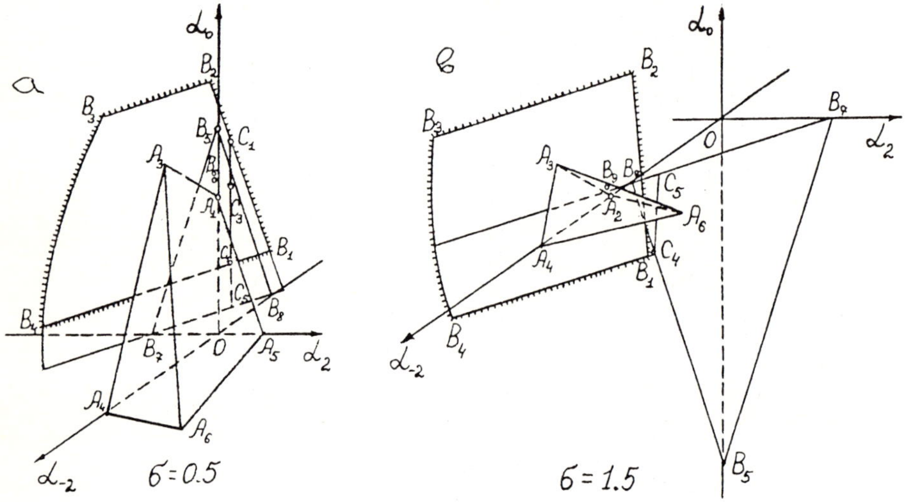

Fig. 2

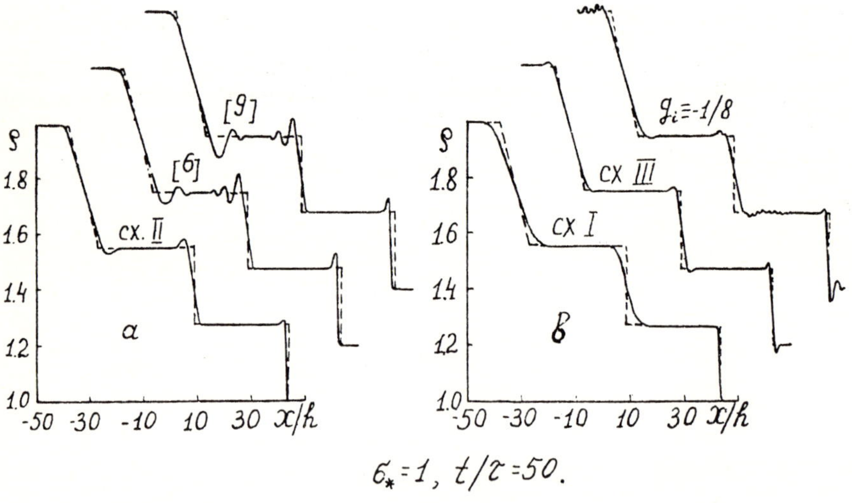

Fig. 3

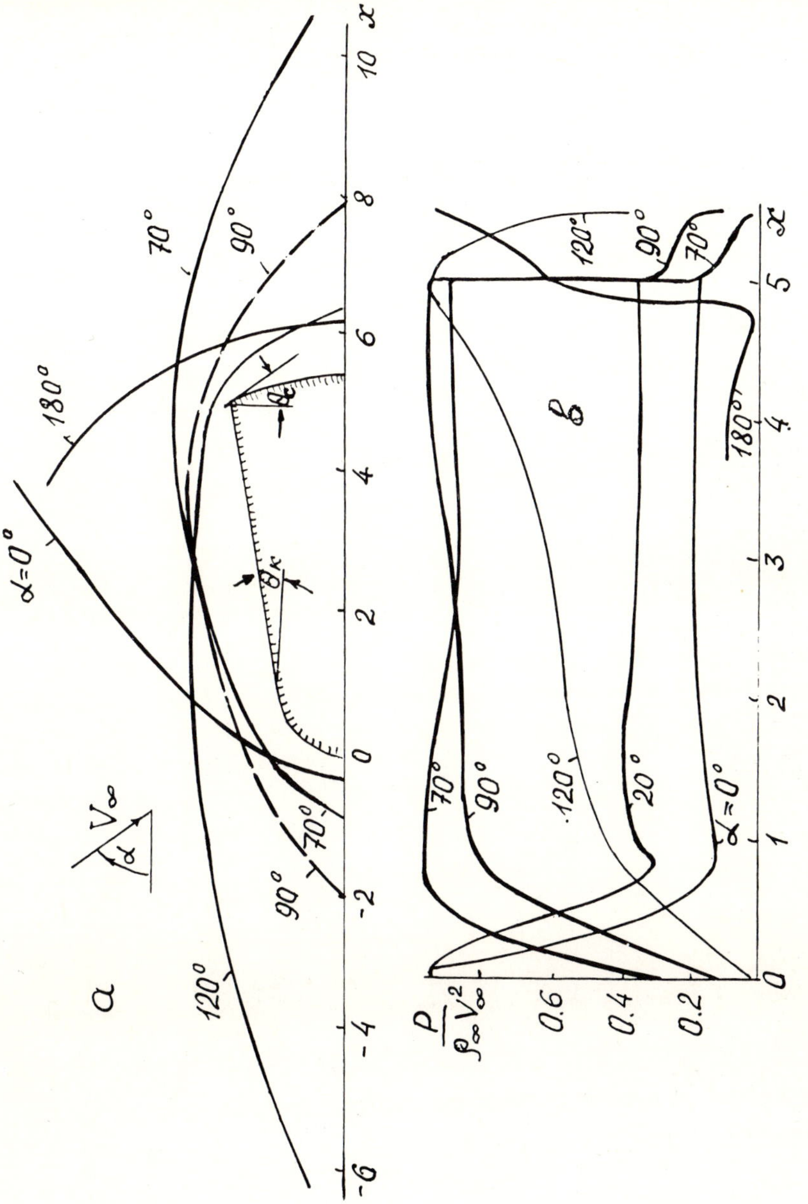

Fig. 4

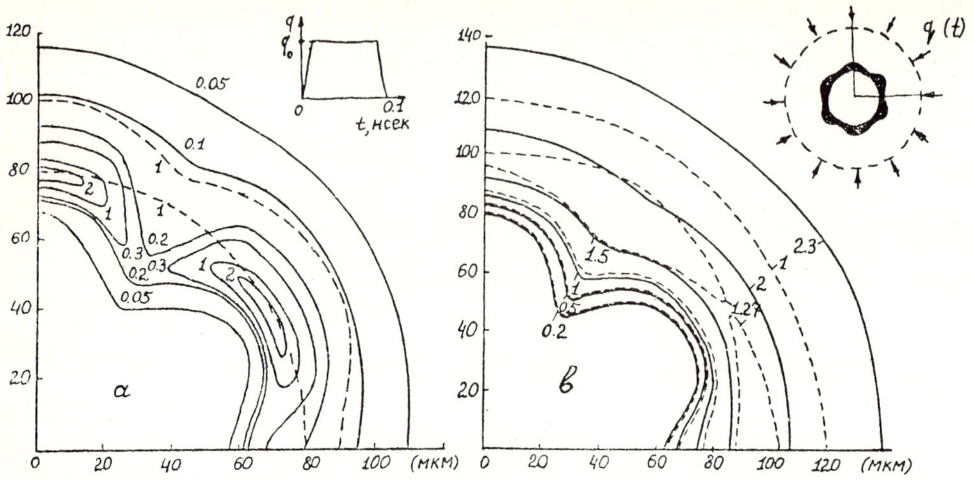

Fig. 5

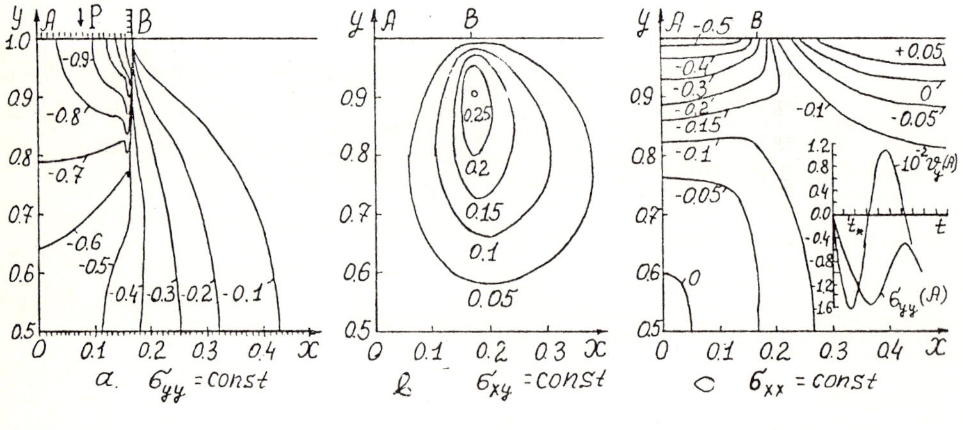

Fig. 6

THE METHOD OF DISCRETE VORTICES IN AEROHYDRODYNAMIC PROBLEMS AND THE THEORY OF MULTIDIMENSIONAL SINGULAR INTEGRAL EQUATIONS

S.M. Belotserkovskii, I.K. Lifanov, M.I. Nisht
N.E.Joukowski Central Aero-Hydrodynamic Institute,
Moscow, USSR

At present the method of discrete vortexes has been greatly developed and has had a wide application. Here by means of numerical calculation there is a transition from continual distribution of the flow parameters and other quantities along space and processes of their changing with time. A continious vortex layer on the body and outside it is simulated by the system of discrete vortexes i, and boundary conditions are seen in a number of designed points j (Fig.1). The continious process of changing the flow parameters (circulations, loading) with time is changed by a staged process.

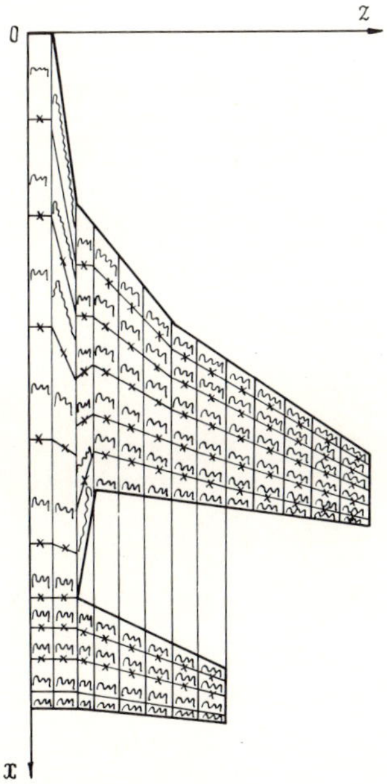

In order to find circulations of discrete vortexes it is enough to have boundary conditions on the wing surface, the conditions closing of vortex systems Chapligyn-Joukowski hypothesis for sharp trailing edges and in nonstationary problems it is enough to have only initial conditions.

The problem is reduced to the solution of the system of linear algebraic equations regarding unknown circulations Γ_i.

The required class of solutions is marked only by the selection of mutual arrangement of discrete vortexes and designed points on the aerodynamic surface:

Figure 1

$$\sum_{i=1}^{n} \Gamma_i w_{yij} = f(x_{0j}, z_{0j}), \quad j = 1, 2, \ldots, n.$$

Here the character of the conversion of the solution into infinity or its restriction on the edge is given a priori and is obtained from the solution of the system of linear algebraic equations automatically. It is supported by numerous computations by means of computer.

The method of discrete vortexes made it possible to solve all main linear subsonic problems of stationary and nonstationary theory of complex wing form and of the vehicle as a whole [1,2,3].

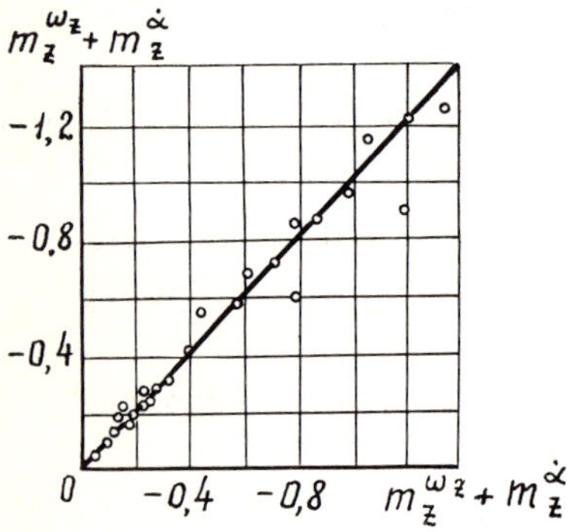

Figure 2

Fig. 2 compares designed and experimental data according to longitudinal damping coefficient of various shape wing in platform. Each point characterizes a wing and straight line coming from the beginning of coordinates at the angle of 45° corresponds to an ideal coincidence of designed and experimental data.

Numerous comparisons show that the results of numerical computations performed by means of the method of discrete vortexes agrees with exact solutions and experimental data both for the wing and for the vehicle [3]. Until recently, however, there was no any mathematical basing of this method.

This work proves that the method of discrete vortexes results in consideration of following equations:

1) for the wing of infinite span:

$$\int_{-b}^{b} \frac{\gamma(x)\,dx}{x_0 - x} = f(x_0),$$

2) for the wing of finite span of complex shape in planform

$$\iint_{G} \gamma_z(x,z) \frac{1}{(z_0-z)^2}\left(1 + \frac{x_0 - x}{\sqrt{(x_0-x)^2+(z_0-z)^2}}\right) dx\,dz = f(x_0, z_0),$$

where G - wing projection on the plane OXZ.

In the process of mathematical basing of the problems given above a new approach was developed for Gouchy studying singular integral equations (s.i.e.) with integrals, i.e. with integrals of the kind:

$$I_1 = \int_{-b}^{b} \frac{\gamma(x)\,dx}{x_0 - x}, \qquad I_2 = \int_{-b}^{b}\int_{-e}^{e} \frac{\gamma(x,z)\,dx\,dz}{(x_0-x)(z_0-z)}.$$

The theory of one-dimentional s.i.e. with integrals of the kind I_1 had been greatly developed in [4] by N.I.Mushelishvily, in [5] by F.D.Gachov and others. It happened due to close connections of such equations with boundary problems for analitical functions of complex variable.

A similar method of investigations of two-dimentional s.i.e. with integrals of the kind I_2 made it possible to study only the equations for tor.

A new approach to s.i.e. was developed on the method of discrete vortexes.[6,7]

At first, we consider canonic s.i.e. by the method consisting of following stages:

1) It is studied quadratic formulas for singular integrals of arbitrary multiplex.

2) With the help of these quadratic formulas the Puankare-Bertran formula is proved for singular integrals of Couchy type of arbitrary multiplex.

3) S.i.e. are changed by the system of linear algebraic equations characteristic for solutions of each class.

4) It is found a function which is a limit for the solution of this system of linear equations.

5) The function thus determined is substituted into initial equation and with the help of Puankare-Bertran formula it is proved that this function is a solution.

By this method exact solutions were obtained for canonic s.i.e. of the first kind at the segment, circle, tor, cylinder and rectangle. In this way not only integrated solutions (in general sense) were obtained but solutions integrated only in the direct meaning according to Couchy. Using the results obtained for canonic s.i.e., it is possible to reduce more complex s.i.e. to the equivalent /in the appropriate class of solutions / Fredholm equations of the second kind.

Linear theory provides reliable results for those regimes of streamlining when kinematic parameters and perturbated velocity are low and linear dependence of aerodynamic loading upon these parameters is maintained. In a number of practically important cases (for instance at separated streamlining) perturbated velocities become commensurable with flight speed and linear methods give great errors. Recently it was widely used the wing with small lengthwise and large sweepback, the wings of complex planform and of a variable sweepback whose streamlining at low angles of attack is accompanied by separation of flow. The method of discrete vortexes made it possible to solve effectively these nonlinear (stationary and nonstationary) problems [8,9].

When studying the regimes of flow separation the escaping of free vortexes from leading edges, the break of surfaces and etc is permissible. Then the requirement of finiteness of velocities (the Chapligin-Joukowski hypothesis)is met. This provides additional conditions for determination of circulation of above mentioned free vortexes from the wing. The position of free vortexes behind the wing is determined from the condition of their motion along the trajectory of liquid particles. The described method makes it possible to determine not only nonlinear aerodynamic characteristics of aerodynamic surfaces but the process of twisting of vortex sheet, its breaking and formation of aerodynamic trail.

Figure 3

Fig.3 shows the vortex structure of aerodynamic trail behind the sheet $\lambda = \infty$ at its separated streamlining under the angle of attack $\alpha = 90^{\circ}$ (Karman way) which was obtained by means of computation at CBM (points show free vortexes escaping from the sheet).

Fig.4 shows vortex structure at separated streamlining of delta wing ($\lambda = 1$, $\alpha = 15^{\circ}$). It is seen that vortex sheet escaping from leading edges is twisted forming steady tight plaits.

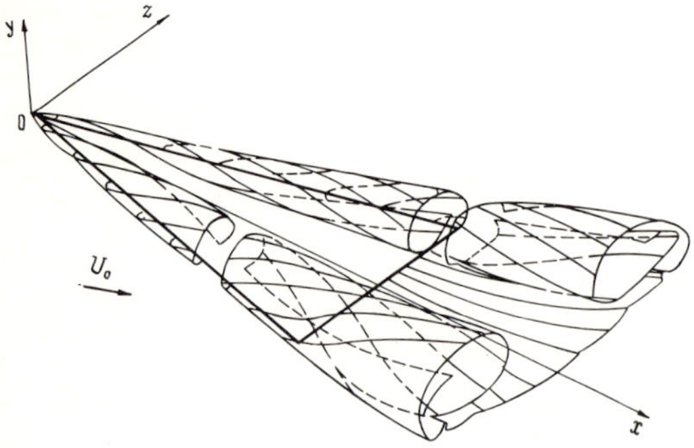

Figure 4

The given data show one indisputable advantage of the method of discrete vortexes, namely: by means of single approach this method makes it possible to solve hydrodynamics problems running from the simplest linear two-dimentional problems to space non-linear ones.

The numerical experiment, comparisons with well-known solutions and experimental data show high effectiveness of these developed approaches. Here it is, of course, important to have exact mathematical demonstration of agreement and correctness of the method of discrete vortexes not only for linear problems but for nonlinear ones as well.

LITERATURE CITED

1. Belotserkovskii, S.M. 1965. Thin aerodynamic surface in subsonic gas flow. "Nauka". Moscow.
2. Belotserkovskii, S.M., Skripach,B.K., Tabachnikov,V.G. 1971. Wing in unsteady gas flow. "Nauka". Moscow.
3. Belotserkovskii, S.M., Skripach, B.K. 1975. Aerodynamic derivatives of vehicle and wing at subsonic speeds. "Nauka". Moscow.
4. Mushelishvily, N.I. 1968. Singular integral equations, "Nauka". Moscow.
5. Gahov, F.D. 1977. Problems of edges. "Nauka". Moscow.
6. Lifanov, I.K., Polonskii, Ya.E. 1975. Proff of the numerical method of discrete vortexes for solving singular integral equations. Prikl. Mathem. and Mech. v.39 N 4. (Appl.Math.Mech. 39 4).
7. Lifanov, I.K. 1978. About singular integral equations with one-dimentional and multiple integrals of Couchy type. Dokl.Akad. Nauk SSSR, v.239 N 2.
8. Belotserkovskii, S.M., Nisht, M.I. 1978. Stalled and unstalled streamlining of thin wings by an ideal fluid. "Nauka". Moscow.
9. Belotsrkovskii, S.M. 1977. Study of the unsteady aerodynamics of lifting surfaces using the computer. Ann.Rev.Fluid Mech. 1977.9.

ANALYSIS OF HIGHER ORDER METHODS FOR THE NUMERICAL SIMULATION OF CONFINED FLOWS

P. BONTOUX, B. FORESTIER and B. ROUX

Institut de Mécanique des Fluides de Marseille
13003 MARSEILLE - FRANCE

ABSTRACT

The compact differencing method developped by Hirsh is used to solve the two dimensional incompressible Navier-Stokes equations. The higher order of accuracy is investigated in the case of the test problem proposed by Pearson. Criterions for the stability of the ADI scheme are derived in terms of the physical parameters, when various types of boundary conditions are involved. Accelerating techniques are tested in order to optimize the convergence of the iterative process. Applications are made in the present paper for the driven cavity problem.

INTRODUCTION

In recent methods, the higher order of accuracy is obtained by using only three points finite difference discretization, when the derivatives of the function are considered as supplementary unknowns. The closure of the system is then ensured with additional relationships, called Hermitian formulae, which relate the values of the function and its two first derivatives at three mesh points. Three points differences have been already developped for a sixth order accuracy by Rubin and Khosla, 1977. The interest is actually centered on the construction of fourth order accurate schemes, as witnessed by the works published by Orszag and Israeli, 1974, Hirsh, 1975, Krause et al, 1976, and Adam, 1977. Peyret, 1978 recently studied the errors corresponding to these various methods for a simple linear differential equation, and showed that they are formally the smallest when implicit hermitian formulae are used for the second derivatives, which is in agreement with our own previous numerical experiments (Bontoux,

Forestier and Roux, 1977). The compact differencing method has then been used extensively as it is also quite readily applicable to the solution of the Navier Stokes equations with an ADI scheme.

COMPACT DIFFERENCING METHOD

The implicit hermitian relationship read as :

$$(u_x)_{i+1} + 4(u_{x_i}) + (u_x)_{i-1} = \frac{3}{h}(u_{i+1} - u_{i-1}) + O(h^4) \tag{1}$$

$$(u_{xx})_{i+1} + 10(u_{xx})_i + (u_{xx})_{i-1} = \frac{12}{h^2}(u_{i+1} - 2u_i + u_{i-1}) + O(h^4) \tag{2}$$

The system includes the second order differential equation under the discretized form :

$$m_{1i}\, u_i + m_{2i}\,(u_x)_i + m_{3i}\,(u_{xx})_i + m_{4i} = 0 \tag{3}$$

When the boundary conditions are known for the function and its derivatives, the (3x3) block tridiagonal system may be solved by using the generalized Thomas algorithm. Usual boundary conditions are however of Dirichlet or Neumann types, and it is then necessary to consider one more additional relation between the values of the function and its derivatives on the boundary and in the field. Such relations which preserve the tridiagonal character of the equations, are only third order accurate. The Padé approximant has been retained :

$$(u_{xx})_i - (u_{xx})_{i+1} + \frac{6}{h}((u_x)_i + (u_x)_{i+1}) + \frac{12}{h^2}(u_i - u_{i+1}) = 0 + O(h^3) \tag{4}$$

GOVERNING EQUATIONS

The present study is devoted to confined flows. The governing equations are considered with the vorticity, ζ, and the stream function, ψ, as dependant variables. The unsteady Navier-Stokes equations are written under the convective form as follows :

$$\zeta_t + u\zeta_x + v\zeta_y = a_1(\zeta_{xx} + \zeta_{yy}) \tag{5a}$$

$$\psi_{xx} + \psi_{yy} = \zeta \tag{5b}$$

where $u = \psi_y$ and $v = -\psi_x$. In the applications where ψ and its first derivative normal to the wall are given on each boundary, the conditions on ψ are overspecified. However, no physical condition exists then for the vorticity ζ_w at the boundaries, and ζ_w must be evaluated in terms of ψ from (5b) as :

$$\zeta_w = \frac{12}{h^2} (\psi_{w-1},\psi_w) + \frac{6}{h}((\psi_n)_{w-1} + (\psi_n)_w) + (\psi_{nn})_{w-1} + O(h^3) \quad (6a)$$

$$\zeta_w = \frac{3}{h^2} (\psi_{w-1}-\psi_w) + \frac{3}{h} (\psi_n)_w - \frac{1}{2} \zeta_{w-1} + O(h^2) \quad (6b)$$

$$\zeta_w = \frac{1}{2h^2} (-\psi_{w-2} + 8\psi_{w-1} - 7\psi_w) + \frac{3}{h} (\psi_n)_w + O(h^2) \quad (6c)$$

A.D.I. SCHEME - NUMERICAL STABILITY

The alternating directions implicit scheme is considered, as it retains at each half step the block tridiagonal character of the finite difference system. The stream function equation is solved as a time dependent problem.

The conventional Fourier analysis predicts the unconditional stability of the ADI scheme, when it is applied at interior points of the field. Computational instabilities however occur in solving N.S. equations (Roache, 1972) and impose a time step restriction of the form $\Delta t/h^2 < a$. The reasons of the lack of convergence have been investigated for the second order scheme, by using an heuristic extension of the Fourier method which involves various types of boundary conditions(Bontoux, Gilly and Roux, 1978). The Fourier solutions of the discretized equations are considered at interior points. The boundary conditions may be expressed as follows from (6b) on the boundary refered by (i,J) :

$$\zeta^n_{iJ} = (- Z^n/2 + 3\ P^n/h^2) \exp\ (I_m(i\theta_x + (J-1)\ \theta_y) + 3u_{iJ}/h \quad (7a)$$

$\psi^n_{iJ} = 0$ (7b), where Z and P are the amplitudes of Fourier solutions. The necessary condition for the suppression of error amplification requires that the modulus of the eigenvalues of the amplification matrix remains less than or equal to unity :

$$\text{Max}_p\ |\lambda_p| \leq 1 \quad (8)$$

where

$$\lambda_p = f\ (u_1 \frac{\Delta t}{h}\ ,\ v_1 \frac{\Delta t}{h}\ ,\ a_1 \frac{\Delta t}{h^2}\ ,\ h) \quad (9)$$

The condition (8) is numerically studied for each set of u_1, v_1, a_1 and h. The corresponding larger value of $\Delta t/h^2$ which verifies (8), is calculated and it determines the heuristic numerical stabitily condi-

Fig.1 : Numerical stability of the ADI scheme (Driven cavity problem)

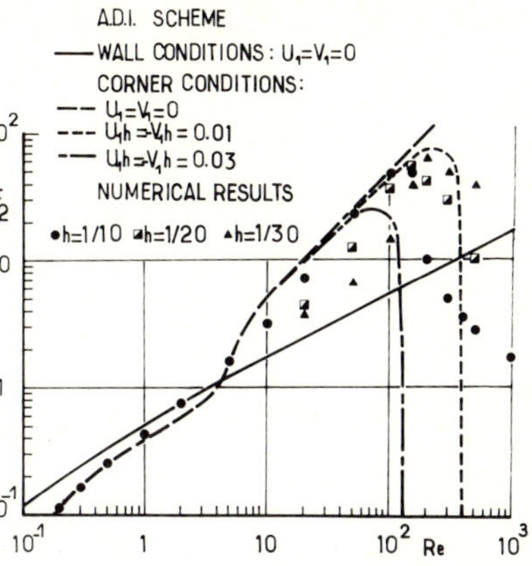

Fig.2 : Influence of the relaxation parameter $\Delta t/2h^2$ on the convergence rate. (Driven cavity problem).

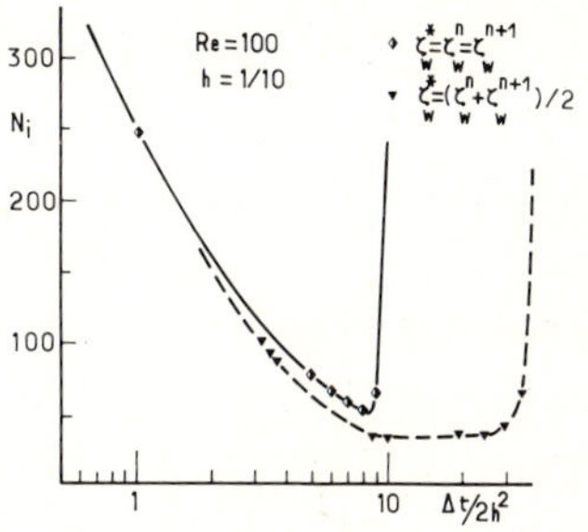

Fig.3 : Accuracy of the solution with the various differencing schemes (Pearson problem).

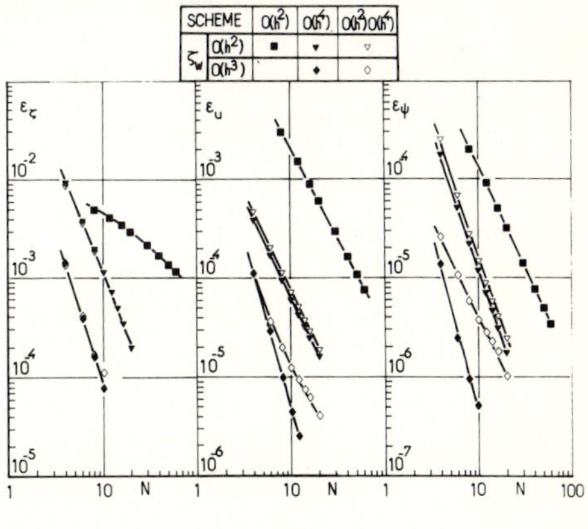

tion. Two cases are to be considered :
- when one mesh point is on the boundary (wall conditions)
- when two mesh points are on the boundaries (corner conditions).
The overall stability is defined theoretically by the most restrictive condition. The analysis has been devèlopped for the driven cavity problem ($a_1 = 1/Re$) (fig. 1) :

- when a_1 is large ($Re < 4$), the corner condition determines the overall stability and gives a quite satisfying prediction of the computational conditions.

$$\Delta t/h^2 < 0.8\, Re^{0.6} \tag{10}$$

- When a_1 is small ($Re > 4$) the wall condition becomes the most restrictive one, but the algorithm is found to converge although the wall condition is no more verified. An empiric upper limit may be deduced from our own numerical experiments as the corner condition. Approximated relationships are :

$$\Delta t/h^2 < (1 + \sqrt{1 + 24\, Re})/6 \quad \text{and} \quad \Delta t/h^2 < Re \tag{11}$$

- the third point to be noted is the influence of the velocity parameters, u_1 and v_1. A cell Reynolds number limit is exhibited by the theoretical study in the vicinity of the corner :
$Re\ |u_1|\ h < 4$, $Re\ |v_1|\ h < 4$. These limits are to be connected to the computational stability results which behavior reveals a sharp decrease when Re varies from 100 to 200. Spatial oscillations are then also encountered in the flow solution ($h = 1/10$).

ACCELERATING TECHNIQUES

When the steady solution is sought, the parameter $\Delta t/h^2$, which is equivalent to a relaxation parameter, has been adjusted to optimize the convergence rate. The optimum conditions are obtained when $\zeta_w^* = (\zeta_w^n + \zeta_w^{n+1})/2$ is used (Fig. 2). The other accelerating technique is the under relaxation of the boundary evaluation ζ_w. In the driven cavity problem, it gives a gain of nearly 15 % on the iterations number.

ACCURACY OF NAVIER-STOKES SOLUTIONS

An exact solution of equations (5a), (5b) proposed by Pearson (1965) for $a_1 = 1$, is used to compare the accuracy of the solutions obtained with the three following methods to solve the N.S. equations on the convective form :

- a second order central difference scheme (SDC)

- the higher order compact differencing scheme (HDC)
- a combined fourth order and second order accurate method (CDC), involving only a compact treatment for ψ equation (Loc, 1977).

The truncation errors are controled by the residuals $\varepsilon_f = \text{Max} \mid f^n - \bar{f}^n \mid / \text{Max} \mid \bar{f}^n \mid$ for ψ, u and ζ variables (Fig. 3). The convergence is defined by an absolute rate of 5×10^{-5} between two successive iterations. The main facts to be noted are :

- with the SDC scheme, the solution is $O(h^2)$ for ψ , u, but O(h) for ζ
- with the HDC scheme, the choice of the boundary evaluation, $O(h^3)$ (6a) or $O(h^2)$ (6c), is determinant for the global accuracy of the solution : the truncation errors are respectively $O(h^{3.1})$ and $O(h^{2.4})$ on ζ; for ψ, the accuracy is then $O(h^4)$ and $O(h^3)$.
- when the CDC scheme is used, the truncation errors for all variables, are quite equivalent in order and magnitude with the errors of the HDC solution, when (6c) is considered. With (6a), the loss of accuracy becomes evident except for ζ. Nevertheless the magnitudes of the errors are substantially smaller than the ones obtained with (6c).

DRIVEN CAVITY PROBLEM

When the coefficient of the diffusive terms is small (Re=400), spatial oscillations are encountered in ζ solution near the driven wall when solving the convective N.S. equations with SDC, HDC or CDC schemes. These wiggles can be damped by reducing the step size h, but it is a very expensive way. Indeed the use of the divergence form results in the damping of the oscillations from Re=400 up to nearly Re=7000 with h=1/35. The second order scheme (SDD) and the combined scheme (CDD) have been written on the divergence form. The respective efficiencies of the different schemes have been experienced on some characteristic quantities : maximum stream function and minimum vorticity on the driving wall.

- At low Re < 10, the solutions of the convective and divergence forms are quite similar, and the highest order scheme (HDC) prevails as for Pearson's problem.
- For Re=100, the results are compared to various ones mentioned in the literature (fig.4). If we consider the convective form, the HDC and CDC schemes achieve respectively results in the 5% accuracy range with approximately 40% and 65% the spatial resolution of the SDC scheme. Rough estimates of the respective CPU times required to obtain this accuracy are 1/6 and 1/2 of the SDC calculation time.

Besides, the treatment on the divergence form provides, a remarkable improvement of the accuracy of the schemes. As shown on fig. 4, the CDD scheme generates then solutions which are much closer to the HDC ones, but with scarce lower efficiency by a ratio of 1.4.

-For higher Re (Re > 400), the SDD and CDD schemes are then the only ones available. The maximum stream function solutions are given in Table 1 for Re=1000. The CDD solutions obtained even with a coarse spatial resolution, are shown to be in rather good agreement with the more accurate solutions proposed in the literature. However it is still necessary to use step size small enough in order to ensure a correct prediction of the recirculating flow near the walls. The fig. 5 shows up that the existence of the upstream secondary vortex at Re=4000 is strongly dependent on the spatial resolution. This vortex is characterized by the length scale L. Comparisons are made in fig.6 between Pan and Acrivos'experiments and some computed results. Up to Re=500 the computed values are not substantially different. They are in rather good agreement with the measurements. When Re > 500, the solution depends on the spatial scheme : the upwind difference scheme seems to provide the best prediction of the decreasing observed by Pan. However, it must not be concluded to the failure of the higher order scheme. Indeed, de Vahl Davis and Mallinson have shown that this agreement is fictitious as the observed decreasing of L is probably due to a 3D effect in the experiments, and they conclude to the failure of the 2D mathematical model for Re >500. The vanishing of L predicted by the upwind difference scheme is only due to false diffusion.

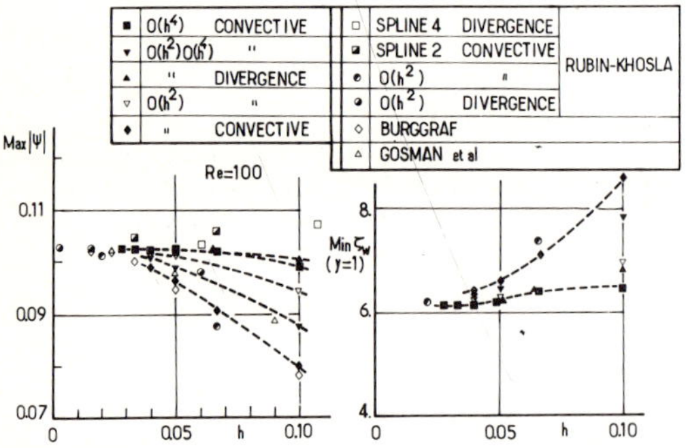

Fig.4 : Efficiency of the various differencing schemes (Driven cavity problem)

DRIVEN CAVITY Re =1000

AUTHORS	SCHEME	FORM	h	$\|\psi\|_{max}$
RUBIN	SPLINE 4	DIV	1/16	0.115
	CENTRAL	"	1/16	0.080
	"	"	1/64	0.114
NALLASAMY	UPWIND	CONV	1/50	0.098
BOZEMAN	"	DIV	1/50	0.081
LOC-DAUBE	UPWIND	CONV	1/20	0.115
	$O(h^2)O(h^4)$	"	1/35	0.114
PRESENT RESULTS	CENTRAL	DIV	1/20	0.112
	$O(h^2)O(h^4)$	"	1/35	0.113
	CENTRAL	"	1/20	0.085
	$O(h^2)$	"	1/35	0.104

Table 1 : Maximum stream function solutions of the driven cavity problem at Re = 1000

CONCLUSION

The problem of Pearson and the driven cavity problem have been used to emphasize the higher order of accuracy provided by the compact differencing method, when solving the N.S. equations. When the coefficient of the diffusive terms is small enough, the higher order method does not avoid the wiggles problem, and, the use of the divergence form is still necessary. Then, the combined $O(h^2)$-$O(h^4)$ accurate method is of a practical interest because the $O(h^2)$ treatment of the vorticity equation is more readily applicable to the divergence form. The use of a third order evaluation of the vorticity at the wall is necessary to maintain the accuracy everywhere. More physical applications of the method have been already made, they concern the natural convection in a solar cell (Roux et al, 1977) and the influence of an oscillating pressure gradient on the flow through a stenotic obstruction (Forestier et al., 1977).

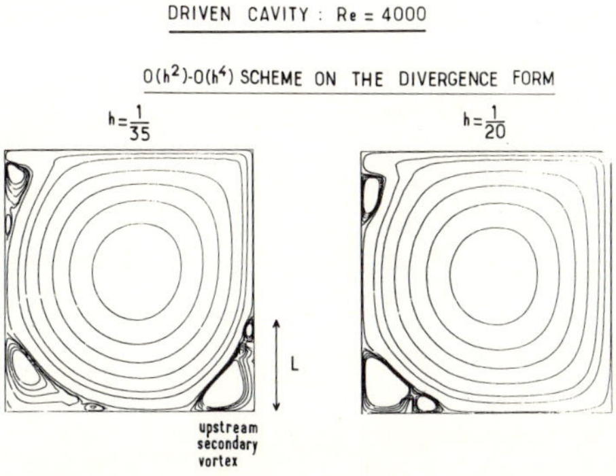

Fig.5 : Influence of the spatial resolution on the stream-function solution

	AUTHORS	DIFFERENCE	FORM	h
▽	BURGGRAF	CENTRAL	CONVECTIVE	1/40
△	DE VAHL DAVIS	"	DIVERGENCE	1/30
□	NALLASAMY	UPWIND	CONVECTIVE	1/50
◇	BOZEMAN	CENT.&UPW.	DIVERGENCE	1/50
●	PRESENT RESULTS	$O(h^4)$	CONVECTIVE	1/35
■		$O(h^2)$ $O(h^4)$	DIVERGENCE	1/35
♦		"	"	1/20

	EXPERIMENTS
◆	PAN ACRIVOS

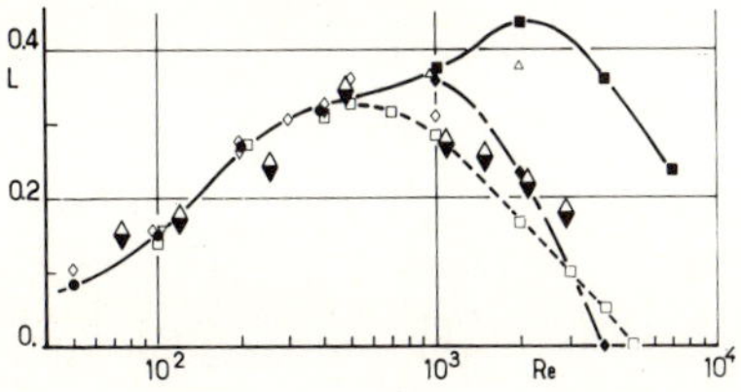

Fig.6 : Influence of the Reynolds number on the upstream-vortex size

REFERENCES

ADAM, Y. (1977), J.C.P., 24, 10-22.

BONTOUX, P., FORESTIER, B., and ROUX, B., 3e Congrès Français de Mécanique, Grenoble, (1977), J. Méca. Appl. Vol.2, N°3, (1978).

BONTOUX, P., GILLY, B. and ROUX, B., (1978), C.R.A.S., Paris, 286, A, 349-352.

BOZEMAN, J.D., and DALTON, C.,(1973), J.C.P., 12, 348-363.

BURGGRAF, O.R., (1966), J.F.M., 24, 113-151.

DE VAHL DAVIS, G., and MALLINSON, G.D., (1976), Comp. and Fluids, 4, 29-43.

FORESTIER, B., et al, (1977), I.N.S.E.R.M., EUROMECH 92, Compiègne, Cardiovascular and pulmonary dynamics, Vol.11, pp.89-102.

GOSMAN, A.A., et al, 1969, Heat and Mass Transfer in Recirculating Flows, Ac. Press, London.

HIRSCH, S.R., (1975), J.C.P., 19, 90-109.

KRAUSE, E., et al, 1976, Comp. and Fluids, 4, 2, 77-92.

LOC,T.P., and DAUBE, O.,(1977), C.R.A.S., 284, A, 1241-1244.

NALLASAMY, M., and PRASAD, K.K., (1977), J.F.M., 79, 2, 391-414.

ORSZAG, S.A., and ISRAELI, M., (1974), Ann. Reviews of Fluid. Mech., 6, 281-318.

PAN, F., and ACRIVOS, A., (1967), J.F.M., 28, 643-655.

PEARSON, C.E., (1965), J.F.M., 28, 611-622.

PEYRET, R., (1978), Conf. Int. sur les Méthodes Numériques pour les Ecoulements Laminaires et Turbulents, O.N.E.R.A., T.P. 1978-29.

ROACHE, P.K. (1972), Computational Fluid Dynamics, Hermosa, Albuquerque.

ROUX, B. et al, (1977), 3ème Congrès Français de Mécanique, Grenoble, (to be published in Numerical Heat Transfer).

RUBIN, S.G., and KHOSLA, P.K., (1977), J.C.P., 24, 217-244.

NUMERICAL SIMULATIONS OF TWO-DIMENSIONAL PLASMA FLOWS

K.V. Brushlinsky, V.V. Savel'ev
Institute of Applied Mathematics, USSR Acad.of Sciences,
Moscow

Numerical simulations and computer methods for plasma flows in coaxial channels of plasma accelerators and magnetoplasma compressors are presented in this paper. As a theoretical basis of the plasma description some different hydrodynamical models are used: the classical magnetohydrodynamics (MHD), the MHD with the Hall effect, a three-component hydrodynamics. Results of the solution of some problems on flow properties, connected with processes in electrode layers and the dynamics of heavy ions admixture, are adduced.

Numerical simulations and computer methods for plasma flows in ducts of plasma accelerators and magnetoplasma compressors are considered. The channel is formed by two coaxial metallic electrodes (fig. I-4). The electric current running between the electrodes produces an azimuthal magnetic field, which accelerates plasma in the direction of the duct axis from the left to the right.

The basic mathematical model uses two-dimensional (axisymmetric) time-dependent magnetohydrodynamics equations, taking into account the viscosity, heat - and electric conductivities and radiation with coeffitients, depending on the temperature. The model includes also the Hall effect, i.e. it takes into account a finite difference between the velocities of the ion and electron plasma components. The MHD - equations with the Hall effect may be considered as a simpliest two-component plasma simulation. It gives in particular a possibility to describe such effects of plasma flows in a duct, for which the polarity of electrodes is important.

A time-dependent problem is numerically solved in a curvilinear region of cylindrical coordinates (r , z) corresponding to the channel shape. Initial conditions create a plasma motion in the z -direction, and the boundary conditions at the left boundary simulate a continuous plasma inflow at the duct entry. Under these conditions the flow regime in the duct stabilizes as a rule and becomes steady-state. As a result of computations properties of such a flow were investigated and in particular there were found some properties, that have been unknown before. The mathematical model, the computer method and a review of results obtained in computations are published in the papers [I, 2]. One of the most essential results of this series of computations is the discovery of steady-state self-compression flows. These are flows, in which a high compression and heating of plasma are observed in a narrow region at the axis of a duct with a shortened central electrode (fig. I, 3, 4). They became the basis of elaboration of the so-called magnetoplasma compressor (MPC) [3]. A typical plasma density distribution in the MPC channel is shown at fig. I, where the lines of constant density are represented [4].

The next step of computations is connected with an analysis of processes in electrode layers which are observed in experiments with large values of discharge current [3]. In MPC channels with solid electrodes there are a large anode discontinuity of the electric potential and a current slip. These singularities may be removed if the anodes are consisted of pivots, so that some part of substance enters into the duct through the anode (ion current transfer). However the further increasing of the discharge current leads to a flow regime with a large cathode potential discontinuity and a current slip. In order to simulate these effects in computation, we substitute the usual hydrodynamic boundary condition $v_n = 0$ at the electrodes by the condition $\rho v_n = f(z)$, where v_n is the normal velocity, ρ is the plasma density, $f(z)$ is a given function of the coordinate z at the electrode. Such a boundary condition gives the possibility to simulate the process of substance flow through the electrodes and to regulate this process by means of the function $f(z)$.

The main computation results in this model are: I) when the plasma inflows in the duct through the anode, the singularities, which have been observed earlier in flows with the Hall effect [I], disappear; 2) the plasma outflow through the cathode produces a current slip, current vortices and a sharp change of the potential near the cathode. At the fig. 2 there are shown the distribution of electric current between the electrodes and the dependence of the potential on the radius in a middle cross-section of the channel, obtained in computations. They are in good agreement with experiments [5]. The last result permits to make a conclusion, that for removing the above cathode layer effects it is necessary to prevent a possible plasma rarefaction near the cathode.

The further investigations are devoted to the problem on the plasma flow refinement from admixture of heavy ions getting in the MPC channel. A three-component plasma model is suggested: the plasma consists of electrons and two kinds of ions - light and heavy. This model uses the well-known transport equations, applied to the three plasma components, and the Maxwell equations for the electromagnetic field. In order to keep the hydrodynamical type of the plasma description, it is necessary to neglect the electron inertia $m_e d\vec{v}_e/dt$ and the displacement current $\partial E/c\partial t$ [I]. Then, on purpose to simplificate the problem, the particles interactions, corresponding to all dissipative processes, are excluded from the equations, and the energy equation of each plasma component is replaced by the adiabatic pressure dependence on the concentration with the exponent $\gamma = 5/3$. The numerical solution method for this problem uses a difference scheme which is analogous to the one, used in [I] for solving the MHD-equations.

The computations carried out in this model have shown that in the duct space between the electrodes an effective separation of ions by electromagnetic forces takes place. Heavy ions (an impurity) are pressed to the cathode (central electrode) and they prevent from the compression of the light plasma component. This fact is demonstrated at the fig.3, where the lines of constant concentrations n_1, n_2 of the light and heavy components are shown. The compression flow properties may be restored, if an axial hole (divertor) is made in the cathode (fig. 4). The impurity outflows from the MPC channel through the divertor, passing on the whole the compression zone. Computations of the plasma flow in the duct with a divertor have confirmed this conclusion. One can see at the fig. 4, that the main plasma component is compressed now at the axis, likewise in the case of the pure plasma in a more simple model [4] (fig. I), and the impurity concentration decreases essentially. The separation and refinement processes depend on flow parameters. The impurity the more strongly reduces the light component compression, the more is the ratio of concentrations n_2/n_1 at the duct entry. If this ratio is small, then the n_1 distribution is like the one, presented at the fig. I. The purity of the main plasma component in the compression zone increases with the increasing of the relative heavy component pressure (for example, due to the temperature increasing). At last, the separation of ions is the more effective, the more essential is the Hall effect, i.e. the more is the value of the dimensionless exchange parameter [I] , defining the Hall effect influence in this flow model.

Thus, the three-component plasma model and the computations, carried out in a complicated channel geometry, confirm the conclusions on the separation of the main ions and an impurity by electromagnetic forces and on the possibility of the plasma refinement in the MPC by means of a divertor. The results are in agreement with qualitative estimates and experimental data. They have permited also to establish a dependence of the plasma separation and refinement processes on different factors and flow parameters.

R E F E R E N C E S

I. К.В. Брушлинский, А.И. Морозов. Расчет двумерных течений плазмы в каналах. В сб. "Вопросы теории плазмы" п/ред. М.А. Леонтовича, вып. 8, стр. 88-I63, М.Атомиздат, I974.

2. K.V. Brushlinsky. Numerical simulation of two-dimensional plasma flow in channels. Comp. Meth. in Appl. Mech. and Engineering, vol. 6, pp.293-308, I975.

3. А.И. Морозов. О процессах в магнитоплазменном компрессоре. Физика плазмы, т.I, № 2, I79-I9I, I975.

4. К.В. Брушлинский, А.И. Морозов, В.В. Палейчик, В.В. Савельев. Расчет компрессионных течений плазмы в коаксиальных каналах. Физика плазмы, т.2, № 4, 53I-54I, I976.

5. А.К. Виноградова, К.В. Брушлинский, А.И. Морозов, В.В. Савельев. О природе прикатодного скачка в магнитоплазменном компрессоре (МПК). Письма в ЖТФ, т.I, вып. I4, 67I-673, I975.

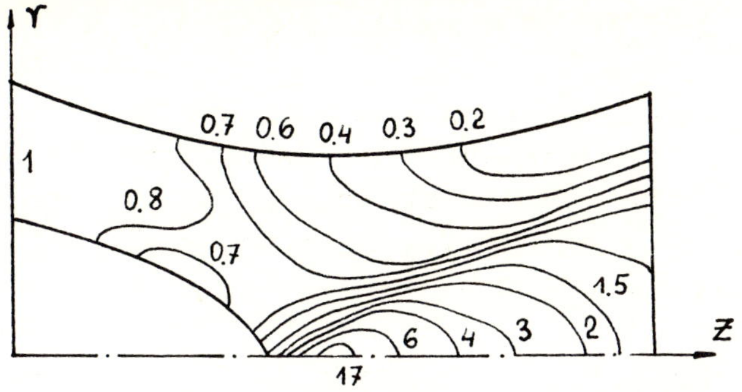

fig. 1

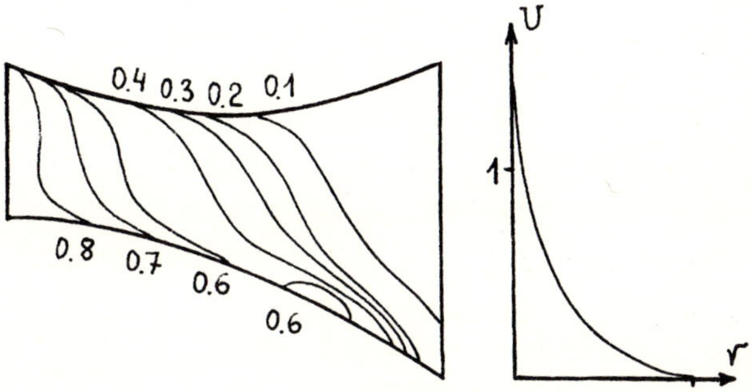

fig. 2

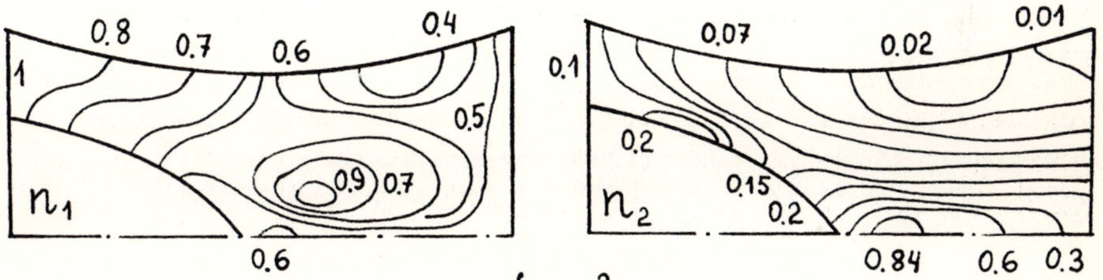

fig. 3

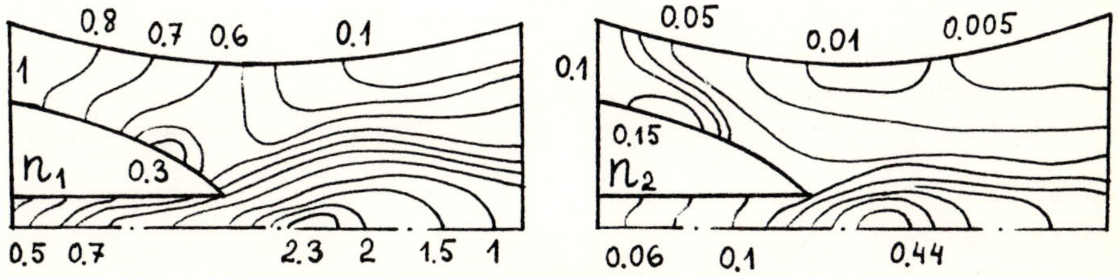

fig. 4

NUMERICAL SIMULATION OF TURBULENT FLOWS WITH A THREE-DIMENSIONAL VORTEX-IN-CELL METHOD

Oscar Buneman, Benoit Couët
Stanford University

Anthony Leonard
NASA-AMES Research Center

The vortex-in-cell method, while employing an Eulerian mesh, does not solve the Navier-Stokes equations directly over this mesh, but it traces the motion of vortex filaments using the velocity field which these filaments themselves create. The velocity field is not calculated directly by Biot-Savart's law of interaction as in[1,2], but by creating a mesh-record of the vorticity field, then integrating a Poisson equation to get the stream function and generating a mesh-record of the velocity field. The vortex filaments are then stepped forward in time.

Vortex pushing methods--as distinct from Navier Stokes methods--have a history of success in two dimensions[3,4,5,6]. In three dimensions[1,2], one of us has applied it to a small number of simple vortex filaments, but at considerable cost because of the time required to sum all the mutual Biot-Savart type interactions between the many elements in all the filaments.

The "cell" or "mesh" method speeds up the calculation of the interactions and allows the three dimensional vortex pushing method to be applied to a space densely filled with vortex filaments, each filament being resolved in fine detail along its length. The techniques of making optimal use of the mesh for evaluating interactions between different fluid elements are the same as those used by plasma physicists for calculating particle interactions[7,8,9].

In creating a code for the evaluation of local flow fields due to a family of vortex filaments we have therefore taken over the principles and architecture of a code developed for magnetic field evaluation in plasma simulations[10]. Each time step in our program consists of calculating the contributions of each element around each vortex filament to the vorticity field $\vec{\omega}(x,y,z)$. Information on $\vec{\omega}$ is kept in three-dimensional arrays over a grid in x,y,z-space, one such array for each component. The vector stream-function $\vec{\Psi}$ is then obtained by solving Poisson's equation $\nabla^2\vec{\Psi} = -\vec{\omega}$, and the velocity field is obtained as $\vec{v} = \text{curl}\,\vec{\Psi}$. Each vortex element on each vortex filament is then advanced through $\vec{v}\delta t$, evaluated locally from the grid record of $\vec{v}$ components.

Inversion of the Laplacian operator ∇^2 and curling are performed in Fourier-transform space, where one has $\vec{v} = -i\vec{k} \wedge \vec{\omega}/k^2$ for the harmonic which goes like $\exp(i\vec{k}\cdot\vec{r})$. The deposition of contributions to $\vec{\omega}$ into its discrete table is done by quadratic spline interpolation, and so is the evaluation of the local $\vec{v}$ from the tabulation of the $\vec{v}$-components. Indeed, we do not tabulate grid values of $\vec{\omega}$ and $\vec{v}$ directly, but spline coefficients, and in fitting splines to data, we choose those which give the best mean-square fit to each Fourier harmonic[11].

One finds that the table of such spline coefficients of $\vec{v}$ can be generated directly from the Fourier harmonics of $\vec{v}$ if, instead of back-transforming these harmonics directly, one first multiplies them by the factor

$$\left(\frac{2}{k_x}\sin\frac{k_x}{2}\right)^3 \Bigg/ \left(1 - \sin^2\frac{k_x}{2} + \frac{2}{15}\sin^4\frac{k_x}{2}\right)$$

and then transforms them. There is one such factor for each dimension (i.e. with k_y and k_z in place of k_x). The same factor is used when the vorticity harmonics are related to the table of spline coefficients of vorticity, so that the factor occurs twice overall in the process of generating the velocities from the vorticity contributions.

A set of vortex filaments with infinitesimal cross section represents a vorticity field which has singularities on each filament. Mathematically, such filaments described by the space curves $\vec{r}_1(\xi), \vec{r}_2(\xi), \ldots$ with circulations $\Gamma_1, \Gamma_2, \ldots$ generate a vorticity field given by:

$$\vec{\omega}(\vec{r}) = \sum_i \Gamma_i \oint \delta(\vec{r} - \vec{r}_i(\xi)) \frac{d\vec{r}_i}{d\xi} d\xi$$

where ξ is a particle coordinate along each filament. Using a grid with spacing h, and Fourier transforming, eliminates the singularities since only harmonic numbers up to π/h are recognized and the vortices are thus automatically broadened. It amounts to introducing a filter for $\vec{\omega}$:

$$\vec{\omega}(\vec{r}) \longleftarrow \iiint G(\vec{r}-\vec{r}') \, \vec{\omega}(\vec{r}') \, d^3r'$$

where "G" now indicates the transverse profile of the vortex filaments. We control G by applying a filter in $\vec{k}$ space which is almost gaussian and thus represents an almost gaussian vortex profile--as it would be after some average life time under viscous vortex broadening and inviscid vortex stretching. It can therefore be said that we are solving the Navier-Stokes problem in a given range of scales.

The shape factor G must be applied again when one calculates the evolution of each space curve in time from the continuous velocity field $\vec{v}(\vec{r})$ in that:

$$\frac{\partial \vec{r}_i(\xi,t)}{\partial t} = \iiint G(\vec{r}_i - \vec{r}') \, \vec{v}(\vec{r}',t) \, d^3r'$$

In practice, this means applying the square of the transform of G as a filter in $\vec{k}$ space. (Our $\vec{k}$-space factor is not exactly a gaussian: it has to come to a strict zero at $|k| = \pi/h$. We use a cubic spline profile in $\vec{k}$-space which translates to a sinc^4 profile in configuration space, with negligible halos.)

Compared with "vortex pushing" by direct Biot-Savart interaction, we observe that the computational effort in the new method goes up directly as the number of vortex elements, rather than quadratically. We derived an estimate of the CPU time per time step to move a set of vortex filaments made of M nodes. In general, CPU seconds/time step = $0.34 + \frac{M}{5000}$. It is therefore well suited to studying large numbers of vortex filaments or rings and can be looked upon as a method for simulating continuous vorticity.

The convenience of fast Fourier transforms, (exploited here for both the purpose of numerically performing calculus and controlling shaping or filtering) is available only when boundary conditions are imposed over rectangular domains. We have chosen periodicity in all 3 dimensions for our first studies. Our grid is a modest 16 by 16 by 16, but extension to 32^3 and even 64^3 is quite feasible on present-day computers (plasma simulations, involving rather more data handling, have been done over a 64^3 grid on an IBM 360/91 machine in a time-sharing environment by one of us[10]).

All calculations were done on a CDC 7600. A first test was done on a single vortex ring of radius $R = 4$ about the z-axis. Its center is initially located at (8,8,8) in our mesh and thereafter moves along the z-axis. The circulation is $\Gamma = 2$. Figures 1 and 2 show pictures of the velocity field in the middle of the cells, taken at $t = 0$. The ring propagated at constant velocity along the z-axis. Slight distortions of the ring became visible at $t \approx 200$ and grew slowly, presumably due to the presence of image rings.

Next, we studied the Taylor-Green vortex system, ideally suited to triple periodicity conditions[12,13]. The initial velocity field we used is

$$
\begin{aligned}
U(x,y,z) &= 2 \cos(x) \sin(y) \sin(z) \\
V(x,y,z) &= - \sin(x) \cos(y) \sin(z) \\
W(x,y,z) &= - \sin(x) \sin(y) \cos(z)
\end{aligned}
$$

Therefore, the initial vortex lines are the curves $x = \text{const}$, $\sin(y) \sin(z) = \text{const}$. For our purposes, we represent our vorticity field by 144 vortex filaments, each broken into an average of 24 nodes. Figures 3 and 4 show pictures of the velocity field in the middle of the cells, taken at $t = 0$. Figures 5 and 6 show views of the system in its initial state. Figure 7 shows a still after 20 steps, Figure 8 after 60 steps.

The results show that (i) the vortex filaments themselves do not remain planar and (ii) there is a considerable tendency towards disordering and convoluting. Because of the coarseness of our mesh, the effective range of scales for these simulations is rather small, but with a 32^3 or 64^3 simulation we will be able to resolve turbulent flow characteristics in better detail. In particular, one would like to look at the energy spectra.

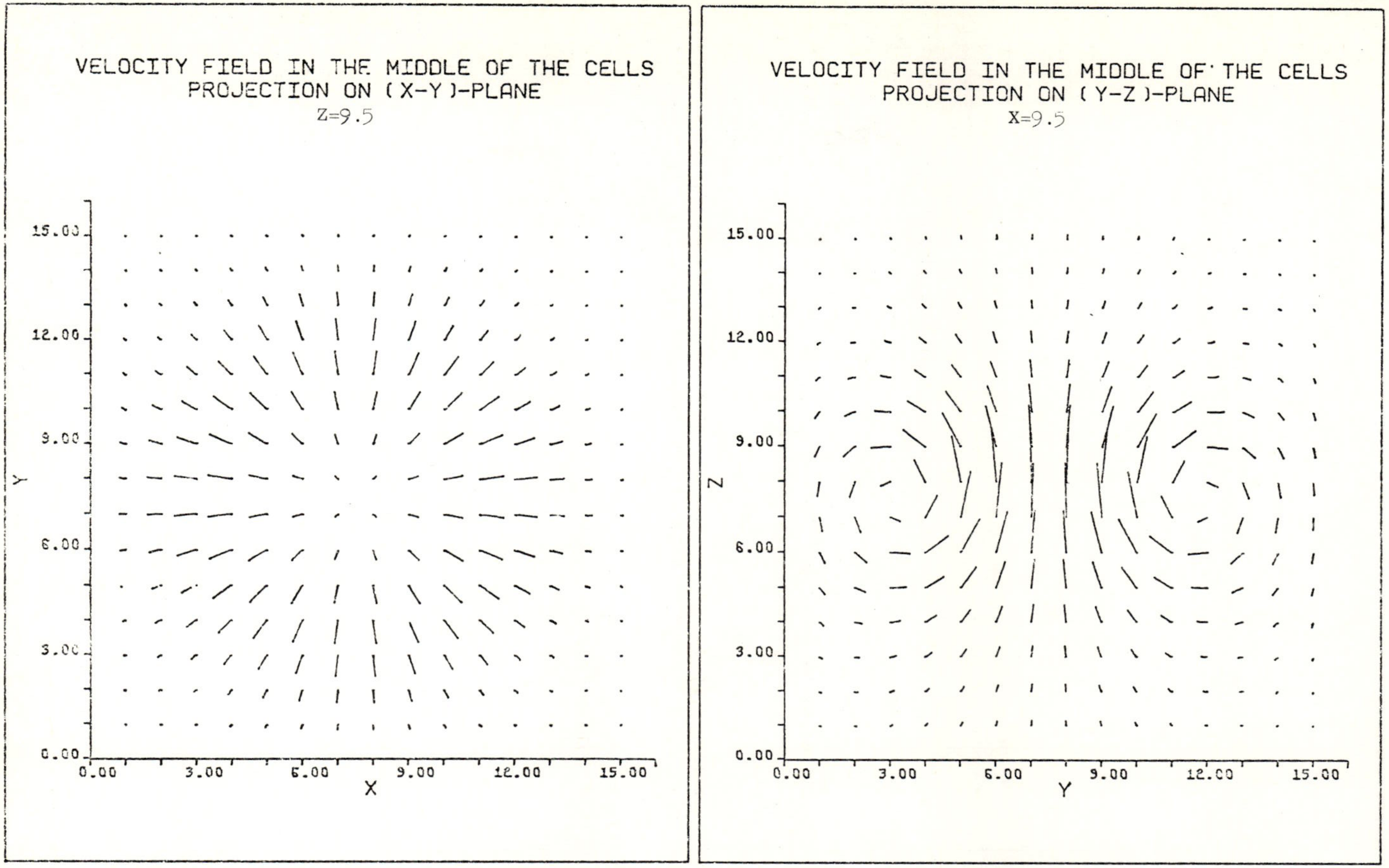

Figures 1 and 2: Single vortex ring, t = 0 .

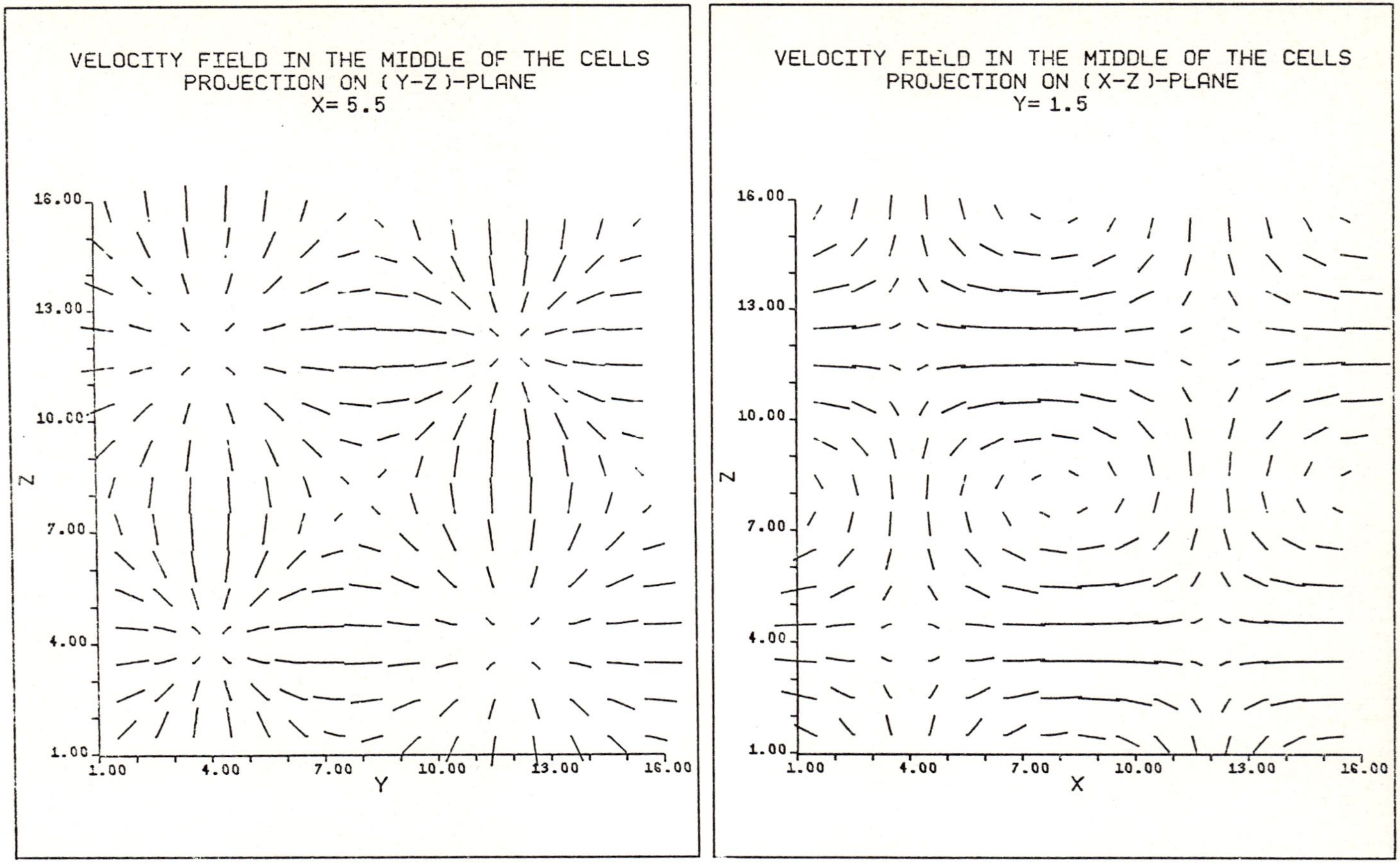

Figures 3 and 4: Taylor-Green vortex system, $t = 0$.

Figure 5: Tilted y-z plane at t = 0 .

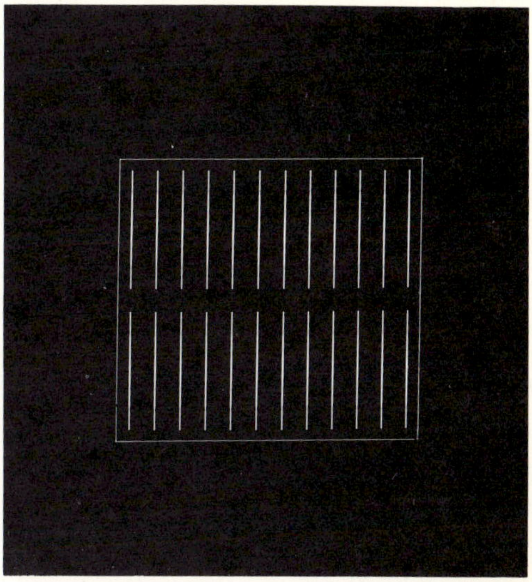

Figure 6: x-z plane at t = 0 .

Vortex Filaments for Taylor-Green System

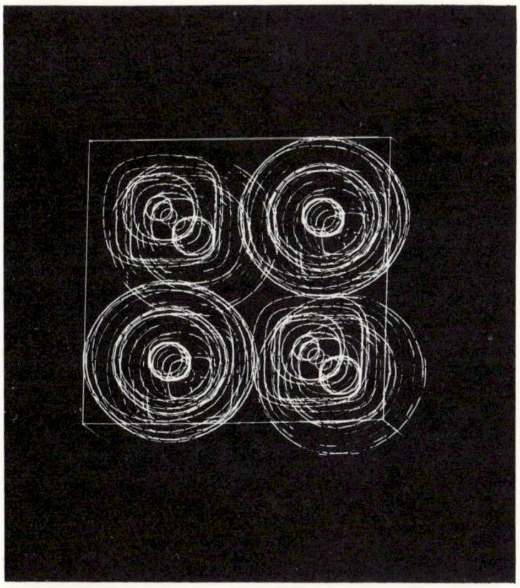

Figure 7: Tilted y-z plane at t = 0.6 .

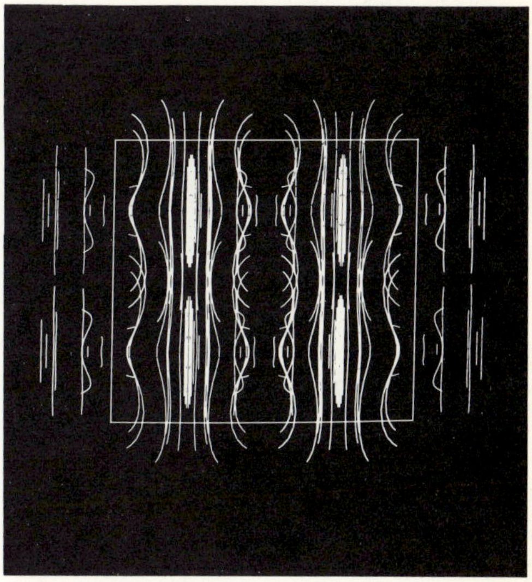

Figure 8: x-y plane at t = 1.8.

REFERENCES

1. A. Leonard, "Numerical Simulation of Interacting, Three-Dimensional Vortex Filaments", Lecture Notes in Physics 35, Proc. 4th Intl. Conf. on Numerical Methods in Fluid Dynamics, 245-250, Springer-Verlag, Berlin (1975).

2. A. Leonard, "Simulation of Three-Dimensional Separated Flows With Vortex Filaments", Lecture Notes in Physics 59, Proc. 5th Intl. Conf. on Numerical Methods in Fluid Dynamics, 280, Springer-Verlag, Berlin (1976).

3. A. J. Chorin, J. Fluid Mechanics 57, 785 (1973).

4. J. P. Christiansen, "Numerical Simulation of Hydrodynamics by the Method of Point Vortices", J. Comp. Physics 13, 363-379 (1973).

5. J.A.L. Thomson and J.C.S. Meng, "Studies of Free Buoyant and Shear Flows by the Vortex-in-cell Method", Lecture Notes in Physics 35, Proc. 4th Intl. Conf. on Numerical Methods in Fluid Dynamics, 403-416, Springer-Verlag, Berlin (1975).

6. R. S. Rogallo, "Proc. AIAA 2nd Comp. Fluid Dyn. Conf.", AIAA, 67 (1975).

7. J. P. Christiansen and K. V. Roberts, Papers 40 and 52, Proc. of Comp. Physics Conf., Culham Lab. (1969).

8. B. Alder, S. Fernbach, N. Rotenberg, Eds., "Method of Computational Physics" 9, Academic Press, New York (1970).

9. S. S. Wang, "Grid-Insensitive Computer Simulation of the Kelvin-Helmholtz Instability and Shear Flow Turbulence", Ph.D. thesis SU-IPR Report No. 710, Stanford University (1977).

10. O. Buneman, "The Advance From 2-D Electrostatic to 3-D Electromagnetic Particle Simulation", Comp. Physics Comm. 12, 21-31 (1976).

11. O. Buneman, "Variationally Optimized, Grid-Insensitive Vortex Tracing", Lecture Notes in Physics 35, Proc. 4th Intl. Conf. on Numerical Methods in Fluid Dynamics, 111-116, Springer-Verlag, Berlin (1975).

12. S. A. Orszag, "Numerical Simulation of the Taylor-Green Vortex", Computing Methods in Applied Science and Engineering, R. Glowinski and J. L. Lions, eds. Springer, New York (1974).

13. G. I. Taylor and A. E. Green, "Mechanism of the Production of Small Eddies from Large Ones", Proc. Roy. Soc. A158, 499-521 (1937).

MATHEMATICAL AND NUMERICAL ASPECTS OF DISCRETE KINETIC THEORY

H. Cabannes, R. Gatignol, D. Leguillon
Université Pierre et Marie Curie
4 place Jussieu - 75005 Paris (France)

Using discrete kinetic theory Boltzmann's equation is replaced by a system of partial differential equations. For this system and for certain models the global existence of the initial value problem has been proved. For a three dimensional model with 14 velocities we have investigated steady Couette flow, and for plane models with 4 velocities both steady and unsteady Couette flows. Numerical computations have been carried out.

The discretization of space velocities in kinetic theory of gases allows us to replace Boltzmann's equation by the system [1,2]

$$\text{(1)}\qquad \frac{\partial N_i}{\partial t} + \vec{u}_i \cdot \vec{\nabla} N_i = \frac{1}{2} \sum_{jkl} A_{ij}^{kl} (N_k N_l - N_i N_j) \qquad (i = 1,2,\ldots p)$$

where the coefficients A_{ij}^{kl} denote positive or zero constants, and the vectors $\vec{u}_i$ constant velocities; the unknown functions $N_i(\vec{x},t)$ denote the densities at the point $\vec{x}$, and at time t, of molecules with velocity $\vec{u}_i$. The replacement of Boltzmann's equation by the system (1) makes it possible for us to prove a global existence theorem [3], to study the steady [4,5] and unsteady [6] Couette flow and to compute numerically the solution of the shock tube problem [7] .

I - For the initial value problem [3] , we wish to find the solution of the system (1) which at the time t = 0 takes the given values :

$$\text{(2)}\qquad N_i(\vec{x},0) = N_{oi}(\vec{x}) \qquad (i = 1,2,\ldots p)$$

a - When the functions $N_{oi}(\vec{x})$ are continuously differentiable, the problem has a local solution. That means that there exists a positive number δ_o such that, in the interval $0 < t < \delta_o$, the problem (1),(2) has one and only one solution. When the initial values satisfy the inequalities $0 < N_{oi}(\vec{x}) < K_o$, one can take $\delta_o = (A\,K_o)^{-1}$ with

$2\,\dot{A} = p^2 \operatorname{Sup} A_{ij}^{kl}$, and the local solution has the following properties: 1°) the densities $N_i(\vec{x},t)$ are non negative, 2°) if the initial values are independant of one of the space variables, the densities are also independant of this space variable, 3°) if the initial values are periodic with period $\vec{\pi}$, the densities are periodic in $\vec{x}$ with the period $\vec{\pi}$.

b - For certain particular models we can show that the local solution is bounded ($N_i(\vec{x},t) < K$) by a constant K which depends only on the initial values; global existence (for $t > 0$) then results. A bound K has been built for the model with fourteen velocities, obtained by joining the center of a cube to the centers of the faces and to the vertices. This is constructed in the case when the initial values $N_{oi}(\vec{x})$ depend only on the abscissa x, and when the number of the molecules initially inside a cylinder bounded by the collisional cross section S,

$$\int_{-\infty}^{\infty} \left\{ \sum_{i=1}^{14} N_{oi}(\xi) \right\} S\, d\xi = \alpha_o ,$$

is small enough. The existence of such a bound is established by integration of the conservation equations over triangles in the x,t plane; these triangles have as edges the axis of the abscissae and some characteristics of the system (1) . We also obtain the inequality

$$\operatorname{Sup} N_i(x,t) < 5\, K_o + \frac{4}{3}\, \alpha_o \operatorname{Sup} N_i(x,t)$$

which proves the result stated if $\alpha_o < \frac{3}{4}$.

c - When the initial values $N_i(\vec{x})$ depend only on the abscissa x and are periodic with period π , the H-theorem of Boltzmann leads to the inequality (3) [3]

$$(3) \qquad \int_{-\infty}^{\infty} \left\{ \sum_{i=1}^{14} \tilde{N}_i(\xi,t_1)\, d\xi \right. < \frac{56\ K_o}{1 - \operatorname{Log} \frac{2L}{\pi}}$$

where

$$\begin{cases} \tilde{N}_i(x,t_1) = N_i(x,t_1) & \text{for } X - L \leqslant x \leqslant X + L \\ \tilde{N}_i(x,t_1) = 0 & \text{for } |X - x| < L \end{cases}$$

In the case of the fourteen velocities model we deduce global existence from inequality (3) when the initial values are periodic and then establish global existence for arbitrary bounded initial values.

The global existence of the solution has also been established in the case of the boundary values corresponding to the shock tube problem [3,7]. The solution has then been calculated numerically by a fractional step method, and an example is illustrated in figure 1 .

II - a - The steady Couette flow problem has been solved analytically [5] for the above 14 velocities model, when the walls are parallel to one of the faces of the cube. The moduli of velocities have the values c or $c\sqrt{3}$, and the temperature T is a macroscopic variable independant of the mean velocity $\vec{u}$:

$$3\,k\,n\,T = \mu\,n\,(c^2 - \vec{u}^2) + 2\,\mu\,c^2 \sum_{i=1}^{8} N_i$$

in which k denotes Boltzmann's constant, μ, the mass of molecules, N_i, the densities of molecules with velociy of modulus $c\sqrt{3}$, and n, the total density. For a gas of this type there exist 27 types of possible collisions, 12 of which occur between molecules having velocities of different modulus. This model has permitted us to demonstrate, not only the effect of slip velocity on the walls, but also the effect of slip in temperature. The mean velocity is not yet a linear function of the distance to the walls. The presence of molecules having a velocity parallel to the walls, molecules which in the reality constitute a set of measure zero, has a consequence that the slip velocity effect does not tend to zero with Knudsen number (limiting case of hydrodynamics); this paradoxical result disappears when we drop the four velocities parallel to the walls.

b - The steady Couette flow problem has also been solved for a plane four velocities model with the most general boundary conditions [4] . The exact solution has been obtained and compared with the solution of the Navier-Stokes equations of the model; in both solutions the anisotropic character of the medium appears. Naturally, the solution of the Navier-Stokes equations with the boundary conditions corresponding to the adherence of the fluid at the walls differs from the solution of the kinetic equations. However both solutions have the same limit when the Knudsen number approaches zero. On the other hand, the solution of the Navier-

Stokes equations with boundary conditions which take account of the slip velocity effect at the walls coïncides with the solution of the kinetic equations. The effect of slip velocity on the walls is of the order of the Knudsen number.

c - Unsteady Couette flow has been treated numerically for a plane, four velocities model [6]. Initially the gas is at rest and so are the two walls, which are parallel; one of the walls is then set in motion and moves parallel to itself. The molecules on each wall are diffusely reflected and we have studied the motion of the gas and its limit when time increases indefinitely. In terms of dimensionless variables we have to solve the following problem

$$(4) \qquad \frac{\partial n_j}{\partial \tau} - \cos\left(\theta + j\frac{\pi}{2}\right)\frac{\partial n_j}{\partial \eta} = n_{j+1}\, n_{j+3} - n_j\, n_{j+2}$$

$$(5) \qquad n_j(0,\eta) = \frac{1}{4}, \qquad n_j(\tau,\eta) > 0 \qquad (j = 1,2,3,4)$$

$$(6) \qquad \text{for} \quad \eta \pm \frac{\delta}{2}, \qquad v = 0$$

$$(7) \qquad \text{for} \quad \eta = \frac{\delta}{2}, \qquad n_3 = n_4$$

and for $\eta = -\frac{\delta}{2}$,

$$(1 - 2\sin\theta\, u_w - u_w^2 \cos 2\theta)\, n_1 = (1 + 2\cos\theta\, u_w + u_w^2 \cos 2\theta)\, n_2$$

in which u_w (velocity of the lower wall) and θ (orientation of the velocities of molecules) are given; u_w is a function of time, θ is a constant; v denotes the component perpendicular of the walls of the mean velocity; η is the distance to the plane equidistant from the walls.

$$\left(\sum_{j=1}^{4} n_j\right) v = -\sum_{j=1}^{4} n_j \cos\left(\theta + j\frac{\pi}{2}\right)$$

The component parallel to the walls of the mean velocity will be denoted by u .

The numerical solution of the system (4)-(7) has been computed by a fractional step method with $2\,u_w = \tanh\tau$, $\delta = 1$ and different values of θ . For $\theta = \frac{\pi}{4}$ the differential system to solve is linear and we obtain $v = 0$. In figure 2 are represented the curves $u(\eta,\tau)$ for different values of τ ; the steady profile corresponding to a wall with the velocity $u_w = 0.5$ is practically reached for $\tau = 8$ (8 times the mean time of flight). For $\theta \neq \frac{\pi}{4}$, the system is non linear and the mean velocity is not parallel to the walls; the component v is not zero, as a result of collisions. In figure 3, the curves $v(\eta,\tau)$ are represented for different values of τ . As time is increased, the point where the component v is maximum oscillates from the lower wall to the upper wall, and this maximum value approaches zero when the time increases indefinitely. New computations corresponding to different values of the velocity u_w are now being carried out and the results will be compared with those obtained by Chu [8].

References

[1] Gatignol R. Lecture Notes in Physics, 36, 1975, Springer

[2] Godunov S. and Sultangazin U. Uspeki Math. Nauk, 26, n°3, 1971

[3] Cabannes H. Journal de Mécanique, 17, 1-22, 1978

[4] Gatignol R. The Physics of Fluids, 20, 2022-30, 1977

[5] Cabannes H. Journal of Fluid Mechanics, 76, 273-87, 1976

[6] Gatignol R. Eleventh Symposium on Rarefied Gas Dynamics, 1978

[7] Leguillon D. Comptes-Rendus Ac. Sci. Paris, 285 A, 1125-28, 1977

[8] Chu C.K. Fifth Symposium on Rarefied Gas Dynamics, p. 589, 1967

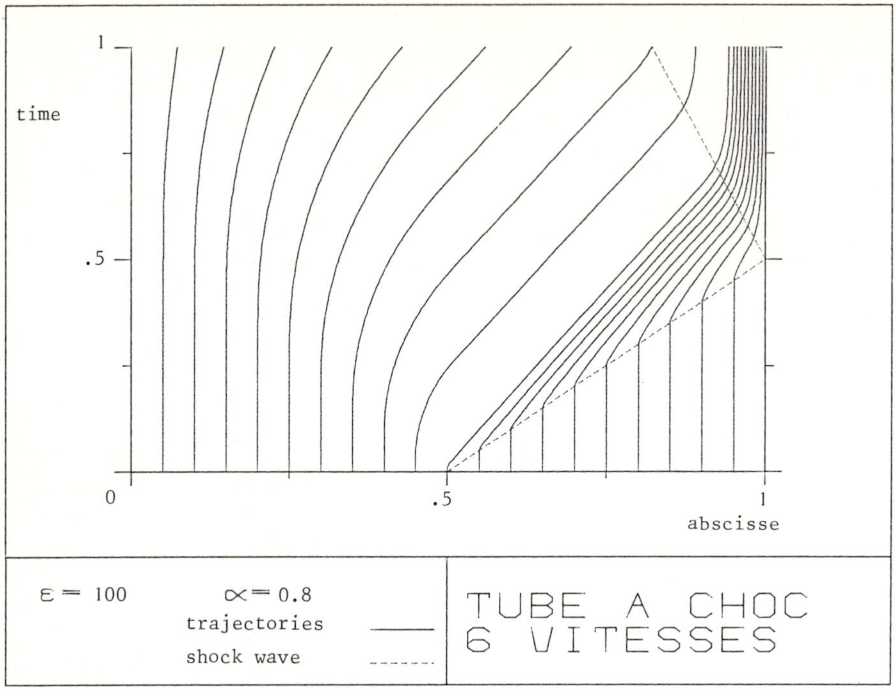

Fig. 1 - Shock tube problem [7]

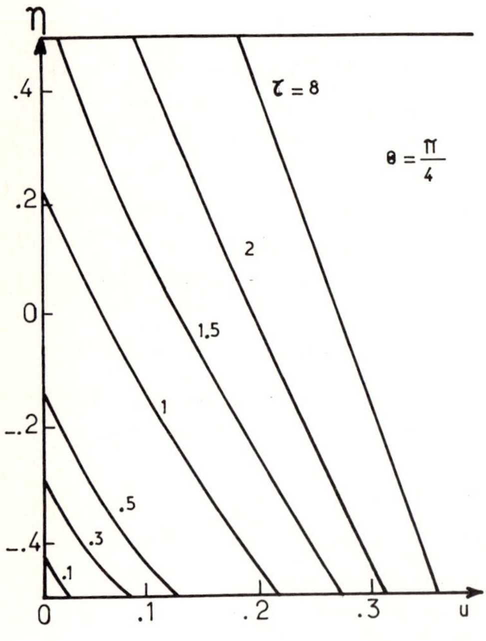

Fig. 2 - Mean velocity $u(\eta,\tau)$

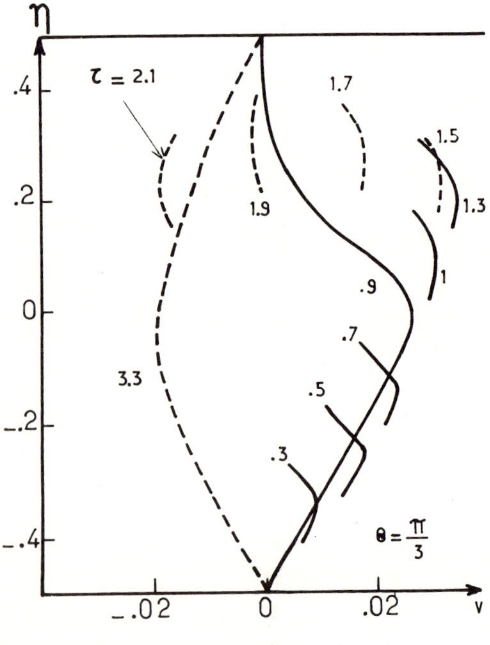

Fig. 3 - Mean velocity $v(\eta,\tau)$

RESOLUTION NUMERIQUE D'UNE EQUATION DE CONSERVATION PAR UNE APPROCHE VARIATIONNELLE

J.J. Chattot[**], J. Guiu-Roux[***], J. Laminie[***]

RESUME

L'utilisation de schémas semi-implicites ou implicites pour la résolution d'équations aux dérivées partielles du second ordre gouvernant le mouvement d'un fluide parfait en régime subsonique et transsonique, a contribué à l'efficacité et la popularité des méthodes de relaxation. Pour des systèmes du premier ordre et en particulier pour les équations d'Euler, ces schémas implicites n'ont pas permis jusqu'à présent d'aboutir à des méthodes aussi efficaces. L'approche proposée pour traiter des systèmes du premier ordre consiste à utiliser une formulation variationnelle du type moindres-carrés conduisant à un système équivalent d'équations aux dérivées partielles du second ordre. Cette méthode est illustrée sur un problème modèle de résolution de l'équation de conservation de la masse dans une tuyère, dont le champ de vitesse est donné. Ceci constitue une première étape en vue de la résolution des équations d'Euler stationnaires. Des résultats obtenus par une méthode de différences finies et une méthode d'éléments finis sont présentés. Des comparaisons détaillées ont été rendues possibles par le choix d'un même maillage pour les deux méthodes de discrétisation.

INTRODUCTION.

Pour le calcul d'écoulements stationnaires de fluide parfait en régime subsonique et transsonique, les méthodes de relaxation ont connu un essor considérable depuis l'introduction par Murman et Cole [1] d'un schéma de discrétisation semi-implicite mixte et d'une méthode de relaxation par colonne. La méthode développée initialement en théorie

** Office National d'Etudes et de Recherches Aérospatiales.
*** Université Paris-Sud - 91405 Orsay, (France)

des petites perturbations transsoniques a été étendue par de nombreux auteurs à l'équation complète du potentiel. Les propriétés de stabilité et de convergence très favorables de ces méthodes sont la conséquence d'un bon conditionnement des matrices associées aux systèmes algébriques obtenus après discrétisation du problème. Pour cette raison aussi ces méthodes peuvent être "accélérées" par l'utilisation de schémas implicites de type factorisation approchée comme le préconise Ballhaus [2]. Dans tous ces cas l'équation traitée est une équation aux dérivées partielles secondes.

Pour les systèmes d'équations aux dérivées partielles premières les méthodes de relaxation n'ont pas encore donné des résultats aussi satisfaisants que pour l'équation du potentiel. La raison tient au fait qu'il est difficile pour des dérivées premières de trouver des schémas de discrétisation qui conduisent à des matrices bien conditionnées. Steger et Lomax [3] ont utilisé des schémas à trois points décentrés, et un algorithme d'échange très particulier pour la relaxation par colonne. Cette méthode, appliquée aux équations d'Euler stationnaires dans une approximation équivalente à l'équation du potentiel mais utilisant les variables (ρ, u, v), faisait jouer un rôle priviligié à la direction coïncidant avec la colonne. Plus récemment, Beam et Warming [4] ont présenté un schéma implicite pour résoudre les équations d'Euler instationnaires. Il semble que le caractère implicite du schéma ne suffise pas à lever la restriction imposée par le critère de Courant-Friedrichs-Levy (CFL) s'il n'est pas accompagné d'un bon conditionnement de la matrice qui lui est associée.

La présente approche consiste à transformer le système du premier ordre en un système équivalent du second ordre au moyen d'une formulation variationnelle du type moindres-carrés.

Le système du second ordre ainsi obtenu peut être discrétisé au moyen de schémas centrés bien adaptés au traitement de problèmes aux limites. Cette méthode a été appliquée à la résolution des équations d'Euler quasi-monodimensionnelles représentant l'écoulement dans une tuyère très élancée [5]. Des configurations bloquées et non-bloquées ont été calculées.

Nous présentons ici une application à la résolution de l'équation de conservation de la masse dans une tuyère bidimensionnelle, dont le champ de vitesse est donné. Ceci constitue une première étape en vue de la résolution des équations d'Euler stationnaires.

1. FORMULATION VARIATIONNELLE.

Soit Ω un domaine simplement connexe, de frontière $\Sigma = \partial\Omega = \Sigma_1 \cup \Sigma_2$.

Soit $\vec{q} = (u,v)$ un champ de vitesse donné sur $\bar{\Omega}$.

On se propose de calculer la masse volumique en tout point du domaine de façon à vérifier l'équation de conservation de la masse :

(1) $$\frac{\partial\rho u}{\partial x} + \frac{\partial\rho v}{\partial y} = 0.$$

Cette équation pour ρ est hyperbolique. Les caractéristiques sont les lignes de courant qui sont spécifiées par la donnée de $\vec{q}$. Un problème bien posé pour ρ consiste à formuler un problème aux valeurs initiales :

soit Σ_1 et Σ_2 les parties de la frontière Σ telles que : $\vec{q}.\vec{n} \leq 0$ sur Σ_1, $\vec{q}.\vec{n} \geq 0$ sur Σ_2, $\vec{n}$ est la normale unitaire extérieure à Ω

(2) $$\rho(x,y) = \rho_1(x,y) \quad \text{si} \quad (x,y) \in \Sigma_1$$

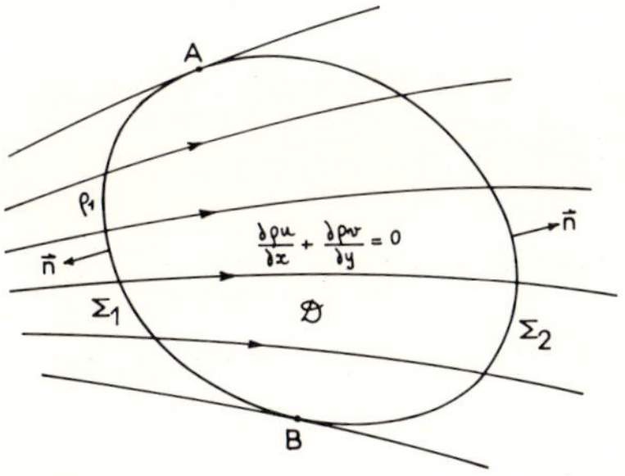

Fig. 1 – Problème de Cauchy pour la masse volumique.

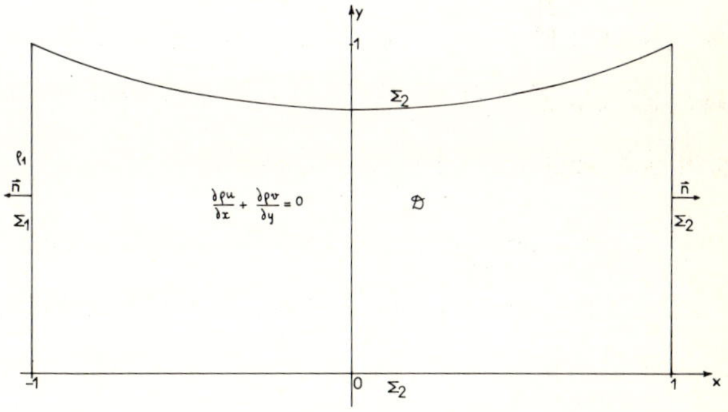

Fig. 2 – Conservation de la masse dans une tuyère.

(3) Soit $$I = \int_\Omega \left\{\frac{\partial\rho u}{\partial x} + \frac{\partial\rho v}{\partial y}\right\}^2 dxdy.$$

La formulation variationnelle consiste à rechercher ρ qui minimise I :

(4) $$\operatorname*{Inf}_{\rho} I.$$

La recherche du minimum de I équivaut à annuler l'intégrale :

$$\int_{\Omega} \left\{ \frac{\partial \rho u}{\partial x} + \frac{\partial \rho v}{\partial y} \right\} \left\{ \frac{\partial \tau u}{\partial x} + \frac{\partial \tau v}{\partial y} \right\} dxdy = 0, \tag{5}$$

pour toute fonction test τ suffisamment régulière et nulle sur Σ_1. On peut vérifier que ceci revient à résoudre une équation aux dérivées secondes dans Ω et à satisfaire une équation aux dérivées partielles premières sur Σ_2 :

$$-\int_{\Omega} \tau \left\{ u \frac{\partial}{\partial x} + v \frac{\partial}{\partial y} \right\} \left\{ \frac{\partial \rho u}{\partial x} + \frac{\partial \rho v}{\partial y} \right\} dxdy + \int_{\Sigma_2} \tau \rho \vec{q}.\vec{n} \left\{ \frac{\partial \rho u}{\partial x} + \frac{\partial \rho v}{\partial y} \right\} ds = 0, \quad \forall \tau. \tag{6}$$

Il faut noter que pour une configuration de tuyère (figure 2), les points singuliers A et B (figure 1) sont étirés sur les parois supérieures et inférieures de la tuyère où l'écoulement est tangent. Ces lignes de courant frontière font partie de Σ_2.

2. METHODE DE DIFFERENCES FINIES.

Le domaine est discrétisé en cellules de forme trapézoïdales par l'introduction d'un maillage régulier engendré par une transformation géométrique du plan (ξ,η) sur le plan (x,y). Les équations (6) sont écrites et discrétisées dans les coordonnées curvilignes (ξ,η). Les dérivées partielles premières $(\frac{\partial \xi}{\partial x},\ldots)$ et secondes $(\frac{\partial^2 \xi}{\partial x^2},\ldots)$ sont évaluées numériquement à partir des noeuds du maillage comme il est indiqué dans [6].

Pour les points intérieurs, les schémas utilisés pour approcher les dérivées partielles secondes sont des schémas centrés, précis au second ordre :

$$\left(u \frac{\partial^2}{\partial \xi^2} \rho u\right)_{i,j} = u_{i,j} \frac{\rho u_{i+1,j} - 2\rho u_{i,j} + \rho u_{i-1,j}}{\Delta \xi^2}, \quad O(\Delta \xi^2) \tag{7}$$

$$\left(u \frac{\partial^2}{\partial \xi \partial \eta} \rho v\right)_{i,j} = u_{i,j} \frac{\rho v_{i+1,j+1} - \rho v_{i-1,j+1} - \rho v_{i+1,j-1} + \rho v_{i-1,j-1}}{4\Delta \xi \Delta \eta}, \quad O(\Delta \xi + \Delta \eta)^2 \tag{8}$$

avec des formules correspondantes pour les autres termes. Au total, la discrétisation de l'équation du second ordre conduit à un schéma à 9 points.

Pour les points-frontières de Σ_2, l'équation du premier ordre est discrétisée au moyen de schémas décentrés et précis du premier ordre. On peut montrer que le décentrement doit être effectué à l'opposé de la vitesse pour assurer un bon conditionnement à la matrice. Par exemple, pour la situation représentée sur la figure 3 nous avons utilisé les schémas $(u > 0,\ v > 0)$:

$$(9) \qquad \left(\frac{\partial \rho u}{\partial \xi}\right)_{i,j} = \frac{\rho u_{i,j} - \rho u_{i-1,j}}{\Delta \xi}\ ,\ O(\Delta \xi)$$

$$(10) \qquad \left(\frac{\partial \rho v}{\partial \eta}\right)_{i,j} = \frac{\rho v_{i,j} - \rho v_{i,j-1}}{\Delta \eta}\ ,\ O(\Delta \eta)$$

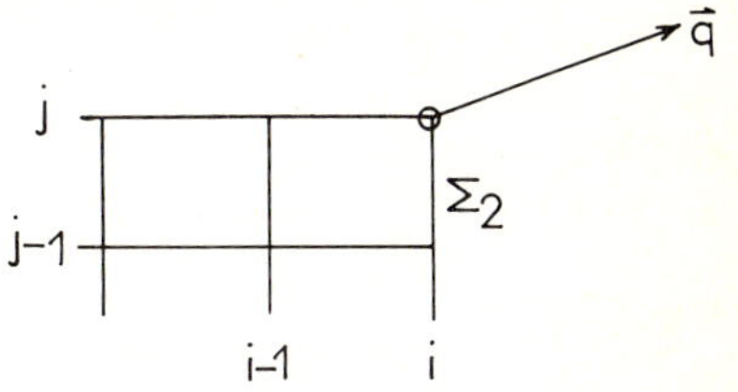

Fig. 3 – Discrétisation de la condition aval.

L'utilisation de schémas décentrés indique que l'on retrouve le caractère hyperbolique du problème près des frontières aval.

La méthode utilisée pour résoudre le système d'équations pour ρ est une méthode de relaxation par ligne ou par colonne. Le facteur de relaxation est pris égal à l'unité. Typiquement 400 à 1000 itérations permettent d'obtenir la convergence complète de la méthode, ce qui est comparable à une résolution par relaxation d'un problème elliptique avec des conditions aux limites de type Neumann.

3. METHODE D'ELEMENTS FINIS.

Soient u et v donnés dans Ω et :

$$V(\Omega) = \{\rho \in L^2(\Omega) \backslash \operatorname{div} \rho\vec{q} \in L^2(\Omega),\ \vec{q} = (u,v)\}$$

$$K = \{\rho \in V(\Omega) \backslash \rho = \rho_1 \quad \text{sur } \Sigma_1\}$$

$$K_o = \{\rho \in V(\Omega) \backslash \rho = 0 \quad \text{sur } \Sigma_1\}.$$

On cherche ρ réalisant :

$$(11) \qquad I(\rho) = \operatorname*{Inf}_{\tau \in K} I(\tau),$$

ρ vérifie alors (5), $\forall\ \tau \in K_o$, (cf. [7]).

Bien que les quantités ρ, u, v présentent en général des discontinuités, on a choisi d'utiliser des éléments finis continus, et en particulier les éléments les plus simples sur des quadrilatères, notés habituellement "élément Q_1", pour lesquels les fonctions de base sont de la forme $a+bx+cy+dxy$ (a, b, c, d constantes) sur chaque élément. Pour plus de détails consulter les références [8], [9] et [10].

Dans une première approche u et v sont discrétisés de la même façon que ρ

sur la base habituelle $\{\varphi_i\}_{i\in J}$ des éléments Q_1. Les résultats obtenus ainsi sont excellents dans les cas des champs de vitesse réguliers, mais lorsque le champ de vitesse est discontinu par exemple lorsqu'un choc droit sépare deux régions d'écoulement uniforme, les conditions de saut ne sont pas satisfaites. Par analogie avec les différences finies on peut qualifier un tel schéma de "non-conservatif". Une explication possible est que cette formulation demande trop de régularité aux fonctions ρ, u et v.

Dans une seconde approche on utilise l'approximation de type Q_1 pour les produits ρu et ρv. On résout le problème suivant :

- Problème Ph :

Soient u_i, v_i, ρ_i les valeurs respectives de u, v, ρ aux noeuds de la discrétisation. On désire trouver ρ_h tel que :

$$\rho_h u = \sum_{j\in J} \rho_j u_j \varphi_j , \qquad \rho_h v = \sum_{j\in J} \rho_j v_j \varphi_j$$

et vérifiant

$$(12) \qquad \int_\Omega \left(\frac{\partial(\sum_{j\in J} \rho_j u_j \varphi_j)}{\partial x} + \frac{\partial(\sum_{j\in J} \rho_j v_j \varphi_j)}{\partial y}\right)\left(\frac{\partial u_\ell \varphi_\ell}{\partial x} + \frac{\partial v_\ell \varphi_\ell}{\partial y}\right) dx\, dy = 0 \qquad \forall \varphi_\ell,\ \ell\in J.$$

La résolution du problème Ph se ramène simplement à la résolution d'un système linéaire dont la matrice est bien conditionnée. Les intégrales qui apparaissent dans l'équation (11) sont évaluées au moyen d'une formule de Gauss à quatre points. Le système est résolu par une méthode de relaxation par point.

4. RESULTATS.

Le maillage utilisé pour les méthodes de différences finies et d'éléments finis est présent sur la figure 4.

Les essais ont été effectués avec deux types de champs de vitesse : un champ de vitesse continu à divergence nulle et un champ discontinu simulant la présence d'une onde de choc dans le divergent. Le champ de vitesse à divergence nulle est obtenu par superposition des champs produits par deux tourbillons disposés symétriquement par rapport à l'axe Ox et d'intensités opposées.

Pour le champ de vitesse continu les résultats obtenus par la méthode de différences finies et la première méthode d'éléments finis sont présentés sur les figures 5 et 6. La solution exacte correspond à $\rho = 1$. On peut noter une erreur relative 5.10^{-2} pour les différences finies et de 5.10^{-3} pour la méthode d'éléments finis.

Dans le cas de la simulation d'un choc parabolique dans le divergent (figure 4), la méthode des différences finies donne pour le calcul du débit sortant de la tuyère une faible perte de 3.10^{-2} par rapport au débit entrant (figure 7). Un résultat comparable a été obtenu par la seconde méthode d'éléments finis (figure 8), et la conservation du débit est assurée à 1.10^{-2} près.

CONCLUSION.

Une approche variationnelle pour résoudre des équations de conservation du premier ordre a été appliquée à la résolution de l'équation de conservation de la masse dans une tuyère. Des discrétisations du type différences finies et éléments finis ont été utilisées. Les deux méthodes ont donné des résultats satisfaisants aussi bien dans le cas d'un écoulement continu que dans le cas d'un champ de vitesse discontinu, simulant une onde de choc. Une excellente précision est obtenue avec la méthode des éléments finis. On peut noter également que celle-ci ne nécessite pas un traitement particulier près de la frontière aval (en différences finies des schémas décentrés sont utilisés en ces points) ce qui confère une grande simplicité à cette approche.

REFERENCES.

[1] Murman, E.M. et Cole, J.D.- Calculation of Plane Steady Transonic Flows, AIAA Journal, Vol. 9, n° 1, 1971.

[2] Ballhaus, W.F., Jameson, A. et Albert, J.- AIAA Paper n° 77-634, AIAA Computational Fluid Dynamics Conference, Albuquerque, N.M., U.S.A., Juin 1977.

[3] Steger, J.L. et Lomax, H.- Generalized Relaxation Methods Applied to Problems in Transonic Flow, Proc. II Int. Conf. Num. Meth. Fluid Dynamics, M. Holt editor, Springer-Verlag, 1971.

[4] Beam, R.M. et Warming, R.F.- Numerical Calculations of Two-dimensional Unsteady Transonic Flows with Circulation, NASA TN D-7605, Février 1974.

[5] Chattot, J.J.- Relaxation Approach to the Steady Euler Equations in Transonic Flow, AIAA Paper n° 77-636, AIAA Computational Fluid Dynamics Conference, Albuquerque, N.M., U.S.A., Juin 1977.

[6] Chattot, J.J., Coulombeix, C. et Tomé, C.- Calculs d'écoulements transsoniques autour d'ailes
La Recherche Aerospatiale, à paraître, 1978.

[7] Temam, R. et Ekeland, I.- Analyse Convexe et Problèmes variationnels. Dunod, 1974.

[8] Ciarlet, P.G.- Numerical Analysis of the Finite Element Method. Cours, Montréal, 1971.

[9] Raviart, P.A.- Cours, Paris, 1972.

[10] Strang, G. et Fix, G.J.- An Analysis of the Finite Element Method, Prentice-Hall.

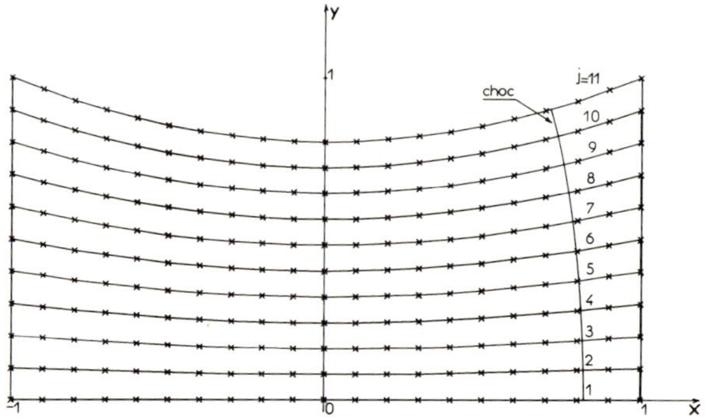

Fig. 4 — Maillage.

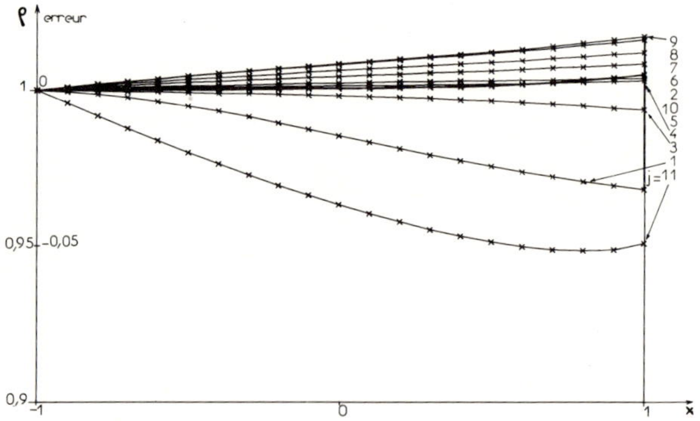

Fig. 5 — Champ de vitesse à divergence nulle.
Méthode des différences finies.

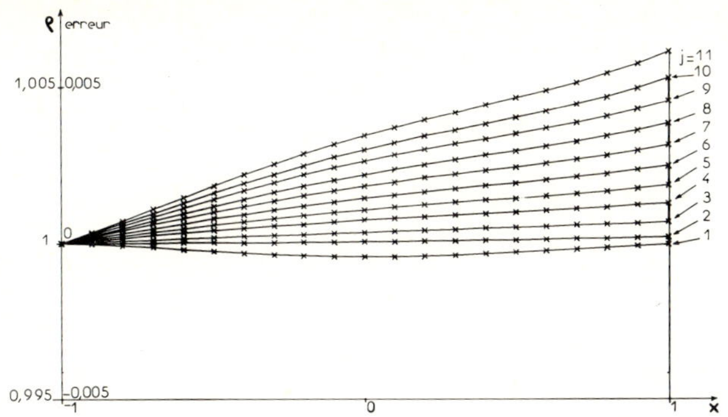

Fig. 6 – Champ de vitesse à divergence nulle. Méthode des éléments finis.

Fig. 7 – Simulation d'un choc. Méthode des différences finies.

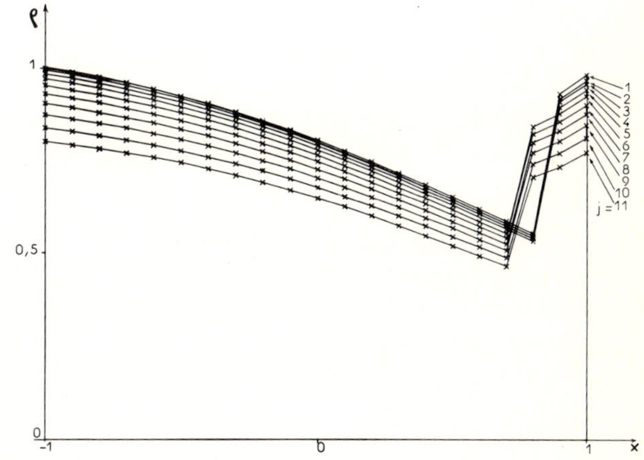

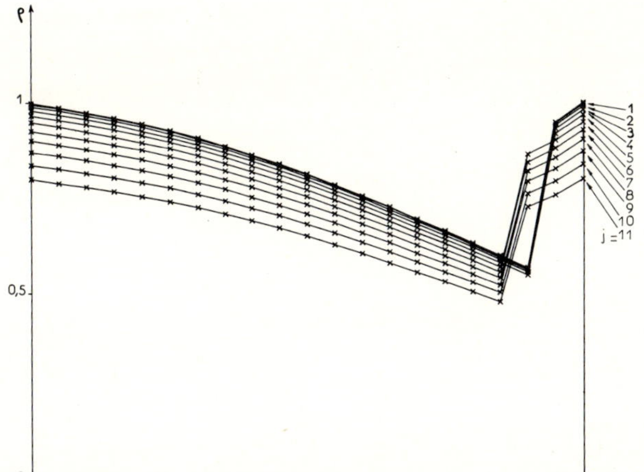

Fig. 8 – Simulation d'un choc. Méthode des éléments finis.

ERRORS IN FINITE DIFFERENCE SOLUTIONS OF NAVIER-STOKES EQUATIONS

by

Sin-I Cheng

Department of Mechanical and Aerospace Engineering
Princeton University
Princeton, New Jersey U.S.A.

I. INTRODUCTION

This paper is concerned with the accuracy of numerical solutions with coarse meshes of nonlinear partial differential equations, such as the Navier-Stokes. By presuming smooth, convergent difference approximations, the author carried out some first differential analyses[1] of the computational errors and recommended an upper error bound of 0.03 $Re\Delta x^2$ for second order accurate conservative difference schemes, computed at $Re\Delta x \lesssim 2$. In most computational solutions of complex problems, the $Re\Delta x$ is appreciably larger and the solutions are oscillatory. Neither the local truncation error nor the suggested upper bound provide any meaningful measure of the errors of the computed solution. Indeed, the computed solutions at successive mesh refinements often become "worse" and suggest possible nonuniform convergence of the sequence as $\Delta x \to 0$. Therefore we solve analytically the difference formulations of some model problems to learn about the roles of various physical and computational parameters, including the mesh Reynolds number $Re\Delta x$. The analytic results are verified by computational solutions of the model equations with various time dependent algorithms.

II. MODEL ANALYSIS

The following model problems with known exact steady state differential solutions are chosen for study.

1. Model 1: The Burgers' equation

$$u_t + uu_x = \frac{1}{Re} u_{xx} \tag{1}$$

with specified boundary values

$$u(x=0) = 0, \quad u(x=1) = -1 \tag{2}$$

2. Model 2: The same Burgers' equation with upstream boundary value and downstream extrapolation

$$u(x=-1) = 1 \quad \frac{\partial u}{\partial x}(x=1) = 0 \tag{3}$$

3. Model 3: The one-dimensional gas dynamic equations

$$\begin{cases} \rho_t + (\rho u)_x = 0 \\ (\rho u)_t + (\rho u^2)_x + p_x = \frac{1}{RE} u_{xx} \\ p = \Lambda \rho^r \end{cases} \tag{4}$$

with upstream boundary values and downstream extrapolation conditions

$$u(x=-1) = 1, \quad \rho(x=-1) = 1$$

$$\frac{\partial u}{\partial x}(x=1) = 0, \quad \text{and} \quad \frac{\partial \rho}{\partial x}(x=1) = 0 \; . \tag{5}$$

The steady state solution of (4) and (5) represents a jump from an upstream uniform flow to a downstram uniform flow with velocity Λ', which is the real solution of u from the Hugoniot relation

$$u + \Lambda/u^r = 1 + \Lambda \tag{6}$$

r is the polytropic index defining the thermodynamic process of transition and Λ is the initial data upstream.

Equations (1) and (4) are discretized according to the algorithm:

$$u_{xx} \sim (U_{j+1} - 2U_j + U_{j-1})/\Delta x^2 \tag{7a}$$

$$uu_x = (u^2/2)_x \sim \left[\gamma U_j(U_{j+1}-U_{j-1}) + (U_{j+1}^2 - U_{j-1}^2)\right]/(2+\gamma)2\Delta x \tag{7b}$$

which renders the resulting difference formulations strictly "conservative", i.e., summable without residue in the interior of the domain of summation.[2] Thus the Stokes theorem can be applied to the difference form to relate the boundary conditions explicitly with any interior point values. The difference form of Eq. (1) in the steady state is

$$U_{j+1}^2 + U_j(U_{j+1}-U_{j-1}) - U_{j-1}^2 = \frac{2(2+\gamma)}{Re\Delta x}\left[U_{j+1}-2U_j+U_{j-1}\right] \tag{8}$$

The difference form of Eqs. (4) with discretization parameter γ_1 for the continuity and γ_2 for the momentum equation is

$$2\left[(\rho U)_{j+1} - (\rho U)_{j-1}\right] + \gamma\cdot\left[\rho_j(U_{j+1}-U_{j-1}) + U_j(\rho_{j+1}-\rho_{j-1})\right] = 0$$

$$2\left[(\rho U^2)_{j+1} - (\rho U^2)_{j-1}\right] + \gamma_2\left[U_j\left\{(\rho U)_{j+1} - (\rho U)_{j-1}\right\} + (\rho U)_j(U_{j+1}-U_{j-1})\right]$$
$$+ \Lambda\left[(\rho_{j+1})^r - (\rho_{j-1})^r\right] = \frac{2(2+\gamma_2)}{RE\Delta x}\left[U_{j+1}-2U_j+U_{j-1}\right] \quad (9)$$

The downstream extrapolation conditions are expressed as

$$U_J = J_{J-1} \quad \rho_J = \rho_{J-1}\text{, etc.} \qquad (10)$$

There are other algorithms that lead to strictly conservative difference formulations for complicated nonlinear terms.

Real solutions of each model system of equations are obtained analytically as follows:

1. Obtain exact solution(s) of the system for special values of $Re\Delta x$ (or $RE\Delta x$) such as $R\hat{e}\Delta x$, $R\overset{*}{e}\Delta x$, ... $R\tilde{e}\Delta x$

2. Identify the region of rapid variation (or jump) in the exact solution in which the fully nonlinear difference relations must be taken.

3. Obtain linearized small perturbation solutions of the difference equations system about the exact solution in its region of small variation, generally including the boundary. The linearized solution serves to relate the boundary values to those function values where the linear region joins the nonlinear (or jump) region in the interior.

4. Sum up all the difference relations over the entire domain to relate the boundary values.

5. Obtain all the real sets of solutions of the function values from as many algebraic equations deduced in 2 through 4.

6. Asymptotic results for finite but large $Re\Delta x$ are studied with or without explicit solution(s) of the nonlinear system of algebraic equations.

For practical range of $Re\Delta x > 10$, the one term asymptotic results give simple functional dependence of the error of the "exact difference solution" compared with the "exact differential solution". Such an analysis is illustrated in Ref. 3 for Model 1 and detailed in Ref. 4 for the other cases. The analytical results have been verified by digital computations with the time dependent schemes of Brailovskaya,[5] of Victoria-Widhopf,[6] of Peyret-Viviand,[7] and of Cheng-Allen,[1] all with the same conservative steady state equation (8). All the computed results under the same steady state criterion agree with one another and with the analytic estimates to three significant figures. As shown in Fig. 1, the errors of computed results for negative values of γ increases to some maximum at $Re\Delta x \sim 3$ and then reaches some minimum at larger $Re\Delta x$. The solutions in different ranges of $Re\Delta x$ are illustrated in Fig. 2. Similar results for Model 2 and Model 3 are given in Figs 3 and 4 and in Figs. 5 to 7, respectively. When the time dependent scheme like MacCormack is used, which does not reduce to (8) in the steady state, the error increases monotonically as $\sim 0.06\ Re\Delta x^2$ without plateau or extrema as $Re\Delta x$ increases (Fig. 1). The presence of error minimum at suitable large $Re\Delta x$ values may be a characteristic of the strictly conservative difference formulation.

III. ANALYTIC INFERENCES

The following qualitative inferences pertinent to practical computations may be obtained from the results of the above model study:

1. It is generally sufficient to compute with 8 - 10 mesh points within the characteristic length of the problem when $Re\Delta x$ is in the appropriate range. Many more mesh points do not improve the accuracy of the computed results at the same $Re\Delta x$.

2. There exist for a certain class of difference formulation critical mesh Reynolds numbers ($\widehat{Re}\Delta x$, $\overset{*}{Re}\Delta x$, etc.) at which computations will yield smooth but abrupt jump solutions with one or more mesh points within the jump. They can be excellent approximate solutions at coarse mesh, although solutions with minimal error occur at slightly different values of $Re\Delta x$. For suitable choices of the algorithm, the trough of the error curve around $\overset{*}{Re}\Delta x$ can be flat so that computations in a broad range of values of $Re\Delta x$ can yield results as accurate as what might be expected from smooth solutions at $Re\Delta x \sim 1$.

3. The magnitudes of these critical $Re\Delta x$ can be determined a priori for the model problems. They depend on the γ's representing the higher order details of the difference algorithms. The value of $\overset{*}{Re}\Delta x$ depends more importantly on the discretization parameter (γ_1) for the continuity equation than on γ_2 of the momentum equation (Fig. 6). The thermodynamic variables in the polytropic relation in (4) appears unimportant.

4. With proper boundary formulation (such as Model 1), computations at $Re\Delta x \lesssim 1$ (or $< \widetilde{Re}\Delta x$) yield smooth and accurate results with an upper bound of $\sim 0.03\ Re\Delta x^2$ as is determined from the first differential analysis.[1] The sequence appears to converge to the correct solution as $Re\Delta x \to 0$. Computational results at $Re\Delta x > \widetilde{Re}\Delta x$ are oscillatory and may or may not exhibit minimum error at some critical $Re\Delta x > \widetilde{Re}\Delta x$.

5. With uncertain boundary formulation (such as downstream extrapolation in Models 2 and 3), smooth computed solutions will be obtained at $Re\Delta x < \tilde{Re}\Delta x$ but they give wrong Hugoniot or jump magnitudes. The sequence of solutions can "converge" to the wrong solution as $Re\Delta x \to 0$. The oscillatory computed solutions at $Re\Delta x > \tilde{Re}\Delta x$ appears to diverge as $Re\Delta x$ decreases toward $\tilde{Re}\Delta x$. But, at or near the suitable critical $Re\Delta x$ under proper choices of γ's, they can provide excellent "asumptotic approximations" with small maximal errors; quite acceptable in practice.

The above inferences are drawn from the study of only the one-dimensional model equations, and discretized only in strictly conservative difference formulations. They are not expected to apply if the difference formulation is non-conservative since Stokes' theorem is crucial to the analysis. They are, however, hoped to hold without serious modifications for more complex equation systems in multi-space dimensions. For the latter purpose, we shall only present some numerical evidences.

IV. TWO DIMENSIONAL COMPUTATIONS WITH NAVIER-STOKES EQUATIONS

The propagation of an oblique, planar shock wave in a uniform supersonic stream at $M = 2$ and its eventual reflection from an inviscid wall have been computed at various mesh Reynolds numbers.[9] The simplest algorithm with $\gamma_1 = \gamma_2 = 0$ (Cheng-Allen) was adopted for discretizing the Navier-Stokes equations with downstream extrapolation conditions. Fig. 8 shows the computed results of a shock with pressure ratio $p_2/p_1 = 1.910$. If the 1-D model results should be assumed to apply to the components normal to the shock, the estimate of the critical Reynolds number $(Re\Delta n_1)crit$ based on the upstream velocity U_{1n} will be $4/\left(1 - \frac{U_{2n}}{U_{1n}}\right)$, according to the procedure in Ref. 2, p. 181. The critical Reynolds number $(Re\Delta x1)crit.$ based on the mesh size Δx in the x-direction will be 15.2 for the above case. We then observe in Fig. 8 that smooth solutions are obtained with $(Re\Delta x)/(Re\Delta x1)crit < 1$. Transition from smooth to oscillatory solutions takes place gradually when $(Re\Delta x1)/(Re\Delta x1)crit \sim 1$. The reflection of the above shock from an inviscid wall is shown in Fig. 9. The reflected shock has a pressure ratio $p_3/p_2 = 1.885$ with $(\tilde{Re}\Delta x1) = 22.1$. The same inference concerning the transition of the smooth to oscillatory solutions is observed. Fig. 10 shows the computed pressure profiles for a weak shock with $p_2/p_1 = 1.096$ and $(\tilde{Re}\Delta x1) = 122$. Fig. 11 gives the variation of the error norms with $Re\Delta x1$. The occurrence of an error minimum of $\lesssim 3\%$ appears when $(Re\Delta x1)/(Re\Delta x1)crit$ is between 3.1 and 6.2. This ratio compares with 2 as is shown in Fig. 7 for the case $\gamma_1=\gamma_2=0$ for the one dimensional gas dynamics model. In view of the fact that the case $\gamma_1=\gamma_2=0$ departs substantially from the optimal curve shown in Fig. 6 and that the model is in one rather than two space dimensions, the agreement is encouraging. How well the analytic inference outlined in the previous section may hold and to what extent these inferences may have to be modified when more complex problems are solved computationally with the Navier-Stokes equations, clearly needs further study.

ACKNOWLEDGMENTS

The above investigation has been sponsored by the Office of Naval Research, U. S. Navy under Contract No. N00014-75-C-0376. The author is grateful of the continued interest and support of this research. The use of the IBM 360-91 computer, which is partially supported by the National Science Foundation and Princeton University is also gratefully acknowledged.

REFERENCES

[1]Cheng, Sin-I, "Accuracy of Difference Formulation of Navier-Stokes Equations", The Physics of Fluids, Supplement II, December 1969.

[2]Cheng, Sin-I, "A Critical Review of Numerical Solution of Navier-Stokes Equations", Lecture Notes in Physics, No. 41. Progress in Numerical Fluid Dynamics, edited by J. Wirz, Springer Verlag, 1975.

[3]Cheng, Sin-I and Shubin, Gregory, "Computational Accuracy and Mesh Reynolds Numbers", to appear in J. Comp. Physics.

[4]Shubin, Gregory, "One-Dimensional Gas Dynamic Modeling and Computational Accuracy", Ph.D. Thesis, Princeton University, 1977. A shortened version is submitted to the Journal of Computational Physics.

[5]Brailovskaya, I. Y., "A Difference Scheme for Numerical Solution of the Two-Dimensional, Nonstationary Navier-Stokes Equations for a Compressible Gas", Soviet-Physics-Diklady, Vol. 10, No. 2, 1965.

[6]Victoria, K. J. and Widhopf, G. F., "Numerical Solution of the Unsteady Navier-Stokes Equations in Curvilinear Coordinates: The Hypersonic Blunt Body Merged Layer Problem". Lecture Notes in Physics, Vol. 19, Springer Verlag, 1973.

[7]Peyret, R. and Viviand, H., "Calcul de l'Econlement d'un Fluide Visquenx Compressible Autour d'un Obstacle de Forme Parabolique". Lecture Notes in Physics, Vol. 19, Springer Verlag, 1973.

[8]MacCormack, R. W., "The Effect of Viscosity in Hypervelocity Impact Cratering", AIAA Paper No. 69, 1969.

[9]Messina, Neale A., "A Computational Investigation of Shock Waves, Laminar Boundary Layers, and their Mutual Interaction", Ph.D. Thesis, Princeton University, 1977.

CAPTIONS OF FIGURES

Fig. 1. Maximum error E_∞ as a function of mesh Reynolds number $Re\Delta x = \frac{\Delta U \Delta x}{\nu}$ for Burgers' Model with downstream boundary value.

Fig. 2. Type of solutions of Burgers' model with downstream boundary value.

Fig. 3. Type of solutions of Burgers' model with downstream extrapolation.

Fig. 4. E_∞ vs $Re\Delta x$ for Burgers' model with downstream extrapolation.

Fig. 5. Type of solutions of 1-D gas dynamics model.
(a) jump next to downstream boundary $k = J-2$.
(b) jump at intermediate point $k < J-2$.

Fig. 6. Optimal discretization parameters for 1-D gas dynamic model.

Fig. 7. E_∞ vs $Re\Delta x$, for 1-D gas dynamics model $r = 1.4$, $k = J-1$.

Fig. 8. Pressure field of an oblique shock with $p_2/p_1 = 1.910$ in Mach 2 supersonic flow.

Fig. 9. Pressure field of shock reflection from inviscid wall $p_2/p_1 = 1.910$, $p_3/p_2 = 1.885$.

Fig. 10. Pressure field of an oblique weak shock with $p_2/p_1 = 1.096$.

Fig. 11. Error norms as a function of $Re\Delta x$ for shock calculations.

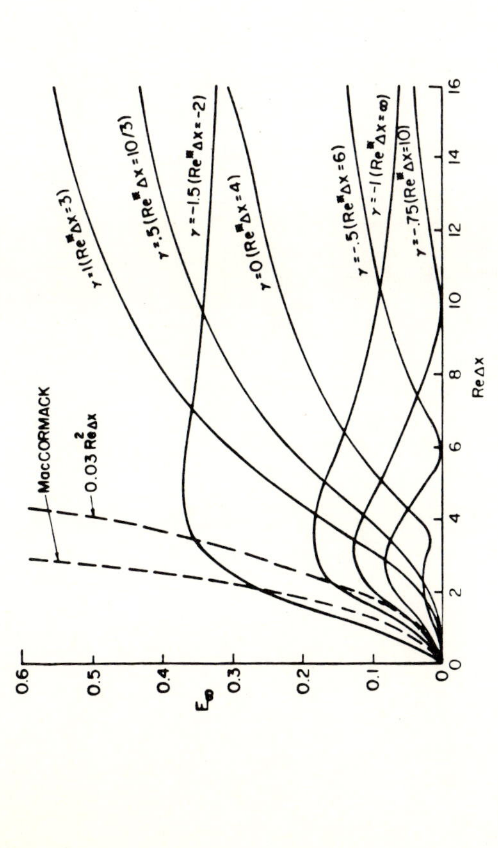

FIGURE 1

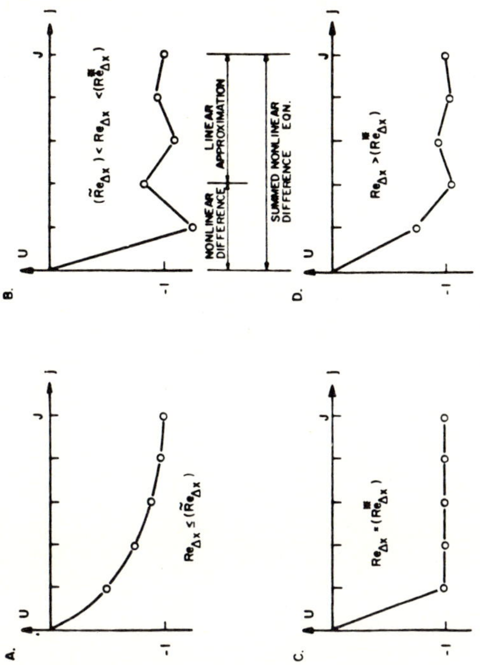

FIGURE 2

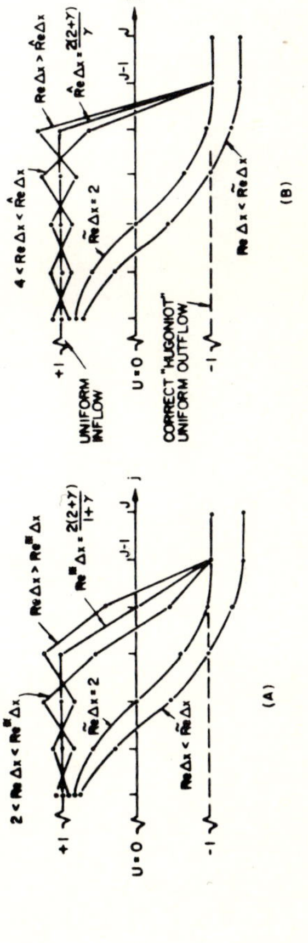

FIGURE 3

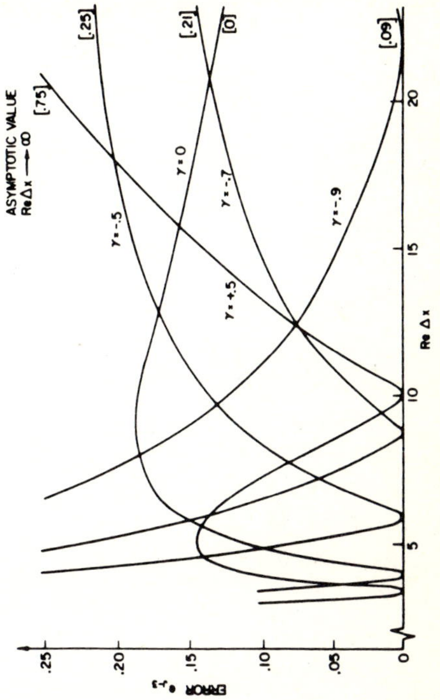

FIGURE 4

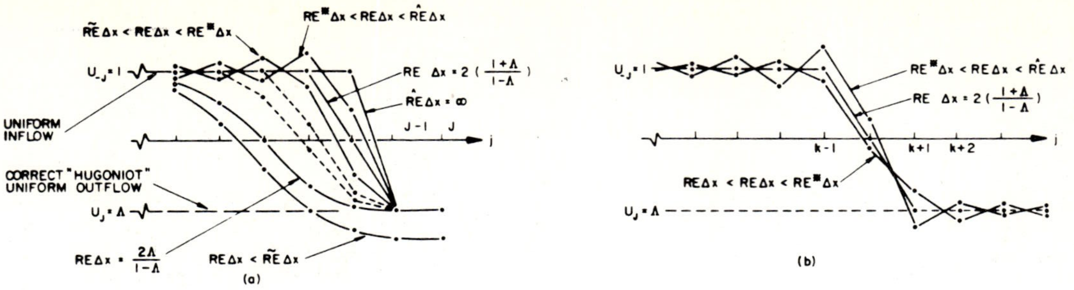

FIGURE 5

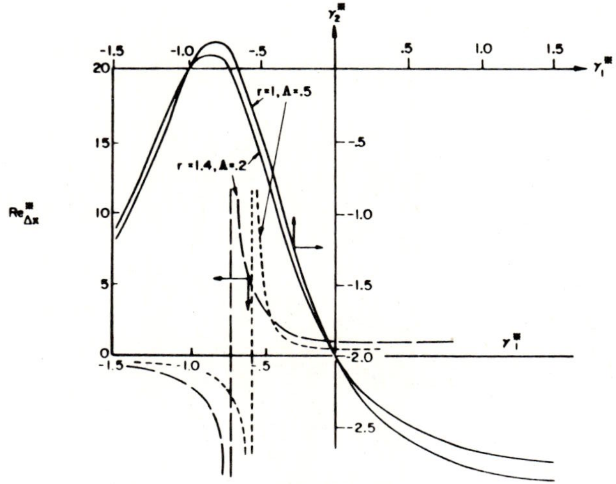

FIGURE 6

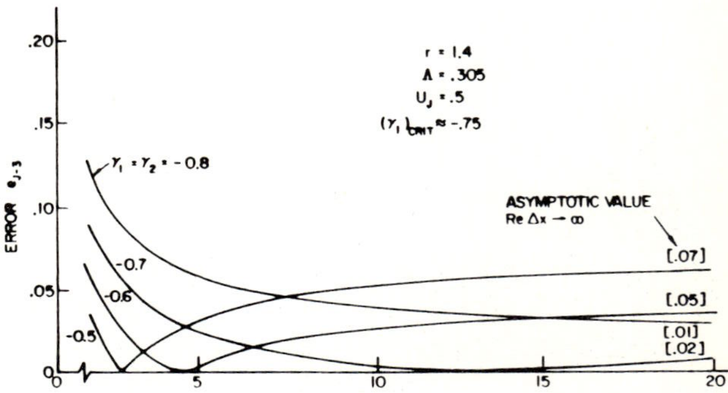

FIGURE 7

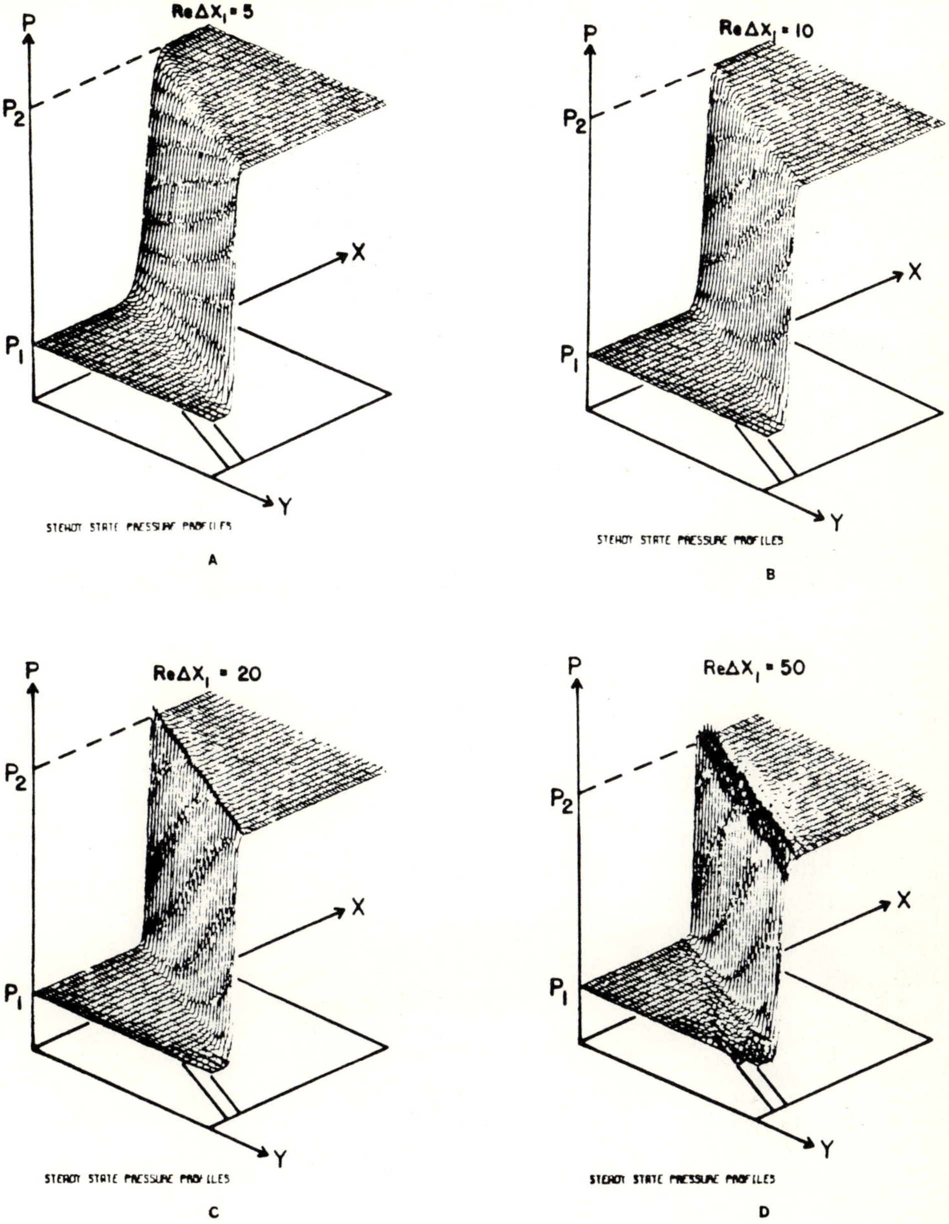

Re∆X1 = 5
P
P2
P1
X
Y
STEADY STATE PRESSURE PROFILES
A
Re∆X1 = 10
P
P2
P1
X
Y
STEADY STATE PRESSURE PROFILES
B
Re∆X1 = 20
P
P2
P1
X
Y
STEADY STATE PRESSURE PROFILES
C
Re∆X1 = 50
P
P2
P1
X
Y
STEADY STATE PRESSURE PROFILES
D

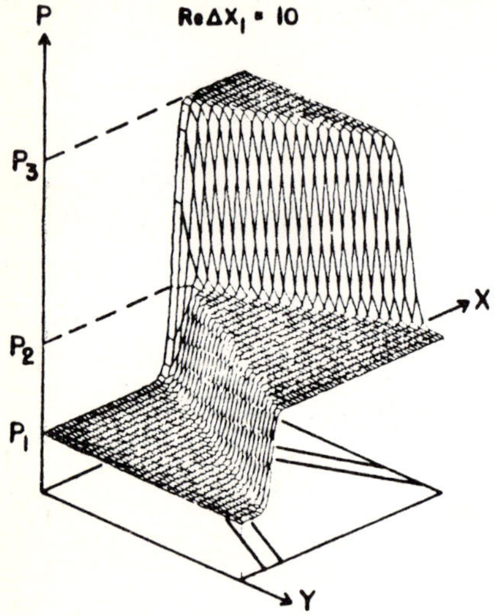

STEADY STATE PRESSURE PROFILES

A

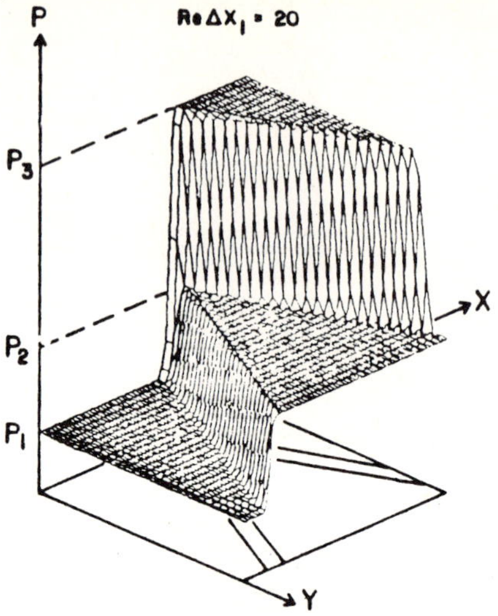

STEADY STATE PRESSURE PROFILES

B

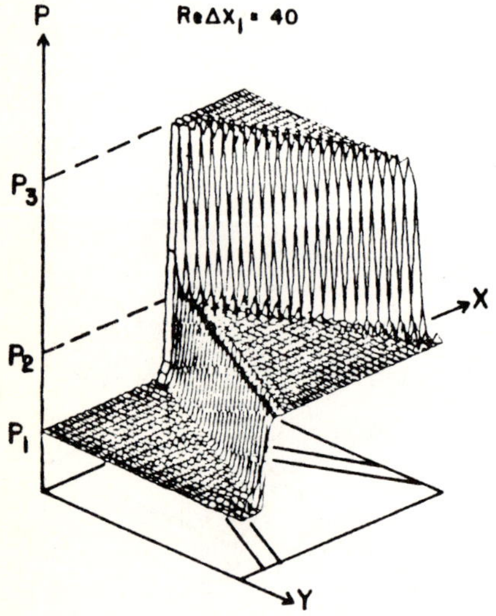

STEADY STATE PRESSURE PROFILES

C

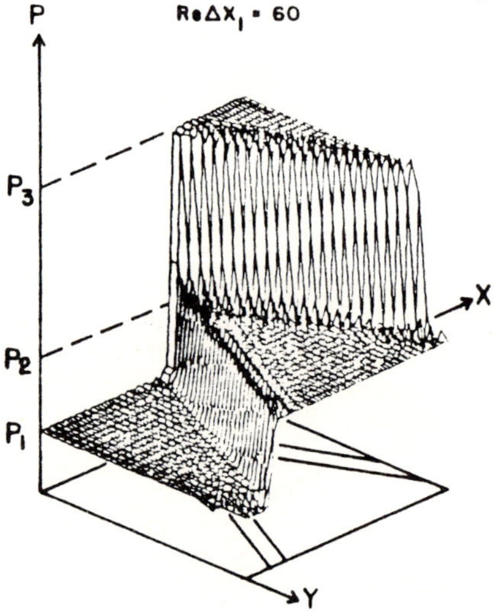

STEADY STATE PRESSURE PROFILES

D

$(Re\Delta X)_1 = 1505.$

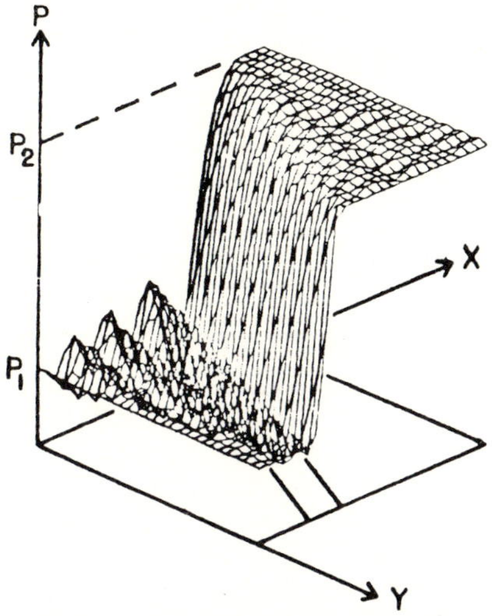

STEADY STATE PRESSURE PROFILES

A

$(Re\Delta X)_1 = 752.5$

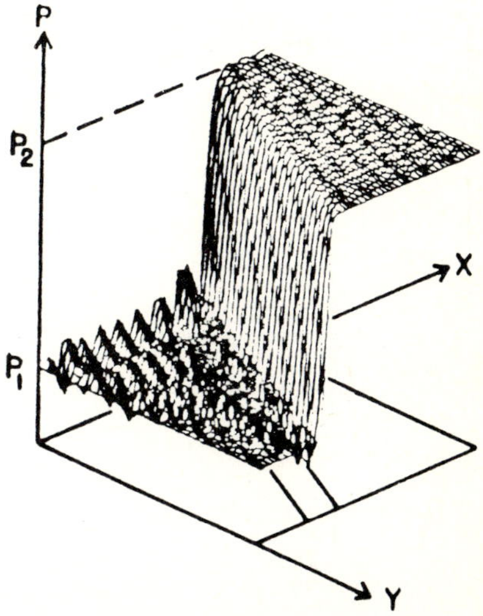

STEADY STATE PRESSURE PROFILES

B

$(Re\Delta X)_1 = 376.25$

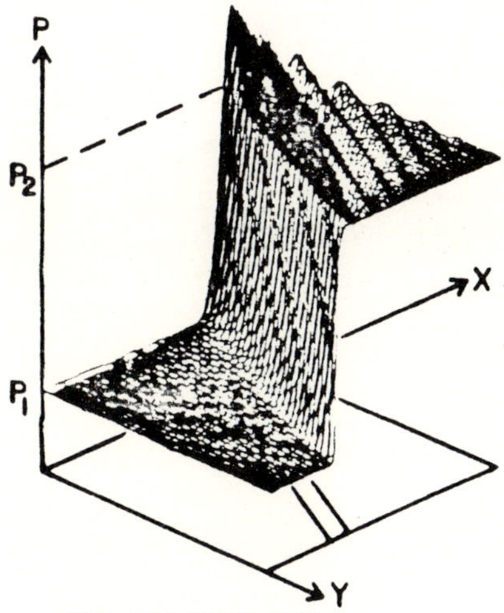

STEADY STATE PRESSURE PROFILES

C

$(Re\Delta X)_1 = 188.125$

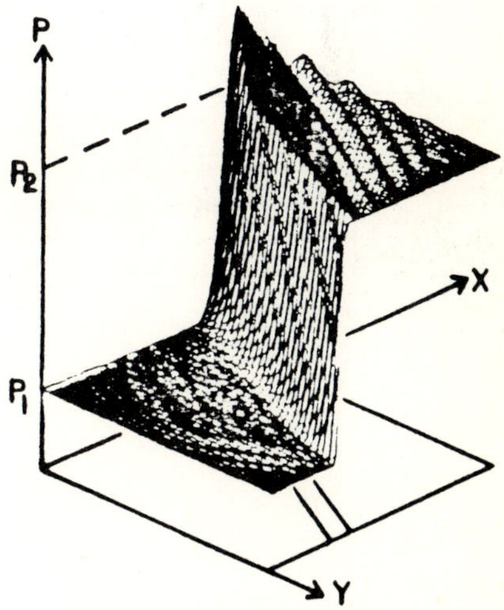

STEADY STATE PRESSURE PROFILES

D

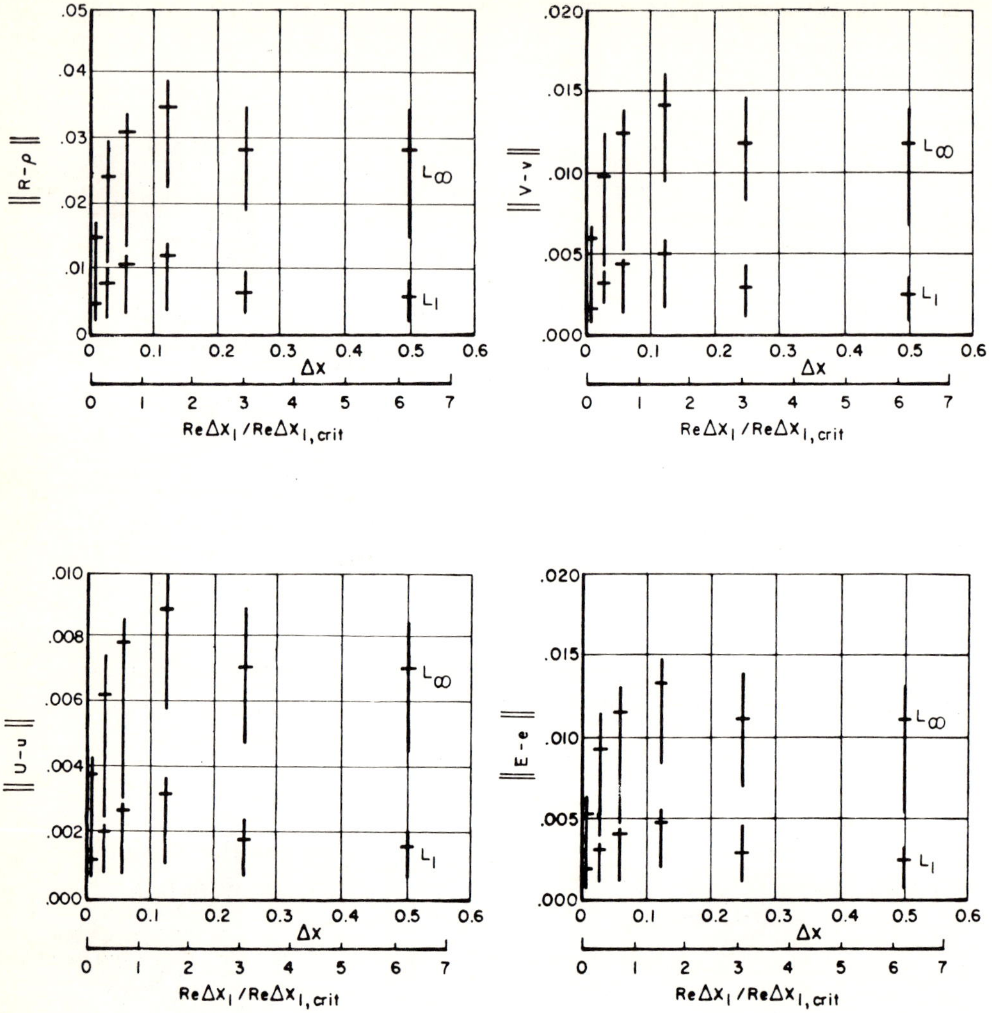
	R-ρ	
	V-v	
	U-u	
	E-e	
L∞
L1
Δx
ReΔx1 / ReΔx1,crit

A FREE BOUNDARY PROBLEM IN HYDRODYNAMIC LUBRICATION INCLUDING SURFACE TENSION.

C. Cuvelier

Delft University of Technology

Dept. of Mathematics

Delft, Netherlands.

Summary

The aim of this paper is to study hydrodynamic partial lubrication in a journal bearing where the surface tension on the cavity-fluid interface is taken into account. For solving this problem we present two methods. When the surface tension σ is equal to zero we formulate the problem as a variational inequality of elliptic type which can be solved numerically by standard methods. For $\sigma \geq 0$ we formulate the problem as a problem of optimum design. For some values of σ we observe the so called fingering phenomenon which is typical for partial lubrication in a journal bearing and of great engineering interest.

* * *

I THE REYNOLDS MODEL FOR PARTIAL LUBRICATION.

The generally accepted differential equation for hydrodynamic lubrication is the so-called Reynolds equation for the pressure p, formulated by Reynolds. His theory is limited to full oil films of positive pressure and does not hold true when film rupture occurs due to cavitation.

It can be proved that under some conditions, for which we refer to CAMERON [1] and FATINO, FRENE, GODET [4], the general equations describing the motion of the lubricant (i.e. conservation of mass, energy and momentum) reduce to the Reynolds equation.

The Reynolds equation is an equation of elliptic type and for the unique solvability we have to give boundary conditions of Dirichlet or Neumann type for the pressure. This mathematically well posed boundary value problem may predict negative values of the pressure in some parts of the oil film. Physically this is impossible and we observe the formation of cavitation bubbles filled with the vapour phase of the oil. This phenomenon is called partial lubrication and the main problem is that the position of the cavities is not known a priori (free boundary problem). In a simplified model (cf. CAMERON [1]), we can say that on the interface of the liquid and the cavitation zones the following transition conditions hold

$$(1.1) \qquad p - p_c = -\frac{\sigma}{R} \qquad\qquad \frac{\partial p}{\partial n} = 0$$

where p denotes the pressure in the lubricant, p_c the pressure in the cavity, σ the surface tension of the liquid, R the radius of curvature of the cavity and n is the unit vector normal to the boundary of the region of cavitation.

In the case of a journal bearing the unknown function is the pressure distribution p in a thin film of lubricant, with viscosity μ, contained in the narrow clearance between two cylinders C_1 (radius r_1) and C_2 (radius r_2) of equal length L, open at both sides to atmospheric pressure (which we put equal to zero for simplicity). The axes of C_1 and C_2 are parallel though distinct and the motion of C_2 is a rotation around its axis with constant angular velocity ω.

The Reynolds equation has the following form

$$(1.2) \qquad -\frac{\partial}{\partial x_1}\left(h^3 \frac{\partial p}{\partial x_1}\right) - \frac{\partial}{\partial x_2}\left(h^3 \frac{\partial p}{\partial x_2}\right) = -\frac{6\mu UD}{c^2}\frac{\partial h}{\partial x_1}$$

where $x = \{x_1, x_2\} \in \Omega =]0,1[\times]0,\frac{L}{D}[$, $D = 2\pi r_2$, $h = 1+\varepsilon \cos 2\pi x_1$ = film thickness and $U = \omega r_2$

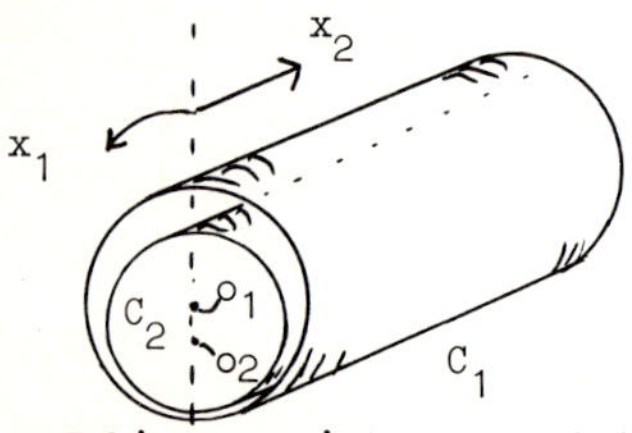

c = radial clearance = $r_1 - r_2$
ε = eccentricity = $\frac{e}{c}$, $0 \leq \varepsilon < 1$
$e = |O_1 - O_2|$ = distance between the central axis.

Taking now into account the possibility of cavitation (with $p_c = 0$ for simplicity), the free boundary problem can be stated as follows:

(1.3) Find a function p (pressure) and a subset Ω_c (cavity region) of Ω such that

$$-\frac{\partial}{\partial x_1}\left(h^3 \frac{\partial p}{\partial x_1}\right) - \frac{\partial}{\partial x_2}\left(h^3 \frac{\partial p}{\partial x_2}\right) = -\frac{6\mu UD}{c^2}\frac{\partial h}{\partial x_1} \quad \text{in } \Omega \setminus \Omega_c$$

$$p = 0 \text{ in } \Omega_c \text{ , } p > 0 \text{ in } \Omega \setminus \Omega_c$$

$$p = 0 \text{ on } \partial\Omega$$

$$p = -\frac{\sigma}{R} \text{ , } \frac{\partial p}{\partial n} = 0 \text{ on } \partial\Omega_c$$

Notice that the condition "$p = 0$ on $\partial\Omega$" implies that there is an oil groove at $x_1 = 0$. We also can take boundary conditions of periodic type.

II $\sigma = 0$. VARIATIONAL INEQUALITY

Setting $f = -\frac{6\mu UD}{c^2}\frac{\partial h}{\partial x_1}$, we make the following assumption:

$$\Omega_c \subset \{x \in \Omega | \ f(x) \leq 0\}$$

which amounts to saying that the cavities only occur in the divergent part of the journal bearing. With this assumption we can formulate problem (1.3) as a variational inequality (cf. CIMATTI [2], DUVAUT, LIONS [3]):

$$(2.1)\quad \left| \begin{array}{l} \text{Find } p \in K = \{q \in H_0^1(\Omega) \mid q \geq 0\} \text{ such that} \\ a(p,q-p) \geq (f,q-p)_{L_2} \qquad q \in K. \end{array} \right.$$

where $a(p,q) = \sum_{i=1}^{2} \int_\Omega h^3 \frac{\partial p}{\partial x_i} \frac{\partial q}{\partial x_i}\, dx$, $(f,q)_{L_2} = \int_\Omega f\, q\, dx$

Problem (2.1) is equivalent to the following minimization problem:

$$\left| \begin{array}{l} \text{Find } p \in K \text{ such that} \\ J(p) = \inf_{q \in K} J(q), \text{ where } J(q) = a(q,q) - 2(f,q)_{L_2} \end{array} \right.$$

By standard methods we can prove existence and uniqueness of a solution $p \in K$ and even that $p \in C^1(\bar{\Omega})$.

For a numerical resolution of the problem we discretize the functional J by means of a finite difference method and solve the discrete problem with a point-overrelaxation method with projection (cf. GLOWINSKI, LIONS, TREMOLIERES [6]). The following figures give an idea of the behaviour of the solution:

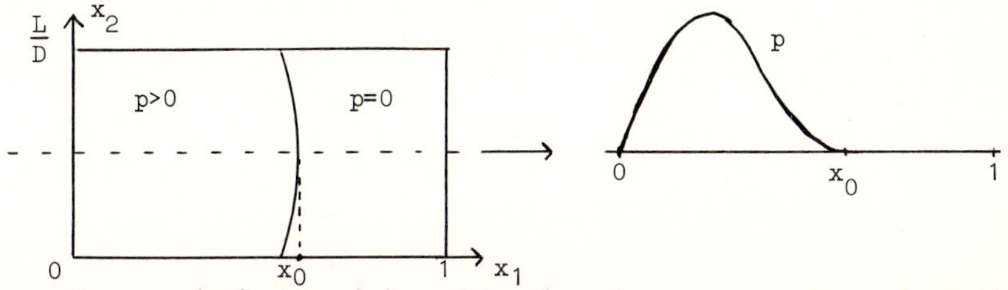

Some characteristic quantities of physical interest are the load W, given by

$$W = (W_H^2 + W_V^2)^{\frac{1}{2}}\ ,\ W_H = -D^2 \int_\Omega p \sin 2\pi x_1\, dx,\ W_V = D^2 \int_\Omega p \cos 2\pi x_1\, dx$$

the attitude angle ϕ, given by

$$\mathrm{tg}\, \phi = \frac{W_H}{W_V}$$

and the Sommerfeld number S, given by

$$S = \frac{4\pi W c^2}{\mu U D^2 L}$$

For a typical situation ($\frac{L}{D} = 1$, 42×13 points in Ω, $\frac{\mu U D}{c^2} = 100$) we can compare the values of ϕ, ψ (= $360^{\circ} * x_0$) and S (column VI in table 1) with the corresponding values of CAMERON [1] (column CAM.).

III OPTIMUM DESIGN

We now take into account the surface tension of the lubricant: $\sigma \geq 0$. Problem (1.3) can be formulated as an optimal control problem of a distributed parameter system where the control is a part of the boundary. Such a problem is called a problem of optimum design. The formulation is the following:

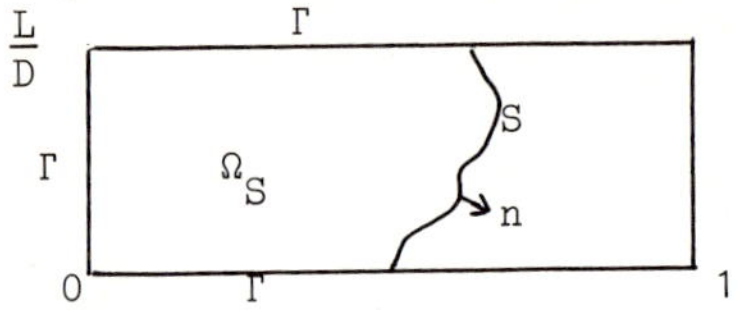

(3.1) Find a boundary $S \in \mathcal{S}$ (= class of admissible boundaries) such that

$$E(S) = \inf_{S' \in \mathcal{S}} E(S')$$

where E is given by

(3.2) $$E(S) = \int_S \left|p + \frac{\sigma}{R}\right|^2 ds, \quad R = R(S) \text{ radius of curvature of } S$$

and $p = p(S)$ by

(3.3) $$\begin{cases} -\dfrac{\partial}{\partial x_1}\left(h^3 \dfrac{\partial p}{\partial x_1}\right) - \dfrac{\partial}{\partial x_2}\left(h^3 \dfrac{\partial p}{\partial x_2}\right) = f & \text{in } \Omega_S \\ p = 0 \text{ on } \Gamma \qquad \dfrac{\partial p}{\partial n} = 0 \text{ on } S \end{cases}$$

We calculate the variation $\delta E = E(S') - E(S)$ of E where

$$S = \{x(s) \mid s \in [0,\tfrac{L}{D}]\} \text{ and } S' = \{x(s) + \alpha(s)\, n(s) \mid s \in [0,\tfrac{L}{D}]\}$$

Using the techniques of PIRONNEAU [8] and LIONS [7] we obtain

(3.4) $$\delta E = 2\int_S \alpha\left[\left(\frac{\partial}{\partial n}\left(\frac{\sigma}{R}\right) + \frac{p + \frac{\sigma}{R}}{R}\right)\left(p + \frac{\sigma}{R}\right) + f\psi + \sum_{i=1}^{2} h^3 \frac{\partial p}{\partial x_i}\frac{\partial \psi}{\partial x_i}\right] ds + o(\alpha)$$

where ψ is the adjoint state defined by

(3.5) $$\begin{cases} -\dfrac{\partial}{\partial x_1}\left(h^3 \dfrac{\partial \psi}{\partial x_1}\right) - \dfrac{\partial}{\partial x_2}\left(h^3 \dfrac{\partial \psi}{\partial x_2}\right) = 0 & \text{in } \Omega_S \\ \psi = 0 \text{ on } \Gamma \qquad h^3 \dfrac{\partial \psi}{\partial n} = p + \dfrac{\sigma}{R} \text{ on } S \end{cases}$$

The numerical algorithm can schematically be written as follows: Choose $S_0 \in \mathcal{S}$. Once $S_0,\dots,S_i = \{x(s) \mid s \in [0,\frac{L}{D}]\}$ are known, we use, in order to apply a first order finite element method, an automatic triangulation subroutine which generates the internal nodes from the boundary nodes. We calculate $p = p(S_i)$ from (3.3) and $\psi = \psi(S_i)$ from (3.5). Then choosing

$$\alpha = -\left[\left(\frac{\partial}{\partial n}\left(\frac{\sigma}{R}\right) + \frac{p + \frac{\sigma}{R}}{R}\right)\left(p + \frac{\sigma}{R}\right) + f\psi + \sum_{i=1}^{2} h^3 \frac{\partial p}{\partial x_i}\frac{\partial \psi}{\partial x_i}\right]_{S_i}$$

we see that for small ρ

$$E(S_{i+1}(\rho)) \leq E(S_i), \text{ where } S_{i+1}(\rho) = \{x(s) + \rho\alpha(s)\, n(s) \mid s \in [0, \tfrac{L}{D}]\}$$

Next we compute $\bar{\rho}$ such that $E(S_{i+1}(\bar{\rho})) = \inf_{\rho} E(S_{i+1}(\rho))$ and finally we set $S_{i+1} = S_{i+1}(\bar{\rho})$.

The number of elements used is equal to 2 * 20 * 7 = 280 and some results for $\sigma = 0$ are tabulated (column OC).

Experimentally it is found by FLOBERG [5] that for some values of $\sigma > 0$ the cavitation happens in such a way that long, fingerlike cavities form. These fingers are regularly spaced in the direction of the flow. For $\sigma = 1$, $\frac{L}{D} = 1$, $\frac{\mu UD}{c^2} = 100$ and $\varepsilon = 0.9$ the numerical solution for S is qualitatively in agreement with these experiments.

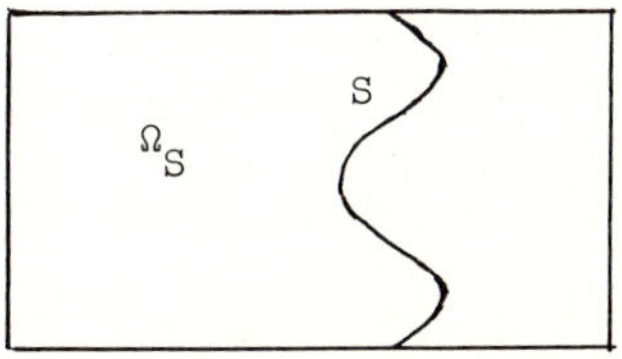

A quantitative study of this stability problem, which is of engineering interest, will be the subject of a forthcoming paper.

For problems of hydrodynamic lubrication with free boundary described by the Stokes equations we refer to CUVELIER [9].

* * *

$\frac{L}{D} = 1$	ϕ			$\psi = 360 * x_0$			S^{-1}		
	VI	OC	CAM	VI	OC	CAM	VI	OC	CAM
=0.9	152.8	152.9	153.5	188.9	193.0	191	0.0177	0.0211	0.0188
0.7	135.7	135.9	136.0	193.7	202.0	199.	0.0722	0.0817	0.0775
0.5	123.1	123.1	123.1	201.1	206.9	205.	0.1683	0.1849	0.180
0.3	111.7	111.8	111.5	202.7	210.8	210.	0.3757	0.4003	0.398
0.1	100.6	100.5	100.5	208.7	212.9	213.5	1.3207	1.3638	1.33

table 1.

* * *

REFERENCES

[1] A. CAMERON. Principles of lubrication. London. Longmans e.a. (1966).

[2] G. CIMATTI. On a problem of the theory of lubrication governed by a variational inequality. Appl. Math. and Opt. 3. p. 227-242 (1977).

[3] G. DUVAUT, J.L. LIONS. Les inéquations en mécanique et en physique. Paris Dunod. 1971.

[4] B. FANTINO, J. FRENE, M. GODET. Conditions d'utilisation de l'équation de Reynolds en mécanique des films minces visqueux. C.R.A.Sc. t 272. p. 691-693. (1971).

[5] L. FLOBERG. Cavitation in lubricating oil films. In 'Cavitation in real liquids'. New York. Elsevier. p. 138-146 (1964).

[6] R. GLOWINSKI, J.L. LIONS, R. TREMOLIERES. Approximation des inéquations de la mécanique et de la physique. Paris. Dunod. 1976.

[7] J.L. LIONS. Contrôle des systèmes gouvernés par des équations aux dérivées partielles. Paris. Dunod. 1968.

[8] O. PIRONNEAU. Sur les problèmes d'optimisation de structure en mécanique des fluides. Thèse à l'Université de Paris VI. 1976.

[9] C. CUVELIER. Free boundary problems in Stokes flows. to appear.

APPROXIMATION VISCOSITY AND STABILITY OF DIFFERENCE SCHEMES

Yu.M. Davydov

Computation Center
USSR Academy of Sciences
Moscow, USSR

In the present paper the problems of stability of nonlinear differences schemes with the consideration of the approximation of boundary conditions by difference relations are investigated with the help of Yanenko-Shokin method of differential approximations [1-2 et.al.]. Matrix connections are obtained between different kinds of approximation viscosity. The "coarse particle" method [3-6 et.al.] is extended to calculate space three-dimensional problems.

The method of differential approximations is quite efficient and is widely applied for the investigation of numerical algorithms by many authors [7-20 et.al.]. The detailed survey of papers dealing with the analysis of the stability of difference schemes by means of differential approximations is given in [21].

Some authors in the investigation of stability of gas dynamics difference schemes use matrix non-negativity (as a rule, diagonal terms)of approximation viscosity. However the matrices of approximation viscosity in various papers are treated differently depending upon the appearance of the initial system of differential equations

$$\rho_t+(\rho u)_x=0,\ (\rho u)_t+(p+\rho u^2)_x=0,\ (\rho E)_t+[(p+\rho E)u]_x=0, \qquad (1)$$

$$\rho_t+(\rho u)_x=0,\ \rho u_t+(p+\rho u^2)_x-(\rho u)_x u=0,\ \rho E_t+[(p+\rho E)u]_x-E(\rho u)_x=0, \qquad (2)$$

$$\rho_t+(\rho u)_x=0,\ (\rho u)_t+(p+\rho u^2)_x=0,\ (\rho J)_t+(\rho uJ)_x+pu_x=0, \qquad (3)$$

$$\rho_t+(\rho u)_x=0,\ \rho u_t+(p+\rho u^2)_x-u(\rho u)_x=0,\ \rho J_t+\rho uJ_x+pu_x=0 \qquad (4)$$

and the choice of basic functions: $W=(\rho, u, E)$, $\widetilde{W} = (\rho, \rho u, \rho E)$, $W^* = (\rho, u, J)$, $\widetilde{W}^* = (\rho, \rho u, \rho J)$. Therefore, the results in these papers are incomparable and they do not give necessary information. Let us establish connections between various kinds of approximation viscosity. Let us designate system (1) by $L_1(W)=0$ or $L_2(\widetilde{W}) = 0$; system (2) by $L_3(W) = 0$; system (3) by $L_4(W^*) = 0$; or $L_5(\widetilde{W}^*) = 0$, system (4) by $L_6(W^*) = 0$. The parabolic form of the second differential approximation with respect to Δx and the first one with respect Δt, is written as

for system (1) $\begin{cases} L_1(W) = aW_{xx} + b\,W_{xxx} + \Delta_1 , & (5) \\ L_2(\widetilde{W}) = \mu\,\widetilde{W}_{xx} + \eta\,\widetilde{W}_{xxx} + \Delta_2 , & (6) \end{cases}$

for system (2) $L_3(W) = \nu\,W_{xx} + \theta\,W_{xxx} + \Delta_3 ,$ (7)

for system (3) $\begin{cases} L_4(W^*) = a^* W^*_{xx} + b^* W^*_{xxx} + \Delta_4 , & (8) \\ L_5(\widetilde{W}^*) = \mu^* \widetilde{W}^*_{xx} + \eta^* \widetilde{W}^*_{xxx} + \Delta_5 , & (9) \end{cases}$

for system (4) $L_6(W^*) = \nu^* W^*_{xx} + \theta^* W^*_{xxx} + \Delta_6 ,$ (10)

where Δ_i ($i = 1,\ldots\ldots, 6$) are terms proportional to $\Delta x, \Delta x^2, \Delta t$, not containing W_{xx} or W_{xxx}, $\widetilde{W}_{xx}$ or $\widetilde{W}_{xxx}$ etc., respectively. Matrices b, η, θ, b^*, η^*, θ^* are introduced in order to take account of space derivatives, which is especially important in regions with large gradients. Making the coefficients equal and the derivatives unchanged in (5)-(10), ((5) and(6) etc.) we shall obtain unknown connection formulas. For compactness of writing we shall formally define 6-component vector-columns

$$a_b = \begin{Bmatrix} a_1 \\ a_2 \\ a_3 \\ b_1 \\ b_2 \\ b_3 \end{Bmatrix}, \mu_\eta = \begin{Bmatrix} \mu_1 \\ \mu_2 \\ \mu_3 \\ \eta_1 \\ \eta_2 \\ \eta_3 \end{Bmatrix}, b_a = \begin{Bmatrix} a^1 \\ a^2 \\ a^3 \\ b^1 \\ b^2 \\ b^3 \end{Bmatrix}, \theta_\nu = \begin{Bmatrix} \nu^1 \\ \nu^2 \\ \nu^3 \\ \theta^1 \\ \theta^2 \\ \theta^3 \end{Bmatrix}, a^*_b = \begin{Bmatrix} a^*_1 \\ a^*_2 \\ a^*_3 \\ b^*_1 \\ b^*_2 \\ b^*_3 \end{Bmatrix}, \mu^*_\eta = \begin{Bmatrix} \mu^*_1 \\ \mu^*_2 \\ \mu^*_3 \\ \eta^*_1 \\ \eta^*_2 \\ \eta^*_3 \end{Bmatrix}, b^*_a = \begin{Bmatrix} a^{1*} \\ a^{2*} \\ a^{3*} \\ b^{1*} \\ b^{2*} \\ b^{3*} \end{Bmatrix}, \theta^*_\nu = \begin{Bmatrix} \nu^{1*} \\ \nu^{2*} \\ \nu^{3*} \\ \theta^{1*} \\ \theta^{2*} \\ \theta^{3*} \end{Bmatrix},$$

where a numerical superscript denotes lines and a subscript-columns of respective matrices. Then, the formulas of transition between the matrices of approximation viscosity are

$$\begin{cases} a_b = A\,\mu_\eta , \\ \mu_\eta = A^{-1} a_b , \end{cases} \begin{cases} b_a = B\,\theta_\nu , \\ \theta_\nu = B^{-1} b_a , \end{cases} \begin{cases} a^*_b = {}^*A\,\mu^*_\eta , \\ \mu^*_\eta = {}^*A^{-1} a^*_b , \end{cases} \begin{cases} b^*_a = {}^*B\,\theta^*_\nu , \\ \theta^*_\nu = {}^*B^{-1} b^*_a , \end{cases}$$

where

$$A = \left\| \begin{matrix} 1 & u & E & 0 & 3u_x & 3E_x \\ 0 & \rho & 0 & 0 & 3\rho_x & 0 \\ 0 & 0 & \rho & 0 & 0 & 3\rho_x \\ 0 & 0 & 0 & 1 & u & E \\ 0 & 0 & 0 & 0 & \rho & 0 \\ 0 & 0 & 0 & 0 & 0 & \rho \end{matrix} \right\|, \quad B = \left\| \begin{matrix} 1 & 0 & 0 & 0 & 0 & 0 \\ u & 1 & 0 & 0 & 0 & 0 \\ E & 0 & 1 & 0 & 0 & 0 \\ 0 & 0 & 0 & 1 & 0 & 0 \\ 0 & 0 & 0 & u & 1 & 0 \\ 0 & 0 & 0 & E & 0 & 1 \end{matrix} \right\|,$$

$$^{*}A = \left\| \begin{matrix} 1 & U & J & 0 & 3u_x & 3J_x \\ 0 & \rho & 0 & 0 & 3\rho_x & 0 \\ 0 & 0 & \rho & 0 & 0 & 3\rho_x \\ 0 & 0 & 0 & 1 & u & J \\ 0 & 0 & 0 & 0 & \rho & 0 \\ 0 & 0 & 0 & 0 & 0 & \rho \end{matrix} \right\|, \quad ^{*}B = \left\| \begin{matrix} 1 & 0 & 0 & 0 & 0 & 0 \\ u & 1 & 0 & 0 & 0 & 0 \\ J & 0 & 1 & 0 & 0 & 0 \\ 0 & 0 & 0 & 1 & 0 & 0 \\ 0 & 0 & 0 & u & 1 & 0 \\ 0 & 0 & 0 & J & 0 & 1 \end{matrix} \right\|.$$

For brevity , there are not given here connection formulas between the matrix elements of approximation viscosity "with asterisks" and "without asterisks". In a similar way we can obtain other connection formulas.

We shall discuss, in brief, the properties of difference schemes of the "coarse particle" method. Various physico-mechanical problems are solved by the "coarse particle" method [3-6, 22 et.al.]: sub-, trans- and supersonic flows around bodies of complex form, external and internal flows using both Euler and Navie-Stokes equations, heterogeneous flows, high speed elastic-plastic collisions of metals, jet flows, break-away flows with the formation of reverse-circular zones; there are studied problems of hydrodynamic stability, the interaction of laser radiation with matter, the extension of a plasma corona, flows in tubes, cosmophysical problems; there are calculated flows of a dissociative and ionized gas, flows with radiation, diffraction and explosion problems and so on. A short survey of two-dimensional (plane and axisymmetrical) problems solved by the "coarse particle" method is presented in [22].

At present, an apparatus of three-dimensional fractional cells is developed and a rather wide class of space three-dimensional problems is studied. In particular, there is solved the problem of a flow around a three-dimensional parallelepiped at arbitrary angles of attack α and inclination β. As an example, we consider here the case $\alpha = \beta = 0$, when it is sufficient to carry out calculations in the 1/4 part of a region calculated. Fig. 1 shows the front part of the region calculated with an element of a calculation mesh. A curvilinear three-dimensional surface shaded with quasivertical lines corresponds to the position and shape of a shock wave in a flow around a three-dimensional parallelepiped represented by a solid line. There is given, for comparison, a surface shaded with horisontal lines- a shock wave for a two-dimensional step having the same height and length as a three-dimensional body. The Mach number of an incoming flow in both cases is M = 2.84. Figs. 2-4 show in sections perpendicular to the incoming flow projections of velocity vectors on these sections; figs. 2, 3 correspond to a region of a shock layer in front of the body, fig. 4 - to a break-away zone behind the body. One can clearly see qualitatively three-dimensional effects. In front of the body (figs. 2, 3) we observe the "extension" of a flow and distinctly trace the position of a shock wave as the boundary of an undisturbed flow.

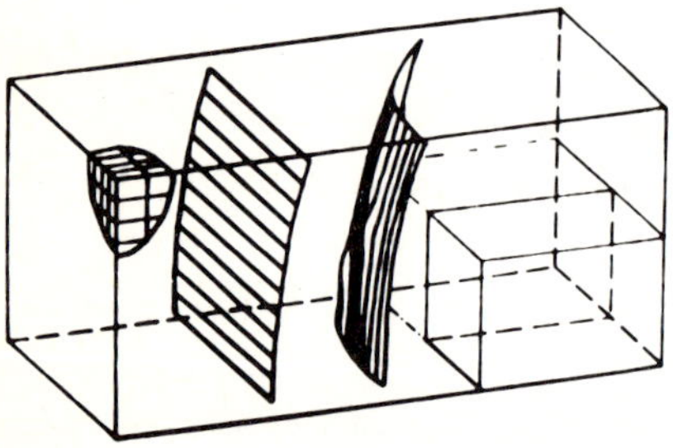

Fig. 1

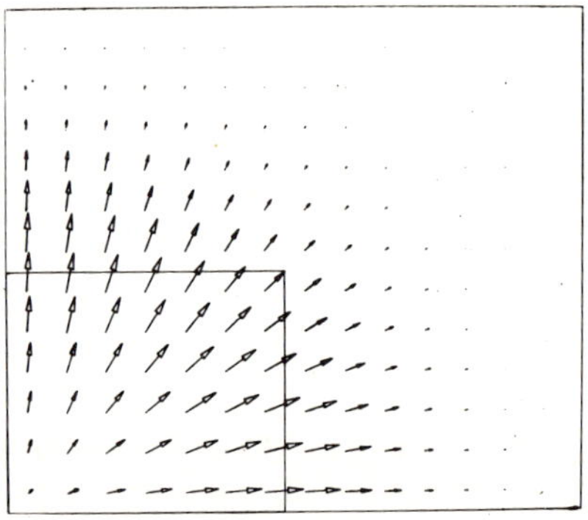

Fig. 2

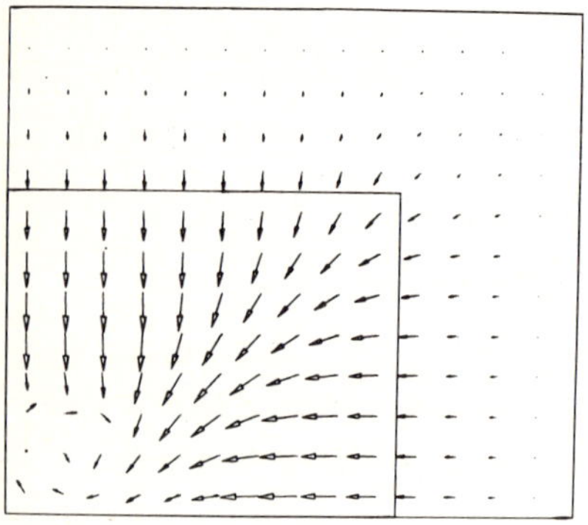

Fig. 4

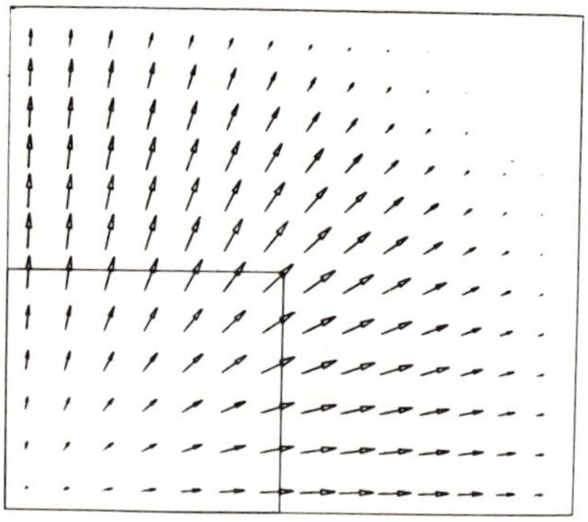

Fig. 3

Far from the body an undisturbed region is not great (fig. 2), while approaching a front surface the shock wave departs farther (fig. 3). The flow becomes practically parallel above a rather long lateral surface. A break-away is observed in the wake of the body. Fig.4 shows a transerve "cut" of a break-away region. Velocity vectors in an external flow "converge" at a break-away zone, a separation contact surface is crearly seen. Note the presence of an interesting three-dimensional effect: vorticity in the region of a reverse-circular flow in a transverse direction. Thus, in spite of symmetric homogeneous initial data a non-symmetric solution is observed: a spiral departure of a vortex film from the rear part of the body.

The difference schemes of the "coarse particle" method were investigated in detail by means of the method of differential approximations. Multiparametric (some tens of parameters) difference schemes generalizing the "coarse particle" method were considered. The effect of splitting on stability was investigated [9]. The concept of differential approximations of boundary conditions was introduced. The matrices of approximation viscosity were regarded both for full and fractional cells taking account of injection and other physical factors on the surface of a body in a flow as well.

As an example, we present one of the conditions of stability for full cells

$$a_{11} = 0.5u\Delta x - 0.25\, u_x \Delta x^2 + 0.5\,\Delta t\,(p_\rho - u^2) \geqslant 0 .$$

Consider differential approximations for nonuniform meshes. At first, let us obtain the difference scheme of the "coarse particle" method with ΔM of the first order accuracy on a non-uniform mesh.

We shall introduce a nonuniform with respect to space (but fixed in time) mesh $\{x_i\}$, where the difference $x_{i+1} - x_i$ is taken as a mesh parameter $h_{i+1/2}$. As to a mesh in time $\{t^n\}$ we shall suggest that $t^{n+1} = t^n + \Delta t^n$ where Δt^n can be variable (it can depend upon the number n of a time layer). The system of equations of a Euler stage is approximated as follows:

$$\tilde{u}_i^n = u_i^n \frac{p_{i+1}^n - p_{i-1}^n}{h_{i+1/2} + h_{i-1/2}} \cdot \frac{\Delta t}{\rho_i^n}, \quad \tilde{E}_i^n = E_i^n - \frac{(p_{i+1}^n + p_i^n)(u_{i+1}^n + u_i^n) - (p_i^n + p_{i-1}^n)(u_i^n + u_{i-1}^n)}{2(h_{i+1/2} + h_{i-1/2})} \cdot \frac{\Delta t}{\rho_i^n} .$$

The difference equations on the Lagrangeon and final stages are given in the form

$$\rho_i^{n+1} \varphi_i^{n+1} - \rho_i^n \tilde{\varphi}_i^n + \Delta t \langle\!\langle \rho u \varphi \rangle\!\rangle_i^n = 0 ,$$

$$\langle\langle \rho u \varphi \rangle\rangle_i^n = \frac{\left[(\rho\tilde{\varphi})_{i+1}^n + (\rho\tilde{\varphi})_i^n\right](\tilde{u}_{i+1}^n + \tilde{u}_i^n) - \left[(\rho\tilde{\varphi})_i^n + (\rho\tilde{\varphi})_{i-1}^n\right](\tilde{u}_i^n + \tilde{u}_{i-1}^n)}{2(h_{i+1/2} + h_{i-1/2})} +$$

$$+ \frac{1}{2(h_{i+1/2} + h_{i-1/2})} \cdot \left\{ \frac{f_{i-1/2}}{h_{i-1/2}} \left[(\rho\tilde{\varphi})_i^n - (\rho\tilde{\varphi})_{1-1}^n\right] \cdot \left| \tilde{u}_{i-1}^n + \tilde{u}_i^n \right| - \right.$$

$$\left. - \frac{f_{i+1/2}}{h_{i+1/2}} \left[(\rho\tilde{\varphi})_{i+1}^n - (\rho\tilde{\varphi})_i^n\right] \cdot \left| \tilde{u}_i^n + \tilde{u}_{i+1}^n \right| \right\}, \quad \varphi = (1, u, E).$$

$\langle\langle \rho u \varphi \rangle\rangle_i^n$ designates the difference approximation of a differential operator $(\rho u \varphi)_x$; $f_{i+1/2} = f(h_{i+1/2})$, where $f(x)$ is a continuously differentiated function defined at $\min_i h_{i+1/2} \leqslant x \leqslant \max_i h_{i+1/2}$. For the sake of approximation it is necessary that $f = O(\max_i h_{i+1/2})$. With $h_{i+1/2} = f_{i+1/2} = h = \text{const}$ we obtain a conventional difference scheme of the "coarse particle" method with ΔM^n of the first order accuracy on a uniform mesh.

The first differential approximation with respect to Δx and a zero one with respect to Δt of the difference scheme of the "coarse particle" method on a nonuniform mesh is written as

$$\begin{cases} \rho_t + (\rho u)_x = \frac{1}{2}(f|u|\rho_x)_x , \\ (\rho u)_t + (p + \rho u^2)_x = \frac{1}{2}(f|u|(\rho u)_x)_x , \\ (\rho E)_t + \left[(p + \rho E)u\right]_x = \frac{1}{2}(f|u|(\rho E)_x)_x . \end{cases}$$

The terms approximation viscosity proportional to Δx are equal to $(f\ u\ (\rho\varphi)_x)_x/2$ where $\varphi = (1, u, E)$ for equations of continuity, impulse, energy. If we take $f(x) = \alpha x$ ($\alpha = \text{const} > 0$) , then "diffusion coefficients" will be proportional to a local dimension of a cell. In case $f(x) = s = O(\max_i h_{i+1/2})$,"diffusion coefficients" will depend only upon a local velocity, i.e., approximation viscosity will be the same (in case of the equality of velocity modules) both for big and small cells. In a general case we can take $f = f(\rho, u, E, x)$ and function f can be chosen so as to obtain the best solution.

Let us proceed to the investigation of stability on the boundaries of a calculation region. The type of boundary conditions and their difference approximation strongly affect the stability of the algorithms used. In the "coarse particle" method [3-6] boundary conditions are given with the help of layers of fictitious cells

[3] and a technique of fractional cells [4] is used on boundaries of complex configuration. A theoretical analysis confirmed by numerous calculations reveals that approximation does not get worse if simpler formulas for full cells [3] are used on the boundaries, thus, keeping to the formulation of boundary conditions for fractional cells suggested in [4].

We shall consider a system of gas dynamics equations (1) written in a divergent form. Let difference boundary conditions be given so that at any point of the boundary $u = O(\Delta x^2)$. As an example, let us consider the conditions of non-penetration on a body.

The difference boundary conditions for (u,E) are constructed with the consideration of splitting relative to physical processes. We shall regard a one-dimensional case. A linear relative dimension of a fractional cell along axis X is denoted by ℓ, then the length of a fractional cell equals $\ell \cdot \Delta x$. The boundary of a body is situated to the right from a fractional cell. Subscript i refers to a fractional cell and $i+I$ – to a boundary cell.

The boundary conditions for gas dynamic variables ρ, u, E are determined on a full step as follows:

$$\rho^n_{i+I} = \sum_{k=0}^{m_\rho} \theta^\rho_k \rho^n_{i-k} \qquad \left(\sum_{k=0}^{m_\rho} \theta^\rho_k = I \right),$$

$$u^n_{i+I} = -\sum_{k=0}^{m_u} \theta^u_k u^n_{i-k} \qquad \left(\sum_{k=0}^{m_u} \theta^u_k = I \right), \quad (12)$$

$$E^n_{i+I} = \sum_{k=0}^{m_E} \theta^E_k E^n_{i-k} \qquad \left(\sum_{k=0}^{m_E} \theta^E_k = I \right).$$

The boundary conditions on a fractional ($n + 1/2$) step are given in a similar way:

$$\tilde{u}^n_{i+I} = -\sum_{k=0}^{m_{\tilde{u}}} \theta^{\tilde{u}}_k \tilde{u}^n_{i-k} \qquad \left(\sum_{k=0}^{m_{\tilde{u}}} \theta^{\tilde{u}}_k = I \right),$$

$$\tilde{E}^n_{i+I} = \sum_{k=0}^{m_{\tilde{E}}} \theta^{\tilde{E}}_k \tilde{E}^n_{i-k} \qquad \left(\sum_{k=0}^{m_{\tilde{E}}} \theta^{\tilde{E}}_k = I \right). \quad (13)$$

Expanding eq. (12) in a Taylor series with respect to a mesh parameter x in the neighbourhood of point x, we obtain differential approximations of difference boundary conditions (12)

$$(\alpha_\rho - I)\rho_x + \frac{I}{2}(\beta_\rho - I)\rho_{xx}\Delta x = O(\Delta x^2),$$

$$2u + (\alpha_u + I)u_x\Delta x + \frac{I}{2}(\beta_u + I)u_{xx}\Delta x^2 = O(\Delta x^3), \quad (14)$$

$$(\alpha_E - I)E_x + \frac{I}{2}(\beta_E - I)E_{xx}\Delta x = O(\Delta x^2).$$

Later, there are also required relations

$$\sum_{k=0}^{m_{\tilde{\psi}}} \theta_k^{\tilde{\psi}} \psi_{i-k}^n = \psi + \alpha_{\tilde{\psi}} \psi_x \Delta x + \frac{I}{2} \beta_{\tilde{\psi}} \psi_{xx} \Delta x^2 + O(\Delta x^3) \ , \ \psi = (u, E). \quad (15)$$

It is not difficult to see that in eqs.(I4)-(I5) there are used designations

$$\alpha_{\xi} = -\sum_{k=0}^{m_{\xi}} k\theta_k^{\xi} \ , \quad \beta_{\xi} = \sum_{k=0}^{m_{\xi}} k^2 \theta_k^{\xi} \ , \quad \xi = (\rho, \ E, \ u, \ \tilde{E}, \ \tilde{u}) \ .$$

From a physical boundary condition

$$u_{\text{on the body}} = u(x + (\ell - I/2)\Delta x) = 0 \quad (16)$$

it follows

$$2u + (2\ell - I) u_x \Delta x = O(\Delta x^2). \quad (17)$$

Comparing coefficients in eqs. (14.2) and (17) we have

$$2 + \alpha_u = 2\ell \ . \quad (18)$$

In this case a difference boundary condition (12.2) approximates a physical boundary condition (16) with accuracy $O(\Delta x^2)$.

Let us introduce differential approximations for considering fractional cells given above. Relative to boundary conditions (12) we shall suppose that $\alpha_\rho \neq I$, $\alpha_E \neq I$. This means that $\rho_x = O(\Delta x)$, $E_x = O(\Delta x)$, but not $\rho_x = O(\Delta x^n)$, $E_x = O(\Delta x^n)$, where $n \geq 2$.

As an example, we shall derive the differential approximation of a continuity equation for the case when a flow runs into a cell.

The continuity equation has the form

$$(\rho_i^{n+I} - \rho_i^n)\Delta x = \Delta M_{i-I/2}^n - \Delta M_{i+I/2}^n \ ,$$

where

$$\Delta M_{i+I/2}^n = \frac{I}{2} \rho_i^n (\tilde{u}_i^n + \tilde{u}_{i+I}^n)\Delta t \quad \text{and} \quad \tilde{u}_i^n = u_i^n - \frac{\Delta t}{2\Delta x} \frac{p_{i+I}^n - p_{i-I}^n}{\rho_i^n} \ .$$

Then

$$(\rho_i^{n+I} - \rho_i^n)\frac{\Delta x}{\Delta t} = \frac{I}{2}\rho_{i-I}^n \left[u_i^n + u_{i-I}^n - \frac{\Delta t}{2\Delta x \rho_{i-I}^n}(p_i^n - p_{i-2}^n) - \right.$$

$$\left. - \frac{\Delta t}{2\Delta x \rho_i^n}(p_{i+I}^n - p_{i-I}^n) \right] - \frac{I}{2}\rho_i^n \left[u_i^n - \frac{\Delta t}{2\Delta x \rho_i^n}(p_{i+I}^n - p_{i-I}^n) - \right. \quad (19)$$

$$\left. - \sum_{k=0}^{m_{\tilde{u}}} \theta_k^{\tilde{u}} \left(u_{i-k}^n - \frac{\Delta t}{2\Delta x \rho_{i-k}^n}(p_{i-k+I}^n - p_{i-k-I}^n)\right) \right] .$$

First, consider convective terms (19)

$$\frac{I}{2}\rho^n_{i-I}(u^n_{i-I}+u^n_i)-\frac{I}{2}\rho^n_i(u^n_i-\sum_{k=0}^{m_{\tilde{u}}}\theta^{\tilde{u}}_k u^n_{i-k})=$$

$$=\Delta x\left\{-(\rho u)_x-\frac{I}{2}(\alpha_u-\alpha_{\tilde{u}})\rho u_x+\rho_{xx}\left[u\frac{\Delta x}{2}-u_x\frac{\Delta x^2}{4}(I+\frac{\beta_\rho-I}{\alpha_\rho-I})+O(\Delta x^3)\right]+O(\Delta x)\right\}. \tag{20}$$

Here Δ_ρ are terms not containing ρ_{xx}. Taking account of eq. (14.2) the coefficient with ρ_{xx} in eq. (20) is written as

$$-(\alpha_u+2+\frac{\beta_\rho-I}{\alpha_\rho-I})u_x\frac{\Delta x^2}{4}+O(\Delta x^3). \tag{21}$$

Consider now terms appearing due to splitting

$$-\frac{I}{2}\rho^n_{i-I}\left[\frac{\Delta t}{2\Delta x\rho^n_{i-I}}(p^n_i-p^n_{i-2})+\frac{\Delta t}{2\Delta x\rho^n_i}(p^n_{i+I}-p^n_{i-I})\right]+\frac{\rho^n_i}{2}\left[\frac{\Delta t}{2\Delta x\rho^n_i}(p^n_{i+I}-p^n_{i-I})-\right.$$

$$\left.-\sum_{k=0}^{m_{\tilde{u}}}\theta^{\tilde{u}}_k\frac{\Delta t}{2\Delta x\rho^n_{i-k}}(p^n_{i-k+I}-p^n_{i-k-I})\right]=\Delta x\left\{\rho_{xx}\left[C^2\frac{\Delta t}{2}(\frac{\beta_\rho-I}{\alpha_\rho-I}+I+\right.\right.$$

$$\left.\left.+\sum_{k=0}^{m_{\tilde{u}}}k\theta^{\tilde{u}}_k)+O(\Delta x)\Delta t\right]+O(\Delta t)\right\}. \tag{22}$$

Consider a time term

$$\rho^{n+I}_i-\rho^n_i=\Delta t\left[\rho_t+\frac{\Delta t}{2}\rho_{tt}+O(\Delta t^2)\right]. \tag{23}$$

From the initial system (1) we have

$$\rho_{tt}=(p+\rho u^2)_{xx}=\rho_{xx}C^2+\Delta_\rho+O(\Delta x^2), \tag{24}$$

where $C=\sqrt{p_\rho}$ is an isothermic sound velocity.

Thus, taking account of eqs. (20)-(24) the differential approximation of a continuity equation will be written as

$$\rho_t+(\rho u)_x+\frac{I}{2}(\alpha_u-\alpha_{\tilde{u}})\rho u_x=\rho_{xx}\left\{-u_x\frac{\Delta x^2}{4}(\alpha_u+2+\frac{\beta_\rho-I}{\alpha_\rho-I})+C^2\frac{\Delta t}{2}\left[\frac{\beta_\rho-I}{\alpha_\rho-I}+\sum_{k=0}^{m_{\tilde{u}}}k\theta^{\tilde{u}}_k\right]+\right.$$

$$\left.+O(\Delta x^3+\Delta x\Delta t+\Delta t^2)\right\}+O(\Delta x+\Delta t).$$

Differential approximations for different conditions and different flow directions are derived in an analogous way.

Considering the positiveness of the coefficients of "diffusion" with ρ_{xx}, u_{xx}, E_{xx} (diagonal terms of the matrix of approximation viscosity) in equations of continuity, impulse and energy, respectively, as a stability criterion, we obtain the following stability conditions:

1) A flow runs into a cell

$$\alpha_u + 2 + \frac{\beta_\rho - I}{\alpha_\rho - I} + \Omega W_\rho > 0 \,,$$

$$(\alpha_u + I)\left[I + \frac{I}{2}(\beta_{\tilde u} - \beta_u)\right] + 2(2-\gamma) + \frac{I}{2}(\alpha_{\tilde u} - \alpha_u)(\beta_u + I) - \gamma\Omega > 0 \,,$$

$$\alpha_u + 2 + \frac{\beta_E - I}{\alpha_E - I} + (\gamma - I) + \Omega W_E > 0 \,.$$

2) A flow runs out of a cell.

$$\alpha_{\tilde u} + (I + \alpha_u - \alpha_{\tilde u})\frac{\beta_\rho - I}{\alpha_\rho - I} + \Omega W_\rho > 0 \,,$$

$$\frac{3}{2}(\alpha_u - \alpha_{\tilde u})(\beta_u + I) - 2(I + \beta_u - \beta_{\tilde u})(I + \alpha_u - \alpha_{\tilde u}) + (\alpha_u + I)\left[I + \frac{3}{2}(\beta_u - \beta_{\tilde u})\right] + 2(\gamma - I) - \gamma\Omega > 0 \,,$$

$$(I + \alpha_u - \alpha_{\tilde u})(I + \alpha_{\tilde E} - \alpha_E)\frac{\beta_E - I}{\alpha_E - I} + \alpha_{\tilde u}(I + \beta_{\tilde E} - \beta_E) - (\alpha_u + I)(\alpha_{\tilde E} - \alpha_E)\frac{\beta_E - I}{\alpha_E - I} - (\gamma - I) + \Omega W_E > 0 \,.$$

3) A flow runs into a cell across both boundaries.

$$2 + \alpha_u - \alpha_{\tilde u} + (I + \alpha_{\tilde u})\frac{\beta_\rho - I}{\alpha_\rho - I} + \Omega W_\rho > 0 \,,$$

$$2(I + \beta_u - \beta_{\tilde u})(I + \alpha_u - \alpha_{\tilde u}) - \frac{3}{2}(\alpha_u + I)(\beta_u - \beta_{\tilde u}) - (\beta_u + I)\left[I + \frac{3}{2}(\alpha_u - \alpha_{\tilde u})\right] + 2(2 - \gamma) - \gamma\Omega > 0 \,,$$

$$2 + \alpha_u - \alpha_{\tilde u}(I + \beta_{\tilde E} - \beta_E) + (\gamma - I) + \left[I + \alpha_{\tilde u}(I + \alpha_{\tilde E} - \alpha_E)\right]\frac{\beta_E - I}{\alpha_E - I} + \Omega W_E > 0 \,.$$

4) A flow runs out of a cell across both boundaries.

$$(\alpha_u + I)\frac{\beta_\rho - I}{\alpha_\rho - I} + \Omega W_\rho > 0 \,,$$

$$-(\beta_u + I)\left[I + \frac{I}{2}(\alpha_{\tilde u} - \alpha_u)\right] + 2(\gamma - I) + \frac{I}{2}(\alpha_u + I)(\beta_u - \beta_{\tilde u}) - \gamma\Omega > 0 \,,$$

$$(\alpha_u + I)\frac{\beta_E - I}{\alpha_E - I} - (\gamma - I) + \Omega W_E > 0 \,.$$

Here

$$\frac{I}{\Omega} = |u_x| \frac{\Delta x^2}{2C^2 \Delta t}, \qquad W_\rho = \frac{\beta_\rho - I}{\alpha_\rho - I} + \sum_{k=0}^{m_{\tilde{u}}} k\theta_k^{\tilde{u}},$$

$$W_E = -(\gamma - I) + \frac{\beta_\rho - I}{\alpha_\rho - I} + \sum_{k=0}^{m_{\tilde{u}}} k\theta_k^{\tilde{u}}.$$

With the help of the method of differential approximations "boundary" conditions of stability for fractional cells of the "coarse particle" method in the presence of injection and sink of the flow as well as for other physical conditions on a body are obtained in [9]. The conditions of stability on open boundaries of a calculation region with the consideration of an undisturbed flow and a free outflow are also investigated by means of the above approach.

The method of differential approximations is employed successfully for near-boundary points (situated from the boundary at a small number of knots) as well, where the stability criteria obtained are different from respective conditions both for internal and boundary points of the mesh.

It should be noted, that the apparatus of differential approximations and the analysis of stability according to a "dissipative coefficient" can be insufficiently accurate in regions with large gradients. In this case, it is reasonable to make use of the differential-difference representation of "dissipative coefficients", which allows us to exactly determine these coefficients not only in a linear but in a nonlinear case.

REFERENCES

1. N.N. Yanenko, Yu.I. Shokin. Doklady Akad.Nauk SSSR, 1968, 182, N 4, 776-778. (Russian)

2. N.N. Yanenko, Yu.I. Shokin. Phys. Fluids, 1969, 12, N 12, part II, II-28÷II-33.

3. O.M. Belotserkovskii, Yu.M. Davydov. Zh.Vych. Mat. i Mat. Fiz., 1971, 11, N 1, 182-207. (Russian)

4. Yu.M. Davydov. Zh. Vychisl. Mat. i Mat. Fiz., 1971, 11, N 4, 1056-1063. (Russian)

5. O.M. Belotserkovskii, Yu.M. Davydov. Lect. Notes in Phys., 1973, 19, N 2, 25-32.

6. O.M. Belotserkovskii, Yu.M. Davydov. Acta Astronautica, 1974, 1, N 11-12, 1467-1474.

7. O.M. Belotserkovskii, Yu.M. Davydov. Sb."Probl. Prikl. Mat. i Mekh." M., Nauka, 1971, 145-155. (Russian)

8. Yu.M. Davydov. Sb. "Trudy MFTI, Ser. Aerofiz., Prikl. Mat.", M., MFTI, 1971, 79-92. (Russian)

9. Yu.M. Davydov, V.P. Skotnikov. Issledovanie Drobnykh Yacheek v Metode "Krupnykh Chastits". M., V Ts AN SSSR, 1978. (Russian)

10. F.H. Harlow. Los Alamos Scient. Lab., Rep. N LAMS-2452, 1960.

11. R.F. Warming, B.J. Hyett. J. Comp. Phys., 1974, 14, N 2, 159-179.

12. A. Lerat, R. Peyret. Lect. Notes in Phys., 1975, N 35, 251-256.

13. A. Lerat, R. Peyret. Computers and Fluids, 1974, 2, N 1, 35-52.

14. J. Roseman, G. Zwas. J. Comp. Phys., 1975, 19, N 2, 229-235.

15. C.W. Hirt. J. Comp. Phys., 1968, 2, N 4, 339-355.

16. B.J. Daly. Math. Comp., 1963, 17, N 84, 346-360.

17. W.F. Ballhaus, H. Lomax. Lect. Notes in Phys., 1975, N. 35, 57-63.

18. W.C. Rivard, T.D. Butler, O.A. Farmer, P.J. O'Rourke. Los Alamos Scient. Lab., Rep. N LA-5426-MS, 1973.

19. A.A. Amsden. Los Alamos Scient. Lab., Rep. N LA-3466, 1966.

20. P.J. Roache. J. Comp. Phys., 1972, 10, N 2, 169-184.

21. Yu.M. Davydov, V.P. Skotnikov. Differentsialnye Priblizhenya Raznostnych Shem. M., V Ts AN SSSR, 1978. (Russian)

22. Yu.M. Davydov. Sb. "Pryamoye Chislennoe Modelirovanie Techenii Gaza", M., V Ts AN SSSR, 1978. (Russian)

A STUDY ON CURVILINEAR COORDINATES AND MACRO - ELEMENTS FOR MULTIPLY CONNECTED FLOW FIELDS

A. Di Carlo, R. Piva, G. Guj
University of Rome
Rome, Italy

ABSTRACT

A computational model for the solution of 2-D Navier-Stokes equations for incompressible fluids, particularly suitable for multiply connected flow fields is developed, exploiting the peculiar features of F.E. and F.D. methods in a two-level discretization technique. Numerical generation of body fitted curvilinear coordinates is considered at the first discretization level (macro-element mapping). The standard MAC scheme is extended to distorted grids, by assuming velocity contravariant components as field variables within a formulation of the problem in general tensor form. A low-order F.E. analog of MAC scheme, based on different interpolation functions for the two velocity components, is also proposed.

1. INTRODUCTION

1.1 Generalities on the problem. Flows of incompressible viscous fluids in multiply connected domains occur in many engineering applications, where the fluid motion about solid bodies (e.g. blade cascades in turbomachines or tube arrays in heat exchangers), is of primary interest.

The numerical solution of Navier-Stokes equations for such problems still presents several challenging difficulties mainly connected to:

a) The presence of recirculating regions which requires locally a finer discretization mesh;
b) The multiple connectivity of the computational domain which excludes the adoption of the most simple numerical procedures;
c) The irregualr shape of outer and inner boundaries typical of most relevant technical problems, which implies the use of a geometrically versatile numerical scheme.

The above points, all together, lead to the adoption of non uniform, distorted computational grids, either if a F.D. or a F.E. method is considered.

1.2 Finite Element Models. The treatment of complicated and disconnected boundaries is a routine matter even with the most elementary F.E. models, whereas involving complex and inaccurate procedures with standard F.D. techniques. It should be realized however that such F.E. capability derives from the model equations being assembled by systematically ignoring any possible regularity of the mesh pattern. That is, the most general node/element connection is a priori considered, as though the mesh were consisting of elements different from each other in shape and dimension.

It hence appears that when a certain topological or even metric mesh regularity is available - as it is the case in most fluid dynamics applications - the above capability is not only wasted but turns into costly drawback giving rise to length, idle computations [1]. The model we propose in the present paper - based on a two level discretization, is concerned with maintaining this F.E. versatility only at the proper level, thus avoiding the loss of efficiency resulting from the standard F.E. technique.

F.E. models experience also accuracy problems strictly connected with the choice of field variables and related interpolation schemes. The interest is here focused on the primitive variable formulation [2,3]. Such accuracy problems are essentially independent of mesh regularity and completely analogous to the ones encountered in F.D., as it will be discussed later.

1.3 Finite Difference Models. With the development of efficient numerical methods for generating boundary fitted curvilinear coordinates [4,5], the traditional limitation of F.D. to regular meshes is removed. Through the use of suitable mappings F.D. can compete with the metric versatility of F.E. and any problem defined on a complicated domain, becomes approachable, in principle, by a regular F.D. mesh constructed on a regular domain in the transformed plane. However the above technique based on the mapping of the entire domain, whichever its connectivity, onto a single rectangle seems to experience increasing difficulties for large number of bodies included in the flow field. To further generalize this mapping technique, the present method proposes the use of an assemblage of several rectangles, each of them being the map of a portion of the physical domain. The most typical feature of F.E. philosophy is in such way recovered.

With regard to accuracy problems in primitive variable formulation, the question of variables location has found a satisfactory solution starting from the MAC scheme [6] developed for rectangular grids. However, generalization of this method to distorted grids, as the one developed [7] in connection with atmospheric flows, are not yet employed in general fluid dynamics numerical simulations.

From the above arguments it can be deduced that comparable versatility and difficulties are presently encountered in F.D. and F.E. methods and it seems worthwhile to undertake a combined study of both methods.

2. PROPOSED MODEL

2.1 Subdivision into Macro-elements. To be specific, let consider the triply connected domain depicted in fig. 1a, which exhibits some general features relevant for the following discussion. Notice, in particular, that the outer boundary and one of the two inner boundaries, namely the body B contour, present a singularity. Fig. 1a also shows a subdivision into seven curvilinear quadrilateral macro-elements, each of which can be smoothly mapped onto a rectangle, as illustrated in fig. 1b, where some inter-element connections have been indicated with a line in order to rapresent with out distortion the transformed macro-elements in a plane.

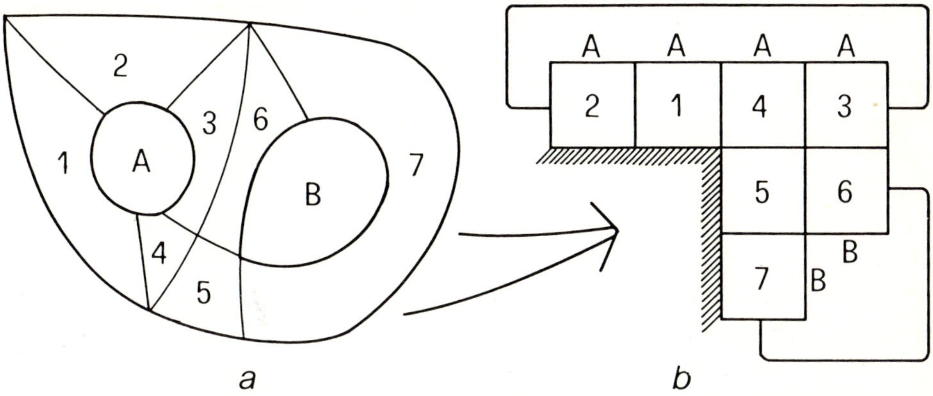

Fig. 1: (a) Subdivision of the physical domain into macro-elements;
(b) Macro-elements in the transformed plane.

Although the subdivision of fig. 1 is only an example, it is appearent that the separate mapping of different macro-elements into rectangles gives more freedom than the mapping onto a single rectangle, for the description of multiply connected fields. Due to the modular character of the m.e. subdivision the extension to a large number of internal bodies does not generate, from a qualitative point of view, increasing difficulties. Furthermore, a suitable m.e. choice may always avoid that a singularity on the physical boundaries be smoothed on the transformed boundaries or viceversa, thus yelding everywhere a one-to-one mapping. Obviously the local failure of one-to one correspondence affects the accuracy of the numerical solution of the flow field more or less, depending on the boundary conditions (e.g. slip or no-slip). The described subdivision constitutes the first of the two discretization levels, while the second one, internal to each m.e., is performed together with the mapping on the transformed plane.

2.2 Numerical Mapping and Field Discretization. Each macroelement is mapped onto a rectangle, by a transformation based on boundary fitted curvilinear coordinates. Being x,y the cartesian coordinates and ζ,η the curvilinear coordinates in the physical plane (fig. 2), such transformation is obtained by the numerical solution of the following system of quasi-linear elliptic partial differential equations:

$$\begin{aligned} \alpha x_{,\zeta\zeta} - 2\beta x_{,\zeta\eta} + \gamma x_{,\eta\eta} &= 0 \\ \alpha y_{,\zeta\zeta} - 2\beta y_{,\zeta\eta} + \gamma y_{,\eta\eta} &= 0 \end{aligned} \qquad (2.1)$$

where:

$$\alpha = (x_{,\eta})^2 + (y_{,\eta})^2 \; ; \quad \beta = x_{,\zeta}\, x_{,\eta} + y_{,\zeta}\, y_{,\eta} \; ; \quad \gamma = (x_{,\zeta})^2 + (y_{,\zeta})^2 \qquad (2.2)$$

and a comma denotes differentiation, with boundary conditions given by the assigned functions $x(\zeta,\eta)$ and $y(\zeta,\eta)$ along the m.e. boundary. These boundary conditions assure, obviously, only the continuity of the transformation functions $x=x(\zeta,\eta)$, $y=y(\zeta,\eta)$. If continuity of derivatives is also desired, the above conditions must be relaxed, and an iteration involving adjacent macro-elements is needed (the practical usefullness of this strong continuity appears questionable).

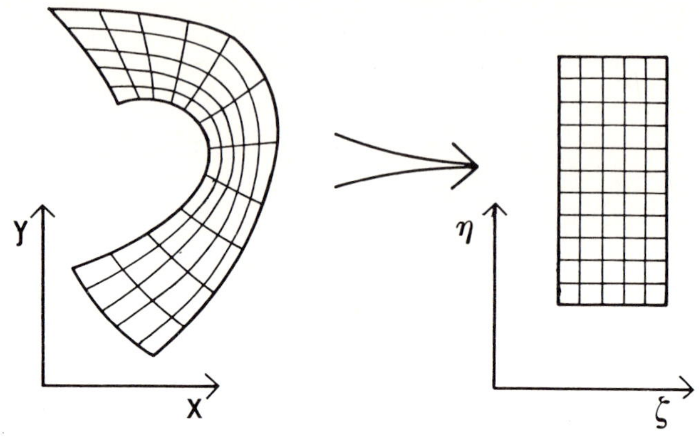

Fig. 2 - Macro-element in physical and transformed plane -

Equations 2.1, which generate a non-orthogonal coordinate system (more flexible than orthogonal systems for general application) may be efficiently solved through an iterative F.D. procedure [5]. The numerical solution is performed in a uniform rectangular mesh, defined on the transformed plane. The same mesh which is subsequently employed for the solution of the fluid dynamics equations, constitutes the second discretization level. At this inner level, it seems natural the adoption of a standard F.D. scheme. If a F.E. formulation is preferred, in order to ease the generation of the matching conditions between m.e.'s, it should anyhow be exploited the mesh regularity by direct construction of recurrent difference equations, thus avoiding the repeated use of local matrices [1].

2.3 Macro-Elements Matching. The solution of a m.e. inner field requires the velocity parameters connected to its contour to be assigned as boundary values. In turn such parameters play the role of unknowns in the matching of macro-elements, the equations being given by the momentum conservation equations in a reference system associated with the common boundary.

The incompressibility constraint, while determining (indirectly through the local mass conservation equation) the pressure inside each m.e. to within a reference value, allows such reference pressures (except one) to be determined through an integral formulation of the mass conservation. Namely for each m.e. an equation can be written, stating the total mass flux through its boundary to be zero. If n is the number of m.e. only (n-1) of these flux eqs. are independent and correspondingly (n-1) m.e. pressures can be determined in terms of the n-th pressure which remains the unique reference pressure for the entire flow field. As a general remarke, it has to be noticed that the above described matching eqs. have to be constructed through a F.E. assembling procedure due to the topological irregularity of the m.e. grid.

3. DISCRETE FORMULATIONS

3.1 Field Variables. The numerical solutions of Navier-Stokes eqs. for two dimensional flows of incompressible fluid are generally given in terms of velocity and pressure, as primary dependent variables, or stream function and velocity. The former (primitive variables) formulation is here preferred - beside its straightforward extension to three dimensions - because the treatment of boundary conditions is direct and easier, especially in multiply connected fields. Moreover, the most delicate numerical problems being connected with the incompressibility constraint and the computation of pressure, it seems more interesting to consider the formulation in which such difficulties explicitly appear.

3.2 Finite Difference Formulations. The various F.D. schemes based on primitive variables, differentiate essentially according to the variable location in the mesh points. Three difference schemes are depicted in fig. 3 for a square grid.

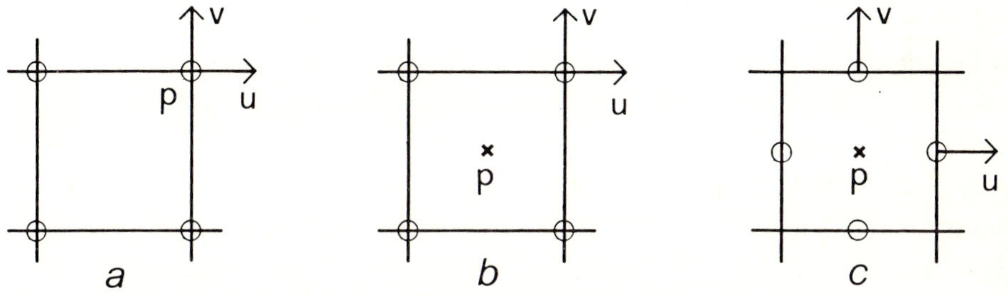

Fig. 3 - Variables location in F.D. schemes -

Scheme (a) with both velocity components and pressure defined at all corners produces spurious oscillations (of wave lenght two times the mesh size) either in velocity and in pressure fields.

With scheme (b), where the pressure is displaced to center of the cell, only checkerboard pressure oscillations remain. Nevertheless a scheme of this kind keeps to be used for distorted grids [8,9], seemingly in order to avoid a more sophisticated definition of velocity components.

Finally with scheme (c), well known as MAC scheme [6], where the velocity components are defined at the cell midsides, no oscillations are produced. The extension of MAC scheme, which locates differently the two velocity components, to distorted grids, is less obvious than the others and requires a deeper understanding of its peculiar features (see sect. 4.1).

3.3 Finite Element Formulations. The F.E. schemes in primitive variables (fig. 4) differentiate according to the interpolation functions adopted for the different variables, exhibiting a completely analogous behaviour with F.D. schemes previously analyzed.

Scheme (a) emploies the same interpolation functions (e.g. piecewise linear) for all variables, while scheme (b) uses a lower (e.g. piecewise constant) interpola tion function for pressure. Both schemes give oscillations, the first one in u,v,p, the second one only in p [2,3].

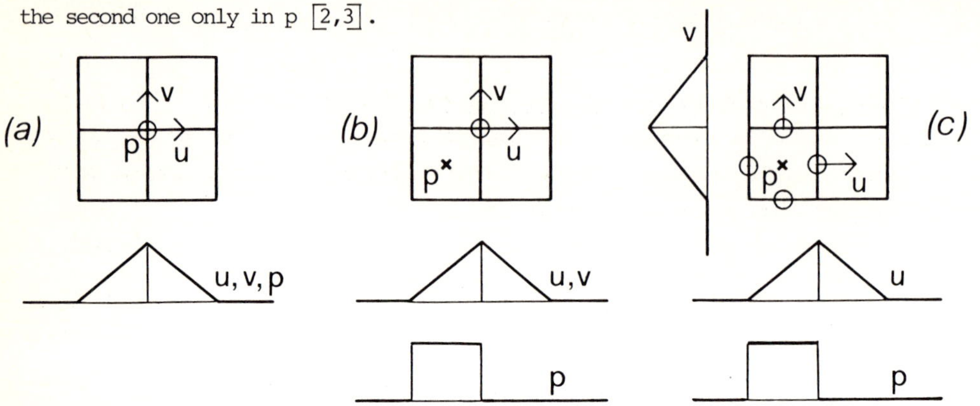

Fig. 4 - F.E. schemes and related interpolation functions -

A F.E. analog of MAC scheme has not yet been presented in the literature. It can be accomplished, as schematically illustrated in fig. 3c,by differentiating the interpolation function of the two velocity components (e.g.: u p.w. linear in x, constant in y; v p.w. linear in y, constant in x; p p.w. constant). Being the tangential velocity component discontinuous across an inter-element boundary, the shear stress has to be continuous and therefore independently interpolated (e.g. p.w. linear). This new F.E. scheme, that we present here concisely and whose capabilities are under investigation, eliminates all spurious oscillations. The extension to general (curvilinear) quadrilateral elements implies the same problems discussed for F.D. schemes.

4. MATHEMATICAL FORMULATION IN GENERAL CURVILINEAR COORDINATES

<u>4.1 MAC Scheme in General Curvilinear Coordinates.</u> The rationale of the standard MAC scheme is that the velocity variables, located at the cell midsides, are strictly connected to the mass fluxes across the cell sides. This feature has to be maintained, when moving to a curvilinear quadrilateral cell (fig. 5).

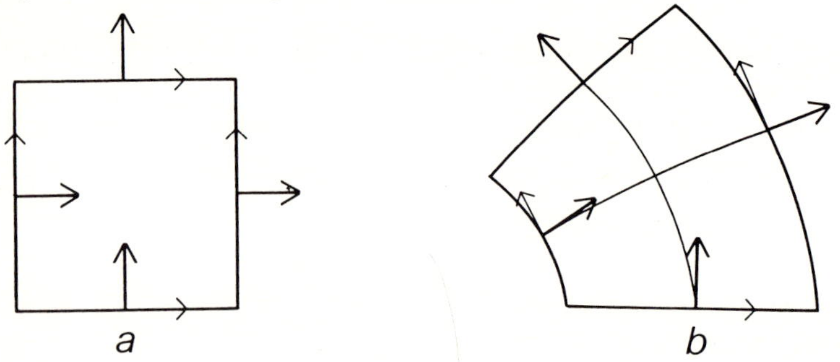

Fig. 5 - Velocity variables definition in MAC scheme -

At this purpose the contravariant velocity components must be considered and as sumed as variables, as shown in fig. 5b [7]. As a consequence the Navier-Stokes eqs. have to be written in general tensor form, as illustrated in the following subsection.

4.2 Navier-Stokes Equations in General Tensor Form. Let denote by $v^i (i=1,2)$, p, τ^{ij} the contravariant velocity components, pressure and contravariant deviation (viscous) stress components. The mass and momentum conservation eqs., following the notations of Flügge [10], can be expressed in the form:

$$v^i|_j = 0 \tag{4.1}$$

$$\rho\dot{v}^i + \rho(v^i v^j)|_j = -p|^i + \tau^{ij}|_j \tag{4.2}$$

where:

$$v^i|_j = v^i{}_{,j} + v^k \Gamma^i_{kj} \tag{4.3}$$

is the covariant derivative of velocity, a comma denoting partial differentiation and Γ^i_{kj} the Christoffel symbol of the second kind;

$$p|^i = g^{ij}\, p|_j = g^{ij}\, p_{,j} \tag{4.4}$$

is the contravariant derivative of pressure and g^{ij} are the contravariant components of the metric tensor;

$$\tau^{ij}|_j = \tau^{ij}{}_{,j} + \tau^{ik}\,\Gamma^j_{kj} + \tau^{kj}\,\Gamma^i_{ki} \tag{4.5}$$

is the covariant derivative of viscous stress.

The constitutive viscous equation is given by:

$$\tau^{ij} = \mu(v^i|^j + v^j|^i - g^{ij}\, v^k|_k) \tag{4.6}$$

where $v^i|^j = g^{jk} v^i|_k$ is the contravariant derivative of velocity.

It is worthwhile to make explicit the connection between the above introduced symbols with the transformation functions $x(\zeta,\eta)$ and $y(\zeta,\eta)$ of section 2.2. Namely, the metric tensor is given by:

$$g^{11} = \frac{\alpha}{J^2}\ ; \quad g^{12} = g^{21} = -\frac{\beta}{J^2}\ ; \quad g^{22} = \frac{\gamma}{J^2} \tag{4.7}$$

where $J = x_{,\zeta} y_{,\eta} - x_{,\eta} y_{,\zeta}$ (4.8)

represents the Jacobian of the transformation; and α,β,γ are defined by eqs. (2.2).

The Christoffel symbols are defined in terms of the first and the second derivatives of the transformation functions, as follows:

$$\Gamma^k_{ij} = g^{kl}\, x^m{}_{,l}\, x^n{}_{,ij}\, \delta_{mn} \tag{4.9}$$

where δ_{mn} is the Kronecker symbol and x^1, x^2 stay for x and y, and ζ^1,ζ^2 for ζ,η. By substituting eqs. (4.7 - 4.9) into eqs. (4.1 - 4.6), we obtain after straightforward, yet tedious manipulations:

for the continuity equation $\frac{1}{J}\left[(Ju)_{,\zeta} + (Jv)_{,\eta}\right] = 0$ (4.10)

where u,v stay now for v^1, v^2;

for the ζ momentum equation:

inertia term $\rho\dot{u}$

convective term

$$+ \rho/J\left[(Juu)_{,\zeta} + (Juv)_{,\zeta}\right]$$

$$+ \rho/J\left[y_{,\eta}(\alpha_{,\zeta\zeta}\, uu + 2x_{,\zeta\eta}\, uv + x_{,\eta\eta} vv)\right.$$

$$\left. - x_{,\eta}\,(y_{,\zeta\zeta}\, uu + 2y_{,\zeta\eta}\, uv + y_{,\eta\eta}\, vv)\right] \quad =$$

pressure term $= -\frac{1}{J^2}(\alpha\, p,_\zeta - \beta\, p,_\eta)$

viscous term $+ 1/J\left[(J\tau^{11}),_\zeta + (J\tau^{12}),_\eta\right]$

$$+ 1/J\left[y,_\eta(x,_{\zeta\zeta}\tau^{11} + 2x,_{\zeta\eta}\tau^{12} + x,_{\eta\eta}\tau^{22}) - x,_\eta(y,_{\zeta\zeta}\tau^{11} + 2y,_{\zeta\eta}\tau^{12} + y,_{\eta\eta}\tau^{22})\right] \qquad (4.11)$$

where: $\tau^{11} = \mu/J^3\left[J\alpha u,_\zeta - 2J\beta u,_\zeta + J,_\zeta\alpha u - J\alpha,_\zeta u + J,_\eta\alpha v - J\alpha,_\eta v - J\alpha v,_\eta\right]$ (4.12)

and analogously for the other two components.

As stated before the viscous stress components are, in F.E. schemes, independently interpolated. Correspondingly in the F.D. scheme the viscous stress field is discretized by locating τ^{12} at the cell corners and τ^{11}, τ^{22} at the center of the cell, together with pressure.

5. CONCLUDING REMARKS

The present technique, exploiting some peculiar features of F.E. and F.D., seems to be an appropriate blending of the two methods, particularly suitable to analyze multiply connected flow fields. This approach is presently being tested on some sample problems, by adopting a time dependent integration technique, which avoids for an accurate choice of the time steps, the necessity of an iteration procedure between the two levels of discretization, bringing to the steady state through a false transient. The search of the true transient would require, at each time step, few iterations between the two solution levels.

REFERENCES

1) Whiteman, J.R., "Lagrangian Finite Element and Finite Difference Methods for Poisson Problems", ISNM 27 Birkhauser-Verlag, 1975.

2) Piva, R. and Di Carlo, A., "Numerical Techniques for Convection/Diffusion Problems" The Mathematics of Finite Elements and Applications II, Academic Press, 1977.

3) Huyakorn, P.S., Taylor, C. et al., "A Comparison of Various Mixed-Interpolation Finite Elements in the Velocity-Pressure Formulation of the Navier-Stokes Equations", Computers and Fluids, vol. 6, pagg. 25-35, 1978.

4) Thames, F.C., Thompson, J.F. et al., "Numerical Solutions for Viscous and Potential Flow about Arbitrary Two-Dimensional Bodies Using Body-Fitted Coordinate Systems", J.C.P., vol. 24, n. 3, July 1977.

5) Thompson, J.F., Thames, F.C., Mastin, C.W., "TOMCAT - A Code for Numerical Generation of Boundary-Fitted Curvilinear Coordinate System on Fields Containing Any Number of Arbitrary Two-Dimensional Bodies", J.C.P., vol. 24, n.3, July 1977.

6) Harlow, F.H., Welch, J.E., "Numerical Calculations of Time-Dependent Viscous Incompressible Flow of Fluid with Free Surface", The Physics of Fluids, vol. 8, 1965.

7) Tzvi Gal-Chen, "On the Use of a Coordinate Transformation for the Solution of the Navier-Stokes Equations", J.C.P., vol. 17, pagg. 209-228, 1975.

8) Hirt C.W., Amsden A.A., Cook J.L. "ALE Method for all Flow Speeds" J.C.P., 14, 1974.

9) Pracht W.E. "Calculating 3-D Fluid Flows at all Speeds with . . ." J.C.P., 17, 1975.

10) Flügge W. "Tensor Analysis and Continuum Mechanics", Springer-Verlag, 1972.

SEPARATING, INCOMPRESSIBLE, LAMINAR BOUNDARY-LAYER FLOW OVER A SMOOTH STEP OF SMALL HEIGHT

D. Dijkstra

Dept. of Appl.Math., Twente University of Technology, Enschede, The Netherlands.

1. INTRODUCTION.

The triple-deck model developed by Stewartson [1] and Messiter [2] for the description of the local interaction of a laminar boundary-layer with the outer flow, has proved to be a powerful instrument for the treatment of a non-uniformity appearing in a boundary-layer (Stewartson [3]). The success of the theory when applied to the flat plate trailing edge is well-known (Jobe and Burggraf [4], Chow and Melnik [5], Veldman and Van de Vooren [6]).Another important example where a triple-deck structure arises, was put forward by Sychev [7] in his theory on imcompressible separation for large Reynolds numbers. In his paper Sychev formulated the problem but he did not present a (numerical) solution; specifically, the question of regularity (or singularity) of the solution at the separation point was not considered.

Recently, an approximate solution to the Sychev problem has been presented by Smith [8]. It appears that the only stage (in Smith's method of solution) where an approximation is made, is the treatment of the reverse flow for which the Flügge-Lotz and Reyhner approach is adopted. Smith pays much attention to the asymptotic structure of the solution and he advances strong arguments for the uniqueness of a solution to the Sychev problem. The remaining difficulty seems to be the reverse flow region, which is infinite in extent (on the scale of the triple deck) and requires information from downstream infinity.

In the present paper we will try to find an answer to the following questions.

(i) If separation occurs within the lower deck of an incompressible triple-deck, will it be regular or singular?

(ii) Can we develop a numerical method (to solve the lower deck equations) which treats reverse flow in a more satisfactory way than by means of the Flügge-Lotz and Reyhner approach?

To answer these questions a special geometry is chosen in order to provoke separation within the lower deck of a Stewartson - Messiter triple-deck. In fact we will consider incompressible boundary-layer flow over a smooth step of small height. The dimensions of the step are such that triple-deck theory can be applied to resolve the flow structure near the step. The step height enters in the problem as a parameter and it appears that above a certain value of the step height a finite separation bubble occurs.

In the early stage of this investigation the governing equation was discretized with the Crank-Nicolson method but near separation oscillations were found in the numerical results. Later on the approach given by Carter and Wornom [9] was adopted: the Crank-Nicolson scheme was replaced by the 3-point backward (3B) scheme and no oscillations occurred. In this paper strong arguments will be advanced in favor of the 3B-scheme.

2. FORMULATION AND GOVERNING EQUATIONS.

Consider a smooth step of small height h^* at a distance L^* downstream of the leading edge of a flat plate (fig. 1). The symbol * denotes a dimensional quantity and the Reynolds number is

$$Re = U_\infty^* L^*/\nu = \varepsilon^{-8}. \tag{2.1}$$

In terms of the variables

$$(x_1, y_1) = (x^*, y^*)/L^*, \quad \psi_1 = \psi^*/(L^* U_\infty^*) \tag{2.2}$$

the Blasius stream function at $x_1=0$ is given by

$$\psi_1 = \varepsilon^4 B(\varepsilon^{-4} y_1) \sim \varepsilon^{-4} \lambda y_1^2/2 \quad , \quad \lambda = B''(0) = .332057 \tag{2.3}$$

where B denotes the Blasius function. Near the wall the leading term in (2.3) represents the unperturbed flow ($h^*=0$) near the origin 0 and it serves as the oncoming flow for the lower deck of a Stewartson-Messiter triple deck structure, which developes near 0 in case $h^* \neq 0$, provided that $h^* = 0(\varepsilon^5 L^*)$ or less. The shear constant λ will be removed from the problem and the variables will be scaled according to lower deck requirements (see e.g. Smith [10]):

$$\begin{aligned} x_2 &= \lambda^{5/4} \varepsilon^{-3} x_1 , \quad y_2 = \lambda^{3/4} \varepsilon^{-5} y_1 , \quad \psi_2 = \lambda^{\frac{1}{2}} \varepsilon^{-6} \psi_1 \\ h^* &= 2\delta \lambda^{-3/4} \varepsilon^5 L^*, \quad p^* = p_\infty^* + \rho^* U_\infty^{*2} \lambda^{\frac{1}{2}} \varepsilon^2 P \end{aligned} \tag{2.4}$$

The numerical factor 2δ represents the numerical step height and the shape of the step is taken as (fig. 2)

$$y_2 = F(x_2) = \delta(2-\text{erfc}(x_2)) \tag{2.5}$$

This wall shape is smooth and it flattens very fast if we leave the origin $x_2 = 0$, so that away from $x_2 = 0$ the wall is a flat plate again.

The problem in the lower deck reads (fig. 2)

$$\frac{\partial(\psi_2, u_2)}{\partial(y_2, x_2)} = -\frac{dP}{dx_2} + \frac{\partial^2 u_2}{\partial y_2^2} \quad , \quad u_2 = \frac{\partial \psi_2}{\partial y_2} \tag{2.7.1}$$

$$y_2 = F(x_2) \;:\; \psi_2 = \partial\psi_2 / \partial y_2 = 0 \tag{2.7.2}$$

$$y_2 \to \infty \;:\; \psi_2 \sim y_2^2/2 + y_2 A(x_2) \tag{2.7.3}$$

$$x_2 \to \infty \;:\; \psi_2 \sim y_2^2/2 \tag{2.7.4}$$

where the displacement A is related to the pressure P via the elleptic condition

$$\pi A'(x_2) = \int_{-\infty}^{\infty} \frac{P}{\xi - x_2} d\xi \qquad (2.7.5)$$

The pressure is eliminated from (2.7.1), transposition is applied and the leading term subtracted from the stream function:

$$\xi = x_2,\ \eta = y_2 - F(x_2)\ ,\ \psi_2(x_2,y_2) - \eta^2/2 = \psi(\xi,\eta) \qquad (2.8)$$

In terms of the shear perturbation Ω there results

$$U \frac{\partial\Omega}{\partial\xi} + V \frac{\partial\Omega}{\partial\eta} = \frac{\partial^2\Omega}{\partial\eta^2} = \frac{\partial^4\psi}{\partial\eta^4} \qquad (2.9.1)$$

$$U = \eta + u,\ \Omega = \partial u/\partial\eta\ ,\ u = \partial\psi/\partial\eta\ ,\ V = -\partial\psi/\partial\xi \qquad (2.9.2)$$

$$\eta = 0 : \psi = \partial\psi/\partial\eta = 0,\ P'(\xi) = \partial\Omega/\partial\eta \qquad (2.9.3)$$

$$\eta \to \infty : u \to A(\xi) + F(\xi)\ ;\ \xi \to -\infty : \psi \to 0\ ,\ \Omega \to 0 \qquad (2.9.4)$$

$$\pi A'(x) = \int_{-\infty}^{\infty} \frac{P(\xi)}{\xi - x} d\xi. \qquad (2.9.5)$$

Note that the wall shape F now enters in the problem via (2.9.4).

3. NUMERICAL METHOD.

3.1 The grid.

Stretching functions have been used for resolution:

$$\xi = \xi(s) = \frac{22}{15}\ \frac{s + 4s^3}{1 - s^2}\ ,\ \eta = \eta(t) = 4\ \frac{t}{1-t} \qquad (3.1)$$

The effect of the transformation on ξ is depicted in fig.3. The computational domain in the (s,t)-plane is taken as $-s_B \le s \le s_B$, $0 \le t \le 1$ and uniform steps in s and t are used. The streamwise domain of eq. (2.9.1) is cut off at $s_B = 13/16$ ($\xi_B = 12.7659$) while the lateral coordinate η runs from 0 to ∞. For $|\xi| > \xi_B$ asymptotic results have been used where necessary. Three grids have been used, Coarse, middle and fine respectively. The grid "coarse" has stepsizes 1/16 and the other grids have the stepsizes halved in both directions. The grid "fine" contains 105 x 65 points.

3.2 The global iteration.

The parabolic / elliptic problem (2.9) is iteratively solved. One global step consists of a sweep through the boundary-layer (2.9.1), which yields the pressure, followed by a computation of the Hilbert integral (2.9.5), which yields a new displacement A. This set-up is the one developed by Jobe and Burggraf [4] and underrelaxation is applied with factor.1. A completely different method is being developed by Veldman [11] which seems to be promising and can be overrelaxed in certain cases.

3.3 The Hilbert integral.

The integral (2.9.5) is split up in portions over the ranges $|\xi| \leq \xi_B$ and $|\xi| \geq \xi_B$. The infinite tails are calculated with asymptotic results for P, while over the range $|\xi| \leq \xi_B$ the singularity is subtracted and the remaining continuous integrand (with value P' at ξ=x) is treated with Simpson's rule in the coordinate s.

3.4 The boundary layer routine.

In each mesh point at a grid line corresponding with a boundary layer station three variables (ψ, u, Ω) appearing in (2.9) are simultaneously computed. Central differencing on the lateral derivatives $\partial/\partial\eta$ is used while $\partial/\partial\xi$ is replaced by its 3-point backward approximation. The variables u and ψ are calculated with

$$u(\xi,\eta) = u(\xi,\infty) - \int_\eta^\infty \Omega(\xi,\bar{\eta})\, d\bar{\eta} \;,\; \psi(\xi,\eta) = \int_0^\eta u(\xi,\bar{\eta}) d\bar{\eta} \qquad (3.2)$$

The integrations are performed with the trapezoidal rule in t given by (3.1), where ψ is obtained by outward integration while u is found integrating inward. The condition u = 0 at η = 0 serves to determine the wall shear $\Omega(\xi,0)$ and the pressure gradient follows from (2.9.3). The non-linearity is treated by means of Newton iteration with tolerance 10^{-6}.

3.5 Treatment of backflow.

Three modes have been used to cope with the case U<0:

1) 3B-Flare, or Flügge-Lotz and Reyhner approximation:
   set $U\partial\Omega/\partial\xi = 0$ if $U \leq 0$.
2) 3B-Ignore, where we simply march in the main stream direction as if there were no backflow.
3) 3B-Back, where the backflow is treated properly using old values for $U\partial\Omega/\partial\xi$. This method was suggested by Williams [12].

4. RESULTS, DISCUSSION AND CONCLUSIONS.

The quality of the numerical method was investigated by solving a linearized version of the problem (2.9). The linearization was performed about the main stream or equivalently about δ = 0. The resulting problem coincides with (2.9) except that in (2.9.1) U is replaced by η and v by 0. The linearized problem was solved analytically with Fourier transformation. This analysis is comparable to the one used by Smith [10] for a related problem. With the exact solution to the linearized problem available, the next step was to compute it numerically with the method of sec.3. A comparison between numerical results and exact value at ξ = 0 is made in table 4.1 and some results for the linearized case have been plotted in figs. 4 and 5.

grid	num.res.	$O(h^2)\to 0$	$O(h^3)\to 0$	ex. value
	$\tau_\ell(0)$			
coarse	0.40952			
middle	0.41118	0.41173		
fine	0.41170	0.41187	0.41189	0.41190
	$-A'_\ell(0)$			
coarse	0.56281			
middle	0.56094	0.56032		
fine	0.56053	0.56039	0.56040	0.56041
	$P_\ell(0)$			
coarse	0.11195			
middle	0.11260	0.11282		
fine	0.11275	0.11280	0.11280	0.11260

Table 4.1 Numerical results at $\xi = 0$ for linearized problem compared with exact value.

Except for a slight residual error in the pressure it is seen that the results may be improved by Richardson extrapolation and we conclude that the numerical method produces results which behave as expected.

Next, the non-linear problem (2.9) was solved numerically for various δ. The onset of separation was found at $\delta = 6.4$ where the wall shear curve became tangent to the ξ-axis. For $\delta >> 6.4$ the curve moved further downward in a regular manner and it was concluded that not a trace of the Goldstein separation singularity was present. In other words the answer to the query (i) raised in sec. 1 is that separation is regular, at least in the present problem.

Finally, the wall height 2δ was pushed further up to 30 and a closed separation bubble appeared for values of 2δ of that order. The behaviour of wall shear and pressure over the range of the bubble is presented in figs. 6,7 and 8 and it is seen that plateau-regions in pressure and wall shear are built up with much larger gradients near reattachment. This behaviour qualitatively agrees with that calculated by others for closed bubbles (Carter and Wornom [9], Briley and McDonald [13]).

A strong argument in favor of the 3-point backward difference approximation to the streamwise derivative is now put forward. The results for $\delta = 12$ produced with the mode 3B-Ignore (sec. 3.5) were found to be graphically identical with the outcome of the mode 3B-Back (figs. 6 and 7). In addition the magnitude of the backflow (minimum (U) = -0.20) was found to be the same. With the mode

3B-Ignore the diagonal element associated with (2.9.1) in the system of algebraic equations is affected in an unfavourable way for those mesh points where $U < 0$, but the results are virtually identical to the outcome of the mode 3B-Back. With the latter mode the diagonal element is not unfavourably affected in points where $U < 0$. This means that the 3B discretization has very strong stability properties and that the flow in the bubble is mainly controlled by the elliptic Hilbert integral boundary condition (2.9.5). The same conclusion was drawn by Veldman [11] for a small bubble within a classical boundary layer.
A final word about the Flare-approximation (sec. 3.5). It was found that Flare produces results which can be used as a first approximation to the flow field in the bubble but significant deviations occur near reattachment where the reverse flow has maximum magnitude.
It is believed that the method presented in this paper may be valuable for other separating lower deck flows but first of all limitations of the modes Flare and Ignore should be demarcated.

ACKNOWLEDGEMENTS.

The author is indebted to Prof. K. Stewartson for suggesting the subject and for many valuable discussions. He also acknowledges the help and support by Dr. S.N. Brown, Mr. P.G. Williams, Dr. E.F.F. Botta, Dr. A.E.P. Veldman and Dipl. math. P.K.H. Gragert. The main part of the work was done while the author was visiting University College, London on a combined grant from the British Science Research Council and the Netherlands Organization for the Advancement of Pure Research (Z.W.O.).

REFERENCES.

1. Stewartson, K., Mathematika 16, 106-121, 1969.
2. Messiter, A.F., Siam J. Appl. Math. 18, 241-257, 1970.
3. Stewartson, K., Advances in Appl. Mech. 14, 145-239, 1974.
4. Jobe, C.E. and Burggraf, O.R., Proc. Roy. Soc. A340, 91-111, 1974.
5. Chow, R. and Melnik, R.E., Proc. 5th. internat.conf. on num.meth.fl.dyn. 135-142, 1976.
6. Veldman, A.E.P. and van de Vooren, A.I., Proc. 4th internat.conf.on num.meth. fl.dyn., 423-430, 1974.
7. Sychev, V.V., Mech. of Liq. and Gases, Ann Arbor, Michigan, nr.3, 47-59, 1972.
8. Smith., F.T., Proc. Roy. Soc. A356, 443-463, 1977.
9. Carter, J.E. and Wornom, S.F., Aerodyn. Anal. requiring adv. comp., Nasa SP-347, Langley Res. Virginia, 125-150, 1975.
10. Smith, F.T., J.F.M., 57, 803-824, 1973.
11. Veldman, A.E.P., private communication, 1978.
12. Williams, P.G., Proc. 4th. internat.conf. on num. meth. fl. dyn., 445-451,1974.
13. Briley, W.R., and Mc.Donald, H., J.F.M. 69, 631-657, 1975.

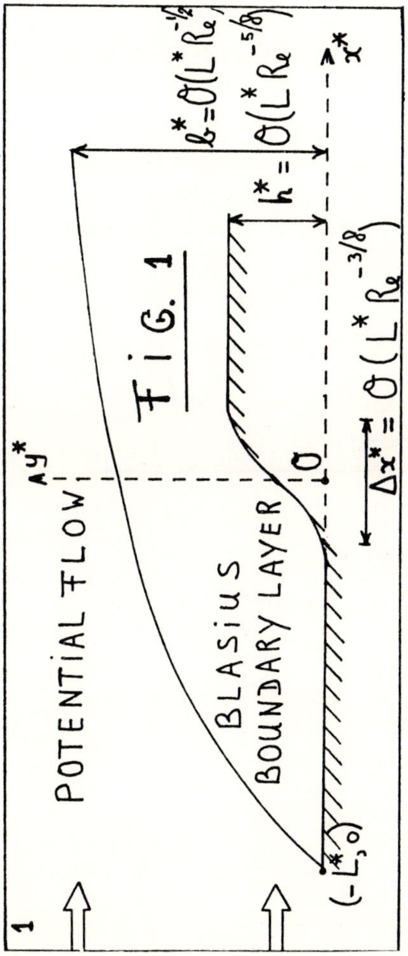

FIG. 1

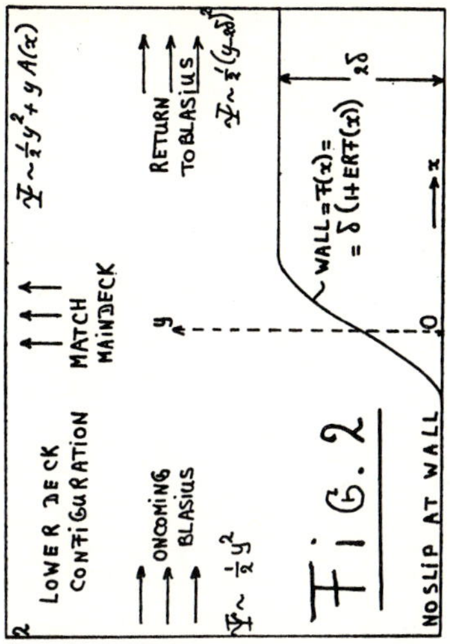

FIG. 2

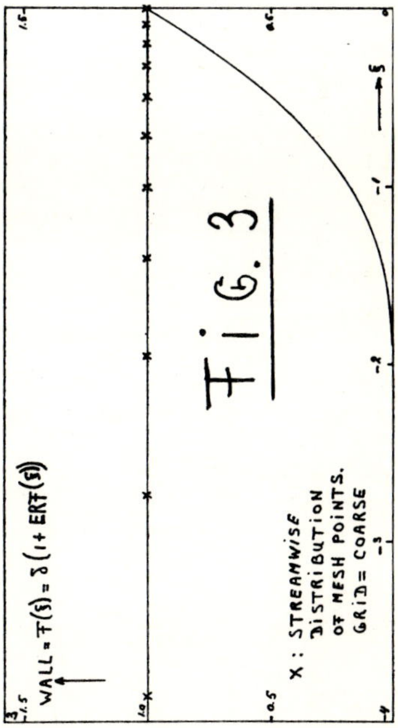

FIG. 3

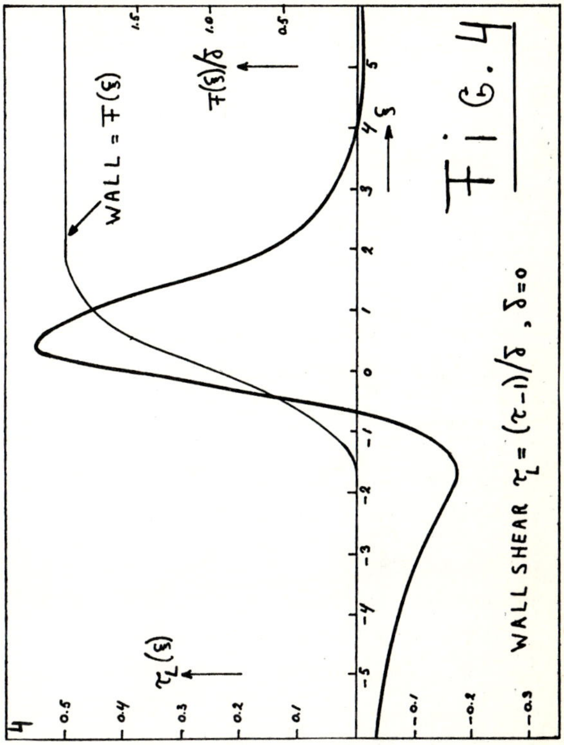

FIG. 4

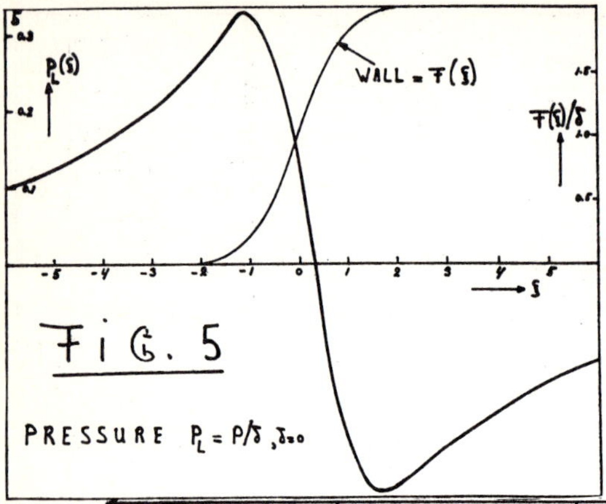
WALL = F(ξ)
F(ξ)/δ
Fig. 5
PRESSURE $P_L = P/\delta$, δ=20

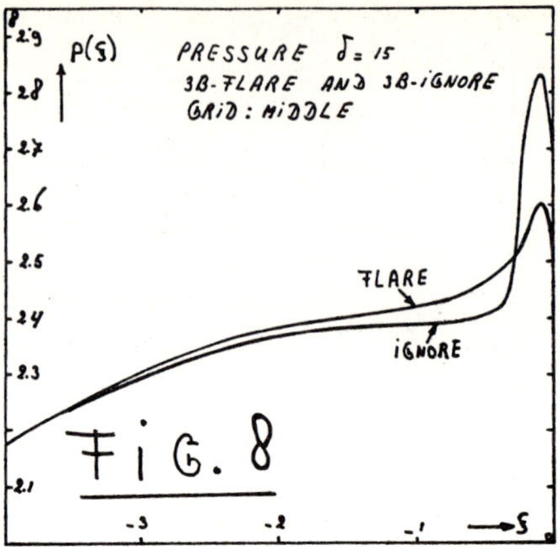
P(ξ)
PRESSURE δ = 15
3B-FLARE AND 3B-IGNORE
GRID: MIDDLE
FLARE
IGNORE
Fig. 8

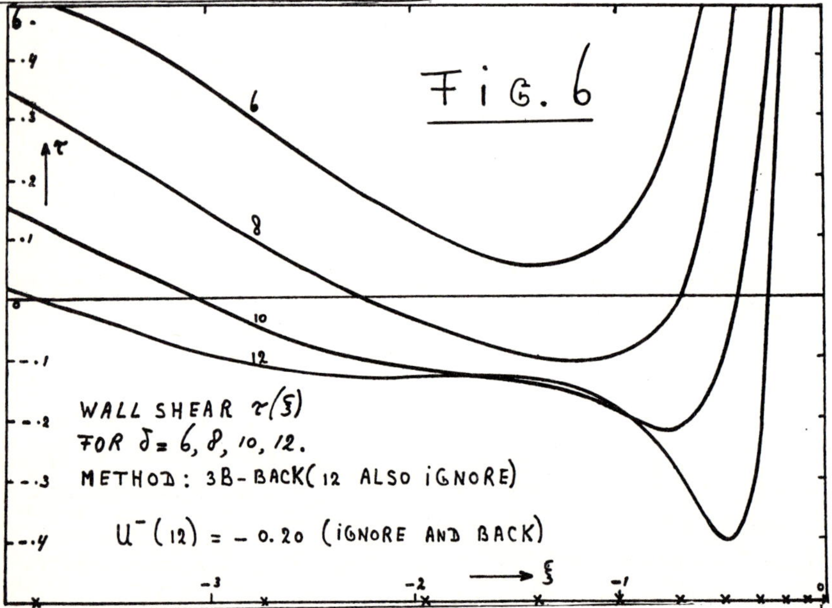
Fig. 6
WALL SHEAR τ(ξ)
FOR δ = 6, 8, 10, 12.
METHOD: 3B-BACK (12 ALSO IGNORE)
U⁻(12) = − 0.20 (IGNORE AND BACK)

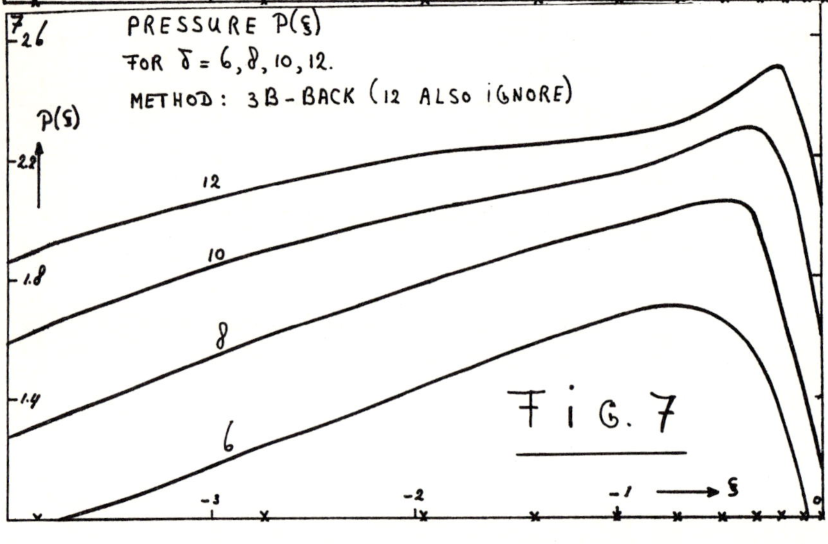
PRESSURE P(ξ)
FOR δ = 6, 8, 10, 12.
METHOD: 3B-BACK (12 ALSO IGNORE)
P(ξ)
Fig. 7

ON CERTAIN SOLUTIONS OF THE NON- STATIONARY EQUATIONS FOR ROTATING FLOW by:

D. Dijkstra*, H. Schippers+ and P.J. Zandbergen*.

*Twente University of Technology, Department of Mathematics Enschede, The Netherlands.

+Mathematical Centre, Amsterdam.

Summary.

In this paper results are presented for three different time dependent problems. In the first place it is shown that the new stationary solutions as constructed in [4] and [5] are unstable.

In the second place the present analysis gives limit cycle solutions in the region where no stationary solutions exist i.e. $-0,16054 < s < -1,4351$. These results are at variance with those reported earlier in [2].

In the third place an interesting phenomenon has been demonstrated for the flow between two disks. It appears that in certain cases there is a transition Reynolds number above which the flow "remembers" the initial sense of rotation.

1. *Introduction.*

In the last years a number of papers on rotating flow above an infinite disk or between two disks have appeared [1], [2], [3], [4], [5], which have considerably added to the knowledge available and thus stimulated further research on a number of still open questions.

In [4] and [5] it has been shown that for $-0,16054 < s < 0,074526$* a second solution branch of the Navier-Stokes equations for the single disk problem can be constructed, and that also more branches will occur. The nature of the solutions pertaining to these branches has been given for $s = 0$ in [5].

A question which arises is whether or not these solutions are stable. In this paper we will consider this question by using a time dependent method.

Once the method is available it is interesting to see what will happen when intruding the region $-0,16054 < s < -1,4355$ where no stationary solutions exist.

In [2] there have been presented results for this case, but there appears some doubt on the validity of these results.

In the mean time the second author used essentially the same method when analyzing the flow between two infinite disks, where one of the disks is at rest and the other oscillates in its plane with a given frequency. It occurs that an interesting phenomenon is present with respect to the influence of the initial condition depending on the Reynolds number and the ratio of angular velocity and frequency ε. Some results are contained in [6].

First we will present the equations and some details on the numerical approach. Then we will **succesively cover** the problems as described above. Necessarily the presentation must be rather sketchy, but we hope to give a more comprehensive

---

* s is the ratio between the angular velocities of the fluid and of the disk.

account of our investigations in a forthcoming paper.

2. *Basic equations.*

By applying the similarity approach of Von Karmann the velocities (u,v,w) in a cilindrical coordinate system (r,ϕ,z) can be written as

$$(2.1)\quad u = \Omega r \frac{\partial F}{\partial x}(x,\tau),\; v = \Omega r\, G(x,\tau),\; w = -2(\nu\Omega)^{1/2} F(x,\tau)$$

where $x = z\left(\frac{\Omega}{\nu}\right)^{1/2}$ and $\tau = \Omega t$

In that case the Navier-Stokes equations reduce to two partial differential equations for the unknown functions F and G.

$$(2.2)\quad F_{x\tau} = F_{xxx} + 2FF_{xx} - F_x^{\;2} + G^2 - K(\tau)\;;\; G_\tau = G_{xx} + 2FG_x - 2GF_x$$

The function $K(\tau)$ is related to the pressure. For the single disk problem there holds $K(\tau) = G^2(\infty,\tau)$, for the two disk problem the function K is determined by applying boundary conditions.

The appropriate boundary conditions are for the single disk problem.

$$(2.3)\quad F = \frac{\partial F}{\partial x} = 0, G = 1 \text{ at } x = 0 \quad ; \quad \frac{\partial F}{\partial x} = 0, G = s \text{ for } x \to \infty$$

For the two disk problem there holds.

$$(2.4)\quad F = \frac{\partial F}{\partial z} = 0, G = \cos\frac{\tau}{\varepsilon} \text{ at } x = 0 \quad ; \quad F = \frac{\partial F}{\partial z} = G = 0 \text{ at } x_n = 1\left(\frac{\Omega}{\nu}\right)^{1/2} = (R_e)^{1/2}$$

where 1 is the distance between the two disks.

Of course we also need initial conditions but these will be specified for the problems we consider in the different paragraphs.

3. *The numerical approach.*

For solving the equations (2.2) we use a finite difference approach which is essentialy an adaptation of the method as given in [4]. The time derivative is discretized by using the well-known formula

$$(3.1)\quad g_\tau = \frac{3g_{k+1} - 4g_k + g_{k-1}}{2\Delta\tau}$$

The system which arises after applying this formula is known as the three point backward method. The stability properties are very good whereas the accuracy is of second order both in time and space using central differencing in space. In case of the flow between two disks the procedure of Kubicek as given in [3] is used to determine the function K. See also [6].

In most of the calculations as presented here we have taken $r = \frac{\Delta\tau}{(\Delta x)^2} = 5$ The accuracy in some cases has been checked by taking half the time step and half the space step. Using then Richardson extrapolation accurate results can be obtained.

This has not been achieved in all the cases as presented here, since our first purpose here was to obtain an insight in the overall behaviour of the solutions.

4. *Investigation of the stability of second and third branch solutions.*

In order to set an answer to the question whether or not the new-stationary solutions in [4] and [5] are stable we have calculated a number of cases. The first case considered is that of a solution on the second branch for s = -0,1. We perturb the solution by 1% for $\frac{\partial F}{\partial z}$ and calculated the appropriate F. Since we need two sets of values, namely at t = 0 and at t = $\Delta\tau$, we set both these values equal to each other and calculate the resulting function as a function on time. From fig. 1 it is quite clear that the overall behaviour is that at first a large peak occurs, followed by one in the opposite direction. Thereafter the solution converges in a very quiet way to the ultimate values of the solution on the first branch in s = -0,1.

The calculations have been made for a number of ranges in x. In this case we compared a range of 36 with that for 72 and even calculated the results for a range of 108. In all cases it appeared that the behaviour is much alike. The peaks appeared at the same place but were much increased for the larger ranges. Nevertheless the convergence to the final solution is always the same, although some smaller peaks may occur before the convergence to the final solution sets in. To get an insight into the behaviour of the solution as a function of x in fig. 2 there have been given curves for the function G for several time stations.

It should be remembered that the results for smaller time- and space steps differ numerically somewhat, but the trends remain the same.The following case is that of a solution on the third branch for s=0.The range had to be taken 100 in this case because of the fact that this solution for s=0 exhibits large "wiggles" in the profiles for G and F_x. Essentially the same character presents itself as in the former case. Again the large peaks,followed by some smaller ones as indicated in fig.3 and in the end a monotonous convergence to the solution on the first branch. Hence it seems quite clear that the solutions for the new branches are all unstable.

Of course it is worthwhile to obtain some more insight into the nature of the solutions presented by applying analytical methods. To do this we consider a time dependent asymptotic approximation for large values of x.

We set $F = f_\infty + \bar{f}(\tau,x)$, $g = s + \bar{g}(\tau,x)$ (4.1)

where $\bar{f}$ and $\bar{g}$ are considered to be small with respect to f_∞. The resulting equations are (4.2)

$$\bar{f}_{tx} = \bar{f}_{xxx} + 2f_\infty \bar{f}_{xx} - 2s\bar{g}$$

$$\bar{g}_t = \bar{g}_{xx} + 2f_\infty \bar{g}_x - 2s\bar{f}_x$$

This is a set of linear homogeneous equations which should be solved for the conditions

$\bar{f}_x = \bar{g} = 0$ for x = 0 and for $x = x_m$ to be able to compare this analysis with our computational results. This hence constitutes an eigen value problem.

By assuming that $\bar{f}_x = C_f e^{\lambda t} e^{\mu x}$ and $\bar{g} = C_g e^{\lambda t} e^{\mu x}$

we find $(\lambda - \mu^2 - 2f_\infty \mu)^2 + 4s^2 = 0$ From this it follows that

(4.3) $\mu = -f_\infty \pm \sqrt{f_\infty^2 + \lambda \pm 2is}$ It turns out that the boundary conditions can be fulfilled if there holds (4.4) $x_m \sqrt{f_\infty^2 + \lambda \pm 2is} = ik\pi$

$$\text{Hence} \quad (4.5)\ \lambda = -f_\infty^2 - \frac{k^2\pi^2}{x_m^2} \mp 2is$$

From this formula there follows that in general the convergence to the final values oscillates with a half period of $\frac{\pi}{2s}$, at least close in the vicinity of the final solution, for it follows also from (4.5) that the first few eigenfunctions for x_m rather large-all are damped in the same manner.

There is a very good agreement between this rough calculation and the results from the computations as follows from table 1

		s = -0,10	s = -0,15
Half period	eq.(4.5)	15,7	10,47
	Comput.	15,8	10,6

				s = 0
Decrease in amplitude	eq.(4.5)	0,2562	0,9106	0,1436
	Comput.	0,3018	0,8935	0,1358

It is clear from this table, which gives the decrease in $f(\infty)$ over a characteristic length in time that the damping is diminishing towards the bifurcation point at s = -0,16054. As can be expected it is quite easy to show from the linearized form of eqs (2.2) and using appendix A of ref. [4] that the damping is exactly egual to zero right at the point of bifurcation.

5. *Excursions into a prohibited region.*

Of course it is very interesting to see what happens, when we proceed with our calculations in the region where no stationary solutions exist. In [2] there is referred to unpublished results for s = -10, -4 and -2, all lying in the region where a stationary solution exists. As is stated there, there are limit cycle solutions for s = -10, whereas the solutions for s = -4 and s = -2 could not be obtained. The situation there is analogous to the case s = -1,0 which is studied in [2]. As is argumented there, the numerical evidence points to behaviour which is "explosive" in time. If τ_E is the critical time, then there should hold

(5.1) $F(\infty,\tau) \approx \frac{c}{(\tau_E-\tau)^2}$. Essential is that not only $F(\infty,\tau)$ becomes singular, but also $F(x,\tau)$.

Interesting of course is to see what our calculations reveal for the case $s = -1{,}0$. It is clear from these calculations that there is no "explosive" behaviour, but that again the large peaks occur which somewhat damp away in time. The calculations have been performed with a range of 60 which is equal to the range as used by Bodonyi and Stewartson. The stepsize is $\Delta\tau = 0{,}1$ and $\Delta x = 0{,}1$. To investigate the effect of $\Delta\tau$, there have been made calculations with different stepsizes i.e. $\Delta\tau = 0{,}05$, $\Delta\tau = 0{,}02$ and $\Delta\tau = 0{,}01$. The overall behaviour in all cases is essentially the same and it turns out that a stable limit cycle is reached gradually. The first few cycles have been given in figure 4 for $\Delta\tau = 0{,}01$. The quantitative values depend rather heavily on the step size and even for $\Delta\tau = 0{,}01$ the solution has not settled down as can be inferred from a Richardson type extrapolation.

To cover up the whole region, we also made computations for other cases. Also in these cases a limit cycle is reached.
In the case $s = -1{,}55$, which lies in the so called Bödewadt region ($\sigma = -0{,}6452$), where stationary solutions exist a strange behaviour occurs. At first the solution looks like a limit cycle, but one which is apparently converging to a stationary solution. However this solution is not the Bödewadt solution for $s = -1{,}55$, but it is a solution for $s = +1{,}55$ over most of the range of calculation. Only in a limited region near to x_m there is a kind of boundary layer where the flow changes from the situation for $s = +1{,}55$ to the boundary conditions at x_m which prescribe $s = -1{,}55$. Of course further investigations are needed to explain this result.
We feel therefor that contrary to the statements in [2] the present analysis gives rise to the conclusion that there exist limit cycle solutions in the "prohibited region" and that there where stationary solutions exist, the time dependent solutions converge towards stationary values.
There should be made here a remark: in the starting phase the profiles of F_x and G remain constant over part of the range as a function of x but not as a function of time. It is possible to describe their behaviour by an exact solution of the time dependent Navier-Stokes equations. This solution is in terms of two arbitrary constant α and β.

$$(5.2)\quad F_x = 1/2\, s \frac{\sin 2\, s\, (\tau+\alpha)}{\sin^2 s(\tau+\alpha) + \sinh^2 \beta s} \qquad G = 1/2\, s \frac{\sinh 2\, \beta s}{\sin^2 s(\tau+\alpha) + \sinh^2 \beta s}$$

A further remark that should be made at the end of this paragraph is the following. Essentially we use as conditions at $x = x_m$, conditions which would prevail at a plate which rotates which a given tangential velocity and which is porous for flow normal to the plate. As can be seen from the stability analysis in paragraph 4, this introduces eigenfunctions which are certainly not the eigenfunctions belonging to the problem for $x_m \to \infty$. This fact has only a limited influence as long as the quantities F_x and G_x are zero over an appreciable region in the surrounding of $x = x_m$. Now certainly where the large peaks occur this is no longer true, and hence what happens there in the calculations certainly is not what happens for $x_m \to \infty$.

Of course this fact also hampers a proof of the occurrence - or not-occurrence - of the singularity suggested in [2].

There are two possible ways to overcome this difficulty. The first one is to take much longer ranges, but then the calculations become extraordinary expensive. The second one is trying to derive asymptotically correct boundary conditions. At present this last possibility is being investigated.

6. *Some results for the two disk problem.*

We now consider the flow between two disks, one of the disks being at rest, while the other one rotates oscillatory, such that $G = \cos 0{,}1\ \tau$. As is clear from the conditions stated earlier the flow is dependent on the Reynoldsnumber $R_e = \frac{l^2\Omega}{\nu}$.

To illustrate an interesting phenomenon, which probably is related to the occurence of more solutions in the stationary case, we consider the following situations $R_e = 196$ $R_e = 230$ $R_e = 400$.

We give no full details of the method for which we refer to [6]. It is clear from fig. 5, where we present the tangential velocity G halfway the disks, that for $R_e = 196$ the initial conditions fade away and that we get a symmetrical chracter with respect to the sign of G. For $R_e = 400$ it is quite clear that a totally different behaviour is present. The flow "remembers" the original sense of rotation set in at tha start and developes a profile for G which is asymmetrical and maintains the original sense of rotation for the bulk of the fluid.

The result for $R_e = 230$ indicates that this is a case which is very close to the transition Reynolds number. From the figure it is clear that there the solution damps away to the symmetrical profile, but with a very low damping. In this case the initial conditions are almost maintained.

So we have here evidence of the case where a Stewartson type of profile is present for $R_e \leq$ ca. 230, and a Batchelor type of solution for $R_e \geqslant$ ca. 230 and where the initial conditions determine the ultimate sense of the tangential velocity of the bulk. Of course this phenomenon should be considered also from an experimental point of view.

Acknowledgement: The authors are grateful for the efforts of Mr. E.H. Derks in preparing this paper.

References.

[1] N.D. Nguyen, J.P. Ribault and P. Florent. Multiple solutions for flow between coaxial disks. Journal of Fluid Mechanics, 68 (1975), 369-388.

[2] R.J. Bodonyi and K. Stewartson. The unsteady laminar boundary layer on a rotating disk in a counter rotating fluid. Journal of Fluid Mechanics, 79(1977), 669-688.

[3] M. Holodnick, M. Kubicek and V. Hlavneck. Computation of the flow between two rotating coaxial disks. Journal of Fluid Mechanics 81 (1977), 689-699.

[4] P.J. Zandbergen and D. Dijkstra. Non-unique solutions of the Navier-Stokes equations for the Karman swirling flow. Journal of Engineering Mathematics, 11 (1977) 167-188.

[5] D. Dijkstra and P.J. Zandbergen. Some further investigations on non-unique solutions of the Navier-Stokes equations for the Karman Swirling flow. Paper presented at the 13th Biennial Fluid Dynamics Symposium, September 1977 Olstyn, Poland.

[6] H. Schippers. Time dependent flow due to an infinite rotating disk. Memorandum 214 Department of Applied Math. Twente University of Technology, Enschede, The Netherlands.

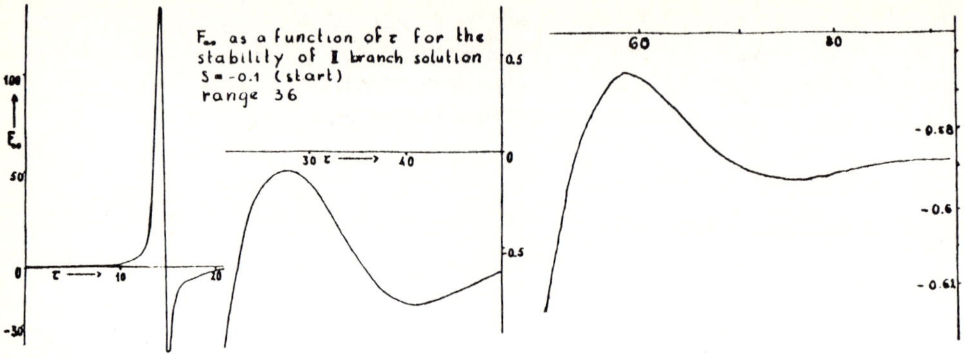

Fig. 1.

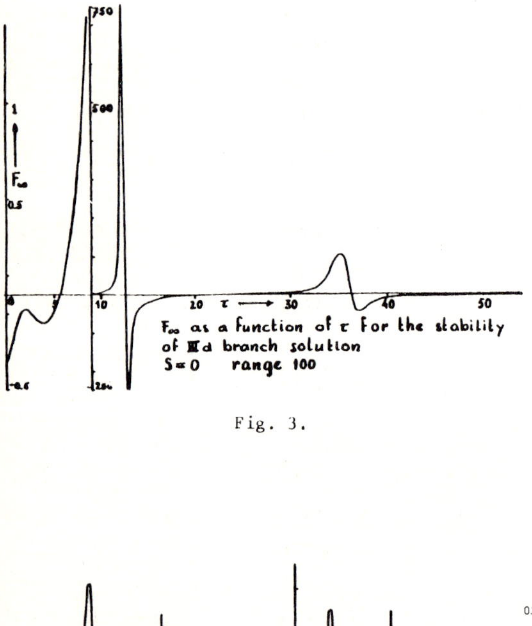

Fig. 3.

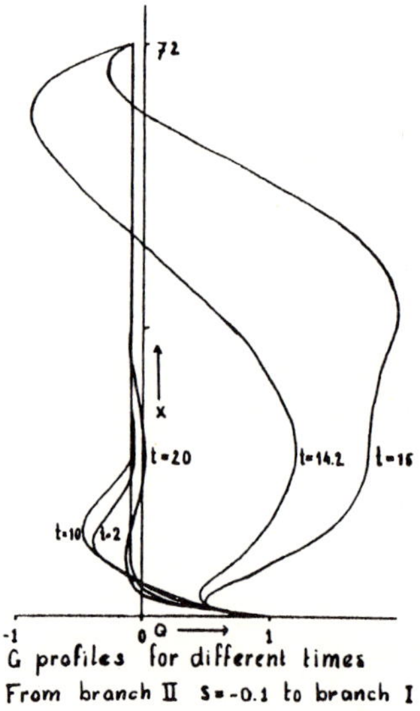

Fig. 2.

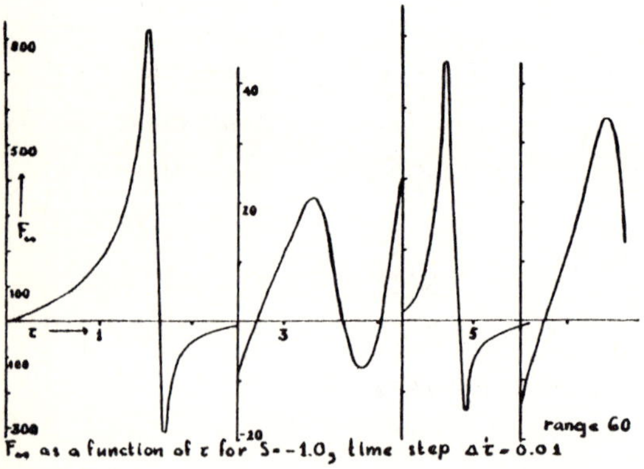

Fig. 4.

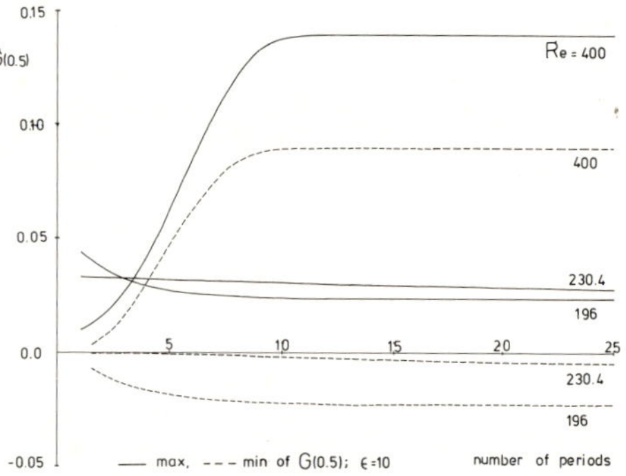

Fig. 5.

A UNIFICATION OF UNIDIRECTIONAL FLOW APPROXIMATIONS

by

Peter R. Eiseman
Scientific Research Associates, Inc.
Glastonbury, Connecticut, USA

INTRODUCTION

In the past decade a number of authors (Refs. 1-9) have numerically solved certain problems (incompressible, compressible, subsonic, supersonic, etc.) with the application of marching methods to an approximation of the three-dimensional Navier-Stokes equations. In each case, an approximation was based upon the assumption that the flow was primarily unidirectional. A motivation for this assumption was derived from the experimentally observable fact that without abrupt changes in geometry a high Reynolds number flow is dominated by only upstream conditions. Here, small disturbances in the primary direction tend to decay quickly. Consequently, a suitable approximation can and has been assumed to produce a well-posed system of governing equations from which a stepwise integration can successfully be applied to a given set of initial conditions. The rate of success with numerical computations tends to support this assumption. As in boundary layer theory, the approximations were obtained by an examination of the relative order of magnitude of each term on an equation by equation basis. Unlike boundary layer theory, these estimates only neglected the diffusive flux in the assumed primary direction. The net result was that the approximate governing equations contained the information which was necessary to accurately model secondary flow phenomena. In addition the approximations, although similar in concept, were each different; the primary differences came from the chosen system of conservation laws (e.g., the transport of momentum, vorticity, energy, etc.) and from the geometric properties associated with each particular problem. These differences also point to the fact that each of the previous approximations were too specialized to obtain an approximation which is valid for all cases where a primary flow direction exists.

The physical properties, associated with a primary flow direction, depend upon the fundamental constitutive relationship of a fluid. For this reason, the approximation presented herein is a direct approximation of the stress tensor which is clearly a statement about the constitution of the fluid. Consequently, such an approximation will be independent of the choice of basic conservation laws governing transport phenomena. Moreover, the approximation will be performed in a manner which preserves the tensor character of the stress. Consequently, the approximation will also be independent of coordinates. In this manner, the limitations of the previous approximations can be overcome. With the added generality, it should be noted that some terms of negligible size will be retained. However, benefits from the added generality will outweigh by far the expense of including these terms. Such benefits will be reflected in the wider range of applicability for a specific algorithm and in the quality of an approximation which can easily be increased because of the added flexibility.

FRAME FIELDS

A fundamental part of the approximation is the creation of a certain frame field. However, before this is examined in detail, some general observations on the applications of frame fields should be mentioned. Specifically, the basic governing equations in fluid mechanics can be written entirely in terms of frame fields. In addition to the desired approximation frame, there are derivative frames, solution frames, and a coordinate frame. In a number of special cases, a frame field has been selected to align solution variables, derivatives, or equations with a given direction of interest. In Jameson's (Ref. 10) generalization of the type-dependent differencing due to Murman and Cole (Ref. 11), a specific derivative frame was employed. This frame, known as the Frenet frame, consisted of vector fields that are tangent, normal, and binormal to the streamlines. The result was called the "rotated difference molecule". This same frame was also used for both a derivative and a solution frame by Lakshminarayana and Horlock (Ref. 12) who applied it to the vorticity equation (for both stationary and rotating systems). Their analysis consisted of successive applications of dot products and differentiation rules for the frame field; the differentiation rules consisted of the Frenet formulas along with other formulas for the normal and binormal directions. The results were a sequence of interesting generalizations of the previous formulations for secondary flow. In other cases, when general curvilinear coordinates were directly used to obtain the transformed equations (in terms of either contravariant or normalized contravariant components), a coordinate frame has been employed. This frame consists of vectors tangent to the respective directions of the coordinate curves. In certain applications, however, it has been found that a solution frame consisting of Cartesian directions is preferable to the coordinate frame (Refs. 13 and 14). In the applications it was referred to as the application of "Cartesian velocity components". Here, the advantages were both a lower required level of differentiability for the coordinates and a better alignment with a far field flow direction. However, in a numerical solution to a problem in curvilinear coordinates, the coordinate frame must always be retained since ultimately it is used for purposes of discretization. That is, the application of derivative and solution frames does not eliminate the use of the coordinate frame. Moreover, the coordinate frame can be used as a local reference frame. In a similar vein, the development of the approximation frame (and the associated approximation) is independent of the other frames and can be referenced to the coordinate frame. In particular, other frames can be applied simultaneously with the approximation frame. For example, different solution, derivative, and approximation frames can be used simultaneously.

THE VISCOUS STRESS APPROXIMATION

To suitably approximate the stress tensor, only the viscous part will be considered since it is the diffusive fluxes which are to be approximated. For subsonic cases, the remaining part of the stress tensor (the pressure field) must be treated carefully. This includes cases where there are both subsonic and supersonic regions. On the viscous stress, the approximation frame is used to determine which parts are to be removed; hence, which parts should contribute little to a flow with an assumed primary direction. To start the construction, this direction is assumed to be given in the form of a vector field $\vec{\alpha}_3$ which identifies a primary direction at each spatial point. Then $\vec{\alpha}_3$ can be extended to an orthogonal triple of vectors $\vec{\alpha}_1$, $\vec{\alpha}_2$, $\vec{\alpha}_3$ to form a frame at each spatial point. This extension must be accomplished in a smooth enough fashion over the whole flow field so that at least one continuous derivative can be taken. The required differentiability occurs because of the requirement to differentiate the components of the stress tensor as they would occur in the conservation laws for momentum, vorticity, or energy. From the construction, a smooth field of orthogonal frames is obtained such that $\vec{\alpha}_3$ is aligned with the assumed primary flow direction and $\vec{\alpha}_1$, $\vec{\alpha}_2$ span orthogonal transverse planes. Altogether, this is the desired approximation frame in which a differential viscous stress cube can be formed (Fig. 1) at each spatial point. The resulting components of viscous stress on a cube surface are either aligned with or are orthogonal to the primary direction. Consequently, the force balances represented by the Navier-Stokes equations are effectively separated into three mutually exclusive directions so that approximations in any given direction do not directly affect other directions. That is, a force in any one of the directions does not project nontrivially onto the other two remaining directions. If the equations of motion were written for the isolated cube, then the stress components would only contribute to the force balance in their respective directions. However, in the primary direction $\vec{\alpha}_3$, the viscous contribution σ^{33} is expected to add little to the strong convective forces; hence, this contribution is ignored, and as a result, the α_3-direction momentum balance is not elliptic in character. In addition, for the $\vec{\alpha}_1$ and $\vec{\alpha}_2$ directions the contribution of the viscous shearing stresses σ^{31} and σ^{32}

are also small relative to convective forces; hence, they also are ignored, and as a result, the respective momentum balances each are not elliptic in character. By contrast, the symmetric counterparts σ^{13} and σ^{23} each add to the force balance in the primary direction, and since they are the primary mechanism for the viscous retardation of the flow they are significant; hence, they cannot be neglected. Due to the orthogonality of the approximation frame, the force balances are mutually exclusive. Moreover, by joining existing cubes along transverse faces, effectively longer and longer viscous stress cubes can be formed. On a differential level, this is possible since the total assumption is that viscous forces on transverse faces are negligible. As a conceptual device, these forces can be considered on a fiber-like object which is aligned with the assumed primary direction. From this viewpoint, only internal viscous forces within the fiber are neglected. That is, the fiber has no stiffness; therefore, the only balance against the convective forces is due to the shearing stress along its boundary. This is particularly appropriate when the fiber is in the boundary layer since a no-slip condition causes the fluid to decelerate from viscous forces and come to rest at the walls (see Fig. 1).

In the approximation frame $\vec{\alpha}_1$, $\vec{\alpha}_2$, $\vec{\alpha}_3$ the full viscous part of the stress tensor σ is given by the equation

$$\sigma = \sigma^{ij}\,\vec{\alpha}_i \otimes \vec{\alpha}_j \tag{1}$$

in which the approximation is given by the assumption that the components σ^{3j} are negligible relative to the convective forces for $j = 1,2,3$. In the equation the summation convention of summing like indices has been applied. Consequently, the approximate stress tensor $\tilde{\sigma}$ is algebraically expressed by

$$\tilde{\sigma} = \tilde{\sigma}^{ij}\vec{\alpha}_i \otimes \vec{\alpha}_j = (1 - \delta_i^3)\,\sigma^{ij}\vec{\alpha}_i \otimes \vec{\alpha}_j \tag{2}$$

where δ_i^3 is the kronecker delta which is defined to be unity for equal indices and zero otherwise. Unlike the original tensor from Eq. 1, this new tensor in Eq. 2 is not symmetric. However, when it is inserted into the given equations of motion, the desired approximate set of governing equations is obtained. Such equations are not elliptic and can be solved by a foward marching procedure. In cases involving subsonic flow, the elliptic character which has been removed must be replaced in the form of some _a priori_ knowledge about the flow. This typically has been done with a special treatment for the pressure field (Refs. 8, 15-17).

THE FORM OF THE APPROXIMATION

Before the approximation as presented in Eq. 2 can be applied, it must be put into a form which is consistent with the chosen system of conservation laws governing the transport phenomena. Within the conservation laws, the stress tensor is usually expressed in component form where the components are the coefficients of the stress in a tensor product basis determined by some frame field $\vec{\beta}_1$, $\vec{\beta}_2$, $\vec{\beta}_3$. This field is typically chosen to be the solution or derivative type as previously discussed. To achieve the desired consistency, the stress components must be taken from the β-frame, transformed into the α-frame, approximated in the α-frame by Eq. 2, and then transformed back into the β-frame so that the results can be used. To start, let w^{ij} be the coefficients of the viscous part of the stress tensor in the β-frame, let $\vec{\alpha}_j = A^i_j\vec{\beta}_i$ be the transformation into the α-frame, and let $\vec{\beta}_i = B^k_i\vec{\alpha}_k$ be the inverse transformation back into the β-frame. The transformations denoted by A and B are technically referred to as elements of a Lie Group which is the structural group for a given bundle of frames (Ref. 18). In terms of this group, the transformation into the α-frame is given by

$$w^{ij}\vec{\beta}_i \otimes \vec{\beta}_j = w^{ij}B^r_i B^s_j\,\vec{\alpha}_r \otimes \vec{\alpha}_s \tag{3}$$

where the multilinearity of the tensor product has been used. Within the tensor product basis determined by the approximation frame α, an application of the approximation from Eq. 2 yields the result

$$(\widetilde{w^{ij}B_i^r B_j^s})\,\vec{\alpha}_r \otimes \vec{\alpha}_s = (1-\delta_r^3)\, w^{ij}B_i^r B_j^s\,\vec{\alpha}_r \otimes \vec{\alpha}_s$$

$$= \sum_{r=1}^{2} w^{ij}B_i^r B_j^s\,\vec{\alpha}_r \otimes \vec{\alpha}_s \tag{4}$$

The transformation back into the β-frame will finish this process. The details are given by the sequence of equalities.

$$(\widetilde{w^{ij}B_i^r B_j^s})\,\vec{\alpha}_r \otimes \vec{\alpha}_s = \sum_{r=1}^{2} w^{ij}B_i^r A_r^m B_j^s A_s^k\,\vec{\beta}_m \otimes \vec{\beta}_k$$

$$= \sum_{r=1}^{2} w^{ij}B_i^r A_r^m \delta_j^k\,\vec{\beta}_m \otimes \vec{\beta}_k \tag{5}$$

$$= \sum_{r=1}^{2} w^{ik}B_i^r A_r^m\,\vec{\beta}_m \otimes \vec{\beta}_k$$

$$= (\delta_i^m - B_i^3 A_3^m)\, w^{ik}\vec{\beta}_m \otimes \vec{\beta}_k$$

This process is summarized in Fig. 2. The coefficient from the last equality is the desired approximation of the viscous stress since it is in a form that can be directly inserted into a given system of conservation laws which govern the fluid motion.

APPLICATIONS

Several applications of the approximation shall be now be mentioned. In each case the central part is the selection of a vector field which adequately defines a primary flow direction. Since the most accurate approximation would result from the velocity field itself, this is a natural selection. One procedure would be to directly use the Frenet frame for the streamlines. This is also the closest approach to the secondary flow equations as presented by Lakshminarayana and Horlock (Ref. 12). However, the Frenet frame is not always well-defined. Certainly problems would arise when the streamlines inflect or have local regions of uniform flow. Each of these possibilities would cause the Frenet frame to suddenly change orientation or lose the distinction between unique normal and binormal directions. As a result, the desired continuity of the frame would be lost in such cases. Consequently, a preferable method would be to use the velocity vector field and to construct a differentiable orthogonal frame around it. This could be chosen to coincide with the Frenet frame in certain regions. Alternatively, the primary direction vector field could be selected as some approximation to the viscous velocity field. For example, the directions from an inviscid potential flow solution could be used. Often, the potential flow directions are just the tangent vectors to the streamline family of coordinate curves in a potential flow coordinate system. Moreover, other primary directions can be selected as tangent vectors to some given family of coordinate curves in some other system of coordinates. When this is done, the approximation is independent of the remaining coordinate directions. In particular, a transverse face of the stress cube (Fig. 1) is not aligned with the tangent plane to the transverse coordinate directions. In a number of cases, a suitable system is given by the tube-like coordinates which were developed by this author in Ref. 15. The associated transformation is given by

$$\vec{x} = \vec{\gamma}(t) + L(r,\theta,t)\left[\hat{n}(t)\cos\phi(\theta,t) + \hat{b}(t)\sin\phi(\theta,t)\right] \tag{6}$$

where $\vec{\gamma}$ is a space curve centerline, $\hat{n}$ and $\hat{b}$ are normal and binormal to $\vec{\gamma}$, L is a length factor which is a measure of radial distance, ϕ is an angular distribution, $(y^1, y^2, y^3) = (r, \theta, t)$ are the curvilinear variables, and $\vec{x} = (x^1, x^2, x^3)$ is the vector of Cartesian variables. The metric data for the tube-like coordinates are given in Fig. 3 and with more detail in Ref. 15. Moreover, if the centerline were to be considered as a bounding surface, then the tube-like coordinates would be a system of the type given by Eq. 15 in Ref. 19.

The natural tangent field to the curves with constant r and θ determines the desired approximation to the primary direction. From this field, the nonzero elements of a suitable A-matrix (in Eq. 5) are found to be $A_1^1 = -g_{33}$, $A_1^3 = g_{31}$, $A_3^3 = 1$, and $A_2^1 = g^{12}$ where g_{ij} and g^{ij} are the respective elements of the metric tensor and its inverse (Ref. 18). The elements of the B-matrix (of Eq. 5) are found by taking the inverse of the A-matrix. In a general curvilinear coordinate system with the derivative and solution frames taken as the coordinate frame, the Navier-Stokes equations are presented in Ref. 19 (pp. 325-328) both in expanded and in conservation law form. By direct substitution of the approximate viscous stress (Eq. 5), the unidirectional approximation for the steady state conservation of momentum is given by

$$\frac{\partial}{\partial y^j}\left\{\left[\rho v^i v^j + g^{ij}p + (\delta_m^j - B_m^3 A_3^j)\, w^{mi}\right]\frac{\partial x^s}{\partial y^i}\sqrt{g}\right\} = 0 \tag{7}$$

where ρ is density, p is pressure, $g = \det(g_{ij})$, and v^i is a contravariant component of velocity. Due to spatial variations in the metric, the contravariant components of velocities are, in some cases, not well suited to numerical computations. The metric is a measure of the lengths and angles within the coordinate frame, and since the contravariant components are coefficients of velocity within this frame, its spatial variation would add error to a numerical solution procedure. A trivial correction would be to normalize each of the frame vectors to unit length. This tact, however, would only remove variations along coordinate curve directions but would not help when severe angular variations occur. With tube-like coordinates, a better choice is evident. Let $\vec{e}_i$ denote the natural vectors $\partial/\partial y^i$ of the coordinate frame. Then choose the solution frame defined by $\vec{\beta}_1 = g^{1i}\vec{e}_i/\sqrt{g^{11}}$, $\vec{\beta}_2 = \vec{e}_2/\sqrt{g_{22}}$, and $\vec{\beta}_3 = \vec{e}_3/\sqrt{g_{33}}$. It is easy to check that each $\vec{\beta}_i$ is of unit length, that $\vec{\beta}_2$ and $\vec{\beta}_3$ respectively, point along the θ and t directions, and that $\vec{\beta}_1$ is orthogonal to the plane spanned by $\vec{\beta}_2$ and $\vec{\beta}_3$. With the assumption of mild axial variations in geometry (which is needed for the basic approximation), the spatial variations in $\vec{\beta}_3$ would be small. By the construction of $\vec{\beta}_1$, the possibly large angular variations between the r and θ directions in the $\vec{e}_i$-frame are removed. Moreover, when the frame is located at a boundary point, $\vec{\beta}_2$ and $\vec{\beta}_3$ span the tangent plane to the boundary at the given point. Consequently, this choice of solution frame is decoupled from the outward r-direction but is otherwise attached to the directions along constant r surfaces so that the assumed primary direction (which coincides with $\vec{\beta}_3$) is closely followed. By contrast, with the motivation of previous successes in other problems (e.g., Refs. 13-14), had a Cartesian solution frame been chosen, then the alignment with the assumed primary direction would not have been as good. The Cartesian frame would, of course, be orthonormal and have no spatial variations. Once the solution frame $\vec{\beta}_1$, $\vec{\beta}_2$, $\vec{\beta}_3$ is chosen as a linear function $\vec{\beta}_i = C_i^j\vec{e}_j$ of the coordinate frame $\vec{e}_1$, $\vec{e}_2$, $\vec{e}_3$, the velocity vector becomes $\vec{v} = v^i\vec{e}_i = v^i(C^{-1})^j_i\vec{\beta}_j \equiv u^j\vec{\beta}_j$ where the u^j are the components in the solution frame. The relationship between the contravariant components and the components in the solution frame is then given by $u^k = C^k_j u^j$. Upon substitution into Eq. 7, the approximate momentum equations are obtained with the new solution frame. A parallel analysis is also valid for both the energy and the vorticity equations.

With tube-like coordinates and contravariant velocity components, viscous flows have been obtained by numerically marching a solution of the approximate equations along a duct with variable cross sectional and axial geometry (Refs. 16-17). The approximate equations consisted of a continuity equation ($\partial(\rho v^i\sqrt{g})/\partial y^i = 0$) and a system of momentum equations (Eq. 7). Since constant total temperature was assumed, there was no energy equation. However, there were further modifications to include an eddy viscosity model of turbulence. This work was supported by NASA Lewis Research Center under Contracts NAS3-17522 and NAS3-19856.

REFERENCES

1. S. V. Patankar and D. B. Spalding, Internat. J. Heat Mass Transfer 15 (1972), 1787-1806.

2. L. S. Caretto, R. M. Curr, and D. B. Spalding, Comput. Meth. Engrg. 1 (1973), 39-51.

3. S. V. Patankar, V. S. Pratap, and D. B. Spalding, J. Fluid Mech. 62 (1974), 539.

4. S. G. Rubin and T. C. Lin, J. Computational Phys. 9 (1972), 339.

5. T. C. Lin and S. G. Rubin, J. Fluid Mech. 59, pt. 3 (1973), 593.

6. W. S. Helliwell and S. C. Lubbard, Computers and Fluids 3 (1975), 83.

7. J. V. Rakich and S. C. Lubbard, NASA SP-347 (1975).

8. W. R. Briley, J. Computational Phys. 14 (1974), 8-28.

9. H. McDonald and W. R. Briley, J. Computational Phys. 19 (1975), 150-178.

10. A. Jameson, Comm. Pure Appl. Math 27 (1974).

11. E. Murman and J. D. Cole, AIAA Journal 9 (1971).

12. B. Lakshminarayana and J. Horlock, J. Fluid Mech. 59 (1975), 97-115.

13. J. F. Thompson, F. Thames, and C. Mastin, J. Computational Physics 15 (1974), 299-319.

14. A. Rizzi and H. Bailey, Proc. 5th ICFD, 347-352.

15. P. R. Eiseman, H. McDonald, and W. R. Briley, United Technologies Research Center Report R75-911737-1 for NASA Lewis Research Center Contract NAS3-17522 (1975).

16. P. R. Eiseman, R. Levy, H. McDonald, and W. R. Briley, NASA CR 3029 (1978).

17. R. Levy, H. McDonald, and W. R. Briley, 6th ICFD.

18. R. Bishop and R. Crittenden, Geometry of Manifolds, Academic Press (1963).

19. P. R. Eiseman, J. Computational Phys. 26 (1978), 307-338.

VISCOUS STRESS CUBE

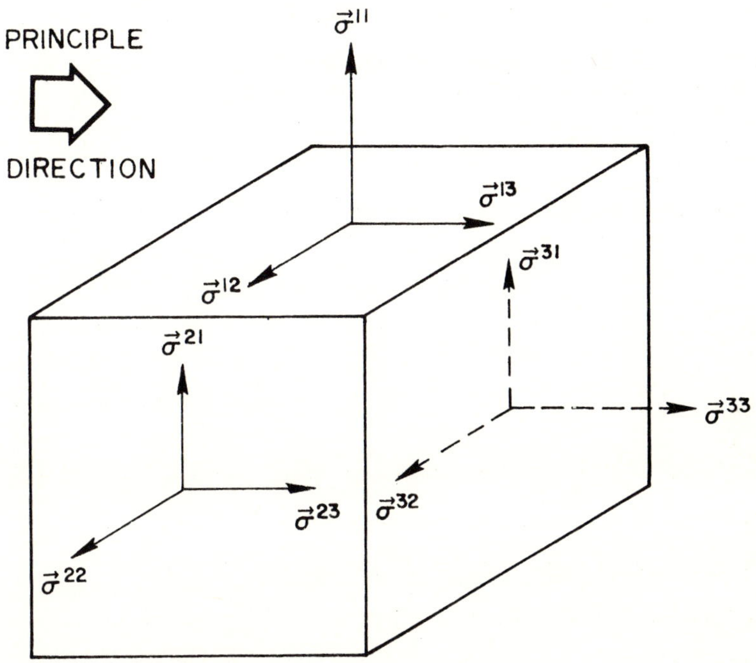

$\sigma = \sigma^{ij} \vec{a}_i \otimes \vec{a}_j$

APPROXIMATION = $\alpha^{3i} = 0$ FOR $i = 1, 2, 3$
RESULTING TENSOR IS NON-SYMMETRIC
THE EFFECTIVE DIFFERENTIAL
ELEMENT IS A FIBER-LIKE THING

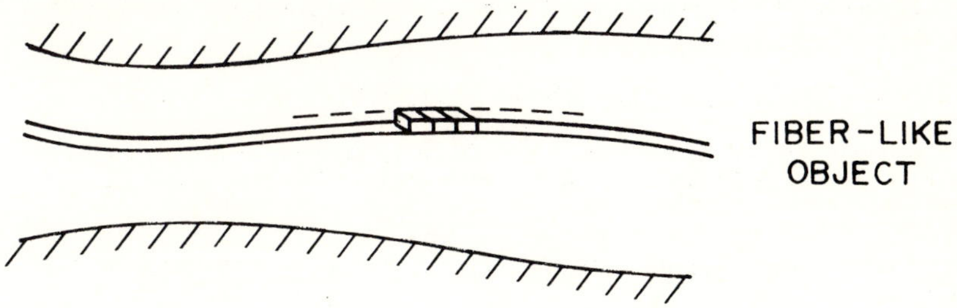

FIGURE 1

THE VISCOUS STRESS APPROXIMATION SCHEME

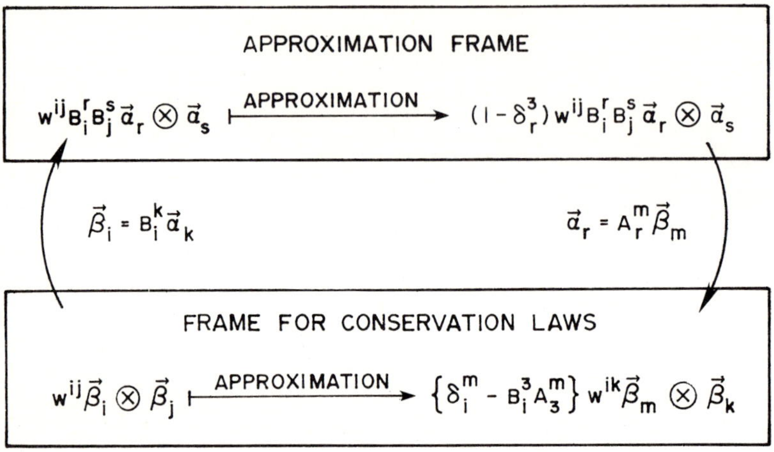

FIGURE 2

TUBE - LIKE COORDINATES

$$(x^1, x^2, x^3) = \vec{\gamma}(t) + L(\theta, r, t)\left\{\hat{n}(t)\cos\phi(\theta,t) + \hat{b}(t)\sin\phi(\theta,t)\right\}$$

METRIC

$$g_{\theta\theta} = \left(\frac{\partial L}{\partial\theta}\right)^2 + \left(L\frac{\partial\phi}{\partial\theta}\right)^2 \qquad g_{\theta r} = \frac{\partial L}{\partial\theta}\,\frac{\partial L}{\partial r}$$

$$g_{rr} = \left(\frac{\partial L}{\partial r}\right)^2 \qquad g_{\theta t} = \frac{\partial L}{\partial\theta}\,\frac{\partial L}{\partial t} + L^2\frac{\partial\phi}{\partial\theta}A$$

$$g_{tt} = B^2 + \left(\frac{\partial L}{\partial t}\right)^2 + (LA)^2 \qquad g_{rt} = \frac{\partial L}{\partial r}\,\frac{\partial L}{\partial t}$$

$$\sqrt{g} = L\frac{\partial L}{\partial r}\,\frac{\partial\phi}{\partial\theta}B$$

$$A = \frac{ds}{dt}\tau + \frac{\partial\phi}{\partial t}$$

$$B = \frac{ds}{dt}(1-\kappa L\cos\phi)$$

FIGURE 3

A FINITE ELEMENT FORMULATION SUITABLE FOR SUBSONIC AND TRANSONIC FLOW

Clive A.J. Fletcher
University of Sydney
Australia

INTRODUCTION

Finite element methods have been applied to flow problems, with the greatest success, where the governing equations are elliptic in character; for such flows finite element methods are competitive with most finite difference methods. For problems like transonic inviscid flow, where for a small part of the region the governing equations are hyperbolic in character, finite element methods, based on a conventional interpretation of the method of weighted residuals, lack the flexibility to allow special treatment of the locally supersonic region.

The current formulation considers a modified Method of Weighted Residuals, in which an additional weighting function, $T(x,y)$, is included in the form

$$\iint T_i(x,y) \cdot W_i(x,y) \cdot R(x,y)\, dx\, dy = 0, \quad i = 1, n, \tag{1}$$

$T_i(x,y)$ is defined on a local basis and contains free parameters that are chosen so that in the supersonic region the correct lack of upstream influence is achieved. The current formulation is suitable for solving the flow past aerofoils in a subsonic freestream for which embedded regions of supersonic flow occur.

Most previous finite element treatments of subsonic, inviscid flow e.g. [1,2] have been in terms of the velocity potential or stream function. Since the governing equations (in this formulation) are highly nonlinear, a large number of triple summations must be evaluated at each step of the iteration. This increases the computation time and leads to a relatively dense stiffness matrix when compared to a finite difference formulation of the same problem. If the governing equations (for the velocity potential) are cast in a variational form it is straightforward to include all the terms associated with compressibility as the source term of a Poisson equation. However the iterative solution of the Poisson equation becomes less efficient and fails at a local Mach number just above unity. Since the pressure distribution at the body surface is usually required the solution for the velocity potential must still be numerically differentiated before the pressure can be obtained. For the above reasons the present formulation is based on the full, inviscid equations of motion in primitive variable form.

FINITE ELEMENT FORMULATION

The governing equations for the primitive variables are written in conservation form as

$$(\rho u)_x + (\rho v)_y = 0 \tag{2}$$

$$(\rho u^2 + P)_x + (\rho uv)_y = 0 \tag{3}$$

$$(\rho uv)_x + (\rho v^2 + P)_y = 0. \tag{4}$$

The pressure is related to the density by the isentropic relationship

$$P = k\rho^{\gamma}. \tag{5}$$

Boundary conditions for equations (1) to (3) are that there is no flow through the body surface and that the farfield conditions (u,v,ρ,P) are specified a large distance from the body. Examination of equations (1) to (3) indicates that the equations are linear in the groups of variables and all have the same form

$$\frac{\partial F}{\partial x} + \frac{\partial G}{\partial y} = 0. \tag{6}$$

The form of equation (6) is taken advantage of in the analytic representations introduced at the element level. Thus

$$\begin{Bmatrix} \rho u \\ \rho v \\ \rho u^2 \\ \rho uv \\ \rho v^2 \\ P \end{Bmatrix} = \sum_j N_j(x,y) \cdot \begin{Bmatrix} (\overline{\rho u})_j \\ (\overline{\rho v})_j \\ (\overline{\rho u^2})_j \\ (\overline{\rho uv})_j \\ (\overline{\rho v^2})_j \\ \overline{P}_j \end{Bmatrix} \tag{7}$$

where $N_j(x,y)$ is a Serendipity rectangular shape function appropriate to the jth node and the overbar indicates the nodal value.

The technique of representing groups of variables rather than single variables has been used before in applying the Orthonormal Method of Integral Relations to supersonic boundary layer flow [3]. However this is the first time such a technique has been applied as part of a finite element formulation. An order-of-magnitude analysis [4] indicates that each group of variables should have the same order of representation. This is in contrast to the situation for incompressible, viscous flow at low Reynolds number in which a second-order velocity representation is usually combined with a first-order pressure presentation [5]. In the current investigation introduction of a Galerkin formulation led to a singular Jacobian and this may have been connected with using the same representation for ρu, ρv as for P.

Representation of groups of variables rather than single variables requires less algebraic manipulation and a lower order numerical scheme. Thus it requires less computation time. It may be noted that the resulting algebraic equations will be linear in the nodal groups of variables. However there are more unknown groups than equations and in relating the groups to the solution variables ($\overline{\rho}_i$, $(\overline{\rho u})_i$ and $(\overline{\rho v})_i$) the nonlinearity reappears.

Substitution of equations (7) into equations (2) to (4) produces the residuals

$$R^{(1)} = \sum_j \frac{\partial N_j}{\partial x} \cdot (\overline{\rho u})_j + \sum_j \frac{\partial N_j}{\partial y} \cdot (\overline{\rho v})_j$$

$$R^{(2)} = \sum_j \frac{\partial N_j}{\partial x} \cdot (\overline{\rho u_j^2} + \overline{P}_j) + \sum_j \frac{\partial N_j}{\partial y} \cdot (\overline{\rho uv})_j \tag{8}$$

and
$$R^{(3)} = \sum_j \frac{\partial N_j}{\partial x} \cdot (\overline{\rho uv})_j + \sum \frac{\partial N_j}{\partial y} \cdot (\overline{\rho v_j^2} + \overline{P}_j) \ .$$

As indicated in the introduction the Method of Weighted Residuals is applied in a generalised form

$$\iint T_i(x,y) \cdot W_i(x,y) \cdot R(x,y)\, dx\, dy = 0, \tag{9}$$

where W_i is the traditional weighting function and T_i is an auxiliary weighting function introduced to permit the lack of upstream influence in the supersonic region to be correctly modelled. As a consequence the shock that terminates the supersonic region may be expected to appear in the solution without the need for shock-fitting.

Both the Galerkin and least-squares formulations were applied to this problem but the Galerkin formulation combined with Newton's method failed to converge for a freestream Mach number greater than 0.32 [4]. The Method of Weighted Residuals will be developed here in terms of the least-squares formulation. Equation (9) becomes

$$\iint T_i(x,y) \cdot \frac{\partial R}{\partial \overline{q}_i}(x,y) \cdot R\, dx\, dy = 0 \tag{10}$$

The choice of T_i suitable for transonic flow will be considered for a one-dimensional, first-order model problem. Thus if the relevant part of equation (10) is

$$\int T_i(x) \cdot \frac{\partial N_i(x)}{\partial x} \left\{ \ldots \frac{\partial N_{i-1}}{\partial x} \cdot \overline{F}_{i-1} + \frac{\partial N_i}{\partial x} \cdot \overline{F}_i + \frac{\partial N_{i+1}}{\partial x} \cdot \overline{F}_{i+1} \ldots \right\} dx,$$

it is necessary to show that with a suitable choice for the auxiliary weighting function, T_i, the resulting algebraic equation formed at the ith node does not depend on $\overline{F}_{i+1}$. T_i is chosen to have the form

$$T_i(x) = 1 - a \cdot x \cdot (1 + x) \quad \text{for } -1 \leqslant x \leqslant 0$$
$$= 1 - a \cdot x \cdot (1 - x) \quad \text{for } 0 \leqslant x \leqslant 1$$

where $x = -1$ corresponds to the $(i-1)^{th}$ node

$x = 0$ corresponds to the ith node

and $x = 1$ corresponds to the $(i+1)^{th}$ node

Evaluation of the various terms in equation (10) leads to

$$- \left[1 + \frac{a}{6}\right] \cdot \bar{F}_{i-1} + 2 \cdot \bar{F}_i - \left[1 - \frac{a}{6}\right] \cdot \bar{F}_{i+1} \tag{11}$$

Thus setting a = 6 removes any influence of the $(i+1)^{th}$ node on the behaviour at the ith node. The extension of the above to two dimensions and quadratic shape functions leads to eight disposable constants per element. The treatment parallels the application of an "upwind-difference" scheme for the Galerkin formulation [6]. The disposable constants are chosen on a local basis to ensure no upstream influence in the region where the flow is locally supersonic.

For the subsonic region it is convenient to set T = 1 (a = 0) and use a conventional least-squares formulation. Equation (10) applied to equation (8) then becomes

$$\iint \left\{ \alpha_1 \cdot \frac{\partial R^{(1)}}{\partial \bar{q}_i} \cdot R^{(1)} + \alpha_2 \cdot \frac{\partial R^{(2)}}{\partial \bar{q}_i} \cdot R^{(2)} + \alpha_3 \cdot \frac{\partial R^{(3)}}{\partial \bar{q}_i} \cdot R^{(3)} \right\} dxdy = 0, \tag{12}$$

where α_1, α_2 and α_3 are scalars that may be used to adjust the significance of the various equations and $\bar{q}_i = \{(\overline{\rho u})_i, (\overline{\rho v})_i, \bar{p}_i\}$. Substitution of equation (8) into (12), and evaluation of the integrals, produces the following algebraic equations

$$S_i^{(m)} = \sum_j \left\{ r_{ij}^{(m)} \cdot (\overline{\rho u})_j + s_{ij}^{(m)} \cdot (\overline{\rho v})_j + t_{ij}^{(m)} \cdot (\overline{\rho u^2})_j + x_{ij}^{(m)} \cdot (\overline{\rho uv})_j \right.$$

$$\left. + y_{ij}^{(m)} \cdot (\overline{\rho v^2})_j + z_{ij}^{(m)} \cdot \bar{P}_j \right\} = 0, \quad m = 1, 3, \quad i = 1, n. \tag{13}$$

The coefficients, $r_{ij}^{(m)}$ etc., are algebraic functions of $\{a_{ij}, b_{ij}, c_{ij}, d_{ij}, u_i, v_i$ and $\rho_i\}$. a_{ij}, b_{ij}, c_{ij} and d_{ij} are algebraic coefficients which arise from the following integrations

$$\begin{aligned} a_{ij} &= \iint \frac{\partial N_i}{\partial x} \cdot \frac{\partial N_j}{\partial x} \, dx\, dy \\ b_{ij} &= \iint \frac{\partial N_i}{\partial y} \cdot \frac{\partial N_j}{\partial y} \, dx\, dy \\ c_{ij} &= \iint \frac{\partial N_i}{\partial x} \cdot \frac{\partial N_j}{\partial y} \, dx\, dy \\ d_{ij} &= \iint \frac{\partial N_i}{\partial y} \cdot \frac{\partial N_j}{\partial x} \, dx\, dy \,. \end{aligned} \tag{14}$$

Equations (14) have been evaluated in conjunction with an isoparametric formulation. The actual expressions for $r_{ij}^{(m)}$ etc. are given elsewhere [4].

Two techniques for the solution of equation (13) have been tried. Firstly it is a desirable feature of the least-squares formulation that it leads to a diagonally dominant stiffness matrix. By considering the dependence of $S_i^{(m)}$ on the diagonal term q_i it is possible to write down the following iterative algorithm of the relaxation type

$$q_i^{\nu+1} = q_i^{\nu} - \lambda \cdot \left[\frac{\partial S_i^{(m)}}{\partial q_i}\right]^{-1} \cdot S_i^{(m)}(\bar{q}^{\nu}). \tag{15}$$

λ is a scalar that can be used to increase the rate of convergence. The algorithm has the advantage of converging from any reasonable starting value and requiring relatively little storage since $[\partial S_i^{(m)}/\partial q_i]$ is a scalar and trivial to invert. However the iterative scheme is rather slow. Newton's method can also be applied to equation (13) which leads to the formula

$$\bar{q}^{\nu+1} = \bar{q}^{\nu} - \left[J_{\nu}\right]^{-1} \cdot S_i^{(m)}(\bar{q}^{\nu}), \tag{16}$$

where J_{ν} is the Jacobian, $\partial S_i^{(m)}/\partial \bar{q}$. Newton's method has the property of quadratic convergence but requires a large central memory to invert the Jacobian even when sparse storage is used [7]. Also the radius of convergence shrinks as the number of unknown, n, is increased [8]. For the flow about an aerofoil with a coarse grid (90 nodal unknowns) a comparison indicated that Newton's method was approximately four times faster than the relaxation method. However it was necessary to apply the relaxation method for ten steps before Newton's method was applied.

RESULTS

Numerical solutions have been obtained for the flow past a circular cylinder, an elliptic cylinder and two representative aerofoils. All the results presented have been obtained for subcritical flow with the relaxation method described above.

For the flow about the circular cylinder a polar grid has been introduced and an isoparametric formulation [7] has been utilised to relate this to a cartesian grid. The farfield boundary condition has been obtained by applying a Prandtl-Glauert coordinate transformation and using complex variable theory.

Results for a freestream Mach number of 0.4 are shown in Figure 1. The finite element solutions were obtained with 91 elements and 829 nodal unknowns. The surface velocity distributions are seen to be in good agreement with the series solutions of Imai [9] and Lush and Cherry [10].

Numerical solutions for the flow about a 2:1 elliptic cylinder at a freestream Mach number of 0.5 are shown in Figure 2. These results were obtained with 91 elements and 829 nodal unknowns. The numerical solution indicated subcritical flow everywhere. The results are compared with the Method of Lines [11] and the agreement is good. Solutions obtained with the Method of Integral Relations [12] indicate that the flow for this problem is just supersonic at the shoulder of the elliptic cylinder.

Results have been obtained for a 6% circular-arc aerofoil for various subcritical Mach numbers. The grid used for this case was obtained by mapping the region external to the aerofoil onto a rectangle in such a way that the internal grid lines are approximately parallel to and perpendicular to the streamlines. The pressure distribution for a freestream Mach number of 0.71 is shown in Figure 3. The finite element results were obtained with 102 elements and 868 nodal unknowns. The farfield boundary conditions for the numerical results shown in Figures 3 and 5 have been obtained by introducing a Prandtl-Glauert coordinate transformation and applying thin-aerofoil theory to the distorted body. The computational results shown in Figure 3 are compared with experimental results due to Knechtel [13]. For the smooth aerofoil results the boundary layer would have been laminar at least to the 50% chord point. For the aerofoil with roughness near the lending edge the boundary layer would have been turbulent throughout. The computational results are in good agreement with the experimental results for the smooth aerofoil, although it should be noted that no correction has been made for the displacement thickness effect.

Computational results have also been obtained for the flow about a NACA-0012 aerofoil. Since this aerofoil is blunt it has a forward stagnation point and hence a rapidly changing solution in the region adjacent to the stagnation point. To ensure that the elements do not become too distorted a grid is introduced (Figure 4) that follows the aerofoil surface and permits a larger density of nodes adjacent to the front stagnation point.

Results for a freestream Mach number of 0.4 are shown in Figure 5. These results were obtained with 88 elements and 767 nodal unknowns in the flowfield. For comparison experimental results due to Amick [14] and computational results due to Emmons [15] are also shown. The finite element results simulated an aerofoil in an unconstrained freestream. In contrast the results of references [14] and [15] shown in Figure 5 are for uncorrected wind-tunnel flow. If a wind-tunnel blockage correction were applied to these results the maximum negative pressure coefficient would be reduced by 4-5%. This would produce closer agreement between the finite element and finite difference solutions. The more negative pressure coefficient produced by the finite difference solution (compared with the experimental results) is consistent with the failure to take into account the displacement thickness effect.

CONCLUSION

A primitive variable, least-squares finite element formulation is proposed for inviscid, compressible flow. A unique feature of the formulation is the analytic representation of groups of terms rather than individual terms. Solutions of high accuracy have been obtained for two-dimensional, subcritical flow and the extension to transonic flow with the aid of a generalised least-squares formulation is not expected to create any special difficulties.

REFERENCES

[1] Periaux, J. "Three Dimensional Analysis of Compressible Potential Flows", I.J.N.M.E. 9, 775-833, 1975.

[2] Labrujere, Th.E. and van der Vooren, J. "Finite Element Calculation of Axisymmetric, Subcritical, Compressible Flow", TR-74162U, N.A.L., 1974.

[3] Fletcher, C.A.J. and Holt, M. "Supersonic Viscous Flow over Cones at Large Angles of Attack", J. Fluid Mech., 74, 561-591, 1976.

[4] Fletcher, C.A.J. "Subsonic, Inviscid Flow by the Finite Element Method", TR-1858(W), W.R.E., 1977.

[5] Olson, M.D. and Tuann, S.Y. "Primitive Variables versus Stream Function Finite Element Solutions of the Navier-Stokes Equations",

Proc. 2nd Int. Sym. F.E.M in Flow Problems, St. Margherita, Italy, June 1976.

[6] Heinrich, J.C. and Zienkiewicz, O.C. "Quadratic Finite Element Schemes for Two-Dimensional Convective Transport Problems", Dept. Civil Eng. Rep. C/R/284/76, University of Swansea, 1976.

[7] Fletcher, C.A.J. "The Application of the Finite Element Method to Two-Dimensional, Inviscid Flow", TN-1606 (WR&D), W.R.E., 1976.

[8] Rheinboldt, W.C. "On the Solution of Large, Sparse Sets of Nonlinear Equations", TR-324, Univ. Maryland, 1974.

[9] Imai, I. "On the Flow of a Compressible Fluid Past a Circular Cylinder", Proc. Phys. & Math. Soc., Japan, 23, 180-193, 1941.

[10] Lush, P.E. and Cherry, T.M. "The Variational Method in Hydrodynamics", J. Mech. App. Math., 9, 6-21, 1956.

[11] Jones, D.J., South, J.C. and Klunker, E.B. "On the Numerical Solution of Elliptic Partial Differential Equations by the Method of Lines", J. Com. Phys., 9, 496-527, 1972.

[12] Belotserkovskii, O.M. and Chushkin, P.K. "The Numerical Solution of Problems in Gas Dynamics", in Basic Developments in Fluid Dynamics (ed. M. Holt), 1, Academic Press, 89-126, 1965.

[13] Knechtel, E.D. "Experimental Investigation at Transonic Speeds of Pressure Distributions over Wedge and Circular-Arc Aerofoil Sections and Evaluations of Perforated Wall Interference", NASA TN-D-15, 1959.

[14] Amick, J.L. "Comparison of the Experimental Pressure Distribution on an NACA-0012 Profile at High Speeds with that Calculated by the Relaxation Method", NACA TN-2174, 1950.

[15] Emmons, H.W. "Flow of a Compressible Fluid Past a Symmetrical Airfoil in a Wind Tunnel and in Free Air", NACA TN-1746, 1948.

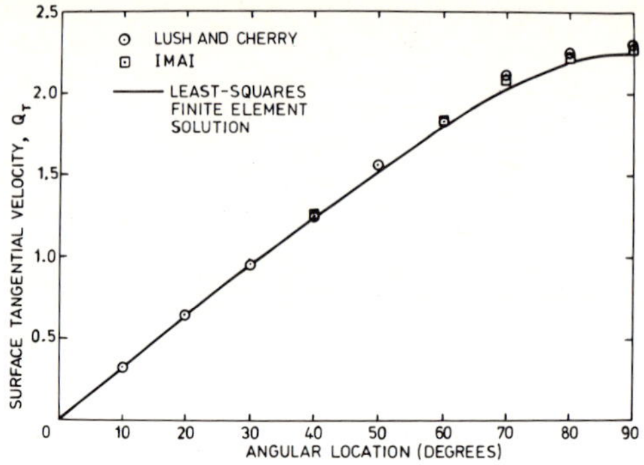

Figure 1 Surface velocity variation for the flow about a circular cylinder at $M_\infty = 0.40$

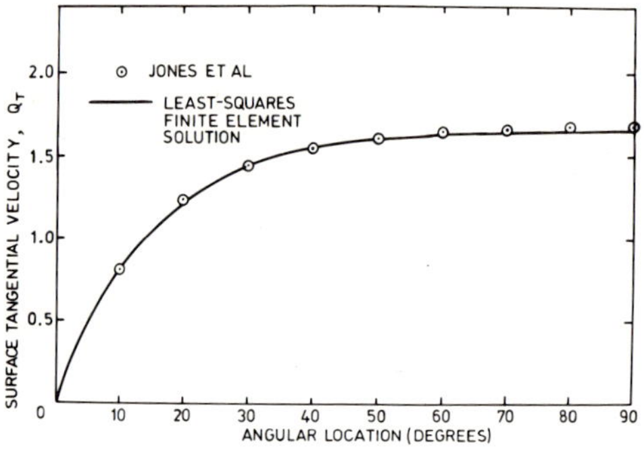

Figure 2 Surface velocity variation for the flow about a 2:1 elliptic cylinder at $M_\infty = 0.50$

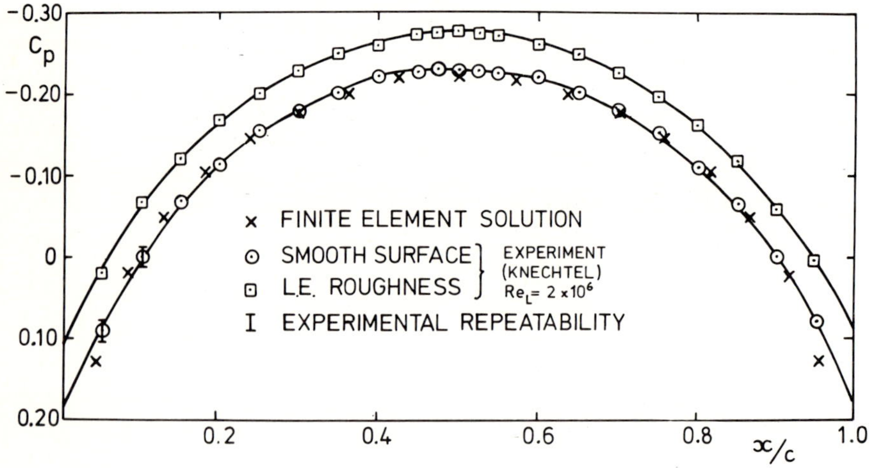

Figure 3 Pressure distribution on a 6% circular-arc aerofoil at $M_\infty = 0.71$

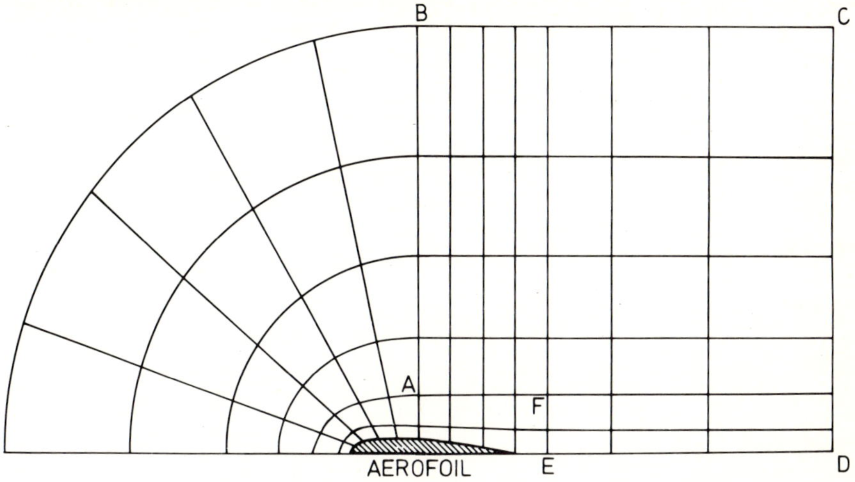

Figure 4 Schematic of grid used for the NACA-0012 aerofoil

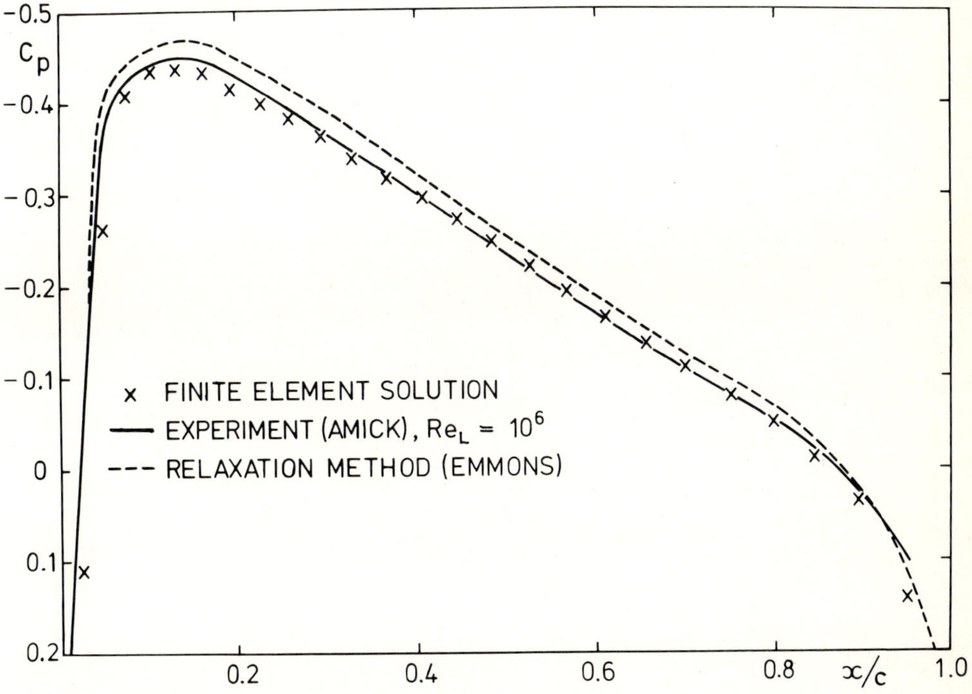

Figure 5 Pressure distribution on a NACA-0012 aerofoil at $M_\infty = 0.40$

TESTS OF COMPUTATIONAL ALGORITHMS FOR INVISCID HYPERBOLIC FLOWS

K.Förster

Institut für Aerodynamik und Gasdynamik der Universität Stuttgart

Stuttgart, Bundesrepublik Deutschland

1. Acknowledgments

This paper evaluates the results of a workshop on boundary algorithms for inviscid hyperbolic flows which was organized by the author at the end of February 1977 under the auspices of the GAMM-Fachausschuß für Numerische Methoden in der Strömungsmechanik, and supported by the Universität Stuttgart and the Deutsche Forschungsgemeinschaft. It constitutes the joint efforts of L.Zannetti (Torino), C.Weiland (Köln), L.Theilemann (Stuttgart), V.Rusanov (Moskva), K.Roesner (Karlsruhe), M.Pandolfi (Torino), T.deNeef (Delft), G.Moretti (NewYork), L.Karlsen (Stockholm), H.Frühauf (Stuttgart) and K.Förster (Stuttgart). Engaged in discussions were further H.Cabannes (Paris), E.Krause (Aachen), C.Morawetz (NewYork), W.Prosnak (Warszawa) and I.Teipel (Hannover).
A detailed report containing the original papers by the authors of the first group above will appear as a Workshop Report in the series "Notes on Numerical Fluid Mechanics" in the Vieweg-Verlag, Wiesbaden, in the fall of 1978.

2. Introduction

The general theme of the workshop was the testing of algorithms for inviscid hyperbolic flows in the most rigorous manner possible. So the first task was to select proper test problems which were sufficiently non-trivial and, on the other hand, could be calculated exactly or at least with sufficiently high (and known) precision.
As an example of continuous, homentropic flow the well-known Ringleb Flow was chosen. Regarded as steady, this flow in its supersonic part provides initial values as well as boundary conditions for well-posed two-dimensional Cauchy problems which are not as trivial as, for instance, the quasi one-dimensional source-sink flow or the Prandtl-Meyer expansion which were tried before as test problems (see for instance Abbett /1/). We used it to test some algorithms for 2D steady supersonic flow.
Regarded as unsteady (with time derivatives theoretically zero) the Ringleb Flow provides initial values and boundary conditions for 2D unsteady flows, covering subsonic, transonic and supersonic problems, all of which were regarded and discussed in the workshop. In this paper, however, instead of listing all cases treated, we will try to work a selection of results into a comparison of methods, and moreover tackle some questions concerning the singular nature of transonic Ringleb Flow

together with its parallels to supercritical airfoils.

For the testing of discontinuities, no truly two-dimensional analytical solution of the gasdynamic equations is known. For this case a model problem has been developed whose solution - where not analytical - is reduced to the solution of ordinary differential equations with high and controlled accuracy. Here too, an application is given.

3. The Ringleb Flow

Because Ringleb's solution $V\psi = \sin\theta$ of Tchapligyn's hodograph equation is very well known (/2/, /3/), we can restrict ourselves to listing (with reference to the figure) the variables and some working equations:

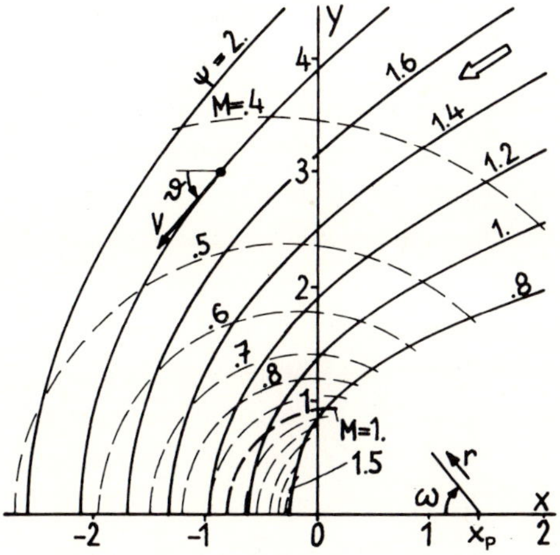

between

V, θ	-	velocity vector
M	-	Mach Number
ρ	-	density
p	-	static pressure
ψ	-	stream function

(nondimensionalized by the stagnation values p_0, ρ_0, $c_0 = \sqrt{1.4\, p_0/\rho_0}$ and an arbitrary length) we have with

$$\gamma := \sqrt{1 - V^2/5}$$

$\rho = \gamma^5$, $p = \gamma^7$, $M = V/\gamma$, $\theta = \arcsin(V\psi)$; and for the coordinates of the physical plane:

$$x = \frac{1}{\rho}\left(\frac{1}{2V^2} - \psi^2\right) + \frac{L}{2} \qquad \text{where} \quad L := \frac{1}{\gamma} + \frac{1}{3\gamma^3} + \frac{1}{5\gamma^5} - \frac{1}{2}\ln\frac{1+\gamma}{1-\gamma}$$

$$y = \pm\frac{\psi}{\rho V}\sqrt{1 - V^2\psi^2}\ .$$

The isotachs are circles given by $\left(x - \frac{L}{2}\right)^2 + y^2 = \left(\frac{1}{2\rho V^2}\right)^2$; they also represent all other isolines because the flow is both homentropic and homenthalpic. To facilitate the various evaluation problems encountered in using the Ringleb Flow for testing purposes, four FORTRAN subroutines were developed giving the properties of the flow depending on the location at a certain streamline or in the physical plane:

Name:	Input:	Output:
KWRING	y , ψ	V , x , ρ
KRINGL	x , y	V , ρ , ψ
PWRING	ω , ψ ; x_p	V , r , ρ
PRINGL	r , ω ; x_p	V , ρ , ψ

The listings of these subroutines are given in the Workshop Report.

Against the Ringleb Flow as a test problem the doubt has been raised that - according to a theorem by C.Morawetz /8/ - this flow is singular and has no neighboring solutions so that some of the workshop problems were not well-posed.

Concerning these objections the following can be stated:

1. The Morawetz Theorem applies to a compressible, inviscid, steady, irrotational, infinite, subsonic flow along a sufficiently smooth wall with an embedded finite (local) supersonic region and it states that the perturbation problem for this case is ill-posed, i.e. that the said type of flow can exist only if inside a certain interval around the point of maximum velocity the wall has a certain singular shape.
2. It follows immediately that all purely super- or subsonic channel- or jet- or mixed (jet along wall) flows derived configurationally from the Ringleb Flow are not struck by this theorem.
3. The Morawetz Theorem does not apply too, to viscous flow, and to quote herself /9/: "If the numerical algorithm admits a shock (say by means of artificial viscosity) so that the entropy condition is satisfied, then one expects the corresponding problem with weak shocks to exist, be unique and probably stable. There will be only one neighboring solution but it will have shocks. In this area there are no theorems, only conjectures, so far ..." . Now at least all finite difference methods contain viscosity terms in a broader sense, simply from the truncation of the Taylor Series, and so they do not solve the inviscid problem but a corresponding pseudo-viscous one. The effect indicated by Prof. Morawetz is nicely demonstrated, for example, by computations of the flow around socalled shockfree supercritical airfoils (the left figure below is composed from calculations done by A.Jameson), and from the workshop papers we can contribute an example where the locally supersonic flow between the Ringleb streamlines $\psi = 1.2$ and 0.8 was computed by a timedependent unsteady method of second order accuracy (last example from chapter 5) and hence with a pseudo-viscosity of third order. The similarities between these two examples are striking.

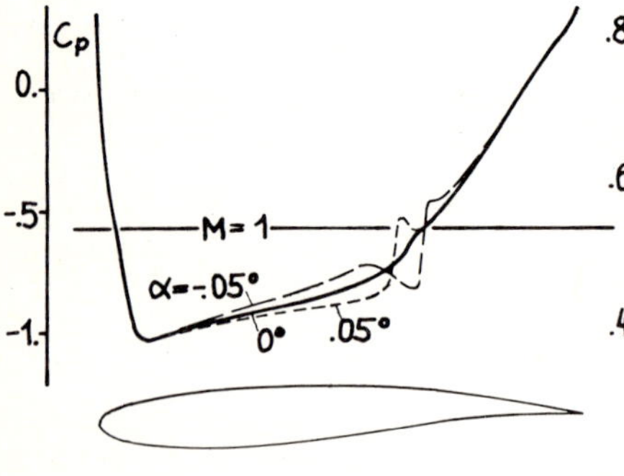

Supercritical airfoil with slightly varied angle of attack

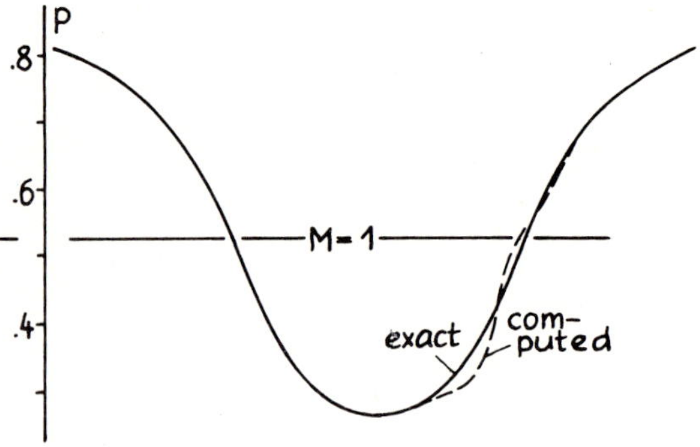

Transonic flow along Ringleb streamline with $\psi = 0.8$

4. Isolated boundary algorithm tests

First some tests of boundary algorithms proper for plane steady supersonic flow and then for plane unsteady flow are described. For these tests, the exact Ringleb values are given as initial values and the errors produced by one single application of the algorithm under test are sought.

The once popular reflection technique was not considered because it is valid only for special configurations such as a straight wall with perpendicular grid lines.

As one promising non-trivial algorithm of second order accuracy that of Abbett /1/, /4/ was tried (Weiland)†. It consists in predicting the new values at the wall by the first step of the well-known MacCormack scheme (onesided, first order accurate finite difference approximation) thus producing a velocity vector which will, in general, not fulfill the tangency condition imposed by the boundary. Then a reflected Prandtl-Meyer wave is assumed such that the velocity vector is turned into the proper direction. In a typical case the error in V turns out to be in the order of 10^{-4}, depending on the (prescribed) grid size. For an algorithm devised to simulate the physical behaviour of the flow (inciding wave taken care of by the onesided predictor, reflected one by the Prandtl-Meyer wave) this is a bit disappointing.

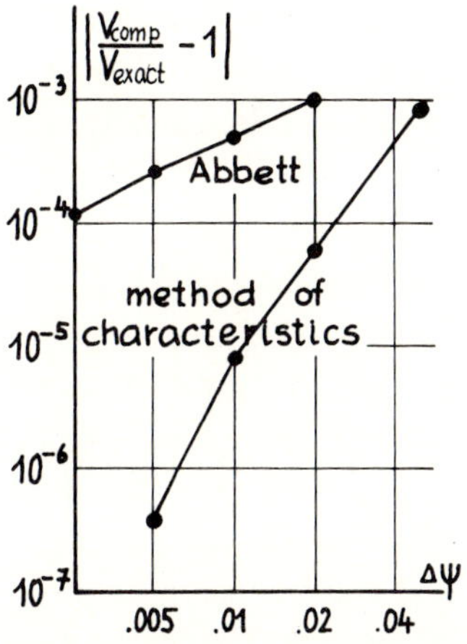

Better results are achieved by the true method of characteristics (Förster) based on the compatibility equation. Compared with Abbett's method the selected algorithm (of second order accuracy, too) gives considerably improved results.

A picture similar to the plane steady flow is exhibited by the unsteady case. As a "simple" algorithm we have here the reference plane method proposed by Moretti and Abbett /5/ some years ago, and a variant with a reference plane sloping along the streamline in the x,y,t-space (Theilemann). The common feature is that the derivatives normal to the reference plane are considered as small forcing terms and thus the threedimensional x,y,t-problem is formally reduced to a twodimensional one in an auxiliary wall-aligned coordinate frame which is solved by the method of characteristics (see figure next page). Both variants differ only with respect to the form of the compatibility equation, so the V-errors are the same in both methods. The p-errors, however, differ and it is interesting to note that obviously the interpolation error in the sloping plane method is much smaller than in the somewhat erratically behaving vertical plane case, though here the errors are about the factor 5 smaller.

† Names in parenthesises indicate the author of the corresponding Workshop paper

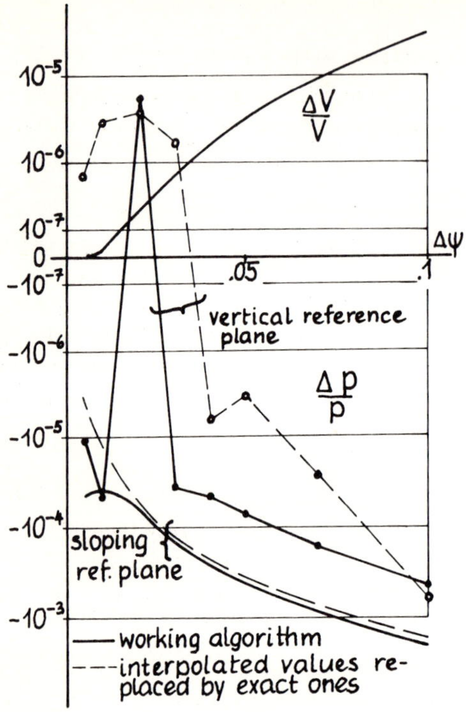

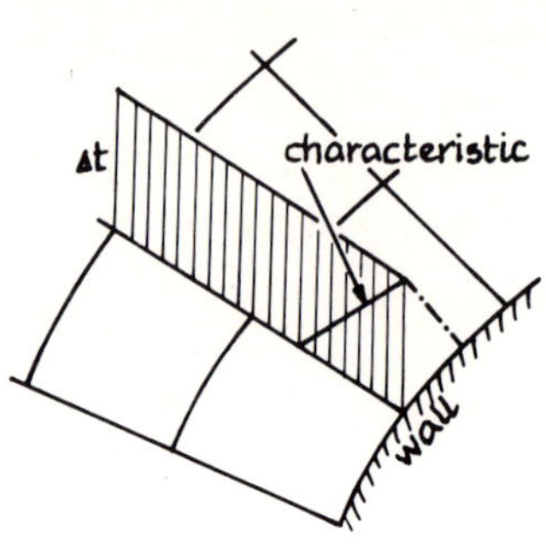

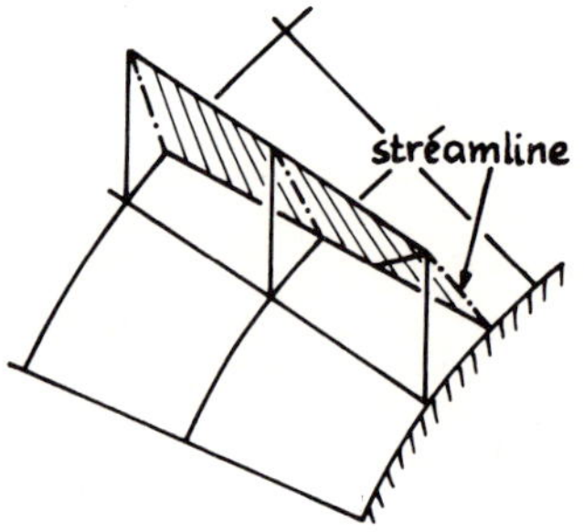

Much better again are the results of the - admittedly also more complicated - three-dimensional method of characteristics (Roesner). Here the ultimate in accuracy is gained by using the true characteristic surfaces for the computation and by carefully selecting the approximation formula for the solid wall boundary. Then for rather large and rather arbitrarily shaped triangles in the plane $t = 0$ which form the basis of the characteristic tetrahedron very high and uniform accuracy can be ob-

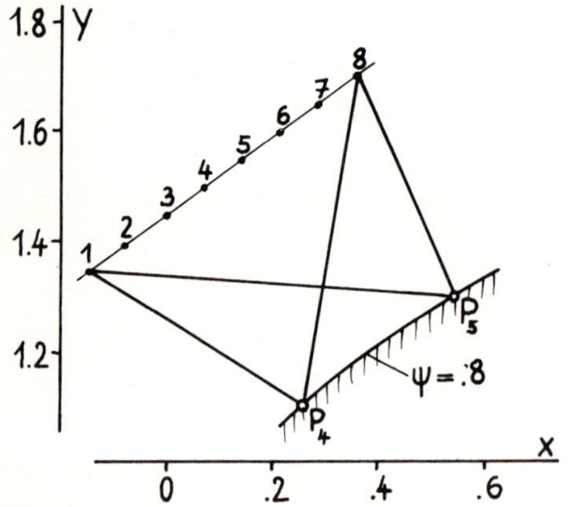

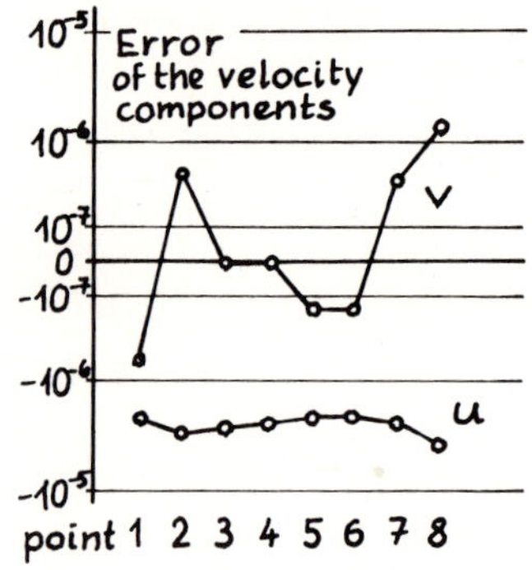

tained. The error diagram refers to a triangle defined by P_4, P_5 and 1 through 8; other triangles yield quite similar pictures.

5. Combined field/boundary tests

Though single-point tests as in the preceding chapter may reveal where an algorithm can be improved in detail, they are no proper means for judging the power of a method in calculating a whole field. This is especially true for "difficult" flows, e.g. transonic ones, but first we will have a look at the simplest case: steady supersonic flow.

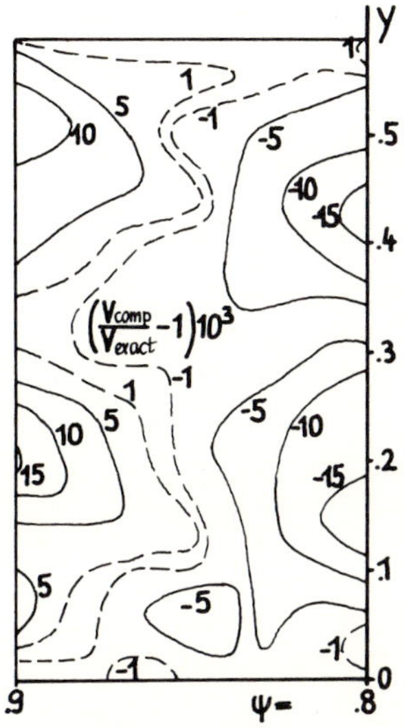

There is first a variant of the well-known Lax-Wendroff-Richtmyer scheme featuring a streamline-aligned coordinate system, and bred for good shock-capturing abilities (Förster). The boundary computation is incorporated into the field point algorithm: two of the three components of the solution vector are computed by linearly extrapolating their derivatives in the marching direction from inside the field to the wall and the third component follows from inserting the boundary condition into the energy equation. According to this rather crude concept which results in a very compact coding we get only a "technical" accuracy with a maximum V-error of 1.5 % for a grid with 10 meshes across the width of the channel. In case the Ringleb Flow is regarded as a jet (prescribing the pressure along the boundary) the errors are much smaller. - Better schemes result if at the wall the compatibility condition from the method of characteristics is used. It must not be expressed in derivatives along characteristic curves as has been done above but can be written with partial derivatives after the coordinates used in the field point calculation (Pandolfi and Zannetti). Inside the field MacCormack's algorithm was used. The figure gives examples for two different mesh sizes, the coordinate frame was a polar one.

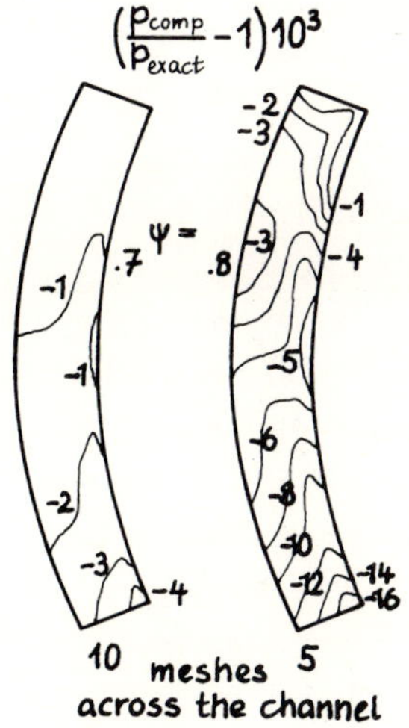

A further gain in accuracy is obtained if the special feature of homentropy is made use of and Riemann invariants are taken as the independent variables (deNeef). In this way, the compatibility condition takes a very simple form and by combining a predictor and a corrector step - both with linear interpolation - second order accuracy is obtained. Using a rather complicated boundary point calculation but only six meshes across the width of the channel the depicted Mach Number error distribution results (left figure next page). It compares favorably with those from the two variants of the MacCormack scheme (righthand predictor/lefthand corrector: central figure, inverse: righthand figure).

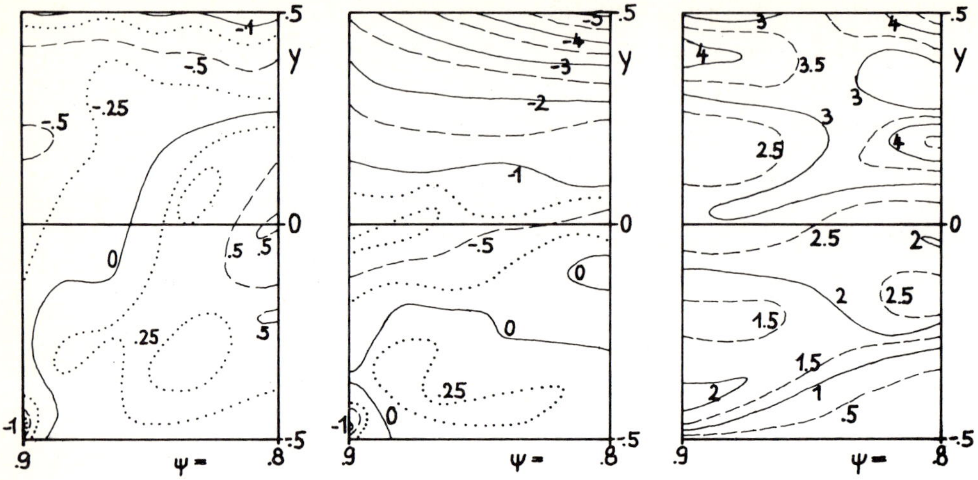

Quite another approach is sought in the BVLR- or "progonka" method† (Weiland). Here the boundary condition at the wall is - running outward along a line y=const - successively fed into the set of implicit difference equations arising from a (nearly) Crank-Nicholson discretization of the gasdynamic equations. At the outer boundary (for which here a characteristic was employed) the resulting equations are supplemented by the boundary condition valid there, and then they are solved for the new set of solution vectors on an inward running sweep; because of the nonlinearity of the equations the procedure must be repeated iteratively. The results are very promising, presumably because of the uniform treatment of field and boundary points. The programming logic, however, is clearly more complicated than in any other finite difference procedure. The best in accuracy can again be expected from the method of characteristics proper. In the following example the inverse variant proposed by Hartree has been used which produces a regular computational grid. In detail: the method is the twodimensional, second order accurate specialization of a threedimensional near-characteristics method after Sauer (Frühauf). With only eight meshes across the width of the channel the errors just peak $4 \cdot 10^{-5}$ which is indeed a very good result.

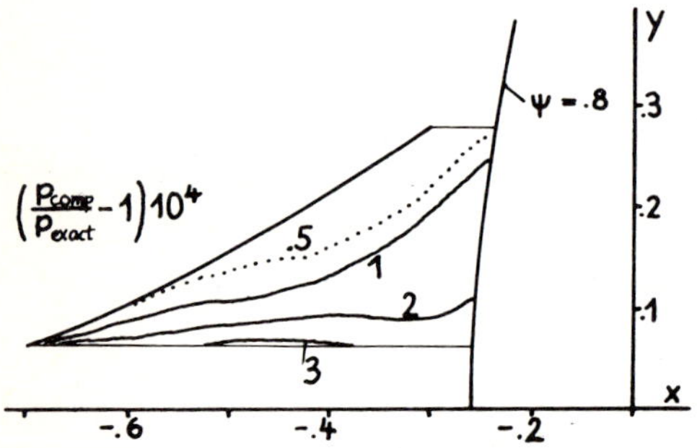

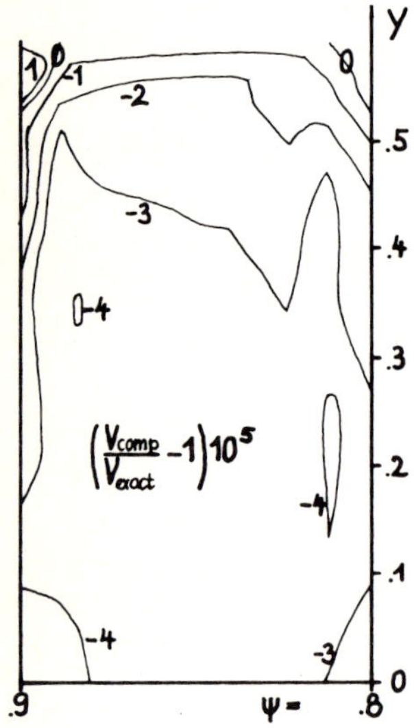

† Due to Babenko, Voskresenskii, Lyubimov and Rusanov /6/

Proceeding now to the difficult problem of unsteady transonic flow the first test to be presented has been done by Rizzi's time-split finite volume technique /7/ (Karlsen). This often-proven method circumvents the differential form of the basic equations and applies the integral conservation laws directly to small rectangular cells

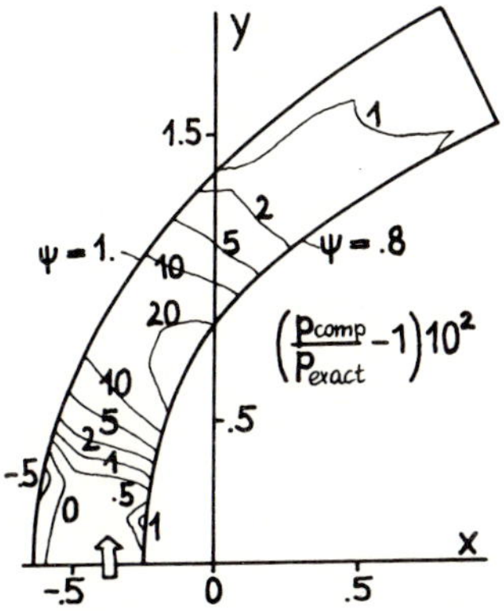

of arbitrary shape. For the example shown it was necessary to introduce a small fourth order damping term to counterbalance the sonic-point minimum of the fourth order dissipative term in the MacCormack scheme which was used to explicitly solve the finite volume flux equations. At the walls the pressure was computed from the normal-momentum equation (which computationally amounts to a Hermitian polynomial perpendicular to the wall) whereas the momentum equation along the wall yielded the velocity. The results, at a first glance, do not seem to be very good. One should consider, however, the relation between the very difficult problem of catching the - physically and mathematically - very sensitive transonic recompression, and the straightforward and simple-to-code programming logic. Furthermore, the boundary algorithm seems not yet to be very refined in the sense developed here and might be the major source of error.

To get better results, a new method had to be developed (Pandolfi and Zannetti). The equations of motion were rearranged to compatibility form in the two possible space-time reference planes, and the spatial derivatives were selected as forward or backward ones according to the slope of the adjoint characteristic. So one could regard

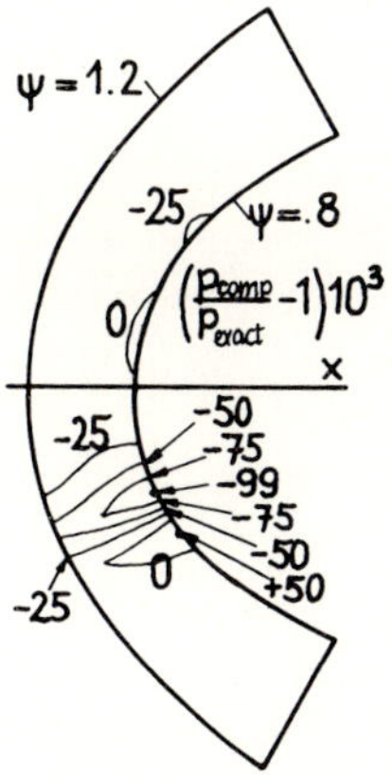

the whole procedure as a variant of the method of characteristics with the compatibility equations expressed in terms of derivatives after the prescribed computational coordinates rather after the characteristic ones. In this way a very stable algorithm was produced. Compared against the Ringleb Flow the p-errors still amount to the order of 10 Percent. Remembering, however, the discussion around the Morawetz Theorem, we should - with all necessary caution - state that here not the singular, inviscid Ringleb solution was approached but a neighboring, viscous flow with a shock (which can clearly be discerned as the steep pressure rise in the last figure in chapter 3). So we should better discriminate between <u>deviations</u> (denoting the difference between the Ringleb Flow and the exact neighboring solution) and true <u>errors</u> (denoting the difference between the exact and the computed neighboring solution). Then the above computations will appear in a much more favorable light than judged from the pure 10 %-difference against the Ringleb Flow.

6. A test case with discontinuous flow (Rusanov)

There is little hope to find a non-trivial problem in true gasdynamics which contains shocks and can be solved analytically or at least with sufficiently high numerical precision. So resort was taken to a mathematically simpler non-physical model which is described in detail in the Workshop Report (a summary is given in the paper by V.Rusanov "A test case for checking computational methods for gas flows with discontinuities" which is to be incorporated into these proceedings). It assumes homentropic flow with an adiabatic exponent of three and employs the Riemann invariants $r,s = u \pm c$. Here we present some results for a case where the initial values and boundary conditions are chosen in such a way as to produce a rightrunning shock inside the field by coalescence of characteristics; this shock is then reflected at the righthand $(z = 0)$ solid wall (the complete FORTRAN program for computing a family of such test cases to an accuracy of 10^{-5} is contained in the Workshop Report). The two diagrams give, for different time levels, the exact solution (left) and one computed by Rusanov's third order scheme /10/.

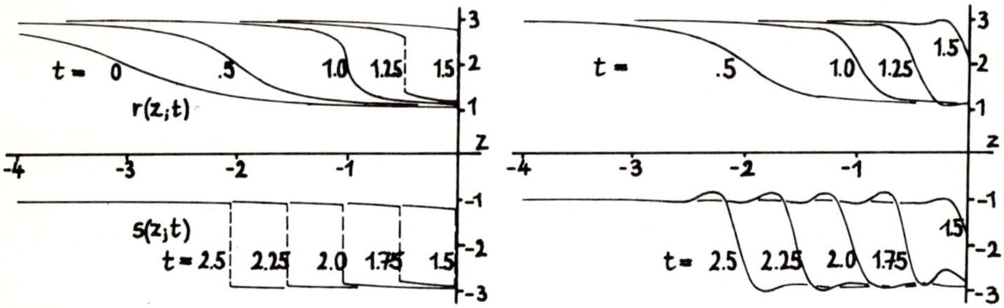

7. References

/1/ Abbett,M., Aerotherm Corp.Rep. 71-41, Mt.View, Cal. 1971

/2/ Ringleb,F., ZaMM 20 (1940), pp.185-198

/3/ Shapiro,A., The dynamics and thermodynamics of compressible fluid flow. Ronald, NewYork 1953, vol.II

/4/ Kutler,P., W.Reinhardt and R.Warming, AIAA-J. 11 (1973), pp.657-664

/5/ Moretti,G. and M.Abbett, AIAA-J. 4 (1966), pp.2136-2138

/6/ Babenko,K., G.Voskresenskii, A.Lyubimov and V.Rusanov, NASA TT F-380 (1968)

/7/ Rizzi,A. and M.Inouye, AIAA-J. 11 (1973), pp.1478-1485

/8/ Morawetz,C., Communs.Pure Appl.Math. 9 (1956) and 10 (1957)

/9/ Morawetz,C., private communication to E.Krause

/10/ Rusanov,V., J.Comput.Phys. 5 (1970), pp.547-571

INITIALIZATION OF THE NAVIER-STOKES EQUATIONS FOR USE IN NUMERICAL SIMULATIONS AND PREDICTIONS OF SEVERE (CONVECTIVE) WEATHER EVENTS

Tzvi Gal-Chen
University of Toronto, Physics Department
Toronto, Ontario, Canada M5S 1A7

1. INTRODUCTION

Considerable progress has been made in the last decade in devising improved techniques for the solution of the Navier-Stokes equations. In many atmospheric science applications, the effective use of these techniques depends, critically, on a complete specification of the initial conditions: hence the term "initialization". An initialization is particularly difficult when interest is focused on severe convective events, characterized by large deviations from hydrostatic balance.

Doppler radars can now measure the three-dimensional kinematics of severe convective storms; the radars cannot, however, measure the temperature and pressure fluctuations.

The purpose of this paper is to outline a procedure whereby the combined use of the momentum equations and the observed wind will permit a unique determination of the temperature and pressure fluctuations.

2. PRINCIPLES OF THE MODEL

a. Governing equations

A special form of the Navier-Stokes equations for low Mach number flows is the so-called "anelastic" approximation (Batchelor, 1953; Ogura and Charney, 1962; Ogura and Phillips, 1962). The continuity and momentum equations of the "anelastic" set may be written in Cartesian coordinates as follows

Continuity equation

$$\partial/\partial x_i(\rho_{0a}u^i) = 0 \tag{2.1}$$

Momentum equations

$$\partial/\partial t(\rho_{0a}u^i) + \partial/\partial x_j(\rho_{0a}u^iu^j) = -\partial p'/\partial x_i - \delta^{i3}\rho' g + \partial\tau^{ij}/\partial x_j + f^i, \tag{2.2}$$

where δ^{i3} is the Kronecker delta and g the constant of gravity. Here the tensorial notation with the summation convection has been used; u^i is the velocity in the x_i direction averaged over a grid of volume $\Delta x\Delta y\Delta z$, where Δx, Δy, Δz are the grid resolution in the x,y,z directions, respectively (i,j,k run from 1 to 3, $x_1\to x$, $x_2\to y$, $x_3\to z$ which is the vertical direction); τ^{ij} is the i,j turbulent stress component; and ρ' and p' are density and pressure deviations from their "basic" hydrostatic (but not necessarily adiabatic) values given by

$$u = v = w = 0 \tag{2.3a}$$

$$\partial p_0/\partial z = \rho_0 g, \; p_0 = p_0(z), \quad \rho_0 = \rho_0(z) \tag{2.3b}$$

$$p_0 = \rho_0 RT_0, \; T_0 = T_0(z) \tag{2.3c}$$

Eq. (2.3c) is the equation of state for ideal gases, R the gas constant for dry air ($\sim 2.8704\times10^2 m^2\ s^{-2}K^{-1}$), and $T_0(z)$ the environmental virtual temperature (i.e., $T\to T + 0.61\ q$, where q is the mixing ratio for water vapor) stratification and is measured easily by conventional radiosondes. The density $\rho_{0a}(z)$ [which is used in (2.1)] corresponds to the hydrostatic adiabatic atmosphere (note that, in general,

$\rho_{0a} \neq \rho_0$), and is given by

$$\rho_{0a} = \rho_{00}(1 - z/H_i)^{c_v/R}, \tag{2.4a}$$

$$H_i = c_p\theta_o/g. \tag{2.4b}$$

There ρ_{00} is the density at z = 0, θ_0 is the potential temperature (constant for adiabatic atmosphere), and c_p and c_v are the specific heats at constant pressure and volume, respectively; for an ideal gas, their values are $c_p = (7/2)R$, $c_v = (5/2)R$. f^i is a general representation for body forces other than gravity.

Thus far, we have not specified explicitly the turbulent fluxes τ^{ij}. Following Deardorff (1975) we assume

$$\tau^{ij} = \rho_{oa}K_m(\partial u^i/\partial x_j + \partial u^j/\partial x_i) \tag{2.5a}$$

$$K_m = 0.12\Delta E^{1/2} \tag{2.5b}$$

$$\Delta = (\Delta x\Delta y\Delta z)^{1/3} \tag{2.5c}$$

$$E = \frac{1}{2}(\overline{u'^2} + \overline{v'^2} + \overline{w'^2}) \tag{2.5d}$$

$$\varepsilon = D(z)0.70E^{3/2}/\Delta. \tag{2.5e}$$

Here K_m is the subgrid eddy viscosity coefficient; Δ is the average grid resolution; Δx, Δy, Δz are the grid resolution in the x, y, z directions, respectively, E is the turbulent energy on scale smaller than the grid resolution; and $\overline{u'^2}$, $\overline{v'^2}$, $\overline{w'^2}$ are the standard deviations from u, v, w, respectively. The overbar indicates an average over the interval

$$(-a,a),\ a = (\Delta x/2,\ \Delta y/2,\ \Delta z/2). \tag{2.6}$$

The reader is reminded that u, v, w are the average velocities over the interval defined by (2.6). In (2.5e) ε is the kinetic energy dissipation, which can also be measured (Frisch and Clifford, 1974). D(z) is a positive empirical function ($\sim$1, away from the surface) designed to partially correct for the various inadequacies of the parameterization.

Because parameterizations of subgrid moist processes require not only a turbulence parameterization but also a microphysical parameterization, we will not make any use of the thermodynamic equation; rather, we will attempt to extract as much information as possible from the dynamical equations.

b. Details of the algorithm and the calculational procedure

b.1 The pressure equation and its implications

Substituting the observed quantities u_i, $\partial u_i/\partial t$ and ε into the horizontal momentum equations [i.e., i = 1,2 in (2.2)] would give a pressure equation of the form

$$\partial p'/\partial x = F \tag{2.7a}$$

$$\partial p'/\partial y = G \tag{2.7b}$$

where F and G are known functions of position. It is known in the theory of partial differential equations that the system (2.7a,b) is overdetermined (Courant and Hilbert, 1962, pp. 1-147). It will have a solution if, and only if

$$\partial F/\partial y = \partial G/\partial x. \tag{2.8}$$

If the measurements were error-free and the turbulent parameterizations exact, then one is assured, assuming the equations of fluid dynamics to be correct, that (2.8) is satisfied. In practice, (2.8) is not satisfied and thus (2.7a,b) does not have a solution in the usual sense. The system can nevertheless be solved in the least-square sense (Gauss, 1809), i.e.,

$$\iint [(\partial p'/\partial x - F)^2 + (\partial p'/\partial y - G)^2] dxdy = \text{Min.} \quad (2.9)$$

This is a standard variational problem (Courant and Hilbert, 1953, pp. 164-274). The resulting Euler equation is a Poisson equation for the pressure

$$\partial^2 p'/\partial x^2 + \partial^2 p'/\partial y^2 = \partial F/\partial x + \partial G/\partial y, \quad (2.10)$$

with the Neumann boundary conditions

$$(\partial p'/\partial x)\cdot n_x + (\partial p'/\partial y)\cdot n_y = F\cdot n_x + G\cdot n_y$$
$$\text{at the boundaries} \quad (2.11)$$

Here n_x and n_y are the direction cosines of the normal to the boundary. The fact that the pressure boundary conditions can be determined from observations, rather than being based on some environmental condition specified almost arbitrarily, is a great asset. It implies that the occurrence of an unavoidable echo-free region will not deteriorate the quality of the pressure inversion in the region where acceptable signals are available.

Inspection of (2.11) reveals that, if a solution to (2.10) subject to the boundary conditions (2.11) exists, it is not uniquely determined. Instead, if $p'(x,y,z_0)$ is a solution at a particular level $z = z_0$, then $p'(x,y,z_0)+c(z_0)$ is also a solution at that level. To remove this spurious component, one can subtract from the pressure its horizontal average.

We have already mentioned that the calculated horizontal pressure gradients will not necessarily balance the horizontal momentum equations; rather, they are a least-square fit to these equations. If indications are that the minimum (i.e., the error) obtained from (2.9) is comparable to the nonlinear terms or the pressure gradients, then one may cautiously adjust the value of the empirical function $D(z)$ [Eq. (2.5e)].

b.2 Deduction of density deviations

Having determined the perturbation pressure fluctuations from the horizontal average, the next task is to determine the density deviations. Performing horizontal averaging on the third equation of motion [i = 3 of (2.2)] one obtains

$$-\partial \langle p' \rangle / \partial z - \langle \rho' \rangle g = \langle H \rangle , \quad (2.12)$$

where H stands for all the other terms of the w equation. By definition, $\langle H \rangle$, $\langle p' \rangle$ and $\langle \rho' \rangle$ are functions of z alone. The equation for the perturbation density deviations from its horizontal average is then

$$-\partial/\partial z(p' - \langle p' \rangle) - (\rho' - \langle \rho' \rangle)g = H - \langle H \rangle . \quad (2.13)$$

As was discussed before, since $p' - \langle p' \rangle$ is uniquely determined, $H - \langle H \rangle$ can be estimated from the radar measurements; thus $\rho' - \langle \rho' \rangle$ can be uniquely determined.

Inspection of (2.12) reveals that the horizontally averaged properties of the convective system are not uniquely determined. Gal-Chen (1978) has indicated how the combined use of an instrumented aircraft, Doppler radars, and Eq. (2.12) might yield this information.

3. THE NUMERICAL EXPERIMENTS

a. General approach

To verify the theoretical predictions of the previous section, data from a numerical model of the planetary boundary layer has been used (Deardorff and Drake, 1974). The model is a special case of the "anelastic" system (2.1), (2.2). It assumes that the environment is unsaturated and the convection is shallow. In that case,

$$\rho' = \rho_{00}\, \theta'/\theta_0 . \tag{3.1}$$

The dynamical equations (2.2) are written as

$$\partial u^i/\partial t + \partial/\partial x_j (u^i u^j) - \partial\tau^{ij}/\partial x_j - f^i = -\partial p'/\partial x_i + (g/\theta_0)\theta'\delta^{i3}, \tag{3.1}$$

where p' is the modified pressure, i.e., $p' \rightarrow p'/\rho_{00}$. Since our discussion thus far is independent of the particular functional form of ρ, the validity of our conclusions for the more complicated cases (e.g., severe storms) is asserted.

The basic strategy can be outlined as follows:

1) Compute the left hand side of (3.1) from the model's prescribed turbulent kinetic energy and winds.

2) Using the proposed algorithm, calculate the pressure and density deviations from their respective horizontal averages.

3) Compare the deduced pressure and density fluctuations with those which were originally produced by the model.

Clearly, if the data set originally satisfied the discretized (finite differences) form of (3.1) and if the same discretization procedure is also used in the proposed algorithm, then there should be no errors. This latter feature is used to ensure that there are no programming errors. By inserting realistically simulated errors into the "observations" the sensitivity of the method to various observational limitations can be assessed.

b. Specifications of the experiments

Twelve numerical experiments have been performed and their detailed analysis has been discussed elsewhere (Gal-Chen, 1978). In this paper only the results of two of these experiments are briefly discussed. In the first experiment (Exp No. 1) the horizontal and vertical extent of the domain where "observations" are available is reduced compared to the full domain size. Thus the pressure boundary conditions are very much different from those imposed by Deardorff's (1974a,b) model; instead, as is discussed in section 2, they are determined from "observations." A situation similar to that of Experiment 1 is expected to happen in reality. In addition the "observed" wind is contaminated by random errors of a magnitude 0.15 m s^{-1}, this accuracy is compatible with the observational capabilities of present Doppler radars. Approximately 40% of the grid points have a velocity with magnitude < 0.60 m s^{-1}; thus in many grid points the relative errors are greater than 25%.

Fig. 1 displays the "observed" versus calculated potential temperature deviations at the vertical level z = 0.45 km. Both the "observed" and calculated temperature have been smoothed following the technique recommended by Gal-Chen (1978). The qualitative agreement is indeed very good. A more quantitative analysis revealed that the rms value of the potential temperature fluctuations is ∿0.08 deg and the rms value of the error is ∿0.01 deg.

In the numerical simulation that produced the "observed" data the time step between two calculations was 4 s. To simulate the nonsimultaneous nature of the radar measurements another data set in which the time interval is 180 s was created. From this relevant quantities at the midpoint t + 90 were obtained. This larger time step still satisfies the consistency conditions

$$\max|\vec{u}|\Delta t/\Delta = O(1), \ \max K_m \Delta t/\Delta^2 = O(1),$$

which is required to ensure that the finite difference scheme approximates the Navier-Stokes equation (Richtmyer and Morton, 1967). This larger time step however violates the stability criteria of this particular explicit finite difference approximation by an order of magnitude and may cause large truncation errors. In addition to this systematic error a random error as in Experiment No. 1 is also added. Fig. 2 displays the observed versus calculated potential temperature deviations at z = 0.45 km. While the qualitative agreement is still very good, some small scale features are different. A more detailed qualitative analysis has revealed that the rms value of the temperature fluctuation is ∿0.08 deg, while the rms value of the error is 0.03 deg, clearly time truncation error has taken its toll.

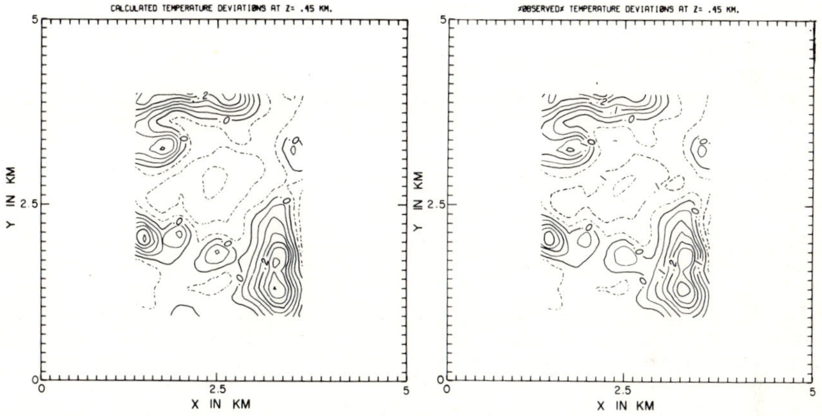

Fig. 1. "Observed" versus calculated temperature deviations for Exp. No. 1

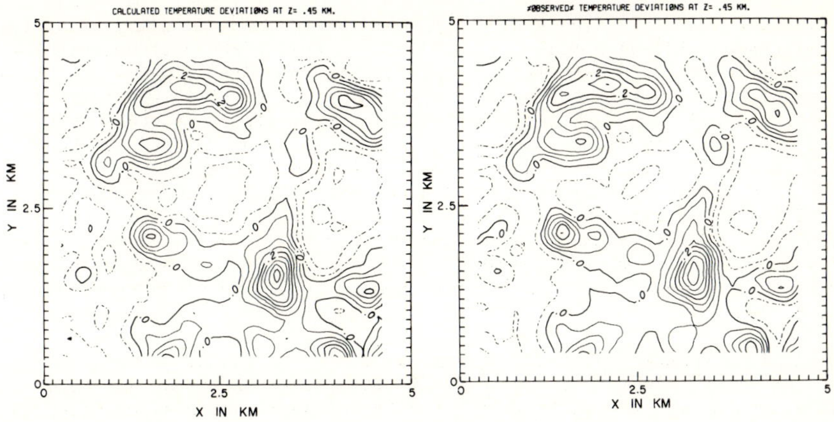

Fig. 2. "Observed" versus calculated temperature deviations for Exp. No. 2

REFERENCES

Batchelor, G. K., Quart. J. Roy. Meteor. Soc. 29, 224-235 (1953).

Courant, R., and Hilbert, D. Methods of Mathematical Physics 1, Intersience, 561 pp. (1953)

Courant, R., and Hilbert, D. Methods of Mathematical Physics 2, Interscience, 830 pp. (1962)

Deardorff, J. W. Bound.-Layer Meteor. 7, 81-106 (1974a)

Deardorff, J. W. Bound.-Layer Meteor. 7, 199-206 (1974b)

Deardorff, J. W. Proc. SESAME Open Meeting, D. K. Lilly, Ed., Boulder, NOAA Environmental Research Laboratories, 251-264 (1975)

Deardorff, J. W., and Drake, M. A. National Center for Atmospheric Research Manuscript (1975). (Available on request from the authors.)

Frisch, A. S., and Clifford, S. F. J. Atmos. Sci. 31, 1622-1628 (1974)

Gal-Chen, T. Mon. Wea. Rev. 106, 587-606.

Gauss, K. F. Theory of the Motion of the Heavenly Bodies Moving about the Sun in Conic Sections (1809; Dover republication, 1963)

Ogura, Y., and Charney, J. G. Proc. Int. Symp. Numerical Weather Prediction, Tokyo, Japan Meteor. Soc., 431-451 (1962)

Ogura, Y. and Phillips, N. A. J. Atmos. Sci. 19, 173-179 (1962)

Richtmyer, R. D. and Morton, K. W. Difference Methods for Initial Value Problems. Interscience, 561 pp (1967)

THE ALGORITHM OF APPROXIMATE CALCULATION OF POTENTIAL-TYPE SINGULAR INTEGRALS AND THEIR APPLICATIONS

T.G. Gegelia, T.Ya. Zviadadze

Institute of Applied Mathematics, Tbilisi University, Tbilisi, USSR

The solutions of many problems of mathematical physics - of hydro- and aerodynamics, of the elasticity theory, of the energy transfer theory, of geophysics and so on - may be expressed with the aid of singular potentials. The term refers to an integral of the type

$$\varphi(x) = \int_S K(x,y) f(y) d_y S \qquad (1)$$

where x and y are points on the n-dimentional space E^n, S a sufficiently smooth m-dimentional manifold lying in E^n (usually $m=n-1$), f a density function, K the kernel of the potential defined, as a rule, on the product $E^n \times S$ and having the following property: when y approaches x, then $K(x,y)$ tends to infinity and the order of infinity coincides with the dimension of the manifold

$$|K(x,y)| \leq C_1 |x-y|^{-m} \qquad (C_1 = const) \qquad (2)$$

With the foregoing assumptions, for $x \in S$ integral (1) does not exist in the ordinary sense of a singular integral but is defined as the principal value.

Thus, The numerical elaboration of the solution of the indicated problems is reduced to the approximate calculation of the singular potential (1).

Here we shall consider boundary value three-dimensional problems of classical elasticity and the same problems with heat stresses taken into account. The representations of the solutions of these problems in form (1) may be found in Kupradze V.D., Basheleishvili M.O., Burchuladze T.V. et al /1/.

Set, for simplicity, S be a plane (boundary value problems for a half-space and contact problems for a space)or a set of parallel planes (boundary value and boundary-contact problems for stratified piecewise homogeneous media) and the elastic medium be transversally-isotropic, S being a plane of isotropy. We introduce a Cartesian coordinate system $Ox_1x_2x_3$. The Ox_1 and Ox_2- axes lie in the plane S . The solutions of the problems in question will then be represented in the form of integral (1), where the kernel K is a 3 x 3 or 4 x 3 matrix whose every element has a singularity of second order while the density f is a three- or four-dimensional vector; in problems of elasticity $\varphi(x)$ is a displacement or stress vector at the point x ; K is the known matrix expressed through elementary functions and depending only on the elastic constants of the medium; f is also the known vector expressed by the boundary and contact conditions of the problem under consideration. Obviously, it may be assumed, without loss of generality, that K and f are scalar functions. That is just what we shall do in what follows.

Since S is the entire plane Ox_1x_2 and therefore contains the neighbourhood of the point at infinity, then, naturally, we need to subject f to some restrictions in this neighbourhood. In addition, f must satisfy some conditions of smoothness, the latter ensuring the existence and continuity of integral (1) and making the cubic formulas applicable with sufficiently good accuracy. These requirements impose, in turn, respective restrictions on the boundary and contact data. It will be assumed that $\forall y \in S$:

$$|f(y)| \leq C_2 |y|^{-\beta}, \tag{3}$$

$$\left|\frac{\partial^j f(y)}{\partial y_i^j}\right| \leq C_3, \qquad i=1,2; \; j=1,2,\ldots,8. \tag{4}$$

Note here that in all the problems mentioned above the kernel K satisfies the conditions

$$\left|\frac{\partial^j K(x,y)}{\partial y_i^j}\right| \leq \frac{C_4}{|x-y|^{j+2}}. \tag{5}$$

In the foregoing assumptions we may choose for any number $\varepsilon > 0$ a number $\tau > 0$, such that

$$\left| \varphi(x) - \int_{C(x',\tau)} K(x,y) f(y)\, d_y S \right| < \varepsilon,$$

where $C(x',\tau)$ is a circle of radius τ centered at the point x' and $x'=(x_1, x_2, 0)$, i.e. it is the projection of the point $x=(x_1,x_2,x_3)$ on the plane Ox_1x_2 For this it is sufficient for τ to satisfy the inequality

$$\tau > |x'| + \left(\frac{2\pi C_1 C_2}{\beta \varepsilon}\right)^{1/\beta} \tag{6}$$

Thus the calculation of integral (1) to within ε is reduced to the calculation of the integral

$$\mathcal{J}(x)=\int\limits_{C(x',\tau)} K(x,y)f(y)\,dy \qquad (7)$$

To calculate integral (7) we employ the numerical cubic formula /2/;

$$\int\limits_{a-h}^{a+h} dy_1 \int\limits_{b-k}^{b+k} F(y_1,y_2)\,dy_2 = kh\Big[F\big(a-\tfrac{h}{\sqrt{3}},\, b-\tfrac{k}{\sqrt{3}}\big)+F\big(a-\tfrac{h}{\sqrt{3}},\, b+\tfrac{k}{\sqrt{3}}\big)+$$

$$+F\big(a+\tfrac{h}{\sqrt{3}},\, b-\tfrac{k}{\sqrt{3}}\big)+F\big(a+\tfrac{h}{\sqrt{3}},\, b+\tfrac{k}{\sqrt{3}}\big)\Big]+R, \qquad (8)$$

where

$$R=\frac{2}{135}kh\left(h^4\frac{\partial^4F(\xi_1,\xi_2)}{\partial\xi_1^4}+k^4\frac{\partial^4F(\xi_1,\xi_2)}{\partial\xi_2^4}\right)-\left(\frac{2}{135}\right)^2k^5h^5\frac{\partial^8F(\xi_1,\xi_2)}{\partial\xi_1^4\,\partial\xi_2^4}, \qquad (9)$$

$$a-h\le\xi_1\le a+h, \qquad b-k\le\xi_2\le b+k\,.$$

Keeping in mind that in our case $F=K(x,\cdot)f$ and estimates (4), (5) are valid we obtain for the error

$$|R|<\frac{2}{135}C_4C_3\frac{kh^5+hk^5}{|x-y|_*^6}+\left(\frac{2}{135}\right)^2C_3C_4\frac{h^5k^5}{|x-y|_*^{10}}, \qquad (10)$$

for $|x-y|_*\le 1$;

$$|R|<\frac{2}{135}C_3C_4\frac{kh^5+hk^5}{|x-y|_*^2}+\left(\frac{2}{135}\right)^2C_3C_4\frac{h^5k^5}{|x-y|_*^2} \qquad (10)$$

for $|x-y|_*>1$,

where

$$|x-y|_*=|x-y|_{|a|-h,\,|b|-k}=\sqrt{(|a|-h-x_1)^2+(|b|-k-x_2)^2+x_3^2}\,.$$

The pitches h and k are chosen automatically according to the following law:

$$\left.\begin{array}{ll} h=k=\delta\,|x-y|_*, & \text{for } |x-y|_*\le 1\\ h=k=\delta\,|x-y|_*^{1/5} & \text{for } |x-y|_*>1\end{array}\right\}, \qquad (11)$$

where δ is an arbitrary positive number; δ is chosen, taking into account the possibilities of the computer and the required accuracy. The calculation begins from rectangle $ABCD$ where

$$A=(0,0),\quad B=(\delta x_3,0),\quad C=(\delta x_3,\delta x_3),\quad D=(0,\delta x_3)\,.$$

Estimate (10) implies that the error R depends on a distance between the points x and y and therefore is different for different rectangeles, but the choice of a variable pitch in agreement with law (11), ensures the same estimate for all rectangles. This estimate has the form

$$|R| < \frac{64}{135} C_3 C_4 \delta^6 \left(1 + \frac{16}{135}\delta^4\right), \qquad (12)$$

As compared with a constant pitch, such a choice of a variable pitch makes the operating time 8-12 times as short as before.

The flow-diagram for all the eight basic problems for a transversally-isotropic elastic half-space is written in the form

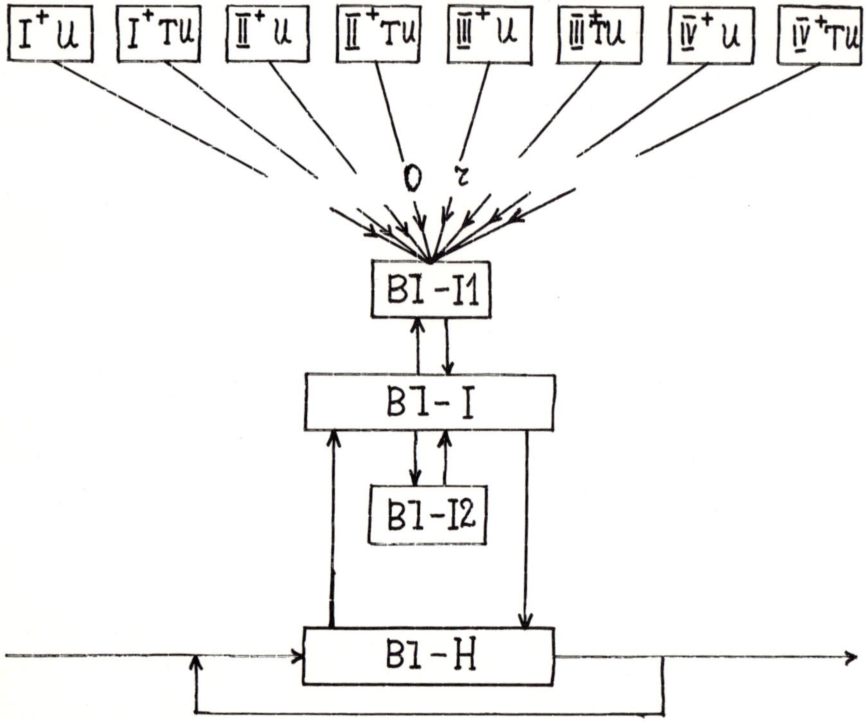

Bl-I is the block for calculating integral (7) with an automatic choice of a pitch;

Bl-I1 is the block for calculation the values of the kernel K at the required point;

Bl-I2 is the block for calculating the values of the density f at the required points;

Bl-H is the block for the initial data (elastic constants, the value of radius r, the coordinates of the point x and others).

We give an example illustrating the accuracy of the results. It is shown in /3/ that magnesium is a transversally-isotropic elastic medium and its elastic constants are taken

$$C_{11} = 5{,}97\,; \quad C_{13} = 2{,}17\,; \quad C_{33} = 6{,}17\,; \quad C_{44} = C_{55} = 1{,}64\,; \quad C_{66} = 1{,}675.$$

Below we give the solution of the first basic problem for this medium when on the boundary the displacement f is given by the formula

$$f_1(y)=f_2(y)=f_3(y)=\begin{cases} 1-y_1^2-y_2^2\ , & \text{for } y_1^2+y_2^2\leq 1,\\ 0\ , & \text{for } y_1^2+y_2^2>1. \end{cases}$$

$\tilde{U}=(\tilde{U}_1,\tilde{U}_2,\tilde{U}_3)$ means the approximate value of the displacement vector calculated by the indicated algorithm while $U=(U_1,U_2,U_3)$ is the exact value of displacement calculated from its exact analitic expression which is constructed elementarily.

x		Displacement calculated by the flow-diagram		Exact values of displacement calculated from its analytic expression	
$x_1=x_2$	x_3	$\tilde{U}_1=\tilde{U}_2$	$\tilde{U}_3$	$U_1=U_2$	U_3
0	1	0,1053904980	0,3317319213	0,1053906828	0,3317321921
0	1/4	0,5177811207	0,7787690631	0,5177811213	0,7778690630
0	1/16	0,8463971013	0,9437952585	0,8463971297	0,9437952678
0	1/64	0,9591461786	0,9859766307	0,9591461717	0,9859766279
0	1/256	0,9896269487	0,9964972943	0,9896269483	0,9964972939
0	1/512	0,9948000526	0,9982489361	0,9948000523	0,9982489357

Above we studied the question of calculating singular potentials

of type (1) when the point x is not a point of the integration set S. The indicated method allows us to approach sufficiently close to the boundary S. We may take x at a distance from S of order 2^{-30} of unity. This is more than enough for the purposes of the elasticity theory and many problems of mathematical physics. But there are cases (some problems of geophysics and energy transfer and other) when it is required to calculate a singular integral. We then use a simple but rather convenient method of calculation. Suppose we need to calculate a singular integral of the type

$$\Psi(z)=\int_S K(z,y)f(y)\,d_yS, \quad z\in S, \tag{13}$$

here K is the kernel defined on $S\times S$ we extend $K(\cdot,y)$ to some neighbourhood $G(S)$ of the set S . Practically such extension is effected directly, from the analytic expression of K. We choose the entire space as $G(S)$. As a result of the extension we obtain a potential of form (1) whose value at the point $z\in S$ is the singular integral (13). Further, we calculate the boundary value of the constructed potential. As a rule, it has the form

$$\lim_{x\to z\in S} \varphi(x)=a(z)f(z)+\Psi(z),$$

where a is the known function. The left-hand side of equality (14) may be replaced by the value of the potential φ at the point x , where x is chosen sufficiently close to the boundary point z . We may give an exact estimate of the error of such approximation.

So, the approximate value of the integral $\Psi(z)$ is expressed by the approximate value of the potential $\varphi(x)$

$$\Psi(z)\approx\varphi(x)-a(z)f(z).$$

In conclusion it should be emphasized that the technique presented here is also directly applicable for the respective problems of thermoelasticity.

References

1. Kupradze V.D., Gegelia T.G., Basheleishvili M.O., Burchuladze T.V. Three-Dimensional Problems of the Mathematical Theory of elasticity, Tbilisi Univ. Press., Tbilisi, 1968.

2. Mikeladze Sh.Ye. The Theory and Practice of Interpolation. Published by the Academy of Sciences of the Georgian SSR, Tbilisi, 1946.

3. Huntington G. Elastic Constants of Crystals, Progress in Physical Sciences (UFN), v.74, iss. 2 - 3, 1961.

AERODYNAMICS AND DYNAMICS OF BODIES WITH MASS LOSS AND SHAPE CHANGE UNDER THE INFLUENCE OF RADIATIVE HEATING

E.A.Gershbein, V.D.Gol'din, S.L.Sukhodol'sky, G.A.Tirsky
Institute of Mechanics, Moscow University,
Moscow, USSR

The reentry of meteorites and cosmic probes with hyperbolic velocities (up to 70 km/sec) into atmospheres of planets is accompanied with their intense convective and radiative heating reaching about several dozens of kilowatts per square centimeter, which results in intense thermochemical destruction of the body's surface (evaporating, melting, burning, erozion) and breaking the whole body when dynamic pressure is sufficiently high.

The meteorites' velocity distribution according to 10933 radio-observations [1] is presented in Fig.1. It is necessary to note that cosmic probes will have alike velocities (40-60 km/sec) during the entry into Jovian atmosphere.

Estimates show [2] that during a time interval of the order of a minute the total mass loss of the body with typical size about 1-3 m occuring with the above reentry velocities becomes comparable to the initial mass of the body and its form, particularly frontal surface, substantially changes. This fact inevitably entails a change of the law of motion as compared to a moving body of constant mass and shape.

On the other hand, velocity of the body significantly affects the character of physico-chemical and gas dynamical processes in a flow near the body and therefore intensity and character of heat transfer to the surface. Under such circumstances the chief mechanism of heat transfer to the body may be either radiative heating or convective one - in laminar, transitional and turbulent flows in a shock layer, - or both depending on velocities, reentry angles of the bodies, their form and size.

Thus whatever character of heat transfer, there appear a complex conjugated problem consisting in the necessity of a coupled solution in every point of the trajectory of radiative gas dynamics equations, describing the flow near the body of arbitrary form with physico-chemical processes, multicomponent diffusion and thermochemical destruction of the body, on the one hand, and of equations of motion of the body with a changing coefficient of drag, middle section and mass, on the other hand. Solution of this problem is necessary for creation a correct quantitative physical theory of meteoric phenomena, for construction of the correct quantitative algorithm of calculation of atmospheric parameters from the law of the motion of the probe, for creation of the theory of ballistic motion and stability of bodies with considerably changing ballistic coefficients at large Reynolds numbers, and for creation of the methods of simulating these complex phenomena in terrestial conditions. There is a very urgent problem to find such forms of bluntness and materials of such thermophysic characteristics which would lead to a minimum change of body's form with mass loss, that is the body's forms would be selfsimilar (in the presence of mass loss) during the motion. For this purpose it is necessary that mass loss in the direction of the body's axis of symmetry would be the same at any distance from axis (the problem of stationary form). At last, the solving of this problems is necessary for choice of optimum reentry trajectories of probes into atmospheres of planets with minimum full weight of necessary thermal protection. In particular this becomes urgent for a probe into major planets because the weight of thermal protection of vehicles of this class becomes one of the important characteristics which determine both structure and configuration of the vehicle on the whole.

Investigation of the formulated problem for such flight conditions, when convective heating in laminar, transitional and turbulent flows in boundary layer near the body play the main role, has been developed for the last ten years to a great extent. But up to this time there is no solution of this problem with intense destruction of the body under the influence of radiative heating.

In this paper a correct formulation of this problem in a hypersonic approximation is given for radiative gas dynamics equations for gas flows in chemical equilibrium; a simple and rathe accurate algorithm of its numerical solution is suggested, taking into account the main physical pricesses: selective radiation and absorption in a shock layer, evaporation of the body's surface and its heat conductivity, reduction of ionization potential at pressures above 10 atm.

Solution of the problem of hypersonic flow past bodies with intensively evaporating surface on the basis of Navier-Stokes equations with real properties of selectively radiating gas taken at every point of the trajectory together with equations of motion of the body in the atmosphere appears difficult even while using modern computers. Because of this fact a reasonable simplification of the problem with acceptable accuracy is of interest. An asymptotic analysis of Navier-Stokes equations shows that a flow in the shock layer with large Reynolds numbers based upon the free stream parameters and those of ablation layer which take place on heat-strained parts of the trajectory during braking in the atmospheres of planets, and in particular in Jovian atmosphere, may be divided into three typical regions. In regions adjoining the shock wave and the surface of the body the effects of the molecular transfer are not essential and they play the main role in intermediate region (in the layer of mixing). In a series of cases while calculating radiative heat flux when chemical components with optical properties greatly different from those of free stream and blown gas are not formed in the layer of mixing, i.e. when the layer of mixing is optically transparent, it may be replaced by a surface of a contact discontinuity.

Thus while formulating the gas dynamics part of the problem a model of two inciscid layers divided by a contact surface is used in this paper. The presence of two small parameters $\varepsilon = \rho_\infty / \rho_s^*$ and $\delta = [(\rho_w^* V_w^{*2})/(\rho_\infty V_\infty^2)]^{1/2}$ in the problem makes possible its essential simplification. As it is well known in the case of hypersonic lows past smooth bodies with impermeable surface without radiative energy transfer a method of boundary layer which was suggested by Chernyi, Freeman, Chester and some other authors may be used for solution of the problem. In such a way solution is found in quadrature for any member of expansion in series in powers of ε . This expansion is verified by comparison with the exact solutions. In the case of selectively radiating gas the solution in quadratures is impossible. Nevertheless the presence of small parameter ε permits to simplify gas dynamics equations and facilitate their solution. Analogous situation takes place in the case of a flow past a body in the presence of injection. Moreover one must mean that the presence of a Newtonian separation point of the front surface in hypersonic theory of flow without injection past such bodies as spheres, ellipsoids of rotation, elliptical cylinders and so on does not allow to find the solution in a whole shock layer. In calculation a flow past bodies such paraboloid, hyperboloid of rotation and so on there is no separation of shock layer and calculation may be made in the whole shock layer.

In this paper an iterational reverse method for obtaining solution in the whole shock layer on the front surface is suggested, which allows, on the one hand, to use simplicity of Newtonian theory of flow and on the other hand to avoid appearance of a separation point. The essence of the method is as follows. The radiative gas dynamics equations are written in orthogonal coordinate system attached to the surface of bow shock (for shock layer) the shape of which is not known before the solution of the problem is reached and with the body's surface (for blown layer) and the terms of the order $O(\varepsilon^2)$ and $O(\delta^2)$ in the equations are neglected. While solving the equations with appropriate boundary conditions the bow shock and contact break surface are iterated, moreover while putting boundary conditions on the body's surface a model of equilibrium evaporation is used.

In order to estimate an accuracy of the solution of the problem of hypersonic flow past blunted bodies by reverse iterational method special calculations of a flow past different bodies were made. In Fig.3 some forms of bow shock are shown. These forms were obtained with the help of reverse iterational method calculating a hypersonic flow past a sphere and a circular cylinder (dotted lines 1 and 2 respectively with constant γ = 1.4). In the same figures for comparison the results of exact calculations of flow past these bodies taken from the paper [12] are marked by a solid line. Small difference between the curves for ε = 0.1667 testifies good accuracy of the method suggested in this paper. Fig.4 testifies the same. It shows pressure distribution along the surface of a sphere (γ = 1.4) received with the help of the present method (dotted line) and that given in paper [12] . A shadded line in Fig.3 corresponds to the pressure distribution according to Busemann. Difference between the solutions decreases when parameters ε has smaller values.

A comparison of asymptotic approach to the problem of strong injection of other gases with numerical ones received by Stulov, Taylor and Masson [7, 8], is shown in Figs 5 and 6.

The inverse iterational method was used for solution of a gas dynamic part of the problem. While putting bounadry conditions on the body's surface a model of equilibrium evaporation is used. Chemical reactions and ionization in the shock layer and in the layer of injected gases are supposed to be equilibrium. A decrease of ionization potential is taken into account.

As under the conditions of this problem the thicknesses of the shock layer and the layer of injected gases are small in comparison with characteristic dimensions of the body, a model of local one-dimensional plane layer is used for the calculation of the radiation transfer. While calculating radiative heat flux and its divergence processes absorption in continuous spectrum and in spectral lines are accounted for. Data on cross sections are taken from papers [9-11] In order to calculate absorption in line spectrum the method of paper [11] is used, in which the line of atoms are united in 9 groups according to this method and each group with covering lines taken into account is replaced by one effective line.

While writing equations of motion of a vehicle along its trajectory in the planet's atmosphere a term corresponding to reactive force which arises due to intensive evaporation of surface heat-shield is evaluated and neglected.

As an example of using this approach to the solving of the problem about reentry of bodies into atmospheres of planets there was considered the problem about reentry into Jovian atmosphere, consisting of 85% H_2 and 15% He, of blunted bodies, in particular of hyperboloid of rotation with a smoothed traling edge, asymptotic angle 80°, curvature nose radius 0.6 m, entry velocity 47.5 km/sec, entry angle -3°, with heat shield, consisting of graphite. The model of isothermal atmosphere is used. In Fig.7 initial form of the body and the form of the body after passing heat-strained part of trajectory are given. In Fig.8 the dotted lines are mass M , velocity V , deceleration $\dot{V}$, the ratio of density of mass flux of an injected gas to the density of mass flux in free stream Q . Drag coefficient C_D , ballistic factor $M/(C_D S)$, real capacity coefficient K_0 showing the part of radiative flux which penetrates through a layer of heat-shield vapour at the line of symmetry (this coefficient equals to a ratio of radiative heat fluxes to the surface of the body and to contact surface), coefficient of mass loss C_m in dependence on time are given in Fig.9. Solid lines in Fig.8 correspond to these values in a case of body entry into Jovian atmosphere with a constant ballistic factor. Note the neglecting the effect of mass loss and shape change while calculating the trajectory of a body in the atmosphere leads in this case of 15% decrease of maximum values of overloads and 12% increase of full mass loss values.

Capacity coefficient K_0 changes in the range 0.48-1.0 during the motion of the body along its trajectory; a minimum capacity coefficient coincides in time with the maximum heat flux to a body. A drag coefficient changes mainly due to change of the shape of a body and to less degree due to the presence of injected gas layer in the region of the flow (thickness of this layer changes with respect to the radiative heat flux to the body).

Fig.10 shows distributions of the ratio of mass injection rates at an arbitrary point of the body and in its critical point with respect to the distance between the point of the body and the axis of symmetry of the body at the tenth, twentith and thirtith seconds of the flight from altitude 210 km above the surface of the planet. Mass loss occures ununiformly along a formative curve of the body and as a result of it the body becomes more blunted and consequently a thickness of shock layer increases. In principle an increase of shock layer thickness with ballistic factor remaining constant leads to an increase of the total mass loss from the surface of the body. In the present case an increase of mass loss does not occur because of decrease of the ballistic factor. In this case a body experiences additional deceleration and this fact influences the value of total radiative energy to a surface of the body during the motion along trajectory more than blunting of the body due to ununiform mass loss from its surface does.

Fig.11 shows surface temperature T_w and capacity coefficient K along the surface of the body with respect to the distance between a point of the body and its axis of symmetry at the thirtith second of flight which corresponds to maximum radiative heat flux to the surface and minimum capacity coefficient at a critical point.

The following figures also correspond to thirtith second of flight. Fig.12 shows the distribution of a radiative heat flux and temperatures across a shock layer with respect to the distance from the bow shock surface at the critical point and the point of the section of the body the radius of which equals 0.5 m. In Figs 13 and 14 distributions of spectral radiative heat flux at a critical point and at a point which belongs to a section of the body with radius 0.5 m are given. The upper lines correspond to contact surface, the lower lines restricting shaded regions correspond to the surface of the body. Solid lines correspond to a portion of the flux in continuous spectrum, dotted lines - to a portion of the flux in the atomic lines. Negative values of the flux in region 6 eV shows a portion of the flux from the shock layer in a continuous spectrum absorbed in the atomic lines in an ablation layer. The layer of blown gaseous products of destruction of graphite heat-shield effectively absorbes in following regions of spectrum: 2-3 eV – C_2 (Swan band), 3-4 eV – C_3 (4050 Å band) and in a region of photoionization continuums of C, C_2, C_3. In spectral regions 0.5-1.6 eV ; 4.1-5 eV ; 6.2-10 eV ablation layer practically does not absorb radiation from shock layer. Radiation in a Balmer series of hydrogen atom giving the main contribution to radiative heat flux from shock layer in a linear spectrum is absorbed in molecular bands in ablation layer.

References

1. Riddell F.R., Winkler H.B. From ICBM to Meteoric Entry. Amer. Roc. Soc., 1961, N 113, 38.
2. Gershbein E.A., Sukhodol'skaya E.Ya., Sukhodol'sky S.L., Tirsky G.A. Abstracts of Reports Presented at the 26th IAF Congress. Lisbon, 1975.
3. Gershbein E.A. Theory of the Hypersonic Viscous Shock Layer at High Reynolds Numbers and Intensive Injection of Foreign Gases. J. Appl. Math. and Mech. (PMM), 1974, N 6, pp.962-972.
4. Chernyi G.G. Introduction to Hypersonic Flow. Acad. Press Inc., New York, 1961.
5. Freeman N.C. On the Theory of Hypersonic Flow Past Plane and Axially Symmetric Bluff Bodies. J. Fluid Mech., 1956, 1, pp.366-387.
6. Chester W. Supersonic Flow Past a Bluff Body with a Detached Shock. J. Fluid Mech., 1956, 1, pp.353-365.
7. Taylor T.D., Masson B.S. Supersonic Flow Past Blunt Bodies with Large Surface Injection. 19th Internat. Astronaut. Congress. N.Y., 1968. Proceedings, v.3 (Propulsion Re-Entry Physics).
8. Stulov V.P. Strong Mass Injection at the Surface of Blunt Body in a Supersonic Flow. Fluid Dynamics (Izvestia AN SSSR, Mekhanika Zhidkosti i Gaza), 1972, N 2.
9. Оптические свойства горячего воздуха. Под ред. Л.М.Бибермана. I970.
10. Каменщиков В.А.,Пластинин Ю.А., Николаев В.М., Новицкий Л.А. Радиационные свойства газов при высоких температурах. М., I97I.
11. Wilson K. Stagnation Point Analysis of Coupled Viscous Radiating Flow with Massive Blowing. NASA CR-1548. 1970.
12. Любимов А.Н., Русанов В.В. Течения газа около тупых тел. М., "Наука", I970.

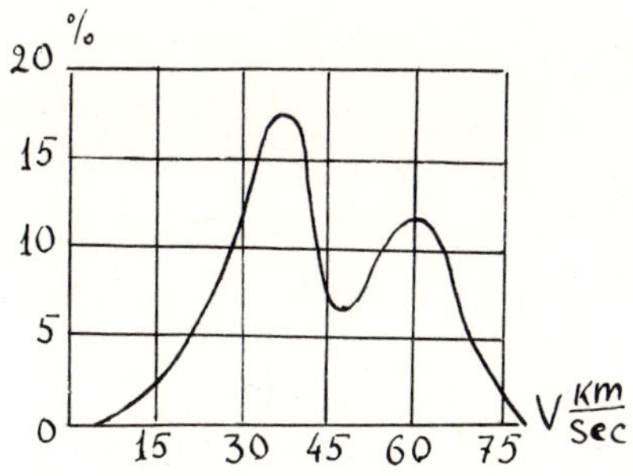

Fig.1

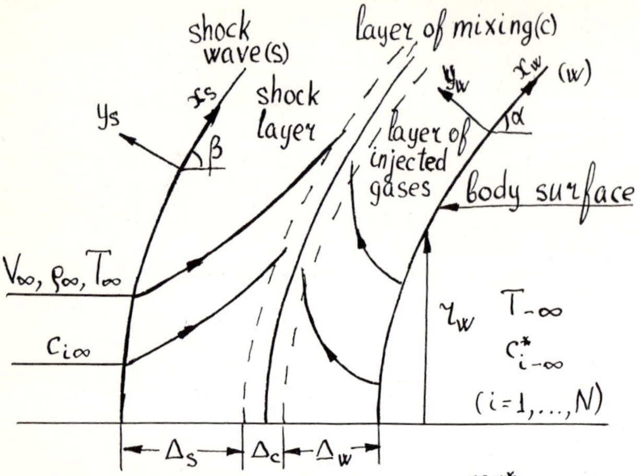

$$\varepsilon \equiv \frac{\rho_\infty}{\rho_s} = 0.05 \div 0.1; \quad \frac{\rho_\infty}{\rho_w} \approx 0.01; \quad \frac{\rho_w^* V_w^*}{\rho_\infty V_\infty} \leq 1.0$$

$$\delta^2 \equiv \Lambda \equiv \frac{\rho_w^* V_w^{*2}}{\rho_\infty V_\infty^2} \lesssim 0.01; \quad Re = \frac{\rho_\infty V_\infty R}{\mu_s^*} = 10^5 \div 10^8$$

$$Re_w = \frac{\rho_w V_w R}{\mu_w^*} = 10^4 \div 10^7$$

$$\frac{\Delta_s}{R} = O(\varepsilon); \quad \frac{\Delta_w}{R} = O(\delta); \quad \frac{\Delta_c}{R} = O\left(\frac{1}{\sqrt{Re}}\right)$$

$$Re^{-1/2} \lesssim Re_w^{-1/2} << \delta \leq \varepsilon << 1$$

Fig.2

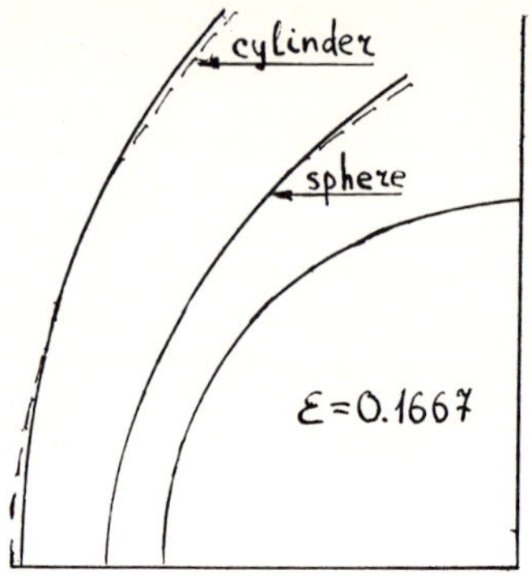

Fig.3

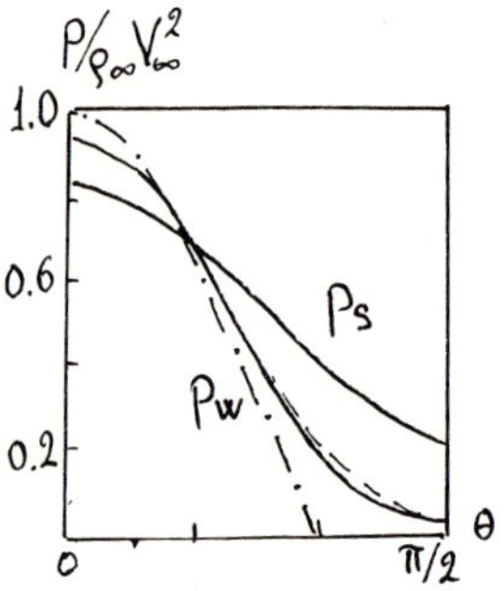

Fig.4

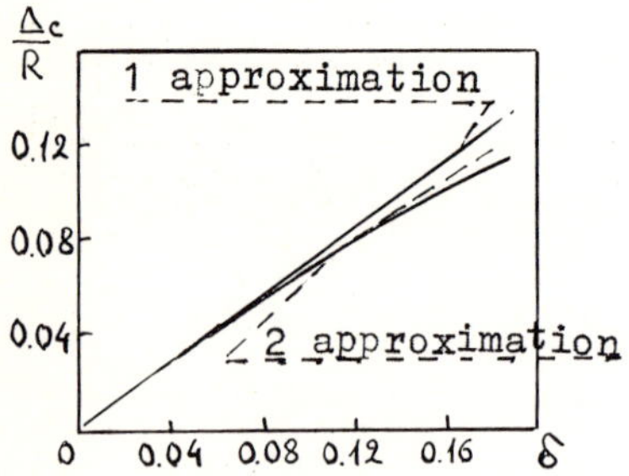

Fig.5. Thickness of a layer of injected gases.
— asymptotic solution [3]
--- numerical solution [8]

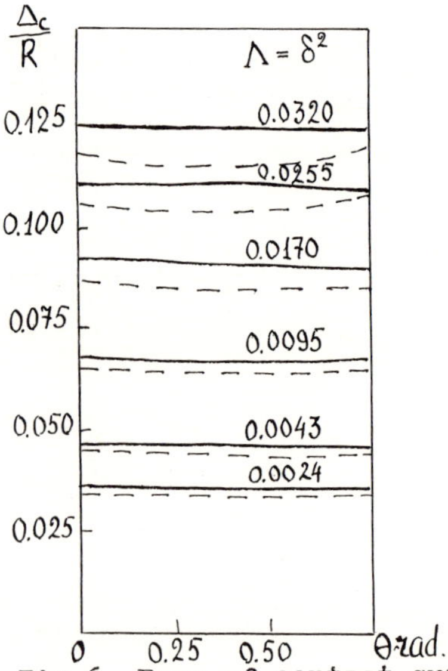

Fig.6. Form of contact surface. Flow past sphere.
$V_w = V_{w0}\cos\theta$, $T_w = const$, $\gamma = 1.4$

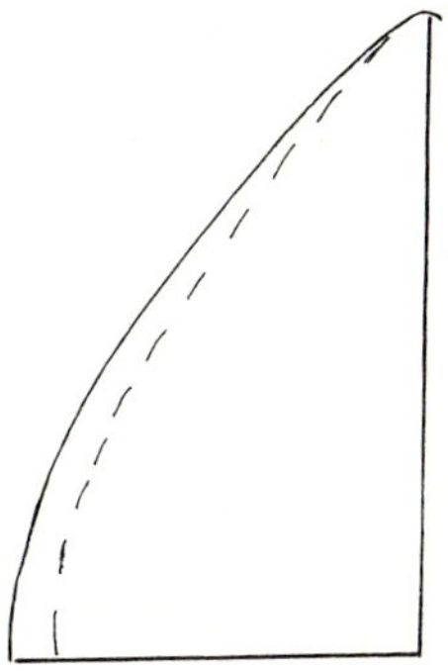

Fig.7

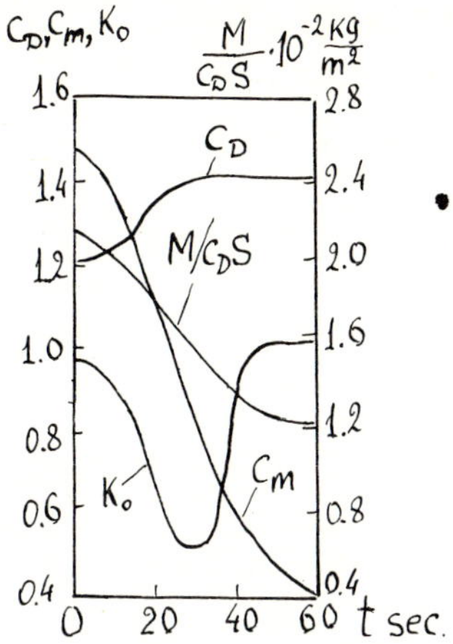

Fig.9

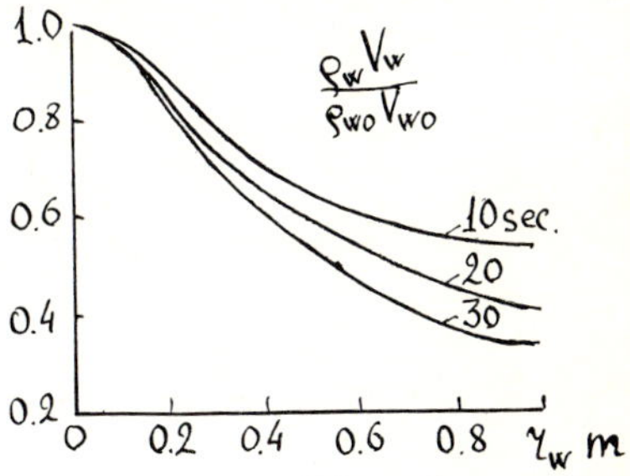

Fig.10

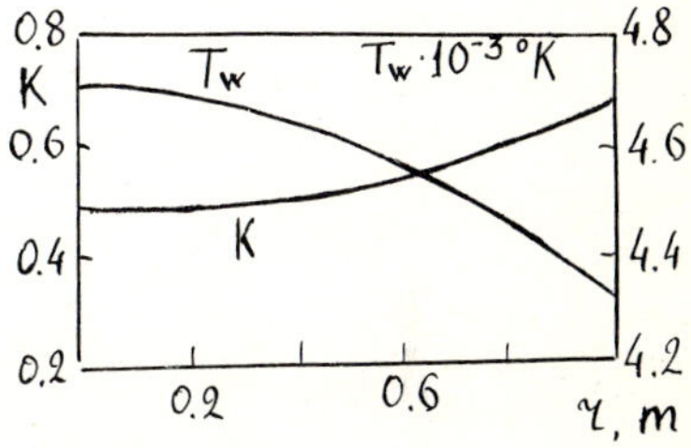

Fig.11

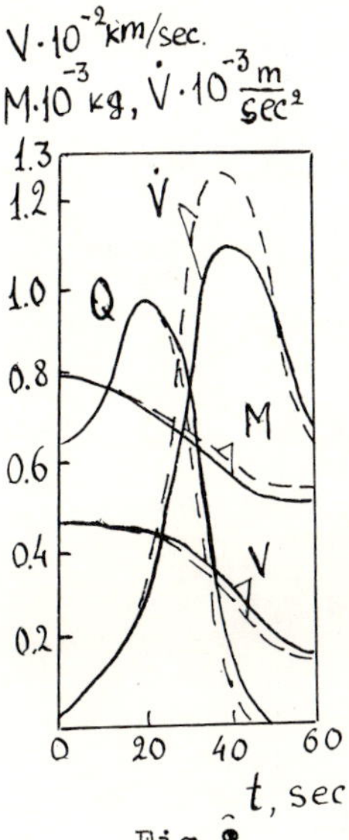

Fig.8

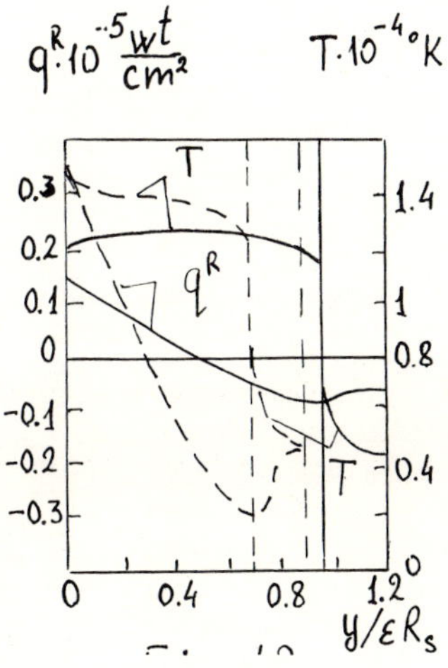

Fig.12

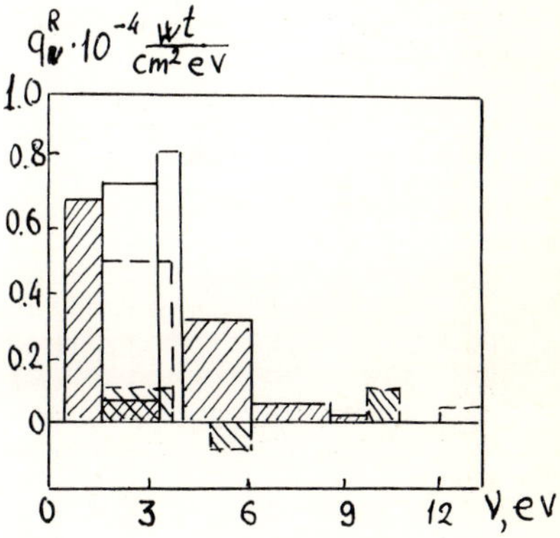

Fig.13

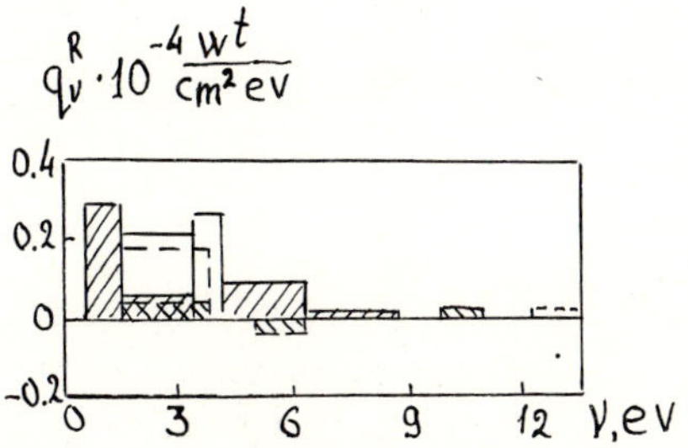

Fig.14

ERGODIC BEHAVIOR OF TWO-DIMENSIONAL INVISCID TURBULENCE

Harland M. Glaz
Naval Surface Weapons Center
White Oak, Silver Spring, Maryland 20910

INTRODUCTION AND THE EQUATIONS OF MOTION

The statistical analysis of solutions of the inviscid two-dimensional Navier-Stokes equations is a subject of great current interest. We study here the ergodic and mixing questions for the dynamical systems obtained through truncation of the Fourier-transformed equations. These systems are integrated numerically; a careful error analysis from the point of view of these statistical questions is presented.

The equations of motion are:

$$(1) \qquad \underline{u}_t + (\underline{u} \cdot \nabla)\underline{u} = -\text{grad } p + \nu \Delta \underline{u} + \underline{F}$$

$$(2) \qquad \text{div } \underline{u} = 0$$

where $\underline{u}$ is the velocity field, p = pressure, ν = viscosity, and $\underline{F}$ = external force. Assuming periodic boundary conditions, the truncated Fourier transform of these equations in the inviscid, $\underline{F} = 0$ case is

$$(3) \qquad u_\alpha(\underline{k},t) = -\frac{i}{2} P_{\alpha\beta\gamma}(\underline{k}) \sum_{\substack{\underline{p},\underline{q}\varepsilon F \\ \underline{p}+\underline{q}=\underline{k}}} u_\beta(\underline{p},t) u_\gamma(\underline{q},t)$$

$$(4) \qquad k_\alpha u_\alpha(\underline{k},t) = 0$$

where summation is indicated for repeated Greek subscripts. F denotes the set of retained Fourier modes for a given truncation (we consider only those sets F which satisfy $\underline{k}\varepsilon F \Rightarrow -\underline{k}\varepsilon F$. This insures that $\underline{u}$ remains real-valued). We have denoted the Fourier transform by $\underline{u}(\underline{x},t) = \sum \underline{u}(\underline{k},t) e^{i\underline{k}\cdot\underline{x}}$ and have chosen the period in such a way that factors of 2π do not appear. The tensor $P_{\alpha\beta\gamma}$ is given by

$$(5) \qquad P_{\alpha\beta\gamma}(\underline{k}) = k_\beta P_{\alpha\gamma}(\underline{k}) + k_\gamma P_{\alpha\beta}(\underline{k})$$

$$(6) \qquad P_{\alpha\beta}(\underline{k}) = \delta_{\alpha\beta} - k_\alpha k_\beta / k^2 .$$

The collection of ordinary differential equations in (3) defines a dynamical system in R^M where M is the total number of independent Fourier mode components $u_\alpha(\underline{k})$. A Liouville Theorem holds for this system. That is,

$$(7) \qquad \sum_{\underline{k}\varepsilon F} \frac{\partial}{\partial u_\alpha(\underline{k},t)} \left(\frac{du_\alpha(\underline{k},t)}{dt} \right) = 0$$

which states that Lebesgue measure is preserved in R^M by the dynamical system. Also, we have conservation of energy and enstrophy (= total vorticity) -

$$(8) \qquad \frac{d}{dt} \sum_{\underline{k}\varepsilon F} u_\alpha(\underline{k},t) u_\alpha(-\underline{k},t) = 0 \ ,$$

$$(9) \qquad \frac{d}{dt} \sum_{\underline{k}\varepsilon F} k^2 u_\alpha(\underline{k},t) u_\alpha(-\underline{k},t) = 0 \ .$$

The proofs of (7) - (9) are trivial. Let S_E and S_Ω denote surfaces of constant energy E and constant enstrophy Ω, respectively. Then, the flow of (3) takes place on the (M-2)-dimensional surface $S = S_E \quad S_\Omega$. Let $\{F_t: S \to S | t \geq 0\}$ denote this flow. One can obtain an F_t-invariant measure, simultaneously for all $t \geq 0$, on S which is absolutely continuous with respect to (M-2)-dimensional Lebesgue measure $d\Sigma$. It is

$$(10) \qquad d\mu = d\Sigma / \ \text{grad } E \quad \text{grad } \Omega \quad \sin\theta$$

where θ is the angle formed by grad E and grad Ω. A derivation of this result may be found in Glaz[7].

Given the F_t-invariance of the measure μ, we can define ergodicity and mixing with respect to μ. A transformation $T: S \to S$ is said to be ergodic if

$$\mu(TA - A) = 0 \Rightarrow \mu(A) = 0 \text{ or } \mu(A) = \mu(M)$$

for every measurable subset A of S. The flow $\{F_t\}$ is ergodic if and only if each $F_t: S \to S$ is ergodic according to the preceding definition. By employing the ergodic theorem, one can obtain the following characterization of ergodicity -

<u>Theorem</u>: The flow $\{F_t\}$ is ergodic if and only if, for each $t \geq 0$ and almost every $(\mu) x \varepsilon S$,

$$(11) \qquad \lim_{T\to\infty} \frac{1}{T} \int_0^T g(F_t x)\,dt = \frac{1}{\mu(S)} \int_S g(x)\,d\mu(x)$$

for every integrable observable $g: S \to R$. Equation (11) states that "time means equal phase means." The flow $\{F_t\}$ is said to be strong mixing if and only if

$$(12) \qquad \lim_{T\to\infty} \int_S g(F_T x) h(x)\,d\mu(x) = \frac{1}{\mu(S)} \int_S g(x)\,d\mu(x) \cdot \int_S h(x)\,d\mu(x)$$

for all integrable $g,h: S \to R$. It is easy to see that mixing implies ergodicity but that the converse is false in general. For proofs and further details concerning these ideas, see Halmos[9] and Sinai[14].

TURBULENCE AND THE MODEL

We consider the relationship between real turbulence and the dynamical systems under study. There are at least four points which should be taken up: the question of 2-D versus 3-D turbulence, realistic boundary conditions versus periodic boundaries (i.e., homogeneous turbulence), the effect of the limit $\nu \to 0$, and the truncated systems versus the untruncated infinite-dimensional system.

In a three-dimensional flow, vortex stretching takes place and is an important mechanism in turbulent energy transfer and dissipation. Because of conservation of vorticity, this phenomenon is excluded in the two-dimensional case; it is not clear whether or not an analogous mechanism exists. Also, in a real turbulent flow, one has a boundary layer in which vorticity creation takes place. This is precluded in our model system by the assumption of periodic boundaries.

The effect of the limit $\nu \to 0$ is not a serious problem in the model, even in the infinite-dimensional form (1) - (2); it can be shown, see Ebin and Marsden[4], that solutions of the equations with $\nu \to 0$ approach those of the inviscid equations in the sense that their Fourier transforms converge uniformly on compacta. In the numerical analysis of finite-dimensional truncations to be presented below, this limit is no problem whatsoever because the results with $\nu = 0$ and with $0 < \nu << 1$ cannot differ substantially in a finite time interval. This leads us to the truncation itself. Of course, it is clear that an extremely large truncation would contain all important physical information. The real question is whether a truncation of reasonable size (i.e., one amenable to a computer analysis) also has this property. We can offer no opinion on this matter.

We shall make the heuristic assumption that equations (3) - (4) model real-world turbulence despite the objections just presented. This allows us to use the language of turbulence in the sequel; the skeptical reader may simply view the system (3) - (4) as an abstract dynamical system with interesting mathematical properties but doubtful physical content.

STATISTICAL PROPERTIES AND COHERENT STRUCTURES

In this section, we discuss the relationship between statistical properties of the dynamical systems described by equations (3) - (4), and physical properties of turbulent flow fields. The main physical features of interest here are the well-defined macroscopic phenomena which are present in most turbulent flows. In particular, we refer to large-scale vortices and intermittency. These coherent structures contrast sharply with the high degree of randomness evidenced in turbulence. It would be of interest to derive a rigorous theory which simultaneously explained these two apparently contrary ideas.

For the dynamical systems (3) - (4) obtained from Fourier analysis of the basic equations (1) - (2), it is standard in the literature to assume a priori that the mixing hypothesis is satisfied. This is of importance in theoretical work because it justifies the interchange of ensemble averages, which can be calculated or approximated, for the physically significant time averages. However, large-scale coherent behavior cannot be derived directly from the mixing hypothesis; indeed, it argues against such behavior. This is so because in a mixing system, states (= velocity fields) not containing such phenomena should have the same probability with respect to the measure μ as those which do contain them. The standard technique to avoid this difficulty is to proceed to the limit as the number of retained modes approaches infinity and to then assume thermodynamic behavior for the limiting system. In our use of the word "thermodynamic," we perceive of each independent Fourier mode component as analogous to an independent degree of freedom in a collection of gas molecules. In the gas molecule situation, it is widely believed that the governing dynamical systems are mixing. Nevertheless, it can be rigorously shown (in simple cases) that thermodynamic concepts such as temperature, pressure, and entropy are well-defined in the limit and are excellent approximations for large-scale systems (i.e., a lot of molecules). Similarly, one might hope that concepts such as macroscopic vortex formation are well defined thermodynamically in the Fourier mode model of turbulence. This would mean that the μ-measure of the set of states containing macroscopic vortices is overwhelmingly large relative to the μ-measure of the whole state space, S. In this way, the regular appearance of large-scale coherent structures can be explained even in the face of random behavior of the underlying system, say mixing or ergodicity.

Unfortunately, the existence of thermodynamic quantities for the Fourier mode model is a matter of conjecture at present. A basic difficulty is that one must contend with two quadratic constants of the motion, total energy and enstrophy, instead of just one - energy - as in gas molecule systems. This immediately implies that equipartition theorems for any single conserved quadratic quantity - e.g., energy - must be false because the dynamics takes place on the surface S, not S_E. The surface S is clearly much more difficult to deal with mathematically than S_E or S_Ω. For further details, see Glaz[7]. There has been interesting analytical and numerical work done on the question of the existence and significance of "negative temperature states," an idea due to Onsager[13]. See Deem and Zabusky[3], Fox and Orszag[5], Montgomery & Joyce[12] for details.

Alternatively, let us suppose that the systems (3) - (4) are not mixing nor even ergodic. It is then somewhat easier to hypothesize theoretical mechanisms which allow for the existence of coherent structures, at least in the case that viscosity is not zero. For, under these circumstances, attractors are possible for the flows governed by equation (3). It is tempting to assume that, for example, the collection of flow fields with large-scale vortex structures is a metastable attracting set for almost all flow fields. We say "metastable," because in case $\nu \neq 0$, all fields tend eventually towards the zero field which is a global attractor. Nontrivial attractors are not possible in the inviscid case because the measure μ given by equation (10) is F_t-invariant.

In view of these considerations, it is of importance to decide the ergodicity and mixing questions for the dynamical systems governed by equations (3) - (4). In this way, the theoretical work in two-dimensional turbulence can proceed on a firm foundation. It is the object of the remainder of this paper to present a feasibility study for this research by means of a careful numerical analysis of low order truncations using techniques which can easily be generalized to the realistic regime of high order truncations.

NUMERICAL METHODS

The ergodic question for a given dynamical system has long been known to be an extremely difficult problem, particularly when the system cannot be integrated in closed form. The mixing question can be equally difficult. The solution of the system (3) - (4) is, of course, not known explicitly, so we must work with an approximate solution. Regarding equations (11) and (12), it is impossible to numerically take the indicated $T \to \infty$ limits, test almost every $x \varepsilon S$, or test all measurable g,h: $S \to R$. Therefore, it is not possible to prove ergodicity or mixing, or their converses, using only numerical techniques. We present our methods and results with this point clearly understood.

Let $\rho(x)$ denote the density of the measure μ, i.e., $d\mu(x) = \rho(x)dx$. The main idea is to carry out the following approximations -

$$(13) \qquad \int_0^T g(F_t x)\,dt \sim \sum_{i=1}^{L} g(F_{t_i} x)\Delta t_i$$

$$(14) \qquad \int_S g(x)\,d\mu(x) \sim \sum_{i=1}^{K} g(x_i)\rho(x_i)$$

$$(15) \qquad \int_S g(F_T x)h(x)\,d\mu(x) \sim \sum_{i=1}^{P} g(F_T x_i)h(x_i)\rho(x_i) \;.$$

In order to integrate the system of first order ordinary differential equations (3), and obtain the flow $\{F_t\}$ thereby, we have used Gear's method - an Adams-Moulton method of variable order. See Hindmarsh[10] for complete documentation. A feature of the method is that local truncation error is a user input, and the time steps, Δt_i, are variable and computed by the routine. In choosing the final time of integration, T, we have taken T to be as large as possible while still maintaining accurate integrations. Accordingly, we tested several initial points by integrating them with the chosen local truncation error (typically, this was 10^{-8} or 10^{-9}) and then reintegrating with a smaller local truncation error. T was taken as that time where the results began to diverge in the second significant figure.

It is clear that the right-hand side of (13) is an approximating Riemann sum for the integral on the left where t_i, Δt_i, and $F_{t_i}x$ are computed using Gear's method. The approximations (14) and (15) were carried out using a straightforward Monte Carlo integration scheme. Doing this obviously involves a parametrization of the surface S so that points can be chosen randomly in the parameter space and then weighted according to the measure μ. The calculations necessary to obtain the parameter space (and those which define the density function ρ) are rather lengthy and will be omitted here for reasons of space. The full details appear in Glaz[7]. In other numerical work of this type, Basdevant and Sadourney[1] and Kells and Orszag[11] both use spherical polar coordinates in velocity space which is somewhat more efficient than the approach used here.

Except for details, this explains how the approximations of (14) and (15) were carried out. The work involved in calculating (15) is quite expensive because it involves a complete integration from t = 0 to t = T for each point selected. Consequently, it was only possible to perform $P = 10^3$ integrations. The Monte Carlo routine itself, (14), is relatively inexpensive and it was possible to take $K = 10^5$. It was found that $K = 10^5$ was adequate for most of the test functions $g: S \to R$ to obtain convergence, but only barely, so $K = 10^3$ was inadequate. This implies that $P = 10^3$ is probably also inadequate.

Given the approximations (13) - (15), we can test various functions $g,h: S \to R$ in equations (11) - (12) and see if we obtain convergence. The obvious question is: How does one measure convergence? This is not at all clear due to the various numerical errors introduced in (13) - (15). Our technique is to employ a comparison between various systems about which we have different types of a priori information. Three different models are used and they can be described by their retained modes:

(16) (I) $K_1 = (0,2)$, $K_2 = (0,1)$, $K_3 = (1,2)$, $K_4 = (1,1)$, $K_5 = (1,0)$
$K_6 = (1,-2)$, $K_7 = (1,-1)$, $K_8 = (2,1)$, $K_9 = (2,0)$, $K_{10} = (2,-1)$,

(II) $K_1 = (1,1)$, $K_2 = (2,1)$, $K_3 = (2,-1)$, $K_4 = (1,-1)$, $K_5 = (3,0)$,

(III) $K_1 = (1,1)$, $K_2 = (2,1)$, $K_3 = (2,-1)$, $K_4 = (1,-1)$,
$K_5 = (1,0)$, $K_6 = (3,0)$.

By work of Hald[8], it was known a priori that system (II) is not ergodic. System (III) differs only slightly from (II); system (I) is considerably more complex and we have no a priori information. We remark that values of E and Ω were chosen with numerical efficiency in mind; there was not enough time to work with a range of values for these two parameters to see if this affects the results.

Finally, we have also compared the results for systems (I), (II), and (III) with the results of a "computer game" described by Chorin[2]. Briefly, this system involves random collisions between point particles in a plane. Spatial coordinates are not retained in phase space - only the two velocity coordinates. This system is almost surely strong mixing and can be used in this regard as a basis of comparison. However, this system contains no time scale since the spatial coordinates are not present; so the comparison must be made with caution. Further details may be found in Glaz[7]. The two "games" used were

$$\text{(17)} \quad \text{(IV)}\ M = 4 \text{ particles},\ E = 1.0,\ A = B = 0\ ,$$

$$\text{(V)}\ M = 8 \text{ particles},\ E = 1.0,\ A = B = 0\ .$$

Here, A and B are the linear momenta which are conserved along with the energy. Thus, the phase space of system (IV) has eight dimensions but only five are independent. Similarly, systems (I) - (III) contain 2M-2 independent directions where M is the number of retained modes (the real and complex parts of each $\underline{u}(\underline{k})$ are independent).

The methodology employed here on the small-scale truncations (I) - (III) can easily be generalized to large-scale models. The results obtained may be viewed as a feasibility study for the more general problem.

RESULTS AND CONCLUSIONS

Several general results can be stated before tabulating the numbers. First, the Monte Carlo integrations converge quickly and are inexpensive to compute. Regarding a larger truncation, the amount of labor per Monte Carlo point is a linear function of the number of retained modes; however, the number of sample points needed for various levels of convergence is not known as a function of number of retained modes. We conjecture that this will not be a serious obstacle for observables of interest.

It is extremely important to recognize that computed approximate trajectories leave the actual trajectory (in the first significant figure) long before the conserved quantities, energy and enstrophy, are in error even in their eighth significant figure. Indeed, one computed trajectory actually consists of portions of 10^2 or 10^3 different exact trajectories if one integrates until the energy is in error. Hence, if one uses conservation of energy and enstrophy as a criterion to choose the final integration time, it is only natural that one sees convergence in equation (11) since two phase space integrations are being compared with each other rather than with a time mean over a single trajectory, as required in the definition. As a consequence, it is easy to conclude that (3) - (4) is ergodic simply as a result of numerical error.

Another general point is that the errors involved in the approximations (13) - (15) are functions of the particular phase observable $g: S \to R$ chosen for study. It is possible to choose observables which force the approximations to converge as quickly (choose a constant function) or as slowly (choose a function with large oscillations or choose a high order moment, x_i^ℓ, where x_i is a coordinate of S) as one desires. This underscores our point that it is impossible to prove ergodicity by a numerical analysis alone. It becomes important to choose observables for study very carefully, both in regards to their numerical stability as well as their physical significance. For example, the second order moments provide the energy and vorticity spectra while being at the same time stable with respect to the numerical study. Most prior studies in this area have concluded that time and phase means are equal for second order moments, and our results agree with this. However, the evidence is much less clear for other observables. The point is that the rate of convergence, as well as the fact of convergence or nonconvergence, is strongly dependent on the observable.

The numerical results are too numerous to be given here in detail. We shall describe a representative sample. Let us first discuss the results for mixing. It turned out that the convergence rates were roughly the same for all five problems, including Problem II, which is not even ergodic. It must be concluded, therefore, that the numerical precision attained with $P = 10^3$ samples is not sufficient to distinguish between mixing and nonmixing behavior in tests based on equation (12).

The results are more encouraging for the study of ergodicity using equation (11); here, different behavior patterns can easily be distinguished. The basic estimate of convergence error is the absolute value of the difference between the ratio of the two sides of equation (11) and unity for many different observables. The vertical of Figure 1 represents an average of this quantity over several such observables. The profiles in time are drawn for each of two starting points for Problems I, II, and III, and one point for Problems IV and V. Note that the final integration times differ for Problems I, II, and III. For Problems IV and V, 5000 discrete steps were run so 100 time steps = 1 second from the point of view of the figures; this has no physical significance. Figure 2 displays the convergence error for six different observables in Problem I (no averaging).

It may be inferred from Figure 1 alone that Problem II is not ergodic since no convergence is exhibited from either starting point. The same may be said for Problem III, but not quite as strongly. Observe how quickly Problems IV and V have converged; nevertheless, they are still converging at $t = 50$, which indicates that convergence errors of 0.05 are not necessarily to be expected solely due to numerical errors. Of course, these problems contain no time scale and their solution involves no truncation error. In any event, we would like to see further convergence in Problem I before concluding that it is ergodic; the final integration time of $T = 30$ has proven insufficient for this problem.

Our basic conclusions are as follows: (1) the basic Monte Carlo integrations can easily be performed; (2) it is possible to distinguish ergodic from nonergodic behavior, but it may be necessary to obtain much more accuracy in the numerical integrations than was the case in this study; and (3) the mixing calculations as performed here are infeasible with existing computer capabilities.

The best advice we can offer workers in this field is to carefully analyze the choice of observables before computing. Since it is not possible to work with all observables, it is important to study only those which will yield numerically reliable results and are physically significant.

REFERENCES

1. C. Basdevant and R. Sadourney, "Ergodic Properties of Inviscid Truncated Models of Two-Dimensional Incompressible Flows," J.F.M., 69, (1975), pp. 673-688.

2. A. J. Chorin, "Numerical Experiments with a Truncated Spectral Representation of a Random Flow," unpublished.

3. G. S. Deem and N. J. Zabusky, "Ergodic Boundary in Numerical Simulations of Two-Dimensional Turbulence," Phys. Rev. Ltr. 27, (1972), pp. 391-399.

4. D. G. Ebin and J. Marsden, "Groups of Diffeomorphisms and the Motion of an Incompressible Fluid," Ann. of Math., 92, (1970), pp. 102-163.

5. D. G. Fox and S. A. Orszag, "Inviscid Dynamics of Two-Dimensional Turbulence," Phys. of Fluids, 16, (1973), pp. 169-172.

6. H. Glaz, "Two Attempts at Modelling Two-Dimensional Turbulence," Turbulence Seminar Berkeley 1976/77, Springer Lecture Notes in Math., #615, 1977, pp. 135-155.

7. H. Glaz, "Statistical Study of Approximations to Two-Dimensional Inviscid Turbulence," Lawrence Berkeley Laboratory Report LBL-6708, 1977.

8. O. Hald, "Constants of Motion in Models of Two-Dimensional Turbulence," Phys. of Fluids, 19, (1976), pp. 914-915.

9. P. R. Halmos, Lectures on Ergodic Theory, (1956), Chelsea Publishing Co., New York.

10. A. C. Hindmarsh, Gear: Ordinary Differential Equation System Solver, 1974, Lawrence Livermore Lab. Report UCID-30001, Rev. 3.

11. L. C. Kells and S. A. Orszag, "Randomness of Low-Order Models of Two-Dimensional Inviscid Dynamics," 1977, preprint.

12. D. Montgomery and G. Joyce, "Statistical Mechanics of 'Negative Temeprature' States," (1973), preprint.

13. L. Onsager, "Statistical Hydrodynamics," Nuovo Cimento, Supp Al, Vol VI, Series IX, (1949), pp. 279-287.

14. Ya. G. Sinai, Introduction to Ergodic Theory, (1976), Princeton Univ. Press.

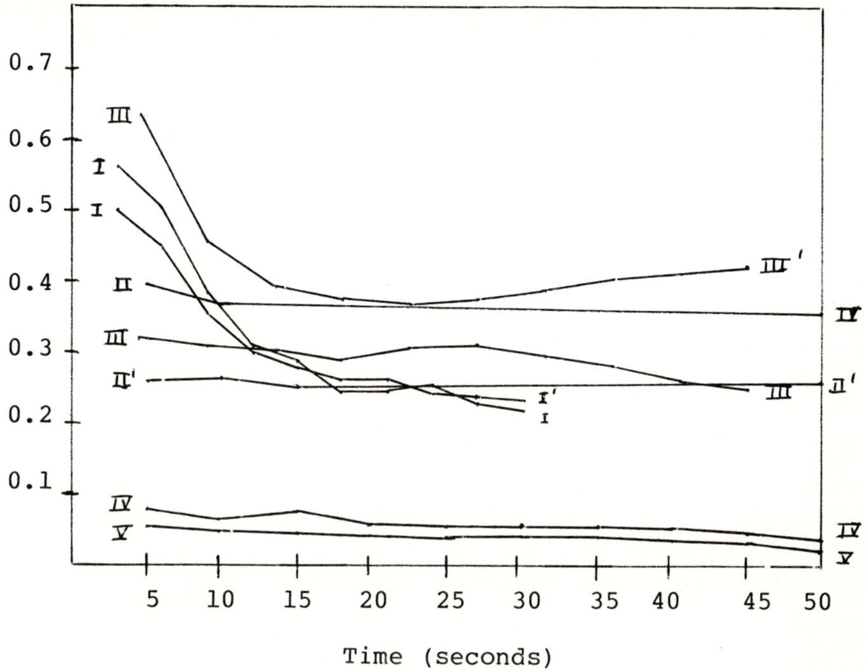

Figure One

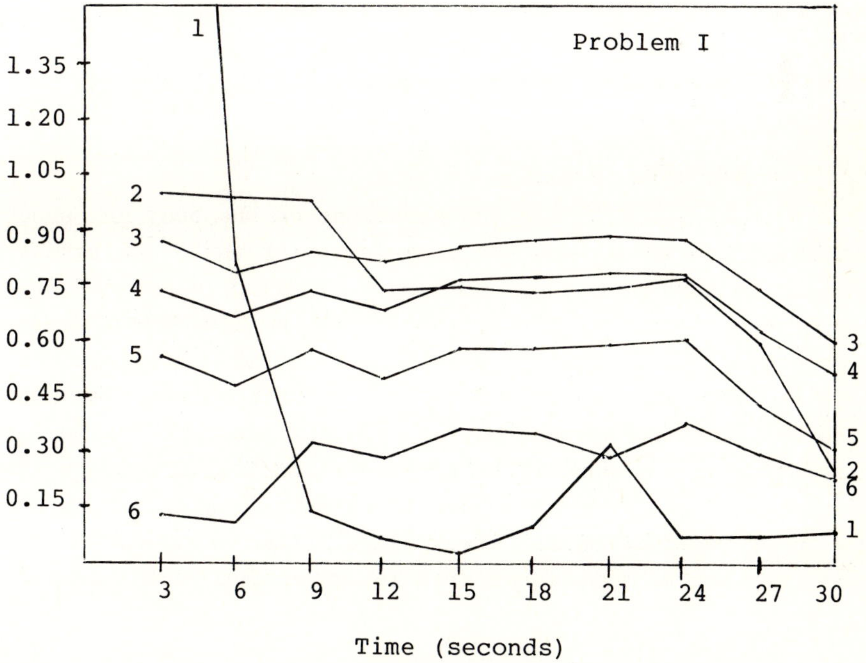

Figure Two

CALCULATION OF THE THREE-DIMENSIONAL SUPERSONIC BLUNT BODY FLOW OF VISCOUS PERFECT GAS AND NONEQUILIBRIUM GAS MIXTURE

Yu.P. Golovachov, V.E. Karyakin, Yu.P. Lunkin, F.D. Popov
A.F. Ioffe Physical-Technical Institute
Leningrad, U.S.S.R.

The numerical methods for calculating the supersonic three-dimensional viscous flows about an axisymmetric blunt body are considered. The stationary solutions are obtained on the base of Navier-Stokes equations by means of the time asymptotic method. At moderate and large Reynolds numbers the viscous terms of these equations which are by the order of magnitude less than $O(1)$ or $O(1/\sqrt{Re})$ in the whole shock layer are ignored [1]. Both simplified sets of differential equations contain all the terms of inviscid flow and boundary layer equations. Velocity vector components are assumed to be equal to zero at the body surface and its temperature is supposed to be constant. Shock slip boundary conditions are used. The flow area is bounded with the surface disposed far enough from the body nose to eliminate the downstream effects on the flow field parameters.

1. Time-dependent curvilinear coordinates (x^1, x^2, x^3) are introdused on the base of spherical coordinate system (r, ϑ, φ):

$$x^1=\vartheta, \quad x^2=(r-r_2)/(r_2-r_1), \quad x^3=\varphi,$$

where $r=r_1(\vartheta)$, $r=r_2(\vartheta,\varphi,t)$ are the equations of the body and shock wave surfaces. To account for the singularity of the coordinate system at $x^1=0$ the mesh points of the computational grid are disposed on the distance of half a step size Δx^1 from the line $x^1=0$. The exact approximation of the elementary solutions (uniform and axisymmetric flows) is required instead of widely used analytical consideration at $x^1=0$. The calculations showed that it was sufficient to obtain a smooth solution in the whole flow area. Realization of this requirement in the finite-difference scheme is especially simple when the tensor form of the initial differential equations is used.

The rectangular computational grid is introduced in the domain considered

$$x^1_k=(k+1/2)\Delta x^1,\quad x^2_l=x^2(l),\quad x^3_m=m\Delta x^3,\quad t^n=n\Delta t,$$
$$k=0,1,\dots,K,\quad l=0,1,\dots,L,\quad m=0,1,\dots,M,\quad n=0,1,\dots$$

Step sizes on the coordinates x^1, x^3 are constant. Step size Δx^2 is variable in order to concentrate the mesh points in the regions where gas dynamic functions vary rapidly. The initial equations are approximated at the point $(k,l+1/2,m)^n$ with the use of implicit finite-difference scheme of the second order accuracy on the spatial coordinates. The values of unknown functions at four mesh points are used to approximate the second derivatives on x^2 (i.e. across the shock layer). First derivatives on x^2 are calculated with the use of two neighbour mesh points. It allows to apply our method for inviscid flow calculation as well.

Covariant velocity vector components U_i and covariant tensor derivatives are evaluated according to tensor analysis formulae, for instance

$$U_i=U_{y\alpha}\frac{\partial y_\alpha}{\partial x^i},\quad \nabla_i U_j=\frac{\partial U_{y\alpha}}{\partial x^i}\frac{\partial y_\alpha}{\partial x^j},\quad \nabla_i T=\frac{\partial T}{\partial x^i},\quad \nabla_2(\eta\nabla_2 U_i)=\frac{\partial}{\partial x^2}(\eta\frac{\partial U_{y\alpha}}{\partial x^2})\frac{\partial y_\alpha}{\partial x^i} \tag{1}$$

where U_y is velocity vector components in the Cartesian coordinate system (y_1,y_2,y_3). Finite-difference representation of the correspondent partial derivatives are (z^n_{klm} is arbitrary grid function)

$$\begin{aligned}
&\partial z/\partial t\sim(z^n-z^{n-1})/\Delta t, && \partial z/\partial x^1\sim(z^n_{k+1}-z^n_{k-1})/(2\sin\Delta x^1)\\
&\partial z/\partial x^2\sim(z^n_{l+1}-z^n_l)/(x^2_{l+1}-x^2_l), && \partial z/\partial x^3\sim(z^n_{m+1}-z^n_{m-1})/(2\sin\Delta x^3)\\
&\frac{\partial}{\partial x^2}(\eta\frac{\partial z}{\partial x^2})\sim\frac{\eta^n_{l+1}(z^n_{l+2}-z^n_l)/(x^2_{l+2}-x^2_l)-\eta^n_l(z^n_{l+1}-z^n_{l-1})/(x^2_{l+1}-x^2_{l-1})}{(x^2_{l+2}-x^2_{l+1}+x^2_l-x^2_{l-1})/2}
\end{aligned} \tag{2}$$

Here and further the subscripts coincident the subscripts of the mesh point $(k,l+1/2,m)$ are omitted.

Finite-difference approximation of the initial equations with the use of formulae analogous (1), (2) results in the following nonlinear algebraic equation at each mesh point $(k,l+1/2,m)^n$

$$A\,X_{l+2}+B\,X_{l+1}+C\,X_l+D\,X_{l-1}=F,\qquad l=0,1,\dots,L-1 \tag{3}$$

where X_l is the column-vector of unknown functions.

Finite-difference approximations of the total continuity equation and momentum conservation equation on the normal to the surface $x^2=$ const only must be retained in the set (3) at $l=0$ to provide a stability. When inviscid flow calculating all the finite-difference equations are to be retained in the set (3) and the corresponuent number of the boundary conditions at the body surface becomes unnecessary.

The iterations are used to solve the nonlinear set of the finite-difference equations at each time step. Newton linearization procedure is applied here for the chemically nonequilibrium flows. At each iteration the solution along the coordinate lines x^1=const is obtained by the elimination method.

Covariant components on x^1 and x^3 are chosen for the velocity and the momentum vectors. It results in decoupling the correspondent momentum conservation equations which solution is obtained independently by the scalar version of the elimination method. Other equations are solved together by the vector variant of the same method. Thus the following iteration for X^n_{klm} vector and the shock front velocity is calculated at each coordinate line x^2. Number of iterations at each time step was usually less than four. Calculations were performing until the time derivatives for all unknown functions in the flow field exceeded the value of 10^{-4}.

Some results for the perfect gas flow about an ellipsoid at angle of incidence $\alpha=10^o$ in regard to the small axis are presented in fig. 1 - 3. The axes length ratio δ =1.5, specific heats ratio $\varkappa$=1.4, Prandtl number Pr=0.75, Mach number M_∞=4, body surface temperature $T_w = 0.15\varkappa M_\infty^2 T_\infty$. Parameters of the computational grid are the following K=8, L=20, M=8. All the distances are referred to the small axis length.

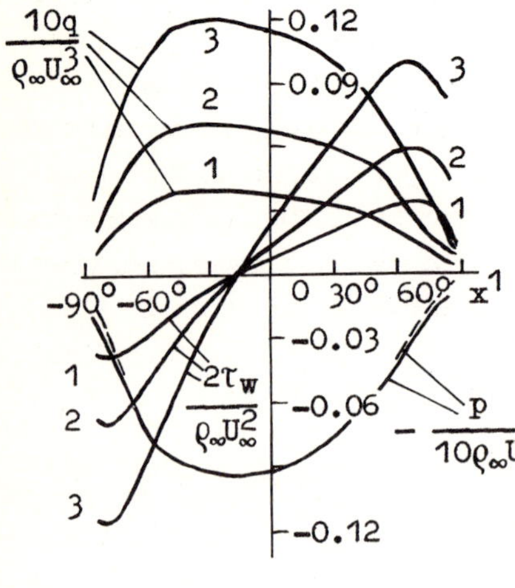

Fig.1

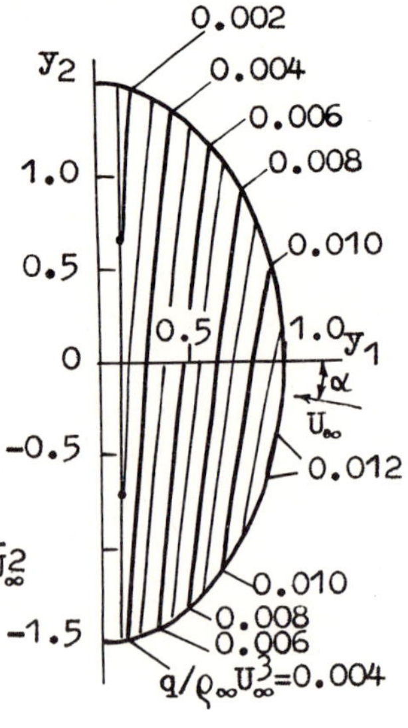

Fig.2

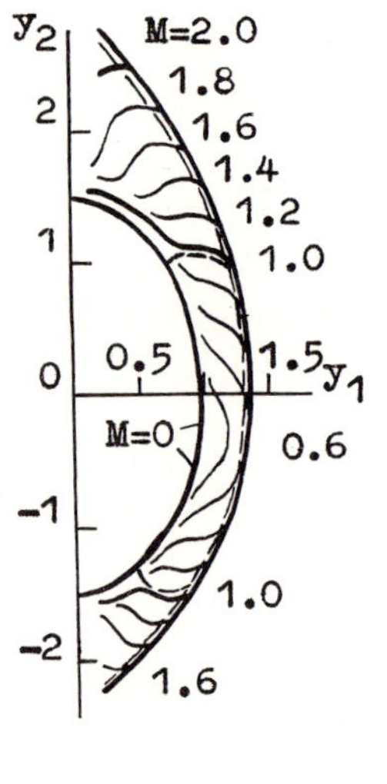

Fig.3

Friction stress and heat flux values at the symmetry plane are plotted in fig.1 for three Reynolds numbers: Re_∞=3333, 1000, 333(curves 1,2,3 correspondingly). Pressure distribution along the body surface for Re_∞=333 is also shown in fig.1. The result of the inviscid flow calculation is shown with the dash line.

The lines of the constant values of heat flux to the body surface (side view) and the lines of the constant values of local Mach number at the symmetry plane are plotted in fig.2,3 for Reynolds number Re_∞= 333. Dash curves in fig.3 show the sonic line and shock front obtained with the use of inviscid flow model. More detailed description of the numerical results is presented in our paper [2].

2.The realization of the above algorithm requires a large computational work. So another method for calculating the three-dimensional viscous flows at small angles of attack is proposed. It based on the power series expansion of the unknown functions when the angle of attack α is considered as the expansion parameter. For the flows about a body of revolution one can obtain that expansion coefficients at α^i are the parts of Fourier serieses on the azimutal coordinate x^3 containing no more terms than i/2+1. According to the symmetry properties on coordinate x^3 they are

$$\begin{aligned} X &= X_{00}(x^1,x^2)+\alpha X_{11}(x^1,x^2)\cos x^3+\alpha^2(X_{20}(x^1,x^2)+X_{22}(x^1,x^2)\cos 2x^3)+\dots \\ X &= \alpha X_{11}(x^1,x^2)\sin x^3 + \alpha^2 X_{22}(x^1,x^2)\sin 2x^3 +\dots \end{aligned} \qquad (4)$$

Vector X_{00} is determined from the solution of the axisimmetric problem. For the following expansion coefficients X_{11}, X_{20}, X_{22} ... the linear sets of two-dimensional equations may be written which are solved one after another. Numerical method described above is applied to solve these equations. In the case considered it requires much less computational work due to the absence of the third spatial coordinate. All the computational work on the series expansions was performed with the computer along with the finite-difference approximation of the initial differential equations.

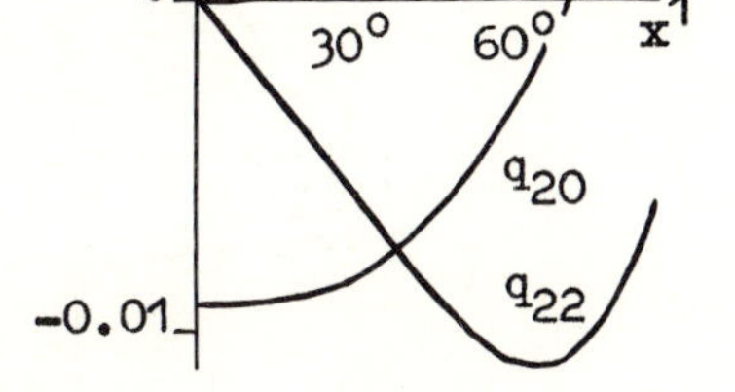

Fig.4

Numerical results for the flow about the ellipsoid at α=10° obtained with only X_{11} term of the expansions (4) differ from the exact three-dimensional solution of part1 less than 5%. Variation of the expansion coefficients for the heat flux on the ellipsoid surface at $Re_{\infty}=333$ is presented in fig.4.

3.The hypersonic carbon dioxide flow about the spherically blunted cones is considered as an example of nonequilibrium flows. The following chemical reactions are taken into account

$$CO_2 + M \rightleftarrows CO + O + M \qquad CO + M \rightleftarrows C + O + M$$

$$O_2 + M \rightleftarrows O + O + M \qquad CO_2 + O \rightleftarrows CO + O_2,$$

where M is an arbitrary particle of mixture. It is supposed that the body surface is absolutely catalytic as regards chemical reactions and its temperature is equal to $2500^{\circ}K$. Constant effective Schmidt numbers $Sc_i=0.5$ are used to account for the multicomponent diffusion. Trans-

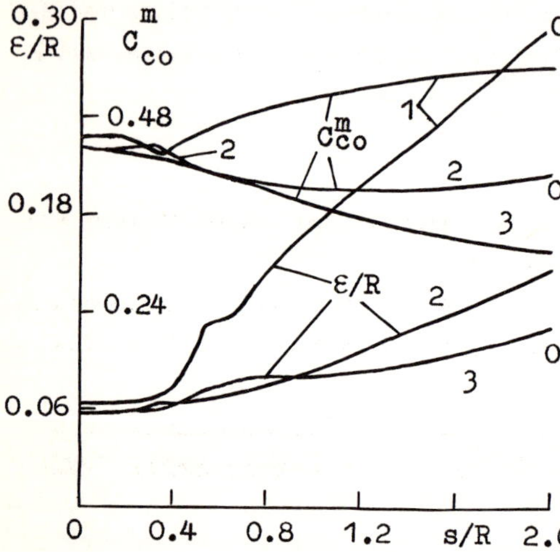

Fig.5

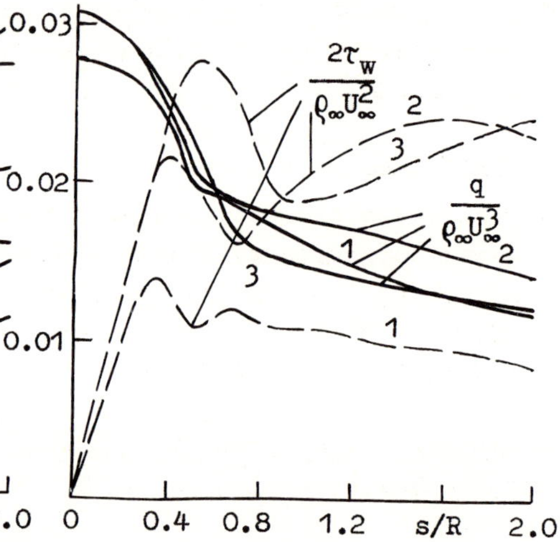

Fig.6

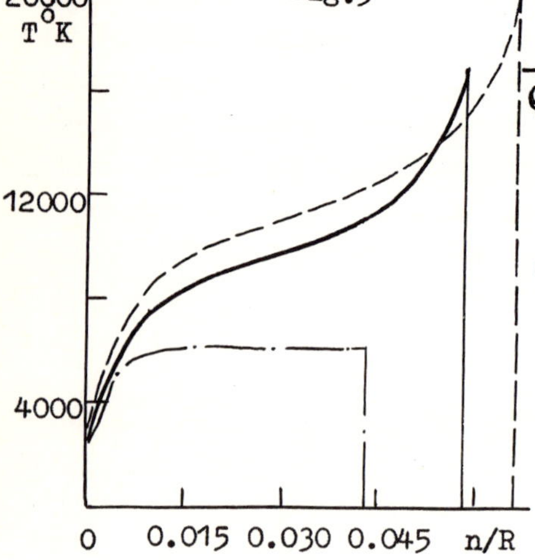

Fig.7

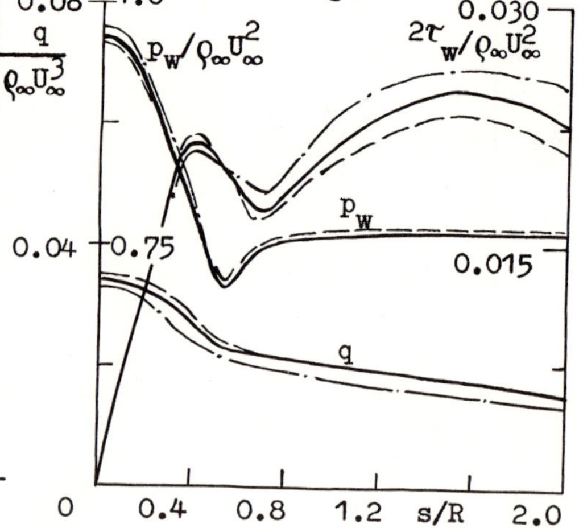

Fig.8

port coefficient are calculated according to kinetic theory formulae [3]. The axisymmetric flow calculations have been performed for the body nose radius R=0.2m and free stream density $\rho_\infty = 0.2 \cdot 10^{-6}$ g/cm^3 with the following computational grid parameters K=12, L=31. Some results for the flight velocity V=6^{km}/s are presented in fig.5,6. Here s is the distance from the stagnation point along the body surface, n is the normal distance from the body surface, C^m_{co} is a maximal value of the carbon dioxide concentration on the lines s=const, curves 1,2,3 corresponds to the cone half-angles 70,60 and 50°. The results for the cone with half-angle value 60° at V=7^{km}/s are plotted in fig. 7,8. Temperature profiles along the symmetry axis are shown in fig.7. Pressure, skin friction and heat flux distributions along the body surface are presented in fig.8. Solid lines show the results for 100% CO_2, dash lines - for the 75%CO_2 - 25%Ar mixture. The results obtained under the local equilibria assumption for 100%CO_2 are shown with dotted curves.

REFERENCES

1. Belotzerkovski, O.M. et al. Numerical Investigation of Modern Gas Dynamics Problems, "Nauka", Moscow, 1974.
2. Karyakin, V.E.; Popov, F.D. Calculation of supersonic three-dimensional viscous and heat-conducting blunt body flow. Ẑ. Vyĉisl.Mat. i Mat. Fiz. 17 (1977), no. 6, 1545-1555.
3. Agafonov, V.B.; Vertushkin, V.K.; Gladkov, A.A.; Polyanski, O.Yu. Nonequilibrium Physical-Chemical Processes in Aerodynamics, "Mashinostroenie", Moscow, 1972.

NUMERICAL SIMULATION OF MHD-PROBLEMS ON THE BASIS OF VARIATIONAL APPROACH

V.M. Goloviznin, A.A. Samarskii,
A.P. Favorskii, T.K. Korshia
Institute of Applied Mathematics, Academy of Sciences
Moscow, USSR

I. Many of modern scientific investigations deal with solution of multi-dimensional MHD-problems. For these problems the finite-difference methods are of the most universal and effective use.

The difference scheme may by considered as the discrete model of a medium. Obviously the scheme of better quality results in better simulation of the medium properties. For example the divergence schemes [I] provide an adequate accuracy of the discontinuous flow computations. The completely conservative schemes [2] may be used successfully even on crude grids.

The differential equations of nondissipative MHD and their most important properties with conservation laws included may be obtained brom variational principles similar to those of Hamilton-Ostrogradskii in classical mechanical systems [3, 4]. The completely conservative finite-difference schemes can be obtained through the analogous approach.

2. As an example we consider the liquid volume G of an idealy conductive adiabatic plasma moving in space. Let Ω_3 be the domain that corresponds to G in Lagrangian coordinates (α, β, γ); $x = x(\alpha,\beta,\gamma,t)$, $y = y(\alpha,\beta,\gamma,t)$, $z = z(\alpha,\beta,\gamma,t)$ the Cartezian coordinates of particles of the medium. The Jacobian $D = D(x,y,z)/D(\alpha,\beta,\gamma) > 0$; $\alpha,\beta,\gamma \in \Omega_3$. The Lagrangian of liquid volume G is

$$\mathcal{L}(t) = \iiint\limits_{\Omega_3} \rho D \left(\frac{\vec{V}^2}{2} - \varepsilon - \frac{\vec{H}^2}{8\pi\rho} \right) d\alpha\, d\beta\, d\gamma \qquad (1)$$

Here $\vec{V} = (u, v, w)$ is the velocity vector, $\vec{H} = (H_x, H_y, H_z)$ the vector of frozen-in magnetic field.

The law of mass conservation is

$$\rho D = \rho_0(\alpha,\beta,\gamma) \qquad (2)$$

The frozen-in condition for magnetic field yields the equations

$$H_x \frac{D(y,z)}{D(\alpha,\beta)} + H_y \frac{D(z,x)}{D(\alpha,\beta)} + \frac{D(x,y)}{D(\alpha,\beta)} H_z = \Phi_{\alpha\beta}$$

$$H_x \frac{D(y,z)}{D(\beta,\gamma)} + H_y \frac{D(z,x)}{D(\beta,\gamma)} + H_z \frac{D(x,y)}{D(\beta,\gamma)} = \Phi_{\beta\gamma} \qquad (3)$$

$$H_x \frac{D(y,z)}{D(\gamma,\alpha)} + H_y \frac{D(z,x)}{D(\gamma,\alpha)} + H_z \frac{D(x,y)}{D(\gamma,\alpha)} = \Phi_{\gamma\alpha}$$

In this paper we consider that all flow parameters depend on only two space coordinates x and y i.e. α and β ,respectively,in Lagrangian coordinates. In this case we can accept

$$\Omega_3 = \Omega_2 \otimes \{0 \leq \gamma \leq 1\}$$

$$\frac{\partial}{\partial z} \equiv 0, \quad \frac{\partial x}{\partial \gamma} = \frac{\partial y}{\partial \gamma} \equiv 0, \quad \frac{\partial z}{\partial \gamma} \equiv 1$$

Equation (2) converts into

$$\rho D = \rho_0(\alpha,\beta) \qquad (2')$$

The relation

$$D\vec{H} = \frac{\partial \vec{r}}{\partial \alpha} \Phi_{\beta\gamma} + \frac{\partial \vec{r}}{\partial \beta} \Phi_{\gamma\alpha} + \frac{\partial \vec{r}}{\partial \gamma} \Phi_{\alpha\beta} \qquad (3')$$

follows from (3).

The time differentiation of (3') gives the equation of induction of magnetic field

$$\frac{d}{dt}(\vec{H}\cdot D) = (\vec{H}\cdot\vec{\nabla}_{\alpha\beta})\vec{V}$$

where

$$\vec{\nabla}_{\alpha\beta} = \left\{ \frac{D(\cdot,y)}{D(\alpha,\beta)}, \frac{D(x,\cdot)}{D(\alpha,\beta)}, 0 \right\} = D\cdot\vec{\nabla}$$

By varying the functional of action $S = \int_{t_0}^{t_1} \mathcal{L}(t)\,dt$ and taking into account additional constraints (2'), (3'), and adiabaticity condition

$$\delta\varepsilon = \frac{p}{\rho^2}\delta\rho \qquad (4)$$

and by setting the first variation δS equal to zero, we obtain the Euler equations

$$\rho D \frac{d\vec{V}}{dt} = -\nabla_{\alpha\beta}\left(P + \frac{H^2}{8\pi}\right) + \frac{\partial}{\partial\alpha}\frac{\vec{H}\Phi_{\beta\gamma}}{4\pi} + \frac{\partial}{\partial\beta}\frac{\vec{H}\Phi_{\gamma\alpha}}{4\pi} \qquad (5)$$

From (4) we obtain the equation for internal specific energy

$$\rho D \frac{d\varepsilon}{dt} = -P\left[\frac{D(u,y)}{D(\alpha,\beta)} + \frac{D(x,v)}{D(\alpha,\beta)}\right]$$

3. We assume that $\Omega_2(\alpha,\beta)$ is a square. Let us introduce the restangular difference grid with the intervals h_α and h_β; ω being the set of its cells and $\bar{\omega}$ the set of its nodes.

Let us approximate the Lagrangian $\mathcal{L}(t)$ by the discrete expression

$$\mathcal{L}_h = \sum_{(i,j)\in\omega} S_{ij}\rho_{ij}\left(\sum_{k\in\omega_1(i,j)} \frac{\vec{V}_k^2}{8} - \varepsilon_{ij} - \frac{\vec{H}_{ij}^2}{8\pi\rho_{ij}}\right) \qquad (6)$$

where $\omega_1(i,j) = \{(i,j),(i+1,j),(i+1,j+1),(i,j+1)\}$ and S_{ij} is the volume of the Lagrangian cell.

The discrete analogs to the law of mass conservation and frozen-in condition are

$$\rho S = m \qquad (2'')$$

$$\begin{cases} H_x S = \Phi_k x_k \\ H_y S = \Phi_k y_k \\ H_z S - \Phi_k z_k = \Phi = const \quad ; \quad k \in \omega_1(i,j) \end{cases} \qquad (3'')$$

The repeated index denotes the summation. The value $\Phi_k = (H_x)_{ij}\frac{\partial S_{ij}}{\partial x_k} + (H_y)_{ij}\partial S_{ij}/\partial y_k$ is the magnetic flux across diagonals of the cell (i,j) in the direction to the k-th node. According to (4) the energy change law is

$$m\frac{d\varepsilon}{dt} = -P\frac{dS}{dt} = -P\left(\frac{\partial S}{\partial x_k}u_k + \frac{\partial S}{\partial y_k}v_k\right) \qquad (4')$$

After the time differentiation frozen-in condition (3'') yields the magnetic induction equation

$$S\frac{d\vec{H}}{dt} = -\vec{H}\frac{dS}{dt} + \Phi_k \cdot \vec{V}_k$$

Varying the action functional with constraints (2")-(4") results in the MHD-equations

$$M\frac{du}{dt}-\left(P+\frac{\vec{H}^2}{8\pi}\right)_K\frac{\partial S_K}{\partial x}+\frac{(H_x)_K\Phi_K}{4\pi}=0$$

$$M\frac{dv}{dt}-\left(P+\frac{\vec{H}^2}{8\pi}\right)_K\frac{\partial S_K}{\partial y}+\frac{(H_y)_K\Phi_K}{4\pi}=0 \qquad (5')$$

$$M\frac{dW}{dt}+\frac{(H_z)_K\Phi_K}{4\pi}=0\ ;\quad K\in Ш_2(i,j)$$

$$M=\sum_{K\in Ш_2(i,j)} m_K/4$$

Here $Ш_2(i,j)=\{(i,j),(i-1,j),(i,j-1),(i-1,j-1)\}$

Schemes (5") are valid in the inner nodes of the grid and accurate to the second-order approximation in h_α and h_β. Simple generalization of the action functional extends (5*) into the boundary nodes. Differential-difference schemes (5*) are completely conservative [4].

4. The difference scheme is deduced from equations (5*) through replacing the time derivative by the finite-difference relations. By time centering the right hand side terms the difference schemes are made completely conservative.

5. The linearization of equations (2")-(5*) produces the differential-difference schemes for MHD-acoustics. The variational approach provides for a space operator of this schemes to be self-adjoint and positive [4]. The scheme stability depends on the implicitness of the scheme. The difference equations are solved by the Newton method.

To calculate the discontinuous flows artificial viscosity is used [4].

6. The variational approach has been used as a basis for numerical algorithms and techniques employed for solving a number of model and applied problems [5, 6, 7].

R E F E R E N C E S

I. Samarskii, A.A. and Popov, Yu.P. Difference schemes of gasdynamics, Moscow, "Nauka", I975.

2. Seliger, R.L. and Whitham, G.B. Variational principles in continuum mechanics, Proc. Roy.Soc., A305, I-25 (I968).
3. Goloviznin, V.M., Samarskii, A.A., and Favorskii, A.P. Variational approach to constructing the finite-difference mathematical models in hydrodynamics, Dokl.Akad.Nauk SSSR, I977, v.235, N 6.
4. Goloviznin, V.M., Korshia, T.K., Samarskii, A.A., and Favorskii A.P. On variation-difference schemes of MHD, Preprint IPM, Akad. Nauk SSSR, N 57, I977.
5. Gasilov, V.A., Goloviznin, V.M., Tishkin, V.F., and Favorskii, A.P., Numerical solution of the model problem on Rayleigh-Taylor instability evolution, Preprint IPM, Akad.Nauk SSSR, N II9, I977.
6. Volkova, R.A., Goloviznin, V.M., Samarskii, A.A., Ulinitch, F.R., and Favorskii, A.P., Calculation of magnetic cumulation.
7. Volosevitch, P.P., Gamalii, E.G., Gulin, A.V., Rozanov, V.V., Samarskii, A.A., Tyurina, N.N., and Favorskii, A.P., Two-dimensional effects during the laser compression of spherical targets, Pis'ma v JETF, v.24, N 5, 283-286 (I976).

NUMERICAL SOLUTION FOR UNSTEADY SEPARATED INVISCID INCOMPRESSIBLE FLOW PAST AN ARBITRARY BODY

V.A. Golovkin, M.A. Golovkin
(USSR)

The problem of the unsteady motion around a variety of bodies with wake followed developing as tangential velocity discontinuities is considered. Numerical and experimental data are compared.

Calculation method for unsteady and separated inviscid incompressible flows have been extensively studied /1-7/. Nevertheless, they have not been sufficiently developed for arbitrary bodies.

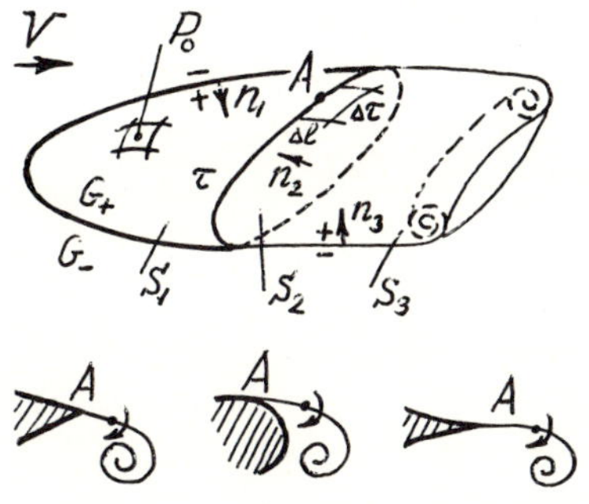

Fig.1

Potential of perturbation velocities vanishing at infinity is known to be defined by the exterior problem, Fig.1:

$\Delta\varphi=0$ in G_-; $\left(\frac{\partial\varphi}{\partial n}\right)=v_{on}-V_n$ on S_-;

$p_+-p_-=0$; $\left(\frac{\partial\varphi}{\partial n}\right)_+=\left(\frac{\partial\varphi}{\partial n}\right)_-$ on S_3,

where v_{on} - speed of points $P_o\in S=S_1\cup S_2$, V - free stream velocity, p - pressure. Evidence dealing with the initial state is a supplement to the problem.

Let $\varphi_-=\varphi_-^{\nu}+F^{\Omega+\gamma}$, where φ^{ν} - double layer potential over S and S_3 with the density ν, $F_-^{\Omega+\gamma}$ - potential known in the domain G_- due both to the vortices filling in the domain G_+ with the intensity $\boldsymbol{\Omega}=2\boldsymbol{\omega}$, where $\boldsymbol{\omega}(t)$ - angular velocity, and to the vortex layer on S with the intensity $\boldsymbol{\gamma}^{\omega}=\mathbf{n}\times\mathbf{v}_S^{\omega}$, where $\mathbf{v}_S^{\omega}$ is the projection of the vector $\mathbf{v}_o^{\omega}=\boldsymbol{\omega}\times\mathbf{r}_o$ on S, with $\mathbf{r}_o$ being position vector of the point P_o. Then there exists no relative fluid motion in the region G_+. Therefore, the flow potential, velocity and pressure on S_- are defined straightforward by the intensity of the hydrodynamic singularities and their derivatives, whereas the boundary condition on S_- is reduced /3, 4/ to the integral equation for $\nu(P_o, t)$, t being time, by transition to the region G_+ :

$$\varkappa \nu(P_o,t)+\int_{S}\nu(P,t)K(P_o,P)dS=Q(P_o,P)-\int_{S_3}\nu_3(P,t)K(P_o,P)dS,$$

where $\varkappa=2\pi$, $K(P_o,P)=\frac{\partial r^{-1}}{\partial n}$ for a three dimensional problem, while $\varkappa=\pi$, $K(P_o,P)=\frac{\partial}{\partial n}\ln\frac{1}{r}$ for two-dimensional one, n - normal at the point $P\in S\cup S_3$, $r=|P_o-P|$, $Q(P_o,t)$ is the known potential.

Integral over S_3 may be thought of as that over the band σ of the width $\Delta\ell$ (ℓ is normal to τ at the point $P'\in\tau$) adjacent to the flow shedding line τ , and that over S_3/σ . $\nu_{\sigma i}(P'_i,t)$ being average value ν_3 in the cell $\Delta\ell\,\Delta_i\,\tau$, $\nu^{(0)}$ - integral equation solution for all i at $\nu_{\sigma i}$ = 0 and $\nu_i^{(1)}$ - one for $j\neq i$ and $\nu_{\sigma i}$ = 1 at ν_{σ_j} = 0, the required density is given by

$$\nu(P_o,t)=\nu_o^{(0)}(P_o,t)+\sum_i \nu_{\sigma i}(P'_i,t)\,\nu_i^{(1)}(P_o,P'_i,t).$$

With the known right-hand side the equations for $\nu^{(0)}$ and $\nu_i^{(1)}$ are Fredholm's integral equations of the second kind (potential theory ones), whose solution approaches are known /8/. Representing their solutions by means of the resolvent, it follows that $\nu^{(0)}$ and $\nu_i^{(1)}$ are derived with the same accuracy as that of the right-hand side terms. Depending on the problem peculiarities and computer capabilities both the Krilov-Bogolubov's method of reducing the equations obtained to a system of linear algebraic ones and iterative approaches are applied for the nonuniform grid (ν = const in the grid cell). Under the above condition the approximate solutions approach the exact ones.

The velocity expression having singularity of the type r^{-1} for $\Gamma_K=\varphi_+-\varphi_-=2\varkappa\nu_K\ (P)\neq 0$, where $P\to A=P_o\in\tau$, and that of $\ln r$ type for $\gamma_{\tau_K}=-\varkappa\nu'_{\ell_K}\neq 0$ at the line τ of the surface S_K (K = 1, 2, 3), Fig.1, then according to the Chapligin--Joukowski condition of the finite fluid velocity near the line τ the condition of the form

$$\Gamma_1-\Gamma_2-\Gamma_3=0\ ,\qquad H=\frac{\partial\Gamma}{\partial t}+\frac{\partial\Gamma}{\partial\tau}\,\frac{\partial\tau}{\partial t}\ ,$$

$\mathbf{n}_3 \equiv \mathbf{n}_1$, $\gamma_{\tau_3} = \gamma_{\tau_1} - \gamma_{\tau_2}$, for $H > 0$; $\mathbf{n}_3 \equiv -\mathbf{n}_2$, $\gamma_{\tau_3} = \gamma_{\tau_2} - \gamma_{\tau_1}$, for $H < 0$ written in the coordinate system moving with the point A, must hold for any t .

Applying the Ostrogradsky-Gaussian formula whose volume integral contains $\operatorname{div} \mathbf{v}$ to the small volume adjacent to τ and limited by S_3 , S_2 and planes normal to τ and ℓ shows that on the smooth portion of S_2 the velocity of the point $A \in \tau$ motion normal to τ coincides with the velocity component of the fluid particle adjacent to the point A. $A \in \tau \subset S$ being angular point, the projection of the fluid velocity with respect to S on a plane normal to τ is zero and one must assume $v_{\ell_2 -} = -\varkappa \nu'_{\ell_2} = \gamma_{\tau_2} \equiv 0$ for $H > 0$ and $\gamma_{\tau_1} \equiv 0$ for $H < 0$.

Note, that the normal coincidence condition should be satisfied. Again, either relation for Γ_i (Thomson's theorem analogue) or that for γ_{τ_i} must be met, then the remain of the conditions will be fulfilled as well. So when representing ν versus $\nu^{(0)}$, $\nu_i^{(1)}$ the conditions for γ_{τ_i} are to be met. For the derivatives entering the continuity equation to be limited near τ (and consequently P'_ℓ as well), the curvature equality condition for lines of intersection of the plane normal to τ with S , S_3 must be satisfied at the point A.

In the point $P \to A$, $P \in S_3$, reference coordinate system we have:

$$v_{\ell_3} = -\frac{|\gamma_{\tau_3}|}{2} \leq 0,$$

$$v_{\tau_3} = \frac{\gamma_{\tau_1} + \gamma_{\tau_2}}{2},$$

$$\gamma_{\ell_3} = \gamma_{\ell_1} - \gamma_{\ell_2},$$

$$\gamma_{\tau_3} = -\sqrt{|2H|}\operatorname{sign} H .$$

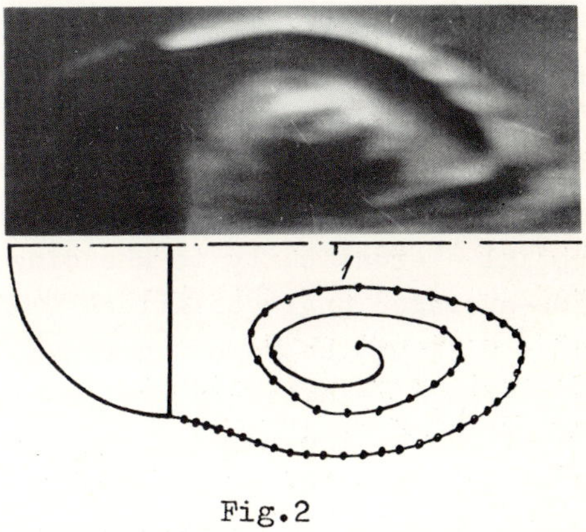

Fig.2

The position $\mathbf{x}$ of the points ν_3 = const of the tangential velocity discontinuity S_3 is defined as $\mathbf{x} = \mathbf{x}_0 + \int_{t_0}^{t} \mathbf{v}\, dt$, where $\mathbf{x}_0 = \mathbf{x}(t_0)$. The vector $\mathbf{v}$ contains the Biot-Savart integral operator taken at the points $\mathbf{x} \in S_3$. The problem of defining the tangential velocity discontinuity is known to be incorrect. Therefore, the regularising operator similar to one

Fig.3

from /9/ is applied, if necessary, at each step of calculation to the solution obtained by presenting the integral in the last expression by the quadrature formula of rectangules. The form of S_3 is slightly dependent on the repetitive use of this operator.

The calculation (Fig.2-4) agrees well with the experimental data obtained by the visualisation of a) separated flow behind a half-cylinder in a highly accelerated free stream (optical vizualization in water tunnel), b) flow in the wake behind a harmonically oscillating airfoil NACA-0015 /10/ ($h = A\cos K\tau$; A = 0.018; K=17; $\tau = \frac{V\cdot t}{b}$, b - chord length), c) flow behind a base of a half-ellipsoid of revolution in highly accelerated free stream /11/.

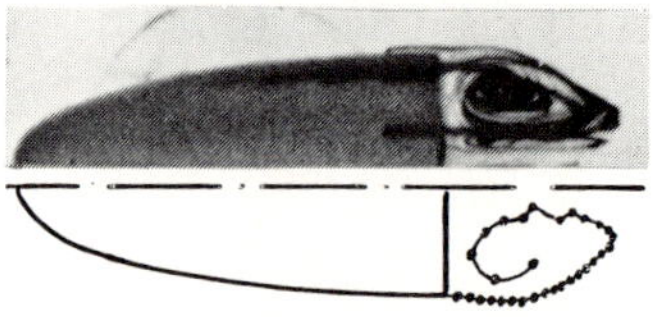

Fig.4

The calculated (the tangential discontinuity adjacent to the trailing edge only) and experimental /12/ normal force coefficients for an airfoil harmonically (A = 0.152, K = 0.134) oscillating across the stream are compared in Fig.5. The effect of the airfoil motion nature on the pressure distribution (tangential discontinuity is shedding at the trailing edge only) at the time-averaged angle of attack $\alpha_o = 5^o$ is illustrated in Fig.6. For K = 0.14 at the moment corresponding to the rearward relative to the free stream airfoil motion at the maximum speed in a period of oscillations three cases are considered: a) sinusoidal oscillations along the free stream (A = 5), b) biharmonical synchronous oscillations (A = 1.19; A_2 = 3.57) and c) asunchronous (A_1 = -1.19; A_2 = 3.57) longitudinal-lateral oscillations. The developed separated flow with the Karman street behind the airfoil as the angle of attack α = 60° set impulsively in motion with cons-

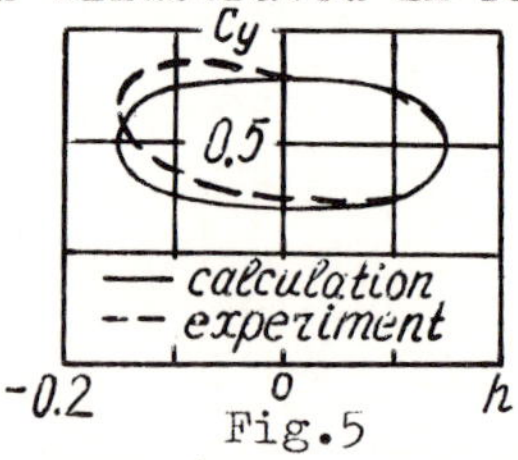

Fig.5

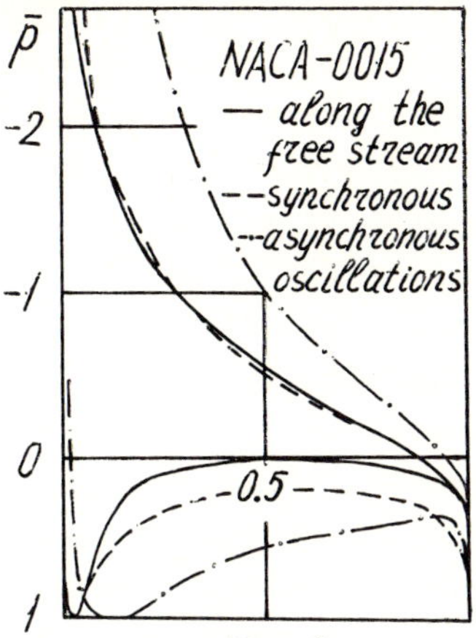

Fig.6

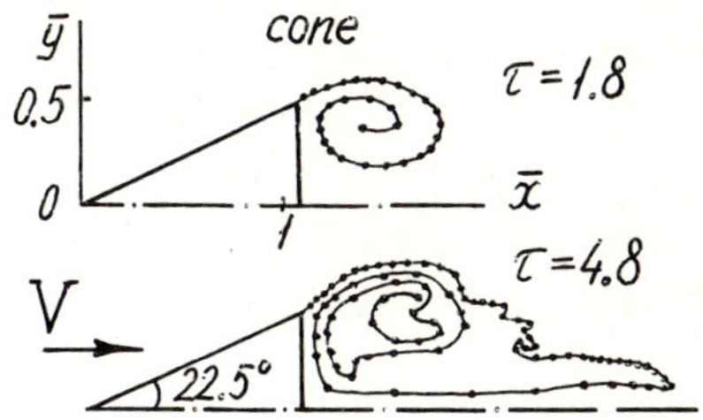

Fig.9

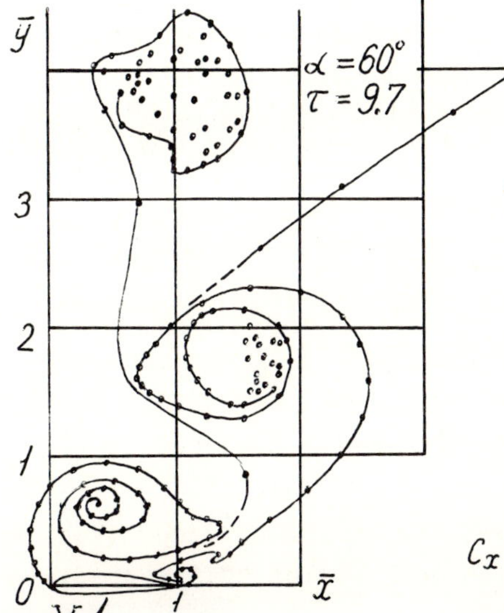
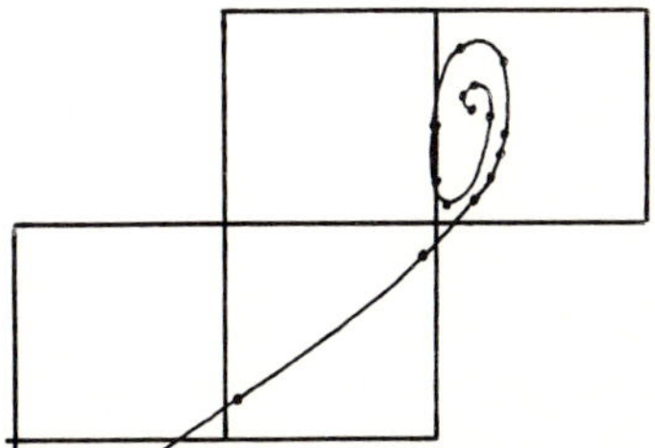

Fig.7

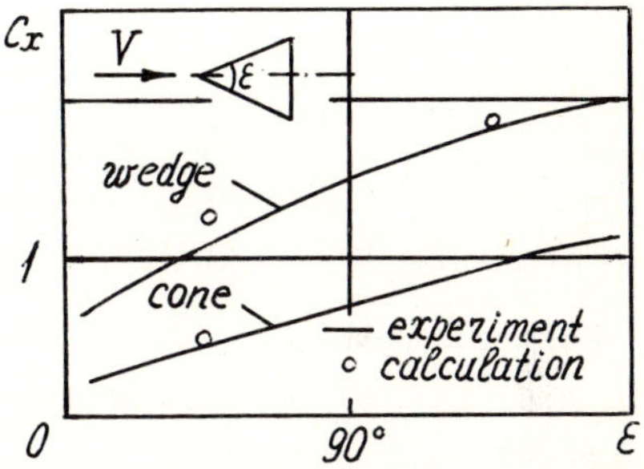

Fig.10

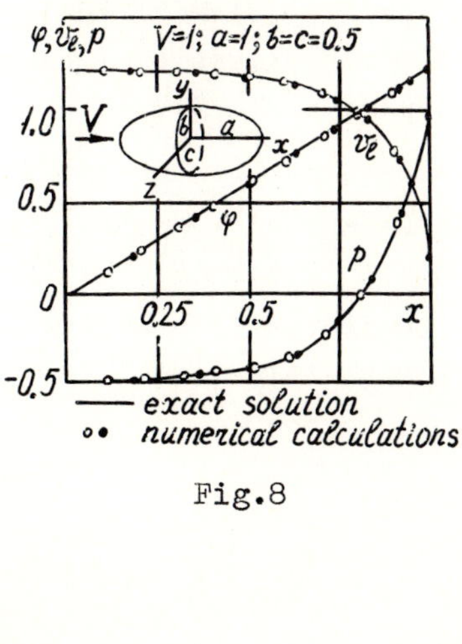

Fig.8

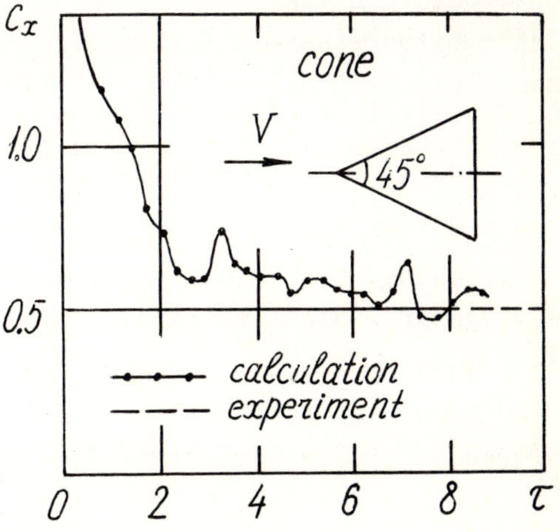

Fig.11

tant velocity is presented in Fig.7.

The calculation of the three-dimensional flow without circulation around an ellipsoid agrees well with the exact solution,Fig.8. The process of developing the axisymmetric separated flow behind a cone set impulsively in constant velocity motion is shown in Fig.9 ($\tau = \frac{V \cdot t}{2R}$, R - mid-section radius). Significant deformation with time of the vortex sheet being helical initially is evident. The time-averaged calculated and experimental drag coefficients C_x are compared, Fig.10, for a cone and wedge with different apex angles ε /13/. The calculated results conform quite closely to the experimental data. Fig.11 shows $C_x(\tau)$ settling for the cone with time after the instanteneous change of the free stream velocity from zero to a constant value at an initial moment, on this case for the axisymmetric flow is considered.

Thus, the examples cited show the satisfactory agreement of the numerical results with the experimental evidence and the usefulness of the solution method.

REFERENCES

1. Ilichov K.P., Postolowsky S.N. - Calculation of the unsteady separated two-dimensional inviscid fluid flow around bodies. - "Izv.Akad.Nauk.SSSR", Mekh.Zhidk.Gasa., N 2, 1972.

2. Belotserkowsky C.M., Nisht M.I. - Nonlinear unsteady theory for the thin wing of an arbitrary plane form. - "Izv.Akad.Nauk.SSSR", Mekh.Zhidk.Gasa, N 2, 1974.

3. Golovkin V.A. - Nonlinear problem of the unsteady flow around an arbitrary aerofoil with a free deformation of the wake. - "Uchen.Zapis.TsAGI", v.III, N 3, 1972.

4. Golovkin M.A. - Method of solution for a problem of the separated inviscid incompressible flow around three-dimensional body, moving arbitrary. - "Uchen.Zapis.TsAGI", v.VIII, N 2, 1977.

5. Sudakov G.G. - Calculation of the separated flow around the cone wings of a small thickness. - "Uchen.Zapis.TsAGI", v. V, N 6, 1974.

6. Marschall F.J., Deffenbaugh F.G. - Separated flow over a body of revolution. - "J.Aircraft", II, v.12, N 2, 1975.

7. Hayashi M., Endo E. - A method for calculating the flow around arbitrary airfoil sections with separation. - "Trans.Japan Soc. for Aeronaut.Sp.Scien.", N 49, 1977.

8. Kantorovich L.V., Krylov V.I. - Approximate methods of the highest analysis. - M.-L. "Phismatgiz", 1962.

9. Molchanov V.F. - A some calculation questions for flows with tangential discontinuities. - "Uchen.Zapis.TsAGI", v.VI, N 4, 1975.

10. Bratt J.B. - Flow patterns in the wake of an oscillating aerofoil. - "ARC R & M", N 2773 (13,001) 1953.

11. Animitsa V.A., Golovkin V.A., Golovkin M.A., Kolkov V.G. - Investigation separated flow around the cutted ellipsoids of revolution with a flat base. - "Uchen.Zapis.TsAGI", v.VIII, N 3, 1977.

12. Liiva J. - Unsteady aerodynamic and stall effects on helicopter rotor blade airfoil sections. - "AIAA Paper"N 68-58, 1968.

13. Fluid-dynamic drag. Hoerner S.F. "Published by the Author", 1965.

DIRECT NUMERICAL METHOD FOR THE VELOCITY PROFILE AND THE FORM OF A LAMINAR JET IN A LIQUID-LIQUID SYSTEM

P. Gospodinov I. Penchev S. Radev

Institute of Mechanics and Biomechanics, Sofia - Bulgaria

ABSTRACT

An exact numerical solution for the velocity profiles and the form of a laminar jet in immiscible Newtonian liquid-liquid systems is obtained. The main difficulty connected with the simultaneous integration of the equations of motion of the jet and of the continuous phase in the presence of a unknown interface is overcome by using an indivisible finite-difference scheme. Thus is avoided any additional iteration for defining the unknown velocities of the points of the interface. The solution of the equations of motion as written in a boundary layer approximation is a function of the following nondimensional numbers: the Reynolds numbers of both phases, the Weber and Froude numbers and the ratio of the densities. The influence of some of these parameters on the jet behaviour is illustrated as well as the influence of the initial velocity profiles at the nozzle exit.

A comparison is made with some known results. Important differences are found to exist between the exact and the approximate velocity profiles and their gradients at the interface. It seems that these differences result from the comparatively inexact description of the boundary layer of the continuous phase when using moment methods. Such a conclusion limits the applicability of the approximate moment solutions to heat and mass transfer problems as well as to the jet stability analysis.

The first numerical solution of the problem of determining the velocity profile and the contraction of the liquid laminar jet is presented in [1] for the case of a gaseous continuous phase. The effects of the continuous phase density and viscosity are taken into account in [5] and a solution is proposed by a moment method. Various moment methods have been also used in other papers (see [3] , [4] etc).

In the present paper the range of applicability of the finite-difference method has been extended over the general case of a jet in liquid-liquid systems. A direct numerical solution of the equations of motion as stated in [5] has been obtained. By the proposed method the solution in [1] can be obtained as a particular case. The main point of the method proposed is the finite-difference formulation of the indivisible boundary problem for both phases. Thus any additional iteration for the points of the interface has been avoided. That is why the computational time is comparable to the time necessary for obtaining the approximate numerical solutions.

EQUATIONS OF MOTION

Consider a vertical stationary axi-symmetric liquid jet that flows with an average axial velocity U_N into another liquid out of a nozzle of a circular cross section of a diameter D_N. We shall suppose that the two liquids in the jet and in the surrounding space are incompressible, Newtonian and immiscible and that the fluid system thus obtained lacks heat - and mass transfer. The geometrical configuration and the cylindrical coordinate systeme used are show in fig. 1.

The equations of motion and the boundary conditions are written in a dimensionless form with the aid of the characteristic scale parameters D_N and U_N. Variables with subscript 1 and 2 refer to the continuous and dispersed phase respectively.

a) Continuous phase. The complete equations of motion of the continuous phase include two dimensionless parameters - Reynolds number $Re_1 = D_N \rho_1 U_N / \mu_1$ (ρ_1 and μ_1 are correspondingly density and dynamic viscosity) and Froude number $Fr = U_N^2 / D_N g$ (g - acceleration of gravity). The analysis of the orders of magnitude shows [5] that the simplification of the boundary layer theory may be applied to those equations when $Re_1 > 100$:

$$(1) \qquad \frac{1}{r}\frac{\partial}{\partial r}(rV_1) + \frac{1}{Re_2}\frac{\partial \varphi_1^{1/2}}{\partial \xi} = 0,$$

$$(2) \qquad V_1 \varphi_1 \frac{\partial \varphi_1}{\partial r} + \frac{\varphi_1^{3/2}}{Re_2}\frac{\partial \varphi_1}{\partial \xi} = \frac{4\varphi_1^{3/2}}{Re_1}\left[\frac{1}{r}\frac{\partial}{\partial r}\left(\frac{r}{2\varphi_1^{1/2}}\frac{\partial \varphi_1}{\partial r}\right)\right],$$

where ($U = \varphi^{1/2}, V$) are the velocity components and $\xi = z / Re_2$ is a new independent variable.

As usually within the boundary layer approximation the presure gradient term is excluded from (2).

b) Dispersed phase. In order to overcome the difficulty due to the presence of the unknown jet surface $r = R(\xi)$, Protean coordinates (ψ, ξ) with ψ being the stream function are used in [1].

According to [1] the equations of motion in a boundary layer approximation with the dispersed phase concerned are of the form

$$(3) \qquad \frac{\partial \varphi_2}{\partial \xi} = N_j + \frac{2}{We(R^2)^{3/2}}\frac{dR^2}{d\xi} + 4\left[\frac{\partial \varphi_2}{\partial \psi} + \frac{r^2 \varphi_2^{1/2}}{2}\frac{\partial^2 \varphi_2}{\partial \psi^2}\right],$$

where $N_j = Re_2 (1-\rho_1/\rho_2)/Fr$ corresponds to a jet issuing downwards and $We = D_N U_N^2/T$ is Weber number with T as a surface tension.

In (3) the radial coordinate of an arbitrary point of the jet is supposed to be a function of ψ and ξ defined by

$$\frac{\partial r^2}{\partial \psi} = \frac{2}{\varphi_2^{1/2}} \quad (4)$$

Equations (3) and (4) form a system of nonlinear partial differential equations which describes the distribution of the axial velocity component within the jet. For determining the radial component the equality

$$V_2 \cdot Re_2 = \varphi_2^{1/2} \frac{\partial r}{\partial \xi} \quad (5)$$

is used.

BOUNDARY CONDITIONS

The equations of motion of both phases must be supplied with boundary conditions, describing the flow along the centerline of the jet and on the outer edge of the continuous phase boundary layer. The conditions describing the character of the interaction between the jet flow and the surrounding fluid must be satisfied on the interface.

The complete set of boundary conditions is listed below:

a) Interface ($\psi = 1/2$)

$$\mu_1 \frac{\partial \varphi_1}{\partial r} = \mu_2 R \varphi_2^{1/2} \frac{\partial \varphi_2}{\partial \psi} \quad (6)$$ (shear stress continuity)

$$p_2 - p_1 = \frac{2}{RWe} \quad (7)$$ (normal stress continuity)

$$\varphi_1 = \varphi_2 \quad (8)$$ (axial and

$$V_1 = V_2 \quad (9)$$ radial component continuity)

b) Jet axis ($\psi = 0$)

$$\frac{\partial \varphi_2}{\partial \psi} = \frac{1}{4} \left[\frac{\partial \varphi_2}{\partial \xi} - N_j - \frac{2}{We(R^2)^{3/2}} \frac{dR^2}{d\xi} \right] \quad (10)$$

c) Outer edge of the boundary layer ($r = r_{max}$)

$$\varphi_1 = 0 \quad (11)$$ (immovable continuous phase)

Note that (7) has been already used in (3) for exluding the pressure gradient. Formally condition (10) follows from (3) upon the assumption that r is equal to zero.

The case of a nonzero velocity of the continuous phase can be also considered by means of condition (11). The function $r_{max}(\xi)$ corresponds to the assumption of a finiteness of the outer boundary layer thickness and appears in (11) beacause of the restriction of finiteness of the numerical mesh.

For the numerical analysis of the flow considered the initial velocity profiles have to be known also. In accordance with the experimental situations of interest a fully developed parabolic profile will be used for the jet flow, i.e.

(12) $$\varphi_2 = 4(1 - 2\psi)$$

or a flat one

(13) $$\varphi_2 = 1.$$

In order to simplify the initial profile problem of the continuous phase imagine that the jet nozzle is surrounded by a circular disk of a sufficiently large radius. Then

(14) $$\varphi_1 \equiv 0 \quad \text{at} \quad \xi = 0.$$

DIRECT NUMERICAL SOLUTION

The equations of motion have been approximated in the sence of the finite-difference method over a suitable grid, shown in fig. 2. The scheme is of the second order of approximation and is implicit in ξ direction. The grid has been constructed in such a way that its lines with a step $\Delta\psi$ along the axis $r = 0$ inside the jet $0 \leq \psi \leq 1/2$ coincide with the streamlines. When the outer layer region is considered these lines with a step Δr are equidistant starting with the line $\psi = 1/2$. The grid lines along ξ with a step $\Delta\xi$ are parallel to the axis r. The values of $\Delta\psi$ and Δr are defined in accordance with the required accuracy of the numerical solution.

An outer iterative procedure has been constructed for determining the unknown radius of the jet. If in the process of the solution an approximation of the jet radius has been obtained then the interface position and the entire flow geometry would be determined.

The task is to find the corresponding velocity distribution after the flow geometry is determined. An inner iterative procedure is used for that purpose because of the nonlinearity of the system of algebraic equations obtained from the approximation of the nonlinear system of partial differential equations.

After a standard linearization the finite-difference versions of eqs. (3) and (2) for the dispersed and continuous phase read:

$$(15) \qquad A_i^{(k-1)}\varphi_{i-1,j}^{(k)} + C_i^{(k-1)}\varphi_{i,j}^{(k)} + B_i^{(k-1)}\varphi_{i+1,j}^{(k)} = F_i^{(k-1)},$$

where k is the number of the internal iterations. The coefficients A_i, B_i, C_i and F_i are known functions of φ, r^2 and V at the knots of the row analysed in the course of the $k-1$ iteration. F_i is at the same time a function of the values of φ on the previous row. Subscripts 1 and 2 in eq.(15) for the continuous and dispersed phase have been omitted.

The finite-difference form of eq. (6) at the interface point of the j-row is analogous to eq.(15). For the first two points of the row eq. (10) is added to eq. (15) and for the last two points of the same row eq. (11) is added. Thus the matrix of the set (15) is tridiagonal and its solution may be found by an elimination method.

The procedure described yields an indivisible boundary problem for both the jet and continuous phase liquids. In such a way any additional iterative procedure for the unknown velocities of the points of the interface is eliminated. The k-approximation of r^2, V and φ for every point of the row analysed is determined with the aid of the k-approximation of the axial velocity distribution. In this way the k-approximation of the j-row is completed. The results obtained are used for calculating the new values of the coefficients A_i, B_i, C_i and F_i. Then the inner iteration is repeated until the prescribed initial accuracy is reached. The velocity distribution calculated for the j-row is used as a zero approximation for the next $j+1$ row ($j > 1$). Eqs. (12) or (13) and (14) yield a value of φ which is used for similar calculations concerning the first row ($j = 1$).

After passing through all the rows along ξ-line one automatically obtains the next approximation for the jet radius $R(\xi)$ by using the solution of eq. (4) with $\psi = 1/2$. The outer iterative procedure described above comes to an end after reaching the previously given accuracy. In this way the final numerical solution of the problem is obtained. The zero approximation for $R(\xi)$ is found by the method of Targ-Shvetz (see [4]), or by means of a function of the type $R(\xi) = 1 - \exp(-c\xi)$ (c is a suitable constant). The iterative process is numerically stable and converges to the exact solution. The rate of convergence does not practically depend on the type of the zero approximation for $R(\xi)$.

ANALYSIS OF THE NUMERICAL RESULTS

The calculated cases are described in table 1 which contains chosen values of all of the five characteristic dimensionless numbers. Case 1 is considered as a basic one. The values of the dimensionless numbers in cases 2 ÷ 4 are identical to those of [5], so that a comparison between the two types of solution could be made. In all the cases a parabolic initial axial velocity profile was used except in the case 3, where a flat one was adopted. Cases 2 ÷ 4 correspond to a jet injected upwards. The calculations were performed with the following steps:

$\Delta\psi = 0.0025$ along ψ line, $\Delta r = 0.00125$ along r line,
$\Delta\xi = 0.005$ along ξ line.

The prescribed accuracy was 10^{-4} with respect to the internal iterations and 10^{-3} with respect to $R(\xi)$. The results obtained differ from each other slightly when doubling $\Delta\xi$, $\Delta\psi$, Δr and this practically confirms the stability of the numerical method.

As for the choice of the r_{max} value in eq. (11) it is conventional practice to balance between the accuracy and the economy of the computational time. Numerical experiments were performed with r_{max} and $2r_{max}$ for different values of Re_2 numbers. Doubling chosen values of r_{max} did not practically affect the numerical results.

Both the jet and the continuous phase streamlines concerning case 1 are shown in fig. 3. In fig. 4 the distribution of the axial velocity is drown for four different cross sections of the jet, including the initial one. The influence of the initial profile and the viscosity of the continuous phase on the jet radius relaxation is ilustrated on fig. 5. A comparison is made with the approximate solution [5]. The axial velocity profile from [5] is compared with the result of the present numerical solution in fig. 6 as well. When compared to [5] the jet radiuses are in good agreement. But the comparison of the velocity profiles for viscous continuous phase (fig. 6) shows considerable difference. The axial velocity gradients on the interface differ from each other in more than 10 % while the surface velocity values - 20 %.

REFERENCES

1. Duda J. L. & Vrentas J. S., Chem.Engng. Sci.,22, 855-869, 1967.
2. Markova M. P. & Shkadov V. Y., Mech. Jidk. i Gaza N°3, 30-37, 1972, (Russian).
3. Meister B. J. & Scheele G. F., A. I.Ch. E.J.,15, 689-699, 1969.

4. Penchev I. F., Radev S. P. & Rakadjiev R. K., Third National Congress of Mechanics, Varna 13-16. IX. 1977, 1977.

5. Yu H. & Scheele G. F., Int. J. Multiphase Flow, 2, 153-169, 1975.

TABLE 1

Case	N_j	We	Re_2	Re_1	μ_2/μ_1	ρ_2/ρ_1	Fr
1	121.47	2.81	660.47	400	0.969	1.600	2.039
2	128.46	5.4	835	1670	1	0.500	6.500
3	128.46	5.4	835	1670	1	0.500	6.500
4	128.46	5.4	835	/	∞	∞	6.500

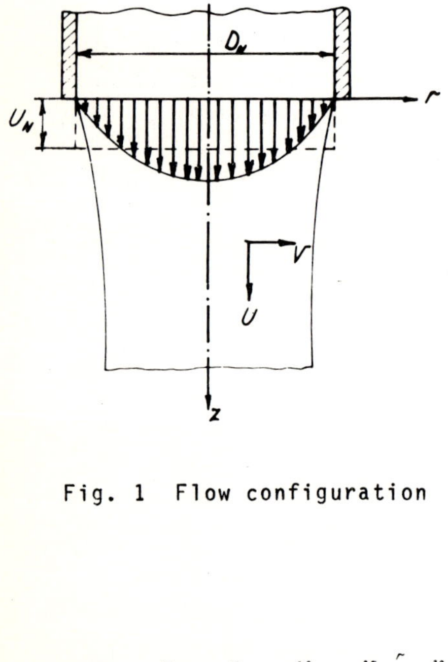

Fig. 1 Flow configuration

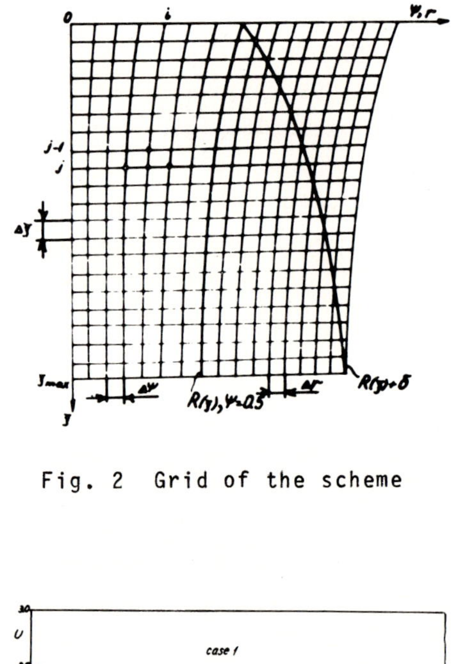

Fig. 2 Grid of the scheme

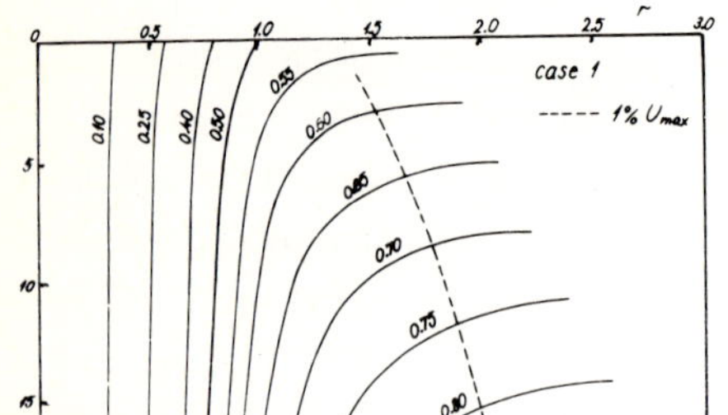

Fig. 3 Stream lines

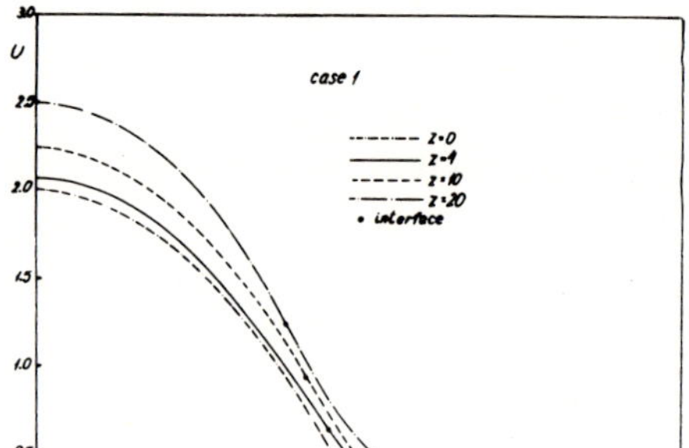

Fig. 4 Axial velocity distribution.

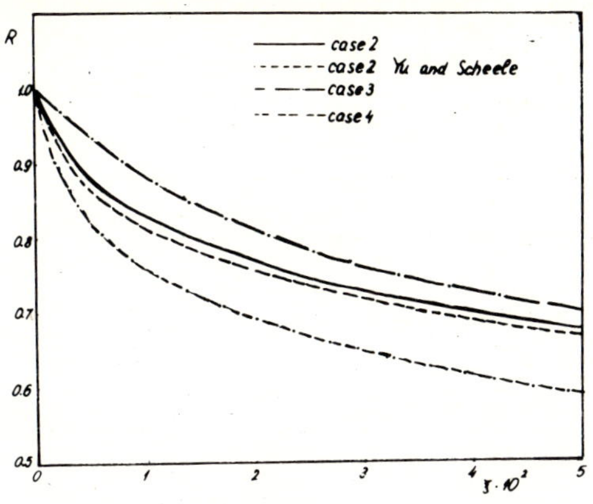

Fig. 5 Comparison between the exact and approximate jet radius

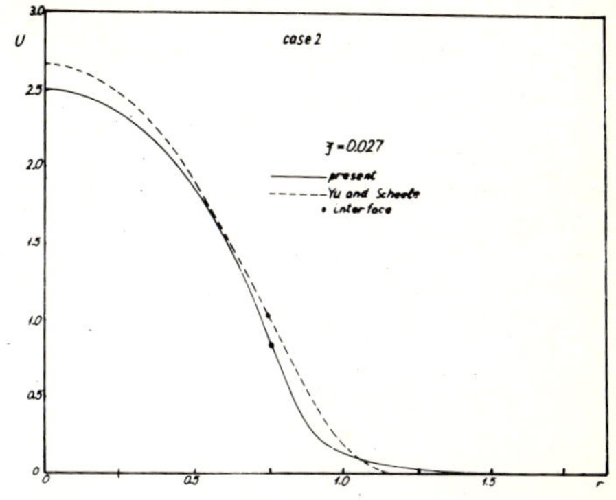

Fig. 6 Comparison between exact and approximate axial velocity distribution

MODEL CALCULATIONS OF SELF-EXCITED OSCILLATIONS IN TRANSONIC FLOW IN A DUCT WITH AN ABRUPT ENLARGEMENT

G. Grabitz

Max-Planck-Institut für Strömungsforschung
Göttingen, Federal Republic of Germany

1. Summary

In this paper, model calculations are given in order to show how a steady transonic flow in a duct with an abrupt enlargement can become unstable and how in this case a disturbance builds up to a limit-cycle oscillation. The theoretical model of the flow incorporates a supersonic expansion area (bounded by straight jet boundaries and a straight normal shock) which is treated with an integral method applied to the gas dynamic equations. The feedback between the base region and the flow area behind the shock is described by various model equations.

From the numerical calculations, model flows are obtained which show nearly the same properties as experimentally observed flow fields. Thus it is possible to explain the formation and mechanism of the oscillations.

2. Problem

The two-dimensional flow field which is considered has the following three main characteristics (Fig. 1): (i) sonic flow (critical pressure p^*) at the smallest cross-section, (ii) expansion fans from the smallest cross section with supersonic velocities and dead-air in the base region at the corners of the duct, (iii) a strong normal shock at the end of the expansion region, followed by a subsonic flow.

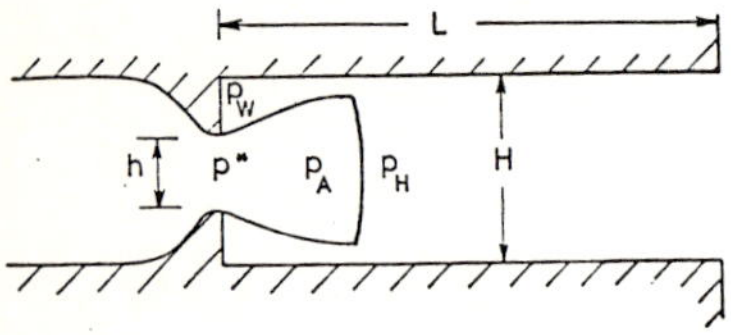

Fig. 1 Schematic diagram of the flow field

Experiments [1, 2] show that the flow field described above can be either steady or unsteady, and that in the latter case self-excited oscillations occur. These oscillations exist if the base pressure p_W is greater than a certain value, and if the cross-sectional area ratio (h/H) is in the appropriate range for a given length if duct L. Typical amplitudes are 10 % of the stagnation pressure.

3. Idealised flow field and basic equations

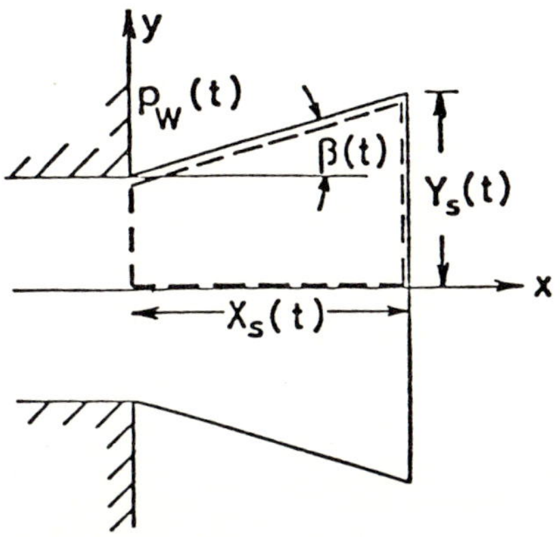

Fig. 2 Idealised flow field

The geometry of the idealised two-dimensional flow field is shown in Figure 2. The shock wave and supersonic flow boundaries are considered to be straight lines. The basic equations in the calculation comprise the equations of continuity, momentum in the transverse and axial directions, and momentum along the jet boundary. Additionally, the relation between the shock position X_S and the shock velocity U_S is introduced in the form $dX_S/dt = U_S$. The equations of continuity and momentum are used in the integral form and applied to the flow field within the dashed line in Figure 2.

4. Approximations used in the equation of the supersonic flow field

The basic equations contain integrals of the form

$$\int_0^{X_S(t)} \int_0^{Y(x,t)} f(x, y, t)\, dy\, dx \qquad (1)$$

The integrands are approximated by linear functions in both the y direction (from the axis of the flow to the flow boundary Y(x, t)) and in the x direction (from the nozzle to the shock wave $X_S(t)$).

With these approximations the mentioned integrals can be easily evaluated, and ordinary differential equations thus derived from the basic equations. The expansion region is then approximately described by five ordinary differential equations in time with six unknown variables (position and velocity of the shock wave and flow variables at certain points of the flow field).

It is not yet known how the pressure p_H in the flow area downstream of the shock wave is related (through the boundary layer) to the base pressure p_W. Different model equations have therefore been developed and tested by numerical calculations for this feedback between p_H and p_W. Two of these model equations will be discussed here (section 6 and 7).

5. Initial flow field (steady solution)

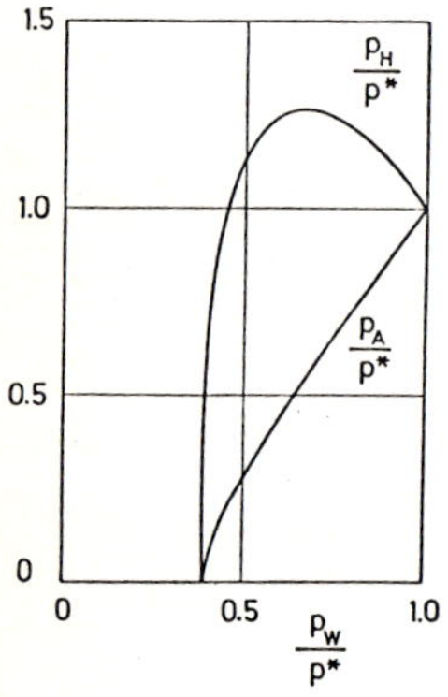

Fig. 3 Steady solutions

A steady initial flow field is assumed, into which a a disturbance is introduced in order to distort the steady solution. The steady solutions are obtained from the algebraic equations which are derived from our approximate differential equation system (Section 4) by cancelling the time derivatives. In Figure 3 the pressure upstream (p_A) and downstream (p_H) of the shock are plotted against the dead-air pressure p_W (p^* is the critical pressure).

6. Feedback Model A

In model A the relation $p_W(p_H)$ is considered to be incorporated in the following differential equation

$$d^2 p_W/dt^2 = (p_G - p_W)\ \omega_o^2 \tag{2}$$

and it is assumed that p_G is an equilibrium pressure related to p_H by $p_G = Kp_H$. The constant K is determinated by the necessity that $p_G = p_W$ for the steady flow case. Hence, the value K can be obtained from Figure 3. This model is motivated by experiments [1] which show that the oscillation frequency is mainly determined by the length L of the duct, with associated resonator frequency ω_o. This frequency is modified influenced by the flow system. The model equation should not be mistaken for an equation for a linear oscillator; in fact the whole system of equations remains non-linear. Equation (2) completes the differential equation system mentioned in section 4. For numerical calculation, the Runge-Kutta-Method is used. An initial value of p_W is chosen, and associate values of the other variables are established which satisfy the steady conditions (Section 5). A disturbance is introduced into the unsteady system of equations by artificially changing one of the variables from its equilibrium value. As a result of this disturbance, there is either an attenuation of the variable parameters or an amplification. Numerical calculations based on model A show the following results: (i) damped oscillations occur if the dead-air pressure ratio $p_W/p^* < k$ $(k \approx 0.55)$, (ii) if $p_W/p^* > k$ then a disturbance in

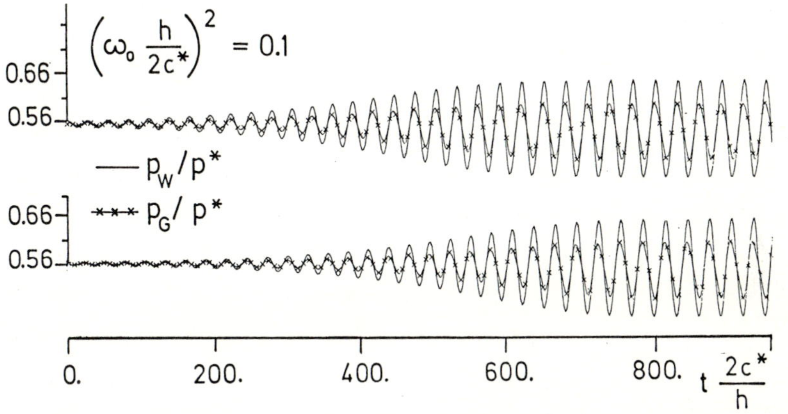

Fig. 4 Variation of p_W and $p_G(=Kp_H)$ with time, caused by disturbing the initial steady flow (a) the base pressure, (b) the shock position (c^* is the critical sound velocity).

the steady conditions leads to an increasing and then a constant amplitude. Figure 4 shows two examples of the pressure oscillations of the air in the base region. The upper trace was obtained by introducing a disturbance in the base pressure p_W, whereas in the lower trace it was the position X_S of the shock wave that was disturbed: the same oscillation results in both cases. This result is a typical example of a limit cycle oscillation. The resulting amplitude is in good agreement with the experiment values.

7. Feedback Model B

Model B assumes a secondary flow in the dead-air region, caused by the expansion and contraction of the supersonic flow during a cycle of the oscillation. The driving pressure difference is again $p_G - p_W$, where p_G is in this case the pressure near the corner of the expansion region. The advantage of model B is that the duct height, and hence the area ratio h/H is introduced into the calculation (see Figure 5). The secondary flow between the expansion region and the duct is described by a one-dimensional incompressible continuity equation and by the Bernoulli equation for unsteady flow:

$$\left(\frac{H-h}{2} - x\tan\beta\right) v(x,t) = \frac{x^2}{2\cos^2\beta}\frac{d\beta}{dt}\,, \quad \int_0^{X_S} \frac{\partial v}{\partial t}\,dx + \frac{v^2(X_S,t)}{2} + \frac{p_G}{\rho_W} = \frac{p_W}{\rho_W} \tag{3}$$

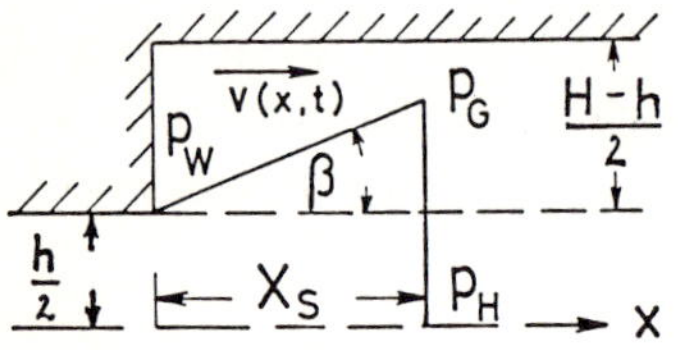

Fig. 5 Explanation of variables

where v(x, t) is the velocity in the dead-air (see Figure 5), and $p_G = Kp_H$ as before.

The following results were obtained from the numerical calculation from model B: (i) as with model A for values of $p_W/p^* < k$, where $k \approx 0.55$, the oscillations were attenuated, (ii) for values of $p_W/p^* > k$ a disturbance created an oscillation with increasing amplitude which was eventually limited to a constant value.

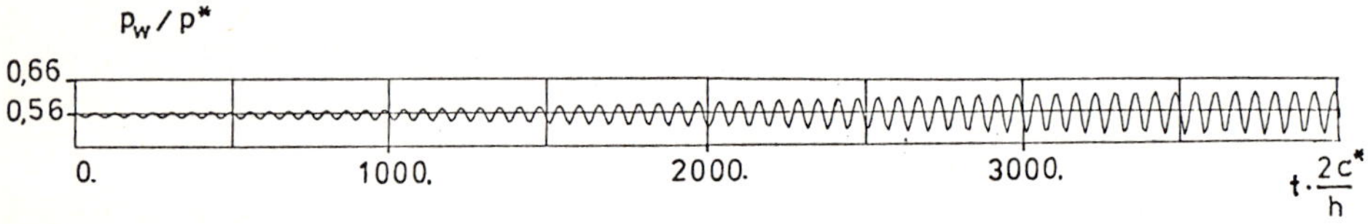

Fig. 6 Onset of pressure oscillations as calculated with model B (p^* critical pressure, area ratio h/H = 0. 5).

An example of the kind of oscillation obtained with model B is shown in Figure 6. The constant peak-to-peak amplitude which is reached at the end of the pressure trace is about 10 % of the critical pressure p^*: this agrees well with the experimental results. The most interesting result obtained with model B is that the calculated frequency now depends only on the flow and the geometrical variables. The two curves in Figure 7 illustrate how the calculate frequencies depend upon the area ratio (h/H) for nozzle heights of 6 mm and 20 mm. The points represent experimental values [1] and show that the calculated and experimental frequencies are in the same range.

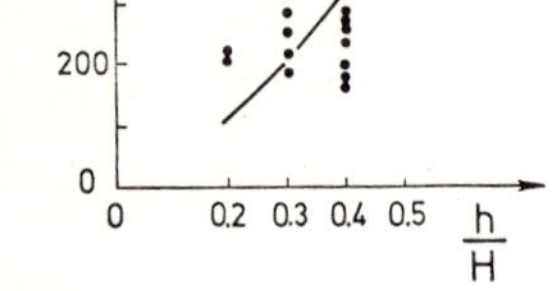

Fig. 7 Frequency of oscillation as a function of area ratio of the enlargement. (Curves for two nozzle heights h calculated using model B; solid circles are experimental points [1])

8. Discussion of the oscillation mechanism

The experimental and theoretical investigations lead to the following idea about the oscillation mechanism and the feedback loop: if the pressure is decreasing in the dead-air region the supersonic jet will expand, and the pressure upstream of the shock wave may decrease. Up to a certain value a lower pressure p_A in front of a shock wave causes a higher pressure p_H downstream of the shock (negative response). A larger pressure p_H behind the shock tends to increase the pressure p_W in the dead-air, as the two regions are connected through the boundary layer. The base pressure p_W is closely related to the pressure p_A in front of the shock wave, and hence a feedback loop is formed with a certain internal time delay. This feedback system works only in the case of a negative response. This idea of a negative response is incorporated in Fig. 2, which is obtained from the solution of the steady flow equations. The limit-cycle oscillations occur for values of $p_W/p^* > 0.55$, which is in the area of negative response.

Although different in formulation, the two theoretical models considered are not contradictory. Model A only uses the equations of gas dynamics to define the supersonic flow in the expansion region which terminates in a normal shock wave. The downstream resonator or filter is characterized by a length L and corresponding resonant frequency ω_o. Model B, with the same gas equations, introduces flow conditions defined by the geometry of the supersonic flow field and the duct geometry, but the length L does not enter. It has been shown experimentally [1] that the length L is an important parameter influencing the value of the frequency of the oscillations for a given area ratio. This influence does not, however, contradict the basic theoretical concept of supersonic flow with shock wave as the real generator, where the downstream part of the duct acts as a resonator or amplifies the amplitude of the generator. The calculated frequency could be used to predict the length of resonator which is required to produce maximum pressure amplitude.

A more detailed description of this work is contained in [2], which gives many experimental results. I thank the other authors of paper [2], particularly Dr. G. E. A. Meier for many helpful discussions.

9. References

[1] J. S. Anderson, W. M. Jungowski, W. J. Hiller and G. E. A. Meier, Flow oscillations in a duct with a rectangular cross section, J. Fluid Mech., 79, 769-784, 1977

[2] G. E. A. Meier, G. Grabitz, W. M. Jungowski, K. J. Witczak and J. S. Anderson, Oscillations of the Supersonic Flow Downstream of an Abrupt Increase in Duct Cross-Section, Mitt. MPI Strömungsf. u. Aerodyn. Versuchsanst., Göttingen, 65/1978

AN ORTHOGONAL FINITE ELEMENT METHOD FOR TRANSONIC FLOW CALCULATIONS

Ch. HIRSCH
Vrije Universiteit Brussel
Dept. Fluid Mechanics
Brussels, Belgium

G. WARZEE
Université Libre de Bruxelles
Dept. Stress Analysis
Brussels, Belgium

ABSTRACT

A new method for the numerical integration of the time dependent Euler equations by Finite Elements is presented based on the introduction of orthogonal shape functions. This leads to a diagonal mass matrix. In certain cases, as linear triangular elements and bilinear quadrilateral elements with parallel sides, the obtained scheme is equivalent to a Finite Volume formulation. For a general mesh, the method automatically generates schemes for arbitrary geometries. Examples of applications are presented.

1. INTRODUCTION

The solution of the steady transonic flow around airfoils or cascades of blades by the F.E. Method poses several problems. Solutions with the full potential flow equation or the small disturbance approximations have been attempted recently and some of the difficulties and possible solutions are discussed in ref. [1]. The main difficulties arise because of the mixed character of the equations and of the fact that the Finite Element Method is best suited for elliptic problems since the basic equations have to be formulated as an integral functional where the solution is given by the stationary state. Difficulties appear when the second variation is not constant sign which is the case in the hyperbolic (supersonic) region of the flow domain. An interesting approach has been developed for the full potential flow equation based on an optimal control formulation in [2] and encouraging results have been obtained for a single airfoil [2] and for cascades [3].

However the potential formulation is limited to weak, isentropic shocks and therefore more general solutions can only be obtained through the numerical integration of the time-dependent Euler equations. Applying a Finite Element, Galerkin method to these equations with finite differences in time and FE in space, an implicit system of first order differential equations in time is obtained with a very large mass matrix which has, at least formally, to be inverted.

This leads to excesive computation time in problems where only the steady state solution is wanted. Besides, the simplest time integration schemes lead to highly dissipative schemes preventing a good definition of shocks. This is shown on some simple examples.

A new method is presented, called the "Orthogonal Finite Element method" in which weight functions orthogonal to the interpolation functions of the finite element discretization are introduced. These orthogonal shape functions, defined locally on an element, are linear combinations of the basic shape functions but are not continuous on inter-element boundaries.

The system of ordinary differential equations obtained with these new functions has a diagonal mass matrix. An analysis of this system shows that it can be made equivalent, in certain cases, to the system obtained with the so-called "finite volume" method where the conservation laws in integral form are directly discretized through a finite difference approach. This is shown to be the case for a general triangulation with linear interpolation functions and for bilinear quadrilateral elements on a mesh with certain regularity properties.

Finally some examples of calculation with a modified Lax scheme are presented for a nozzle.

2. GALERKIN METHOD FOR THE EULER EQUATIONS

The two dimensional Euler equations are written in conservation form

$$\frac{\partial U}{\partial t} + \vec{\nabla}\,\vec{H} = 0 \tag{1}$$

where :

$$U = \begin{Bmatrix} \rho \\ \rho u \\ \rho v \\ \rho E \end{Bmatrix} \qquad H_x = F = \begin{Bmatrix} \rho u \\ \rho u^2 + p \\ \rho uv \\ \rho u\,(E + \frac{p}{\rho}) \end{Bmatrix} \qquad H_y = G = \begin{Bmatrix} \rho v \\ \rho uv \\ \rho v^2 + p \\ \rho v\,(E + \frac{p}{\rho}) \end{Bmatrix} \tag{2}$$

The method of weighted residuals applied to (1) with weight functions W, leads to

$$\int_S \frac{\partial U}{\partial t} \cdot W\, dS + \int_S \vec{\nabla}\,\vec{H} \cdot W\, dS = 0 \tag{3}$$

and a Finite Element-Galerkin approximation with shape functions N_i and weight functions N_j leads to the system of differential equations

$$M_{ij}\,\frac{\partial U_j}{\partial t} + \vec{H}_j\,\vec{K}_{ij} = 0 \tag{4}$$

where

$$M_{ij} = \int_S N_j\, W_i\, dS \qquad \vec{K}_{ij} = \int_S \vec{\nabla} N_j \cdot W_i\, dS \tag{5a}$$

$$U = \sum_j U_j\, N_j \qquad W_i = N_i \tag{5b}$$

The mass matrix M has non-zero elements at all nodes connected geometrically with node j and the system (4) is of implicit nature. Therefore inversion of M will generally be a time consuming task.

For instance, the one-dimensional equation

$$\frac{\partial U}{\partial t} + \frac{\partial F}{\partial x} = 0 \tag{6}$$

with linear shape functions, leads to the system

$$\frac{U'_{i+1} + 4\,U'_i + U'_{i-1}}{6} + \frac{\Delta t}{\Delta x}\;\frac{F_{i+1} - F_{i-1}}{2} = 0 \tag{7}$$

The scheme obtained with a forward time difference is unstable while the backward time differencing gives an unconditionally stable scheme as is easily seen by a linearized Von Neumann stability analysis.

One obtains, with

$$w_i^{n+1} = U_i^{n+1} - U_i^n$$

$$w_{i+1}^{n+1}(1+\alpha_{i+1}) + 4\,w_i^{n+1} + (1-\alpha_{i-1})\,w_{i-1}^{n+1} = -\frac{3\,\Delta t}{\Delta x}(F_{i+1}^n - F_{i-1}^n) \tag{8}$$

where $\alpha_1 = \frac{3\,\Delta t}{\Delta x}\left(\frac{\partial F}{\partial U}\right)_i^n$

A Taylor expansion shows that this scheme has a second order dissipation, leading to a strongly smeared shock (see fig. 1), although it is not strictly dissipative in the sense of Kreiss and therefore strong non-linear post-shock oscillations will appear, e.g. for Burgers equation.

3. AN "ORTHOGONAL" FINITE ELEMENT METHOD (OFE)

It would be desirable to avoid the large mass matrix M, obtained after assembly, while keeping the advantages of the FE approach, in particular with regard to the arbitrary geometries and computational meshes which can be handled with a FEM as easily as a regular mesh system.

In order to obtain a diagonal mass matrix within the weighted residual approach equation (5), one could define continuous weight functions W_i orthogonal to the shape functions N_j in the whole domain. This corresponds to the conjugate shape functions, introduced by Oden [4], which are non-zero over the whole domain while each shape function N_j has only non-zero values within few elements. Moreover the definition of these conjugate shape functions requires the inversion of the full, assembled mass matrix, [4], leading to conjugate functions which are strongly dependent on the geometry, in the sense that the addition of a single element will change the values of these functions every where. Hence, one of the main advantages of FE is lost in the sense that one doesn't work anymore with locally defined functions.

Therefore if one is willing to sacrify the interelement continuity of these conjugate functions and define locally orthogonal shape functions N_j^* through

$$\int_{S_e} N_i\, N_j^*\, dS_e = 0 \qquad \text{for } i \neq j \tag{9}$$

where the integration is performed on a single element the mass matrix would become diagonal on each element and remain so after assembly.

Within one element, the orthogonal shape function N_j^* is defined through

$$\int_{S_e} N_i\, N_j^*\, dS = \mu^{(j)}\,\delta_{ij} \tag{10}$$

Since the shape functions N_i form a complete set in an element one has

$$N_j^*(x,y) = C_{jk}^{(e)}\, N_k(x,y) \tag{11}$$

where the C_{jk} are constants. Introducing (11) in (10) one obtains

$$C_{jk}^{(e)}\, M_{ki}^{(e)} = \mu^{(j)}\,\delta_{ij} \tag{12}$$

and the matrix C is given by the inverse of the local mass matrix $M^{(e)}$

$$[\, C^{(e)} \,] = \lceil\, \mu \,]\,[\, M^{(e)} \,]^{-1} \qquad (13)$$

where $[\,\mu\,]$ is the diagonal matrix of the normalization coefficients.

The dimensions of these square matrices are equal to the number of nodes in each element and therefore inversion of $M^{(e)}$ is an easy task.

3.1. Linear Elements in One Dimension

Refering to fig. 2, the linear shape functions for the element between nodes (i - 1) and i, element (1), are

$$N_{i-1}^{(1)} = \frac{x_i - x}{h_{i-1}} = (1 - \xi) \; ; \; N_i^{(1)} = \frac{x - x_{i-1}}{h_{i-1}} = \xi \qquad (14)$$

where ξ is the local coordinate and the following orthogonal shape functions are obtained (fig. 1)

$$N_{i-1}^{(1)*} = \mu_{i-1} \frac{2}{h_{i-1}^2} \cdot [-3\,(x - x_i) - h_{i-1}\,] = \mu_{i-1} \frac{2}{h_{i-1}} (-3\,\xi + 2) \qquad (15)$$

$$N_i^{(1)*} = \mu_i \frac{2}{h_{i-1}^2} [\,3\,(x - x_i) + 2\,h_{i-1}\,] = \mu_i \frac{2}{h_{i-1}} (3\,\xi - 1) \qquad (16)$$

Generally, for an arbitrary mesh these functions will not be continuous at the interelement boundaries. However in the linear case, choosing

$$\mu_{i-1} = \mu_i = h_{i-1}$$

will lead to continuity of the orthogonal shape functions.

Applying the weighted residual method with the orthogonal shape functions to the linear equation (6), one obtains after assembling the two equations for node i.

$$\frac{dU_i}{dt} + \frac{F_{i+1} - F_{i-1}}{(h_{i-1} + h_i)} = 0 \qquad \text{for } \mu_i^{(1)} = h_{i-1} \text{ and } \mu_i^{(2)} = h_i \qquad (17)$$

or $$\frac{dU_i}{dt} + \frac{F_i - F_{i-1}}{2\,h_{i-1}} + \frac{F_{i+1} - F_i}{2\,h_i} = 0 \quad \text{for } \mu_i^{(1)} = \mu_i^{(2)} = 1 \qquad (18)$$

These are finite difference formula's for an arbitrary mesh spacing which are automatically generated through the OFE method.

3.2. Linear Elements in Two Dimensions

Considering a two-dimensional domain with linear triangular elements (fig. 3) the shape functions are [4].

$$N_i = \frac{1}{2\,\Delta} (a_i + b_i\,x + c_i\,y) \qquad i = 1, 2, 3 \qquad (19)$$

where a_i, b_i, c_i are coordinate dependant and Δ is the area of the triangle.

The orthogonal shape functions become

$$N_1^* = \frac{3\,\mu_1}{\Delta}(3\,N_1 - N_2 - N_3) \qquad N_2^* = \frac{3\,\mu_2}{\Delta}(-\,N_1 + 3\,N_2 - N_3) \qquad (20)$$

$$N_3^* = \frac{3\,\mu_3}{\Delta}(-\,N_1 - N_2 + 3\,N_3)$$

and are discontinous at the interelement boundaries except for

$$\mu_1 = \mu_2 = \mu_3 = \Delta \qquad (21)$$

An interesting observation is that an application of the OFE method with the normalization (21) to the Euler equations (1), lead after straighforward calculation to the equations

$$S\,\frac{dU_{i,j}}{dt} + \sum_{k=1}^{6} \bar{F}_k\,\Delta\,Y_k - \sum_{k=1}^{6} \bar{G}_k\,\Delta\,X_k = 0 \qquad (22)$$

where the summation on k extends to the 6 sides of the hexagonal domain of fig. 3. The bars on F and G indicate the arithmetic average of the values at the two nodes of each side and $\Delta\,X_k$, $\Delta\,Y_k$ are the differences in the Y and X coordinates of these nodes, S is the surface of the hexagon.

Equation (22) then clearly corresponds to the Finite Volume discretization of the integral Euler equations applied to the hexagonal domain S of fig.3.

$$\int_S \frac{dU}{dt}\,dS + \oint_C (F\,dy - G\,dx) = 0 \qquad (23)$$

A similar analysis for quadrilateral elements fig. 4 and normalisation $\mu_i = S^{(e)}$ where $S^{(e)}$ is the area of the element, shows that for a geometry with parallel lines, the OFE method leads to the discretisation of the Finite Volume approach equation (23), but applied to the volume indicated in heavy lines in fig. 4 (*). However this is not the case for an arbitrary mesh.

In this case, fig. 5, application of the OFE method leads to expressions for the spatial derivatives of equation (1) which contain the contribution from all eight nodes surrounding P. The coefficients of the extreme corner nodes, like (i + 1, j + 1) vanish when the mesh lines become parallel.

(*) In this case, the orthogonal shape functions are given by

$N_j^* = \mu_j\,(1 + 3\,\xi\xi_j)\,(1 + 3\,\eta\eta_j)$ in local coordinates (ξ, η) where ξ_j, η_j take the values $\pm$ 1. For vertical i lines and arbitrary j lines, one obtains

$N_j^* = \mu_j(1 + \alpha_j\,\xi\xi_j)\,(1 + 3\,\eta\eta_j)$ where α_j is coordinate dependant.

One obtains generally, an equation of the form

$$\frac{dU_i}{dt} + \sum_{k=1}^{9} (S_k F_k + T_k G_k) = 0 \tag{24}$$

where the summation is over all nodes around P (P included) and the coefficients S_k and T_k are only dependent on the geometry.

For a regular orthogonal mesh, a central difference scheme in space is obtained. Various finite difference schemes in time can be adopted for equation (24) and a modified Lax scheme [5] called the damping surface technique is used for the example shown in the following section.

3.3. Example Calculation

In order to illustrate the method, a transonic nozzle has been tested with arbitrary geometry for the computational mesh. The scheme proposed in [5] has been adopted, although other schemes can be used, for instance by applying the OFE-method to the time-discretized equation analyzed by Beam α Warning [6], in the space factored form. This allows an automatic generation of numerical schemes for arbitrary geometries.

The damping surface technique [5] applied to equation (24) leads to

$$U_{i,j}^{n+1} = \overline{U_{i,j}^{n}} + \sum_{k=1}^{n_{i,j}} (S_k F_k + T_k G_k) - \alpha (\overline{U_{i,j}^{L}} - U_{i,j}^{L}) \tag{25}$$

where $\overline{U_{i,j}^{n}}$ is the average of U on all nodes surrounding (i,j) which contribute to the sum on k ($n_{i,j}$ nodes). The variables at time indicated by L correspond to values which are updated every L time iterations. The coefficient α is close to 1. Fig. 6 shows the Mach number distribution for the nozzle.

4. CONCLUSION

The Orthogonal Finite Element presented in this paper leads to numerical schemes which include as particular cases some Finite Volume formulations as well as generalized Finite Difference schemes. The advantages of Finite Element methods with regard to the arbitrary geometries are kept while the disadvantage of large mass matrices and hence implicit equations with large bands are avoided.

REFERENCES

1 M.M. HAFEZ, C.C. WELLFORD, C.L. MERKE α E.M. MURMAN, "Numerical Computation of Transonic Flows by Finite Element and Finite-Difference Methods", Flow Research Report 70, Jan. 1977, (U.S.A.)

2 R. GLOWINSKI, J. PERIAUX, O. PIRONNEAU, "Use of Optimal Control Theory for the Numerical Simulation of Transonic Flow by the Method of Finite Elements", Proc. of the Fifth International Conf. on Numerical Methods in Fluid Dynamics, Lecture Notes in Physics, Vol. 54, p. 205, 1976

3 Ch. HIRSCH, "Transonic Flow Calculation in Blade Rows with an Optimal Control-Finite Element Formulation", Bat-Sheva Seminar on Finite Elements for Non-Elliptic Problems, Tel-Aviv, July 1977

4 J.T. ODEN, "Finite Elements of Non Linear Continua", Mc-Graw Hill, 1972

5 M. COUSTON, P.W.Mc DONALD, J. SMOLDEREN, "The Damping Surface Technique for Time-Dependent Solutions to Fluid Dynamics", VKI-TN 109, 1975

6 R. BEAM α R. WARMING, J. Comp. Physics 22, 1976, 87-110.

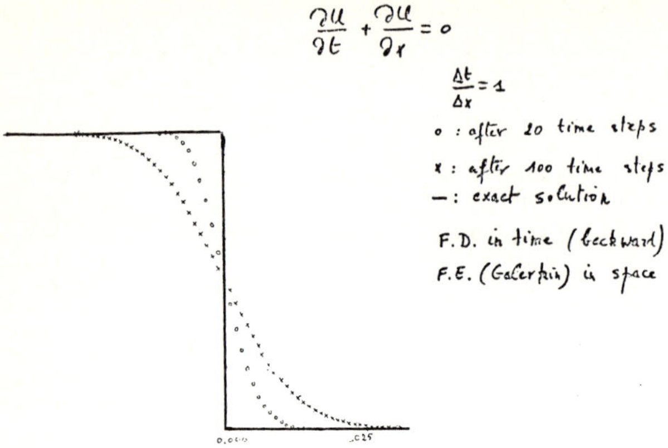

Fig. 1 : GALERKIN Method

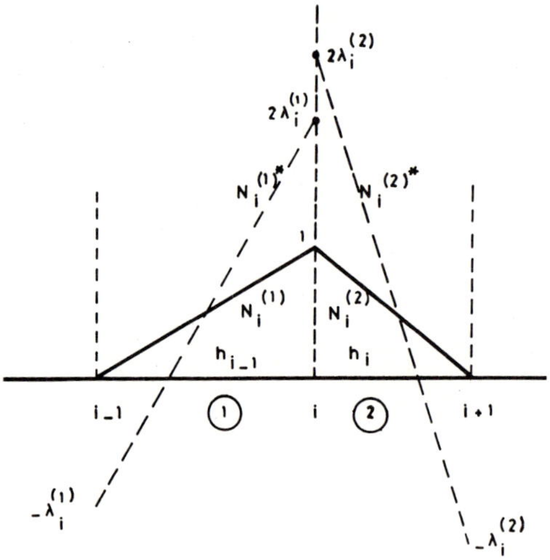

Fig. 2 : Linear Orthogonal Shape Functions

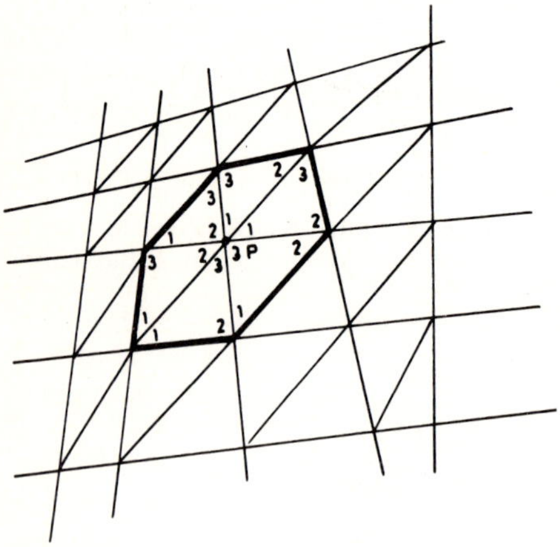

Fig. 3 : Linear triangular elements

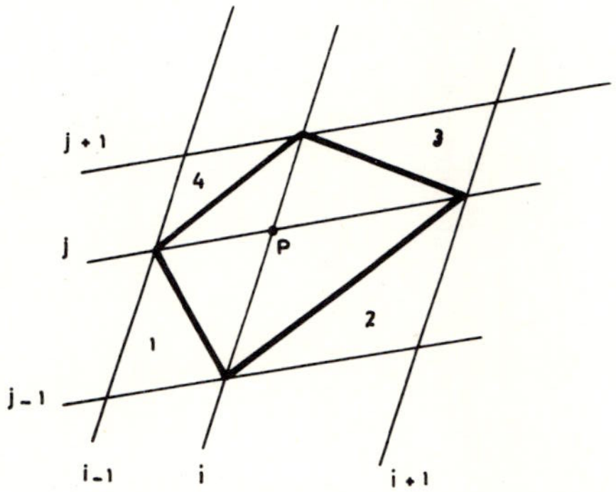

Bilinear quadrilateral elements

Fig. 4

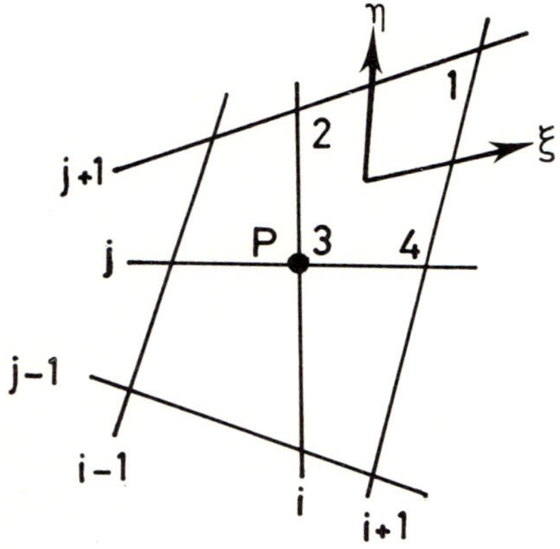

Fig. 5

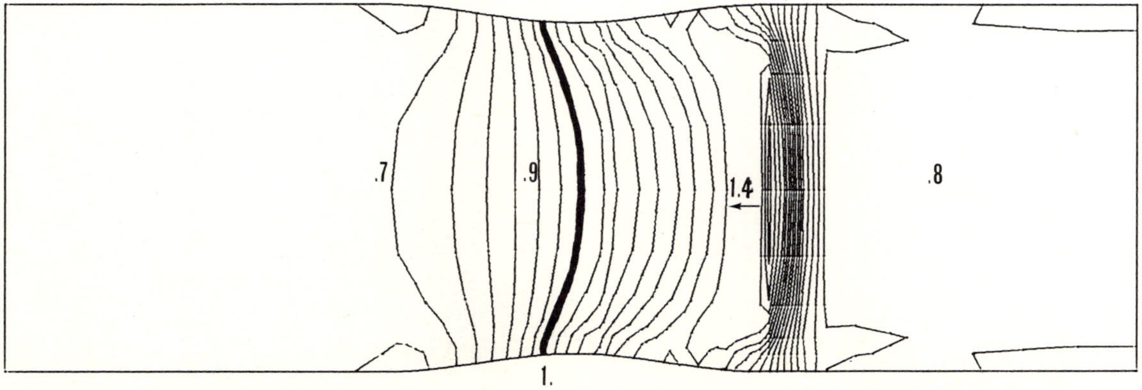

ISOMACH LINES

Fig. 6 : Transonic nozzle calculation

SOLUTION OF TURBULENT TRANSPORT EQUATIONS BY AN ACCURATE NUMERICAL METHOD

by

Richard S. Hirsh*, Douglas M. Friedman* and Tuncer Cebeci**
McDonnell Douglas Corporation, Douglas Aircraft Company
Long Beach, California

INTRODUCTION

The turbulence model of Cebeci and Smith, based on an algebraic eddy-viscosity formulation, has been used with considerable success to compute a wide range of turbulent boundary-layer flows. This model does not, however, allow for the transport of turbulence properties and its applicability is limited to near-equilibrium flows. The range of flow configurations relevant to aircraft applications includes trailing-edge flows and wakes, merging boundary layers, both with strong streamline curvature, shockwave boundary-layer interactions, and regions of separation. These flows include regions which are far from equilibrium and the transport associated with higher turbulence models is required. Higher turbulence models include the single-equation approach of Bradshaw[1,2] supplemented by either an algebraic or differential length-scale equation, two-equation models where turbulence kinetic energy and dissipation rates are considered as dependent variables[2,8,11], and Reynolds stress models involving five and more equations for turbulence transport[2,10,11].

Of major importance to an aircraft manufacturer is the calculation of the flow past wings. Current wing geometries lead to viscous flow regimes which can be broken up into four significant parts. The leading-edge region, where the presence of slots or other leading-edge devices create merging boundary-layer flows; the major portion of the wing, where most of the research concentrates; the trailing edge, where flaps again create merging boundary layers; and the resulting flow in the wake behind the wing.

For flows which can be classified as standard boundary-layer flows, i.e., the boundary layer on most of the wing, and the flow in the far wake, the success of eddy-viscosity models is well established. These are the stock and trade of most engineers concerned with aircraft-type flows. The turbulent-length scale is fairly well fixed in these flows, and the mean velocity distribution contains no relative minima or maxima. For merging boundary layers, the assumption of eddy-viscosity proportional to mean velocity gradient yields unrealistic results when this gradient is small or zero. For anything but symmetric wakes, the length scale is difficult to assign a priori unless recourse to some previously obtained experimental data is available. Higher-order turbulence models overcome both these drawbacks, although they do introduce some (mathematical) problems of their own which will be discussed below. Major modifications to this general pattern such as shock-wave–boundary-layer interactions or separation only tend to make the comments below even more valid.

The major computational requirements for all these flows are that the calculations be performed efficiently and accurately. In some cases, the configuration of the flow itself imposes very severe restrictions on the computational method. For merging boundary layers, the presence of both the wall region and the shear layer between the two merging flows makes grid point placement an important consideration. The finite-difference grid assumes even more significance in the near-wake–trailing-edge region where it has recently been shown to

---

*Senior Engineer/Scientist, Aerodynamics Research
**Chief Aerodynamics Engineer, Research

be possible to advance past the trailing edge and into the near wake through the singularity, by allowing a numerical double structure within the wake[5].

This double structure uses two separate scalings for the boundary-layer depending upon whether one is considering the inner layer, dominated by the singularity at the trailing edge[7], or the outer layer, where the inviscid forces are dominant. In each region a different grid size is required, and in addition a match at some location within the boundary layer must be made to ensure the continuity of the stream function and all its derivatives. This last requires that no differencing across the match point be made in each of the two regions. The Keller Box scheme is ideally suited to such a double structure since it can easily accommodate the different mesh sizes required in each region, and does not difference across the interface since it is a two-point-differencing scheme[4]. In addition, the method has been shown many times to be highly accurate and efficient [2,3].

Our goal is to compute complex turbulent flows using transport equations and the Box scheme. However, prior to investigating these difficult flows, experience needs to be gained concerning the behavior of the combination of the numerical technique and the particular turbulence model chosen. The aim of this paper is a basic study of the use of transport equation models for boundary-layer flows. The transport equations are solved by incorporating them into the framework of the accurate and efficient Keller-Cebeci procedure for solving the boundary-layer equations[2,3]. The test cases will be drawn from the standard data gathered into the 1968 Stanford Conference Proceedings[6] for the calculation of turbulent boundary layers.

TRANSPORT EQUATIONS

Of the many models which consider the transport of turbulence quantities, the two equation models with k, the turbulence kinetic energy, and ε, the rate of turbulence dissipation, as variables have enjoyed the most success. These models, basically developed at Imperial College, have a curious history. Originally developed for a few specific nonequilibrium boundary-layer problems[2,5] and then used with dramatic impact to calculate free shear layers[9], their usage seems to have been restricted recently solely to Navier-Stokes solutions where recirculation is an important aspect of the flow[11]. Although much success has been achieved, there are still large gaps in the documented properties of these models on the simpler flows where they should work at least as well as the algebraic models. Questions as to the predictive capability in pressure gradients, both favorable and adverse; the modelling of the wall layer to reduce the computational domain; how to estimate initial profiles for given experimental data; the "universality" of constants used previously; etc., need to be investigated for boundary layers before confidence can be gained in the use of these models to the more complex flows encountered on today's aircraft.

Governing Equations

The following equations form the basis for the k-ε model to be studied:

$$u \frac{\partial u}{\partial x} + v \frac{\partial u}{\partial y} = \frac{\partial}{\partial y}\left[(\nu + \nu_t)\frac{\partial u}{\partial y}\right] + u_e \frac{du_e}{dx} \tag{1}$$

$$u \frac{\partial k}{\partial x} + v \frac{\partial k}{\partial y} = \frac{\partial}{\partial y}\left[\left(\nu + \frac{\nu_t}{\sigma_k}\right)\frac{\partial k}{\partial y}\right] + S_k \tag{2}$$

$$u \frac{\partial \varepsilon}{\partial x} + v \frac{\partial \varepsilon}{\partial y} = \frac{\partial}{\partial y}\left[\left(\nu + \frac{\nu_t}{\sigma_\varepsilon}\right)\frac{\partial \varepsilon}{\partial y}\right] + S_\varepsilon \tag{3}$$

$$\frac{\partial u}{\partial x} + \frac{\partial v}{\partial y} = 0 \tag{4}$$

where

$$\nu_t = C_\mu \frac{k^2}{\varepsilon}, \qquad S_k = \nu_t \left(\frac{\partial u}{\partial y}\right)^2 - \varepsilon, \qquad S_\varepsilon = C_1 \frac{\varepsilon}{k} \nu_t \left(\frac{\partial u}{\partial y}\right)^2 - C_2 \frac{\varepsilon^2}{k} \tag{5}$$

and the five "universal" constants have been taken to be

$$\sigma_k = 1.0, \qquad \sigma_\varepsilon = 1.3, \qquad C_\mu = 0.09, \qquad C_1 = 1.55, \qquad C_2 = 2.0 \tag{6}$$

which are the values given by reference 8. There was some question as to whether these constants were a consequence of the modelling of the turbulent phenomena, or represented the best compromise with the data as computed by the particular numerical scheme chosen. Therefore, no changes were made in their specification when used with the highly accurate Box method to be discussed below. The results will show that the constants are not a function of the numerical method, but of the physics.

Boundary Conditions

To complete the specification of the problem, both boundary and initial conditions are required. The conditions in the freestream are relatively easy to obtain. The most obvious of these is that the velocity equal the edge velocity, i.e.

$$y \to \delta: \qquad u = u_e \tag{7a}$$

However, the value of k, and certainly ε, are not usually measured and reported for experimental tests. Therefore, we must rely upon our model to make the edge values consistent. Writing the limiting values of equations (2) and (3) at the boundary-layer edge, where all y-gradients are zero, yields

$$u_e \frac{dk_e}{dx} = -\varepsilon_e \tag{7b}$$

$$u_e \frac{d\varepsilon_e}{dx} = -C_2 \frac{\varepsilon_e^2}{k_e} \tag{7c}$$

These two equations can be solved in finite-difference form along with the solution of eqs. (1) - (3), but they also have an analytic solution given by

$$k_e = k_o \left[1 - (1 - C_2) \frac{\varepsilon_o}{k_o} \int_0^x \frac{dx}{u_e}\right]^{1/(1-C_2)} ; \quad \varepsilon_e = \varepsilon_o \left[1 - (1 - C_2) \frac{\varepsilon_o}{k_o} \int_0^x \frac{dx}{u_e}\right]^{C_2/(1-C_2)} \tag{8}$$

where the subscript zero refers to the initial data station, to be discussed below. We currently use eqs. (8) in our solution.

The remaining conditions to be imposed are either conditions directly at the wall, or conditions which are valid in the fully turbulent portion of the boundary layer, at $y^+ > 50$, say. It is well known that calculating right to the wall requires an extremely fine grid to define the viscous sublayer and so for computational efficiency our boundary conditions were imposed in the turbulent flow, away from the wall at $y = y_1$, corresponding to $y^+ \simeq 50$. Four relations are required which may be regarded as specifying u, v, k, ε at the inner boundary, although there is actually a strong coupling between all the variables.

The law of the wall is used to specify u as

$$u_1 = \frac{u_\tau}{\kappa}\left(\ln \frac{u_\tau y_1}{\nu} + c\right) \tag{9a}$$

where the friction velocity u_τ is obtained by assuming that the shear in the sublayer varies linearly, i.e.

$$\frac{\tau_1}{\rho} = \frac{\tau_w}{\rho} + \alpha y_1 = u_\tau^2 + \alpha y_1 \tag{9b}$$

The value of τ, can be obtained from mixing-length theory

$$\frac{\tau_1}{\rho} = (\kappa y_1)^2 \left[\left(\frac{\partial u}{\partial y}\right)_{y=y_1}\right]^2 \tag{9c}$$

and α is determined by integrating the momentum equation (1) over the sublayer to give

$$\alpha = 0.3\,\frac{du_1^2}{dx} - \frac{d}{dx}\left(\frac{u_e^2}{2}\right) \tag{9d}$$

The result of combining eqs. (9a) - (9d) is a highly nonlinear boundary condition which not only involves u at the boundary, but also its derivative, $\partial u/\partial y$.

The relation for the normal velocity, v, is obtained by using the continuity equation, (4), and assuming that, in the sublayer, u is only a function of y^+. This yields

$$v_1 = -\frac{u_1}{u_\tau}\frac{du_\tau}{dx}\,y_1 \tag{10}$$

where u_τ is obtained from (9b) - (9d) above. Again, this boundary condition represents a highly nonlinear relation between v, and u and its derivative.

The third boundary condition comes from the experimentally determined relation between the shear and the turbulence energy, and the definition of the turbulent viscosity

$$\tau = \nu_t \frac{\partial u}{\partial y} = 0.3k \tag{11}$$

Using eq. (5) for ν_t gives, at $y = y_1$,

$$k_1 \left(\frac{\partial u}{\partial y}\right)_{y=y_1} = \frac{0.3}{C_\mu}\,\varepsilon_1 \tag{12}$$

The final condition assures consistency between the mixing length and k-ε models at the edge of the sublayer. Using eq. (9c) in eq. (11) gives:

$$\varepsilon_1 = \frac{C_\mu}{(0.3)^2} \cdot \frac{(0.3k_1)^{3/2}}{\kappa y_1} \qquad (13)$$

It appears that in attempting to save computational effort by only integrating to the edge of the sublayer, we have increased the complexity of the boundary conditions by a comparable amount. However, as will be seen, the use of the Box method of solution easily accommodates such coupled nonlinear conditions as (9), (10), (12), (13).

Initial Conditions

For the parabolic problem posed by eqs. (1) - (3), initial data must be provided at the first station for the solution to proceed. In most cases, the experimental data provided by any test is not adequate to use for a finite-difference method requiring values at specified grid points. Thus, interpolation and smoothing techniques are usually required. Rather than attempt to do this for a large number of cases, we have chosen to use the measured values of boundary-layer properties as input to an empirical formula which fits measured data very well over a large range of flows.

Coles' law-of-the-wall/law-of-the-wake formula, modified to give a zero gradient at the edge of the boundary layer is[3]

$$u = \frac{u_\tau}{\kappa}\left[\ln\left(\frac{u_\tau y}{\nu}\right) + C + \Pi\left(1 - \cos\frac{\pi y}{\delta}\right) + \left(\frac{y}{\delta}\right)^2\left(1 - \frac{y}{\delta}\right)\right] \qquad (14)$$

As shown in reference 3, this can be used along with expressions for δ^* and θ to form a set of three relations for the five unknowns c_f, R_{δ^*}, R_θ, δ and Π Supplying c_f and R_θ from experiment allows the solution for the other three, and hence specifies the velocity by (14).

Since we are dealing with a transport model, in addition to the velocity, initial profiles are also necessary for k and ε. We have available the relation given by eq. (11) with ν_t given by the Cebeci-Smith eddy-viscosity formula[2,3], which is a two-layer model, and since ν_t is discontinuous at the matchpoint, very undesirable features are introduced into the evaluation of the k profile. An alternate method is suggested in reference 2 where equation (1) is written as

$$u\frac{\partial u}{\partial x} + v\frac{\partial u}{\partial y} = -\frac{1}{\rho}\frac{\partial p}{\partial x} + \frac{1}{\rho}\frac{\partial \tau}{\partial y} \qquad (15)$$

The pressure gradient is then set equal to zero and eq. (15) is integrated across the boundary layer to get an expression for τ. We have, however, included the effect of nonzero pressure gradient, and used eq. (14) for the u profile in the integration of eq. (15) to get a consistent evaluation of τ. When used in conjunction with eq. (11) this generates the initial profile required for k. Once τ and u have been generated, the definition of turbulence dissipation can be used to get the initial profile for ε, i.e.

$$\varepsilon = \tau\frac{\partial u}{\partial y} \qquad (16)$$

NUMERICAL METHOD

Equations (1) - (4) and their associated boundary conditions are solved using the Keller-Cebeci Box method for boundary layers . The initial step of this procedure is to recast the governing equations (1) - (4) as a first-order system. To do this, the continuity equation is satisfied by introducing a stream function, f:

$$u = \frac{\partial f}{\partial y}, \qquad v = -\frac{\partial f}{\partial x}$$

Then, eqs. (1) - (4) become (where primes indicate differentiation with respect to y, and v and p have changed meanings)

$$u = f'$$

$$v = u'$$

$$w = k'$$

$$p = \varepsilon'$$

$$u\frac{\partial u}{\partial x} - v\frac{\partial f}{\partial x} = (b_1 v)' + u_e\frac{du_e}{dx} \qquad (17)$$

$$u\frac{\partial k}{\partial x} - w\frac{\partial f}{\partial x} = (b_2 w)' + S_k$$

$$u\frac{\partial \varepsilon}{\partial x} - p\frac{\partial f}{\partial x} = (b_3 p)' + S_\varepsilon$$

This represents a system of seven equations for seven unknowns. We have the correct number of boundary conditions, and the remaining initial conditions are easily generated. Profiles for v, w and p are generated by differentiating the initial profiles already obtained for u, k and ε. The value of the stream function, f, is obtained by integrating u, accounting for the value of the edge of the sublayer by using the excellent correlation of u for wall flows developed by Whitfield[12].

Equations (17) are now centered on a finite-difference grid, and two-point differences are used to create a coupled, nonlinear system of finite-difference equations.

The boundary conditions (9) and (12) which involve $\partial u/\partial y$ can now be written in terms of $v(=u')$. Instead of the need to use one-sided differences, which would be the case with standard finite-difference schemes, a major advantage of the box method is its ability to include derivatives (solved as an unknown of the problem) in consistent relations at the boundary.

All the nonlinear terms in eqs. (17) are handled by a Newton iteration scheme with the following exceptions. The nonlinear viscosity coefficients b_1, b_2, b_3 and the source terms S_k and S_ε, are lagged one iterative cycle. Previous treatments of turbulent boundary layers using the Box method have found that this treatment of b does not deteriorate the convergence process[2,3]. The source terms were not treated exactly because a correct linearization eliminated many of the zero entries in the resulting coefficient matrix, and it was felt this would slow the solution procedure. All the boundary conditions are treated exactly by Newton's method. No ill effects could be detected from this linearization.

The linearization produces a block tridiagonal system for the seven unknowns, f, u, v, k, w, ε, p. This coupled system is solved in a very efficient manner by an extension of the block elimination method (which accounts for all the zeros in the matrices) discussed in reference 3, to systems with seven unknowns. Some previous treatments of transport models have had difficulties when the equations were only weakly coupled, and specific measures had to be taken to ensure a strong link between the variables[10,11]. Use of our procedure automatically accounts for coupling between all of the variables.

RESULTS AND DISCUSSION

The computer program incorporating the k-ε model and the Box Method was tested on numerous experimental configurations documented in the 1968 Stanford Conference. Representative results are presented here in figures 1-3. All constants necessary to specify the turbulence quantities were fixed at their recommended values, and never changed.

The results presented in figure 1 show the calculated skin friction compared with the experimental results of three equilibrium boundary layers. Figure 1a is for a mildly favorable pressure gradient (1300); figure 1b is a flat-plate flow (1400); figure 1c is a slightly adverse gradient (2500). In all cases the agreement between the calculations and experiment is very good. Note that for the favorable gradient, the calculations fall below the measurements, and for the adverse gradient, the predictions are high. This trend was consistently observed in all calculations.

More detailed comparisons are presented in figures 2 and 3. Figure 2 gives c_f, δ^* and θ calculations for another favorable gradient equilibrium flow (2700). Again the comparisons are very good with the δ^* and θ predictions being somewhat better than c_f. Figure 3 shows comparisons for an easily calculated nonequilibrium adverse gradient flow heading towards separation (4400). The agreement is excellent except close to separation where again the c_f prediction is high.

Thus, the k-ε model predicts these flows well, with the formulation described. A comparison with the mixing length model shows little difference, although the highly refined mixing length of Cebeci-Smith[2,3] may still be marginally better. However, the consistent behavior of the k-ε model for pressure gradient flows tends to indicate that possibly a minor modification in the wall law or the turbulence constants would account for the differences. One would then expect that in highly adverse gradients, far from equilibrium, the k-ε transport model to be superior. This is currently being investigated.

REFERENCES

1. Bradshaw, P., Ferriss, D.H. and Atwell, N.P.: Calculation of Boundary-Layer Development Using the Turbulent Energy Equation. JFM, Vol. 28, p. 593, 1967.

2. Cebeci, T. and Smith, A.M.O.: Analysis of Turbulent Boundary Layers. Academic Press, New York, 1974.

3. Cebeci, T. and Bradshaw, P.: Momentum Transfer in Boundary Layers. McGraw-Hill/Hemisphere Publishing Corp., Washington, 1977.

4. Cebeci, T., Stewartson, K. and Williams, P.G.: On the Response of a Stagnation Boundary Layer to a Change in the External Velocity. To be published in SIAM J. Num. Anal., 1977.

5. Cebeci, T., Thiele, F., and Stewartson, K.: On Near-Wake Laminar and Turbulent Shear Layers. To be published, 1978.

6. Coles, D. and Hirst, E.A.: Computation of Turbulent Boundary Layers — 1968, AFOSR-IFP-Stanford Conference, Vol. 2, Thermosciences Division, Stanford University, Stanford, 1969.

7. Goldstein, S.: Concerning Some Solutions of the Boundary Layer Equations in Hydrodynamics. Proc. Camb. Phil. Soc., Vol. 26, p. 1, 1930.

8. Jones, W.P. and Launder, B.E.: The Prediction of Laminarization with a Two-Equation Model of Turbulence. Int. J. H. M.T., Vol. 15, p. 301, 1972.

9. Launder, B.E., et al: The Prediction of Free Shear Flows — A Comparison of the Performance of Six Turbulence Models. In Proceedings of the Conference on Free Turbulent Shear Flows, NASA-SP-321, Vol. 1, 1973.

10. Naot, D., Shavit, A. and Wolfshtein, M.: Numerical Calculation of Reynolds Stresses in a Square Duct with Secondary Flow, Wärme-Stoffübertrag, Vol. 7, p. 155, 1974.

11. Pope, S.B. and Whitelaw, J.H.: The Calculation of Near-Wake Flows. JFM, Vol. 73, p. 9, 1976.

12. Whitfield, D.L.: Turbulent Flow Near a Wall Based on a Turbulent Kinetic Energy Equation. Paper presented at DEA Meeting, Dayton, Ohio, 1976.

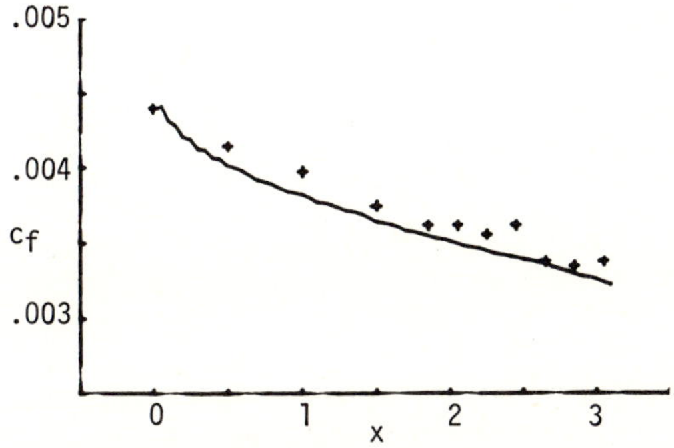

(a) Case 1300 — Favorable gradient

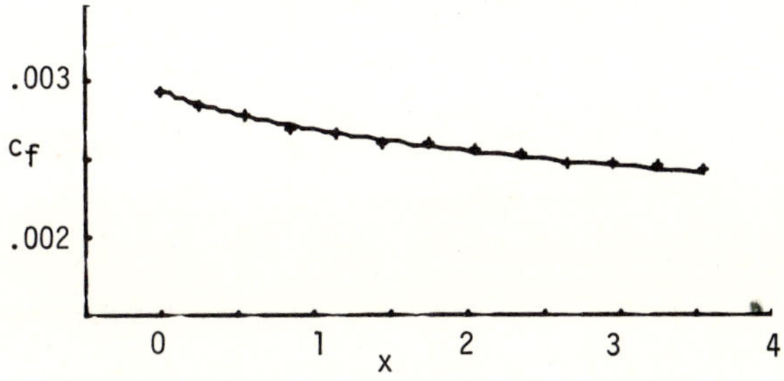

(b) Case 1400 — Flat-plate, zero gradient

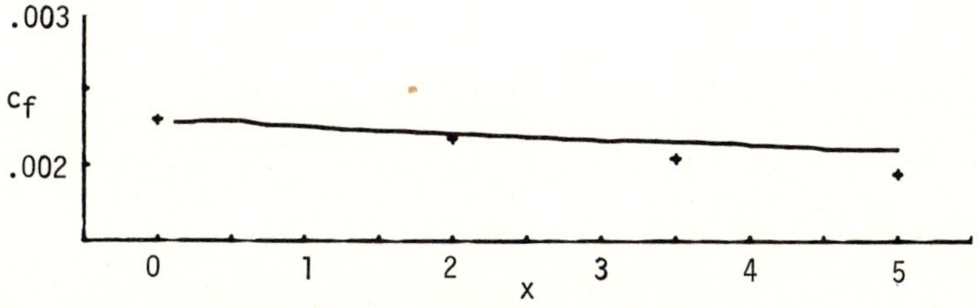

(c) Case 2500 — Adverse gradient

Figure 1. Skin-friction predictions for various pressure gradient flows.

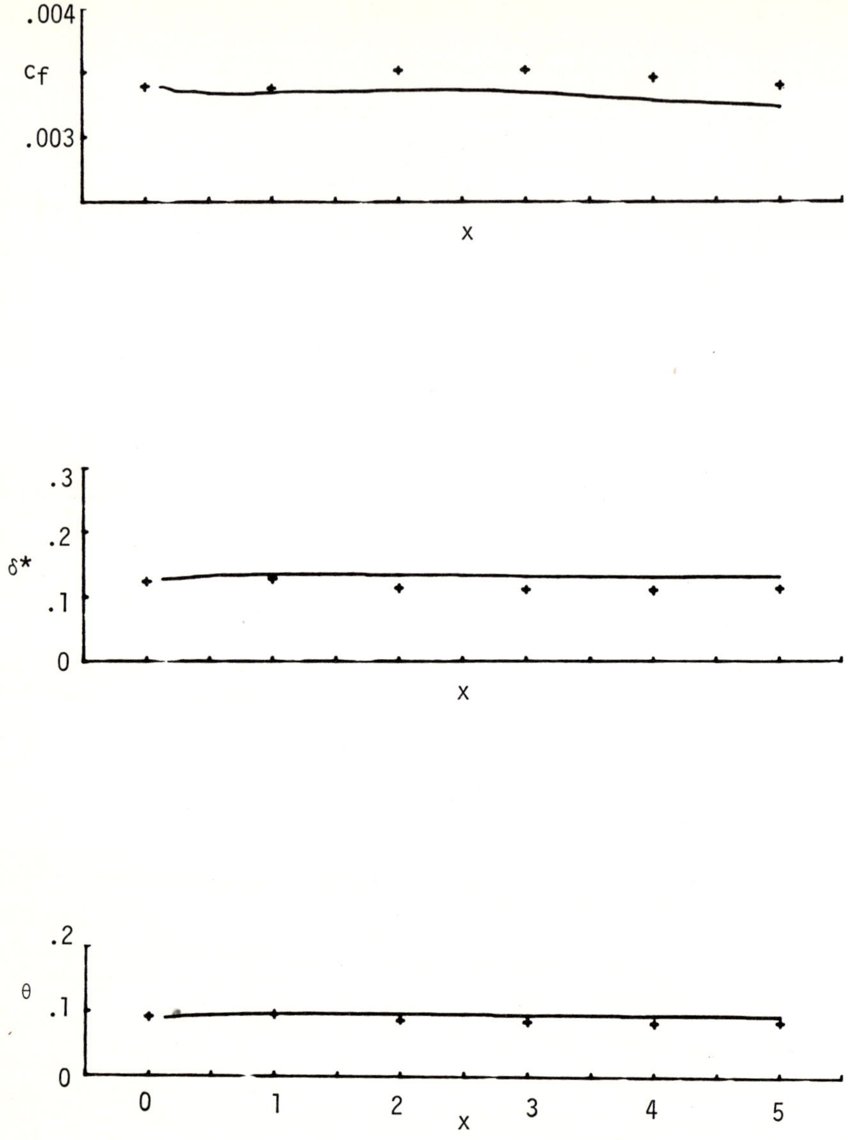

Figure 2. Comparison of predicted boundary-layer parameters for case 2700 (favorable gradient)

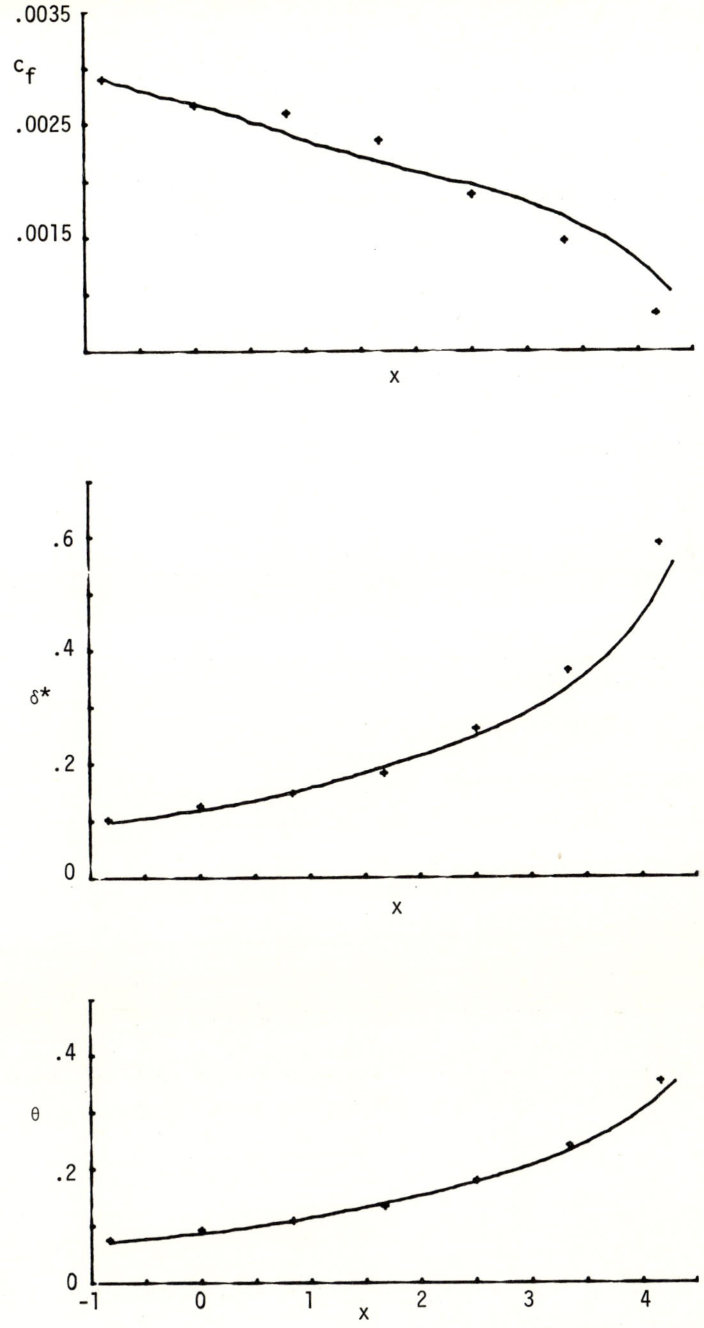

Figure 3. Comparison of predicted boundary-layer parameters for case 4400 (adverse gradient)

A NEW NUMERICAL METHOD FOR SOLVING

EXTERIOR LINEAR ELLIPTIC PROBLEMS*

by A. JAMI and M. LENOIR **

Ecole Nationale Supérieure de Techniques Avancées

32 Boulevard Victor 75015 PARIS FRANCE

We present a method for dealing with a wide variety of *exterior linear problems* involving an elliptic operator. Assuming that we know a *GREEN function* of the operator, we derive a *variational formulation* of the problem in a *bounded domain*. The problem is then discretized and solved through the *Finite Element Method*.

For the sake of simplicity we shall consider a particular but *non-trivial* problem

THE (2-D) WAVE RESISTANCE PROBLEM

A body is moving horizontally with a constant velocity V_∞ under a free surface (FS), see figure 1. We assume that the fluid is incompressible and inviscid, and that the flow is steady and irrotationnal. The *linearized problem* set for the *absolute velocity potential* ϕ in a *relative reference frame* is

$$(P_1)\quad \begin{cases} \nabla^2 \phi = 0 \text{ in } (\check{\Omega}), & (1) \\ \dfrac{\partial\phi}{\partial n} = 0 \text{ on } (S), & (2) \\ \dfrac{\partial\phi}{\partial y} + \dfrac{V_\infty^2}{g}\dfrac{\partial^2\phi}{\partial x^2} = 0 \text{ on } (FS_0), & (3) \\ \phi \to \Phi = V_\infty x \text{ when} & \\ x \to -\infty \text{ or } y \to -\infty , & (4) \end{cases}$$

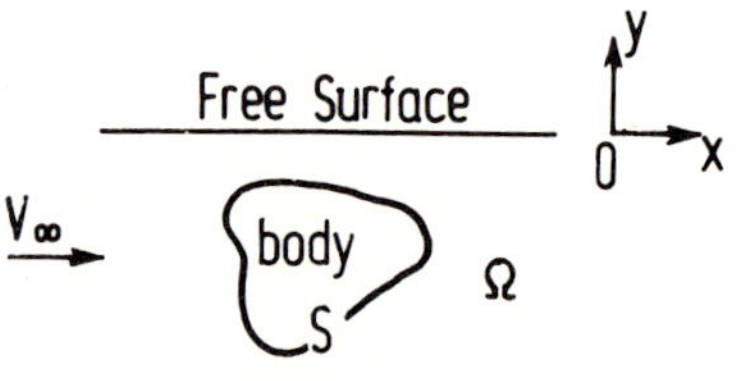

Fig:1

---

* This research was supported by the Direction des Recherches, Etudes et Techniques under contracts n° 76/1044 and 76/1167.

** ERA n° 070.664 Hydrothermodynamique of the Centre National de la Recherche Scientifique, PARIS.

in which g is the gravitational acceleration ;

(1) is the *LAPLACE's equation* in the *unbounded domain* $(\check{\Omega})$,

(2) the *slipping condition* on the body,

(3) the *linearized free surface* (NEUMANN-KELVIN) condition set on $\{y = 0\}$, which is the unperturbed free surface location, note that it involves a tangential second derivative ;

(4) means that there is no perturbation of the *uniform flow potential* Φ upstream and down below.

Moreover, because of energy considerations, equations (1) through (4) induce *finite amplitude waves downstream.*

This problem has already been solved using various methods : HAVELOCK [1] uses a *superposition of multipoles* to represent the potential outside of a circle.

The recent *Localized Finite Element Method* [2], [3] needs a basis of eigenfunctions which is not known in the infinite depth case.

The *singularity distribution method* [4] needs an intricate analytical preliminary work which is not tractable for high degree approximation; this method has already been applied in the (3-D) case but does not achieve a satisfactory accuracy.

As the domain $(\check{\Omega})$ is unbounded, (P_1) cannot be solved through Finite Elements.

PROBLEM FOR ϕ RESTRICTED TO A BOUNDED DOMAIN.

Let ϕ be a solution of (P_1) and G a GREEN function of the LAPLACE's operator which satisfies (3) and vanishes upstream and down below (see [5]); using the third GREEN's identity one derives the *Integral Representation* formula :

$$(I\ R) \quad \forall\ M \in (\check{\Omega}),\ \phi(M) = \Phi(M) + \int_S \phi(P) \frac{\partial G}{\partial n_p} (M,P)\ ds_p \ .$$

Let us point out that (I R) only involves an integration over (S).

Consider an *arbitrary curve* (Σ) lying in $(\check{\Omega})$ and without any intersection with (S); we shall now derive a problem for ϕ in the enclosed domain (Ω) (figure 2);

then for each point of (Σ), (I R) can be expressed as

(5) $\quad \phi_{|\Sigma} = \Phi_{|\Sigma} + A\,\phi_{|S}$

Using (5) as a "DIRICHLET condition" on (Σ), ϕ is obviously solution of (P_2) in the bounded domain (Ω).

$$(P_2)\quad \begin{cases} \nabla^2 \phi = 0 \ \textit{in}\ (\Omega) \\ \dfrac{\partial\phi}{\partial n} = 0 \ \textit{on}\ (S) \\ (5)\ \textit{on}\ (\Sigma) \end{cases}$$

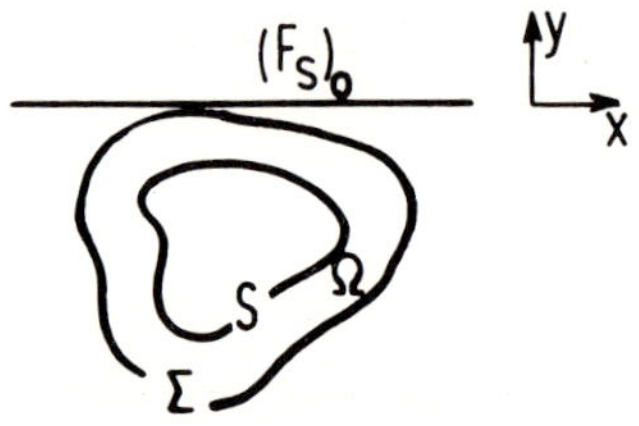

Fig:2

In order to solve (P_2) using Finite Elements, we must derive a

VARIATIONAL FORMULATION

As it is usual when dealing with DIRICHLET problems, we shall introduce a *linear and continuous extension operator E :*

f being defined on (Σ) only, E f is defined over the whole domain (Ω). E satisfies the following properties :

a) $E\,f\,(M) = f\,(M) \quad \forall\, M \in (\Sigma)$,

b) $E\,f\,(M) = 0 \quad \forall\, M \in (S)$.

We now introduce a new unknown function ϕ' as

(6) $\quad \phi' = \phi - E\,\phi_{|\Sigma}$.

Thanks to a), b) and using (I R) we can write

(7) $\quad \phi = \phi' + E\,A\,\phi'_{|S} + E\,\Phi_{|\Sigma}$

Let $H^1(\Omega)$ be the usual SOBOLEV's space, $V = \{w \in H^1(\Omega) \mid w_{|\Sigma} = 0\}$ and

(8) $\quad a(u, v) = \int_\Omega (\vec{\nabla} u \,|\, \vec{\nabla} v)\, dw$.

Then by using (P_2) and (7), we obtain a variational formulation for ϕ' :

(P_3)

Find ϕ' *in* V , *such that*, $\forall\ \psi \in V$,
$a(\phi', \psi) + a\,(E\,A\,\phi'_{\mid S}\,, \psi) = -\,a\,(E\,\phi_{\mid \Sigma}\,, \psi)$.

When, as in figure 3, and for the problem dealt with, $\partial\Omega$ contains a portion (L) of (FS_0), we have

$$(9) \qquad a\,(u,v) = \int_\Omega (\vec{\nabla}\, u \mid \vec{\nabla}\, v)\, dw - \frac{V_\infty^{\ 2}}{g} \int_L \frac{\partial u}{\partial x}\, \frac{\partial v}{\partial x}\, dx\ .$$

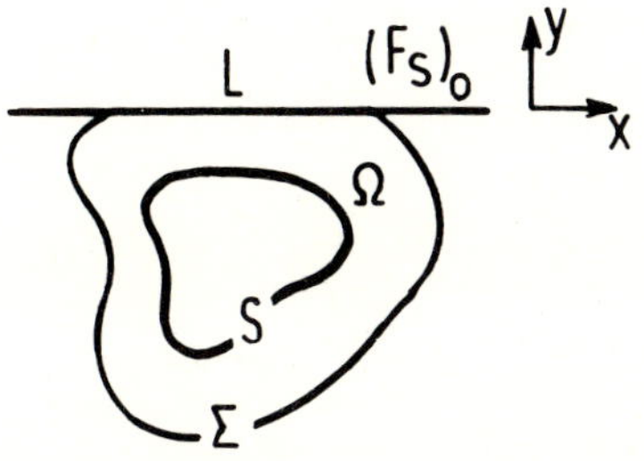

Fig:3

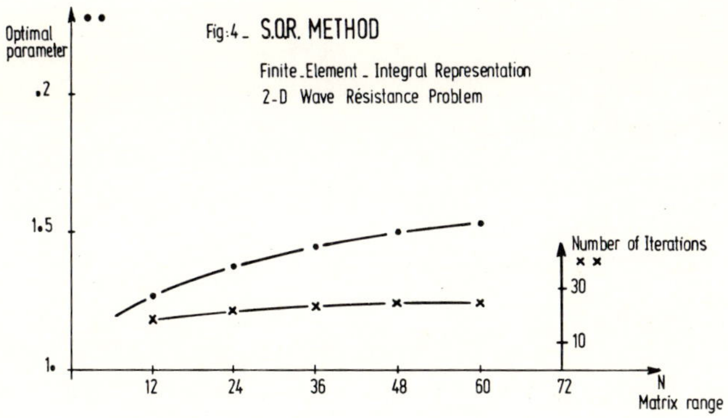

RESULTS AND COMMENTS

The F.E. discretization of (P_3) related either to (8) or (9) is straightforward ; it has been given in [6] for such a model case. The corresponding *rigidity matrix is non-symmetric*, but in those cases, a Successive Over Relaxation scheme works quite well ; moreover as shown on figure 4 numerical experiments allow a good prediction of the optimal relaxation parameter and of the corresponding number of iterations.

The arbitrary location of the boundary (Σ) allows the choice of a *narrow domain* (Ω) surrounding the body ; therefore a discretization involving *only one layer* of elements, and thus very few degrees of freedom, proves quite sufficient in many cases. As discussed in [7] , isoparametric eight-node rectangular elements have been used.

Figure 5 shows the wave resistance of an immersed circular cylinder of diameter D against the FROUDE number ($F_D = V_\infty / \sqrt{gD}$).

Fig: 5 . Wave Resistance (2-D)

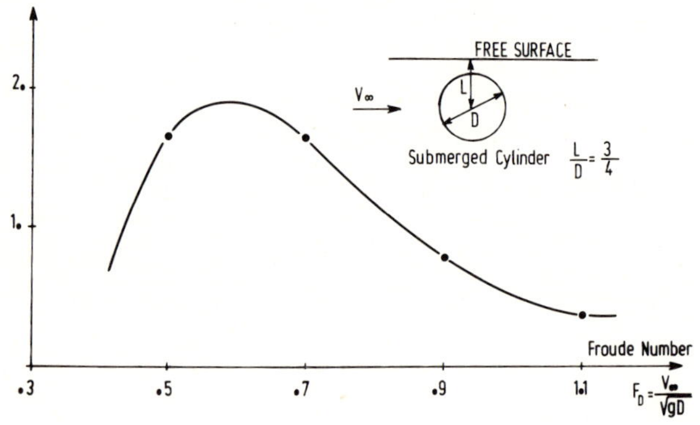

On figure 6 the precision of our method and of the singularity distribution method [4] are compared. The values are plotted against the number of unknowns on (S) which is the fundamental parameter for evaluating the computing time of both methods.

Fig 6_ Wave_Resistance (2-D) $F_D = 0.5$ $\frac{L}{D} = \frac{3}{4}$

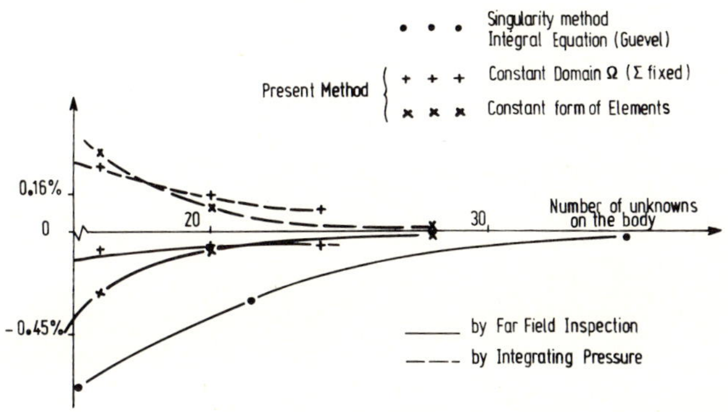

CONCLUDING REMARKS

This method is an appreciable improvement for the solution of the problems of hydrodynamics. Some other (2-D) problems have been worked out, e.g.

harmonic motion of a ship,

HELMHOLTZ equation in unbounded domains.

Moreover some more intricate problems are under development such as

(3-D) harmonic motion and wave resistance of a ship,

(2-D) diffraction on an island in the shallow water case with non constant depth.

Note that, on the one hand, our method is not restricted to elliptic operators, for instance time-dependant problems could be dealt with; on the other hand, similar ideas have been recently used [8] for dealing with corner-type singularities.

REFERENCES

[1] T.H. HAVELOCK *"The Forces on a Circular Cylinder Submerged in a Uniform Stream"*. Proceedings of the Royal Society of London, A, Vol. 157, p 526-534 (1936).

[2] K.J. BAI, R.W. YEUNG *"Numerical Solutions to Free Surface Flow Problems"*. Tenth Naval Hydrodynamics Symposium, Session VII, Technical papers, M.I.T. Cambridge (1974).

[3] S. MARCOUYOUX *"Calcul, selon une méthode d'éléments finis localisés, de la résistance de vagues d'un profil immergé dans un écoulement à surface libre, dans l'approximation de NEUMANN-KELVIN"*. Rapport E.N.S.T.A. , n° 087, Paris (1977).

[4] P. GUEVEL, J.P.V. CORDONNIER *"Numerical Solution of the NEUMANN-KELVIN problem by the method of singularities"*. Second International Conference on Ship Hydrodynamics, Berkeley, California (1977).

[5] J.V. WEHAUSEN, E.V. LAITONE *"Surface Waves"* Handbuch der Physik, Vol IX, Berlin, Springer Verlag, p 446-778 (1960).

[6] A. JAMI, M. LENOIR *"Variations numériques sur le thème du mariage ou comment coupler éléments finis et représentation intégrale"*. Journées Eléments finis de Rennes (Mai 1978).

[7] A. JAMI, M. LENOIR *"A Variational Formulation for Exterior Problems in Linear Hydrodynamics"* Computer Methods in Applied Mechanics and Engineering (to appear).

[8] BARDOS, CEA, GRISVARD, LOZI *"Calcul numérique de solutions singulières de problèmes aux limites"* Colloque d'Analyse Numérique, Giens (mai 1978).

THE TRANSFORMATION OF EXTERNAL DISTURBANCES INTO THE BOUNDARY LAYER WAVES

Yu.S.Kachanov, V.V.Kozlov, V.Ya.Levchenko, V.P.Maksimov
Institute of Pure & Applied Mechanics
USSR Academy of Sciences, Novosibirsk 630090

It is well known that the transition to the turbulence in boundary layers is the result of the development of small disturbances, so called Tollmien-Schlichting waves (T-S waves). The problem of T-S wave appearence is of interest in the study of the hydrodynamic stability and the transition. There is a close connection between a free-stream disturbances level and a transition position. However, the mechanisms of the transformation of this external disturbances to eigenoscillations of the boundary layer (T-S waves) are not studied as yet.

It was shown theoretically in [1] that small vortex disturbances of an outer flow did not excite T-S waves in the developed boundary layer. The intensity of that sort of disturbances damps sharply from the outer boundary of the boundary layer to the wall. It was discovered in [2-3] that longitudinal acoustic oscillations of the outer flow do not influence on characteristics of the boundary layer stability and do not lead to the T-S waves generation. However, the appearence of T-S waves was discovered in the region of the leading edge of the flat plate at its cross irradiation by sound waves [4]. The generation of T-S waves in the boundary layer was discovered in [3] in the region of the leading edge as a result of its vibration in cross direction.

This paper describes the theoretical and experimental study of the process of generation of vortex T-S waves in the region of the flat plate leading edge in the presence of different free-stream disturbances.

The theoretical study is based on the numerical solution of the incompressible Navier-Stockes equations linearized relative to a small amplitude of disturbances. External disturbances are given in the form of a cross stationary wave as well as a travelling wave running on the leading edge of the plate. The main flow is determined from the solution of full Navier-Stockes equations, which in variables ψ , ω have the form

$$\frac{\partial \omega}{\partial t} + u\frac{\partial \omega}{\partial x} + v\frac{\partial \omega}{\partial y} = \frac{1}{Re}\frac{\partial^2 \omega}{\partial x^2} + \frac{\partial^2 \omega}{\partial y^2},$$

$$\omega = \frac{1}{Re}\frac{\partial^2 \psi}{\partial x^2} + \frac{\partial^2 \psi}{\partial y^2}, \qquad (1)$$

$$u = \frac{\partial \psi}{\partial y}, \quad v = -\frac{\partial \psi}{\partial x}.$$

The reference scales for x, y, t, u, v are L, $L/\sqrt{Re}$, L/U_∞, U_∞, $U_\infty/\sqrt{Re}$ $(Re = U_\infty L/\nu)$ respectively (see Fig.1). Let us seek a solution for disturbances in the form

$$\omega(x,y,t) = \omega(x,y,t) - \Omega(x,y) = f(x,y)\,e^{-i\bar{\beta}t},$$

$$\psi'(x,y,t) = \psi(x,y,t) - \Psi(x,y) = g(x,y)\,e^{-i\bar{\beta}t}$$

($\bar{\beta} = \beta \cdot L/U_\infty$ - nondimensional frequency). In this case, the equations for complex amplitudes of disturbances ($f = f_r + i f_i$, $g = g_r + i g_i$) will have the form

$$-i\bar{\beta}f + U\frac{\partial f}{\partial x} + V\frac{\partial f}{\partial y} - \frac{\partial \Omega}{\partial y}\frac{\partial g}{\partial x} + \frac{\partial \Omega}{\partial x}\frac{\partial g}{\partial y} = \frac{1}{Re}\frac{\partial^2 f}{\partial x^2} + \frac{\partial^2 f}{\partial y^2}, \qquad (2)$$

$$f = \frac{1}{Re}\frac{\partial^2 g}{\partial x^2} + \frac{\partial^2 g}{\partial y^2}.$$

This equations were solved numerically for two kinds of external disturbances.

In the first case a cross irradiation of the leading edge by sound waves was considered. Cross sound disturbances can be represented qualitatively in the form

$$v_\infty(x,y,t) = v_e(x)\,e^{i(\alpha_s y - \bar{\beta}t)} \qquad (3)$$

If the sound wave length $\lambda_s = 2\pi/\alpha_s \gg Y_2$ (Y_2 - cross size of the integration region) and the sound velocity $C_s \gg U_\infty$, it is possible to consider the disturbing liquid column to be moving across the flow within the boundaries of the integration region as a whole with a practically constant phase. The expression (3) is simplified

$$v_\infty(x,t) = v_e(x)\, e^{-i\beta t} \tag{4}$$

and is used as a boundary condition (Fig. 1, a).

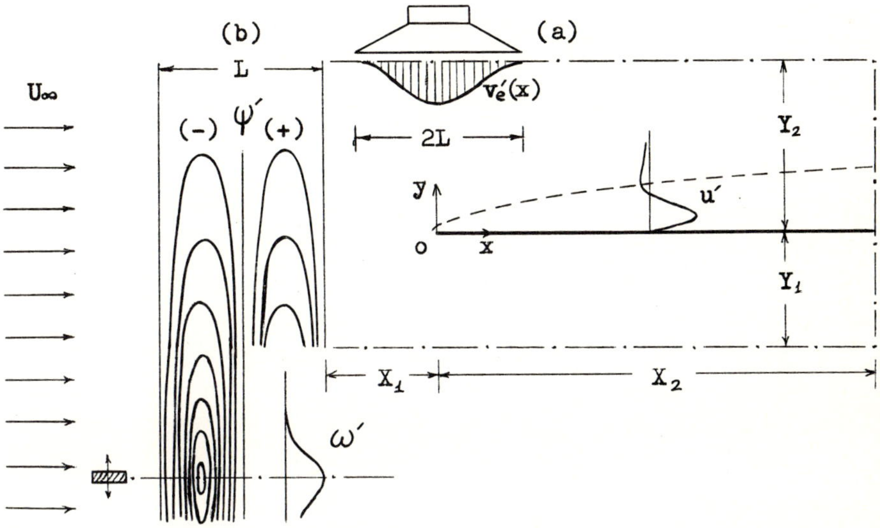

Fig. 1.

Furthermore, the stationary wave (4) can approximate qualitatively small vibrations of the leading edge of the plate. The experimentally determined amplitudes of plate vibrations at the generation of T-S waves [3] were as small as some microns while the extension of the vibrating section of the plate was as large as tens of centimeters. Therefore neglecting changing of the plate curvature, it is possible qualitatively to use flow oscillations relative to the still plate which are concentrated in the region of the leading edge instead of small oscillations of the leading edge relative to the undisturbed flow. The integration region for the system (2) in this case is $-X_1 \leq x \leq X_2$, $0 \leq y \leq Y_2$. The boundary conditions are

$$-X_1 \leq x < 0, \qquad y=0, \qquad \frac{\partial g}{\partial y} = 0, \qquad \frac{\partial f}{\partial y} = 0,$$

$$0 < x \leq X_2, \qquad y=0, \qquad g = \frac{\partial g}{\partial y} = 0 \quad \left(f = \frac{\partial^2 g}{\partial y^2}\right),$$

$$x = X_2, \qquad 0 \leq y \leq Y_2, \qquad \frac{\partial^2 g}{\partial x^2} = -\alpha^2 g, \qquad \frac{\partial^2 f}{\partial x^2} = -\alpha^2 f,$$

$$1 < x \leq X_2, \qquad y = Y_2, \qquad \frac{\partial g}{\partial y} = -\frac{\alpha}{\sqrt{Re}} g, \qquad f = 0,$$

$$-X_1 \leq x \leq 1, \qquad y = Y_2, \qquad g = \varphi(x), \qquad f = \frac{1}{Re} \varphi_{xx},$$

$$x = -X_1, \qquad 0 \leq y \leq Y_2, \qquad g = \varphi(-X_1), \qquad f = 0.$$

The function $\varphi(x)$ determines the form of the external disturbance profile $\varphi_x = v_e'(x)$ (see Fig. 1, a). Reynolds number Re and nondimensional frequency $\bar{\beta}$ are parameters of the problem.

In the second case the interaction between running disturbances and the leading edge of the plate was simulated numerically. In the experiments the disturbances in the free stream were produced by vibrating ribbon. It is known that vibrating ribbon in the free stream originates the disturbance like Karman vortex street, which propagates downstream with the speed of the flow. In the experiments the (nonvortex) part in an outlying district of the kernel of "vortex street" reached the leading edge. This part (of the "vortex street") in integration region is approximated very well by the function

$$\psi'(x, y, t) = e^{-\alpha y/\sqrt{Re}} \, e^{i\alpha(x - t)} \tag{5}$$

if to suppose that the vortex kernel pass below the integration region. Thus, the travelling wave (5) (see Fig. 1, b) was set as the external disturbance in this case. The integration region was $-X_1 \leq x \leq X_2$, $-Y_1 \leq y \leq Y_2$. The boundary conditions were

$$x = X_2, \qquad -Y_1 \leq y \leq Y_2, \qquad \frac{\partial^2 g}{\partial x^2} = -\alpha^2 g, \qquad \frac{\partial^2 f}{\partial x^2} = -\alpha^2 f,$$

$$-X_1 \leq x \leq X_2, \qquad y = Y_2, \qquad \frac{\partial g}{\partial y} = -\frac{\alpha}{\sqrt{Re}} g, \qquad f = 0,$$

$$x = -X_1, \qquad -Y_1 \leq y \leq Y_2, \qquad g = e^{-\alpha y/\sqrt{Re} - i\alpha X_1}, \qquad f = 0,$$

$$-X_1 \leq x \leq X_2, \qquad y = -Y_1, \qquad g = e^{\alpha Y_1/\sqrt{Re} + i\alpha x}, \qquad f = 0.$$

In this case the wave length of running disturbances (Fig. 1, b) was assumed as a reference length L, and $\bar{\beta} = 2\pi$. Therefore this problem has a single parameter - Reynolds number Re.

The equations (2) were solved numerically by an implicit method, so the differential equations were replaced by corresponding difference ones with second-order truncation error in the integration region at the uniform rectangular grid

$$-i\Delta y^2\bar{\beta} f^K_{m,n} + \frac{\Delta y^2}{2\Delta x} U_{m,n}\left(3 f^K_{m,n} - 4 f^K_{m,n-1} + f^K_{m,n-2}\right) + \frac{\Delta y}{2} V_{m,n}\left(f^K_{m+1,n} - f^K_{m-1,n}\right) +$$

$$+ \frac{\Delta y^2}{2\Delta x} R_{m,n}\left(g^{K-1}_{m,n+1} - g^K_{m,n-1}\right) + \frac{\Delta y}{2} S_{m,n}\left(g^K_{m+1,n} - g^K_{m-1,n}\right) =$$

$$= \frac{\Delta y^2}{\Delta x^2}\frac{1}{Re}\left(f^K_{m,n-1} - 2 f^K_{m,n} + f^{K-1}_{m,n+1}\right) + f^K_{m-1,n} - 2 f^K_{m,n} + f^{K-1}_{m+1,n} ,$$

$$\Delta y^2 f^K_{m,n} = \frac{\Delta y^2}{\Delta x^2}\frac{1}{Re}\left(g^K_{m,n-1} - 2 g^K_{m,n} + g^{K-1}_{m,n+1}\right) + g^K_{m-1,n} - 2 g^K_{m,n} + g^K_{m+1,n} ,$$

$$R_{m,n} = -\left(\frac{\partial \Omega}{\partial y}\right)_{\substack{x = n\cdot\Delta x \\ y = m\,\Delta y}} , \qquad S_{m,n} = \left(\frac{\partial \Omega}{\partial x}\right)_{\substack{x = n\cdot\Delta x \\ y = m\,\Delta y}} ,$$

k is the iteration index. Nodal values on grid lines parallel to the y axis are determined by a usual direct method which proceed iteratively in the x direction, what is one iteration loop. The iteration process was ended when most relative variations of all grid variables from one iteration loop to the following had not exceeded the given value $\varepsilon = 10^{-3}$. The coefficients $U_{m,n}$, $V_{m,n}$, $R_{m,n}$, $S_{m,n}$ characterizing the main flow were determined from the solution of equations (1) for the plane semi-infinite plate in the same integration region at $\frac{\partial \omega}{\partial t} = 0$ and corresponding Reynolds number.

The unknown wavenumber α , being a part of boundary conditions, is a function of x -coordinate and it was calculated during the iteration process.

The experiment was performed in the low-turbulence wind tunnel on the flat plate 10 mm thick, 1m wide and 1,5m long. The leading edge was in the form of two conjugated semiellipsises with the ratio of semiaxis 2mm:50mm from working side and 8mm:132mm from opposite one. The turbulence level was no more than 0,04%,. Sine vortex disturbance with small amplitudes (~ 0,04%) propagating with the main flow velocity were excitated in the free stream by means of the vibrating ribbon technique. The transformation of a far part of the voetex street from the vibrating ribbon into T-S waves in the region of the leading edge was studied. The kernel of a vortex street was under the plate in this case. The registration of amplitude and phase fields of the longitudinal component of disturbances and of mean velocities was carried out by the hot-wire anemometer (DISA 55 D 01).

It is also necessary to note that when the vortex street passed over the plate coming into contact with a growing boundary layer, T-S waves were not excited and pulsations intensity damped sharply to the wall. It is in a good agreement with the result of theoretical paper [1] .

The calculations were made in the large range of parameters: a) - $Re = 0,5\cdot10^4,\ 10^4,\ 2\cdot10^4,\ 4\cdot10^4;$ $\bar{\beta} = 0,8,\ 1.32,\ 2;$ b) - $Re = 10^4,\ 2\cdot10^4,\ 4\cdot10^4$.

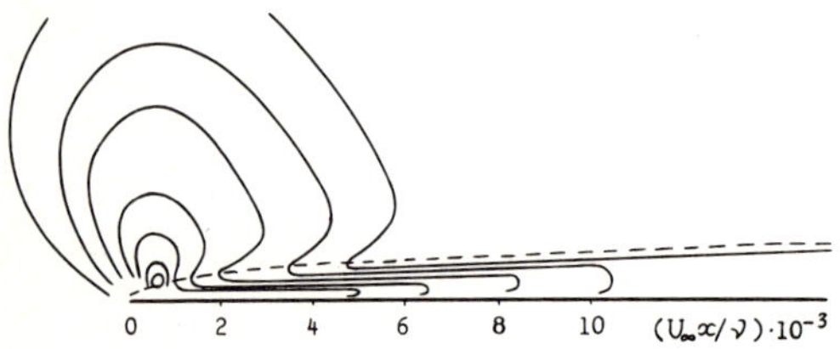

Fig. 2

In Fig. 2 isolines of the rms value of component disturbances field ($U_{rms} = mod(\partial g/\partial y)$) are shown for the case corresponding to Fig. 1a, $Re = 10^4$, $\bar{\beta} = 1,32$. A maximum value of the disturbance intencity near the leading edge (closed curve) was considered to be unity. The rest lines follow in 1/8 in order of decreasing. The corresponding curves for the traveling wave case (see Fig. 1b) are given in Fig. 3. Calculated curves ($Re = 2\cdot10^4$) are represented in the upper part of the figure. The results of experimental measurements are represented in lower part reflected relative to x - axis. The experiments were performed at $U_\infty = 5,6$ m/sec and $\beta = 106$ Hz, corresponding to the Reynolds number $Re = 2\cdot10^4$ at numerical calculations. One can see that both theoretical and experimental result are in a good qualitative agreement. Some differences are explained by a finite thickness of a real plate in experiments and pressure gradient due to it. The most cha-

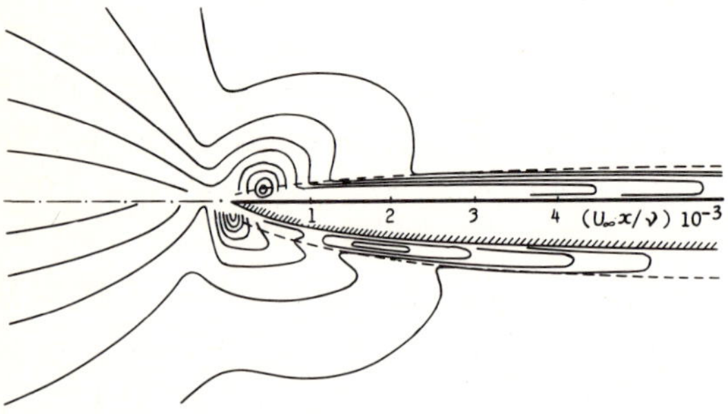

Fig. 3

racteristic detail of the transformation is a localized region of higher intensity of longitudinal component of pulsations in a leading edge region. Large local pressure pulsations were discovered numerically at the same place. The far part of a running vortex street is to the left of this region, the T-S waves generated on the leading edge propagate inside the boundary layer to the right. The disturbance intensity outside the boundary layer decreases sharply. Sizes of the intensity jump region were ~ 1 mm under the experiment conditions.

A mechanism of transformations of free-stream disturbances into boundary layer eigen-waves is seen well from numerical results. The streamlines for different time moments of a half-period are shown in Fig. 4 and Fig. 5 for cases "a" and "b" (see Fig. 1) respectively. The dotted lines represent the outer boundaries of the boundary layers. The development of vortex disturbances in the boundary layer from a moment of their origin can be traced in this figures. The speed of disturbance propagation in the boundary layer changes from $c/U_\infty \sim 0{,}5$ in the leading edge region to $\sim 0{,}35 \div 0{,}4$ downstream. The theoretically calculated profile of u-component of disturbances is in a good qualitative agreement with classical T-S wave profile (see Fig. 1).

It should be noted that processes of wave arising in the cases of stationary and traveling waves (see Fig. 4 and 5 correspondingly) are similar qualitatively. It is associated with the fact that the reason for their appearence is one and the same. The common feature of cases "a" and "b" is the presence of a periodical v-component of velocity in free-stream disturbances. It is corroborated by experimental results when external disturbances had not v-component of pulsations. It was achieved by turning the vibrating ribbon to 90° in the plane perpendicular to the flow direction. The waves in the boundary layer do not arise in this case nor in the case of the irradiation of the plate by longitudinal sound waves [3].

When disturbances are propagating in the boundary layer downstream, the maximum of Urms decreases as it is shown in Fig. 6. It takes place right up to $Re^* = U_\infty \delta^*/\nu = 635$. Then the amplitude begins to increase until $Re^* = 1000$ in the full accordance with the classical linear theory of hydrodynamic stability (dotted line).

It was also discovered that the magnitude of the intensity jump near the leading edge (see Fig. 3) depended on Reynolds number Re both in the theory and in the experiments $\ln(u'_0 / u'_\infty) = k \ln(Re)$ and $k \sim 1$, or $\ln(u'_0 / u'_\infty) = -k \ln(\bar{\beta})$.

Fig.5.

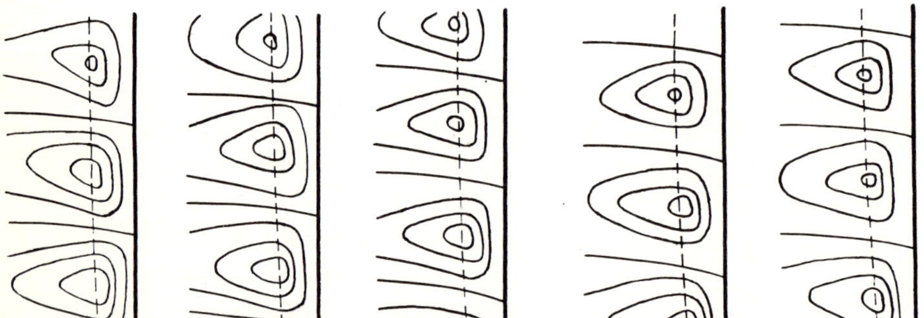

Fig.4.

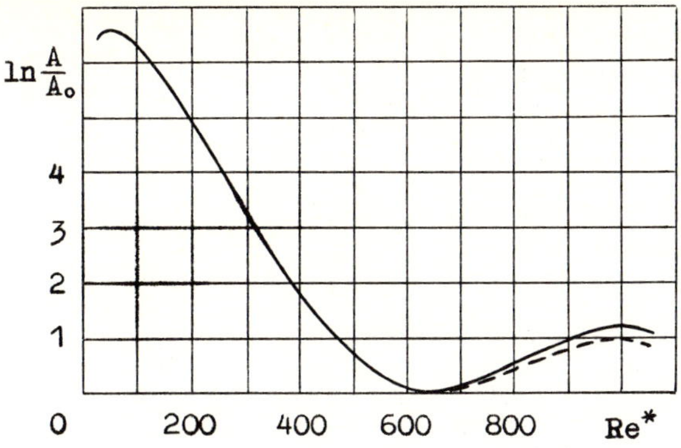

Fig. 6

Thus, the results of numerical simulation of mechanism of the generation of boundary layer waves can explain qualitatively the generation of T-S waves in the leading edge region which is observed in the experiments with imposing of cross vibrations of plate [3], cross sound waves and running vortex [4].

R e f e r e n c e s

1. Rogler H., Reshotko E. SIAM J.Appl.Mech., 1975, vol.28, No.2, p.431.
2. Levchenko V.Ya., Solovyev A.S. Izv. Sib. Otd. Akad. Nauk USSR, 1975, No.13, Vip.3, p.10.
3. Kachanov Yu.S., Kozlov V.V., Levchenko V.Ya. Izv. Sib. Otd. Akad. Nauk USSR, 1975, No.13, Vip.3, p.18.
4. Mechel F., Schilz W. Acustica, 1964, Vol.14, p.325.

THE EFFECT OF STRONG MAGNETIC FIELD ON THE SHIFT FLOW OF VISCOUS INCOMPRESSIBLE ELECTROCONDUCTING FLUID

H.E.Kalis
Latvian State University, Riga, USSR
A.A.Klyukin, Yu.B.Kolesnikov
Physics Institute, Latvian SSR Academy of Sciences, Riga-Salaspils, USSR

In a strong magnetic field a flow of viscous incompressible electroconducting fluid has a number of peculiarities due to the velocity structure deformation and the formation of specific Hartmann boundary layers. The basic difficulty of the construction of mathematical models results from the fact that these peculiarities should be taken into account (see /1/ and literature references in /1/).

In the present paper an attempt is made to overcome problems of this kind which are related to the study of one class of free shift stationary flows, i.e., a plane-parallel and axisymmetrical flow of rotating fluid (analogue of the Lehnert flow /2/) in the presence of external homogeneous magnetic field. In both cases motion is due to an electromagnetic force which arises when electric current with a component normal to the magnetic field induction vector is passing through the field. An analytical solution and grid method elaborated yielded results in agreement with experimental studies on the effect magnetic field upon the velocity structure and the flow stability.

The plane-parallel shift flow

Fluid is in a rectangular cross-section channel ($|x|<L, 0<y<C,$ $|z|<\infty$) with nonconducting walls in the presence of magnetic field $B=(B_0\cos\varphi_0, B_0\sin\varphi_0, 0)$. Using a thermocouple or linear electrodes ($x=\pm a$), or electrodes with finite width ($x\in[a,b]$, $x\in[-b,-a]$) placed at the wall of the channel $y=0$, the electric current is induced in the fluid (Fig.1.1).

The motion is described by a system of time-independent equations with respect to velocity component $u_z=u(x,y)$ and the component of an induced magnetic field strength $H_z=H(x,y)$ /1/:

$$\begin{cases} L_1 u + L_2 u + Ha\left(\cos\varphi_0 \frac{\partial H}{\partial x} + \sin\varphi_0 \frac{\partial H}{\partial y}\right) = 0 \\ L_1 H + L_2 H + Ha\left(\cos\varphi_0 \frac{\partial u}{\partial x} + \sin\varphi_0 \frac{\partial u}{\partial y}\right) = 0, \end{cases} \tag{1}$$

$$L_1 = \frac{\partial^2}{\partial x^2}, \quad L_2 = \frac{\partial^2}{\partial y^2},$$

where Ha is Hartmann number /1/. At the channel walls $u=0$, and $H|_{y=c} = H|_{x=\pm L} = 0, H|_{y=0} = H_0(x)$, where $H_0(x)$ is the function determined by current distribution at the electrodes. For finite electrodes

$$H_0(x) = \begin{cases} -1, |x| \le a \\ -\frac{b-|x|}{b-a}, a \le |x| \le b \\ 0, |x| \ge b \end{cases} \tag{2.1}$$

or more general boundary condition

$$H\Big|_{y=0} = -1(|x| \le a); \quad \frac{\partial H}{\partial y}\Big|_{y=0} = 0(a \le |x| \le b); \quad H\Big|_{y=0} = 0(|x| \ge b). \tag{2.2}$$

In the case of linear electrodes

$$H_0(x) = \begin{cases} -1, |x| < a \\ 0, |x| > a \end{cases} \tag{2.3}$$

Substituting function $S_\pm = H \pm u$ we obtain a corresponding boundary problem. Using a method of variables separation with the account of boundary conditions (2.1) and (2.3) we obtain analytical solutions of the initial problem. For example, in the case of finite electrodes at $\varphi_0 = \pi/2$ solutions are in the following form

$$u(x,y) = sh(\alpha y)\, A(x,y), \quad H(x,y) = ch(\alpha y)\, A(x,y),$$

where $\alpha = \frac{Ha}{2}, \quad A(x,y) = \sum_{k=0}^{\infty} a_k \cos(\lambda_k x)\, R(\alpha,\lambda_k,y), \quad \lambda_k = \frac{2k+1}{2L}\pi,$

$$R(\alpha,\lambda_k,y) = \frac{sh\left((c-y)\sqrt{\alpha^2+\lambda_k^2}\right)}{sh\left(c\sqrt{\alpha^2+\lambda_k^2}\right)}, \quad a_k = -\frac{2}{L(b-a)\lambda_k^2}\left[\cos(\lambda_k a) - \cos(\lambda_k b)\right]$$

The use of asymptotic expansions of integrals of rapidly oscillating functions /3/ yields estimations of solutions at large values of Ha: $u(x,y) \approx -\frac{1}{2}H_0(x), H(x,y) \approx \frac{1}{2}H_0(x)$ Asymptotics is practically attained at $\alpha y \sim 1$, since a Hartmann boundary layer of $O(Ha^{-1})$ thickness is formed.

The fluid expenditure $Q = \int_0^c \int_{-L}^{L} u(x,y)\,dx\,dy$ is determined by the following formula

$$Q=\frac{4}{L}\Big\{-\sum_{K=0}^{\infty}\frac{(-1)^K\lambda_K C_K(\lambda_K)\,ch(\alpha L\cos\varphi_0)}{(\lambda_K^2+\alpha^2\cos^2\varphi_0)^3}\Big[\alpha\cos\varphi_0\cos(\lambda_K a)\,sh(\alpha a\cos\varphi_0)+$$

$$+\lambda_K\sin(\lambda_K a)\,ch(\alpha a\cos\varphi_0)\Big]+\sum_{K=1}^{\infty}\frac{(-1)^K\tilde\lambda_K C_K(\tilde\lambda_K)\,sh(\alpha L\cos\varphi_0)}{(\tilde\lambda_K^2+\alpha^2\cos^2\varphi_0)^3}\times$$

$$\times\Big[\alpha\cos\varphi_0\sin(\tilde\lambda_K a)\,sh(\alpha a\cos\varphi_0)-\tilde\lambda_K\cos(\tilde\lambda_K a)\,sh(\alpha a\cos\varphi_0)\Big],$$

where

$$C_K(q_K)=\alpha\sin\varphi_0-\frac{\sqrt{\alpha^2+q_K^2}\;sh(\alpha c\sin\varphi_0)}{sh(c\sqrt{\alpha^2+q_K^2})},$$

$$q_K=\lambda_K;\ \tilde\lambda_K,\ \tilde\lambda=\frac{K\pi}{L},\ b=a.$$

Using a method of Fourier integral transform we obtained analytical solutions of boundary problems in half-space $y>0$ and between two planes $y=0$, $y=c$ /4/. A boundary problem for(2.1) is also solved using a grid method. By means of inserted grids we constructed the monotone difference scheme of the second order of approximation which takes into account the peculiarities of a boundary layer at large Hartmann numbers. Every equation of the boundary problem with respect to the $S_\pm$ function was approximated by difference equations /5/. In the case of $\varphi_0=\pi/2$ a difference approximation describes exactly Hartmann layer at the wall $y=0$ of $O(Ha^{-1})$thickness. Boundary problem$(\varphi_0=\pi/2)$

$$L_1S_\pm\pm L_2S_\pm\pm Ha\frac{\partial S_\pm}{\partial y}=0,\qquad(0<x<L,\ 0<y<c)$$

$$S_\pm\big|_{y=c}=S_\pm\big|_{x=L}=\frac{\partial S_\pm}{\partial x}\Big|_{x=0}=0;\ S_\pm\big|_{y=0}=H_0(x),$$

is substituted for the difference one

$$\Lambda_1S_\pm+\varkappa\Lambda_2S_\pm\pm Ha(S_\pm)_{\mathring y}=0,$$

where

$$\Lambda_1S=\frac{1}{h_1^2}(S_{i+1,j}-2S_{ij}+S_{i-1,j}),\ \Lambda_2S=\frac{1}{h_2^2}(S_{i,j+1}-2S_{ij}+S_{i,j-1}),$$

$$S_{\mathring y}=\frac{1}{2h_2}(S_{i,j+1}-S_{i,j-1}),\ \varkappa=\frac{Rh}{2}\,cth\frac{Rh}{2},\ S_{ij}=S(x_i,y_j),S=S_+,S_-$$

$$x_i=(i-1)h_1\ (1\le i\le N),\ y_j=(j-1)h_2\ (1\le j\le M),\ L=(N-1)h_1;\ c=(M-1)h_2,$$

$R_h=Ha\,h_2$ is a grid Reynolds number, N,M is a number of basic lines in the direction of axes x, y. To obtain grid functions S_{ij} with a step h_1, h_2 an absolutely stable iteration method by Douglas-Rackford is used /6/. In order to make the results more exact in the flow core we used inserted grids which permitted to avoid artificial viscosity of the order of $O(Hah_2^2)$. An approximate solution is made more exact using the grid ω_{2h} according to formula

$$S_h^* = \gamma_1 S_h + \gamma_2 S_{2h} , \tag{3}$$

where

$$\gamma_1 = (R_h^{-1} \, th R_h - 1)/(0{,}5 \, th R_h / th \frac{Rh}{2} - 1), \quad \gamma_2 = 1 - \gamma_1 ,$$

S_h, S_{2h} are the solutions of finite difference equations by means of variable directions method with grids ω_h, ω_{2h} taking into account linear interpolation. It is seen that $\varkappa \to 1, \gamma_1 \to \frac{4}{3}, \gamma_2 \to -\frac{1}{3}$ at $R_h \to 0$ (in central differences the scheme is monotone if $R_h < 2$), and $\varkappa \to R_{h/2}, \gamma_1 \to 2, \gamma \to -1$ at $R_h \to \infty$ (a monotone scheme with one-sided directions of differences with the first order of approximation).

In the case of boundary condition (2.2) an approximation of the initial system of differential equations (1) follows from the difference equations. Boundary conditions (2.2) were approximated by the one-sided differences of the second order of precision, and the difference equations obtained were realized at $\varphi_0 = \pi/2$ in sequence by variable directions method. A method of harmonic analysis permits to show the stability of the iteration process provided

$$\frac{\tau Ha}{h_2} < 2 + \sqrt{8} .$$

<u>Axisymmetrical shift flow of rotating fluid</u>

The fluid is in a cylindrical vessel with nonconducting walls ($0<r<L, 0<z<C, 0\leq\theta\leq 2\pi$) in the presence of magnetic field parallel to the axis (Fig.2.1). At the planes $z=0$ and $z=C$, perpendicular to the field there is a couple of circular electrodes used for current distribution in the fluid.

The motion of the fluid is described by a system of equations (1) for velocity azimuthal components $u_\theta = u(r,z)$ and magnetic field strength $H_\theta = H(r,z)$. Here $y=z$, $\varphi_0 = \pi/2$, $L_1 = \frac{\partial}{\partial r}(\frac{1}{r}\frac{\partial}{\partial r}(r\cdot))$, $L_2 = \frac{\partial^2}{\partial z^2}$. At the axis of the vessel $u|_{r=0} = H|_{r=0} = 0$. At the walls $H|_{r=L} = 0$, $H|_{z=0} = H_0(r)$, $H|_{z=C} = aH_0(r)$, a is equal, larger or smaller than zero. Function $H_0(r)$ likewise in the case of a plane-parallel fluid is determined from current distribution on the electrodes. For electrodes of finite width, in particular, it is written assuming that input current density is constant. Analytical solutions $u(r,z), H(r,z)$ are obtained using Fourier-Bessel expansion of function $H_0(r)$. Solutions of corresponding boundary problems in the half-space and between two planes $(z>0, 0<z<C)$ are found by the method of Bessel in-

tegral transform. In the case of finite electrodes a boundary problem is also solved by means of the grid method according to schemes similar to those in section I for case (2.1).

For the indicated kind of fluid experimental research was conducted with an experimental set reported in /7/. The high of cylindrical vessel was 7 cm and that of the mean radius of the rotating layer 3.5 cm. The evtectic alloy indium-gallium-tin was used as a working medium. Azimuthal and radial components of the mean and pulsed velocity were measured with a conduction anemometer.

Results and discussion

In fig.1.1 for $\varphi_0 = \pi/2$ the mean velocity profiles of the plane-parallel flow (sectionl) are shown for different Hartmann numbers. The calculated velocity fields in this case as well as for $\varphi_0 = \pi/4$ (Fig.1.2,1.3) indicate that with growing Ha the flow tends to be concentrated between electrodes with decreasing gradients in the direction of magnetic field. If the magnetic field orientation is directed along axis (Fig.1.4) we have two opposite flows with a pronounced shift zone between them. The homogeneity of the flow is growing in the direction of the field with Ha increasing and the shift zone width reduces.

Fig. 2.2-2.5 represent investigation results on the axisymmetrical flow (section 2) obtained by mathematical methods as well as experimentally. At $a = 0$ it is seen in fig.2.1 and 2.2 that with Ha growing the flow is concentrated between two electrodes, and at $Ha \geqslant 400$ it becomes actually homogeneous in the direction of axis z. At $a > 0$ for numerical calculations (Fig. 2.3, 2.4) we have two opposite flows with a separating plane perpendicular to the applied magnetic field.

It is known that /2/ if a magnetic field effects the velocity profile (and perturbation), a hydrodynamic stability of the flow is either growing or decreasing. For the class of the shift flows under consideration excluding the case mentioned in section 1, (Fig.1.4) characteristic is an increase in hydrodynamic stability under the action of magnetic field if the velocity profile in the radial direction is not considerably changing. An experimental stability curve for axisymmetric flow is shown in fig.2.5. Instability area is above the curve and an area of stability is under the curve.

The basic and obviously the only reason of the flow destabilization is deformation of the mean flow with appearance of extreme points in velocity profiles. The presence of these points results in a decreasing flow stability and generation of very intensive perturbations. Here the loss of the flow stability is accompanied by a stabilizing effect of a magnetic field directly on the perturbations. With growing magnetic field the latter effect is less essential due to the reconstruction of the flow into a plane one in the plane perpendicular to the field. This kind of flow is not effected by the field, so it does not influence the development of the two-dimensional perturbations in a plane perpendicular to the field.

The measurement of the perturbation structure showed that at $Ha \geq 200$ perturbations arising are two-dimensional, they represent a chain of vortices going in sequence along a flow with axes parallel to an external magnetic field. It should be noted that the number of vortices can be both even and uneven according to the flow regime and Hartmann parameter.

Strongly unstable flows also arise in magnetic field at the electroconductivity unhomogeneity and its boundaries. In paper /8/ an anologue of the flow considered in a prizmatic tube with rectangular cross-section was studied. It was shown that with growing magnetic field the deformation of the main flow also occurs with the formation of the M-shaped velocity profile which is generating perturbations having a high intensity level. Here the mean and pulsed structures in a strong enough magnetic field acquire a pronounced two-dimensional character.

For a problem considered in a strong magnetic field in regime beyond critical transition to the two-dimensional turbulent flow is anticipated /9/.

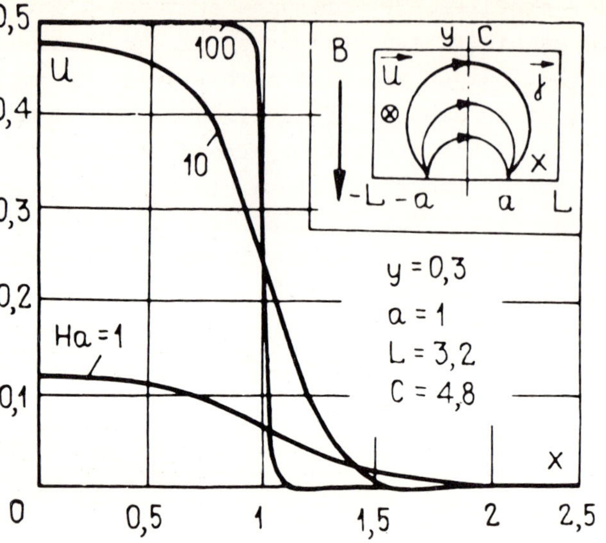
0,5
0,4
0,3
0,2
0,1
u
100
10
Ha = 1
B
y
C
u
γ
x
-L
-a
a
L
y = 0,3
a = 1
L = 3,2
C = 4,8
x
0
0,5
1
1,5
2
2,5

Fig. 1.1

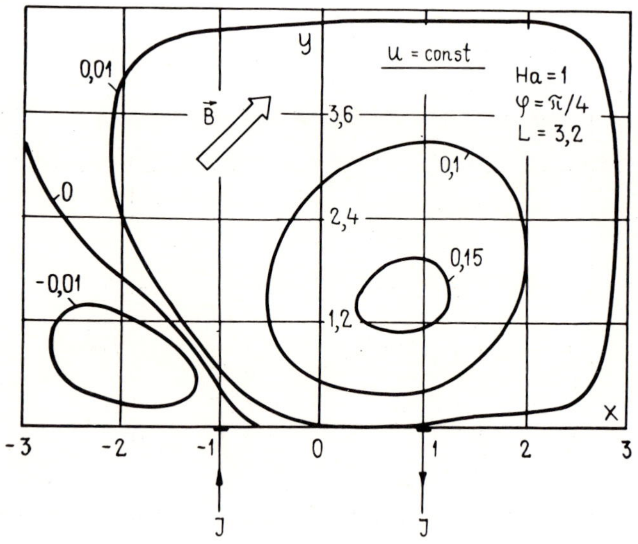
y
u = const
Ha = 1
φ = π/4
L = 3,2
0,01
B
3,6
0,1
0
2,4
0,15
-0,01
1,2
x
-3
-2
-1
0
1
2
3
J
J

Fig. 1.2

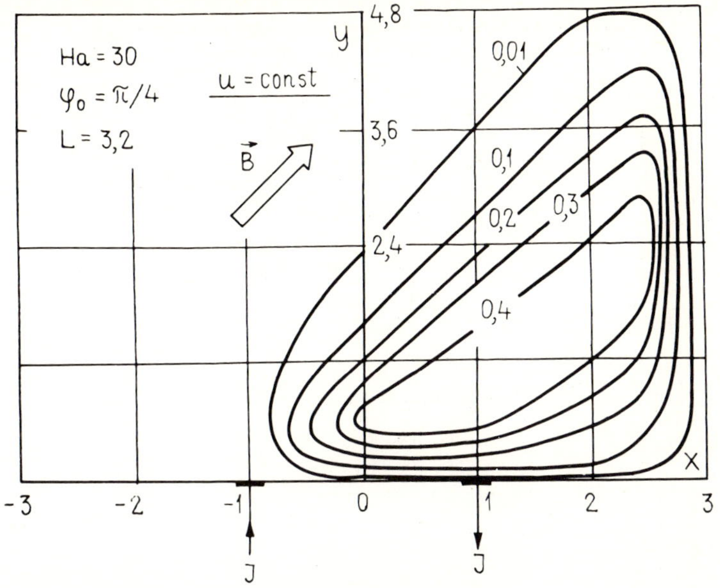

Ha = 30
φ₀ = π/4
L = 3,2
u = const
B
y
x
4,8
3,6
2,4
0,01
0,1
0,2
0,3
0,4
-3
-2
-1
0
1
2
3
J
J

Fig. 1.3

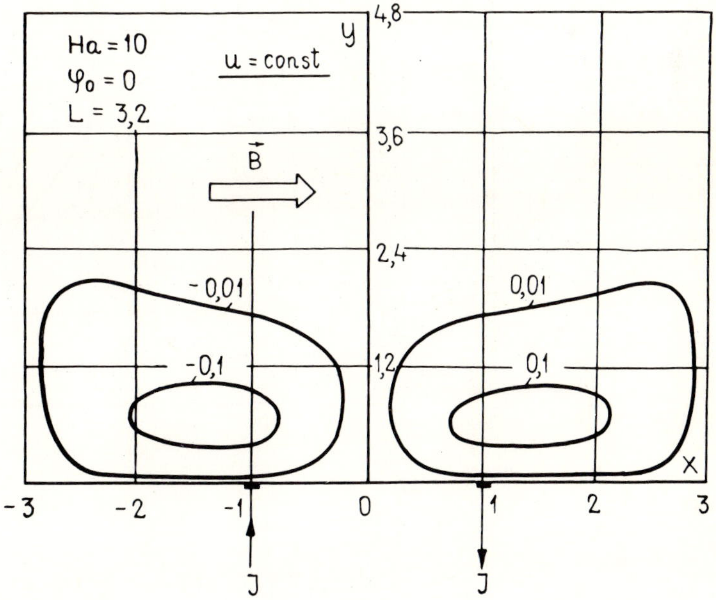

Ha = 10
φ₀ = 0
L = 3,2
u = const
B
y
x
4,8
3,6
2,4
1,2
-0,01
-0,1
0,01
0,1
-3
-2
-1
0
1
2
3
J
J

Fig. 1.4

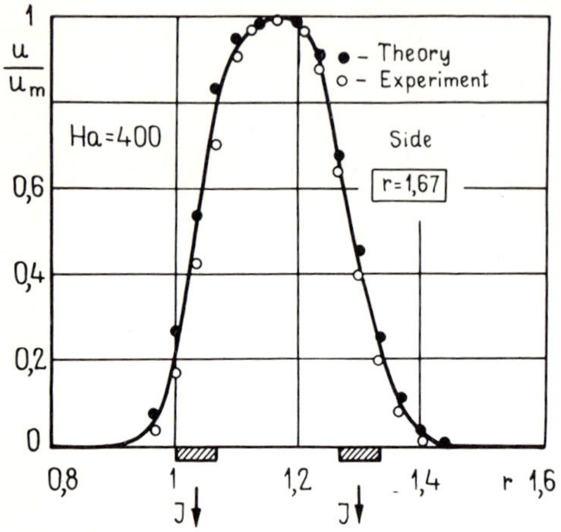

Fig. 2.1

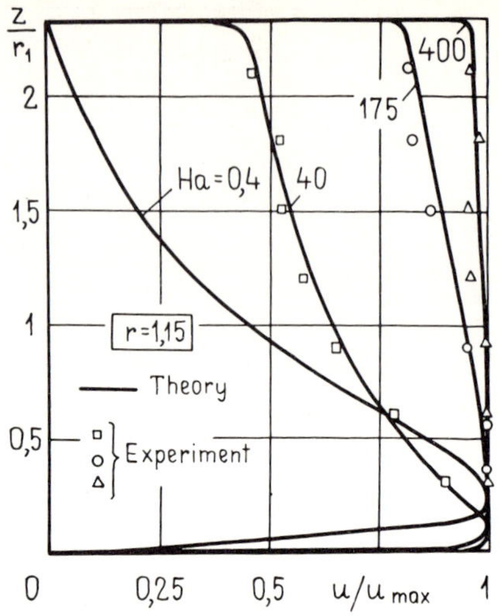

Fig. 2.2

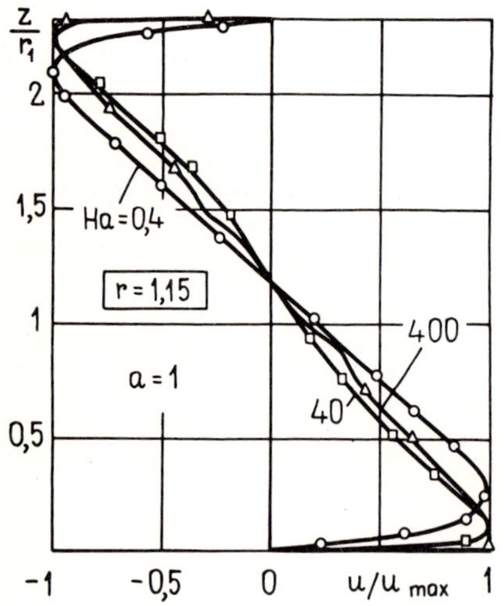

Fig. 2.3

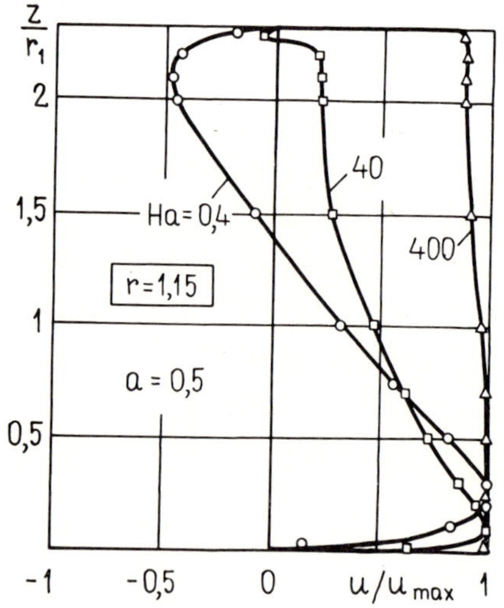

Fig. 2.4

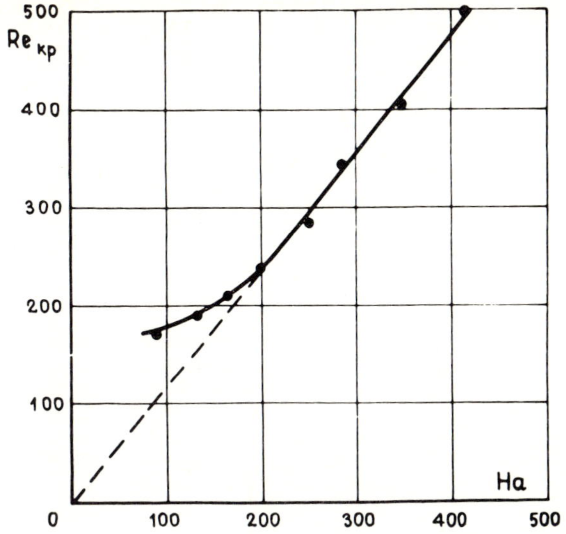

Fig. 2.5

References

1.Hunt J.C.R., Stewartson K. J.Fluid Mech. 38,part 2, 225-236 (1969)
2.Lehnert B. Proc.Roy.Soc. A233, 299-307 (1955)
3.Riekstin E.J. Latv. Math. Ežegodnik 16, Riga, Zinatne, 89-92 (1975)
4.Kalis H.E. Magnitnaya Gidrodinamika 2, 65-67 (1978)
5.Ilyin A.M. Matemat. Zametki 6, 2, 234-237 (1969)
6.Janenko N.N. Metod drobnih šagov reshenija mnogomernih zadach matematicheskoi fiziki. Novosibirsk, Nauka (1967)
7.Levin V.B. The 9th Riga meeting on MHD. Riga, 35 (1978)
8.Kolesnikov Yu.B. Magnitnaya Gidrodinamika,3, 32-35 (1972)
9.Vocish A.D., Kolesnikov Yu.B. Magnitnaya Gidrodinamica, 1, 35-38 (1977)

A LAGRANGIAN METHOD FOR THE NUMERICAL SOLUTION OF THE EULER EQUATIONS FOR TRANSONIC FLOWS

Goetz H. Klopfer
National Research Council Research Associate, Ames Research Center, NASA
presently with
Nielsen Engineering and Research, Mountain View, CA, USA

INTRODUCTION

One of the major problems in computational transonic aerodynamics is the accurate, simple, and efficient treatment of boundary conditions. Considerable effort has been expended in this area, and some promising methods have been developed. For example, three-dimensional transonic small disturbance codes are now used routinely in the U.S. for the design of aerodynamic configurations, Ballhaus (1977). In this approach the wing boundary condition is imposed in terms of airfoil slopes applied on flat mean surface approximation to the wing. Considerable simplification results at the expense of accuracy.

For a more accurate treatment in which boundary conditions are satisfied on the wing surface, some form of coordinate transformation is required. This presents two problems: 1) generation of a suitable body-oriented coordinate transformation with a proper distribution of mesh points, and 2) solving the governing flow field equations in this complicated coordinate system. Some progress has been made in addressing the first problem, grid generation, for two-dimensional flows by Thompson et al. (1974) and others. However, extension of this procedure to three-dimensional and unsteady flows is a formidable task. A number of methods have been developed to address the second problem, solving for the flow field variables once the grid has been generated. Examples are the finite volume method developed by MacCormack (reviewed by Ballhaus, 1977) and recently extended by Jameson (1977) to the treatment of potential flows; the implicit method of Steger (1977); and the Euler-Lagrange technique of Harlow and Amsden (1974).

In the present paper, a Lagrangian approach is developed for the solution of steady and unsteady transonic flows about airfoils. The advantage of this procedure is that it provides a simple solution to the two problems mentioned above. Unlike the methods of Harlow et al. (1974), the computation is done in the Lagrangian plane and thus the mesh is automatically generated as part of the solution procedure. Another advantage of the present method is that the initial distribution of fluid particles is arbitrary, and particles can be clustered near the stagnation streamline for better resolution of the aerodynamic forces. The secular mesh distortion, which usually destroys the accuracy of Lagrangian computations, does not occur for transonic flows except at stagnation points.

GOVERNING EQUATIONS

Differential Equations

The equations describing the conservation of mass, momentum, and energy for an inviscid, compressible ideal gas in Lagrangian coordinates are

$$U_\tau + F_a + G_b + H = 0 \tag{1}$$

where

$$U = \begin{bmatrix} J\rho \\ J\rho u \\ J\rho v \\ J\rho E \\ x \\ y \end{bmatrix} \quad ; \qquad F = \begin{bmatrix} 0 \\ y_b p \\ -x_b p \\ p(y_b u - x_b v) \\ 0 \\ 0 \end{bmatrix}$$

$$G = \begin{bmatrix} 0 \\ -y_a p \\ x_a p \\ p(-y_a u + x_a v) \\ 0 \\ 0 \end{bmatrix} \quad ; \quad \text{and} \quad H = \begin{bmatrix} 0 \\ 0 \\ 0 \\ 0 \\ -u \\ -v \end{bmatrix}$$

and ρ, p, u, v, and E are the density, pressure, Cartesian velocity components, and the energy ($E = p/\rho(\gamma-1) + (u^2 + v^2)/2$) of a fluid particle located at (x,y) in the fixed Cartesian coordinates system and identified by the Lagrangian coordinates (a, b). The Jacobian J is the volume of a fluid particle in the Lagrangian system,

$$J = \frac{\partial(x,y)}{\partial(a,b)} = x_a y_b - x_b y_a \ .$$

The proper conservation form of the equations are obtained by transforming the gas dynamic equations in Cartesian coordinates (x,y,t) into Lagrangian coordinates (a,b,τ) by Viviand's procedure (1974). The transformation is defined by the last two equations of system (1) and $\tau = t$.

The solution of the first equation of system (1) is trivial; i.e., ρJ = constant which states that the mass of a fluid particle remains constant for all time. Thus system (1) represents a system of five equations to be solved for the five unknowns u, v, E, x, and y. The density ρ is derived from the Lagrangian derivatives of x and y.

The major difficulty with the Lagrangian system (1) is that it becomes rapidly stiff with any type of mesh distortion. This can be seen from the eigenvalues of the Jacobian matrices $\partial F/\partial U$ and $\partial G/\partial U$; they are $(0, \pm cJ^{-1}(x_b^2 + y_b^2)^{1/2}, 0, 0, 0)$ and $(0, \pm cJ^{-1}(x_a^2 + y_a^2)^{1/2}, 0, 0, 0)$, respectively, where c is the sonic speed. Mesh distortion causes the metrics x_a, y_a, x_b, and y_b to take on large disparate values and thus render system (1) stiff.

Boundary Conditions

Three types of boundaries are considered: impermeable wall, inflow, and outflow. In the present paper the impermeable wall condition is imposed on the far field and stagnation particle path. For simplicity only symmetrical nonlifting bodies will be considered so that the stagnation particle paths are known a priori. Stagnation points which are the only severe source of the Cartesian mesh distortion are eliminated by cusping the aerodynamic bodies at the stagnation points. With this device the mesh rezoning procedures required by most classical Lagrangian methods are not necessary.

The boundary particle paths are prescribed in the Cartesian space by

$$b(x,y,t) = y - f(x,t) = \text{constant} \qquad (2)$$

where the direction of the Lagrangian coordinate 'a' is streamwise. For the inviscid flows considered in this paper the impermeable wall condition requires that the fluid velocity normal to the surface must coincide with the normal surface velocity. The flow tangency condition becomes

$$v = f_t + uf_x \tag{3}$$

The tangential momentum equation is

$$(\rho Ju)_\tau + \frac{1}{(1 + f_x^2)} \{(y_b - x_b f_x)p\}_a + \frac{\rho J f_x}{(1 + f_x^2)} \{f_t + 2uf_{xt} + u^2 f_{xx}\} = 0 \tag{4}$$

and the energy equation along the boundary particle path is

$$(\rho JE)_\tau + \{(y_b - x_b f_x)pu\}_a - f_t \{(x_b p)_a - (x_a p)_b\} = 0 \tag{5}$$

At the inflow boundary, U is prescribed according to uniform inflow. At the outflow the "pseudo-compressibility" device developed by Orzag (1976) is employed. In the present paper it simply means linear extrapolation of U to the outflow boundary and then artificially reducing the density of the outflow particles. It can be shown that errors in prescribing the outflow condition cannot propagate upstream and thus remain in an outflow boundary layer. This procedure seems to work well for subsonic, supersonic or even mixed subsonic-supersonic outflow.

NUMERICAL METHOD

The numerical solution of system (1) is obtained by an explicit hyperbolic equation solver. Unfortunately most of the standard numerical schemes have been devised for Eulerian systems and do not work well for the Lagrangian system. The methods of Harlow, et al. (1974) are not applicable to the present system. After much experimentation (ref. 6) it was found that the Flux Corrected Transport (FCT) method of Boris and Book (1976) yielded the most accurate results. It is a monotone preserving, second order accurate, explicit scheme, utilizing time-step splitting, the Courant-Frederick-Levy condition for stability is CFL ≤ 1 and for monotonicity is CFL $\leq \sqrt{3}/2$. Up to now, the major drawback of the FCT method has been the proper inclusion of boundary conditions.

In the present paper the boundary conditions are satisfied as follows. The entire flow field including the boundary points is advanced to the new time step using the FCT method to solve system (1) only. Since FCT is essentially a five-point scheme, two fictitious points are required beyond the boundary points. These are obtained by linear extrapolation from the interior flow field. In this step no attempt has been made to satisfy any of the boundary conditions. The points along the boundary particle paths are then updated by solving equations (2-5) with a fully implicit (backward Euler) scheme. At the same time the inflow and outflow boundary points are updated.

RESULTS

To demonstrate the flexibility of the present method, several sample calculations are presented. They include a one-dimensional Riemann problem, oblique shock and Mach reflection off a straight wall, and transonic flow over a parabolic arc airfoil.

The initial conditions of the Riemann problem are depicted in fig. 1a. Three calculations are shown to demonstrate the differences between the Eulerian and Lagrangian approach and the effects of two different numerical algorithms for a Lagrangian calculation. The results show that the optimum generalized Lax-Wendroff scheme (Lerat and Peyret, 1974) gives poor results; large dispersion errors follow the shock and expansion waves. The Lagrangian approach exhibits much lower amplitudes of the expansion wave oscillations. The final frame of fig. 1 shows a Flux Corrected Transport solution for the Lagrangian approach. Most of the non-physical dispersion waves are eliminated.

The results of an oblique shock wave and a Mach reflection off a straight wall are shown in figs. 2 and 3. In both cases the solution is started with uniform flow in a straight channel and then the ramp angle is smoothly increased from 0° to $\tan^{-1}(.2)$ over one time unit. Discontinuities in the surface slopes and curvatures are smoothed out over a small distance (5 to 10% of the ramp length). Steady state results are given in fig. 2b for the oblique shock. Unsteady results are shown in fig. 3b for the Mach reflection problem in terms of pressure contours and Lagrangian particle positions. The important point to note is that shock waves are captured over 2 cells and slip surfaces over 3 or 4 cells even in cases where the shocks are oblique to the grid lines.

The transonic parabolic arc airfoil results are given in figs. 4 and 5. The first case is a parabolic arc airfoil in a free stream Mach number of 0.785. The thickness of the airfoil varies as shown on the top of fig. 4a. After two time units (chord lengths traveled at free stream speed) the final thickness of $\delta = 0.10$ is reached. The mid-chord ($x = 0.525$) pressure coefficient is shown as a function of time in fig. 4b. The time evolution of the pressure is shown in fig. 4c. The present method shows good agreement with the low frequency small disturbance approximation of Ballhaus and Lomax (1974) except at early and late times. The differences at early times are due to the low frequency small disturbance approximation, i.e. it cannot pick up piston wave effects resulting from the thickening airfoil surface.

The final figure shows the same airfoil at a higher free stream Mach number $M = 0.83$. Comparison is made with an Euler solution obtained by an implicit scheme (Steger, 1977). The present method results are not yet at steady state, but can be seen to be approaching the steady state results of Steger.

CONCLUSION

The main purpose of the present study was to show that a Lagrangian approach can simultaneously solve the problem of grid generation and the complication of solving the body-oriented transformed equations. The significant results are that the present method can solve the compressible flow field around arbitrary aerodynamic configurations for both steady and unsteady flows for a wide range of Mach numbers. In principle the method can be easily extended to three-dimensions. Although the present method is solved by an explicit method, implicit methods can also be employed.

REFERENCES

1. W. F. Ballhaus (1977), Numerical Methods in Fluid Dynamics, Chap. 3, ed. Wirz and Smolderen.

2. W. F. Ballhaus and H. Lomax (1975), Proc. of the 4th Int'l Conf. on Num. Methods in Fluid Dynamics, pp. 57-63.

3. J. P. Boris and D. L. Book (1976), Methods in Computational Physics, Vol. 16, Chap. 11, Academic Press, New York.

4. S. Z. Burstein (1964), AIAA J. 2, 12, 2111-2117.

5. F. H. Harlow and A. A. Amsden (1974), J. Comp. Physics 16, 1-19.

6. G. H. Klopfer (1978), NASA TM-78456.

7. S. A. Orzag (1976), Proc. of the 5th Int'l Conf. on Num. Methods in Fluid Dynamics, pp. 32-51.

8. A. Lerat and R. Peyret (1975), Proc. of the 4th Int'l Conf. on Num. Methods in Fluid Dynamics, pp. 251-256.

9. J. L. Steger (1977), AIAA 10th Fluid Dynamics and Plasma Physics Conference, Albuquerque, New Mexico.

10. J. F. Thompson, F. C. Thames, and C. W. Mastin (1974), J. Comp. Physics 15, pp. 299-319.

11. H. Viviand (1974), La Recherche Aerospatiale, No. 1, 65-66.

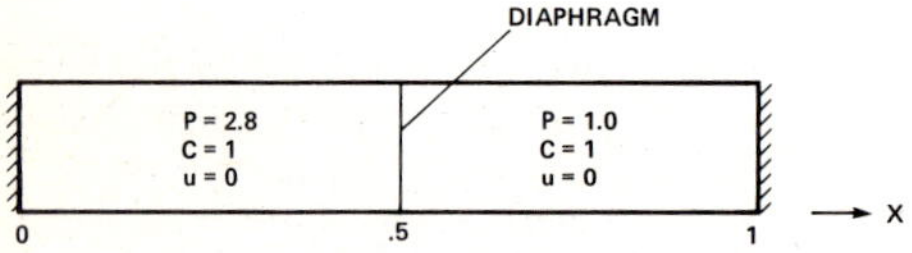

Schematic

(a)

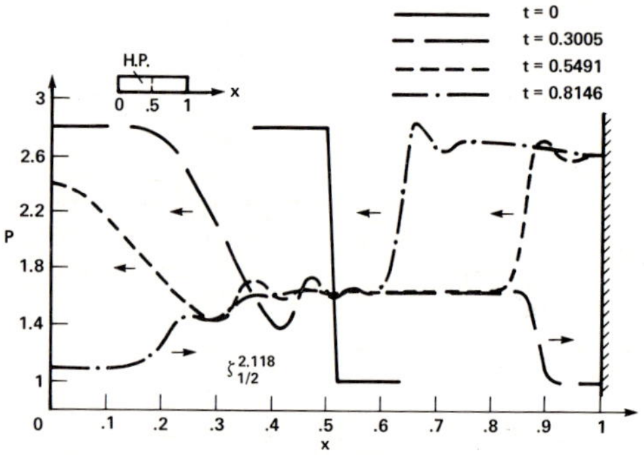

Eulerian approach - solved by optimum generalized Lax Wendroff scheme (Lerat & Peyret 1974)

(b)

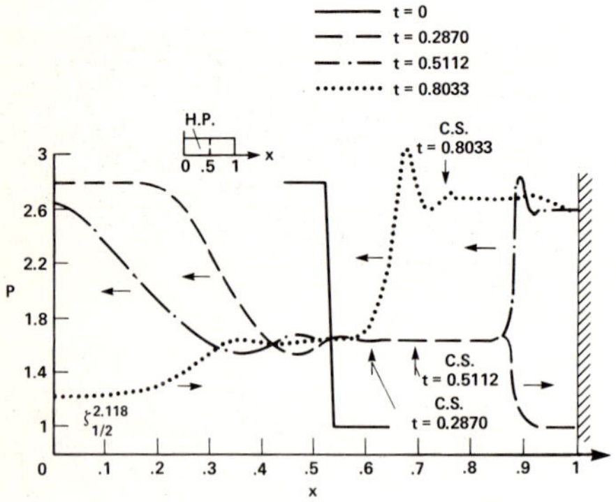

Lagrangian approach - solved by optimum generalized Lax-Wendroff scheme (Lerat & Peyret 1974)

(c)

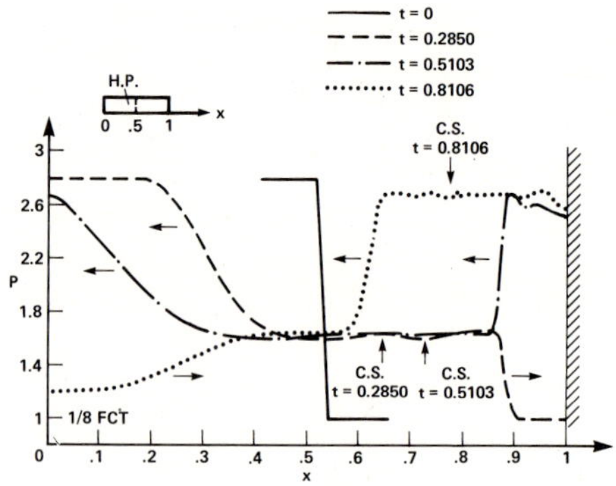

Lagrangian approach - solved by FCT scheme (Boris & Book 1976)

(d)

Figure 1. One-Dimensional Riemann Problem , 51 points.

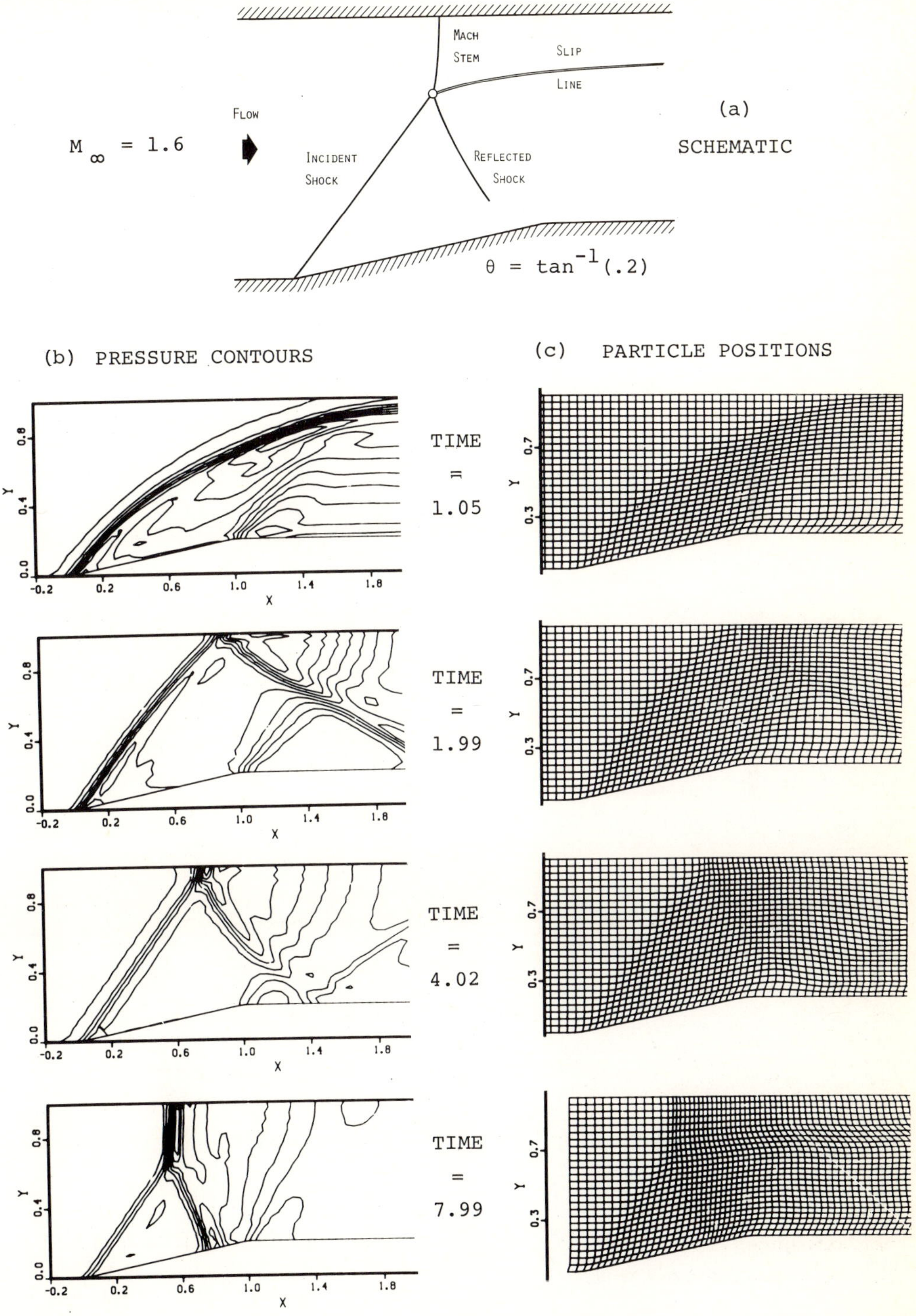

Figure 3. Mach Stem Reflection

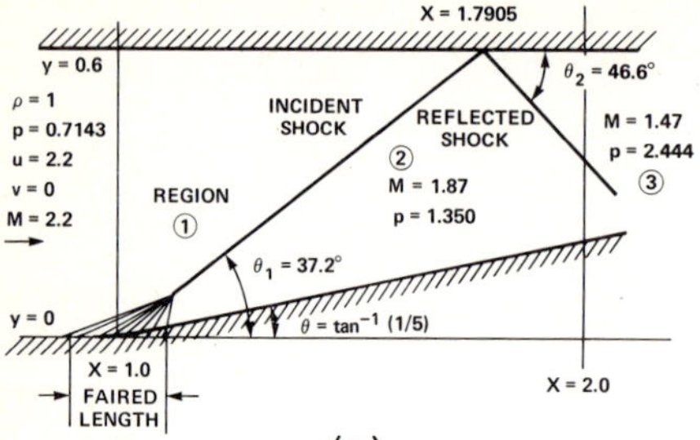

(a)

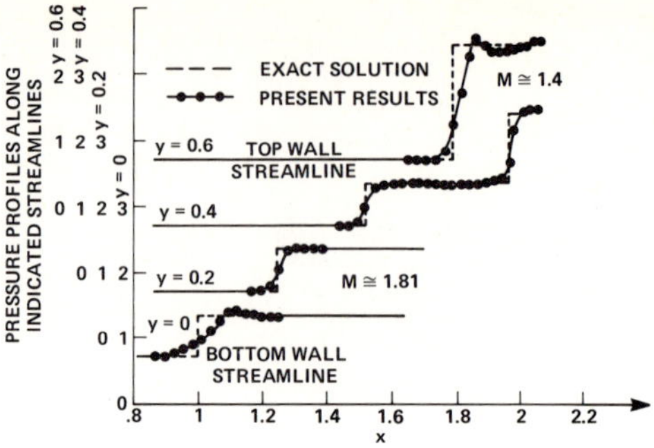

(b)

Figure 2. Oblique Shock Wave Reflection

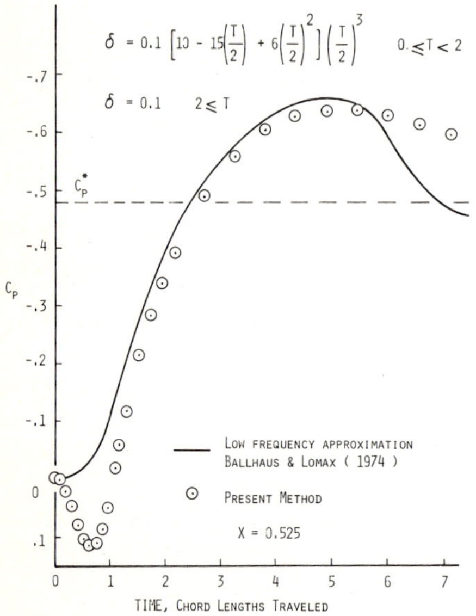

(a) TIME HISTORY OF MIDCHORD PRESSURES ON A PARABOLIC ARC AIRFOIL, $M_\infty = 0.735$

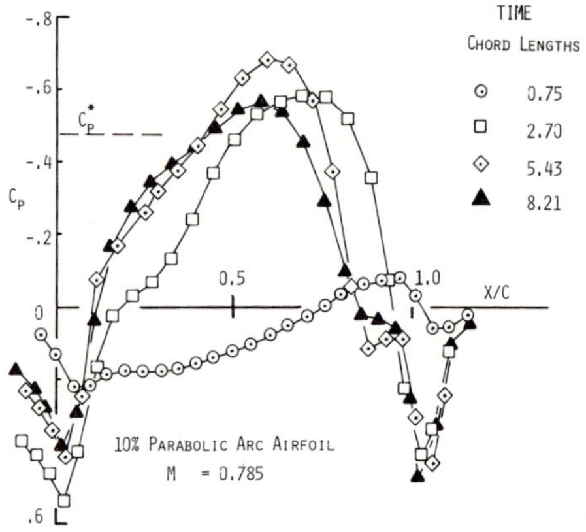

(b) TIME EVOLUTION OF PRESSURE COEFFICIENT

Figure 4. Unsteady Transonic Flow

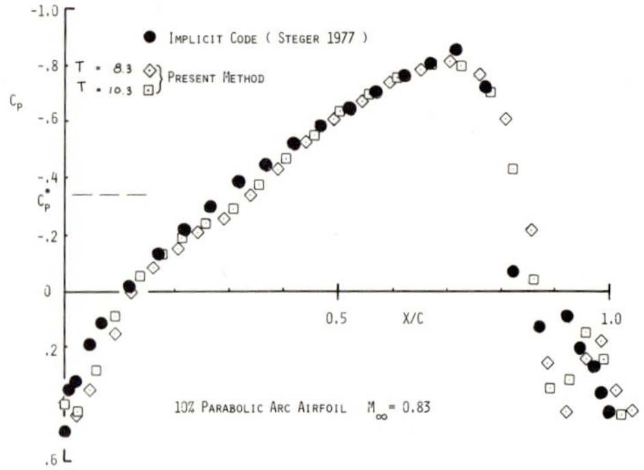

Figure 5. Steady Transonic Flow

ON COMPUTATIONAL MODELLING OF THE TUNGUSKA CATASTROPHE

V.P.Korobeinikov, B.V.Putjatin, P.I.Chushkin
and L.V.Shurshalov

Steklov Mathematical Institute and Computing
Centre of the USSR Academy of Sciences,

Moscow

1. Introduction

Trajectory and energetic parameters of the Tunguska meteorite may be found from numerical solution of an inverse problem based on the gasdynamical model of the Tunguska catastrophe [1]. Here such ground forest destructions are determined which correspond to the real forest flattening in the catastrophe region. For that a direct three-dimensional non-stationary gasdynamic problem on a model explosion of a semi-infinite cylindrical charge with variable specific energy distribution along its axis in the real inhomogeneous atmosphere should be repeatedly computed with shock waves reflection from the ground. Because of complexity of this problem the authors have worked out a whole system of approaches [1-4] combining exact numerical and approximate analytical methods, gradually improving and extending the model taking into consideration more and more delicate effects and constantly checking the conformity of the model to the phenomenon under study.

The following parameters of the Tunguska meteorite were determined as the result of the inverse problem solution [3]: the height of the final burst $H_0 = 6.5$ km, the trajectory inclination angle $\alpha = 40^{\circ}$, the energy of the final explosion $E_0 = 10^{23}$ erg, the specific energy of the ballistic wave $E_1 = 1.4 \cdot 10^{17}$ erg/cm . The summary value of these two energies found only on the basis of the forest destructions is equal to 9.5 Mton. The total trotyl equivalent of the phenomenon must somewhat exceed this value. The latest estimates [5] based on available seismograms and barograms give its quantity equal to 12.5 ± 2.5 Mton.

In solution [3] some important factors were not considered. Being included in the calculation they may somewhat change the above mentioned figures. Among these factors are: complex non-one-dimensional character of the flow, dissociation, ionization and radiation at high temperatures, non-isothermity and wind gradient in the real atmosphere, energy release in upper atmosphere layers during the meteorite flight. In this paper significance of these factors is analysed.

2. Two-dimensional problem on spherical and cylindrical explosion in the inhomogeneous atmosphere

The consideration of this problem is reasonable because conditionally shock waves of the Tunguska phenomenon may be divided into two parts: one from the final quasi-spherical explosive decomposition and the other from quasi-cylindrical ballistic wave. Here we take into account both separately and jointly high temperature thermodynamic properties of air and non-isothermal properties of the atmosphere. The governing equations in cylindrical coordinates r, z are written below in the common denotations

$$\rho\, du/dt + \partial p/\partial r = 0 \quad , \quad \rho\, dw/dt + \partial p/\partial z = -\rho g$$

$$d\rho/dt + \rho(\partial w/\partial z + \partial u/\partial r + \nu u/r) = 0 \qquad (1)$$

$$\partial E/\partial t + \partial[w(E+p)]/\partial z + \partial[u(E+p)]/\partial r + \nu u(E+p)/r =$$

$$= -\rho g w \quad , \quad E = \varepsilon + \rho(u^2 + w^2)/2$$

The caloric equation of state for air was taken in the form $\varepsilon = p/(\gamma - 1)$ similar to the case of the perfect gas but with variable quantity $\gamma = \gamma(p, \rho)$ [6]. Convenient and very accurate analitic fits drew up by the authors were used to represent the table data for the standard undisturbed atmosphere conditions, namely, for the pressure p, the density ρ and the gravity g.

For the solution of the problem the well-known finite-difference Godunov's method [7, 8] was applied. The computational grid used was moving and non-uniform with the shock front treated as its boundary line. The radial size of the network cells was diminishing towards the shock wave and was automatically handled to the flow [9]. For uniform approximation along the angular coordinate Θ the network angle divisions were uniformly distributed and the basic divergent form of the equations was written somewhat non-standard. For the continuity and energy equations it was common [8], while for the two momentum equations it was taken as it had been proposed for all the equations in [10]. Such compromise choice provides better accuracy near the axis of symmetry and not worse than in usual scheme [8] fulfilment of the conservation laws.

Use of the equation of state for real air leads to some complication of the numerical algorithm in the method [7, 8] in comparison with the perfect gas case, mainly because of more complex computational procedure for decomposition of arbitrary discontinuity at the cells boundaries and at the shock front [11]. However, the algorithm for the real gas can be easily reduced to the one for the perfect gas without losses in accuracy.

Some numerical results for a spherical explosion with the energy $E_0 = 10^{23}$ erg and the height $H_0 = 8$ km (these values are close to those which might have taken place for the Tunguska event) are shown in Fig. 1-3.

In Fig. 1 the relative shock overpressure Δp_B is depicted versus the shock radius R_B. The lower curves correspond to the lowest shock wave point, the upper to its highest point. Four different models are considered. The results for the non-isothermal atmosphere and real air are shown by the solid curves, while for the isothermal exponential atmosphere and perfect gas with $\gamma = 1.4$ by the dashes. The circles and triangles correspond to intermediate cases of the isothermal atmosphere - real air and the standard atmosphere-perfect gas models. These results illustrate the separate influence of the real atmosphere and real air properties on the shock wave intensity. They reveal that for the downward going shock the high temperature thermodynamics of air is more important than the non-isothermal properties of the atmoshpere (provided that the chosen exponential atmosphere is close to the real atmosphere near the ground surface).

In Fig. 2 for the standard atmosphere the fall of the maximum temperature T_{max} in the central part of the explosion region is drawn as a function of time $\tau = t/t_*$, $t_* \approx 3.5$ sec, for both the perfect gas and the real air models. Though in this example temperatures are relatively low, the difference between the results corresponding to these two models is considerable. For the perfect gas T_{max} is higher at the beginning of the explosion but then it drops rapidly. For the real air T_{max} is lower at small τ because of dissociation and ionization but it is droping slower due to the recombination processes going with the release of energy.

Analysis shows that the Tunguska meteorite energy estimates [3] may somewhat (to within 15-20%) increase if one takes into account the high-temperature thermodynamic properties of air. On the contrary, the non-isothermal nature of the atmosphere is not substantial in this sense.

Some results of accuracy evalution are presented below. We compare data for 8 different networks characterized by a parameter M (the total number of cells is proportional to M^2). In this problem it is possible to obtain additional information by computing a one-dimensional explosion in a homogeneous medium with the help of the two-dimensional finite-difference scheme, the results then being compared with the known standard solutions. In Fig. 3 the value Δp_B at some sufficiently distant R_B for a one-dimensional explosion calculated by the two-dimensional scheme is compared with one-dimensional numerical data [12] (Δp_B = 0.211 , horizontal line). The convergence to zero of the function δ_M which is the difference between two successive values of Δp_B is also demonstrated here for both the one-dimensional (dash - dotted line) and two-dimensional (dashed line) explosions. These and other data show the same qualitative character of convergence for one - and two-dimensional problems computed by the two-dimensional procedure. In both cases the convergence is monotonic, rather slow and the limit of Δp_B is approached from lesser values. Because of more complex flow structure the two-dimensional results for each M are farther away from the limit than the one-dimensional ones. But the convergence speed for the two-dimensional case is considerably faster. For this reason sufficiently fine networks provide approximately the same accurancy in these two cases. This fact can help to choose the network in accordance with the accuracy demanded.

3. Quasi-one-dimensional problem on explosion with radiation in the inhomogeneous atmosphere

The influence of radiation was evaluated using the solution of the problem mentioned in the heading. The gasdynamic equations along a ray at an angle Θ with the vertical was employed in the Lagrangian form

$$\vartheta = 1/\rho = R^{\delta-1}\,\partial R/\partial m \;, \qquad u = \partial R/\partial t$$
$$\partial \varepsilon/\partial t + p\,\partial\vartheta/\partial t = Q \;, \quad \partial u/\partial t + R^{\delta-1}\,\partial p/\partial m = g\cos\Theta \qquad (2)$$
$$p = p(T,\vartheta) \;, \quad \varepsilon = \varepsilon(T,\vartheta)$$

The index δ is equal to 1, 2, 3 correspondingly for plane, cylindrical or spherical symmetry, m is the mass coordinate, Q is the specific radiative heat supply, all the other denotations are common. The ambient atmosphere was supposed to be radially exponential.

The radiation equations determining Q were considered in the diffusion limit and the "grey" gas model [13] was used, namely,

$$Q = 4\pi \kappa \left(I_0 - \sigma T^4/\pi\right) \vartheta$$
$$\partial I_0/\partial R + 3\kappa I_1 = 0 \, , \quad \partial\left(I_1 R^{\delta-1}\right)/\partial R + \kappa\left(I_0 - \sigma T^4/\pi\right) R^{\delta-1} = 0 \qquad (3)$$

where I_0 , I_1 were the first two coefficients in the Legendre expansions of the radiative intensity.

The finite - difference method [14] with an artificial viscosity was employed to solve gasdynamic equations (2). The flux variant of the double-sweep method [15] was used to integrate the radiation transport equations. The time steps were limited by three conditions: 1) the usual Courant condition; 2) the atrificial viscosity condition [14]; 3) the demand that the internal energy alteration at a time step should not exceed 1%.

The equations of state for air and the absorbtion coefficient were taken from tables [16] and [17] correspondingly. The absorbtion coefficient was interpolated between the Plank and Rosseland averages with provision for the local optical thickness.

Some results of the calculations for the downward shock wave propagation are presented in Fig. 4-7. We considered both spherical and cylindrical explosions which were simulated by expansions of the corresponding volumes of compressed hot gas. In the spherical case the basic parameters were: $E_0 = 10^{23}$ erg, $H_0 = 6.5$ km, the initial sphere radius $r_0 = 0.2$ km. Three values of the initial temperature $T_0 = 10000$, 11000, 18000^{0}K inside the sphere were taken.

In Fig. 4 the relative share of energy E lost due to radiation is depicted as a function of time t for three T_0 . In the considered range of T_0 this share is significant and increases with T_0 .

The relative overpressure Δp_B in dependence on the shock radius R_B for $T_0 = 18000^0$K is presented in Fig. 5 (the solid curve - with radiation, the dashed curve - without radiation). At $R_B = 3$ km the shock intensity in the radiating gas is almost twice lower than in the case without radiation.

For the same conditions in Fig. 6 the temperature distributions along the radius R at the time $t \approx 1.1$ sec are drawn. Abrupt temperature gradient take place near the explosion centre and near the contact surface. The temperature rise at the centre is caused by the secondary converging shock. The rise is gradually flattening with time in the radiating gas but is retaining if radiation is not taken into consideration.

A calculated example of the cylindrical explosion is given in Fig. 7 for the following initial data: $E_1 = 1.4 \cdot 10^{17}$ erg/cm , $H_0 =$ 13 km , $\alpha = 40^o$, $r_0 = 40$ m , $T_0 = 17000^oK$. Temperature distributions at some time moments (from 0.05 to 0.7 sec) are presented. The cylindrical explosion has the same qualitative features as the spherical one. Only the share of energy lost by radiation is smaller here.

Thus, in course of the Tunguska event a considerable amount of energy might have been lost due to radiation. Consequently, the energy estimates [3] of the Tunguska phenomenon may increase some 10 - 20 % due to this process. Naturally, the results obtained here demand further improvement. In particular, the non-one-dimensional property of the flow and the selective character of the radiation transfer should be included.

4. Model ballistic calculations

To estimate the amount of energy released in upper layers of the atmosphere during the Tunguska cosmic body flight and to reveal general features of such an energy release, some simple model ballistic computations were carried out. The hypersonic flight of spherical bodies with the constant ballistic parameter B ($B = m/c_x S$, m is the body mass, S is the body mid-section area, c_x is the drag coefficient) was calculated. In some testing cases the ballistic parameter B was taken in accordance with experimental data. The specific ballistic energy $E_1 = \frac{1}{2} c_x \rho V^2 S$ and its integral quantity E_{1c} released along the trajectory were determined.

The integral energy E_{1c} is shown in Fig. 8 as the function of the height H for a number of bodies with various values of the ballistic parameter B . The body density is taken equal to 0.1 g/cm^3, the trajectory inclination angle $\alpha = 40^o$. In all the cases we demand that the specific ballistic energy at the final point of the trajectory be equal to $1.4 \cdot 10^{17}$ erg/cm , that corresponds to solution [3] . In the plot the larger entry velocities of the body correspond to the smaller values of B . As is seen the energy release along the passive portion of the trajectory at heights above 20 km (where the ballistic wave is weak and does not throw down the forest) may achieve 1-3 Mtons. To determine the total trotyl equivalent of the phenomenon the energy release for the passive portion of the trajectory must be added to the summary value of the explosion and ballistic energies corresponding to the active portion of the trajectory. It is also interesting that the ballistic energy distribution along the active portion of the trajectory for small values of B has a maximum at a low height and can be approximated here by an appropriate mean distribution.

References

1. V.P.Korobeinikov, P.I.Chushkin, L.V.Shurshalov. Astr. Acta 17, № 4/5 (1972).
2. V.P.Korobeinikov, P.I.Chushkin, L.V.Shurshalov. Izv. AN SSSR, Mekh. Zhidk. Gasa, № 3 (1974).
3. V.P.Korobeinikov, P.I.Chushkin, L.V.Shidlovskaya, L.V.Shurshalov. Lecture Notes in Physics 59 (1976).
4. V.P.Korobeinikov, P.I.Chushkin, L.V.Shurshalov. Zh. Vych. Mat. Mat. Fiz. 17, № 3 (1977).
5. A.Ben-Menahem. Phys. Earth and Planetary Interiors 11 (1975).
6. H.L.Brode. Phys. Fluids 2, № 2 (1959).
7. S.K.Godunov. Mat. Sb. 47, № 3 (1959).
8. S.K.Godunov, A.V.Zabrodin, G.P.Prokopov. Zh. Vych. Mat. Mat. Fiz. 1, № 6 (1961).
9. L.V.Shurshalov. Dokl. AN SSSR 230, № 4 (1976).
10. B.S.Masson, T.D.Taylor, R.M.Foster. AIAA J. 7, № 4 (1969).
11. L.V.Shurshalov. 13th Int. Conf. Advan. Probl. Methods in Fluid Mechanics, Sept. 5-10, 1977, Poland.
12. D.E.Okhotsimskii et al. Tr. Mat. Inst. Steklova 50 (1957).
13. G.I.Marchuk, V.I.Lebedev. Numerical Methods in the Transport Theory, Moscow, Atomizdat (1971).
14. H.L.Brode. Astr. Acta 14, № 5 (1969).
15. L.M.Degtyarev, A.P.Favorskii. Zh. Vych. Mat. Mat. Fiz. 9, № 1 (1969).
16. N.M.Kuznetsov. Thermodynamic Functions and Shock Adiabats for Air at High Temperatures. Mashinostroenie (1965).
17. L.V.Avilova et al. Optical Properties of Hot Air. Moscow, Nauka (1970).

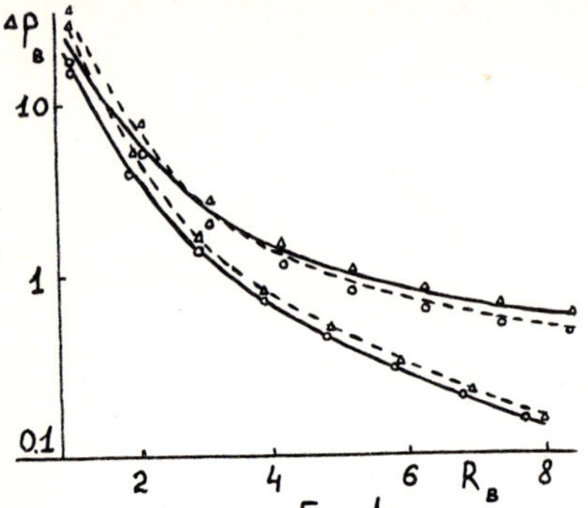

Fig. 1

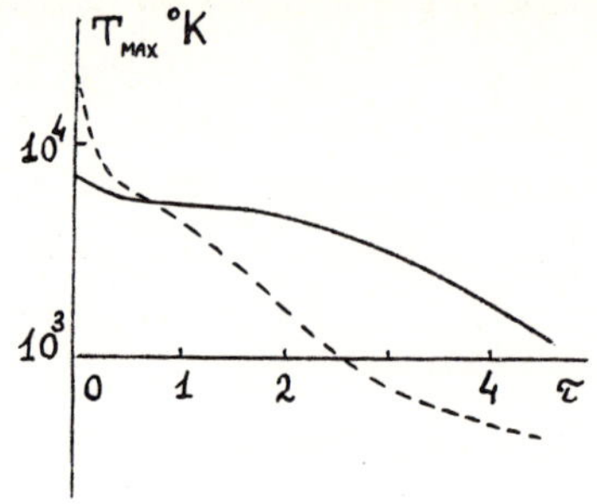

Fig. 2

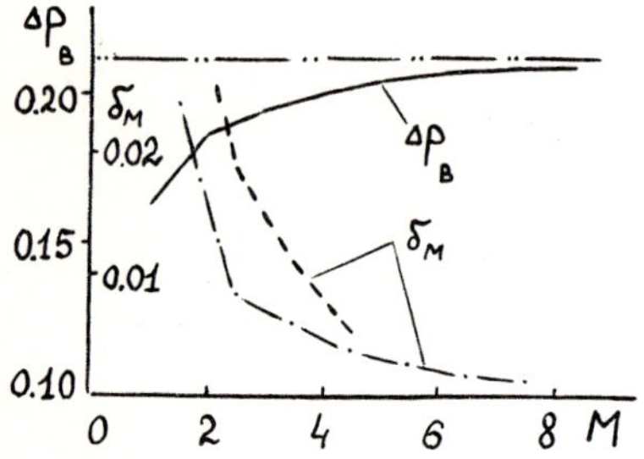

Fig. 3

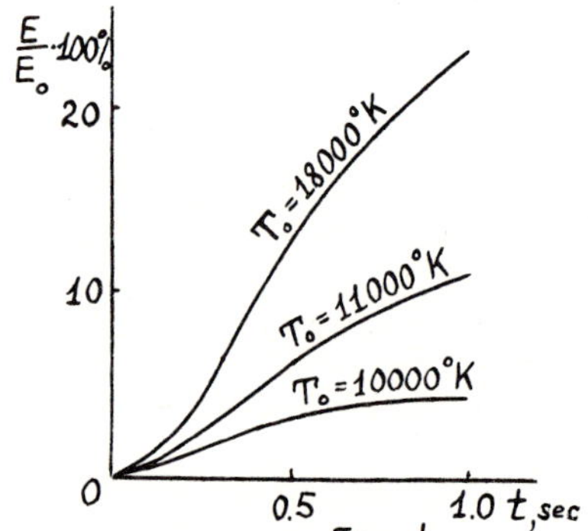

Fig. 4

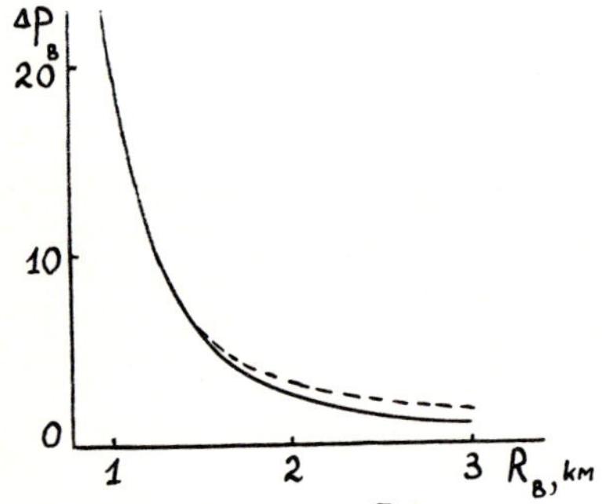

Fig. 5

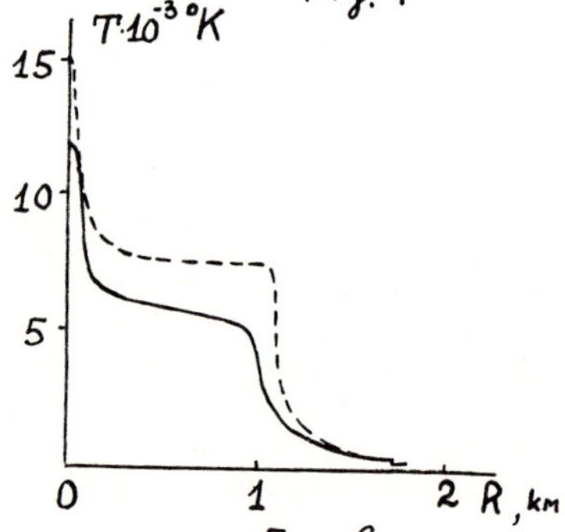

Fig. 6

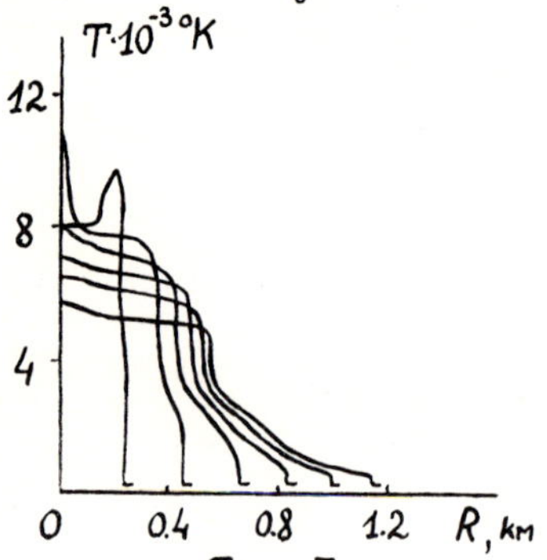

Fig. 7

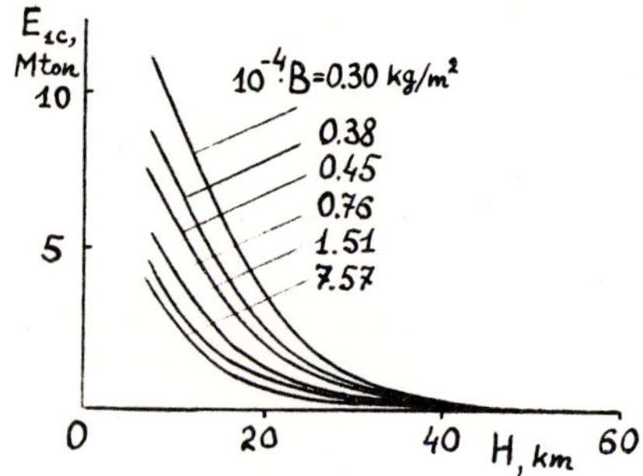

Fig. 8

NUMERICAL SOLUTION OF TRANSONIC FLOW THROUGH A CASCADE WITH SLENDER PROFILES

K.Kozel,J.Polášek,M.Vavřincová
Department of Applied Mathematics
Faculty of Mechanical Engineering
Prague,Czechoslovakia

The paper describes a method for the solution of a plane steady inviscid transonic flow through a cascade with thin and low cambered profiles.A weak solution of boundary value problem is defined and the difference problem approximating the differential problem is formulated.The difference problem is solved by a succesive line relaxation method.Some results of transonic flow through cascade with double circular arc symmetrical airfoils are compared with experiment (published at Göttingen 1975).Next numerical results show a change of the flowfield in described cascade with the change of the angle of attack,upstream Mach number,stagger angle and numerical test of influence of boundary layer effects.

I.Differential problem

Consider plane transonic cascade flow with thin and low cambered profiles.This flow can be described by small disturbance theory,the governing equation of which is the Cole's equation for the perturbation velocity potential

$$[K-(\varkappa+1)\varphi_x]\varphi_{xx}+\varphi_{\tilde{y}\tilde{y}}=0\,,\quad K=(1-M_\infty^2)/\delta^{2/3},\quad \tilde{y}=\delta^{1/3}\cdot y, \tag{1}$$

or in conservation form

$$[K\varphi_x-\tfrac{1}{2}(\varkappa+1)\varphi_x^2]_x+[\varphi_{\tilde{y}}]_{\tilde{y}}=0. \tag{1a}$$

A linearized condition at the profile surface is as follows

$$\varphi_{\tilde{y}}(x,\tilde{y})=F'_{u,e}(x)-\frac{\alpha}{\delta}\,,\quad x\in\langle -1+il_1,\,1+il_1\rangle,\quad \tilde{y}=i\delta^{1/3}l_2,\quad i=0,\pm1,\pm2,\dots \tag{2}$$

where δ is the thickness ratio, α is the angle of attack or inlet velocity angle with x-axis.The body shape is given by $y=\delta.F_{u,e}(x)$.

In our case the field of velocity vector $\overline{W}(U,V)$ is periodical with a period $\bar{l}(l_1,l_2)$, $l_1=|\bar{l}|\cos\beta$, $l_2=|\bar{l}|\sin\beta$, $\tilde{l}_2=l_2\delta^{1/3}$ where β is the stagger angle and $tg\tilde{\beta}=\delta^{1/3}tg\beta$ (see fig. 1).

We can consider a solution in one period $\Omega^{(i)}$ only

$$\Omega^{(i)}\equiv\{(x,\tilde{y});\ x\in(-\infty,\infty),\ \tilde{y}\in\langle i\tilde{l}_2,\,(i+1)\tilde{l}_2\rangle\},\quad i=0,\pm1,\pm2,\dots$$

or in some finite period

$$\Omega_G^{(i)}\equiv\{(x+il_1,\,\tilde{y}+i\tilde{l}_2),\ (x,\tilde{y})\in\Omega_G^{(0)}\},$$

$$\Omega_G^{(0)}\equiv G_1G_2G_3G_4$$

$$G_1[-\tilde{x},0],\quad G_2[\tilde{x},0]$$

$$G_3[\tilde{x}+l_1,\tilde{l}_2],\quad G_4[-\tilde{x}+l_1,\tilde{l}_2],$$

because of the disturbances vanish quickly at sufficient distance from the profiles.

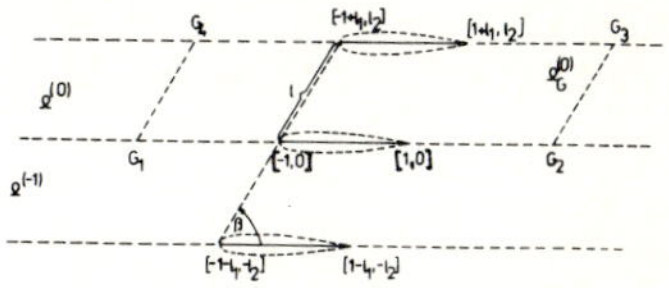

Fig. 1.
Cascade geometry in (x,y) coordinate system

Now we shall describe a class of functions where we shall find a weak solution of our transonic cascade problem(TC-problem).

Let the next conditions for the function $\varphi(x,\tilde{y})$ are fulfilled:

1) $\varphi(x,\tilde{y})$ is continuous in $\Omega^{(i)}$ or $\Omega_G^{(i)}$,

2) $\varphi_x(x,\tilde{y})$, $\varphi_{\tilde{y}}(x,\tilde{y})$ could be discontinuous only along a curves or in

some points $\Omega^{(i)}$ or $\Omega_q^{(i)}$. The set of the curves and the points has finite number of elements in every bounded subdomain $D \subset \Omega^{(i)}$. A bounded limits $\varphi_x(x\pm 0, \tilde{y})$, $\varphi_{\tilde{y}}(x\pm 0, \tilde{y})$ from different sides of that curves exist. Functions φ_x, $\varphi_{\tilde{y}}$, φ_{xx}, $\varphi_{x\tilde{y}}$, $\varphi_{\tilde{y}\tilde{y}}$ are continuous in other points of region $\Omega^{(i)}$, $\Omega_q^{(i)}$.
3) $|\varphi_x| < L$, $|\varphi_{\tilde{y}}| < L$ (L-positive real number) in all points of continuity of φ_x, $\varphi_{\tilde{y}}$. Then function $\varphi \in K(\Omega^{(i)})$ or $\varphi \in K(\Omega_q^{(i)})$.

Consider

$$\Gamma = -\oint_c \varphi_x dx + \varphi_{\tilde{y}} d\tilde{y} \tag{3}$$

where $c \subset \Omega^{(i)} \cup \Omega^{(i-1)}$ is closed and piecewise smooth curve enclosing the slit $\tilde{y} = i\tilde{l}_2$, $x \in \langle -1+il_1, 1+il_1 \rangle$.

Now we can define weak solution of our TC-problem:

Let $\varphi \in K(\Omega^{(0)})$ or $\varphi \in K(\Omega_q^{(0)})$ fulfils relation

$$\oint_d [K\varphi_x - \tfrac{1}{2}(\varkappa+1)\varphi_x^2] d\tilde{y} - \varphi_{\tilde{y}} dx = 0 \tag{4}$$

along every closed and piecewise smooth curve $d \subset \Omega^{(0)}$ or $d \subset \Omega_q^{(0)}$ and the next conditions:

a) $\varphi_x = \varphi_{\tilde{y}} = 0$, $x \to -\infty$ or $(x, \tilde{y}) \in \overline{G_1 G_4}$, (5)

b)
$$\varphi_x(x, 0+) = \varphi_x(x+l_1, \tilde{l}_2 - 0),$$
$$\varphi_{\tilde{y}}(x, 0+) = \varphi_{\tilde{y}}(x+l_1, \tilde{l}_2 - 0), \quad |x| > 1, \tag{6}$$

c)
$$\varphi_{\tilde{y}}(x, 0+) = F_u'(x) - \frac{\alpha}{\delta},$$
$$\varphi_{\tilde{y}}(x+l_1, \tilde{l}_2 - 0) = F_e'(x) - \frac{\alpha}{\delta}, \quad |x| \le 1,$$

$F_{u,e}(x)$ is a piecewise smooth function, $F_u(x) \ge F_e(x)$. (7)

d) $\varphi_x = K_1$, $\varphi_{\tilde{y}} = K_2$ for $x \to \infty$ or $(x, \tilde{y}) \in \overline{G_2 G_3}$. K_1, K_2 are unknown constants. (8)

Remark: a) If a case $\Gamma \neq 0$ is considered, the Joukovski's condition, that the flow leaves the trailing edge smoothly, must be applied. We consider this condition in (6) for $x > 1$ and in the properties of class $K(\Omega^{(0)})$. The Joukovski's condition chooses a value of circulation Γ fulfilling relation (6) for $x > 1$. $\Gamma \neq 0$ and definition $K(\Omega^{(0)})$ require

$$\Gamma = \varphi(x, 0+) - \varphi(x+l_1, \tilde{l}_2 - 0), \quad x \ge 1. \tag{9}$$

b) Relations
$$\langle K\varphi_x - \tfrac{1}{2}(\varkappa+1)\varphi_x^2 \rangle (d\tilde{y})_s - \langle \varphi_{\tilde{y}} \rangle (dx)_s = 0,$$
$$\langle \varphi_x \rangle (dx)_s + \langle \varphi_{\tilde{y}} \rangle (d\tilde{y})_s = 0, \quad \langle \varphi_x \rangle = \varphi_x(x+0, \tilde{y}) - \varphi_x(x-0, \tilde{y}), \tag{10}$$

which are fulfilled along a shock wave, immediately follow from (4) and vortex-free condition in integral form.

c) Consider region $\Omega_q^{(0)}$. Integration (4) along $G_1 G_2 G_3 G_4$ with using (5), (6), (7), (8) and relation (3) for circulation Γ give two algebraic equations for K_1, K_2 along $\overline{G_2 G_3}$. Two solutions for K_1, K_2 exist, we want only solution that for $\Gamma = 0$ gives $K_1 = 0$, $K_2 = 0$, then

$$K_1 = (\varkappa+1)^{-1}\{(K + \cot g^2\tilde{\beta}) - \sqrt{(K + \cot g^2\tilde{\beta})^2 + 2(\varkappa+1)\Gamma(l_1)^{-1}\cot g^2\tilde{\beta}}\}, \tag{11}$$

$$K_2 = -\{\Gamma(l_1)^{-1} + K_1\}\cot g\tilde{\beta}; \quad \Gamma \le -l_1[K + \cot g^2\tilde{\beta}]^2 \cdot [2(\varkappa+1)\cot g\tilde{\beta}]^{-1}. \tag{12}$$

d) Our boundary value problem is some periodical modification of Neuman's problem. We find only φ_x, $\varphi_{\tilde{y}}$. Let φ is a solution, then $\varphi + C$ is solution too (C is a constant). Then we consider $\varphi(-\bar{x}, 0)$ = const. and (5), (6), (8), (10) give

$$\varphi(x, 0+) = \varphi(x+l_1, \tilde{l}_2 - 0), \quad x < -1, \tag{13}$$
$$\varphi(x, 0+) = \varphi(x+l_1, \tilde{l}_2 - 0) + \Gamma, \quad x \ge 1. \tag{14}$$

II. Difference problem

Consider region $\Omega_q^{(0)}$ and transformation $(x, \tilde{y}) \to (x, t)$, $\bar{t}^0(\cos\tilde{\beta}, \sin\tilde{\beta})$ in $(x, \tilde{y})$ coordinate system. Then

$$\varphi_{\tilde{y}\tilde{y}} = (\cot g^2\tilde{\beta})\,\varphi_{xx} - 2\frac{\cos\tilde{\beta}}{\sin^2\tilde{\beta}}\,\varphi_{xt} + \frac{1}{\sin^2\tilde{\beta}}\,\varphi_{tt}\,.$$

Consider grid points $P_{ij} = P(x_i, t_j)$ in new coordinate system $x_i = -\bar{x} + ih_1$ $(i = 0, 1, 2, \dots, M)$, $t_j = jh_2$ $(j = 0, 1, 2, \dots, N)$.

Following two difference schemes had been used:

α) The difference approximation of the nonlinear term of (1) depends on whether $E_{ij} = [K - (x+1)(2h_1)^{-1}(\varphi_{i+1,j} - \varphi_{i-1,j})] > 0$ or $E_{ij} < 0$. For $E_{ij} > 0$ we used central differences

$$\varphi_x = (2h_1)^{-1}(\varphi_{i+1,j} - \varphi_{i-1,j}) + O(h_1^2),\quad \varphi_{xx} = (h_1)^{-2}(\varphi_{i+1,j} - 2\varphi_{ij} + \varphi_{i-1,j}) + O(h_1^2),$$

for $E_{ij} < 0$ we used backward differences(α_1)

$$\varphi_x = (2h_1)^{-1}(\varphi_{ij} - \varphi_{i-2,j}) + O(h_1),\quad \varphi_{xx} = (h_1)^{-2}(\varphi_{ij} - 2\varphi_{i-1,j} + \varphi_{i-2,j}) + O(h_1)$$

or(α_2)

$$\varphi_x = (2h_1)^{-1}(5\varphi_{ij} - 8\varphi_{i-1,j} + 3\varphi_{i-2,j}) + O(h_1^2),\quad \varphi_{xx} = (h_1)^{-2}(2\varphi_{ij} - 5\varphi_{i-1,j} + 4\varphi_{i-2,j} - \varphi_{i-3,j}) + O(h_1^2).$$

For $\varphi_{\tilde{y}\tilde{y}}$ we used central differences in elliptic($E_{ij} > 0$) and in hyperbolic($E_{ij} < 0$) point

$$\varphi_{xx} = (h_1)^{-2}(\varphi_{i+1,j} - 2\varphi_{ij} + \varphi_{i-1,j}) + O(h_1^2),$$

$$\varphi_{tt} = (h_2)^{-2}(\varphi_{i,j+1} - 2\varphi_{ij} + \varphi_{i,j-1}) + O(h_2^2),$$

$$\varphi_{xt} = (4h_1h_2)^{-1}(\varphi_{i+1,j+1} - \varphi_{i+1,j-1} - \varphi_{i-1,j+1} + \varphi_{i-1,j-1}) + O(h_1^2, h_2^2).$$

In hyperbolic point we used some modification of rotated difference scheme because it is necessary for fulfilling stability condition in the grid points where $(E_{ij} < 0) \wedge (\cot g\tilde{\beta} > \sqrt{(x+1)\varphi_x - K})$. For boundary condition $\varphi_{\tilde{y}} = (\sin\tilde{\beta})^{-1}(-\varphi_x\cos\tilde{\beta} + \varphi_t) = F_u'(x) - (\alpha/\delta)$, $x \in \langle -1, 1\rangle$, $\tilde{y} = 0$

we used following approximation

$$\varphi_x = (2h_1)^{-1}(\varphi_{i+1,j} - \varphi_{i-1,j}) + O(h_1^2),\quad \varphi_t = (2h_2)^{-1}(-3\varphi_{ij} + 4\varphi_{i,j+1} - \varphi_{i,j+2}) + O(h_2^2).$$

In difference equation for points lying at $\tilde{y} = 0$, $|x| > 1$ we used condition of periodicity (13),(14) for points with $\tilde{y} < 0$. In the points with $\tilde{y} = \tilde{\ell}_2$, $|x| > 1$ we used only condition of periodicity (13),(14). Downstream conditions (11),(12) are used in difference scheme.

β) The second difference scheme is conservative and is derived using conservation law in integral form (4) for each cell(fig.2a). The approximation (4) along boundary of mentioned cell gives

$$\{[K\varphi_x - \tfrac{1}{2}(x+1)\varphi_x^2]_{i+\frac{1}{2},j} - [K\varphi_x - \tfrac{1}{2}(x+1)\varphi_x^2]_{i-\frac{1}{2},j}\}\,h_2\sin\tilde{\beta} +$$

$$+ \{-[\varphi_{\tilde{y}}]_{i+\frac{1}{2},j} + [\varphi_{\tilde{y}}]_{i-\frac{1}{2},j}\}\,h_2\cos\tilde{\beta} - \{[\varphi_{\tilde{y}}]_{i,j-\frac{1}{2}} - [\varphi_{\tilde{y}}]_{i,j+\frac{1}{2}}\}\,h_1 = 0.$$

Considering different approximation $[K\varphi_x - \frac{1}{2}(x+1)\varphi_x^2]_{i+\frac{1}{2},j}$ for $E_{ij} \gtrless 0$ and $[K\varphi_x - \frac{1}{2}(x+1)\varphi_x^2]_{i-\frac{1}{2},j}$ for $E_{i-1,j} \gtrless 0$, we obtain the difference scheme that uses the same differencing for nonlinear term as Murman [6] and other terms are the approximation $\varphi_{\tilde{y}\tilde{y}}$ (multiplied by some constant) as in previous scheme (α). Near profiles we used approximation (4) along boundary cell(fig.2b).

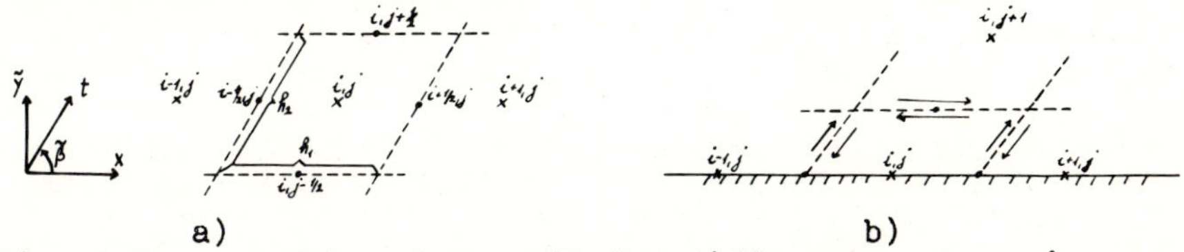

Fig. 2: Inner and boundary cell for difference scheme in conservation form in $(x, \tilde{y})$ coordinate system

III. Iteration procedure

The described system of difference equations is solved by successive line relaxation method. The values of coeffcients E_{ij} are fixed in every point and iteration. We go through region $\Omega_q^{(o)}$ from left to right in the positive direction of x-axis. The values $K_1(y)$, $K_2(y)$

and γ are changed every tenth iteration. A convergence of the iteration process depends on the choice of suitable relaxation parameter ω .Von Neumann test based on local Fourier analysis requires f.e. for scheme (α):

1) $\omega \in (0,2)$, if $E_{ij} > 0$ in every point in the line $x = x_i$ in (x,t) coordinate system, $|x_i| > 1$,

2) $\omega < 2[1 + a\cos\tilde{\beta}]^{-1}$, $a = h_2/h_1$,for the points in the line $x = x_i$, $-1 \le x_i \le 1$ if $E_{ij} > 0$ in every point in the line $x = x_i$,

3) $\omega = 1$ if $E_{ij} < 0$ in some point in the line $x = x_i$.

In real iteration process we used as a suitable value of parameter ω in the second case $1 \le \omega \le 2[1 + a\cos\tilde{\beta}]^{-1} - 0{,}25$ .A convergence of the iteration process is controlled using pressure coefficients c_p for points lying on the "profiles".

IV. Results

We shall now present some numerical results of the flow through cascade with double circular arc symmetrical 8% airfoils.

A)We shall mention some cases with $\beta = \pi/4$, $\delta = 0{,}08$, $p/c = 1$ and different values M_∞ , α .Some results were compared with experimental data from the Institute of Thermomechanics Czechoslovak Academy of Sciences.Our upstream Mach numbers are slightly higher(about 0,02-0,03) then those measured in experiment.The angles of attack were considered nonzero, $\alpha \in \langle 0^\circ ; 2{,}5^\circ \rangle$.

Fig. 3

a)Experiment : $M_\infty = 0{,}813$; $\alpha = 0^\circ$; $\beta = 45^\circ$; $p/c = 1$
Calculation: $M_\infty = 0{,}835$; $\alpha = 1{,}3^\circ$; $\beta = 45^\circ$; $p/c = 1$

b)Experiment : $M_\infty = 0{,}832$; $\alpha = 0^\circ$; $\beta = 45^\circ$; $p/c = 1$
Calculation: $M_\infty = 0{,}856$; $\alpha = 2{,}5^\circ$; $\beta = 45^\circ$; $p/c = 1$

c)Experiment : $M_\infty = 0{,}849$; $\alpha = 0^\circ$; $\beta = 45^\circ$; $p/c = 1$
Calculation: $M_\infty = 0{,}876$; $\alpha = 0{,}5^\circ$; $\beta = 45^\circ$; $p/c = 1$

d)Experiment : $M_\infty = 0{,}863$; $\alpha = 0^\circ$; $\beta = 45^\circ$; $p/c = 1$
Calculation: $M_\infty = 0{,}892$; $\alpha = 1^\circ$; $\beta = 45^\circ$; $p/c = 1$

Comparison of our numerical results with similar experimental results is given in figures 3a-d.White strips in the interferogrammes are places with constant density that changes from strip to strip by the same constant value.The numerical results are mapped using curves M=const. ,where the difference between curves equals $\Delta M = 0{,}05$.It is possible to say that differences of 2,5 strip in the interferogram is approximatelly equal to difference $\Delta M = 0{,}05$ in the mapping of our numerical results.

Using qualitative comparison we can say that agreement in inlet part and in the first part of the channel between profiles is good as we mention localization and shape of the strips and,especially, closed sonic line.In the second part of the channel between profiles we see only qualitative agreement in the shape of the strips and curves M=const.,but there is possible to observe some differences in localization and shape.The differences are probably caused by viscosity effects in boundary layer and in a wake that we see clearly in the interferogrammes.In experiments as well as in numerical results we see that the flow was realized for upstream Mach number slightly higher than Mach number when the sonic line is for the first time closed.

B)Next figures 4a-c show numerical test of the change of transonic flowfield in described cascade with the change of the angle of attack,upstream Mach number and stagger angle(compared with experiment published at Göttingen 1975).The differences in sonic line shape in 4c, $\beta = 35°$ on upper profile are probably caused by viscosity effects in a wake which are not considered in numerical solution.

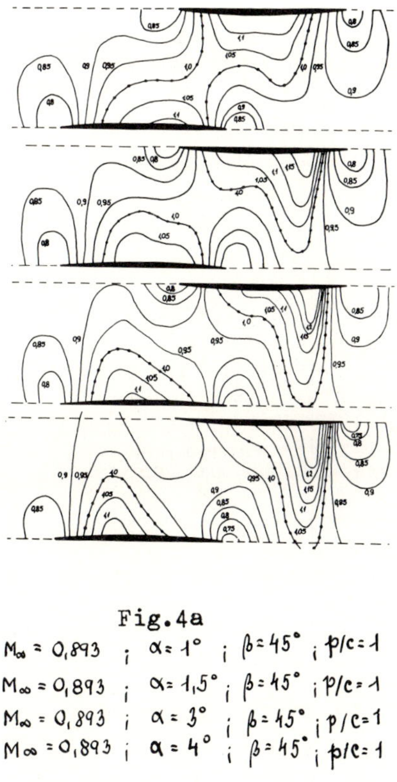

Fig.4a

$M_\infty = 0{,}893$; $\alpha = 1°$; $\beta = 45°$; $p/c = 1$

$M_\infty = 0{,}893$; $\alpha = 1{,}5°$; $\beta = 45°$; $p/c = 1$

$M_\infty = 0{,}893$; $\alpha = 3°$; $\beta = 45°$; $p/c = 1$

$M_\infty = 0{,}893$; $\alpha = 4°$; $\beta = 45°$; $p/c = 1$

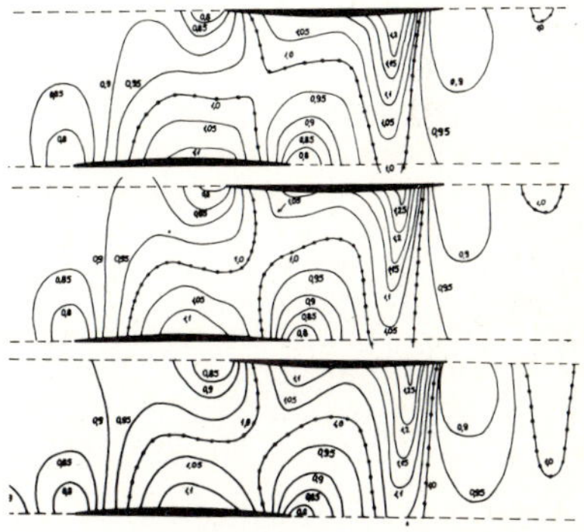

Fig.4b

$M_\infty = 0{,}9$; $\alpha = 1{,}3°$; $\beta = 45°$; $p/c = 1$

$M_\infty = 0{,}91$; $\alpha = 1{,}3°$; $\beta = 45°$; $p/c = 1$

$M_\infty = 0{,}92$; $\alpha = 1{,}3°$; $\beta = 45°$; $p/c = 1$

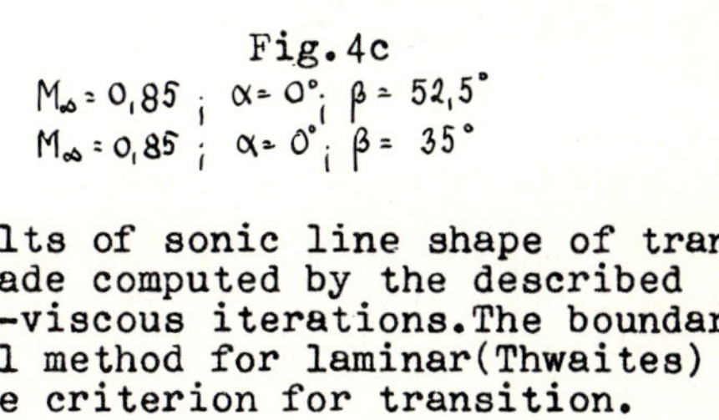

Fig.4c

$M_\infty = 0{,}85$; $\alpha = 0°$; $\beta = 52{,}5°$

$M_\infty = 0{,}85$; $\alpha = 0°$; $\beta = 35°$

C)Figure 5 shows numerical results of sonic line shape of transonic flowfield in a particular cascade computed by the described (inviscid) method and using inviscid-viscous iterations.The boundary layer was computed by simple integral method for laminar(Thwaites) and turbulent(Green) part with simple criterion for transition.

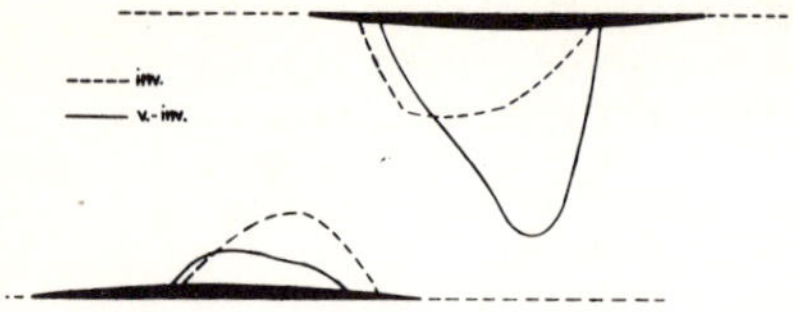

Fig.5

$M_\infty=0,845;\ \alpha=0^\circ;\ \beta = 45^\circ;\ p/c=1$

All presented results were computed by difference scheme ($\alpha 1$) and ($\alpha 2$),the cases (A) were also computed by conservative scheme(β) because the flowfield in figures 4c,4d is near choked regime and we controlled conservative properties of the difference scheme (α). The maximal difference in Mach number between results of the schemes ($\alpha 1$),(β) for first two cases (A) is about 0,005.The greater difference is possible to observe in last two cases (A) in the region of the end of the shock wave on upper profiles,where $|\Delta M| \leq 0,015$.

We used 91x17 or 141x17 grid points,26 along profile surface.For all presented results computer IBM 7040 was used,the process converged approximatelly about 30-300 iterations with suitable choice of initial values.It takes about 6-45 minutes of computer time.

Conclusion

On the basis of mentioned comparisons it is possible to say,that described numerical method correctly mapps the field of transonic cascade flow for considered cases.A localization and shape of the sonic line is in good agreement with experiment.The method calculated also flows(in a good agreement with experiment) with upstream Mach number slightly higher than those where the sonic line closed for the first time.Accuracy of the calculated results proved to be better than expected.

References

[1] Cole,J.D.;Murman,E.M.:AIAA Journal,No 1,1971

[2] Rozhdestvenski,B.L.;Yanenko,N.N.:Systems of quasilinear equations,Moskva 1968(in Russian)

[3] Jameson,A.:Comm.Pure and Appl.Mathem.,No 3,1974

[4] Kozel,K.;Polášek,J.;Vavřincová,M.:Rpt. K-201,No 020,1975 (in Czech)

[5] Dvořák,R.:Symposium Transsonicum II,1975

[6] Murman,E.M.:Symposium Transsonicum II,1975

[7] Kozel,K.;Polášek,J.;Vavřincová,M.:The Journal of Mechanical Engineering,No 2,1978

CZOCHRALSKI BULK FLOW IN THE GROWTH OF GARNET CRYSTALS

W. E. Langlois

IBM Research Laboratory
San Jose, California 95193

1. Introduction

The term "Czochralski bulk flow" was coined [1,2] to describe a slight geometrical idealization of the melt-flow configuration arising in the Czochralski crystal-pulling process. In this process, the growing crystal is lifted very gradually from a crucible of melt, as shown in Figure 1. In practice, the free surface has a meniscus near the growing crystal; to ignore this, and treat the free surface as flat, is the "bulk-flow" idealization.

In order to keep heterogeneities in the finished crystal to a minimum, the crystal grower strives for perfect axial symmetry in the melt velocity and temperature fields. To disperse asymmetries which arise from nonuniform heating or other accidental occurrences, he rotates the crystal, the crucible, or both. The rotation rate is large enough that the crystal extraction rate can be neglected in comparison.

Czochralski growth of liquid semiconductor crystals was investigated in [2]. The simulated flow field exhibited a rather pronounced vertical stratification: regions of steep vertical gradients near the crystal face and crucible bottom were separated by a relatively quiescent region occupying about 80% of the flow field. This result suggested that it might be profitable to stretch the axial coordinate in order to concentrate grid points within the steep gradient regions. That is the approach taken in the present study.

2. Equations of Motion

Our starting point is the Navier-Stokes equations for axisymmetric flow, normalized with respect to the units of time (τ), length (λ), and temperature (θ), with Boussinesq approximation and with viscous dissipation neglected in the energy equation. To take advantage of the expected vertical stratification we introduce a transformation which stretches the axial coordinate, taking

$$z = F(s).$$

The objective is to select a smooth function F such that a finite difference grid with uniform Δs will concentrate grid points in those strata where steep vertical gradients are expected to occur.

To eliminate the pressure, we define a vorticity ω and streamfunction ψ according to

$$\omega = \frac{\partial w}{\partial r} - \frac{1}{F'}\frac{\partial u}{\partial s}, \tag{1}$$

$$u = \frac{1}{rF'}\frac{\partial \psi}{\partial s}, \quad w = -\frac{1}{r}\frac{\partial \psi}{\partial r} \tag{2}$$

It follows from (1) and (2) that the vorticity and streamfunction are related through the diagnostic equation

$$\frac{1}{rF'}\frac{\partial}{\partial s}\left(\frac{1}{F'}\frac{\partial \psi}{\partial s}\right) + \frac{\partial}{\partial r}\left(\frac{1}{r}\frac{\partial \psi}{\partial r}\right) = -\omega . \tag{3}$$

The prognostic equations for the vorticity, azimuthal velocity, and absolute temperature are

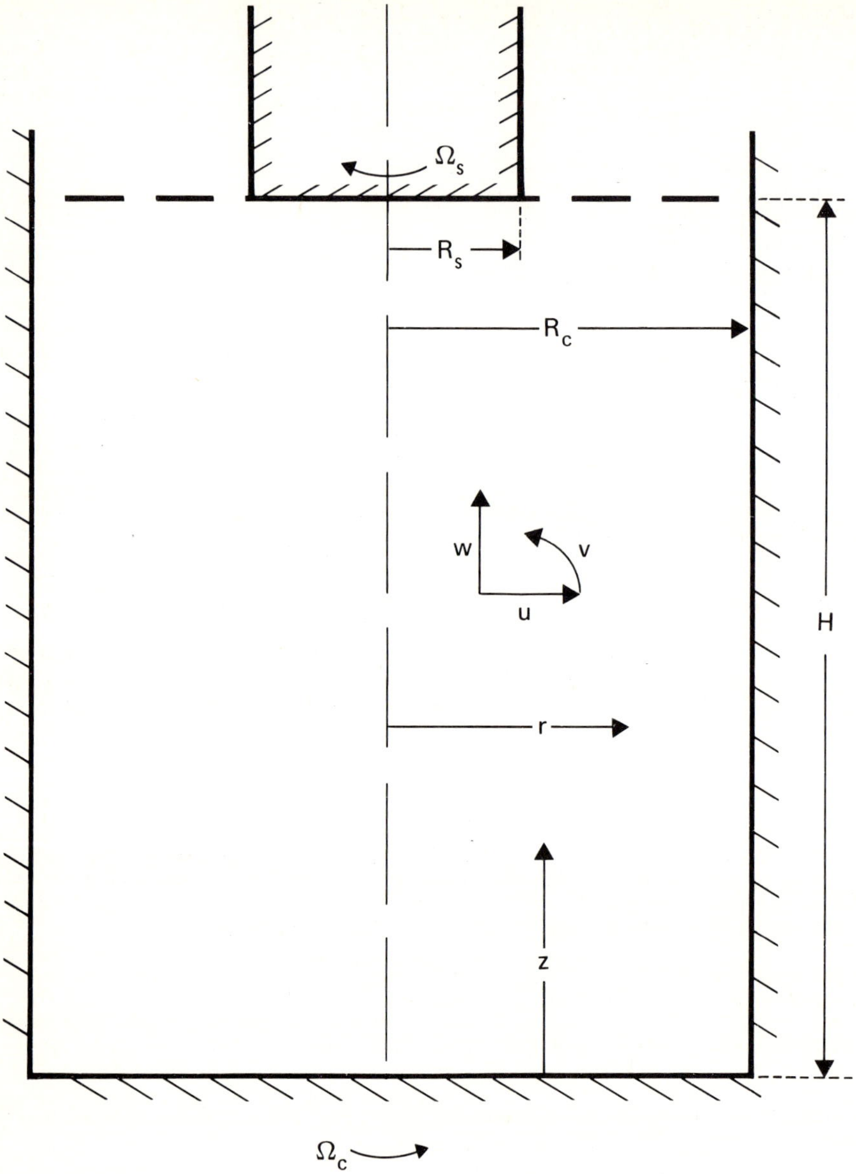

Figure 1. Geometry of the Czochralski technique.

$$\frac{\partial\omega}{\partial t} + \frac{\partial}{\partial r}(u\omega) + \frac{1}{F'}\frac{\partial}{\partial s}(w\omega) + \frac{2v}{F'}\frac{\partial v}{\partial s} = B\frac{\partial T}{\partial r} + E\nabla^2\omega - E\frac{\omega}{r^2},$$

$$\frac{\partial v}{\partial t} + \frac{\partial}{\partial r}(uv) + \frac{1}{F'}\frac{\partial}{\partial s}(wv) + \frac{2uv}{r} = E\nabla^2 v - E\frac{v}{r^2}, \tag{4}$$

$$\frac{\partial T}{\partial t} + \frac{1}{r}\frac{\partial}{\partial r}(ruT) + \frac{1}{F'}\frac{\partial}{\partial s}(wT) = K\nabla^2 T,$$

where

$$\nabla^2 = \frac{\partial^2}{\partial r^2} + \frac{1}{r}\frac{\partial}{\partial r} + \frac{1}{F'}\frac{\partial}{\partial s}\left(\frac{1}{F'}\frac{\partial}{\partial s}\right),$$

and where the dimensionless coefficients B, E, K are related to the volumetric expansion coefficient α, the gravitational acceleration g, the kinematic viscosity ν, and the thermal diffusivity κ according to

$$B = \frac{\alpha g}{\lambda(\tau^2\theta)}, \quad E = \frac{\nu}{\lambda^2/\tau}, \quad K = \frac{\kappa}{\lambda^2/\tau}.$$

Finite difference analogs of (4) are used to advance the vorticity, azimuthal velocity, and temperature fields from one time step to the next. These analogs can be succinctly described, provided we first establish some notation.

To begin, let the symbol f signify any of the dependent variables ω, v, T, or ψ. Further, let $f^k_{i,j}$ denote $f[(i-1)\Delta r,(j-1)\Delta s,t]$ and $f^{k+1}_{i,j}$ denote $f[(i-1)\Delta r,(j-1)\Delta z,t+\Delta t]$. Next, we introduce four operators $\mathcal{A}$, $\mathcal{B}$, $\mathcal{C}$, $\mathcal{D}$(advective, buoyant, centrifugal, diffusive). Except for $\mathcal{B}$, each of these entails an integer parameter n. For $\mathcal{A}$ and $\mathcal{D}$, the value of n may be 0 or 1; for $\mathcal{C}$ it may be 1 or 2.

The advective operator is defined by

$$\begin{aligned}\mathcal{A}(f_{i,j};n) = {} & \\ & f_{i,j} + \frac{\Delta t}{4r_i\Delta r\Delta z_j}\left[\frac{1}{1-n\Delta r/2r_i}\left(\psi^k_{i,j+1}-\psi^k_{i,j-1}+\psi^k_{i-1,j+1}-\psi^k_{i-1,j-1}\right)f^u_{i-1/2,j}\right] \\ & - \frac{1}{1+n\Delta r/2r_i}\left(\psi^k_{i+1,j+1}-\psi^k_{i+1,j-1}+\psi^k_{i,j+1}-\psi^k_{i,j-1}\right)f^u_{i+1/2,j} \\ & + \left(\psi^k_{i-1,j}-\psi^k_{i+1,j}+\psi^k_{i-1,j-1}-\psi^k_{i+1,j-1}\right)f^u_{i,j-1/2} \\ & - \left(\psi^k_{i-1,j+1}-\psi^k_{i+1,j+1}+\psi^k_{i-1,j}-\psi^k_{i+1,j}\; f^u_{i,j+1/2}\right).\end{aligned}$$

In this definition

$$r_i = (i-1)\Delta r, \quad \Delta z_j = \frac{1}{2}\left[F(j\Delta s)-F((j-2)\Delta s)\right],$$

and $f^u_{i-1/2,j}$ denotes either $f_{i-1,j}$ or $f_{i,j}$, depending upon which is upstream, with analogous meanings for $f^u_{i+1/2,j}, f^u_{i,j-1/2}, f^u_{i,j+1/2}$.

The other three operators are less complicated:

$$\mathcal{B}(f_{i,j}) = \frac{B\Delta t}{2\Delta r}(f_{i+1,j}-f_{i-1,j}),$$

$$\mathcal{C}(f_{i,j};n) = \frac{\Delta t}{r^n_i\Delta z_j}v^k_{i,j}(f_{i,j+1}-f_{i,j-1})$$

$$\mathscr{D}(f_{i,j};n) = f_{i,j} + \frac{E^n K^{1-n}\Delta t}{\Delta z_j}\left[\frac{1}{\Delta z_{j+1/2}}(f_{i,j+1}-f_{i,j}) - \frac{1}{\Delta z_{j-1/2}}(f_{i,j}-f_{i,j-1})\right]$$

$$+ \frac{E^n K^{1-n}\Delta t}{(\Delta r)^2}\left[\left(1+\frac{\Delta r}{2r_i}\right)\left(f_{i+1,j}-f_{i,j}\right) - \left(1-\frac{\Delta r}{2r_i}\quad f_{i,j}-f_{i-1,j}\right)\right] - \frac{nE\Delta t}{r_i^2} f_{i,j} ,$$

where

$$\Delta z_{j+1/2} = F(j\Delta s)-F((j-1)\Delta s) .$$

In terms of the four operators, the finite difference prognostic equations are

$$\omega_{i,j}^{k+1} = \mathscr{D}\{\mathscr{A}[\omega_{i,j}^k+\mathscr{B}(T_{i,j}^k)-\mathscr{C}(v_{i,j}^k;1);1];1\} ,$$

$$v_{i,j}^{k+1} = \mathscr{D}\{\mathscr{A}[v_{i,j}^k-\mathscr{C}(\psi_{i,j}^k;2);1];1\} ,$$

$$T_{i,j}^{k+1} = \mathscr{D}\{\mathscr{A}[T_{i,j}^k;0];0\} ,$$

A finite-difference analog of the diagnostic equation (3) is

$$\psi_{i,j}\left[\frac{1}{(\Delta r)^2}\left(\frac{r_i}{r_{i+1/2}}+\frac{r_i}{r_{i-1/2}}\right)+\frac{2}{\Delta z_{j+1/2}\Delta z_{j-1/2}}\right]$$

$$= r_i\omega_{i,j} + \frac{1}{(\Delta r)^2}\left[\frac{r_i}{r_{i+1/2}}\psi_{i+1,j}+\frac{r_i}{r_{i-1/2}}\psi_{i-1,j}\right]+\frac{1}{\Delta z_j}\left[\frac{\psi_{i,j+1}}{\Delta z_{j+1/2}}+\frac{\psi_{i,j-1}}{\Delta z_{j-1/2}}\right] .$$

For most of the work described here, this equation was solved by successive overrelaxation. As soon as a direct solver was developed [3] it was incorporated into the program.

3. Boundary Conditions

The boundary conditions are those used in [1] and [2]: axial symmetry; no-slip velocity condition and specified temperature on all solid surfaces; Stefan-Boltzmann radiation from the melt free-surface, which remains stress-free and almost flat. Incorporating the obvious modifications required by the stretched vertical coordinate, these are:

AXIS $r = 0$ $(i=1)$

$$\psi = 0,\ \omega = 0,\ v = 0,\ \partial T/\partial r = 0 ,$$

CRUCIBLE BOTTOM $s = 0$ $(j=1)$

$$\psi = 0,\ \omega = -\frac{2\psi(r,\Delta s,t)}{r_i(\Delta z_{3/2})^2},\ v = r\Omega_c,\ T = T_c ,$$

CRUCIBLE WALL $r = R_c$ $(i=M)$

$$\psi = 0,\ \omega = -\frac{2}{R_c(\Delta r)^2}\psi(R_c-\Delta r,s,t),\ v = r_c\Omega_c,\ T = T_c ,$$

CRYSTAL FACE $s = s_H$, $r \le R_S$ $(j=N;\ i=1,2,\ldots,i_S-1)$

$$\psi = 0,\ \omega = -\frac{2}{r_i(\Delta z_{N-1/2})^2}\psi(r,s_H-\Delta s,t),\ v = r\Omega_s,\ T = T_s\ ,$$

FREE SURFACE $s = s_H$, $R_S<r<R_c$ $(j=N;\ i=i_S,i_{S+1},\ldots,M-1)$

$\psi = 0$, $\omega = 0$, $\partial v/\partial s = 0$, $\partial T/\partial t = \partial T/\partial t$ from the prognostic equation

$$-2\theta^3\tau\varepsilon\sigma T^4/\lambda\rho c_v\Delta z_{N-1/2}\ .^*$$

In this system: ε is the surface emissivity; the dimensional parameters ρ, c_v, and σ are, respectively the melt density, the melt specific heat, and the Stefan-Boltzman constant; s_H is the value of s at the top surface, i.e., $H = F(s_H)$.

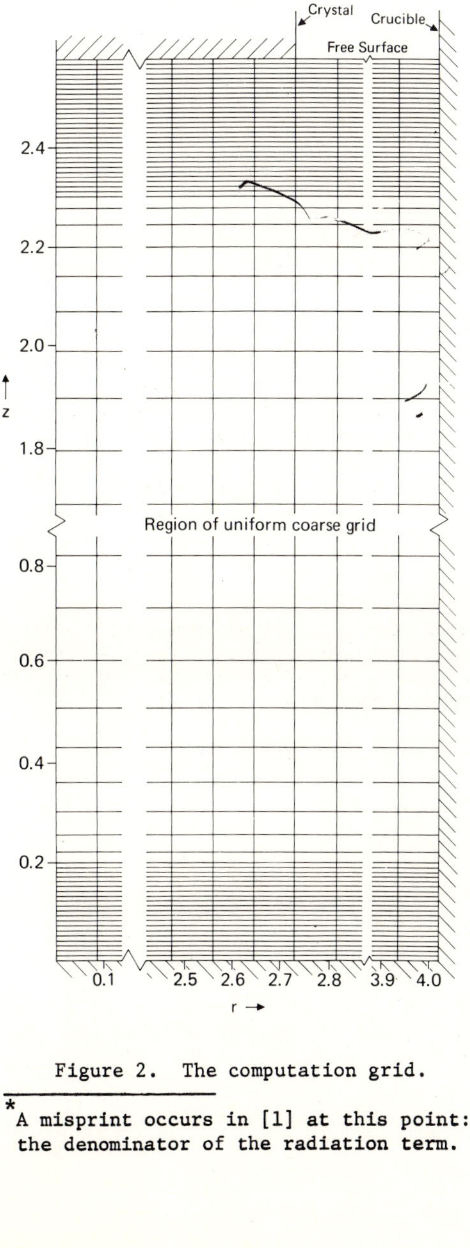

Figure 2. The computation grid.

4. Czochralski Growth of Garnets

Molten garnets differ from liquid semiconductors [2] in several important particulars:

- Viscosities are considerably lower.
- Melting points are significantly higher, so that radiation from the melt free-surface takes on greater importance.
- For operational reasons, the garnet crystal grower usually prefers to avoid the additional complication of rotating the crucible, i.e., $\Omega_c = 0$.

If the measuring system is CGS, a typical setup for growing crystals of $Gd_3Ga_5O_{12}$ (Gadolinium gallium garnet, or GGG) would be represented by the parameter set

$$\begin{aligned} &R_s = 2.794,\ R_c = 4.191,\ H = 2.580 \\ &\Omega_s = 1.6755,\ \Omega_c = 0,\ T_s = 2033,\ T_c = 2200 \\ &B = 0.02646,\ E = 0.00164,\ K = 0.0464 \\ &2\theta^3\tau\varepsilon\sigma/\lambda\rho c_v = 1.026 \times 10^{-12}\ . \end{aligned} \quad (5)$$

The simulation for this case was carried out on a 50×75 grid, with uniform mesh in r-s space. The stretching function F is numerically prescribed in such a way that the grid points are clustered near the top and bottom of the flow field and widely spaced in the interior. To minimize numerical reflection, the vertical spacing between grid points increases by arithmetic progression in seven steps as we move

*A misprint occurs in [1] at this point: the factor Δz was omitted from the denominator of the radiation term.

from the interior into a dense-packed region. With this restriction, specifying $\Delta z_{j-1/2}$=0.01 for j=1,2,...,21 and also for j=48,49,...,75 uniquely determines the grid shown in Figure 2. A double time step was used: the thermal diffusion and radiation must be carried out with a very short time-step Δt_θ for reasons of stability; the remainder of the computation has much less stringent requirements and a larger step Δt can be used. For the parameter set (5) and the grid of Figure 2, the computation proceeds smoothly with Δt=0.002136 and $\Delta t_\theta = \Delta t/5$.

The meridional circulation in Czochralski bulk flow in a stationary crucible is essentially a "contest" between buoyant convection and centrifugal pumping by the rotating crystal. Figure 3 shows that, for the parameter set (5), buoyant convection is a decisive winner. The configuration seems to be steady state, in contradistinction to the liquid semiconductor simulation of [2], which evolved to time-periodic oscillations. Isotherms of the flow are shown in Figure 4.

5. References

1. W. E. Langlois and C. C. Shir: "Digital Simulation of Flow Patterns in the Czochralski Crystal-Pulling Process," Computer Methods in Appl. Mech. and Eng. 12, p. 145-152 (1977).

2. W. E. Langlois: "Digital Simulation of Czochralski Bulk Flow in a Parameter Range Appropriate for Liquid Semiconductors," J. Crystal Growth 42, p. 386-399 (1977).

3. G. H. Golub and W. E. Langlois: "Direct Solution of the Equation for the Stokes Stream Function," to appear in Computer Methods in Appl. Mech. and Eng.

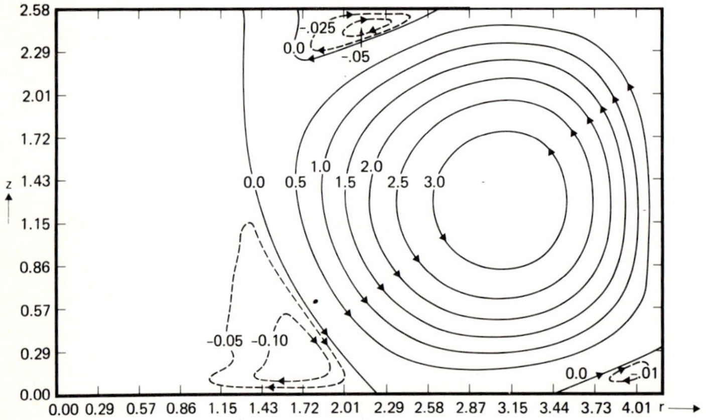

Figure 3. Streamlines of the meridional circulation.

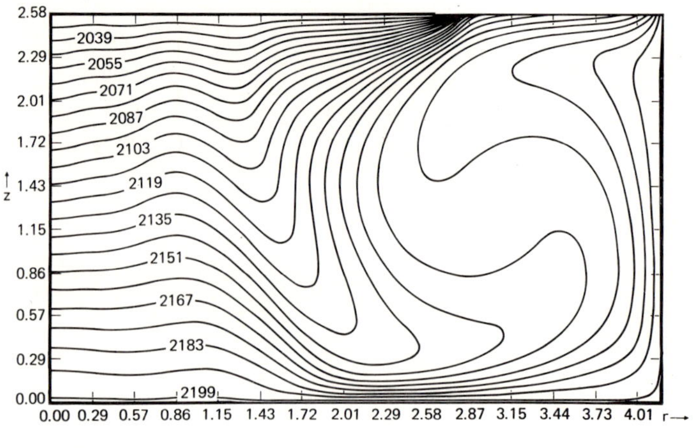

Figure 4. Isotherms.

NUMERICAL SHOCK STRUCTURE AND NONLINEAR CORRECTIONS FOR DIFFERENCE SCHEMES IN CONSERVATION FORM

Alain Lerat

Ecole Nationale Supérieure d'Arts et Métiers, 75013 Paris and Laboratoire de Mécanique Théorique, Université P. et M. Curie Paris.

ABSTRACT : *In this paper the scalar equation $u_t + f(u)_x = 0$ is approximated by three-point difference schemes in conservation form with an order of accuracy p equal to one or two. The partial differential equations approximated by the schemes with (p + 1)-th order accuracy are derived and their shock structures are analytically obtained and found to be close to the numerical shock profiles. By using these results, correction terms are proposed for second order schemes.*

INTRODUCTION

A fruitful method of studying finite difference schemes of an order of accuracy p for a hyperbolic equation is to analyze their "equivalent equation" up to some order q, that is the partial differential equation approximated by the schemes with an order of accuracy q greater than p. This method has given rise to a number of works ; some are related to the stability theory of schemes [1] - [4] , some are related to the problem of parasitic oscillations in the numerical solutions [5] - [10] and others to a variety of purposes [11] - [15] . The main advantage of the method is that it can be applied to nonlinear schemes. Since equivalent equations are derived by Taylor expansions, it might seem that the method is not valid when the solution of the exact problem is discontinuous. However, the study [5] has clearly indicated that parasitic oscillations in numerical shock profiles can be explained by using an equivalent equation. Furthermore, the validity of the equivalent equation method has been proved in [10] for the approximations of discontinuous solutions of a linear equation (see also [9] and [13]). The main purpose of the present paper is to add to the understanding of the validity of the method by determining shock structures of nonlinear equivalent equations.

THREE-POINT SCHEMES IN CONSERVATION FORM

We consider the initial value problem for a scalar conservation law :

$$(1) \qquad u_t + f(u)_x = 0, \qquad u(x,0) = u_0(x), \qquad -\infty < x < \infty .$$

We are interested in weak solutions of (1) and especially in "shocks", discontinuity waves which satisfy the jump relation and also the strict entropy condition. We recall that the "jump relation" [16] is

$$(2) \qquad f(u_l) - f(u_r) = s(u_l - u_r)$$

where s is the speed of propagation of the discontinuity, and u_l and u_r are the states on the left and on the right of the discontinuity, respectively. The strict "entropy condition" [17] can be written as :

$$(3) \qquad \frac{f(u) - f(u_r)}{u - u_r} < s \qquad \text{for } \min(u_r,u_l) < u < \max(u_r,u_l).$$

In order to approximate the weak solutions of (1), we consider a three-point finite difference scheme "in conservation form" [18], [13], i.e.

$$(4) \qquad v_j^{n+1} = v_j^n - \sigma\left[h(v_j^n, v_{j+1}^n) - h(v_{j-1}^n, v_j^n)\right]$$

where $v_j^n = v(j\,\Delta x, n\,\Delta t)$, $\sigma = \Delta t/\Delta x$, Δt and Δx are the time and space increments, and h is a function of two arguments related to f by the condition

$$h(u,u) = f(u) \qquad \text{for all } u.$$

We introduce the Courant number $\eta(u) = \sigma f_u(u)$ where f_u is the first derivative of f, and define $\theta(u) = \sigma[h_u(u,u) - h_v(u,u)]$ where h_u and h_v are the first derivatives of h with respect to its first and second variables, respectively.

Then it can be shown that : (i) scheme (4) is second-order accurate if the functions θ and η^2 are equal or first-order accurate if not, (ii) the linearized form of scheme (4) is L_2 - stable if and only if

$$(5) \qquad \eta^2 \leqslant \theta \leqslant 1$$

Let us now give some examples of schemes in the form (4) useful in what follows.

a) If $h(u,v)=(u-v)/2\sigma + [f(u)+f(v)]/2$, we get the Lax scheme [16] for which $\theta = 1$, and the stability condition (5) becomes $|\eta| \leqslant 1$.

b) If $h(u,v) = f(u)$, we get a one-sided scheme for which $\theta = \eta$, and (5) becomes $0 \leqslant \eta \leqslant 1$.

c) If $h(u,v) = h^{LW}(u,v) \equiv [f(u)+f(v)]/2 - \sigma f_u[(u+v)/2] \cdot [f(v)-f(u)]/2$, we get the Lax-Wendroff scheme [18] for which $\theta = \eta^2$, and (5) becomes $|\eta| \leqslant 1$.

d) If $h(u,v) = h^{LW}(u,v) + (u-v)/8\sigma$, we get the transport stage of the SHASTA algorithm [19] for which $\theta = \eta^2 + 1/4$, and (5) becomes $|\eta| \leqslant \sqrt{3}/2$.

e) Finally if

$h(u,v) = [(\alpha-\beta)f(v)-(1-\alpha-\beta)f(u)]/2\alpha + f\{(1-\beta)u+\beta v-\alpha\sigma[f(v)-f(u)]\}/2\alpha$,

we get the class of $\mathcal{S}_\beta^\alpha$ schemes [6], [8] for which $\theta = \eta^2$, and (5) becomes $|\eta| \leqslant 1$. This class contains all the second-order accurate predictor-corrector schemes. In particular $\mathcal{S}_{1/2}^{1/2}$ is the Richtmyer scheme and $\mathcal{S}_0^1$, $\mathcal{S}_1^1$ are the MacCormack schemes.

SHOCK STRUCTURE FOR FIRST-ORDER SCHEMES

For scheme (4), the equivalent equation up to second-order can be written as :

$$(6) \qquad u_t + f(u)_x = \Delta x\left[d(u)u_x\right]_x$$

where $d = (\theta - \eta^2)/2\sigma$. For a fixed Δx, we seek a "shock structure" of (6), that is a solution of (6) which depends only on $\xi = (x - st)/\Delta x$ and which satisfies the following conditions :

(7) $u \longrightarrow u_l$ (resp. u_r), $\quad u_\xi \longrightarrow 0 \quad$ as $\xi \longrightarrow -\infty$ (resp. $+\infty$)

where u_l, u_r and s are constants. By setting $u = u(\xi)$ in (6) and integrating it with respect to ξ, we obtain :

$$(8) \qquad d(u)u_\xi = f(u) - su + \text{const.}$$

Let us assume that scheme (4) satisfies the stability condition (5) with $\eta^2(u) < \theta(u)$ for $\min(u_r,u_l) \leq u \leq \max(u_r,u_l)$. Then it is easy to show that a shock structure of (6) exists if and only if u_l, u_r and s satisfy the jump relation (2) and the strict entropy condition (3). This shock structure can be analytically determined. In the case where $f(u) = u^2/2$, from (8) and (7) we deduce

$$(9) \qquad \xi_\eta = (\theta - \eta^2)(\eta - \eta_r)^{-1}(\eta - \eta_l)^{-1}$$

where $\eta_r = \sigma u_r$ and $\eta_l = \sigma u_l$. Fig.1 shows good agreement between the solution of (9) and the numerical shock profile for $u_l = 2$, $u_r = 1$, $\eta_l = 1/2$ and for the Lax scheme.

It is interesting to compare the thickness of shock structures of (6) for various first-order accurate schemes. With Δx as unity, this thickness can be defined by $\tau = (u_l - u_r)/\max|u_\xi|$. In the case where $f(u) = u^2/2$, it follows from (9) that :

$$(10) \qquad \tau = (\eta_l - \eta_r) \min_{\eta_r<\eta<\eta_l} \left[(\theta - \eta^2)(\eta - \eta_r)^{-1}(\eta_l - \eta)^{-1}\right]$$

Fig. 2 shows the variation of τ with η_l for $u_l = 2$, $u_r = 1$ and for $\theta = 1$ (Lax scheme), $\theta = \eta$ (one-sided scheme) and $\theta = \eta^2 + 1/4$ (SHASTA transport stage). Only the one-sided scheme has a bounded shock thickness as $\eta_l \longrightarrow 0$. Furthermore it has the smallest shock thickness for any Courant number η_l. Numerical experiments confirm these results.

SHOCK STRUCTURE FOR SECOND-ORDER SCHEMES

When $\theta = \eta^2$, the scheme (4) is second-order accurate and its equivalent equation up to third-order can be written as :

$$(11) \qquad u_t + f(u)_x = \Delta x^2 \left[e_1(u)u_{xx} - e_2(u)b(u)\;(u_x)^2/2 \right]_x$$

where $e_1 = -\eta(1 - \eta^2)/6\sigma$, $e_2 = 1/3 - \eta^2 - \bar{h}_{uv}$, $b(u) = f_{uu}(u)$ and $\bar{h}_{uv}(u) = h_{uv}(u,u)/b(u)$. For example, $\bar{h}_{uv} = 0$ for the Lax-Wendroff scheme and $\bar{h}_{uv} = \left[-\alpha\eta^2 + (2\beta - 1)\eta + \beta(1 - \beta)/\alpha\right]/2$ for $\mathcal{S}_\beta^\alpha$ schemes.

For a fixed Δx, we seek now a "shock structure" of (11), that is a solution of (11) which depends only on $\xi = (x - st)/\Delta x$ and which satisfies the conditions (7) and also : $u_{\xi\xi} \longrightarrow 0$ as $\xi \rightarrow \mp\infty$. We shall suppose that u_l, u_r and s satisfy (2) and (3). By setting $u = u(\xi)$ in (11), integrating it whith respect to ξ and using the conditions at infinity, we obtain :

$$(12) \qquad e_1(u)\, u_{\xi\xi} - e_2(u) b(u)\, (u_\xi)^2/2 = F(u)$$

where $F(u) = f(u) - f(u_r) - s(u - u_r)$. By defining $v = u_\xi$ and considering $v = v(u)$ as the dependent variable, we get :

$$(13) \qquad e_1(u) v v_u - e_2(u) b(u)\, v^2/2 = F(u).$$

By setting $z = v^2/2$, we reduce (13) to the linear equation :

$$(14) \qquad e_1(u) z_u - e_2(u) b(u) z = F(u).$$

Let us now assume that the conservation law (1) is "genuinely nonlinear" that is $b(u) \neq 0$, for all u. Then we may choose the Courant number $\eta = \sigma f_u(u)$ as the independant variable and transform (14) into :

$$(15) \qquad \mathcal{E}_1(\eta) z_\eta + \mathcal{E}_2(\eta) z = \Phi(\eta)$$

where $\mathcal{E}_1(\eta) = \eta(1 - \eta^2)/6$, $\mathcal{E}_2(\eta) = 1/3 - \eta^2 - \bar{h}_{uv}\left[u(\eta)\right]$ and

$$\Phi(\eta) = -F\left[u(\eta)\right]/b\left[u(\eta)\right] .$$

As a result the problem has been reduced to finding a solution of (15) satisfying the boundary conditions :

$$(16) \qquad z(\eta_r) = 0 \qquad \text{and } z(\eta_1) = 0,$$

where $\eta_r = \sigma f_u(u_r)$ and $\eta_1 = \sigma f_u(u_1)$. The strict entropy condition implies that $\eta_r < \eta_1$ and we suppose that $-1 \leq \eta_r$ and $\eta_1 \leq 1$ in order to respect the stability condition (5).

We shall express the solutions of (15) in the case where $\bar{h}_{uv}\left[u(\eta)\right]$ is a polynomial of degree at most 2 in η (this holds for all schemes as yet devised). The general solution of (15) is then

$$(17) \qquad z = K(\eta)\,(1 - \eta)^{\delta_+}\,(1 + \eta)^{\delta_-}\,|\eta|^{-\delta_0}$$

where $\delta_+ = 3\,\mathcal{E}_2(+1)$, $\delta_- = 3\,\mathcal{E}_2(-1)$, $\delta_0 = 6\,\mathcal{E}_2(0)$ and the function K is defined, to within a constant, by its derivative :

$$(18) \qquad K_\eta = 6\,\Phi(\eta)\,|\eta|^{\delta_0}\,\eta^{-1}(1 - \eta)^{-1-\delta_+}\,(1 + \eta)^{-1-\delta_-}$$

For example, let us take $f(u) = u^2/2$. Then we have $\eta = \sigma u$, $b(u) = 1$ and $\Phi(\eta) = (\eta - \eta_r)(\eta_1 - \eta)/2\sigma^2$. Let us now choose the $\mathcal{S}_\beta^\alpha$ scheme corresponding to $\alpha = 1$ and $\beta = 0$ (MacCormack scheme). For this scheme we get $\delta_- = -2$, $\delta_0 = 2$ and $\delta_+ = 1$, so it follows from (17) and (18) that :

$$(19)\quad z = \frac{3(1-\eta)\left[-\frac{1}{3}\eta^3 + \frac{A-3}{2}\eta^2 + (3A-B-5)\eta + C + (5A-3B-7)\mathrm{Log}(1-\eta)\right] + 2(A-B-1)}{\left[\sigma\eta(1+\eta)\right]^2}$$

where $A = \eta_r + \eta_l$, $B = \eta_r\,\eta_l$ and C is an arbitrary constant.

With a single constant C, it is not possible to verify the two boundary conditions (16), except in the case where $\eta_l = 1$ and $0 \leq \eta_r < 1$. In this case, the shock structure is monotonic because, in the phase plane (u,v), the solution-curves of (13) are symmetrical with respect to the u-axis. Fig.3 shows a trajectory in the phase plane corresponding to a shock structure for $\eta_l = 1$ and $\eta_r = 0$. Fig. 4 shows the same shock structure in the physical plane and also the numerical solution (determined for all x). The agreement between the solution of the equivalent equation and the solution of the difference scheme is very good. In the case where $\eta_l < 1$ and $0 \leq \eta_r < \eta_l$, the solution (19) which satisfies $z(\eta_r) = 0$ corresponds to a solitary wave and the numerical solution is an oscillatory shock profile. These solutions are shown in Fig. 5 and 6 for $\eta_l = 1/2$ and $\eta_r = 0$. Let us note that for schemes for which numerical shock profiles are always oscillatory (such the Lax-Wendroff or the Richtmyer scheme) the equivalent equation (11) never has a monotonic shock structure.

CORRECTION TERMS FOR SECOND-ORDER SCHEMES

A second-order accurate scheme generally has oscillatory numerical shock profiles. Our aim is here to add correction terms so that the corrected scheme remains a three-point scheme in conservation form and gives monotonic shock profiles. For the sake of simplicity we shall take $f(u) = u^2/2$. Two cases should be distinguished.

a) In the case where the Courant numbers on each side of the shock are either $0 \leq \eta_r < 1$ and $\eta_l = 1$ or $\eta_r = -1$ and $-1 < \eta_l \leq 0$, it is possible to find correction terms which satisfy the above requirements and which preserve the second-order accuracy. For that, we correct the original scheme in such a way that equivalent equation (11) holds with $e_2(u) = 1/3$. Indeed, for this function e_2, we have $\delta_- = \delta_+ = 1$ and $\delta_0 = 2$, and we can find a solution of (17), (18) which satisfies the boundary conditions (16), for any of the considered values of η_r and η_l. Thus we always obtain a monotonic shock structure. The correction terms are added in the following way. Let h be the function which defines the original scheme in the form (4). Then the corrected scheme is written in the form (4) with the new function $h_2(u,v) = h(u,v) + q_2(u,v)$ where

$$q_2(u,v) = \left[\frac{1}{3} - e_2\left(\frac{v+u}{2}\right)\right](v-u)^2/2 \,. \tag{20}$$

Fig. 7 shows the numerical shock profile of this corrected scheme for $u_l = 2$, $u_r = 1$ and $\eta_l = 1$, with the Lax-Wendroff scheme as the original scheme.

b) In the case where either $\eta_r < 0$ and $\eta_l = 1$ or $\eta_r = -1$, $0 < \eta_l$, or else $-1 < \eta_r < \eta_l < 1$, it is no longer possible to find correction terms which satisfy the above requirements and which preserve the second-order accuracy. So, we look for a corrected scheme of first-order accuracy. Since the shock structure of the one-sided scheme ($\theta = \eta$) has been found to be satisfactory for $\eta > 0$, we shall correct the original scheme in such a way that the equivalent equation (6) holds with $\theta = |\eta|$. Thus the corrected scheme is written in the form (4) with the new function $h_1(u,v) = h(u,v) + q_1(u,v)$ where

$$q_1(u,v) = \left[\sigma\left(\frac{v+u}{2}\right)^2 - \left|\frac{v+u}{2}\right|\right](v-u)/2. \tag{21}$$

Fig.8 shows the numerical shock profile of this corrected scheme for $u_l = 2$, $u_r = 1$ and $\eta_l = 1/2$, with the Lax-Wendroff scheme as the original scheme. This correction should be used only in the vicinity of shocks to preserve the second-order accuracy elsewhere.

REFERENCES

[1] HIRT C. - J. Comput Phys.2 (1968) p.339-355.
[2] YANENKO N. and SHOKIN Yu. - Soviet Math Dokl.9(1968) p. 1153-1155.
[3] SHOKIN Yu. - Proc. Steklov Inst. Math. 122 (1973) p. 67-86.
[4] WARMING R. and HYETT B. - J. Comput. Phys. 14 (1974) p. 159-179.
[5] LERAT A. and PEYRET R. - C.R. Acad. Sc. Paris A-276 (1973) p. 759-762.
[6] LERAT A. and PEYRET R. - C.R. Acad. Sc. Paris A-277 (1973) p. 363-366.
[7] Mc GUIRE G. and MORRIS J. - J. Comput. Phys. 11 (1973) p. 531-549.
[8] LERAT A. and PEYRET R. - Rech. Aérosp. (1975) p. 61-79.
[9] CHIN R. - J. Comput. Phys. 18 (1975) p. 233-247.
[10] HEDSTROM G. - Math. Comput. 29 (1975) p. 969-977.
[11] ZWAS G. and ROSEMAN J. - J. Comput. Phys. 12 (1973) p. 179-186.
[12] FOMIN V., VOROZHTSOV E. and YANENKO N. - Comput.Fluids 4 (1976) p. 171-183.
[13] HARTEN A., HYMAN J. and LAX P. - Comm. Pure Appl. Math. 29 (1976) p. 297-322.
[14] HARTEN A. - Comm. Pure Appl. Math. 30 (1977) p. 611-638.
[15] MAJDA A. and OSHER S. - To appear.
[16] LAX P. - Comm. Pure Appl. Math. 7 (1954) p. 159-193.
[17] OLEINIK O. - Usp. Mat. Nauk (1959). AMS Trans. 2-33 (1963) p. 285-290.
[18] LAX P. and WENDROFF B. - Comm. Pure Appl. Math. 13 (1960) p. 217-237.
[19] BORIS J. and BOOK D. - J. Comput. Phys. 11 (1973) p. 38-69.

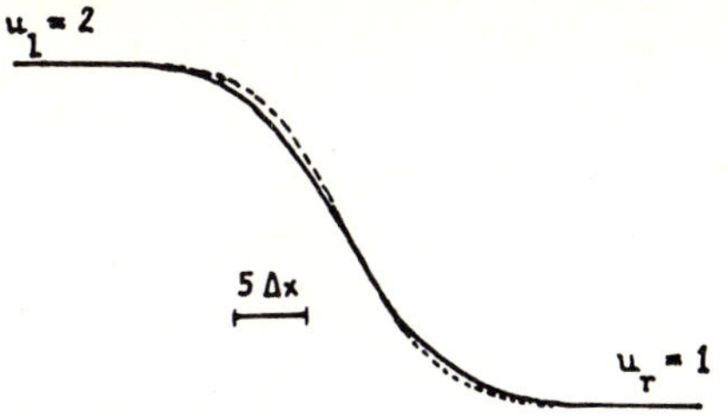

Fig. 1. Shock structure of (6) : —— and numerical shock profile : ------- for Lax scheme with $\eta_1 = 1/2$.

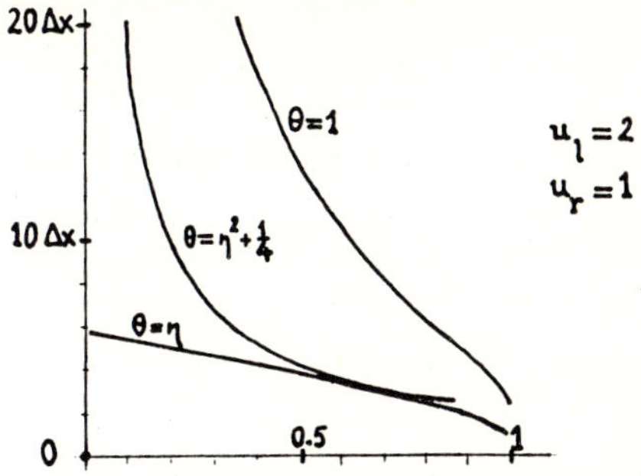

Fig. 2. Shock thickness τ versus η_1

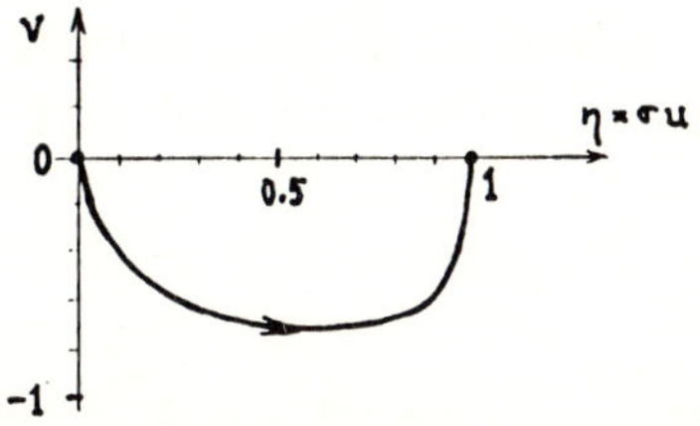

Fig. 3. Shock structure of (11) in the phase plane for MacCormack scheme ($\eta_l = 1$)

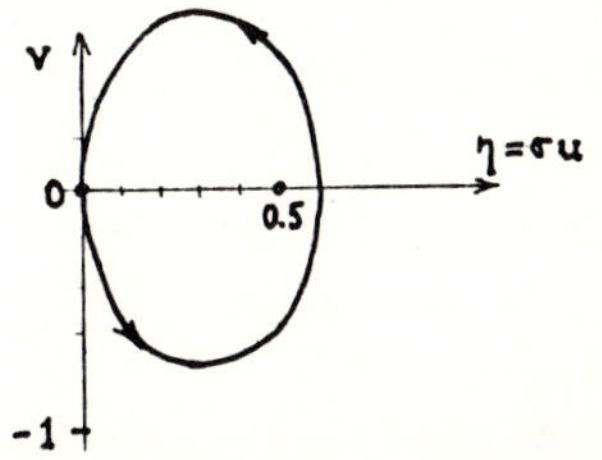

Fig. 5. Solitary wave of (11) in the phase plane for MacCormack scheme ($\eta_l = \frac{1}{2}$)

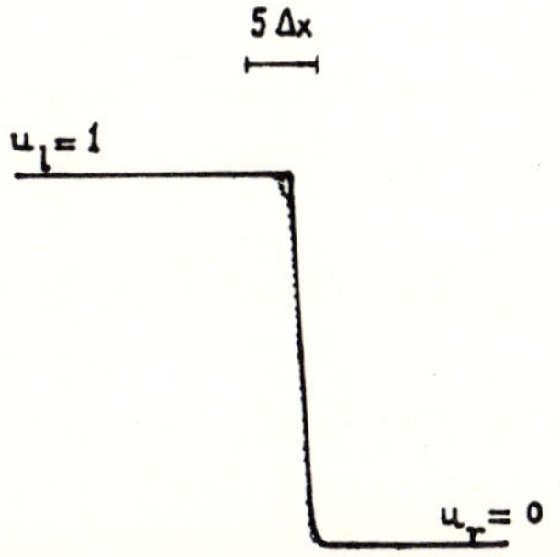

Fig. 4. Shock structure of (11) : —— and numerical shock profile : ------- for MacCormack scheme with $\eta_l = 1$.

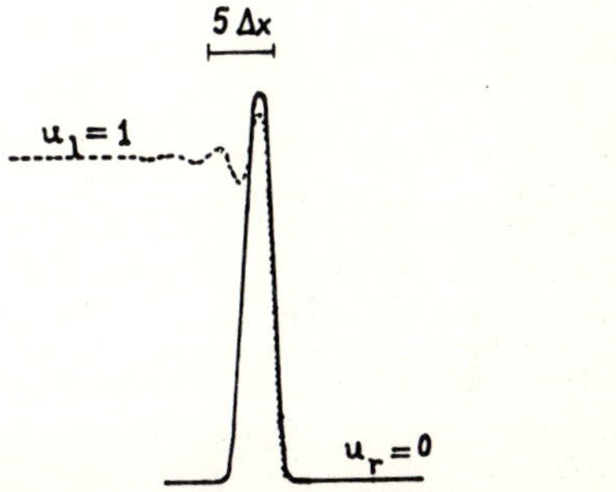

Fig. 6. Solitary wave of (11) : —— and numerical shock profile : ------- for MacCormack scheme with $\eta_l = 1/2$.

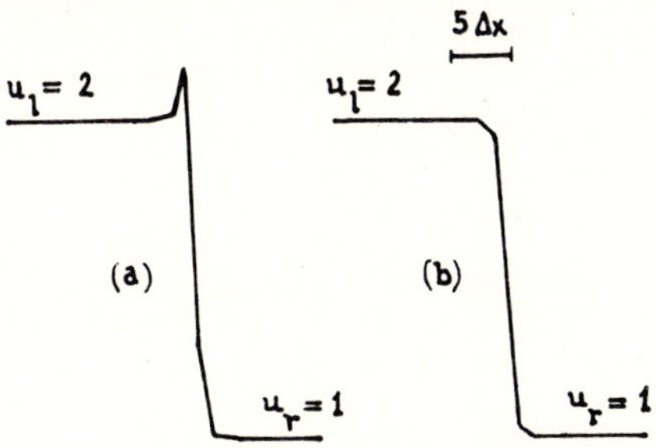

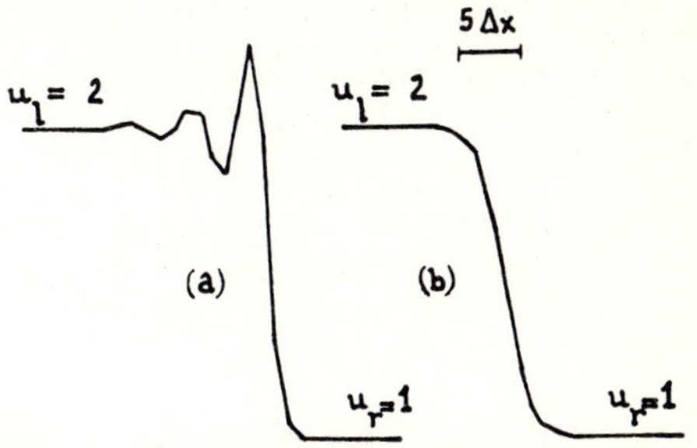

Numerical shock profiles of Lax-Wendroff scheme

Fig. 7. $\eta_l = 1$ (a) without correction (b) with 2nd order correction

Fig. 8. $\eta_l = 1/2$ (a) without correction (b) with 1st order correction

NUMERICAL SIMULATION OF EXPLOSION PHENOMENA TAKING INTO ACCOUNT NON-EQUILIBRIUM PHYSICAL-CHEMICAL PROCESSES

V.A.Levin, V.V.Markov, S.F.Osinkin, J.V.Tunik

Institute of Mechanics of Moscow State University and Steklov Mathematical Institute

Moscow

At present large achievements in development of numerical methods make it possible to investigate complicated gas dynamics processes taking into account many real properties of medium and physical-chemical transformations in it. A specially interesting subject from the scientific and practical point of view is gas dynamics flows which appear under concentrated energy input (under explosion) and go along with chemical reactions and different relaxation processes.

Among the numerical methods of solving gas dynamics problems the splitting method find great application. The theoretical basis of this method was worked out by N.N.Janenko [1]. One of the modifications of the splitting method is wellknown S.K.Godunov's method [2]. The calculation technique of explosive gas dynamics flows was worked out on the scheme basis of this method. The technique takes into account real processes in medium.

The problem of detonation initiation by explosion in hydrogen-chorine mixture has been investigated. The analoguous problem of explosive initiation of detonation in hydrogen-air mixture with model description of chemical reaction processes has been previously considered in detail in [3, 4] in which the worked-out technique was tried.

The system of equations describing one-dimensional non-stationary flow with plane, cylindrical and spherical waves in gas mixture H_2-Cl_2 looks like:

$$\frac{\partial(r^{\nu-1}\rho)}{\partial t}+\frac{\partial(r^{\nu-1}\rho v)}{\partial r}=0$$

$$\frac{\partial(r^{\nu-1}\rho v)}{\partial t}+\frac{\partial(r^{\nu-1}(\rho v^2+p))}{\partial r}=(\nu-1)\,p\,r^{\nu-2} \qquad (1)$$

$$\frac{\partial(r^{\nu-1}(\rho v^2/2+\rho h-p))}{\partial t}+\frac{\partial(r^{\nu-1}\rho v\,(v^2/2+h))}{\partial r}=0$$

The process of chemical interaction is described by the following reactions:

$$Cl_2 + M \underset{K_2}{\overset{K_1}{\rightleftharpoons}} Cl + Cl + M \qquad H_2 + M \underset{K_8}{\overset{K_7}{\rightleftharpoons}} H + H + M$$

$$Cl + H_2 \underset{K_4}{\overset{K_3}{\rightleftharpoons}} HCl + H \qquad HCl + M \underset{K_{10}}{\overset{K_9}{\rightleftharpoons}} H + Cl + M \qquad (2)$$

$$H + Cl_2 \underset{K_6}{\overset{K_5}{\rightleftharpoons}} HCl + Cl$$

Here M denotes the third particle and may specifically denote the particle of deluent for the mixtures deluted by the inert gases. K_j - constants of rates of the corresponding elementary reactions. Their values for the direct reactions (j = 1, 3, 5, 7, 9) were obtained from 5 and the values for the back reactions (j = 2, 4, 6, 8, 10) were determined with the help of equilibrium constants. Here are their expressions:

$$K_1 = 10^{20.96} T^{-3/2} \exp(-57060/RT)$$
$$K_2 = 10^{17.54} T^{-0.77} \exp(1404/T)$$
$$K_3 = 10^{13.8} \exp(-5900/RT)$$
$$K_4 = 10^{13.66} T^{-0.027} \exp(-2472/T)$$
$$K_5 = 10^{15.10} \exp(-2200/RT) \qquad (3)$$
$$K_6 = 10^{13.62} T^{0.307} \exp(-24023/T)$$
$$K_7 = 10^{17.47} T^{-1/2} (1 - \exp(-6000/T)) \exp(-103250/RT)$$
$$K_8 = 10^{15.39} T^{-1} \exp(1710/T)$$
$$K_9 = 10^{13.62} \exp(-81000/RT)$$
$$K_{10} = 10^{11.67} T^{0.42} \exp(12357/T)$$

Dimensions of constants:

$$[K_j] = \begin{cases} cm^3/sec \cdot mole, & j = 1, 3, 4, 5, 6, 7, 10 \\ cm^6/sec \cdot mole^2, & j = 2, 8, 10 \end{cases}$$

If we denote $n_i = \alpha_i/\mu_i$, where α_i - is a mass share of the component i , and μ_i - its molecular weight, than the equations of chemical kinetics will look like:

$$\frac{dn_1}{dt} = \dot{y}_{11} + \dot{y}_{13}$$

$$\frac{dn_2}{dt} = \dot{y}_{22} + \dot{y}_{24}$$

$$\frac{dn_3}{dt} = -2\dot{y}_{11} + \dot{y}_{22} - \dot{y}_{13} + \dot{y}_{35} \qquad \frac{dn_4}{dt} = \dot{y}_{13} - 2\dot{y}_{24} + \dot{y}_{35} - \dot{y}_{22} \tag{4}$$

$$\frac{dn_5}{dt} = -\dot{y}_{22} - \dot{y}_{13} - \dot{y}_{35}$$

Here

$$\dot{y}_{11} = \rho n\,(-K_1 n_1 + \rho K_2 n_3^2)$$

$$\dot{y}_{13} = \rho\,(-K_5 n_1 n_4 + K_6 n_3 n_5)$$

$$\dot{y}_{22} = \rho\,(-K_3 n_2 n_3 + K_4 n_4 n_5)$$

$$\dot{y}_{24} = \rho n\,(-K_7 n_2 + \rho K_8 n_4^2)$$

$$\dot{y}_{35} = \rho n\,(-\rho K_{10} n_3 n_4 + K_9 n_5)$$

The indexes 1, 2, 3, 4, 5 relate resp. to Cl_2, H_2, Cl, H and HCl.

Two equations of this system may be replaced by the final relations, expressing the conservation of atoms of chlorine and hydrogen.

$$\mu_1 n_1 + \mu_3 n_3 + \mu_3 n_5 = 1 - \beta$$

$$\mu_2 n_2 + \mu_4 n_4 + \mu_4 n_5 = \beta$$

where β is constant depending upon the initial composition. Besides we believe the Claiperon's equation

$p = \rho R T n, \; n = \sum_{i=1}^{6} n_i$ to be true, n_6 is a concentration of inert diluent and the full enthalpy is equal to the corresponding sum of partial enthalpies:

$$h = \sum_{i=1}^{6} n_i h_i$$

The equations (1) are written down in terms of non-dimensional variables:

$$\bar{p} = p/\rho_0 Q, \qquad \bar{\rho} = \rho/\rho_0, \qquad \bar{h} = h/Q$$

$$\bar{v} = v/\sqrt{Q}, \qquad \bar{t} = t/t^*, \qquad \bar{r} = r/t^*\sqrt{Q}$$

The line over the non-dimensional variables is omitted and we assume:

$$\rho_0 = 10^{-3}\ г/см^3, \quad Q = 10^7\ кал/кг, \quad t^* = 10^{-7}\ сек$$

Detonation wave is a structure consisting of shock wave and the following chemical reaction zone. On the front side of the shock wave propagating over motionless gas the lows of conservation are true

$$\begin{aligned} \rho_0 D &= \rho_1 (D - v_1) \\ \rho_0 D^2 &= \rho_1 (D - v_1) + p_1 \\ D^2/2 + h_0 &= (D - v_1)^2/2 + h_1 \end{aligned} \qquad (5)$$

where D is the wave velocity and the indexes 0 and 1 correspond resp. to the parameters before and behind the wave. Energy input causing medium motion with detonation wave formation has been simulated by instantaneous increase of temperature and pressure in some domain near by the centre of symmetry of one-dimensional flow under consideration. The difference equations of chemical kinetics have been obtained from differential ones in divergent forms; the initial reaction rates have been calculated by parameters on the known time layer. The partial enthalpy values have been calculated on the basis of linear approximations of its table values [6]. The collapse of arbitrary discontinuity has been calculated without taking into account chemical transformation influence in frozen composition. The whole system of difference equations became explicit and time was limited both by Courant's condition and by the condition of relatively small increase of species concentrations because of chemical reactions. The main difficulty appeared when chosing the optimal difference net. In present case the choice was determined by the presence of narrow zone of intensive chemical transformation behind the initiating shock wave. In the process of counting the mobile difference net connected with the shock wave contained a constant number of knots a part of which always lay in that narrow zone. The optimal net parameters were chosen on the basis of series of numerical experiments and the described technique permitted to calculate with good accuracy the propagation of detonation wave, appearing in the result of instant release of energy. It is well known that for explosive initiation of detonation the quantity of releasing energy should be larger than some critical value [3, 4, 7, 8] , which is an important characteristic of any mixture. The value of critical energy strongly depends on the initial values of its thermodynamical parameters (composition, temperature and pressure).

In the result of calculations these values for all the types of flow symmetry and different initial conditions were determined. The graph in Fig. 1 shows the dependence of critical explosion energy upon the initial mixture pressure, under the rest of the conditions for the spherical symmetry being fixed. The configuration agrees with the theoretical dependence $E \sim P_0^{-2}$, which was earlier obtained by the authors on the basis of analytical analysis 4 .

Fig. 2 shows the dependence of initiation energy upon the value of molar share rH_2 of hydrogen mixture under one and the same initial pressure $P_0 = 0.3$ atm. It was found out that the detonation can't be initiated in mixtures, in which the molar share of hydrogen is less than 15 % or more than 80 % .

S.K.Godunov's method permits to simulate physical processes of another nature. The investigation was held on the possibility of determination of inversion population of carbonic acid gas oscillation levels behind the axis-symmetrical explosive wave propagating over motionless gas mixture $CO_2 + N_2 + He$ [9, 10]. A particle of metionless gas, carried on by the explosive wave, heats and then intensively cools down in non-stationary rarity wave. Under sufficiently quick and effective cooling the inversion of population between two oscillation levels, first antisymmetric and first symmetric, can appear in a particle.

For the description of non-equilibrium processes of oscillating energy transfer and relaxations a known technique based on the difference between the rates of processes was used [11]. In this case great transfer rates of oscillation energy within each type of oscillation permit to introduce non-equilibrium temperatures for each mode - T_1, T_2, T_3, T_4 , and values of mean quantity of quants of certain oscillation type for one molecula, connected with these temperatures $e_i = (\exp \theta_i/T_i - 1)^{-1}$, where θ_i - characteristic temperature of i-th oscillation type. The change of values of e_i or T_i describes interaction between different modes and relaxation of oscillation energy into translational.

Known mathematical model of these processes was used in calculations. It gives the following equations for e_i [11]:

$$
\begin{aligned}
&e_1 = e_2^2 / (e_2 + 1) \\
&\frac{de_2}{dt} = (pQ_{23}\varphi_{23} - p(e_2 - e_2')/\tau')\lambda \\
&\frac{de_3}{dt} = -pQ_{23}\varphi_{23} - \xi_4 p Q_{34}\varphi_{34} \qquad\qquad (6) \\
&\frac{de_4}{dt} = \xi_3 p Q_{34}\varphi_{34} - p(e_4 - e_4')/\tau'' \\
&\lambda = (1+2e_2)^2/(1+6e_2+6e_2^2), \qquad e_i' = e_i(T) \\
&\varphi_{ij} = \begin{cases} e_j(e_i+1)^3 - \exp\left(-\frac{\theta_j - 3\theta_i}{T}\right)(e_j+1)e_i^3, & ij = 23 \\ e_j(e_i+1)\exp\left(\frac{\theta_j - \theta_i}{T}\right) - e_i(e_j+1), & ij = 34 \end{cases}
\end{aligned}
$$

Here indexes 1, 2, 3 relate resp. to symmetrical, deformational, antisymmetrical oscillation modes of CO_2 , 4 - to nitrogen, 5 - to helium. P - pressure, Q_{ij} - transition probabilities under collision of moleculas of i-th and j-th types, which are known functions of temperature T ; $\xi_1 = \xi_2 = \xi_3$ - molar concentration of carbonic acid gas, ξ_4 - of nitrogen, ξ_5 - of helium. In equations (1) the expression for enthalpy in this case has the following form:

$$
h = \frac{\gamma}{\gamma-1}RT + \sum_1^4 g_i Q_i R \xi_i e_i , \quad g_i = \begin{cases} 1 & i \neq 2 \\ 2 & i = 2 \end{cases}
$$

In this case the variables are not concentrations ξ_i but mean values of oscillation energy e_i . Release of energy E_0 was simulated by instant formation of cylindric domain with the size L_0 high pressure P_* and temperature T_* .

In the result of collapse of the desturbed domain the axissymmetrical shock wave propagates over motionless working gas mixture. The processes of inversion formation which we are interested in occur between shock wave and contact surface which seperates energy release products from the working gas. In order to fulfil calculations, relaxation equations as well as gasdynamic equations are reduced to divergent form and are nummerically solved according to the technique described above. In the process of counting the position of surface of contact discontinuity was determined by matching one knot of the mobile calculation net with the contact surface or with the help of determination of the boundary of that part of stream, in which the gas mass equals to the mass of energy release products.

The mobile net was determined in the following way. During the whole calculation period the ratio of lengths of calculation cells on every time layer was fixed. The length of some i-th calculation cell on the next layer was defind as a product of distance of the wave r_s from the axis of symmetry and the i-th number of the fixed row. The choice of the number row was made in such a way that the number of the cells grew near the shock wave and their sizes diminish respectively. The dependence of population inversion levels on determining parameters of the problem P_0, E_0, T_0, L_0/L ($L = \sqrt{E_0/P_0}$) and on conditions for the non-equilibrium temperatures on the shock wave was investigated. Fig. 3 shows the dependence of intensification coefficient under temperatures $T_0 = 293^0K$ and 150^0K on Mach number M_s of wave when other parameters are fixed. Maximal values of the coefficient are observed when Mach numbers range from 1, 2 to 1, 6 and these values are reached in particles carried on by the wave when $M_s \sim 5$. If compared to the particles carried on the earliear stage of flow development ($M_s \gg 5$) they are cooled down to lower temperatures which provide inversion appearance. When the initial temperature decreases the maximal inversion value increases as it is shown in Fig. 3. Fig. 4 shows the parameters distribution when $M_s = 1.27$ for two variants with $T_0 = 293^0K$ and $T_0 = 150^0K$ and for equilibrium conditions on the wave. The inversion zone constitute a significant part of the whole perturbed (or disturbed) domain which increases when the temperature of non-perturbed gas decreases.

The worked-out technique of numerical calculations of gas-dynamic flows where the real physico-chemical processes going on are taken into account, and which is based on Godunov's scheme, may be effectively used when investigating and describing complecated physical phenomena in liquid and gas mechanics.

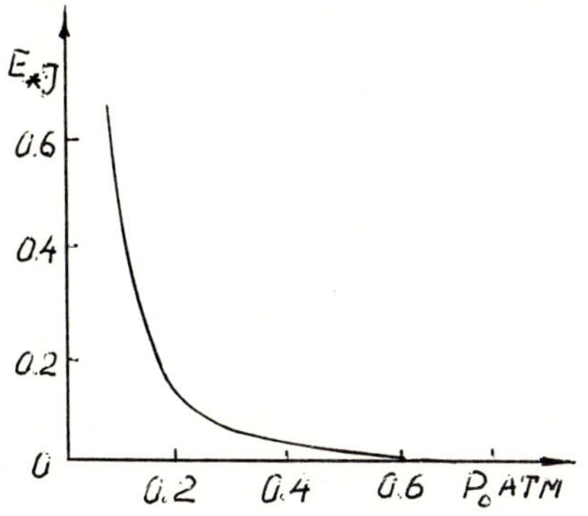

Fig. 1. Dependence of critical energy upon the initial pressure.

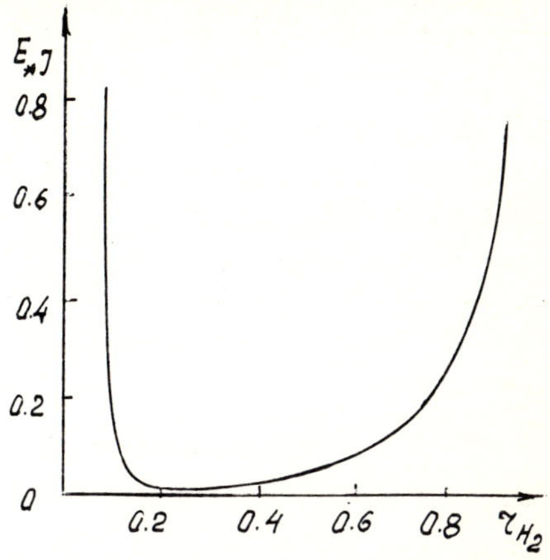

Fig. 2. Dependence of critienergy upon the value of molar share of hydrogen.

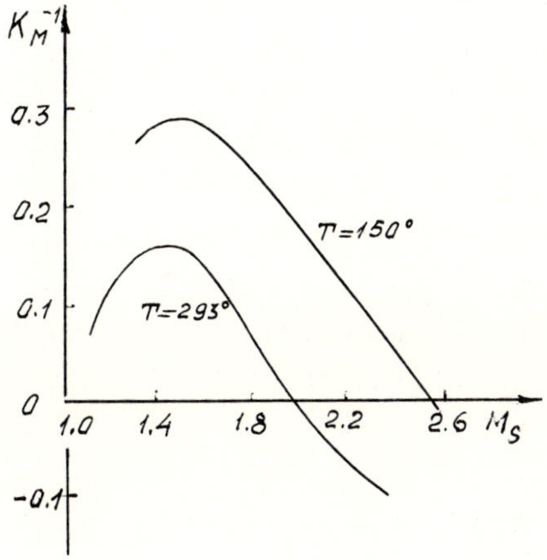

Fig. 3. Dependence of intensification coefficient upon Mach number.

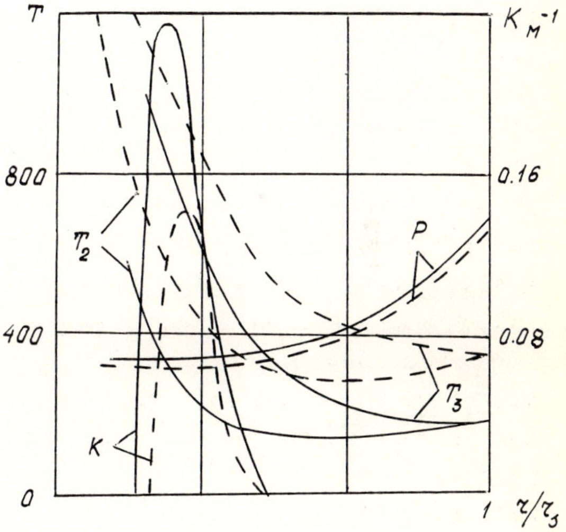

Fig. 4. Distributions of parameters for equilibrium conditions on the wave.

REFERENCES

1. Yanenko N.N. Metod drobnyh shagov reshenija mnogomernyh zadach matematicheskoi fiziki. Novosibirsk, 1967.

2. Godunov S.K., Zabrodin A.V., Prokopov G.P. Raznostnaja skhema dlja dvumernyh nestatsionarnyh zadach gazovoi dinamiki i raschet obtekanija s otoshedshei udarnoi volnoi. Jurnal vychislitelnoi matematiki i matematicheskoi fiziki, 1961, no 6.

3. Levin V.A., Markov V.V. O vozniknovenii detonatsii pri kontsentrirovannom podvode energii. Izvestija AN SSSR, MJG, 1974, no 5.

4. Levin V.A., Markov V.V. Vozniknovenie detonatsii pri kontsentrirovannom podvode energii, Fizika gorenija i vzryva, 1975, no 4.

5. Kondratjev V.N. Konstanty skorosti gazofaznyh reaktsyi, Spravochnik, M., "Nauka", 1970.

6. Gurvich L.V. i dr. Termodinamicheskie svoistva individualnyh veschestv, t. 2, M., Izdatelstvo AN SSSR, 1962.

7. Lee J.H., Knystautas R. Laser spark iginition of chemically reactive gases. AIAA J., 7 (1969), no 2.

8. Korobeinikov V.P. The problem of point explosion in a detonating gas, Astronautica Acta, 1969, no 14.

9. Lee J.H., Bui T.D., Knystautas R. Population inversion in blast waves, Appl. phys. letters, 22 (1973), no 9.

10. Levin V.A., Tunik Yu.V. Koeffitsient opticheskogo usilenija za osesimmetrichnoi udarnoi volnoi, Fizika gorenija i vzryva, 1977, no 3.

11. Losev S.A., Makarov V.N., Pavlov V.A., Shatalov O.G. Issledovanie protsessov v gazodinamicheskom lazere na udarnoi trube bolshogo diametra, Fizika gorenija i vzryva, 1973, no 4.

CALCULATION OF THREE-DIMENSIONAL TURBULENT SUBSONIC FLOWS IN TRANSITION DUCTS*

Ralph Levy, Henry McDonald and W. Roger Briley

Scientific Research Associates, Inc.
Glastonbury, Connecticut 06033, U.S.A.

ABSTRACT

A method for computing three-dimensional turbulent subsonic flow in curved ducts is being developed. A set of tube-like surface oriented coordinates is employed for a general class of geometries applicable to subsonic diffusers with offset bends. The geometric formulation is complex and no previous treatment of this class of viscous flow problems is known to the authors. The duct centerline is a space curve specified by piecewise polynomials. A Frenet frame is located on the centerline at each axial location. The cross sections are described by super-ellipses imbedded in the Frenet frame. Duct surfaces are also coordinate surfaces, which greatly simplifies the boundary conditions. The resulting coordinates are nonorthogonal.

An approximate set of governing equations is employed for viscous flows having strong flow in a primary flow direction. The derivation is coordinate invariant and the resulting equations are expressed in tensor form. These equations are solved by an efficient alternating direction implicit (ADI) method. This numerical method is generally stable and permits solution in difficult geometries using the general tensor formulation.

ANALYSIS

Subsonic flow is known to be governed by equations which are elliptic; that is, by equations which require downstream boundary conditions. Solution by forward marching integration is not appropriate, at least not without some sort of iterative procedure to satisfy the downstream boundary conditions. To circumvent this problem for subsonic flows, so-called "parabolic flow assumptions" are used and it is thus assumed that the pressure field appropriate for irrotational inviscid flow through the passage represents a given, reasonable first approximation to the actual pressure field. The parabolic flow assumptions also presume a primary flow in a predominantly streamwise flow direction and a Reynolds number sufficiently high to permit the assumption of negligible streamwise diffusion. The inviscid axial pressure gradients computed with appropriate downstream boundary conditions are "imposed" upon the flow, much as in conventional boundary layer theory, so as to permit solution by forward marching integration for subsonic flows (cf., Ref. 1). The forward marching integration requires there be no flow reversal in the primary flow direction. For internal flows, the inviscid pressure gradients are corrected for internal flow losses associated with the well-known viscous pressure drop and blockage effects by a process which is consistent with forward marching integration. The imposition of inviscid pressure gradients incorporates *a priori* the elliptic effects associated with a subsonic pressure field without the necessity of solving elliptic equations other than for an inviscid flow.

*This work was sponsored by NASA Lewis Research Center under Contracts NAS3-17522 and NAS3-19856.

In conservation law form and in general curvilinear coordinates, the time average Navier-Stokes equations from Ref. 2 are:

$$\frac{\partial}{\partial y^i}\left(\bar{\rho}\,\bar{v}^i\sqrt{g}\right)=0$$
$$\frac{\partial}{\partial y^j}\left[\left(\bar{\rho}\bar{v}^i\bar{v}^j + g^{ij}\bar{P} + \bar{\omega}^{ij}\right)\frac{\partial x^s}{\partial y^i}\sqrt{g}\right]=0 \tag{1}$$

where the y^i 's are the curvilinear coordinates, the $\bar{v}^i$'s are the curvilinear velocity components, $\bar{\rho}$ is the density, $\bar{\omega}^{ij}$ is the viscous portion of the stress tensor, $\bar{P}$ is the pressure, g^{ij} is the inverse of the metric tensor and g = Det (g_{ij}) and density fluctuations are neglected. In the present application, the stress tensor was modified to neglect streamwise diffusion (Ref. 3) and the pressure was modified by parabolic flow assumptions (Ref. 2).

At present a mixing length model is being used which employs the following eddy-viscosity formulation for the Reynolds stresses:

$$\rho\,\overline{v^i v^j} = \frac{\mu_T}{Re}\,\frac{\partial v^j}{\partial x_i} \tag{2}$$

The mathematical form of the expression for the turbulent viscosity follows that of Beer and Chigier, Ref. 4:

$$\frac{\mu_T}{Re} = \rho \ell^2 (2\bar{\bar{e}}:\bar{\bar{e}})^{1/2} \tag{3}$$

where $\bar{\bar{e}}$ is the mean flow rate of strain tensor

$$\bar{\bar{e}} = 1/2\left[(\nabla\vec{v}) + (\nabla\vec{v})^T\right] \tag{4}$$

The mixing length ℓ is determined from the empirical relationship of McDonald & Camarata (Ref. 5) for equilibrium turbulent boundary layers which can be written

$$\ell(\tilde{y}) = 0.09\,\delta_b \tanh\left[\kappa\,\tilde{y}/(0.09\,\delta_b\right]\cdot\mathcal{D} \tag{5}$$

where δ_b is the local boundary layer thickness, κ is the von Karman constant, taken as 0.43, y is distance from the wall, and $\mathcal{D}$ is a sublayer damping factor defined by

$$\mathcal{D} = p^{1/2}(y^+ - \overline{y^+})/\sigma \tag{6}$$

where P is the normal probability function, $y^+ = \tilde{y}(\tau/\rho)^{1/2}/(\mu/\rho)$, τ is local shear stress, $\overline{y}^+ = 23$, and $\sigma_1 = 8$.

The numerical method presently under development follows the general approach developed by McDonald & Briley (Ref. 6) for laminar supersonic flow in rectangular jets. A detailed discussion of the calculation procedure is not included here, as such a discussion would be lengthy, and discussions of the general approach being followed are given elsewhere (Refs. 6 and 7). The method is based on implicit schemes which are potentially stable for large step sizes. Thus, as a practical matter, stability restrictions which limit the axial step size relative to the transverse mesh spacing and which become prohibitive for locally refined meshes are not a factor in making the calculations. The general approach is to employ an implicit difference formulation and to linearize the implicit equations by expansion about the solution at the most recent axial location. Terms in the difference equations are then grouped by coordinate direction and one of the available alternating direction implicit (ADI) or splitting techniques is used to reduce the multidimensional difference equations to a sequence of one-dimensional equations (Refs. 8 and 9). These linear one-dimensional difference equations can be written in block-tridiagonal or a closely related matrix form and solved efficiently and without iteration by standard block elimination techniques. The general solution procedure is flexible in matters of detail such as the type and order accuracy of the difference approximations and the particular scheme for splitting multidimensional difference approximations. By solving the equations fully coupled and by employing a formal linearization procedure the application of coupled implicit boundary conditions is straightforward. The use of coupled implicit boundary equations played a significant role in the development and implementation of this analysis.

This analysis in its general form has been tested by application to the flows in ducts whose centerlines are specified by piecewise polynomial functions. Cross sections of the ducts in planes normal to this centerline are described by the equation of a superellipse (in local cartesian coordinates):

$$b^a |x|^a + |y|^a = m^a \tag{7}$$

where a is the superelliptic exponent, b is the ratio of major to minor axes (sometimes called the shape factor), and m is the length of the minor axis.

COMPUTATIONS

A series of test calculations has been performed to assist in the development of the analysis and computer program. Flow in the entrance region of a circular straight pipe was computed at a Reynolds number of 500. At an inlet Mach number of 0.01 the results, Fig. 1, compared very favorably with incompressible experimental data, Ref. 10. For a pipe of noncircular cross section at a Reynolds number of 500, the inlet Mach number was progressively increased from 0.01 to 0.1, 0.2 and 0.3 and the expected effects of compressibility were observed. For the case with an entrance Mach number of 0.3 the calculation proceeded until the flow choked, Fig. 1. This behavior corresponds to that expected from solution of the initial value problem posed. Additional calculations using a variety of cross sections have also been performed including elliptical cross sections of shape factors, b, from 1.0 to 2.0 and superelliptical cross sections of superelliptic exponent (a in Eq. 7) from 2.0 to 10.0, shown in Fig. 2. Three of these are shown in Fig. 2.

A series of calculations has been performed for flow through a duct which undergoes an S-shaped bend, as in Fig. 3. These cases were run at a Reynolds number of 500 and an inlet Mach number of 0.1. Four different duct cross sections were used: circular, elliptic with shape factor of 1.5, elliptic with shape factor of 0.667, and superelliptic of exponent 10.0. Although the required inviscid pressure field can be generated by numerous methods, in the S-shaped bend cases considered here a suitable three-dimensional inviscid pressure field was not yet available. The inviscid pressures were generated at each axial station by a highly approximate method based on a free vortex pressure distribution using the local radius of curvature of the centerline. Some sample cross flows near the start of the first bend are presented in Fig. 4. Turbulent flow through the tighter S-shaped bend of Fig. 3 was computed at a Reynolds number of 10^5. Several laminar and turbulent calculations were also performed for transitions from a nearly square to round cross section as shown in Fig. 5.

CONCLUSIONS

A method for computing three-dimensional turbulent subsonic flow in curved ducts has been outlined. A surface-oriented coordinate system is used for a general class of geometries applicable to subsonic diffusers with offset bends. The approximate set of governing equations are applicable to viscous flows having strong flow in a primary flow direction. The formulation is coordinate invariant and the resulting equations are expressed in tensor form. These equations are solved by an efficient alternating direction implicit (ADI) method. This numerical method is generally stable and permits solution in difficult geometries using the general tensor formulation. Further development of the analysis and numerical method appears warranted by the results described here.

REFERENCES

1. Briley, W. R., Numerical Method for Predicting Three-Dimensional Steady Viscous Flow in Ducts. Journal of Computational Physics, Vol. 14, 1974, p. 8.

2. Eiseman, P. R., R. Levy, H. McDonald and W. R. Briley, Development of a Three-Dimensional Turbulent Duct Flow Analysis. NASA CR-3029, 1978.

3. Eiseman, P. R., A Unification of Unidirectional Flow Approximations. Sixth International Conference on Numerical Methods in Fluid Dynamics, 1978.

4. Beer, J. M. and N. A. Chigier, Combustion Aerodynamics. Wiley, New York, 1972.

5. McDonald, H. and F. J. Camarata: An Extended Mixing Length Approach for Computing the Turbulent Boundary-Layer Development. In Proceedings, Stanford Conference on Computation of Turbulent Boundary Layers, Vol. I, Published by Stanford University, 1969, pp. 83-98.

6. McDonald, H. and W. R. Briley: Three-Dimensional Supersonic Flow of a Viscous or Inviscid Gas. Journal of Computational Physics, Vol. 19, No. 2, 1975, p. 150.

7. Briley, W. R. and H. McDonald, Computation of Three-Dimensional Turbulent Subsonic Flow in Curved Passages. United Technologies Research Center Report R75-911596-8, 1975.

8. Douglas, J. and J. E. Gunn, Numer. Math, Vol. 6, 1964.

9. Yanenko, N. N., The Method of Fractional Steps, Springer-Verlag, New York, 1971.

10. Reshotko, E., Experimental Study of the Stability of Pipe Flow, Pt. 1 Establishment of an Axially Symmetric Poiseuille Flow. Jet Propulsion Laboratory Progress Report No. 20-364, October 1958.

ACKNOWLEDGEMENT

The authors are indebted to Mr. B. Anderson of NASA Lewis Research Center for kindly constructing the plots presented in Figs. 2, 3, and 5.

FIGURE I

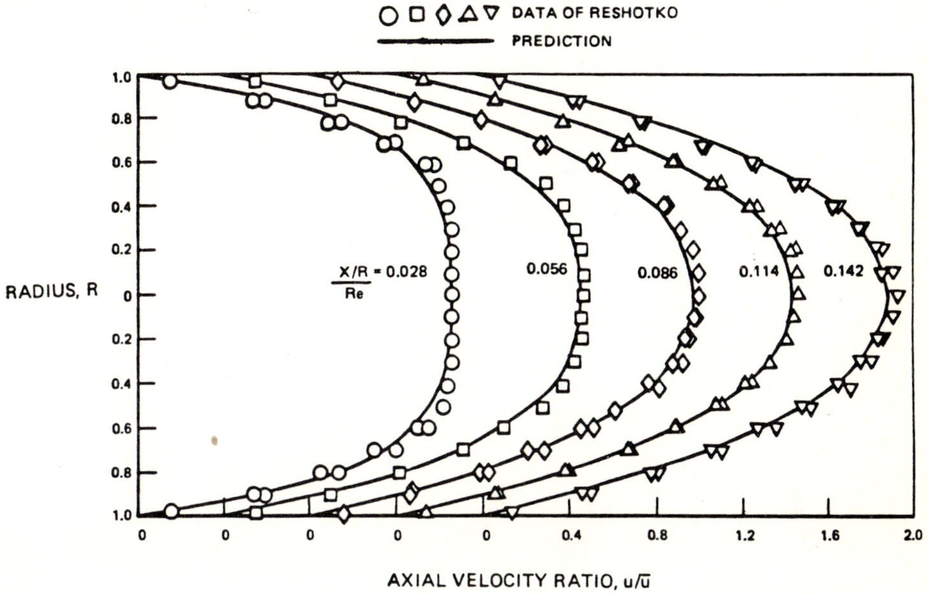

Velocity Profiles in the Entrance Region of a Circular Pipe

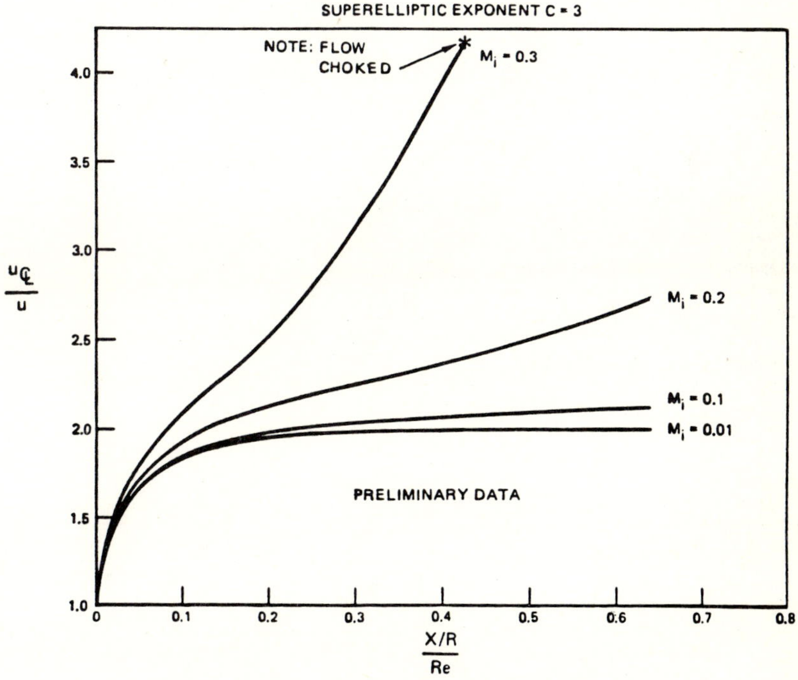

Effect of Mach Number on Centerline Velocity Ratio

FIGURE 2

Entrance Region to an Elliptic Straight Pipe

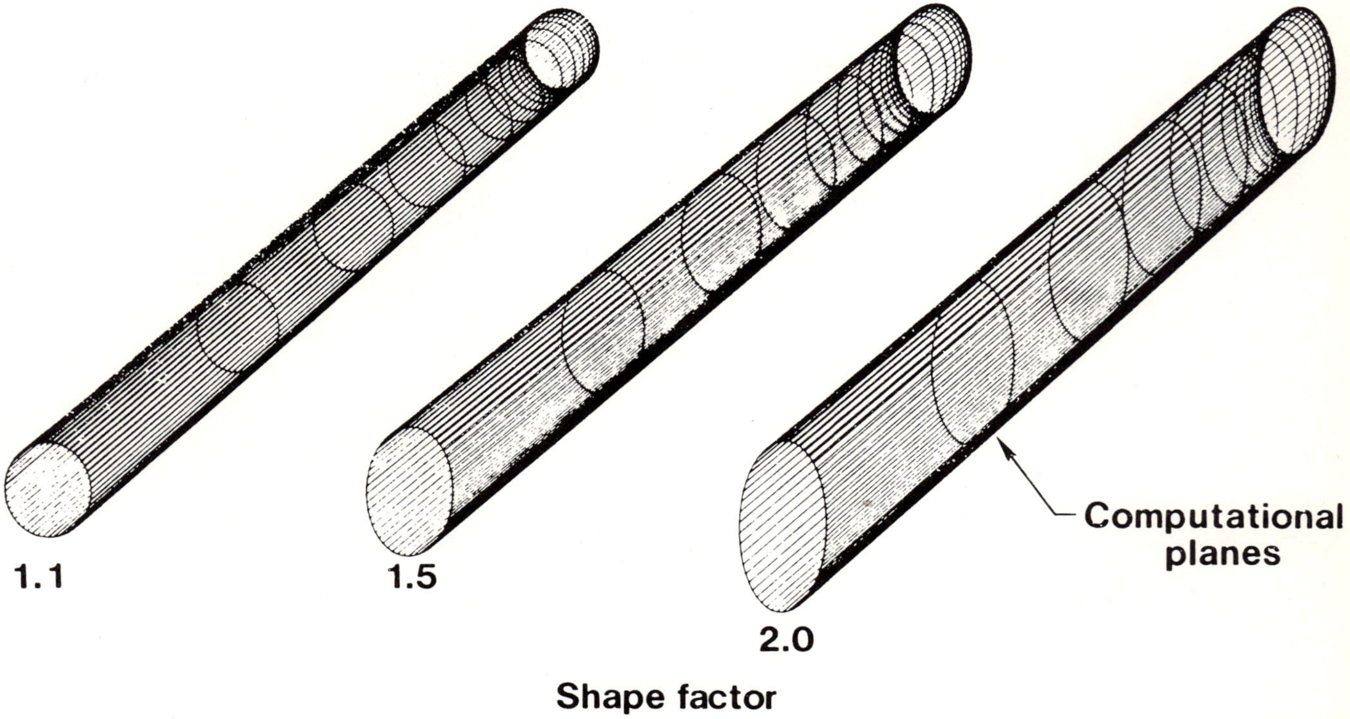

FIGURE 3

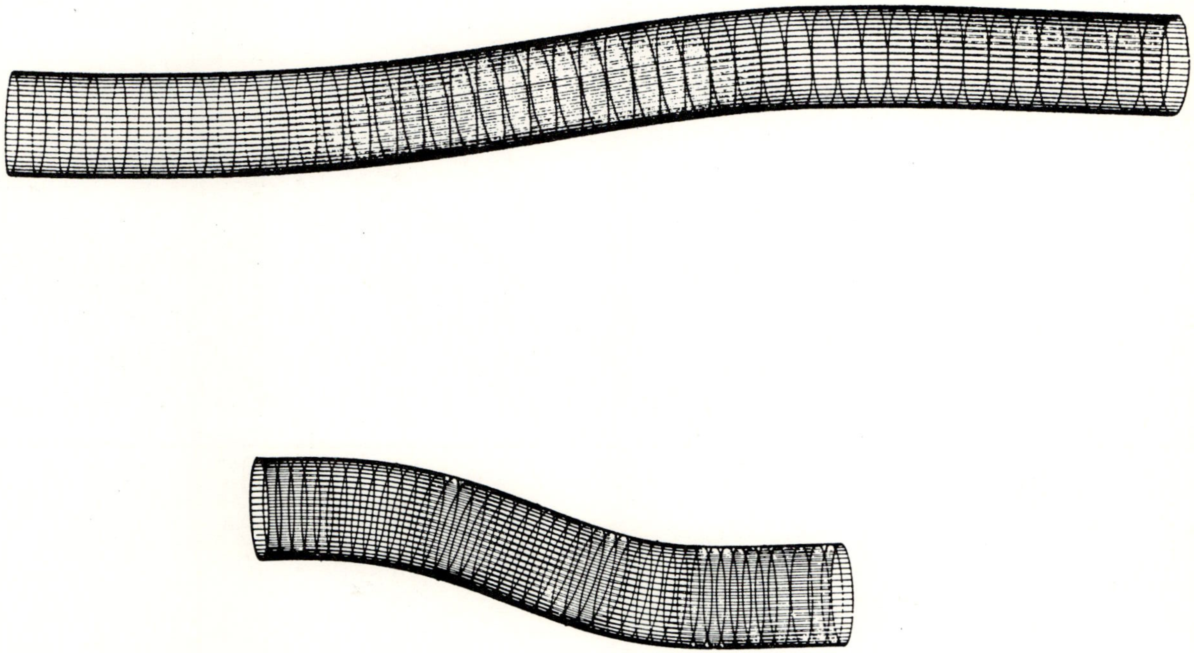

S-SHAPED BENDS WITH CIRCULAR CROSS SECTION

(FLOW LEFT TO RIGHT)

FIGURE 4

CROSS FLOW VELOCITIES IN CURVED DUCTS

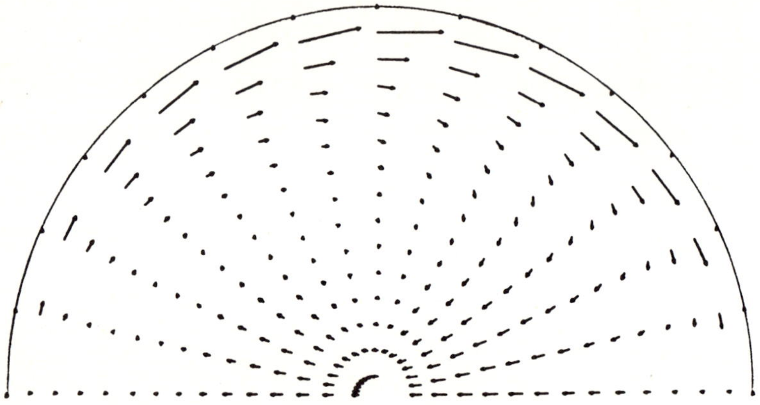

CURVED CIRCULAR DUCT

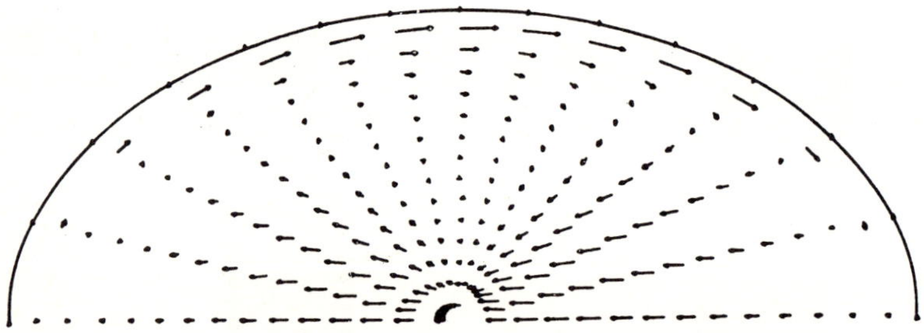

CURVED ELLIPTIC DUCT, SHAPE = 1.5

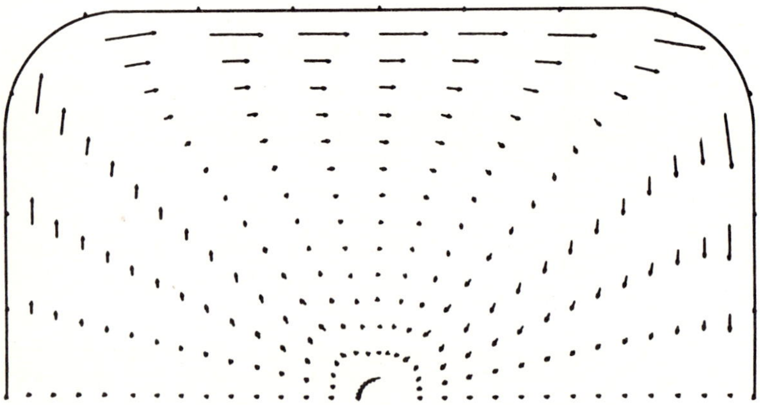

CURVED SUPERELLIPTIC DUCT, EXPONENT = 10

FIGURE 5

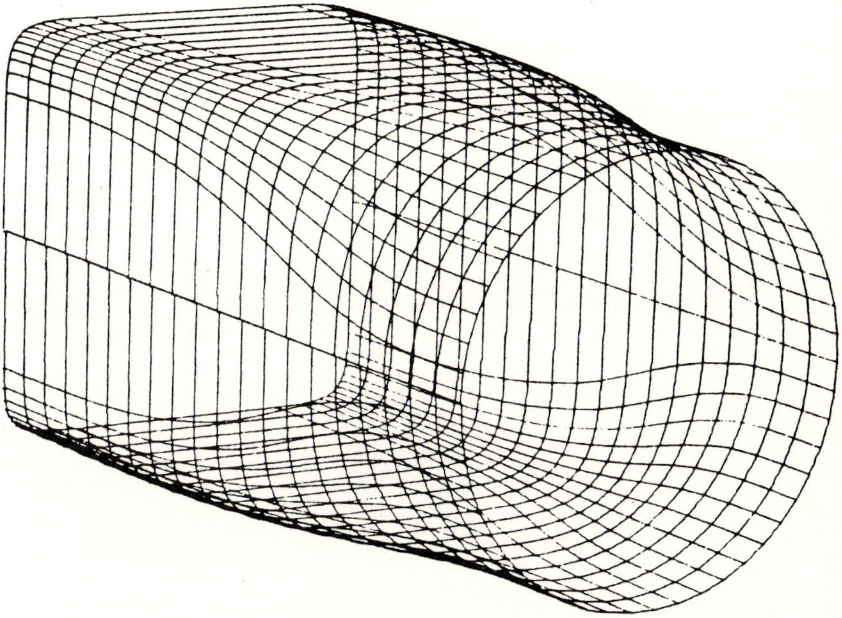

Transition Duct Superellipse Exponent = 10 to Circular

PULMONARY HAEMODYNAMICS

by

J. MACCARIO and R. COLLINS

Université de Compiègne, B.P. 233, 60206 COMPIEGNE (France)

ABSTRACT

Pulmonary hemodynamics is studied in terms of the quasi one-dimensional unsteady nonlinear fluid flow equations which are applied to the approximately 40 generations of branched arterial, capillary and venous distensible vessel segments making up the four lobes of the complete lung. An idealised pressure-area 'tube law' is introduced which provides for varying degrees of vessel collapse. The model predictions agree well with experimental measurements of flow transmission as a function of pulsatile frequency. Pulmonary response is represented schematically in terms of an influence diagram. Pressure pulses are shown to increase in amplitude in the early arterial segments, with the greatest drop occurring across the capillary bed.

INTRODUCTION

In certain respects, the pulmonary circulation constitutes an ideal system with which to demonstrate the application of the principles fluid mechanics to vascular beds in general. The characteristics of the lung most fundamental to fluid dynamics are number, dimensions and material properties of its component vessels, in addition to the architectural pattern (morphology) in which they are connected. Only the mechanical aspects of blood flow in the adult human lung will be investigated, omitting completely the contributions of neural control, the respiratory function of the lung, and the vasomotor effects of oxygen, carbon dioxide, or pharmacologic agents.

The present numerical study of the flow of a viscous incompressible fluid through a network of distensible cylindrical tubes (Fig. 1) with cross-sectional area increasing from stations V to M, and subsequently decreasing from M to O, forms the basis for a fluid dynamical description of the pulmonary circulation. The individual vessel lengths vary inversely with their corresponding cross-sections.

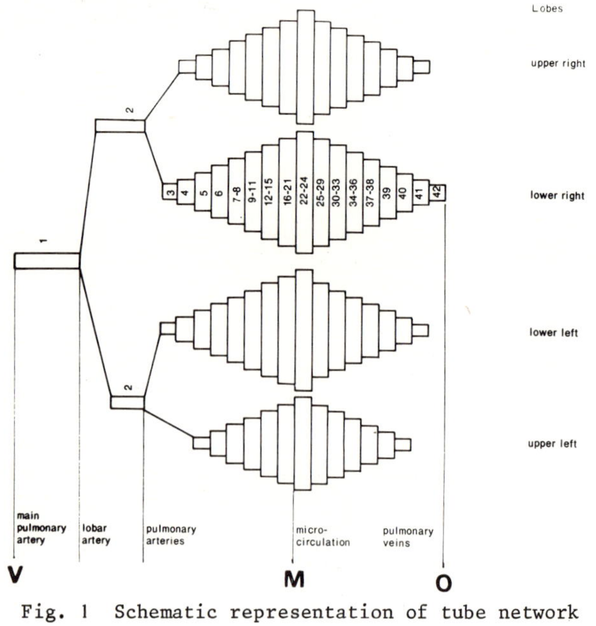

Fig. 1 Schematic representation of tube network

Periodic variations in pressures are specified as boundary conditions at stations V and O. The actual initial conditions are not known a priori. It proves possible to determine initial conditions compatible with the periodic flow boundary conditions through use of a rapidly convergent iterative scheme developed about a nominal rest state.

MATHEMATICAL FORMULATION

The governing flow equations may be reduced to a quasi one-dimensional cross-section-averaged form under the present conditions in which the wave length of propagated waves is much larger than the tube radius (Barnard et al. 1966, Rockwell et al. 1974). The corresponding momentum and continuity equations are expressed respectively as

$$\frac{\partial u}{\partial t} + \frac{\partial}{\partial x}\left(\frac{u^2}{2} + \frac{p}{\rho}\right) = -F \qquad (1)$$

$$\frac{\partial S}{\partial t} + \frac{\partial}{\partial x}(Su) = 0 \qquad (2)$$

where u is the longitudinal fluid velocity averaged over the non-circular cross-section S, and p the fluid pressure. The frictional term F is evaluated approximately on the basis of a quasi-steady laminar boundary layer as

$$F = n\,\frac{8\pi\mu}{\rho} = \gamma\,\frac{u}{S}$$

in terms of the fluid density ρ and the number of vessels n of cross-sectional area S. Milnor (1972) confirms that the velocity profiles are relatively flat under normal pulmonary flow conditions and that there is no gross evidence of turbulence except possibly under certain pathological conditions such as congenital stenosis of the pulmonic valve.

The flow system (1), (2) constitutes 2 equations in three unknowns, and may be completed by an equation of state linking the local cross-sectional area to variations in the transmural pressure, as depicted in Fig. 2.

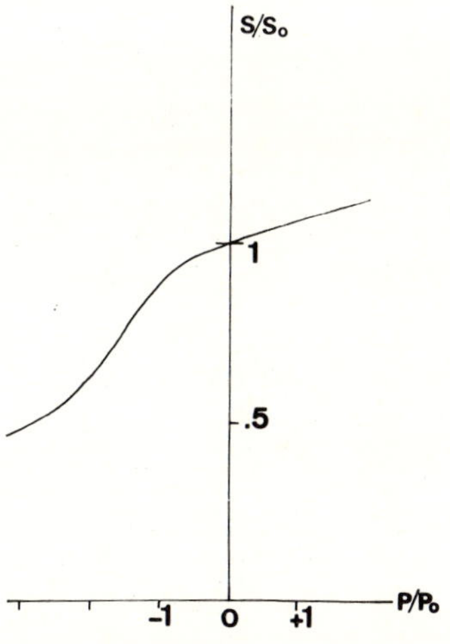

Fig. 2 Relation between transmural pressure and tube cross-section

At branch points, one requires continuity of flow rate and fluid pressure, as standing pressure discontinuities at bifurcations are not physically admissible in view of the elastic response of the tube walls tending to relieve them.

COMPUTATIONAL PROCEDURE

The governing equations are re-cast in characteristic form

$$\frac{dx}{dt} = u \pm c \tag{3}$$

with wave propagation speed

$$c = \left(\frac{S}{\rho} \frac{\partial p}{\partial S} \right)^{1/2}$$

and

$$dp \pm \rho c \, du = \mp \rho c \, \gamma \left(\frac{u}{S}\right) dt \tag{4}$$

The relatively large variation of the derivative $\partial S/\partial p$ throughout the pulmonary branching network leads to numerical instabilities for purely explicit finite difference forms of the characteristic equations. Consequently, it has proved necessary to introduce an implicit numerical procedure for the solution of equations (3) and (4).

Furthermore, the particular properties of the tube law relating transmural pressure to local cross-sectional area lead to variations in the wave speed c of a factor of five. The concomitant changes in the slopes of the characteristics (eq. 3) cause certain regions of the computational net to advance much more slowly than others. The "valleys" so created must be "filled" before proceeding with subsequent calculations of the "fast" regions. Similar numerical difficulties are encountered in the propagation of shock waves in combustible mixtures, and special numerical algorithms must be constructed to deal with them.

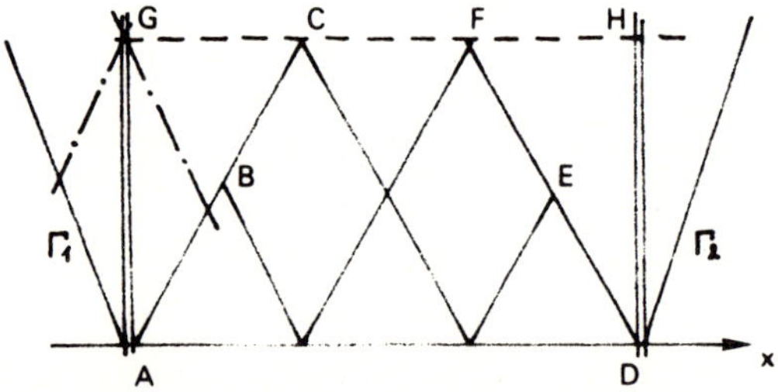

Fig. 3 Computations within a tube segment

Such a procedure has been devised for the pulmonary circulation. It is based on the following three considerations :

i) conservation of the number of computational points within each individual tube segment AD by advancing the time increments in pairs (Fig. 3) to compensate for intermediate staggering.

ii) Computation of u and p at the upstream segment junction G at the new time by back-interpolation from point G (Fig. 3) along the C- characteristic to find its intersection with the C^+ characteristics AB, BC emanating from point A at the previous time step. A similar procedure along the C^- characteristic Γ_1 through point A of the adjacent tube segment leads to the intersection point by back-interpolation along the C+ characteristic through G. Likewise, an identical scheme of interpolation along the characteristic directions permits calculation of the new values of u and p at the downstream junction H.

iii) Levelling of the non-uniform advance of the solution in time to uniform time intervals by automatic computer scanning and discretisation of the various time levels attained following successive paired advances of the time increment (Fig. 4a). The smallest of the resulting time steps Δt is recovered and a grid of time steps is constructed in terms of a series expansion : t, t + Δt, t + $2\Delta t$, ...t + 2^k Δt. Each new computational point is reduced by back-interpolation along the C^+ and C^- characteristics running through it to the next lower time value of the aforementioned grid.

This procedure is subsequently repeated for tube segments having attained times t + 2 Δt (Fig.4b) and t + 4 Δt (Fig 4d) before returning to the intermediate levels t + 3 Δt (Fig 4c). Upon completion of this algorithm all tube segments will have been advanced uniformly to the new common time level t + $4\Delta t$, and the general computation of the characteristics net is resumed.

The numerical computations are somewhat facilitated by selecting a value of Δt 5 % below the smallest time increment apparent in Fig. 4a to assure that forward interpolations will not be necessary. Gradual spacewise skewing of the computational net can be controlled by alternating the directions $C^{\pm}$ along which interpolations are carried out. Local rezoning of the computational network may however be necessary eventually to correct overly-skewed grids. In such cases, new values of u and p are determined by linear interpolations on the new network.

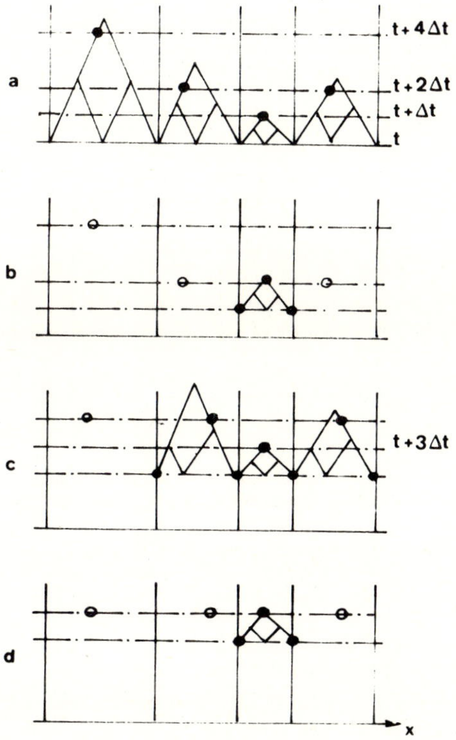

Fig. 4 Full-field computational network

RESULTS

Experiments of Wiener et al. (1966) on the pulmonary circulation of the dog furnish the required morphological and mechanical properties of the branching network, as well as the inlet and outlet boundary conditions on pressure and flow rate for incorporation into the present computational model. Good agreement has been obtained when compared with the experimental results of Maloney et al. (1968) showing the rapid attenuation in the amplitude of transmitted flow oscillations across the lung with increases in pulsatile frequency. Fig. 5 shows such a typical fall-off of the inlet flow rate Q(v), measured at the right ventricle, to its value Q(o) measured in the left atrium at the exit from the lung at a pulsatile frequency of 3 Hz.

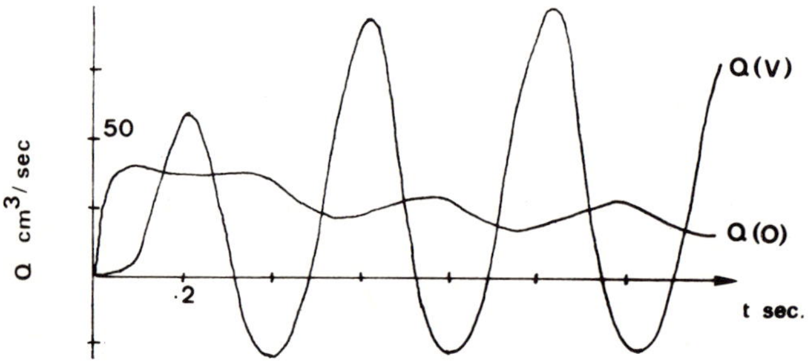

Fig. 5 Flow transmission in the lung

The present computational model however provides a means of studying the mechanics of fluid/vessel interactions within the interior of the lung ; whereas, direct experimental or clinical measurements are limited to the first few inlet and exit generations of branching in the lung.

As an illustration of this capability, Fig. 6 shows the significant weakening of the back flow as one progresses from the arterial segments (Fig. 6a) through the capillary bed (Fig. 6b) to the venous network (Fig. 6c) at a frequency of 3 Hz. In other words, the flow becomes "forward-rectified" simply by the action of the collapsible pulmonary vessels (Fig. 2) which close to inhibit backflow under conditions of adverse pressure gradient. Indeed, such vessels play the role of one-way valves which are otherwise physically absent within the pulmonary circulatory system.

The present model thus effectively opens the "black box" of the lung to mathematical studies of internal flow regulation under various conditions of physical exertion.

CONCLUSIONS

A mechanical flow model of the complete pulmonary circulation has been formulated on the basis of the nonlinear unsteady equations of fluid motion. The introduction of an experimentally verified relationship between transmural pressure and local cross-sectional area, which provides for partial or total collapse of certain vessels, has been demonstrated to be an essential feature of the model. Good agreement has been obtained between predictions of sharp attenuation of flow transmission with pulsatile frequency and the experimental results reported by Maloney et al. (1968) for the isolated dog lung. It has been further shown that even substantial alterations in vessel wall rheology (compliance), without accounting for collapse, cannot produce this observed response.

The nonlinear features of the analysis are important for the proper treatment of wave reflections and the observed steepening of the pressure pulses in the proximal arterial segments of the pulmonary network, The model also predicts no significant differences in the character of forward and retrograde flows, in accordance

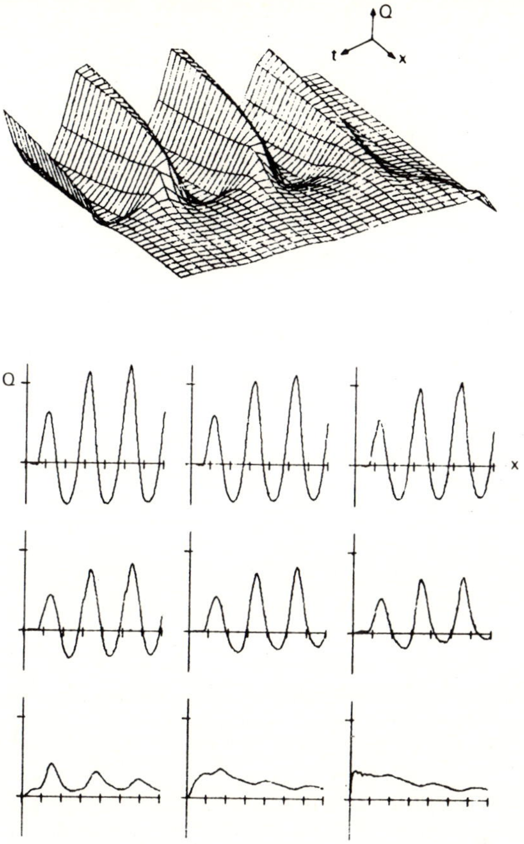

Fig. 6 Flow variations within the lung

with the consensus of experimental findings.

In the interest of simplicity, certain secondary effects, such as momentary losses in stagnation pressure due to possible short-lived bursts of vorticity at bends and branchings in the pulmonary tree during pulsatile flow, have not been included in this preliminary model. The influence of gravity is readily incorporated by adjustment of the pulmonary arterial pressure with height in the lung. Flow in the capillary bed has been treated simply by incorporating the vascular network within the 40-odd generations of the complete pulmonary tree. Computations confirm that the major drop in blood pressure occurs across the capillary bed.

The complex and controversial question of the existence of a spectrum of finite opening times for collapsible vessels (Maloney et al. 1968 a) is only partially answered by the present analysis. Its resolution hinges very critically on the form of the pressure-area law which one adopts. The present relation, which provides good phenomenological agreement with observed pulmonary behaviour, implies a dependence of vessel cross-section on local transmural pressure only. Therefore, changes in cross-sectional area are necessarily synchronous with pressure changes in this approximate formulation.

A definitive conclusion would require application of a simultaneous solution of the fluid and wall equations, without the assumption of such a pre-specified "tube law". An analysis of the dynamics of flow in collapsible tubes (Tedgui and

Collins 1978) proves considerably more complex as it accounts for the significant residual longitudinal tension and local curvature distribution of the vessel walls.

The pulmonary circulation appears to possess a self-regulating control system. It is suggested that mechanical (or even certain vasomotor) effects may act upon the pulmonary arterial pressure level which in turn regulates the instantaneous effective compliance of the collapsible vessel segments, thus determining the character of ensuing wave propagation. Such representations of pulmonary response may eventually be quantified for specific clinical situations, hopefully providing an improved guide to correct diagnosis and therapy of pulmonary disorders.

REFERENCES

BARNARD A.C.L., HUNT W.A., TIMLAKE P. and VARLEY E. (1966) A theory of fluid flow in compliant tubes, Biophysic. J., 6.6, 717-724

MALONEY J.E., BERGEL D.M., GLAZIER J.B., HUGHES J.M.B., WEST J.B. (1968) Transmission of pulsatile blood pressure and flow through the isolated lung, Circulation Res., 23, 11-24

MILNOR W.R. (1972) Pulmonary hemodynamics, in Cardiovascular Fluid Dynamics, ed. D.M. BERGEL, Academic Press, London, New York.

ROCKWELL R.L., ANLIKER M. and ELSNER J. (1974) Model studies of the pressure and flow pulses in a viscoelastic arterial conduit, J. Franklin Institute, 297-5, 405-427

TEDGUI A., COLLINS R. (1978) Flow in collapsible tubes, presented at the Congress on Biomechanics, Paris, Sept. 14-15, 1978.

WIENER F., MORKIN E., SKALAK R. and FISHMAN A.P. (1966) Wave propagation in the pulmonary circulation, Circulation Res., 19, 834-850.

A VORTEX-GRID METHOD FOR BLOOD FLOW THROUGH HEART VALVES

by

M. F. McCracken
Indiana University

and

C. S. Peskin
Courant Institute of Mathematical Sciences
New York University

I. INTRODUCTION

Recently there has been significant progress in developing numerical methods for studying incompressible, two-dimensional fluid flow at high Reynolds number. Chorin [1] has a grid free method involving modified point vortices. However, because vortices must be generated at the boundary at each timestep and the work of computation is proportional to the square of the number of vortices, calculations using his method can be prohibitively expensive except in flow-past-a-body type problems, where the vortices are eventually swept out of the region of computation. Shestakov [6] introduced a hybrid method involving both a mesh and point vortices, but his method is applicable only to the case of flat boundaries. The object of the present work is to study the flow patterns of the mitral valve of the human heart. Because the Reynolds numbers of the human heart are two orders of magnitude higher than the highest of those that can be studied by conventional finite difference methods, we need a high Reynolds number method for solving the incompressible Navier-Stokes equations. Peskin [5] undertook a study of the mitral valve using a finite difference method and we now extend his work to Reynolds numbers in the human physiological range. The boundaries consist of the valve and the heart wall, which we idealize to be infinitely thin. These boundaries move at the local fluid velocity and also exert forces on the fluid. Because of the complicated geometry, existing methods were not applicable to our problem.

The method we developed is described briefly below. It is quite complicated and details of the algorithm may be found in [3]. Like Shestakov's method, ours is a hybrid involving both a computational mesh and point vortices. However, it has several new features. In the first place, we introduce a new algorithm for generating point vortices at the boundary. Chorin's algorithm is based on the slip velocity generated at the boundary in solving the Euler equations. Because the motion of our boundary is not known in advance, his algorithm cannot be used. Instead we show that the boundary forces can be viewed in a natural manner as generating a layer of vortex dipoles along the boundary. Preliminary work on this method of vorticity generation was done by Mendez [4]. To insure that the number of vortices does not grow, after a fixed number of timesteps we spread their vorticity to the mesh. Thus, vorticity is stored in two forms: on the mesh and in the point vortices. The vortices supply high resolution near the boundary and so the mesh need not be refined as the Reynolds number increases.

II. EQUATIONS OF MOTION

The fluid lives on $[0,1] \times [0,1]$ with periodic boundary conditions (or on any region for which a fast Laplace solver is available, see [2]). In non-dimensional units the equations are:

$$(1) \qquad \frac{\partial w}{\partial t} + \nabla \cdot (\vec{u}w) = \frac{1}{R} \Delta w + \hat{z} \cdot (\nabla \times \vec{F})$$

$$(2) \qquad \Delta\psi = w$$

$$(3) \qquad u_1 = -\frac{\partial\psi}{\partial x_2} , \quad u_2 = \frac{\partial\psi}{\partial x_1} ,$$

where $\vec{u}$ is the fluid velocity, w the vorticity, $\vec{F}$ the force, R the Reynolds number, and $\hat{z}$ a unit vector perpendicular to the plane of the motion. Further, we have

$$(4) \qquad \frac{\partial\vec{X}}{\partial t}(s,t) = \vec{u}(\vec{X}(s,t),t)$$

$$(5) \qquad \vec{F}(\vec{x},t) = \int_0^L \vec{f}(s,t)\, \delta(\vec{x} - \vec{X}(s,t))ds ,$$

where $\vec{X}$ is the Lagrangian representation of the boundary, δ the Dirac δ-function, and where $\vec{f}$, the force density, may be computed from $\vec{X}$ (see Peskin [5] for details). In the case where the boundary is not supposed to move, one can also use this method. In that case, the force attracts the boundary to where it should be.

III. GENERATION OF VORTICITY USING THE BOUNDARY FORCES

The boundary forces can be regarded in a natural manner as generating a dipole layer of vortices along the boundary. Below we suppress dependence on time.

$$(6) \qquad \vec{F}(x) = \int_0^L \vec{f}(s)\, \delta(\vec{x} - \vec{X}(s))ds , \quad \text{so that}$$

$$(7) \qquad (\nabla \times \vec{F}) \cdot \hat{z} = \int_0^L \{\nabla \times [\delta(\vec{x} - \vec{X}(s))\vec{f}(s)]\} \cdot \hat{z}ds$$

$$= \int_0^L \nabla\delta(x - X(s)) \cdot \{\vec{f}(s) \times \hat{z}\}\, ds .$$

Taking the dipole moment over an arbitrary region R_1 of fluid, we see that it is

$\int_{R_1 \cap \vec{X}} (-f_2(s), f_1(s))ds$. Thus, $(-f_2(s), f_1(s))$ is the dipole moment density along the boundary. It is natural to discretize this as a layer of point vortex dipoles. Note that they are not oriented normal to the boundary. Adopting this point of view, when the boundary is discretized, at each boundary point, at each timestep, a pair of point vortices is shed into the fluid. Their strengths are respectively

$$\pm \frac{|\vec{f}_k| \Delta t}{dN} ,$$

where $\vec{f}_k$ is the force density at the kth boundary point, N is the number of boundary points, and d is the axis length of the dipole.

IV. THE FINITE DIFFERENCE METHOD

Here we use a method similar to Shestakov [6]. We update according to

$$\left(I + \frac{\Delta t}{2} Q_{x_1}(u_1^n)\right)^{n+1/2} = \left(I - \frac{\Delta t}{2} Q_{x_2}(u_2^n)\right) w^n$$

$$\left(I + \frac{\Delta t}{2} Q_{x_2}(u_2^n)\right) w^{n+1} = \left(I - \frac{\Delta t}{2} Q_{x_1}(u_1^n)\right) w^{n+1/2}$$

$$L\psi^{n+1} = w^{n+1} + w^* \tag{9}$$

$$u_1^{n+1} = -D_{x_2} \psi^{n+1} , \quad u_2^{n+1} = D_{x_1} \psi^{n+1} , \tag{10}$$

where Q_{x_1} and Q_{x_2} are difference operators such that $Q_{x_1}(u_1^n) + Q_{x_2}(u_2^n)$ corresponds to $\nabla \cdot (\vec{u}^n) - \Delta$, where D_{x_1} and D_{x_2} correspond to differentiation in the x_1 and x_2 directions, and where L corresponds to Δ on $[0,1] \times [0,1]$ with periodic boundaries. To solve (9) a fast Laplace solver is used. The vorticity field w^* is the vorticity due to the point vortices. Since vorticity is stored on the mesh and in vortices, the total vorticity is $w + w^*$.

V. CONNECTING THE BOUNDARY AND VORTICES WITH THE MESH

We discretize the boundary into N points located at $\{\vec{X}_k\}_{k=1}^{N}$. At each boundary point we compute the force density $\vec{f}_k$ from the boundary configuration. The way in which this is done is complicated because the stability of the heart valve calculation depends crucially upon correct computation of the forces. We compute them exactly as in Peskin [5] and in this paper we will adopt the point of view that this has already been done. In order to move the boundary points we use the formula

(11) $$\vec{X}_k^{n+1} = \vec{X}_k^n + \Delta t\, \vec{u}^{n+1}(\vec{X}_k^n) ,$$

where $\vec{u}^{n+1}$ is given by (10) on the mesh and interpolated to the point $\vec{X}_k^n$.

Using the boundary forces, point vortices are created as in III. There are 2NK of them located at the points $\{\vec{X}_j\}_{j=1}^{2NK}$, where K is the number of timesteps for which they are retained before being spread to the mesh. A point vortex should be thought of as a δ-function of vorticity. For a detailed discussion, see Chorin [1]. Point vortices move by being convected in each other's velocity fields and by diffusion. We use the formula

(12) $$\vec{Y}_j^{n+1} = \vec{Y}_j^n + \Delta t\, \vec{u}^{n+1}(\vec{Y}_j^n) + \vec{\eta}_j ,$$

where the $\vec{\eta}_j$'s are independent, Gaussian random variables with mean zero and variance $\frac{\Delta t}{R}$. This represents diffusion. We treat convection differently from Chorin. The velocity field for convection is computed through the mesh using (10), instead of summing over the individual vortices. This saves computations, and it provides an automatic smoothing of the vortex singularity. A different smoothing, appropriate for rigid boundaries, was used in [1].

In order to interpolate velocity and spread vorticity, we introduce a function d to approximate the one-dimensional Dirac δ-function:

(13) $$d(x) = \begin{cases} \frac{1}{4}\left(1 + \cos\left(\frac{\pi x}{2h}\right)\right) & |x| < 2h \\ 0 & |x| \geq 2h , \end{cases}$$

where h is the mesh width. To interpolate velocity we use the formula

(14) $$\vec{u}(X_{1k},X_{2k}) = h^2 \sum_{i,j} \vec{u}(ih,jh)\, d(X_{1k} - ih)\, d(X_{2k} - jh) ,$$

where $\vec{X}_k = (X_{1k},X_{2k})$. This is a discretization of

(15) $$\vec{u}(\vec{x}_k) = \int_{\text{fluid}} \vec{u}(\vec{x})\, \delta(\vec{x} - \vec{x}_k)dx .$$

To spread vorticity we use

$$w^*_{ij} = \sum_{k=1}^{L} K_k \, d(Y_{1k} - ih) \, d(Y_{2k} - jh) \quad , \tag{16}$$

where K_k is the strength of the kth vortex, $\vec{Y}_k = (Y_{1k}, Y_{2k})$, and where L is the number of vortices to be spread. Vorticity is spread to the mesh twice in each timestep, once when the oldest vortices are permanently spread to the mesh, and also to calculate w^* , when all the vortices are temporarily spread to the mesh to find the stream function using (9).

VI. SUMMARY OF THE METHOD

Briefly, at each timestep first the forces are updated as in [5]. Then the oldest vortices are spread to the mesh using (16) and new ones are created in their stead using the boundary forces as in III. After that the mesh vorticity and the velocity are updated using (8)-(10). Finally, the boundary points and vortices are moved according to (11) and (12).

VII. CONCLUSIONS

The strength of this method is its generality. It can be used in a wide variety of two-dimensional problems. However, it should be born in mind that the moving of the vortices through the mesh may generate some numerical viscosity at very high Reynolds numbers. A check of this question is currently being undertaken. Our results for the heart-valve problem agree with those of Peskin [5] for low Reynolds number. In our calculations, the mitral valve appears to perform independently of Reynolds number over a range of two orders of magnitude, which is the range of Reynolds numbers found in mammalian hearts. This supports the conjecture that mammalian hearts act independent of Reynolds number.

ACKNOWLEDGEMENTS

Many people have made useful suggestions and contributions to this work. The authors would particularly like to thank A. J. Chorin, D. McQueen, O. Widlund, and E. L. Yellin for interesting conversations; M. Brothers and D. McQueen for helping with the computing; and J. Race for the beautiful job she did typing the manuscript. The research was partially supported by the National Science Foundation under grant NSF MCS-76-07039 and by the National Institutes of Health under research grant HL-17859. Computation was supported by Indiana University and by the U.S. Department of Energy under Contract EY-76-C-02-3077 at the Courant Mathematics and Computing Laboratory of New York University.

REFERENCES

[1] Chorin, A. J. (1972) Springer Lecture Notes in Physics, 19:100-104.

[2] Fischer, Golub, Hald, Leiva, Widlund (1974) Math. Comp., 28.

[3] McCracken, M. and Peskin, C. A vortex method for blood flow through heart valves, to appear 1978.

[4] Mendez, R. (1977) Numerical study of incompressible flow in a region bounded by elastic walls, thesis, Berkeley.

[5] Peskin, C. S. (1977) Numerical analysis of blood flow in the heart, J. Comput. Phys., 25, 220-252.

[6] Shestakov (1975) Thesis, Berkeley, Nat. Tech. Info. Serv., U.S. Dept. of Commerce, UCRL51894.

PICTURES TAKEN FROM A RUN AT 1/10 THE REYNOLDS NUMBER OF THE HUMAN HEART

The upper chamber in the figures below is the atrium, to which blood returns from the lungs; hence, we have a point source of volume in the atrium. The lower chamber is the ventricle; it is separated from the atrium by the mitral valve. The chamber to the left is the outflow track to the aortic valve. Since we study only the filling cycle, that valve is not included. The line segments are velocity vectors normalized to a maximum of one mesh width.

Below Figures 1-4 show the mitral valve opening. In Figures 5-7 it is starting to drift shut under the influence of vortices which form at the tip of the valve leaflets. Figure 8 shows the beginning of atrial systole, in which the atrium contracts, squeezing out a last jet of blood and the valve opens once more. In Figure 9, the valve is closing.

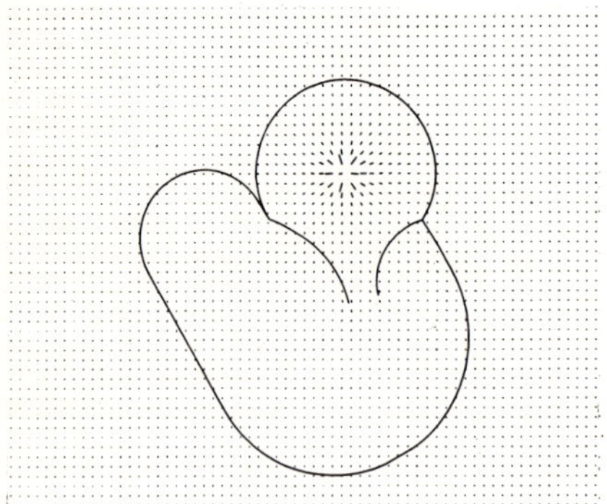

Fig. 1

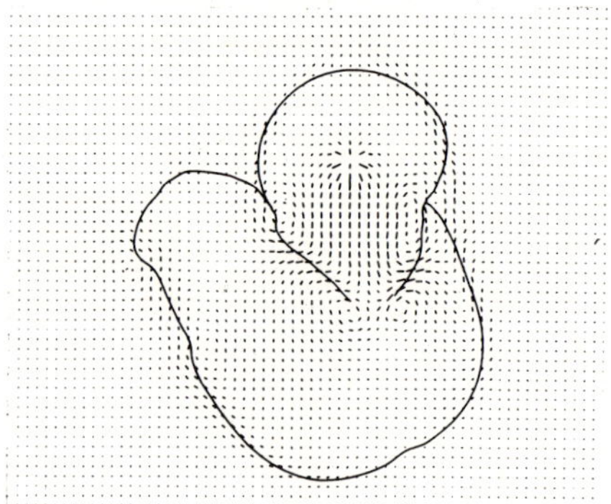

Fig. 2

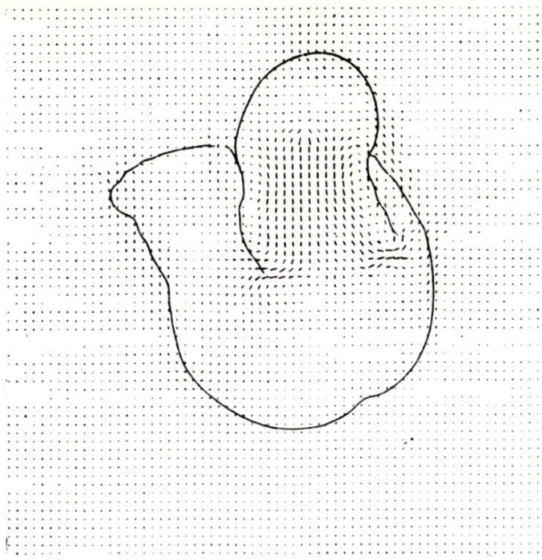

Fig. 3

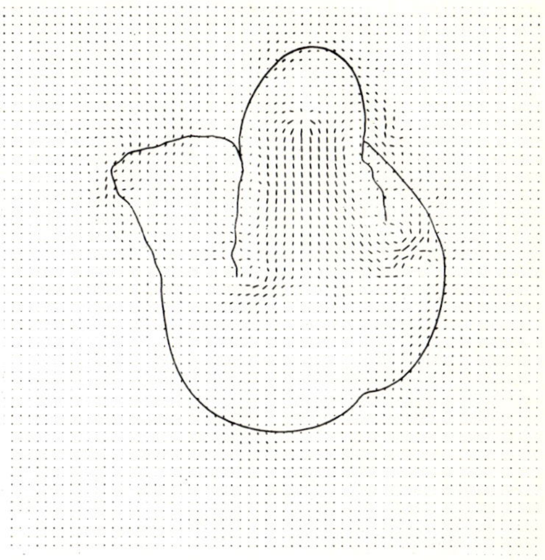

Fig. 4

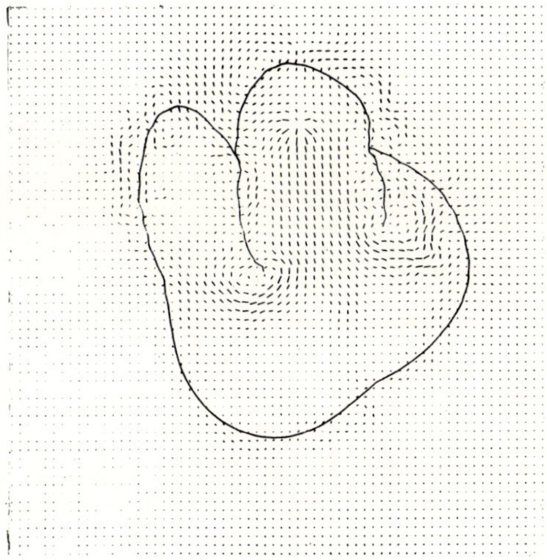

Fig. 5

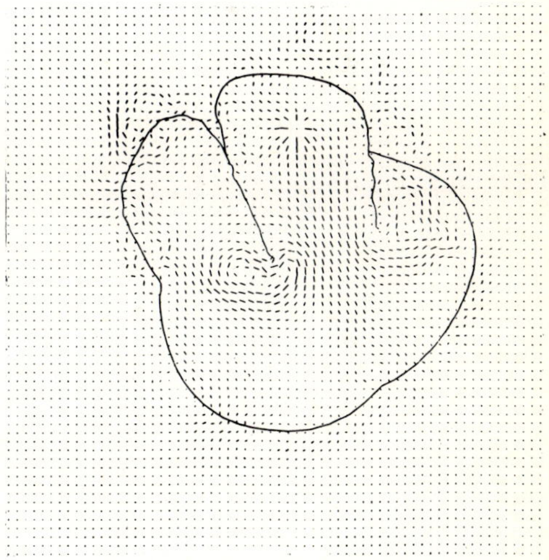

Fig. 6

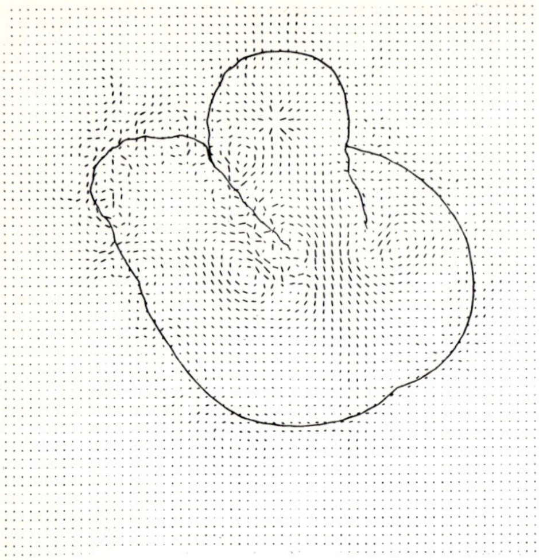

Fig. 7

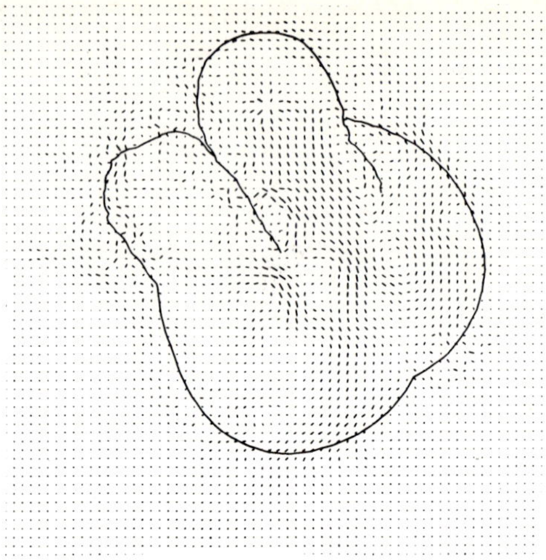

Fig. 8

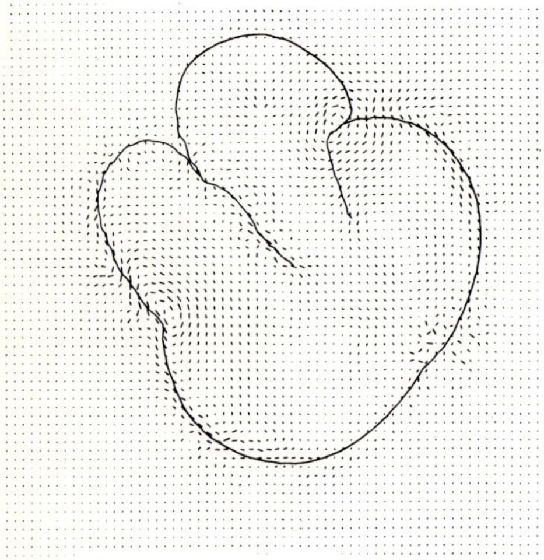

Fig. 9

SUPERSONIC VISCOUS FLOW OVER CONES AT INCIDENCE

Major David S. McRae
Air Force Flight Dynamics Laboratory
Ames Research Center, NASA, Moffett Field, California

and

M. Y. Hussaini*
Ames Research Center, NASA, Moffett Field, California

ABSTRACT

The conically symmetric Navier-Stokes equations are solved by MacCormack's method for the supersonic flow past sharp cones at incidence. To provide closure for the case of turbulent flows, a scalar eddy-viscosity model based on mixing length hypotheses is used. The results are compared with available experimental data.

INTRODUCTION

Previously, the condition of conical symmetry has been applied to the inviscid Euler equations with the goal of obtaining supersonic flow-field solutions for conical bodies. The presence of the temporal derivative term in the equations allows a finite difference integration technique to be marched in time until steady state is reached. Such solutions[1] have been obtained for a restricted range of incidence and Mach number beyond which they encounter numerical instability. Where they exist, these solutions are in disagreement with experimental observations on the lee side of the cone, particularly at moderate and large angles of attack. This disagreement appears to be due to the neglect of the viscous effects which predominate on the lee side as angle of attack is increased. In another study,[2] the lee side viscous effects were modeled by fictitious alteration of the lee surface to yield a pressure distribution in agreement with experiment. Unfortunately, this technique requires that the experiment must always precede the calculation.

In the present analysis the viscous-inviscid interactions are modeled through the inclusion of viscous terms which satisfy the condition of conical symmetry. This assumption implies that the viscous compressible flow over a cone is locally conical. There is experimental evidence[3,4] for the validity of this assumption, particularly for turbulent flows. A recent numerical investigation[5] confirmed the assumption for laminar flow over cones at low values of the hypersonic similarity parameter $\bar{\chi}$. The present results also provide additional support for the adequacy of the assumption.

THE GOVERNING EQUATIONS

The unsteady, compressible, conically symmetric Navier-Stokes equations for a spherical polar coordinate system may be written in conservation form as in Ref. 5:

$$\frac{\partial D}{\partial t} + \frac{\partial F}{\partial \theta} + \frac{\partial G}{\partial \phi} + H' = 0$$

The vectors D, F, G, and H' denote

$$D^T = r^2 \sin\theta(B) \qquad F^T = r\sin\theta(vB - S_2) \qquad G^T = r(wB - S_3)$$

---

*Presently at ICASE, Langley Research Center, NASA.

and

$$H' = r \sin\theta \begin{bmatrix} \rho u \\ -(\rho v^2 - \tau_{22}) - (\rho w^2 - \tau_{33}) + 2(\rho u^2 + kp) - (\tau_{11} + kp) \\ -\cot\theta(\rho w - \tau_{33}) + 3\rho uv - 2\tau_{12} \\ \cot\theta(\rho vw - \tau_{23}) + 3\rho uw - 2\tau_{13} \\ -2u(\rho e + kp) - (\tau_{11} + kp)u - v\tau_{12} - w\tau_{13} \end{bmatrix}$$

where

$$B = [\rho, \rho u, \rho v, \rho w, \rho e]$$

$$S_i = [0, \tau_{i1}, \tau_{i2}, \tau_{i3}, -q_i + u\tau_{i1} + v\tau_{i2} + w\tau_{i3}] \; ; \qquad i = 1,\ 2,\ 3$$

$$k = \frac{\gamma - 1}{2\gamma}$$

and the superscript T indicates the transpose. The velocity components are nondimensionalized with respect to the maximum adiabatic velocity V_{max}, the state variables with respect to their stagnation values, and the time with respect to r/V_{max}; r being the radial distance from the apex of the cone to the spherical surface on which the calculation takes place. The variables ρ, u, v, w and e denote, respectively, the nondimensional density, the velocity components in the (r, θ, ϕ) coordinate system, and the internal energy. The characteristic Reynolds number (Re) is $V_{max} r \bar{\rho}_s / \bar{\mu}$ where $\bar{\rho}_s$ is the stagnation density and $\bar{\mu}$ the coefficient of viscosity. The metric r appears in the equations through the Reynolds number. The nondimensional stress and heat transfer terms are

$$\tau_{12} = \tau_{21} = \frac{1}{Re}\left(\frac{\partial u}{\partial \theta} - v\right)$$

$$\tau_{13} = \tau_{13} = \frac{1}{Re \sin\theta}\left(\frac{\partial u}{\partial \phi} - w \sin\theta\right)$$

$$\tau_{23} = \tau_{32} = \frac{1}{Re \sin\theta}\left[\frac{\partial}{\partial \theta}(w \sin\theta) - 2w\cos\theta - \frac{\partial v}{\partial \phi}\right]$$

$$\tau_{11} = -kp + \sigma_\eta$$

$$\tau_{22} = -kp + \frac{2}{Re}\left(\frac{\partial v}{\partial \theta} + u\right) + \sigma_\eta$$

$$\tau_{33} = -kp + \frac{2}{Re \sin\theta}\left(\frac{\partial w}{\partial \phi} + u \sin\theta + v\cos\theta\right) + \sigma_\eta$$

and

$$\dot{q}_1 = 0\,, \quad \dot{q}_2 = -\frac{1}{2RePr}\frac{\partial T}{\partial \theta}\,, \quad \dot{q}_3 = -\frac{1}{2RePr}\frac{\partial T}{\partial \phi}$$

where

$$\sigma_\eta = -\frac{2}{3Re \sin\theta}\left(2u\sin\theta + \frac{\partial}{\partial \theta}(v \sin\theta) + \frac{\partial w}{\partial \phi}\right)$$

$$Pr = \frac{\bar{\mu} Cp}{\bar{K}} = \text{Prandtl number}$$

C_p = specific heat at constant pressure

$\bar{K}$ = thermal conductivity

TURBULENCE MODEL

For the eddy viscosity of the turbulent flow, simple algebraic relations similar to those of Harris and Morris[6] are assumed. The effective coefficient of viscosity μ and thermal conductivity K are then defined as follows:

$$\bar{\mu} = \bar{\mu}_e + \bar{\mu}_t$$

$$\bar{K} = C_P\left(\frac{\bar{\mu}_\ell}{Pr_\ell} + \frac{\bar{\mu}_t}{Pr_t}\right)$$

where $\bar{\mu}_\ell$ is the molecular viscosity and Pr_ℓ is the Prandtl number associated with the laminar flow (0.72). The turbulent Prandtl number Pr_t is assumed to be 0.9 in the present analysis; and

$$\bar{\mu}_t = \bar{\rho}\,\frac{\ell^2}{r}\left[\left(\frac{\tau_{12}}{\bar{\mu}}\right)^2 + \left(\frac{\tau_{32}}{\bar{\mu}}\right)^2\right]^{1/2}$$

$$\bar{\ell} = 0.09\ r\theta_e\ \tan h\left[\frac{0.4}{0.09}\,\frac{\theta}{\theta_e}\right] D$$

$$D = 1 - \exp\left(-\frac{r\theta}{A}\right)\ , \qquad A = 26.0\,\frac{\bar{\mu}}{\bar{\rho}}\left(\frac{\bar{\rho}}{\bar{\tau}}\right)_w^{1/2}$$

$$\tau_w = \left[(\tau_{12})_w^2 + (\tau_{23})_w^2\right]^{1/2}$$

The specific modifications made to the model for the present study were:

$$\left.\begin{aligned} \bar{\mu}_{t_\theta} &= C_1\bar{\mu}_t \\ \bar{\mu}_{t_\phi} &= C_2\bar{\mu}_t \end{aligned}\right\}\quad \text{Used for Rainbird's data}^7$$

$$\left.\begin{aligned} \bar{\mu}_{t_\theta} &= C_1\bar{\rho}\ell^2\left|\frac{\tau_{12}}{\bar{\mu}}\right| \\ \bar{\mu}_{t_\phi} &= C_2\bar{\rho}\ell^2\left|\frac{\tau_{32}}{\bar{\mu}}\right| \end{aligned}\right\}\quad \text{Used for McElderry's data}^8$$

A relaxation parameter[9] was used in both instances

$$\bar{\mu}'_{t_\phi} = \bar{\mu}_{t_{\phi_o}} + \left(\bar{\mu}_\phi - \bar{\mu}_{\phi_o}\right)\exp(-\Delta\phi/\lambda)$$

$$\lambda = 20\theta_{e_o}\ .$$

The initial point for the relaxation was the mesh point prior to crossflow separation. C_1 and C_2 are nonindependent constants which were used to adjust the levels of $\bar{\mu}_t$ for each coordinate direction. A bar denotes a dimensional quantity and the subscript w indicates evaluation at the wall. The viscous layer edge (θ_e) is defined to be that mesh point nearest to the wall where the modulus of vorticity is less than 0.0025 times the wall value.

THE NUMERICAL METHOD OF SOLUTION

The equation set resulting from the application of the conical symmetry assumption can be solved by time integration on a single spherical surface located at a distance r from the cone apex. A finite difference grid is defined on the spherical surface and MacCormack's[10] well-known finite difference technique is applied.

If the solution $D^i_{j,k}$ is known at time $t = i*\Delta t$ at each point of the computational mesh on the spherical surface (at a distance r from the cone apex), the solution at time $t = (i + 1)*\Delta t$ is calculated at each interior mesh point (θ, ϕ) as follows.[10]

$$\text{Predictor:}\quad \tilde{D}^{i+1}_{j,k} = D^i_{j,k} - \frac{\Delta t}{\Delta\theta}\left(F^i_{j+1,k} - F^i_{j,k}\right) - \frac{\Delta t}{\Delta\phi}\left(G^i_{j,k+1} - G^i_{j,k}\right) - \Delta t H'_{j,k}$$

$$\text{Corrector:}\quad D^{i+1}_{j,k} = \frac{1}{2}\left[D^i_{j,k} + \tilde{D}^{i+1}_{j,k} - \frac{\Delta t}{\Delta\theta}\left(\tilde{F}^{i+1}_{j,k} - \tilde{F}^{i+1}_{j-1,k}\right) - \frac{\Delta t}{\Delta\phi}\left(\tilde{G}^{i+1}_{j,k} - \tilde{G}^{i+1}_{j,k-1}\right) - \Delta t \tilde{H}'^{\,i+1}_{j,k}\right]$$

Auxiliary conditions imposed during the integration were as follows: (1) no slip conditions at the body surface with wall temperature maintained constant at the experimentally determined value; (2) free-stream conditions maintained at the outer edge of the mesh; (3) lateral symmetry of the flow quantities is maintained along the $\phi = 0°$ and $\phi = 180°$ lines; and (4) initial values for all fluid quantities were either free stream or a previous solution at similar conditions.

The use of free-stream conditions at the outer mesh boundary required that the bow shock wave be captured during the integration. To overcome the oscillations which occurred in the vicinity of the shock wave for high Mach number conditions, a technique called normal stress damping was used.[5] This technique involved changing the sign on the normal stress term (σ_η) and increasing the magnitude of this term to provide the necessary damping.

NUMERICAL RESULTS

The validity of any heretofore untried assumption applied to the governing equations can only be determined through comparison of the results obtained by calculation with results of experiment. To test the validity of the present technique, published experimental data were chosen over a Mach number range of 1.8 to 14 for comparison. Brief comparisons of the calculated results with experimental data are presented here for Mach numbers through 8; further comparisons are available in Ref. 5.

For Reynolds numbers corresponding to laminar flow conditions, the well-documented experimental study of Tracy[11] was chosen. These data were collected for a sharp 10° half-angle cone in air at a Mach number of 7.95. The free-stream Reynolds number was 1.25×10^6/ft with measurements taken 4 in. from the cone apex. Other conditions were $P_{t\infty} = 259.3$ psia and $T_{t\infty} = 1360°R$. Calculated and experimental results were obtained at angles of attack from 0° to 24°. Shown in Fig. 1 is the comparison of the experimental and calculated surface pressure at $\alpha = 24°$. The agreement is quite good except on the windward side of the cone, where experimental errors are suspected due to the use of pressure taps which are large in relation to boundary layer thickness. Figure 2 shows the calculated crossflow velocity vectors plotted for the $\alpha = 24°$ case. This figure clearly shows the large lee side vortices found in the experimental study. Also note that two stagnation points exist on the lee plane of symmetry: one associated with the dividing streamline around the vortex, and one "vortical singularity-like" stagnation point which occurs near the edge of the viscous layer. Figure 3 gives a pictorial representation of all flow field features compared with experiment. The calculated features, including the crossflow separation point and imbedded shock wave position, agree quite well with experiment.

For Reynolds numbers at which boundary layer transition has occurred, two experimental cases were chosen for comparison. The first was the experimental study of McElderry,[8] which was made using a 6° half-angle sharp cone at $M = 6.05$ in air. Data were collected at 36.3 in. along the cone at a free-stream Reynolds number of 5.0×10^6/ft. Stagnation pressure and temperature of the flow were 280 psia and 850°R, respectively. Wall temperature was 0.61 $T_{t\infty}$. Figure 4 gives the comparison of calculated and experimental surface pressure at 12° angle of attack. Two calculations are shown, one for laminar flow at $Re = 1.5 \times 10^6$ and one for turbulent flow at $Re - 15.0 \times 10^6$. The surface pressure for the laminar case agrees well until the first experimental separation point is reached. The laminar numerical solution does not separate until $\phi = 163°$ with the experimental primary separation occurring at $\phi = 130°$ and a secondary separation at $\phi = 166°$. The calculation using the turbulence model of the present study shows qualitatively a more accurate behavior of the pressure on the lee side (calculated separation point at $\phi = 130°$), but too low a turbulent boundary layer displacement thickness from approximately $\phi = 45°$ onward results in a lower surface pressure than that obtained in the experiment. The turbulence model was modified for this case by using $|\tau_{12}/\mu|$ in $\mu_{t\theta}$ and $|\tau_{32}/\mu|$ in $\mu_{t\phi}$ rather than the vector sum of τ_{12}/μ and τ_{32}/μ. The values of C_1 and C_2 used were 1.0 and 0.5, respectively.

The second turbulent flow case chosen for comparison is that of Rainbird.[7] The data of interest were collected on a 5° sharp cone in air at a Mach number of 1.8. Measurements were taken 3.94 ft along the cone at a free-stream Reynolds number of 7.5×10^6/ft. Stagnation pressure and temperature were 25 psia and 530°R, respectively, with wall temperature assumed to be T_{t_∞}. Figure 5 gives the comparison of calculation and experimental surface pressure for two sets of values for the constants C_1 and C_2. The dotted line in the figure gives the surface pressure obtained by using 1.0 for both C_1 and C_2. The turbulence model is therefore unmodified except for the use of relaxation in the crossflow component. The resulting pressure distribution does not agree with experiment past the $\phi = 110°$ point. The primary separation line is at $\phi = 140°$ in the calculation and $\phi = 132°$ in the experiment. The secondary separation does not exist in the calculation for this value of C_1 and C_2. The solid line in Fig. 5 gives the results of reducing the magnitude of eddy viscosity by $C_1 = 0.35$ and $C_2 = 0.525$. The surface pressure distribution for these values matched experiment more closely with the calculated primary separation at $\phi = 126°$. The secondary separated region was present in the calculation and the secondary separation line matched experiment at $\phi = 156°$.

A comparison in Fig. 6 of the calculated limiting streamline angle with the experimental surface oil flow angle confirms the more adequate agreement in the separated region for the modified values of eddy viscosity.

CONCLUSION

The solution of the Navier-Stokes equations subject to a conical symmetry assumption is shown to result in entirely adequate agreement with experiment for laminar flow about cones at angle of attack. The present laminar solutions exhibit all of the features found in the experimental case without depending on experiment for input.

It is apparent that the unmodified algebraic eddy viscosity model of the present study will not give agreement with experiment. However, the modifications proposed to the model give quantitative agreement with the data of Rainbird and qualitative agreement with the data of McElderry.

REFERENCES

1. Daywitt, J., Anderson, D., and Kutler, P., AIAA 15th Aerospace Sciences Meeting. 1977, Paper 77-86.

2. Miyazawa, M., Proceedings 10th International Symposium on Space Technology and Science, Tokyo, 1973, pp. 433-442.

3. Stetson, K. F., ARL 71-0127, Aerospace Research Laboratories, Wright-Patterson AFB, Ohio, August 1971.

4. R. F. Feldhun, AIAA J., 9, 1971, pp. 1074-1081.

5. McRae, D. S., AFFDL-TR-76-139, Air Force Flight Dynamics Laboratory, Wright-Patterson AFB, Ohio, 1976.

6. Harris, J. E. and Morris, D. J., Proceedings 4th International Conference on Numerical Methods in Fluid Dynamics, 1974, Lecture Notes in Physics No. 35, pp. 204-211.

7. Rainbird, W. J., AGARD Conference Proceedings No. 30, AGARD Specialists Meeting on Hypersonic Boundary Layers and Flow Fields, London, May 1-3, 1968, pp. 19-1 to 19-19.

8. McElderry, E. D., Jr., AFFDL TM-74-170, Air Force Flight Dynamics Laboratory, 1974.

9. Shang, J. S., Hankey, W. L., and Law, C. H., AIAA J., 14, 1976, pp. 1451-1457.

10. MacCormack, R. W., AIAA Paper 69-354, 1969.

11. Tracy, R. R., California Institute of Technology, Aeronautical Laboratories Memorandum No. 69, 1963.

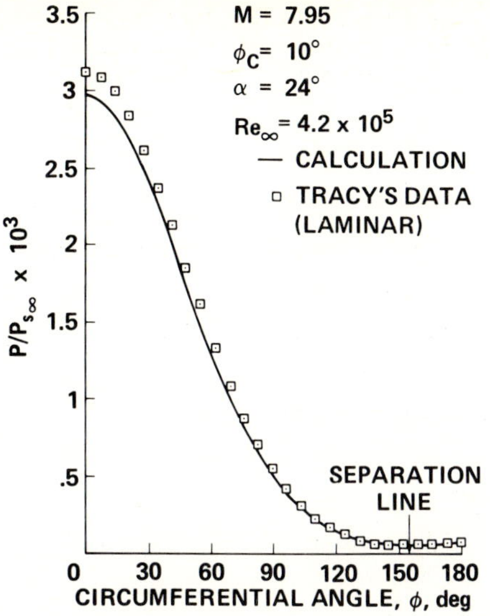

Figure 1. Comparison of calculated and experimental surface pressure at Tracy's[11] conditions, α = 24°.

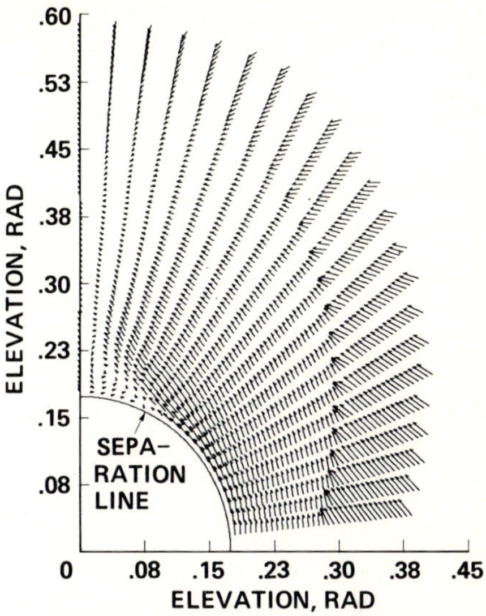

Figure 2. Calculated crossflow velocity vectors as seen on spherical surface at Tracy's[11] conditions, α = 24°.

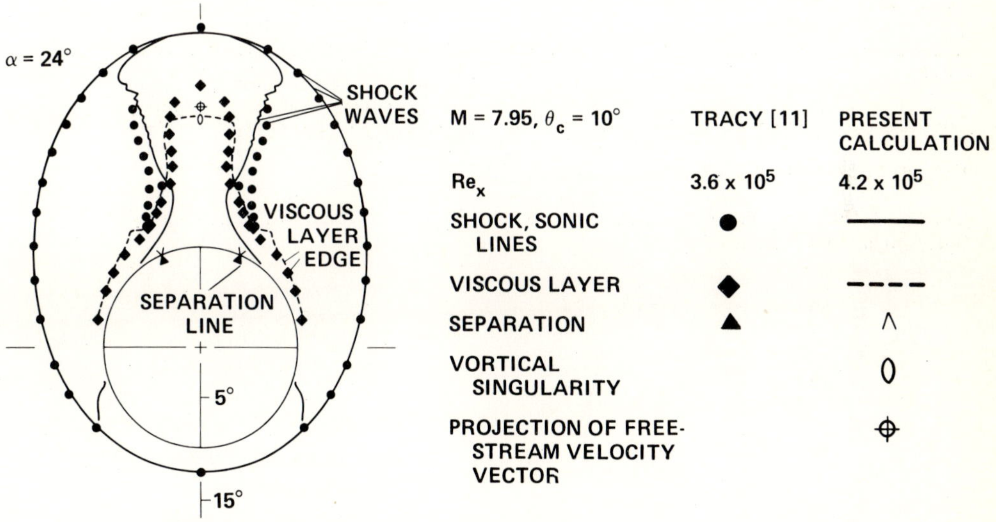

Figure 3. Comparison of calculated and experimental flow-field features at Tracy's[11] conditions, α = 24°.

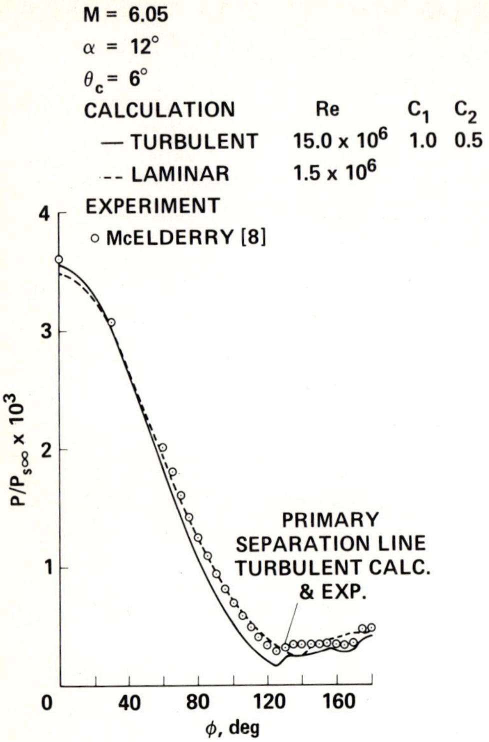

Figure 4. Comparison of calculated and experimental surface pressure at McEderry's[8] conditions, α = 12°.

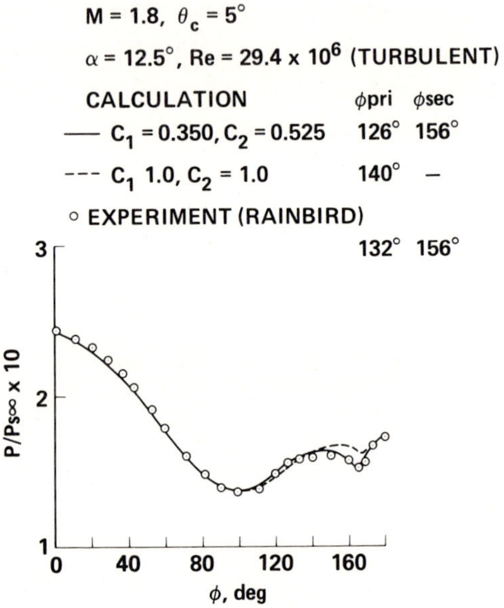

Figure 5. Comparison of calculated and experimental surface pressure at Rainbird's[7] conditions, α = 12.5°.

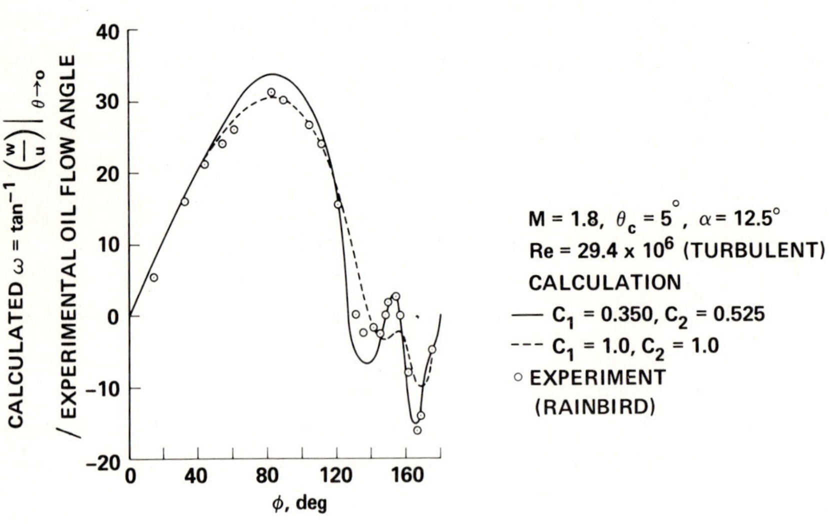

Figure 6. Comparison of calculated surface streamline angle with experimental surface oil flow angle.

CALCULATION OF SUPERSONIC GAS FLOWS ABOUT WINGS

Minaylos A.N.
(USSR)

A numerical method and a routine have been created for calculation of internal and external supersonic flows. Flow fields about wings and other airplane elements have been investigated.

For the numerical computation of three-dimensional perfect gas flows the marching shock capturing finite difference technique using the set of Euler equations is presented /1-2/. Calculations are executed in the longitudinal direction x in Cartesian coordinates x , y , z ; a body surface crosses net cells and is approximated by elements of planes in each cell in accordance with the method of minimum squares. Such approach gives a "strong" conservativeness of the system, and allows to investigate flow fields about any type of bodies using a single program.

The inclusion a new type of bodies into the program demands the introducing of a new block for the calculation of nonstandard cells at the body. These cells are replaced by "equivalent" ones whose volume is equal to that of a standard cell of the Cartesian coordinates at a given computational step.

A steady version of the monotonic finite difference scheme /3/ with a high order approximation in two variables is used. The schemes with nonuniform mesh stencils have been introduced by soviet mathematicians /3-5/; in this study such a scheme has been used for the calculation of the three-dimensional problem for the first time. The scheme gives all wash-out flow discontinuities more accurately /6/ than those of the first order or nonmonotonic schemes (Fig.1 shows the increase of errors of four different methods in L_2 rate for shock capturing calculation in x direction of a slip surface with $\rho_2/\rho_1=3$, $\varepsilon=(1/\rho_1 h_y)\int_0^y|\rho_{calc}-\rho_{exact}|dy$. The monotony permits to use discontinious initial data, to considerably extend the limits of a possible input parameters: the calculations have been executed in such ranges of Mach numbers, angles of attack and yaw: $1,2 \leq M \leq 20$; $0^0 \leq \alpha \leq 40^0$; $0^\circ \leq \beta \leq 15^\circ$.

The condition of zero normal velocity component is carried out on a body surface and it is needed to know only the pressure. It is determined by a calculation of the interference of a body surface element with the supersonic flow which is defined by the extrapolation from nearest mesh cells. The pressure in the vicinity of the subsonic edges is calculated by the extrapolation on a body surface in z direction. Numerical experiments have shown the errors of such extrapolation to have been localized in the vicinity of the edge.

A modification of the method is introduced that eliminates arising "calculated" entropy layers due to a loss of the approximation during "sharp edge capturing" /6/.

The accuracy of the results is checked by "internal" methods (at the 3200 grid points in y , z plane the errors about 2-3%%) and by comparisons with theoretical and measured data of other authors. The efficient method and flexible algorithm made it possible to solve a wide class of problems (during 4 years more than 700 different flow versions). Most results have been numerically received for the first time.

The classification of conical flows about slender wings with delta and trapezoidal planforms has been suggested /7/ depending on (see Fig.2):

a) the kind of the domain of influence,

b) the flow pattern on the low wing surface,

c) the flow scheme over the upper wing surface.

Boundaries of the flow regimes in the space $\theta - \alpha$ ($\theta = \pi/2 - \chi$, χ - edge sweep angle) for a slender delta wing is shown in Fig.3.

The position of the boundary 1-2 is defined by the extrapolation from the region 2 and requires a more precise definition. The condition that the flow with leading edge separation exists under this boundary is introduced on the basis of physical assumption. An analogous but a more rough condition based on the Newtonian theory of thin shock layer is proposed in /8/. The boundary of the separation zone due to the "scheme" dissipation lies under the line 1-2. The boundary condition and the flow separation algorithm /2/ are used to eliminate this difference. Thus, in the solution, the flow inside the vortex core doesn't correspond to a real gas flow (because of the different structure of dissipation tensors). In the regions B III and C III the conical sonic surface doesn't get out a corner point of a body, but supports the slip discontinuity (Fig.4, delta wing). The cross-section of the wing compression side being convex, additional singular straight stream lines may appear on the surface (regime 4 in Fig.2b). In this case the diagram $\theta - \alpha$ may change considerably: for example, regimes 1 and 2 may generally be absent.

The curvature of embedded shocks above a wing limiting a conical subsonic element may be of different sign; they are absent in the region D near a delta wing and a λ -shock formation is possible in the region; it is characteristic of a convex upper surface. Dissipative properties of a finite difference scheme effect the position of the boundaries I-II, III-IV and the dimensions of the separation bubble in the region II; additional studies are necessary to conform these results to a real flow. The separation bubble in the region IV localizes in the vicinity of an edge and, strictly speaking, the conical flow structure is violated. This region corresponds to large angles of attack, and isn't usually realized in experiments of a thick boundary layer.

For rectangular and trapezoidal wings an embedded shock exists in the region D in a gas having passed over a leading edge (Fig.5a); in other cases there are two waves (Fig.5b); in the regime IV the intensive vortex core may divide the shock 2 into two parts(Fig.5c).

Different combinations of regimes on the left and right leading edges are possible in yaw. Versions of such combinations are given in Fig.6.

Accuracy estimates of the aerodynamic characteristics defined by approximate theories are obtained on the examples of delta and rectangular wings of low aspect ratio.

The data of the slender body theory with a separated flow at delta plate edges /9/ are compared with the calculated results ($75° \leq \chi \leq 89°$; $1,3 \leq M \leq 5$) in Fig.7. Each line in Fig.7 corresponds to a pointed value of the theoretical and calculated results discrepancy (in per cents to calculated values). The acceptable accuracy of the theoretical results corresponds to the transonic flow $M < 1,3$.

The functions $C_n = f(\alpha)$, (χ = const and M = const) calculated for delta plates are presented in Messiter parameters /10/ of the Newtonian shock layer theory. These functions have an envelope (Fig.9), and all the results are piled up near this envelope in a narrow "corridor" for the parameter (characterizing density ratio across a wave) less than 0.7. For $\Omega <$ 2 the corridor width is not in excess of 4% of C_n value. Thus, region of the theory applicability for C_n coefficient covers a wide range of this parameter variation.

A number of examples are calculated with the initial data from such ranges: $45° \leq \chi \leq 80°$; $1,5 \leq M \leq 8$; $\alpha \leq 30°$.

These results are presented as well in the hypersonic parameters of Sychev /11/. These parameters are found to be valid in the whole range of the initial data.

The numerical solutions for a number of fully three-dimensional flows are obtained. The effect of the transverse and longitudinal surface curvature and aerofoil and trailing edge thickness on the flow about a delta wing of low aspect ratio. The flow about a delta wing with cranked leading edge is studied. Two types of the flow are obtained for such a wing with symmetrical aerofoil of 4% thickness ratio (Fig.9). The difference between these flow types is defined by the wing nose flow: without a well-developed separation (corresponds to the conical flow regimes I and IV), or with it (regimes II and III). In the last case an embedded λ - shock appears above the separation zone. Behind the edge crank, the vortex spirals are rolling up into cylindrical vortices, the axis of the vortex core is curving, and the embedded shock structure doesn't change (Fig.9). The bow shock is relaxed by the edge expansion in the vicinity of the edge above the wing. A system of new shocks and vortices, which is characteristic of the flow about a rectangular wing edge, arises behind the section A-A (Fig.9) near the wing edge. Fig.10 presents the full scheme of the flow.

Flow and downwash fields for a wedge-type body of finite width /12/ with (Fig.11-12) and without a basic surface crank are calculated. Such bodies are elements of a supersonic inlet at different operating conditions. An angle of attack is measured relative the upper body surface. Sometimes the slip discontinuity, which rolled up into a vortex core, may arise near the body low edge.

REFERENCES

1. Kosych A.P., Minaylos A.N. Tr. TsAGI, 1977, N 1809.

2. Minaylos A.N. Izv.Akad.Nauk SSSR, Mekh.Zhidk.Gasa, 1978, N 1, 78-89.

3. Kolgan V.P. Uchen.Zap.TsAGI, 1972, N 6, 68-77.

4. Fedorenko R.P. Zh.Vychisl.Mat. Mat.Fiz., 1962, N 6, 1122 - 1128.

5. Goldin V.Ya., Kalitkin N.N., Shishova T.V. Zh.Vychisl.Mat. Mat.Fiz., 1965, N 5, 938-944.

6. Minaylos A.N. Zh.Vychisl.Mat. Mat.Fiz., 1977, N 4, 1058 - - 1063.

7. Minaylos A.N. Uchen.Zap. TsAGI, 1977, N 4, 10-17.

8. Squire L.C. Aeronaut.Quart., 1976, XXVII, N 1.

9. Brown C.E., Michael W.H. NACA TN, 1955, N 3430.

10. Messiter A.F. AIAA J., 1963, N 4, 794-802.

11. Sychev V.V. Prikl.Mat.Mekh., 1962, 2, 205-212.

12. Bosnakov S.M., Minaylos A.N., Remeev N.Kh. Uchen.Zap.TsAGI, 1977, N 6, 125-131.

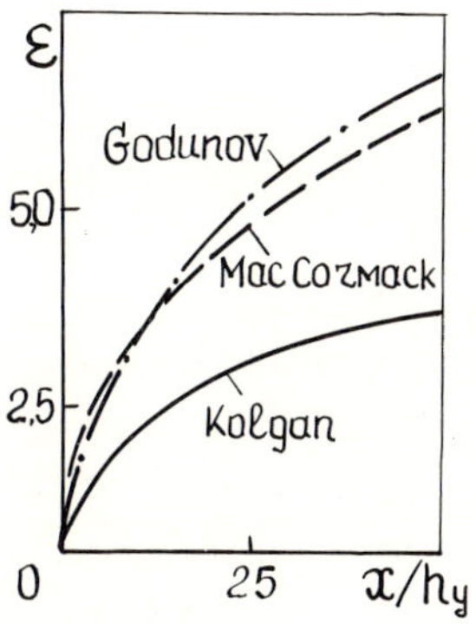

Fig.1

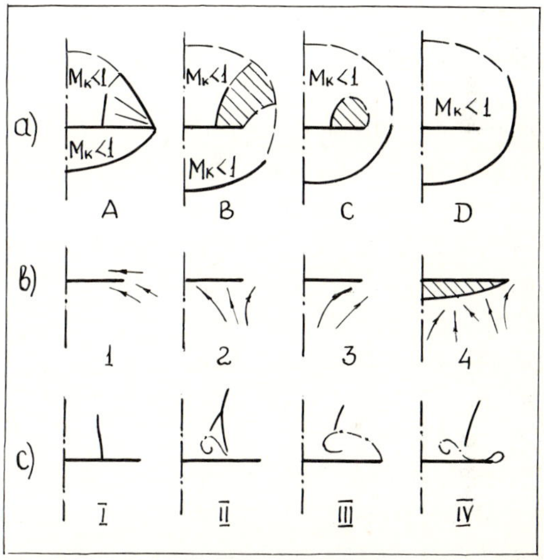

Fig.2

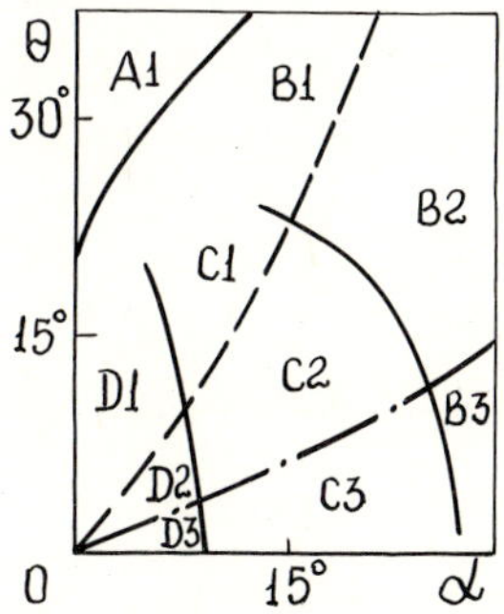

Fig.3

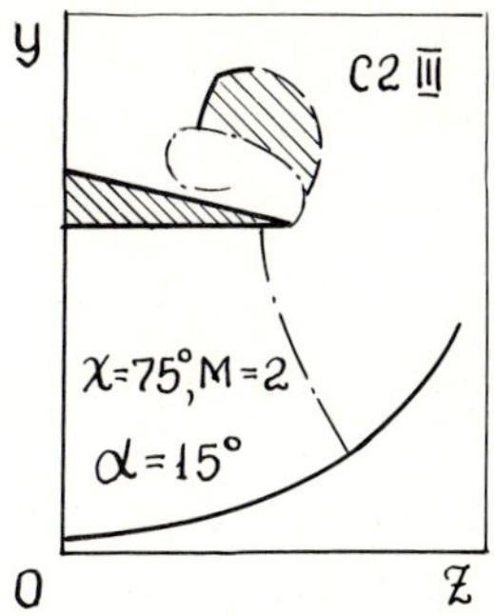

Fig.4

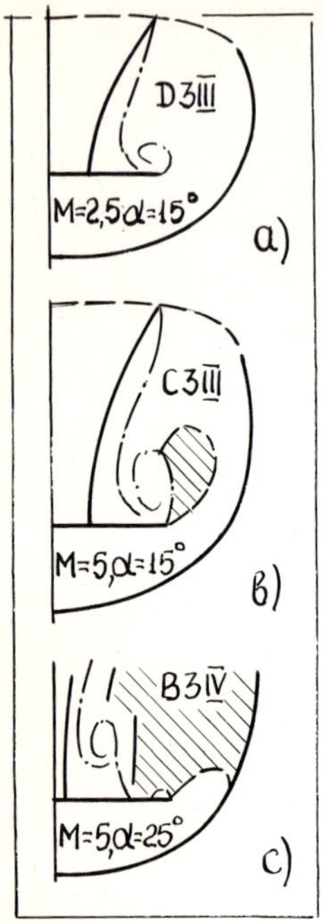

Fig.5

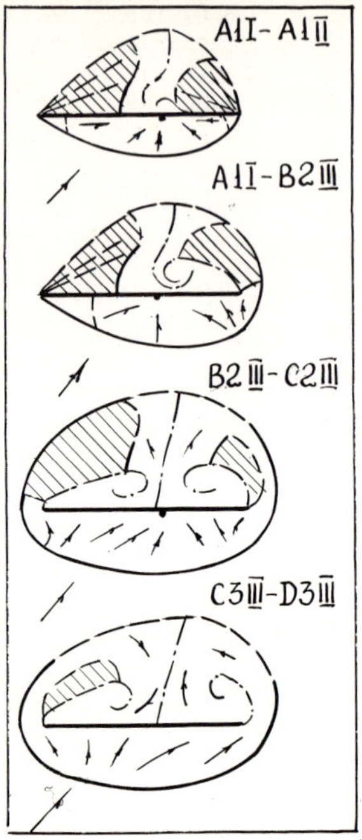

Fig.6

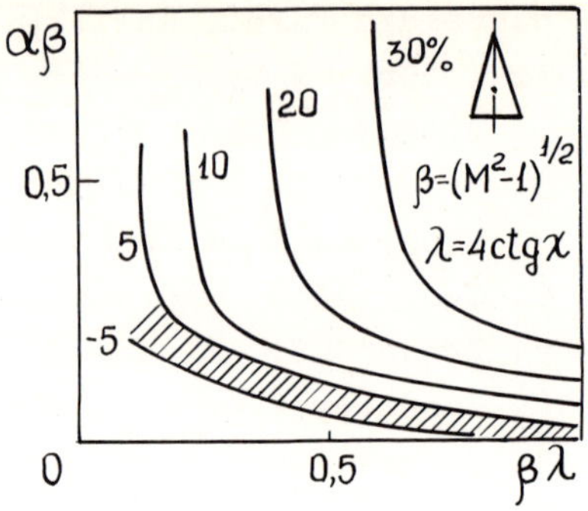

Fig.7

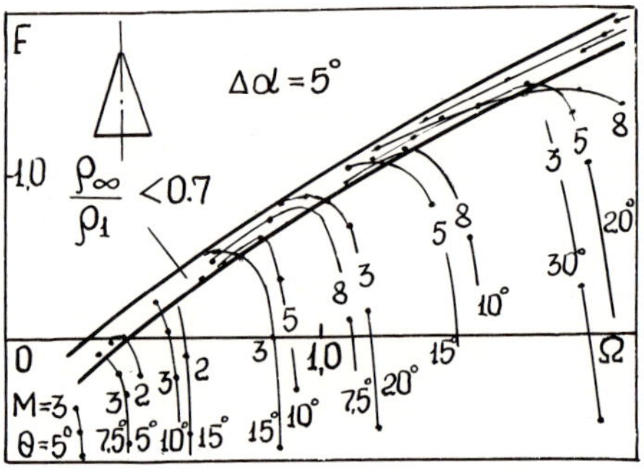

Fig.8

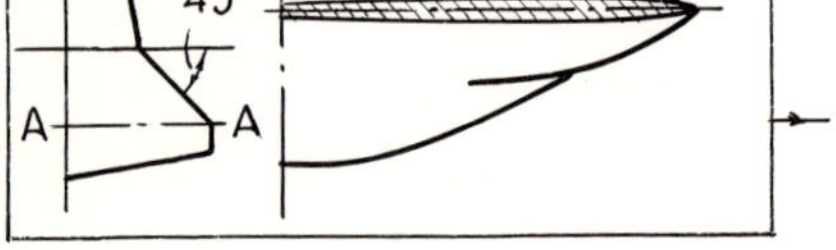

Fig.9

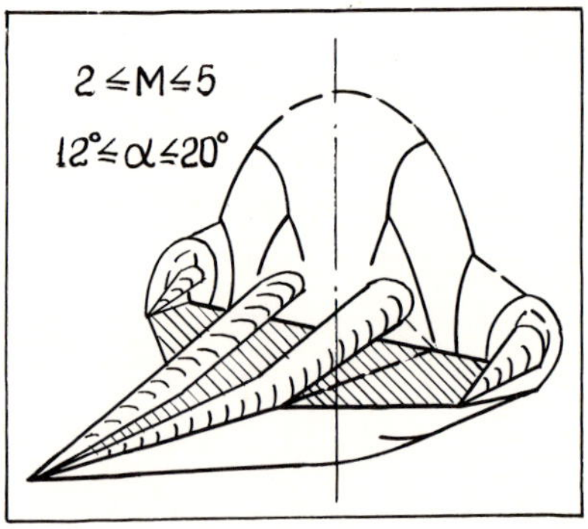

Fig.10

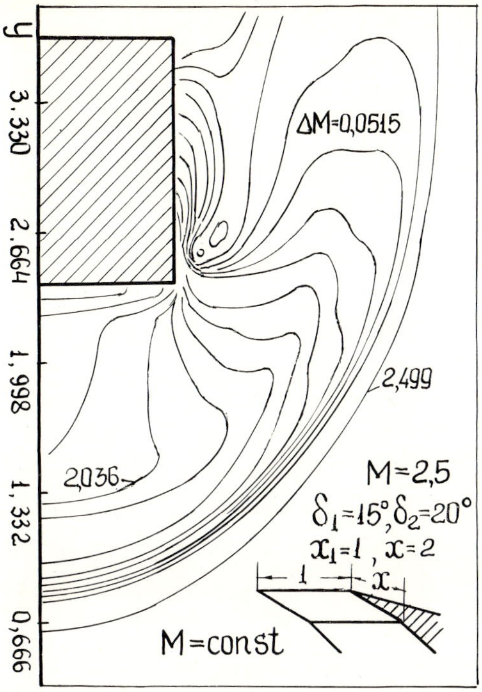
y
3,330
2,664
1,998
1,332
0,666
ΔM=0,0515
2,499
2,036
M=2,5
δ1=15°, δ2=20°
x1=1 , x=2
M=const

Fig.11

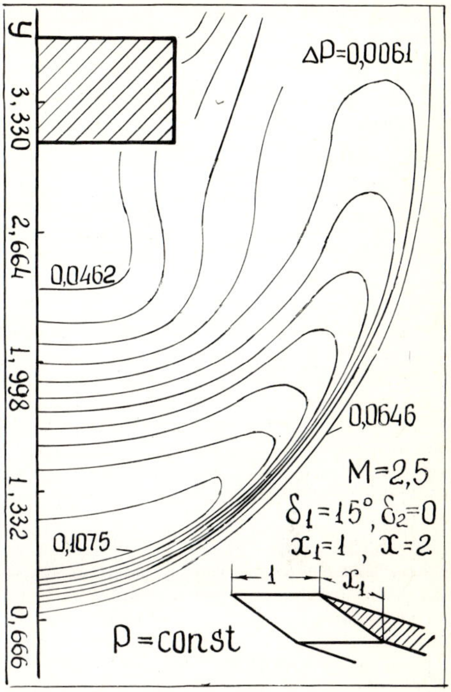
y
3,330
2,664
1,998
1,332
0,666
ΔP=0,0061
0,0462
0,0646
0,1075
M=2,5
δ1=15°, δ2=0
x1=1 , x=2
P=const

Fig.12

LARGE EDDY SIMULATION OF TURBULENT SHEAR FLOWS

by

P. Moin
N. N. Mansour
W. C. Reynolds
J. H. Ferziger

Department of Mechanical Engineering
Stanford University

1. INTRODUCTION AND OVERVIEW

Substantial effort has been devoted to the development of model equations for turbulent flow calculations [1]. These phenomenological models require connections between various statistical properties of the turbulence in order to effect closure of the governing equations. However, the large-scale structures that dominate the important statistical quantities are not universal, and so as yet no truly universal model has been obtained.

Large Eddy Simulation (LES) is an alternative approach that is now being developed. LES seeks to capitalize on the experimental observation that the smaller scales of turbulent motion are more nearly universal and hence more amenable to universal closure models. One then calculates the large-scale motions directly in a time-dependent, three-dimensional computation, which yields one realization of a turbulence time-space history. LES is now in an experimental stage but has promise of becoming a standard tool for computational fluid mechanics with continued improvement in computers.

The objectives of our program have been to develop a solid conceptual framework for LES, to explore its use in first simple and then more complex flows, and to develop new LES algorithms. This paper summarizes the conceptual foundation and concentrates on the numerical methods developed for simulation of the time-developing turbulent mixing layer and turbulent plane Poiseuille flow. The physical aspects of these two simulations will be described in greater detail elsewhere. For more detail see references [9] and [11].

In LES, the field variables (velocity and pressure for incompressible flow) are divided into large-scale and small-scale components,

$$f = \bar{f} + f' \tag{1.1}$$

where the overbars denote the large-scale or "filtered" field and the primes the residual or "subgrid" field. Conceptually, one defines the filtered field by [5]

$$f(\underline{x}) = \int G(\underline{x},\underline{x}')\, f(\underline{x}')\, d\underline{x}' \tag{1.2}$$

The filter function $G(\underline{x},\underline{x}')$ provides the smoothing that is necessary to define the large-scale component of the flow. It will have a parameter, its width, that characterizes the smallest scales of motion retained in the filtered field (the largest scales in the residual field). For homogeneous fields we usually use a Gaussian $G(\underline{x}-\underline{x}')$ with width Δ.

Filtering the Navier-Stokes equations, one obtains the equations for the filtered field,

$$\frac{\partial \bar{u}_i}{\partial t} + \overline{\bar{u}_j \left(\frac{\partial \bar{u}_i}{\partial x_j} - \frac{\partial \bar{u}_j}{\partial x_i}\right)} = -\frac{\partial P^*}{\partial x_i} + \frac{\partial \tau_{ij}}{\partial x_j} + \nu\nabla^2\bar{u}_i \quad , \qquad \frac{\partial \bar{u}_i}{\partial x_i} = 0 \quad . \tag{1.3a,b}$$

Here the term τ_{ij}, which must be modeled, is all the terms containing u_i', i.e., the subgrid-scale Reynolds stress, and P^* is $(P/\rho + \overline{\bar{u}_j\bar{u}_j}/2)$. The form of the convective terms in (1.3a) is chosen because it is easiest to assure the proper inviscid conservations of circulation, momentum, and energy when this form is differenced [2].

The model most commonly used for τ_{ij}, which has some physical bases, is the Smagorinsky model [3],

$$\tau_{ij} = 2\nu_T S_{ij} \quad , \quad \nu_T = (C_s\Delta)^2 \sqrt{S_{ij}S_{ij}} \quad , \quad S_{ij} = \frac{1}{2}\left(\frac{\partial\bar{u}_i}{\partial x_j} + \frac{\partial\bar{u}_j}{\partial x_i}\right) \qquad (1.4)$$

Here ν_T is a local eddy viscosity of the subgrid scale motions, C_s is a constant (the only empirical constant of the LES method), and Δ is the filter width. Note that $\nu_T \to 0$ as $\Delta \to 0$. C_s has been evaluated by matching the computed and experimental rates of energy decay in isotropic turbulence, and has also been calculated directly from a very fine-scale calculation that captured all of the turbulence scales [4].

Filtering produces terms with double overbars (see 1.3a), and these have been handled in two ways. In our earlier work we related $\overline{\bar{u}_i\bar{u}_j}$ to $\bar{u}_i$ and $\bar{u}_j$ using a Taylor's series expansion [5]. In later work we have instead computed the double-bar quantity directly from the bar quantities by Fourier methods [6].

In simulating homogeneous flows, we advance the velocity in time using an explicit algorithm (e.g., the second-order Adams-Bashforth method). The pressure is chosen such that continuity equation is satisfied at the new time step by solving a Poisson equation for pressure, which we do by Fourier methods. This approach is quite efficient and works well, provided care is taken in defining the divergence and gradient difference operators [7].

Our earliest work used a fourth-order central difference scheme for evaluating the spatial derivatives [7]. We have also used a staggered grid method [8], pseudo-spectral methods, and the fourth-order compact scheme of Kreiss [2]; these all work very well for homogeneous flows, with mesh sizes as coarse as $16 \times 16 \times 16$.

Our current work deals with the extension of LES to flows that are inhomogeneous in one spatial dimension, in particular the time-developing mixing layer and channel flow. The LES of these flows required the development of new numerical methods, described herein.

For the mixing layer, a successful LES will have to display the entrainment of irrotational fluid by the vorticity-bearing turbulence. This led us to use the dynamical equations of vorticity rather than the primitive equations for this part of the study. This calculation is discussed in §2.

In the channel flow, one must deal with solid (no-slip) boundaries. Near such a boundary the length scales of the turbulence become highly anisotropic, and viscous effects, which are negligible away from a solid boundary, become very important. This led us to use a grid that is highly nonuniform, very fine near the wall and rather coarse in the center of the channel. The subgrid-scale model was modified to account for the anisotropy of the grid and the importance of viscous effects near the wall.

For the channel flow problem, a new semi-implicit method has been developed. This method avoids the use of the Poisson equation for pressure, and has considerable promise for use in other flows as well as a basis for both finite-difference and spectral methods (see §3). The serious numerical problem that can appear in fully explicit spectral methods [9] is avoided by spectral versions of this method.

All the calculations reported here were carried out on a CDC 7600 at the NASA-Ames Research Center by members of the Stanford team.

2. FREE SHEAR LAYER

The physical problem to be treated is the time-developing shear layer between a uniform stream of velocity $U/2$ as $x_2 \to +\infty$ and another of velocity $-U/2$ as $x_2 \to -\infty$. The vorticity-containing region grows with time as it entrains fluid which was originally non-vortical. Following this growth for long periods of time would require an expanding mesh. The present work used a fixed mesh and hence simulates only a relatively short part of the history of the shear layer.

Since vorticity plays such a central role in this flow, the vorticity equations were chosen as the basis for the calculation. The necessary equations can be found either by filtering the continuous equations for the vorticity or by taking the curl of the filtered equation (1.3a) for the velocity. The filtered vorticity equations are:

$$\frac{\partial \bar{\omega}_i}{\partial t} + \frac{\partial}{\partial x_j}(\overline{\bar{u}_j \bar{\omega}_i} - \overline{\bar{u}_i \bar{\omega}_j}) + \frac{\partial}{\partial x_j} W_{ij} = \nu \nabla^2 \bar{\omega}_i \qquad (2.1)$$

where W_{ij} represents all of the terms arising from the subgrid-scale vorticity. This term has been modeled by

$$W_{ij} = -\frac{\partial}{\partial x_j}(\nu_T \bar{\omega}_i) + \frac{\partial}{\partial x_i}(\nu_T \bar{\omega}_j) \qquad (2.2)$$

which satisfies all of the requirements of a valid subgrid-scale model: antisymmetry, vanishing in an irrotational region, and being definitely energy-dissipative. For the small-scale eddy viscosity, it is natural to use

$$\nu_T = (C_v \Delta)^2 (\bar{\omega}_i \bar{\omega}_i)^{1/2} \qquad (2.3)$$

C_v was evaluated by reference to experiments in isotropic turbulence decay [6].

The problem is to solve Eqn. (2.1), subject to a given initial vorticity distribution and appropriate boundary conditions. The flow is very sensitive to the initial vorticity distribution, and the major focus of the study has been the effect of various initial conditions on the time development of the shear layer. In the directions in which the flow is statistically homogeneous, i.e., in x_1, the streamwise direction, and x_3, the spanwise direction, we used periodic boundary conditions. In the x_2-direction, we imposed the boundary conditions at some large but finite distance from the vorticity-containing layer. Specifically, the boundary conditions on the velocity (which will be needed in order to obtain the velocity field from the vorticity field) are

$$\left.\begin{aligned} \frac{\partial \bar{u}_1}{\partial x_2} &= 0 \qquad (2.4a)\\ \bar{u}_2 &= 0 \qquad (2.4b)\\ \frac{\partial \bar{u}_3}{\partial x_2} &= 0 \qquad (2.4c) \end{aligned}\right\} \quad \text{at} \quad x_2 = 0, L$$

All of the spatial derivatives are computed by the pseudospectral method [10], i.e., by means of Fourier transforms. In the x_1 and x_3 directions, in which the boundary conditions are periodic, this can be done by using the standard complex discrete Fourier transform. For example, to compute $\partial \bar{u}_1/\partial x_1$ we represent

$$\bar{u}_1(x_1,x_2,x_3) = \sum_{k_1} \hat{\bar{u}}_1(k_1,x_2,x_3)\, e^{ik_1 x_1} \qquad (2.5a)$$

Then

$$\frac{\partial \bar{u}_1}{\partial x_1} = \sum_{k_1} (ik_1)\, \hat{\bar{u}}_1(k_1,x_2,x_3)\, e^{ik_1 x_1} \qquad (2.5b)$$

Hence, we first Fourier transform $\bar{u}_1$ in the x_1 direction by FFT, then multiply by ik_1, and then invert to obtain $\partial \bar{u}_1/\partial x_1$. For the last component of the transform, we multiply by zero rather than k_1 in order to maintain conservation [6]. Hereafter, we shall use k' to indicate places where this is done.

If $f(x_2)$ vanishes at $x_2 = 0$ and L, we use the discrete Fourier sine series:

$$f(x_2) = \sum_{n_2} \hat{f}^s(n_2) \sin\left[\frac{n_2\pi}{(N_2-1)h_2} x_2\right] \tag{2.6}$$

whose inverse is

$$\hat{f}^s(n_2) = \frac{1}{(N_2-1)} \sum_{x_2} f(x_2) \sin\left[\frac{n_2\pi}{(N_2-1)h_2}\right] \tag{2.7}$$

Here N_2 is the number of mesh points and h_2 is the mesh size in the x_2-direction. If $\partial f/\partial x_2$ vanishes at $x_2 = 0$ and L, then we use a discrete Fourier cosine series.

The Fourier sine transform can be computed very efficiently by using a modified version of the FFT routine; we call this the fast discrete sine transform (FDST). The FDST and its cosine analog, the FDCT, each require approximately twice the computational time as the FFT.

To compute the derivative of a function $f(x)$ which is expanded in a discrete sine series, we just take its sine transform to form $\hat{f}^s(k_\ell)$ then multiply $\hat{f}^s(k_\ell)$ by k_ℓ and invert the result using a cosine transform. A similar procedure is used for a function which is expanded in a cosine series, except that the roles of the sine and cosine transforms are reversed [11].

To advance in time, a second-order Adams-Bashforth method is used. If $\partial\bar{\omega}_i/\partial t = M_i$, ω_i at time $n+1$ is given by

$$\bar{\omega}_i^{n+1} = \bar{\omega}_i^n + \Delta t \left(\frac{3}{2} M_i^{(n)} - \frac{1}{2} M_i^{(n-1)}\right) \tag{2.10}$$

In our case,

$$M_i = -\frac{\partial}{\partial x_j} \overline{(\bar{u}_j\bar{\omega}_i - \bar{u}_i\bar{\omega}_j)} - \frac{\partial}{\partial x_j} W_{ij} \tag{2.11}$$

The calculation is started with one step by the Euler method.

Having the vorticity field $\bar{\omega}_i$ at time step n, we have to find the velocity field in order to be able to advance in time. To do so, we define a vector potential ψ_k, such that

$$\bar{u}_i = \varepsilon_{ijk} \frac{\delta}{\delta x_j} \psi_k \tag{2.12}$$

The δ's denote the difference approximation to the partial derivatives. ψ_i can be chosen to be solenoidal; i.e.,

$$\frac{\delta}{\delta x_i} \psi_i = 0 \tag{2.13}$$

Taking the discrete curl of Eqn. (2.12) and using Eqn. (2.13), we get

$$\frac{\delta}{\delta x_j} \frac{\delta}{\delta x_j} \psi_i = -\bar{\omega}_i \tag{2.14}$$

The discrete Fourier transform of Eqn. (2.14) is

$$-k^2\psi_i = -\hat{\bar{\omega}}_i \tag{2.15}$$

Thus we divide $\hat{\bar{\omega}}_i$ by k^2 and invert the Fourier transform to obtain ψ_k. The velocities are obtained by taking the curl of the vector potential.

Computations based on the method outlined above were carried out to study the effects of initial conditions on the growth of the shear layer. These simulations revealed that the rate of growth of the layer is very sensitive to the initial field. An initial field consisting of the observed mean field plus random turbulence will not grow properly. However, if the mean field is replaced with a row of large-scale vortices that produce the same mean field, proper growth is observed. The rate of growth is sensitive to the three-dimensional character of the large-scale vortices. Experiments [12] confirm these observations. More details on these physical results will be reported elsewhere (see [11]).

3. CHANNEL FLOW

A. Grid Selection

In the numerical simulation of turbulent channel flow, all the variables are non-dimensionalized by the shear velocity, u_τ, and the channel half width, δ. The Reynolds number based on δ and u_τ is specified as 640, and the calculation therefore imposes a fixed pressure gradient on the flow.

In the vertical direction $(-1 \leq y \leq 1)$, a nonuniform grid spacing is used. The following transformation gives the location of grid points in the vertical direction [13]. For notation simplicity, we occasionally shift from (x_1,x_2,x_3), (u_1,u_2,u_3) to (x,y,z), (u,v,w). We used

$$y_j = \frac{1}{a} \tanh\left[\xi_j \tanh^{-1}(a)\right] \tag{3.1a}$$

with a uniform mesh in ξ,

$$\xi_j = -1 + 2(j-2)/(N-3) \quad , \quad j = 1,2,\ldots,N \tag{3.1b}$$

N is the total number of grid points in the y direction, and the adjustable parameter of the transformation is a $(0 < a < 1)$. We used $a = .98346$, $N = 65$. The location of the first few grid points in the vicinity of the (lower) wall and their corresponding y^+ $(= y_w u_\tau/\nu)$ values are given below for the Reynolds number used

j	y_j	$y_{w_j} = 1 + y_j$	$y^+_{w_j}$
1	-1.002	fictitious external	point
2	-1.0	0.0	0.0
3	- .997219	.00278	1.78
4	- .993983	.00602	3.85
5	- .990220	.00978	6.26
6	- .985847	.01415	9.06

The length L_x and L_z of the computational box in the streamwise (x) and spanwise (z) directions, in which periodic boundary conditions are used, should be long enough to include the important large eddies [9]. Based on the two-point correlation measurements of Comte-Bellot [14], we used $L_x = 2\pi$, $L_z = 4\pi/3$. Due to computer cost and storage limitations, only 16 uniformly spaced grid points were used in each of these directions. The resulting grid resolution is not adequate for complete simulation of the important large-scale field. Consideration of the known fine spacings between flow structures in the vicinity of the wall suggests that at least 32 times as many grid points would be required in the x-z planes to resolve them completely. Nevertheless, we found that this calculation did reveal these structures qualitatively, though not at their proper scale, and many other features are properly predicted [9].

Near the wall we modified the subgrid model to reflect a transition from inertia-dominated subgrid motions to viscous-dominated turbulence. For details see [9].

B. Numerical Differentiation

In the y direction we used central differencing with variable grid spacing, $h_j = y_j - y_{j-1}$ and $j = 2,3,\ldots,64$. The partial derivatives, with the first truncation error term included, are then

$$\left(\frac{\partial f}{\partial y}\right)_j = \frac{f_{j+1} - f_{j-1}}{h_{j+1} + h_j} - \frac{1}{2}(h_{j+1} - h_j)\left(\frac{\partial^2 f}{\partial y^2}\right)_j + O\left(h_j^2\right) \tag{3.2}$$

$$\left(\frac{\partial^2 f}{\partial y^2}\right)_j = 2\left[\frac{f_{j-1}}{h_j(h_j+h_{j+1})} - \frac{f_j}{h_j h_{j+1}} + \frac{f_{j+1}}{h_{j+1}(h_j+h_{j+1})}\right] - \frac{h_{j+1}-h_j}{3}\left(\frac{\partial^3 f}{\partial y^3}\right)_j + O\left(h_j^2\right) \tag{3.3}$$

It can be easily shown [15] that these expressions are second-order accurate in terms of $\Delta\xi$.

The pseudo-spectral method was used for calculation of partial derivatives in the streamwise and spanwise directions. The accuracy thereby obtained in part compensates for the use of a coarse mesh.

C. A Semi-Implicit Scheme

Due to the presence of diffusion terms in the governing equations and the use of a very fine mesh near the boundaries, the time-step requirement of a fully explicit method is severe. To circumvent this difficulty, we devised a semi-implicit algorithm that treats pressure and part of the diffusion terms implicitly in the y direction. The equation of continuity is solved directly. It is important to appreciate that the ideas developed in this section are not restricted to finite spatial differencing or to this particular flow.

Equation (1.3a) may be written as

$$\frac{\partial \bar{u}_i}{\partial t} = H_i - \frac{\partial \bar{P}}{\partial x_i} + \left(C_i \tilde{\nu}_T + \frac{1}{Re_\tau}\right)\frac{\partial^2 \bar{u}_i}{\partial x_2^2} \qquad \text{(no summation)} \tag{3.4}$$

where

$$H_i = -\bar{u}_j\left(\frac{\partial \bar{u}_i}{\partial x_j} - \frac{\partial \bar{u}_j}{\partial x_i}\right) + \frac{1}{Re_\tau}\left(\frac{\partial^2 \bar{u}_i}{\partial x_1^2} + \frac{\partial^2 \bar{u}_i}{\partial x_3^2}\right) + \frac{\partial}{\partial x_1}\left[\nu_T\left(\frac{\partial \bar{u}_i}{\partial x_1} + \frac{\partial \bar{u}_1}{\partial x_i}\right)\right]$$

$$+ \frac{\partial}{\partial x_3}\left[\nu_T\left(\frac{\partial \bar{u}_i}{\partial x_3} + \frac{\partial \bar{u}_3}{\partial x_i}\right)\right] + \left(\frac{\partial}{\partial x_2}\nu_T\right)\left(\frac{\partial \bar{u}_i}{\partial x_2} + \frac{\partial \bar{u}_2}{\partial x_i}\right) + \nu_T' \frac{\partial}{\partial x_2}\left(\frac{\partial \bar{u}_i}{\partial x_2} + \frac{\partial \bar{u}_2}{\partial x_i}\right) + \delta_{i1} + Q_i$$

$$C_i = 1 + \delta_{i2}, \qquad Q_i = \tilde{\nu}_T \frac{\partial}{\partial x_2}\left(\frac{\partial \bar{u}_2}{\partial x_i}\right) d_i \qquad \text{(no summation)}$$

$$d_i = 1 - \delta_{i2}, \qquad \tilde{\nu}_T = \langle \nu_T(x_1,x_2,x_3) \rangle_{x_1,x_3}, \qquad \nu_T' = \nu_T - \tilde{\nu}_T$$

$\langle\ \rangle_{x_1,x_3}$ indicates the average of bracketed quantity in x_1-x_3 plane. The rationale for this decomposition of ν_T will be clear shortly.

For time advancing, we use the Adams-Bashforth method on H_i, and the Crank-Nicolson method [16] on $\partial\bar{P}/\partial x_i$ and $\partial^2\bar{u}_i/\partial x_2^2$, in the right-hand side of Eqn. (3.4). To avoid iteration, we evaluated $\tilde{\nu}_T$ at time step n. Thus, we have

$$\bar{u}_i^{n+1} = \bar{u}_i^n + \Delta t\left(\frac{3}{2}H_i^n - \frac{1}{2}H_i^{n-1}\right) - \frac{\Delta t}{2}\left(\frac{\partial \bar{P}^{n+1}}{\partial x_i} + \frac{\partial \bar{P}^n}{\partial x_i}\right)$$

$$+ \left(\frac{1}{Re_\tau} + C_i\tilde{\nu}_T^n\right)\frac{\Delta t}{2}\left(\frac{\partial^2 \bar{u}_i^{n+1}}{\partial x_2^2} + \frac{\partial^2 \bar{u}_i^n}{\partial x_2^2}\right) \qquad \text{(no summation)} \tag{3.5}$$

Setting

$$\beta_i(x_2) = -\frac{2/\Delta t}{\left(\frac{1}{Re_\tau} + C_i\tilde{\nu}_T^n(x_2)\right)}$$

rearrangement of Eqn. (3.5) yields:

$$\frac{\partial^2 \bar{u}_i^{n+1}}{\partial x_2^2} + \beta_i \bar{u}_i^{n+1} + \beta_i \frac{\Delta t}{2} \frac{\partial \bar{P}^{n+1}}{\partial x_i} = \beta_i \bar{u}_i^n + \beta_i \Delta t \left(\frac{3}{2} H_i^n - \frac{1}{2} H_i^{n-1}\right) - \beta_i \frac{\Delta t}{2} \frac{\partial \bar{P}^n}{\partial x_i} - \frac{\partial^2 \bar{u}_i^n}{\partial x_2^2}$$

(no summation) (3.6)

Finally, we write the continuity equation at time step $n+1$ as

$$\frac{\partial \bar{u}_i^{n+1}}{\partial x_i} = 0 \tag{3.7}$$

We Fourier transform Eqns. (3.6) and (3.7) in x_1 and x_3 directions. This converts the set of partial differential equations (3.6) and (3.7) to a set of ordinary equations for the variables at time step $n+1$, for every pair of Fourier modes k_1, k_3, with $y = x_2$ as the independent variable. In the remainder of this section all the dependent variables are to be interpreted as two-dimensional Fourier transforms of the quantities.

$$\frac{\partial^2 u_1^{n+1}}{\partial y^2} + \beta_1 u_1^{n+1} + ik_1\beta_1 \frac{\Delta t}{2} P^{n+1} = \beta_1 u_1^n + \beta_1 \frac{\Delta t}{2}\left(3H_1^n - H_1^{n-1}\right) - ik_1\beta_1 \frac{\Delta t}{2} P^n - \frac{\partial^2 u_1^n}{\partial y^2} \tag{3.8a}$$

$$\frac{\partial^2 u_2^{n+1}}{\partial y^2} + \beta_2 u_2^{n+1} + \beta_2 \frac{\Delta t}{2} \frac{\partial P^{n+1}}{\partial y} = \beta_2 u_2^n + \beta_2 \frac{\Delta t}{2}\left(3H_2^n - H_2^{n-1}\right) - \beta_2 \frac{\Delta t}{2} \frac{\partial P^n}{\partial y} - \frac{\partial^2 u_2^n}{\partial y^2} \tag{3.8b}$$

$$\frac{\partial^2 u_3^{n+1}}{\partial y^2} + \beta_3 u_3^{n+1} + ik_3\beta_3 \frac{\Delta t}{2} P^{n+1} = \beta_3 u_3^n + \beta_3 \frac{\Delta t}{2}\left(3H_3^n - H_3^{n-1}\right) - ik_3\beta_3 \frac{\Delta t}{2} P^n - \frac{\partial^2 u_3^n}{\partial y^2} \tag{3.8c}$$

$$ik_1 u_1^{n+1} + \frac{\partial u_2^{n+1}}{\partial y} + ik_3 u_3^{n+1} = 0 \tag{3.8d}$$

Thus, for every pair of k_1 and k_3 we have four coupled linear ordinary differential equations with $u_1^{n+1}(k_1,y,k_3)$, $u_2^{n+1}(k_1,y,k_3)$, $u_3^{n+1}(k_1,y,k_3)$, and $P^{n+1}(k_1,y,k_3)$ as unknowns. Note that, with no further complications, one can treat additional terms in Eqn. (1.3a) (e.g., $(1/Re_\tau)(\partial^2/\partial x^2)$, $(1/Re_\tau)(\partial^2/\partial z^2)$, etc.) implicitly.

Note that, in order to avoid evaluating complicated convolution sums, we decomposed ν_T into its planar average $\tilde{\nu}_T(x_2)$ and "fluctuating" component $\nu_T'(x_1,x_2,x_3)$. We used explicit time advancing for $\nu_T'(\partial^2 u_i/\partial y^2)$, whereas $\tilde{\nu}_T(\partial^2 u_i/\partial y^2)$ was treated implicitly. This decomposition of ν_T may not be an optimum one from the standpoint of numerical stability. Other choices are possible. For example, one could decompose ν_T as $\nu_T(x_1,x_2,x_3) = \max_{D}(\nu_T) + \nu_T''(x_1,x_2,x_3)$. This decomposition would be worth exploring.

D. Finite-Difference Formulation and Boundary Conditions

In order to solve Eqns. (3.8) numericall, we used the finite difference operators (3.2) and (3.3) to approximate $\partial/\partial y$ and $\partial^2/\partial y^2$. This gives a set of linear algebraic equations for the Fourier transforms of the dependent variables. The system is of block tri-diagonal form and can be solved efficiently. In order to close the system we must provide a set of boundary conditions, i.e., we have to specify the values of u_1, u_2, u_3, and P at the solid boundaries.

Implementation of velocity boundary conditions poses no problem; we simply set the velocity vector zero on the walls. In order to obtain the pressure boundary conditions, we note that the normal momentum equation (3.8b), at the walls, yields

$$\left[\frac{\partial^2 u_2^{n+1}}{\partial y^2} + \beta_2 \frac{\Delta t}{2} \frac{\partial P^{n+1}}{\partial y}\right]_{y=\pm 1} = -\left[\beta_2 \frac{\Delta t}{2} \frac{\partial P^n}{\partial y} + \frac{\partial^2 u_2^n}{\partial y^2}\right]_{y=\pm 1}$$

This is consistent with the continuous normal momentum equation evaluated at the walls,

$$\left.\frac{\partial P}{\partial y}\right|_{y=\pm 1} = \frac{1}{Re_\tau} \left.\frac{\partial^2 u_2}{\partial y^2}\right|_{y=\pm 1} \tag{3.9}$$

since at $y = \pm 1$, $\beta_2(\Delta t/2) = - R_\tau$. Thus, we formally use Eqn. (3.9) as the pressure boundary condition. However, the algebraic equations require the value of the pressure transform at the wall. For this we use the following expansion.

$$P_2 = P_3 - \frac{h_3}{2}\left(\left.\frac{\partial P}{\partial y}\right|_{j=2} + \left.\frac{\partial P}{\partial y}\right|_{j=3}\right) + O\left(h_3^3\right) \tag{3.10}$$

(Recall that point $j = 2$ is the lower wall.)

Using the finite-difference analog of Eqn. (3.9) for the derivative at $j = 2$ and differencing the derivative at $j = 3$, and using the finite-difference form of the continuity equation at the wall (which determines the value of u_2 at the fictitious external point), we obtain

$$P_2 = \left[\frac{2P_3}{h_3} - \frac{P_4}{(h_3 + h_4)} - \frac{2u_{2_3}}{Re_\tau h_2 h_3}\right] \Bigg/ \left(\frac{2}{h_3} - \frac{1}{h_4 + h_3}\right) \tag{3.11}$$

The velocities u_1 and u_3 at the fictitious external point will not enter into the computation. However, they can be taken as those that force the x_1 and x_3 momentum equations to be satisfied at the boundary. This would be necessary, for example, if one desired to compute the local wall stress. Hence, in this formulation we can regard all of the governing equations as being satisfied at the boundary, as well as in the interior region.

An analogous relation is used for the value of the pressure at the upper wall ($j = N - 1$). Note that the pressure is still indeterminate by a constant, as it should be due to the use of Neumann boundary conditions.

In the case $k_1 = k_3 = 0$, a special solution technique must be used. First, in this special case Eqns. (3.8a) and (3.8c) are independent of each other and of Eqns. (3.8b) and (3.8d). Furthermore, the former two equations are of simple tridiagonal form and can be solved directly to yield $u_1^{n+1}(0,y,0)$ and $u_3^{n+1}(0,y,0)$. Second, the continuity equation and the boundary conditions for u_2 yield

$$u_2(0,y,0) = 0 \tag{3.12}$$

Since pressure is indeterminate by a constant, let

$$\left.P(0,y,0)\right|_{y=-1} = 0 \tag{3.13}$$

Using Eqns. (3.11), (3.12), and (3.13) in conjunction with the finite-difference analog of Eqn. (3.8b), one can solve for $P^{n+1}(0,y_j,0)$, $j = 3,4,\ldots,N-1$.

The pressure boundary conditions required by the difference computation are structured to insure that the normal momentum and continuity equations are satisfied at the wall. In a spectral version of our semi-implicit method, which might for example use finite expansion in Chebychev polynomials to represent the flow variables in the y direction, these equations would be satisfied everywhere to the order of the expansion, and hence in such methods these pressure boundary conditions can not be imposed. However, it can be shown [17] that the spectral version of our semi-implicit scheme requires no wall-pressure boundary conditions, as indeed it should not. Nor would the exact solutions of the continuous equations require any wall pressure boundary conditions.

E. Results

In this section we briefly present some of the computational results. See [9] for details, including the initial field.

The CDC 7600 calculation required approximately 20 sec. per time step. Approximately 2000 time steps were taken, with $\Delta t = 0.001$ for most of the calculation. This is barely a sufficient time for stabilization of the time-average statistics.

Figure 3.1 shows the profiles of $< \overline{uv} >$ and the total Reynolds stress,

$$< -\overline{uv} + \nu_T(\partial\bar{u}/\partial y + \partial\bar{v}/\partial x) >$$

Here $< >$ indicates averaging in both the x-z planes as well as time averaging. This profile indicates that an approximately steady mean velocity is obtained. The average Reynolds stress profile has nearly attained the equilibrium shape which balances the downstream mean pressure gradient in the regions away from the walls. In the vicinity of the walls, viscous stresses are significant and they, together with the Reynolds stresses, balance the mean pressure gradient. Moreover, it should be noted that the subgrid scale contribution to the total Reynolds stress is significant only in the vicinity of the walls [9].

Figure 3.2 shows the streamwise mean velocity profile $< \bar{u} >$. The calculated mean velocity profile shows a distinct logarithmic region, and the agreement with experimental data is satisfactory.

The profile of the streamwise turbulence intensity is shown in Fig. 3.3. Note that the subgrid scale contribution to the total streamwise turbulence intensity is small, and the overall agreement with the experimental data is rather good.

Finally, we mention that the present calculation has reproduced many of the important features of turbulent boundary layers (bursts, streaks, etc.). These results will be presented elsewhere.

4. CONCLUSIONS

We believe that LES shows promise for the future, and that its development should be continued. The methods presented in this paper have been applied to the flows described, and important results concerning the nature of these flows have been obtained. For details see our reports [9, 11] or forthcoming papers.

The financial and intellectual support of this work by the NASA-Ames Research Center is sincerely appreciated.

References

1. W.C. Reynolds, Annual Rev. of Fluid Mech., 8, 183-208 (1976).
2. P. Moin, N.N. Mansour, U.B. Mehta, J.H. Ferziger & W.C. Reynolds, Rept. No. TF-10, Mech. Engrg. Dept., Stanford Univ. (1978).
3. J. Smagorinsky, S. Manage & J.L. Holloway, Mon. Weath. Rev., 93, 727 (1965).
4. R.A. Clark, J.H. Ferziger & W.C. Reynolds, Rept. No. TF-9, Mech. Engrg. Dept., Stanford Univ. (1977).
5. A. Leonard, Adv. in Geophysics, 18A, 237 (1973).
6. N.N. Mansour, P. Moin, W.C. Reynolds & J.H. Ferziger, Proc. Symp. on Turbulent Shear Flows, Penn. State (1977).
7. D. Kwak, W.C. Reynolds & J.H. Ferziger, Rept. No. TF-5, Mech. Engrg. Dept., Stanford Univ. (1975).
8. S. Shaanan, J.H. Ferziger & W.C. Reynolds, Rept. No. TF-6, Mech. Engrg. Dept. Stanford Univ. (1975)
9. P. Moin, W.C. Reynolds & J.H. Ferziger, Rept. No. TF-12, Mech. Engrg. Dept., Stanford Univ. (1978).
10. S.A. Orszag, Stud. in Appl. Math., LI, No. 3, 253 (1972).

11. N.N. Mansour, J.H. Ferziger & W.C. Reynolds, Rept. NO. TF-11, Mech. Engrg. Dept., Stanford Univ. (1978)
12. C. Chandrasuda, R.D. Mehta, A.D. Weir & P. Bradshaw, J. Fluid Mech., 85, 693-704 (1978).
13. U.B. Mehta, AGARD Fluid Dynamics Panel Symp. on Unst. Aerodynamics, Ottawa, Canada (1977).
14. G. Comte-Bellot, Doctoral Thesis, Univ. of Grenoble (1963).
15. F.G. Blottner, Comp. Methods in Appl. Mech. & Engrg., 4, 179-194 (1974).
16. R.D. Richtmyer & K.W. Morton, New York: Interscience, 2nd ed. (1967).
17. P. Moin & J. Kim, to be published (1978).
18. A.K.M.F. Hussain & W.C. Reynolds, J. Fluid Engrg., 97, 568 (1975).
19. J.A. Clark, J. Basic Engrg., 90, 455 (1968).
20. J. Laufer, NACA TN-1053 (1951).

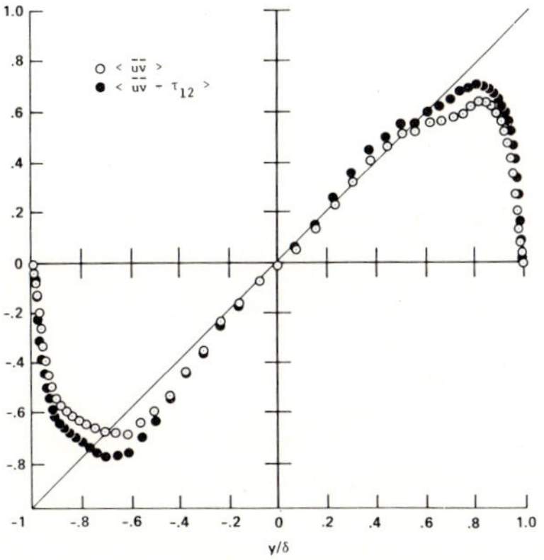

Fig. 4.1. Time-averaged resolvable and total turbulence stress.

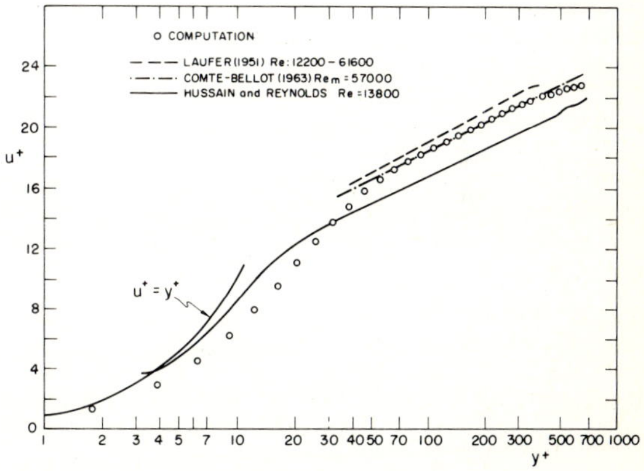

Fig. 4.2. Mean velocity profile. Experimental data of [14], [18], and [20] are included.

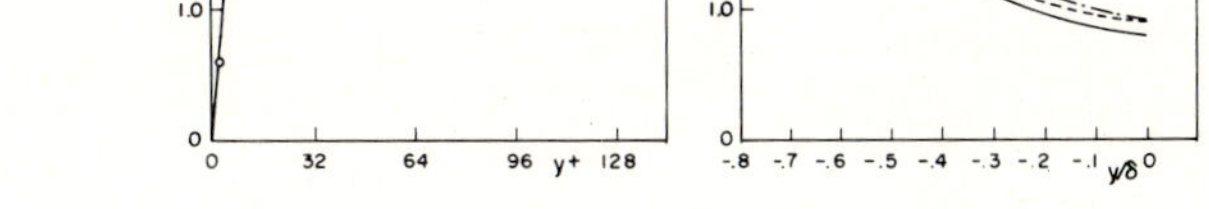

Fig. 3.3. Time-averaged streamwise turbulence intensity in the vicinity of the wall (A) and away from the wall (B). Experimental data of [14], [18], and [19] are included.

NUMERICAL SOLUTION OF THE NONLINEAR STABILITY OF AN IMCOMPRESSIBLE EKMAN BOUNDARY LAYER

by

Hans Bruun Nielsen
Institute for Numerical Analysis
Technical University of Denmark

Hans True
Lab. for Applied Mathematical Physics
Technical University of Denmark

1. Introduction; formulation of the model.

In this paper we concentrate on some special numerical aspects of the problem. For the full discussion of the problem we refer to Iooss et al. (1978) and Nielsen (1977). In this section we shall only briefly review the mathematical model.

A linear theory for the problem results in the differential eigenvalue problem

$$-\mu'' + (\alpha^2+i\alpha Rv)\mu + 2\phi' + i\alpha Ru'\phi = i\alpha Rc\,\mu \tag{1.1a}$$

$$\phi'''' - (2\alpha^2+i\alpha Rv)\phi'' + (\alpha^4+i\alpha^3 Rv+i\alpha Rv'')\phi + 2\mu' = i\alpha Rc(-\phi''+\alpha^2\phi) \tag{1.1b}$$

$$\mu(0) = \mu'(L) = \phi(0) = \phi'(0) = \phi(L) = \phi''(L) = 0 \tag{1.1c}$$

or, formally

$$M_k(\mu,\phi) = i\alpha Rc\cdot N_k(\mu,\phi)\ ,\ k=1,2\ ;\quad \Omega(\mu,\phi) = 0\ .$$

In (1.1) a prime denotes differentiation with respect to the independent variable z, i is the imaginary unit, α the wave number, R the Reynolds' number, $Re(c)$ the phase velocity, and with ε denoting the angle between wave front and a certain x-axis we have

$$u = u(z) = (1-e^{-z}\cos z)\cos\varepsilon + e^{-z}\sin z\cdot\sin\varepsilon \tag{1.2a}$$

$$v = v(z) = -(1-e^{-z}\cos z)\sin\varepsilon + e^{-z}\sin z\cdot\cos\varepsilon\ . \tag{1.2b}$$

We have to determine $(\alpha,\varepsilon) = (\alpha_0,\varepsilon_0)$ such that there is an eigenvalue $i\alpha Rc_1$ for (1.1), for which the growth rate $Im(c_1) = 0$ for the smallest possible value of R. We denote this critical Reynolds' number R_0. Applying bifurcation theory Iooss et al. (1978) show that the flow (1.2) is stable for $R < R_0$ and unstable for $R > R_0$, and a time-periodic stable flow bifurcates from the solution (1.2) and exists for some $R > R_0$ provided the following assumptions can be verified:

H1. The critical eigenvalue $i\alpha_0 R_0 c_1$ is algebraically simple.

H2-H5. $Re(\zeta_{01}) = 0$, $Re(\zeta_{10}) < 0$, $Re(\zeta_{02}) > 0$, $Re(\gamma) < 0$

where (with $(\alpha,\varepsilon,R,c) = (\alpha_0,\varepsilon_0,R_0,c_1)$)

$$\zeta_{01} = 2\pi i\int_0^L\{-u(\bar{\nu}\mu+\bar{\psi}'\phi'+\alpha^2\bar{\psi}\phi) + (\bar{\nu}v'+\bar{\psi}'u')\phi\}dz \tag{1.3a}$$

$$\zeta_{10} = -\frac{i\alpha c}{R} + \frac{2\pi i}{R}\int_0^L\{v(\bar{\nu}\mu+\bar{\psi}'\phi'+\alpha^2\bar{\psi}\phi) + (\bar{\nu}u'-\bar{\psi}'v')\phi\}dz \tag{1.3b}$$

$$\zeta_{02} = -2\pi i\int_0^L\{\bar{\nu}[us-v't+\tfrac{1}{2}(v\mu+u'\phi)] + \bar{\psi}'[ut'-u't+\tfrac{1}{2}(v\phi'-v'\phi)] + \alpha^2\bar{\psi}[ut+\tfrac{1}{2}v\phi]\}dz \tag{1.4}$$

Here overbar denotes complex conjugate, (ν,ψ) are the eigenfunctions of the adjoint problem to (1.1) with the normalization

$$<(\bar{\nu},\bar{\psi}),(\mu,\phi)>\equiv\int_0^L\{\bar{\nu}\mu + \bar{\psi}'\phi' + \alpha^2\bar{\psi}\phi\}dz = \frac{\alpha}{2\pi} \quad , \tag{1.5}$$

and (s,t) are solutions of the boundary-value problem

$$M_k(s,t) - i\alpha Rc\cdot N_k(s,t) = r_k \ , \ k=1,2 \ ; \quad \Omega(s,t) = 0 \tag{1.6}$$

with the right-hand sides given by

$$r_1 = R[(\zeta_{01}+i\alpha u)\mu - i\alpha v'\phi] \ , \quad r_2 = R[(\zeta_{01}+i\alpha u)(-\phi''+\alpha^2\phi)+i\alpha u''\phi] \quad . \tag{1.6a}$$

It is seen, that with $i\alpha Rc = i\alpha_0 R_0 c_1$ the system of differential equations in (1.6) is singular. However, the right-hand sides are orthogonal to (ν,ψ) in the sense that

$$\int_0^L(\bar{\nu}r_1 + \bar{\psi}r_2)dz = 0 \ , \tag{1.7}$$

and unique solutions to (1.6) are defined, when we demand

$$<(\bar{\nu},\bar{\psi}),(s,t)> = 0 \quad . \tag{1.8}$$

Finally, the functional γ is determined from the solutions to two systems of boundary-value problems. These, however, did not present any special numerical problems, and we therefore refer to Iooss et al. (1978) for this problem.

2. Numerical model.

We found it worthwhile to choose the numerical model so that it preserved as many characteristics as possible of the differential subproblems. E.g. if the algebraic eigenvalue problem

$$A \underline{x} = \lambda B \underline{x} \tag{2.1a}$$

is the discretized version of (1.1), then the algebraically adjoint problem

$$\bar{\underline{y}}^T A = \lambda \bar{\underline{y}}^T B \tag{2.1b}$$

should be the discretized version of the adjoint problem to (1.1).

This goal was obtained through the use of a Galerkin approach, and by using the same quadrature formula to approximate the contributions to the Galerkin matrices and the integrals (1.3) we were also able to give a very satisfactory approximation to the singular problem (1.6).

The solutions to the differential problems belong to the space

$$S_{2,4} = \{(f,g) \mid f \in C^2(0,L),\ f(0)=f'(L)=0,\ f(z)\not\equiv 0 \ \wedge \\ g \in C^4(0,L),\ g(0)=g'(0)=g(L)=g''(L)=0,\ g(z)\not\equiv 0\} \tag{2.2}$$

We choose arbitrary functions $(f,g) \in S_{2,4}$, multiply (1.1a) by f, (1.1b) by g, sum, and integrate over $[0,L]$. By use of partial integration and the boundary conditions we obtain

$$a(f,g,\mu,\phi) = i\alpha Rc\cdot b(f,g,\mu,\phi) \tag{2.3}$$

where

$$a(f,g,\mu,\phi) = \int_0^L \{f'(\mu'-2\phi) + f[(\alpha^2+i\alpha Rv)\mu + i\alpha Ru'\phi] + g''\phi'' + g'[(2\alpha^2+i\alpha Rv)\phi'-i\alpha Rv'\phi] + g[(\alpha^4+i\alpha^3 Rv)\phi+2\mu']\}dz \tag{2.3a}$$

$$b(f,g,\mu,\phi) = <(f,g),(\mu,\phi)> \quad . \tag{2.3b}$$

The adjoint problem to (1.1) is treated in the same way, and for any $(f,g)\in S_{2,4}$ we have

$$a(\bar{\nu},\bar{\psi},f,g) = i\alpha Rc\cdot b(\bar{\nu},\bar{\psi},f,g) \tag{2.4}$$

Apart from obtaining a kind of symmetry by using the formulation (2.3,4) rather than (1.1) and its adjoint, we also obtain a reduction in the maximum order of differentiation from 4 to 2 . It can be shown (see e.g. Strang & Fix (1973)) that it is possible to get good approximations to the solution functions by the use of functions in $S_{2,2}$. This permits us to use cubic splines to approximate the functions:

$$f(z) \simeq F(z) = \sum_{j=1}^{n+3} f_j N_j(z) \;, \quad 0 \le z \le L \quad , \tag{2.5}$$

where $\{N_j(z)\}$ are normalized cubic B-splines, defined (see e.g. Cox (1972)) on a set of knots $0 = z_0 < z_1 <\ldots< z_n = L$ with two extra knots at each end. We chose the extra knots to coincide with z_0 and z_n , respectively, and in Nielsen (1977) we showed, that with this choice the boundary conditions $F(0)=F'(L)=0$ are satisfied, if $f_1=0$ and $f_{n+3}=f_{n+2}$, while $F(0)=F'(0)=F(L)=F''(L)=0$ are satisfied, if $f_1=f_2=f_{n+3}=0$ and $f_{n+2}=\theta\cdot f_{n+1}$, where

$$\theta = (z_n-z_{n-1})/(2z_n-z_{n-1}-z_{n-2}) \quad . \tag{2.6}$$

Therefore we approximate $(f,g)\in S_{2,4}$ by $(F,G)\in S^n_{2,2}$, where

$$S^n_{2,2} = \{(F,G) \mid F(z) = \Sigma_{j=2}^{n+2} f_j N_j(z) + f_{n+2}N_{n+3}(z) \;, \quad \{f_j\}\neq\{0\} \wedge \\ G(z) = \Sigma_{j=3}^{n+1} g_j N_j(z) + \theta g_{n+1}N_{n+2}(z) \;, \quad \{g_j\}\neq\{0\}\} \tag{2.7}$$

We see that $(F,G)\in S^n_{2,2}$ can be represented by the 2n-vector of cubic B-spline coefficients

$$\underline{w}_{fg} = [f_2, f_3, g_3, f_4, g_4, \ldots, f_{n+1}, g_{n+1}, f_{n+2}]^T \quad . \tag{2.8}$$

(The reason for this ordering of the coefficients is that it minimizes the bandwidths of the Galerkin matrices A and B in (2.9)).

We now introduce the cubic spline approximations to (f,g) and (μ,ϕ) in (2.3), and approximate the integrals by means of Simpson's formula with function evaluation at the knots and at the midpoints of the knot-intervals. The result can be expressed in the form

$$\underline{w}^T_{fg} \, A \, \underline{w}_{\mu\phi} \simeq i\alpha Rc \; \underline{w}^T_{fg} \, B \, \underline{w}_{\mu\phi} \tag{2.9}$$

where A and B are complex bandmatrices of order $2n$ and bandwidths 15 and 13, respectively. (The difference in bandwidth is caused by the fact that in contrast to the left-hand side integral in (2.3) we do not have "coupling" between f and ϕ or g and μ in the right-hand side integral).

Now (2.3) has to hold for any $(f,g) \in S_{2,4}$, and this implies that (2.9) has to hold for any $\underline{w}_{fg} \in \hat{C}^{2n}$. Therefore the coefficients of the cubic B-spline approximations to μ and ϕ are the components of $\underline{w}_{\mu\phi} = \underline{x}$, an eigenvector of (2.1a), while the corresponding eigenvalue λ approximates $i\alpha Rc$. By means of (2.4) it is easily verified that the corresponding left-hand eigenvector $\bar{\underline{y}}^T$ of (2.1b) will similarly contain the coefficients of the cubic B-spline approximations to $\bar{\nu}$ and $\bar{\psi}$.

After we had found the approximations to the eigenfunctions, we computed the integrals (1.3) using Simpson's formula, and solved equations (1.6) using Galerkin's method: Similarly to the derivation of (2.3) we obtain

$$a(f,g,s,t) - i\alpha Rc \cdot b(f,g,s,t) = \int_0^L (f \cdot r_1 + g \cdot r_2) dz$$

for any $(f,g) \in S_{2,4}$. Using the same procedure as before it is seen, that the coefficients of the cubic B-spline approximations to (s,t) must satisfy a singular system of equations of the form

$$(A - \lambda B)\underline{w}_{st} = C\,\underline{x} \tag{2.10a}$$

where C is a bandmatrix of order $2n$ and bandwidth 15. Corresponding to (1.7) we showed in Nielsen (1977) that the approximations satisfy $\bar{\underline{y}}^T C\,\underline{x} = 0$. Provided λ is a simple eigenvalue (cf. H1), this implies that (2.10a) has an infinity of solutions, and corresponding to (1.8) we define a unique solution by demanding that it should satisfy the orthogonality condition

$$\bar{\underline{y}}^T B\,\underline{w}_{st} = 0 \quad . \tag{2.10b}$$

Finally, Simpson's formula was used again to approximate the integral (1.4).

3. Computational aspects.

In the choice of knots we were guided by the behaviour of the solution functions (cf. fig. 4.2). They vary significantly only for $0 \le z \lesssim 4$, but in order to get good approximations corresponding to large values for L, it is necessary to apply the boundary conditions further out. Given n, the best approximation is obtained if the knots are close for small z, while a larger knot-spacing can be used for large z. In order to avoid the side-effects which may be caused by large jumps in the knot-spacing (cf. Powell (1970)) we decided to gradually widen the know-spacing by constructing the net such that

$$z_{j+1} - z_j = \eta \cdot (z_j - z_{j-1}) \ , \quad j=1,2,\ldots,n-1 \tag{3.1}$$

where $\eta \ge 1$. Given L, n, and η, the knots are uniquely defined.

The two eigenvalue problems (2.1a,b) were solved simultaneously by means of inverse iteration, completed with a generalized Rayleigh quotient for a better determination of λ . See Nielsen (1977) for details.

For the solution of (2.10) we used Gaussian elimination with partial pivoting. Before back substitution we have a system of the form

$$U \underline{w}_{st} = \underline{r} \quad , \tag{3.2}$$

where U is upper triangular. Because of the singularity we must have $u_{2n,2n} = 0$. We found this to be satisfied to the same order of accuracy as we had determined the eigenvalue λ . Further, we found the rank of U to be $2n-1$, thus supporting assumption H1: λ is at least a geometrically simple eigenvalue.

Also $r_{2n} = 0$ must hold, if (3.2) is to have any solution. We found this to be satisfied to the same order of accuracy as we had computed the eigenvector $\underline{x}$. The desired solution, satisfying (2.10b), was found from two linearly independent solutions to (3.2).

4. Results.

With $\eta = 1$ we have been able to verify experimentally that the computed values for $i\alpha Rc$, ζ_{01}, ζ_{10}, and ζ_{02} had errors that behaved like $O(h^4)$, where h is the knot-spacing. This is in accordance with the truncation error of Simpson's formula, and it is the best we can hope for, when using Galerkin's method with cubic splines on a 4th order problem (cf. Strang & Fix (1973)).

The critical parameter values were found to be

$$\alpha_0 = 0.316 \ , \quad \varepsilon_0 = -23^\circ.3 \ , \quad R_0 = 54.2 \quad . \tag{4.1}$$

In the neighbourhood of these values we found that the results for $i\alpha Rc$ and $\{\zeta_{ij}\}$ from the net given by $L = 12$, $n = 24$, $\eta = 1.1$ differed less than 0.3% from the results corresponding to $L \to \infty$, $n \to \infty$, $\eta = 1$.

In fig. 4.1 we show the behaviour of R_c and R_{01} for ε near ε_0 . Here $R_c = R_c(\varepsilon)$ is the minimum value for R for which $Im(c) = 0$ (note, that we also had to determine $\alpha = \alpha_c(\varepsilon)$ for which this minimum was obtained). $R_{01} = R_{01}(\varepsilon)$ is the value for the Reynolds' number, for which $Re(\zeta_{01}) = 0$ (computed with $\alpha = \alpha_0$). We see, that the two curves intersect at the minimum point for $R_c(\varepsilon)$, thus confirming H2. Also assumptions H3-H5 were confirmed by the results obtained with the parameter values (4.1).

In fig. 4.2 we show the computed eigenfunctions (μ,ϕ) corresponding to the parameter values (4.1). The figure shows the modulus and argument of the functions μ and ϕ on complex form.

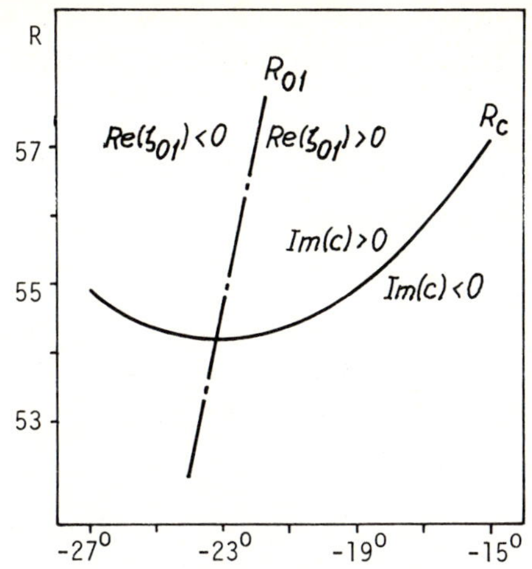

Fig. 4.1. Behaviour near ε_0 .

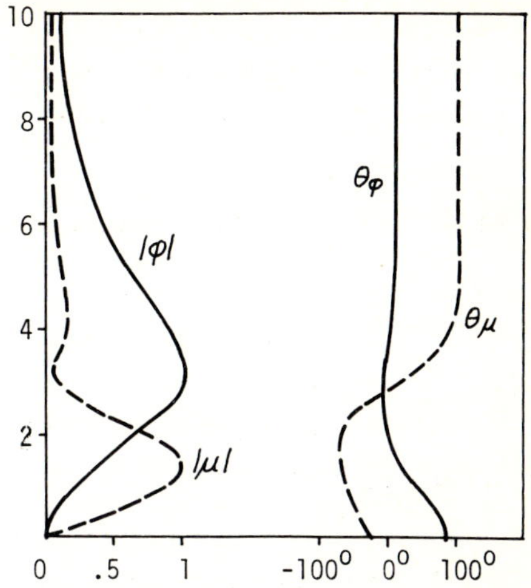

Fig. 4.2. Eigenfunctions (μ,ϕ) .

References.

Cox, M.G. (1972), "The Numerical Evalutation of B-Splines". J.Inst.Maths.Applics. vol. 10, 134-149.

Iooss, G., Nielsen, H.B., and True, H. (1978), "Bifurcation of the Stationary Ekman Flow into a Stable periodic Flow". To appear in Arch.Rat.Mech.

Nielsen, H.B. (1977), "Numerical Solution of a Non-Self-Adjoint Differential Eigenvalue Problem". Report No. NI-77-11, Institute for Numerical Analysis, Technical University of Denmark.

Powell, M.J.D. (1970), "Curve Fitting by Splines in one Variable". In J.G. Hayes (ed.), "Numerical Approximation to Functions and Data". Athlone Press, London.

Strang, G. & Fix, G.J. (1973), "An Analysis of the Finite Element Method". Prentice-Hall, Englewood Cliffs, N.J.

NON-FORWARD-MARCHING METHOD FOR THREE DIMENSIONAL FLOW AROUND SPINNING BODIES

N. Nishikawa
Faculty of Engineering, Chiba University
Yayoi, Chiba, Japan

1. INTRODUCTION

Three dimensional boundary layers around yawed spinning axisymmetric bodies are analyzed by the numerical methods in which the disuse of finite differencing is attained for partial derivatives with respect to one or more independent variables. Great economization in the number of computational grids is realized by extending the local nonsimilarity method[1] to the three dimensional flow or by coupling the method with the method of lines[2];i.e. since these are the nonforward-marching schemes, any initial conditions and piling up of data planes are not necessary. In these methods streamwise derivatives are not discretized, and consequently there appear less requirements of examining the numerical stability and convergence, which cannot be circumvented by marching schemes.

Some types of the two dimensional boundary layers have been analyzed by the local nonsimilarity method, from the proposal[1] to the recent application,[3] taking advantage of its directness due to unnecessity of the solutions up to a concerned location or simplicity in the mathematical expressions. That is, streamwise derivatives are evaluated not by finite differencing expressions but by the simultaneous ordinary differential equations including subsidiary equations derived by differentiation of the original dimensionless equations with respect to streamwise coordinate. Therefore in the local nonsimilarity method the forward marching is not necessary.

The formal extension of the local nonsimilarity method to 3-D flow is "Method I", temporarily named here, in which the circumferential derivatives are estimated by the subsidiary equation reduced by the differentiation of dimensioless basic equations with respect to circumferential coordinate. This method is well applied for the problem for which the solutions at many streamwise locations are not important. After recognizing some inefficiencies of the method in caluculating the developed system to consider higher order derivatives and in knowing the whole flow field, the "Method II" by which these shortcomings can be overcome is proposed. In this Method II, as related to streamwise derivatives, the same treatment as used in Method I is employed, while as for circumferential derivatives finite difference approximation is introduced on the mesh lines which separate the data plane intersecting the streamwise coordinate. Thus in Method II the system of equations is reduced to the simultaneous ordinary differential equations on the lines.

The numerical example by Method I is shown for the flow around the spinning circular cone with angle of attack, in which it is expected that the effect of higher order derivatives appears at a narrow region near the leeside symmetry plane and for which the solutions at many streamwise locations are not required. By the use of Method II the solutions for the incompressible flow around the spinning ellipsoid are obtained for some combinations of parameters such as axis-length ratio, yaw angle, and spinning rate. In Method II the circumferential derivatives are expressed in the finite difference form, since the introduction of many mesh lines is not ineffecient to know the whole flow field, and then the method of lines is applied. The results for the location of separation lines over spinning sphere, or ellipsoid without spin show the reasonable agreement with the previous analyses, though in the present analysis the second order streamwise derivatives are omitted.

2. Examples and Descriptions of the Methods

2.1 Viscous Flow bounded by Conical Shock over Yawed Spinning Cone

The description for Method I is given below for the example of the viscous flow around the cone. The basic equations are essentially the same as those studied by Lin & Rubin[4]. The assumptions to be mentioned specially are followings; the external free stream is conical and the interaction between the viscous layer and inviscid flow[5] is not so large in the region concerned here. The transformed coordinate and the vector potentials for velocity components $U=u/u_e, W=w/w_e$ (u_e=velocity at outer edge) are

$$d\eta=(3PCRe)^{1/2}\rho dy/x \ , \quad \xi=(x/L)^3/3 \ , \quad F=\xi^{\frac{1}{2}}\int U d\eta \ , G=\xi^{\frac{1}{2}}\int W d\eta \qquad (1)$$

where x,y are coordinate along the generator and normal to that, respectively, which are attatched to the cone and do not rotate with the cone and where P,ρ are the dimensionless pressure and density, respectively and $Re=\rho_e u_e x/\mu_e$, assuming $\mu \propto CT$ (C=constant). The reference values are taken from inviscid surface values for zero angle of attack $\alpha=0$ at azimuthal angle $\theta=90^o$. Thus for the case of unity Prandtl number we obtain the dimensionless conservation equations involving the lateral diffusion terms.

$$U_{\eta\eta} + [(U_\eta F_1 - WU_\theta + U_{\theta\theta}/\rho r Re)/\beta + W^2]/3 = \xi(UU_\xi - U_\eta F_\xi) - P_\xi/\rho\gamma M^2) \qquad (2)$$

$$W_{\eta\eta} + [(W_\eta F_1 - WW_\theta + 4W_{\theta\theta}/3\rho r Re)/\beta - UW]/3 = \xi(UW_\xi - W_\eta F_\xi) - P_\theta/\rho\gamma M^2 r \qquad (3)$$

$$P_\eta = M_\infty^2 W^2 \cos\theta_c \beta (P/3Re)^{1/2} \qquad (4)$$

$$T_{\eta\eta} + [T_\eta F_1 - W(T_\theta - P_\theta/r\rho) - 4T_{\theta\theta}/3\rho r Re]/\beta + \xi U P_\xi/\rho - U_\eta^2 - W_\eta^2 - (U_\theta^2 + 4W_\theta^2/3)/\rho r Re]/3 = \xi(UT_\xi - T_\eta F_\xi) \qquad (5)$$

where $F_1=G_\theta/r+(\xi FP_\xi + GP_\theta/r)/2P$, $\beta=\sin\theta_c$, $T=C_pT/u_e^2$, θ_c=cone half angle and where the subscripts η,ξ,θ show the differentiation with respect to η,ξ,θ and

γ,L are specific heat ratio and the cone hight(1 ft), respectively.

The impermeable and adiabatic wall is considered here and then the boundary conditions are written as follows, denoting the spinning rate by ω

$$\eta=0\ ;\ U=F=G=0,\ W=r\omega/u_e,\ T_\eta=0\ ,\quad \eta\to\infty\ ;\ U=U_e(\theta), W=W_e(\theta),\ T=T_e(\theta) \qquad (6)$$

The external boundary conditions are extracted from the numerical data for inviscid conical flows by Babenko et al (BVLR)[6] for $M_\infty \leq 7$, $\alpha \leq 7.5$. Using these data, U_e, W_e are interpolated with sinusoidal functions of θ and the data for T, P, ρ are expressed in cubic expressions.

For the system of equations (2)-(5) the application of Method I is shown through the introduction of the subsidiary equations of $q_\xi, q_\theta, q_{\theta\theta}$ with notation q=U,W, or T. That is, by differentiating equations (2)-(6) with respect to ξ or θ and omitting $q_{\xi\xi}, q_{\theta\theta\theta}$ etc we have

$$q_{\xi,\eta\eta} + \{[q_{\xi\eta}F_1 + q_\eta F_{1,\xi} + A\left(\frac{1}{rRe\rho}\right)_\xi - W_\xi q_\theta]/\beta + B_\xi q + Bq_\xi\}/3$$
$$= (Uq_\xi - U_\eta F_\xi + C) + \xi(U_\xi q_\xi - U_{\xi\eta}F_\xi + C_\xi) \qquad (7)$$

$$q_{\theta,\eta\eta} + \{[q_{\theta\eta}F_1 + q_\eta F_{1,\theta} - W_\theta q_\theta - Wq_{\theta\theta} + \frac{A}{rRe}\left(\frac{1}{\rho}\right)_\theta]/\beta + B_\theta q + Bq_\theta\}/3$$
$$= \xi(U_\theta q_\xi - q_{\theta\eta}F_\xi + C_\theta) \qquad (8)$$

$$q_{\theta\theta,\eta\eta} + \{[q_{\theta\theta\eta}F_1 + 2q_{\theta\eta}F_{1,\theta} + q_\eta F_{1,\theta\theta} - W_{\theta\theta}q_\theta - 2W_\theta q_{\theta\theta} + \frac{A}{rRe}\left(\frac{1}{\rho}\right)_{\theta\theta}]/\beta$$
$$+ B_{\theta\theta}q + 2B_\theta q_\theta + Bq_{\theta\theta}\}/3 = \xi(U_{\theta\theta}q_\xi - q_{\theta\theta\eta}F_\xi + C_{\theta\theta}) \qquad (9)$$

where for $q=U$; $A=U_{\theta\theta}, B=W, C=P_\xi/\rho\gamma M^2$: for $q=T$; $A=4(T_{\theta\theta}-W_\theta^2)/3-U_\theta^2, B=0$

for $q=W$; $A=4W_{\theta\theta}/3, B=-U, C=P_\theta/\rho\gamma M^2 r$. $C=-UP_\xi/\rho-(U^2+W^2-WP_\theta/3r\beta\rho)/\xi$

with

$$\eta\to\infty\ ;\ q_\xi = 0,\ q_\theta=\partial(\text{inviscid sol})/\partial\theta\ ,\ q_{\theta\theta}=\partial(q_\theta)/\partial\theta$$
$$\eta = 0\ ;\ q_\theta = 0,\ U_\xi = T_\xi = 0,\ W_\xi = W/3\xi\ ,\ q_{\theta\theta}=0 \qquad (10)$$

The simultaneous ordinary differential equations (2)-(5) and the derived subsidiary equations such as (7)-(9) can be calculated by the conventional numerical integration program. In present examples the calculated values for the boundary layer edge at $\eta=4\sim8$ were converged to the conditions (6) or (10) with 10^{-3} accuracy after about 30 times iteration. Thus the solutions at the location of x=0.8 ft, for which the finite difference scheme[5] requires the streamwise increments of one hundreds or so, can be obtained by only 4 rows of grids respectively for $q, q_\xi, q_\theta, q_{\theta\theta}$.

In Fig. 1 the obtained temperature profiles are shown and compared with the profile by Lin & Rubin[4] originaly for density. The qualitatively good agreement shown here is reasonable when we consider that in the present formulation the higher order derivatives $q_{\xi\xi}, q_{\theta\xi}$ etc are not taken into account. The effect of yaw angle on the streamwise shear stress is shown in Fig. 2 for a few leeside locations. We can observe that the shear stress near the leeside symmetry plane is affected by α more strongly compared to that at shoulder region of the cone. It is emphasized here that the solutions at a location can be obtained in the CPU time of fifteen seconds or so by the computer HITAC-8800(the same class as IBM-360) and then all the data for completing Fig. 2 can be prepared within 135 seconds CPU time at most. The role of any parameter can be clarified within a few minuits. We shall have to observe Re number dependence[7] of Magnus force to be shown from solutions at wider range of Re than here.

As can be seen from our reduction, the higher order derivatives to be appeared in the subsidiary equations (7),(8),(9) are truncated to enclose the system of equations. Then, if the retention of the truncated terms in the above system is appropriate for the aim of analysis, the subsidiary equations for the retained derivatives should be reduced through the differentiation of already given subsidiary equations. However the effort to solve the reformulated system of more or less 16 ordinary differential equations may not be small. Due to this and because the solutions at many streamwise coordinate values are necessary in some problems to grasp the appearance of drastic behavior of flow, we should introduce the following Method II.

2.2 Incompressible Boundary Layer over the Spinning Ellipsoid

The flow around the ellipsoid with incidence angle has been studied by Wang[8] for the zero spin case by the use of the combination of some types of finite difference expressions. Recently the Box scheme of more simple form was applied for the same case as Wang by Hirsh & Cebeci[9]. Similarily to these studies one of the main matters of interest is the location of separation line, however for the spinning body,then the solutions at relatively many points distributed over the ellipsoid are necessary. Thus the "Method II" introduced later is more advantageous than Method I because the use of subsidiary equations was particularly developed to know a part of flowfield.

The coordinate system for the ellipsoid is shown in Fig.3 where only V_∞ is the dimension-possessing property and U,W are non-dimensional ones referred to V_∞. Z,V are stretched in accordance with the Reynolds number $Re=V_\infty a/\nu_\infty$. Thus we have

$$\frac{\partial U}{h\partial s} + \frac{\partial V}{\partial Z} + \frac{\partial W}{r\partial\theta} - KU = 0 \tag{11}$$

$$\frac{U\partial U}{h\partial s} + V\frac{\partial U}{\partial z} + \frac{W\partial U}{r\partial\theta} - KW^2 = -\frac{\partial P}{h\partial s} + \frac{\partial^2 U}{\partial z^2} \tag{12}$$

$$\frac{U\partial W}{h\partial s} + V\frac{\partial W}{\partial z} + \frac{W\partial W}{r\partial\theta} + KUW = -\frac{\partial P}{r\partial\theta} + \frac{\partial^2 W}{\partial z^2} \tag{13}$$

where $e=[1-(b/a)^2]^{1/2}$ $h=[(1-e^2s^2)/(1-s^2)]^{1/2}$ $K=-\frac{\partial r}{\partial x}\frac{1}{rh}$

and with the boundary conditions: at Z=0, U=V=0, $W=r\omega/V_\infty$ and with notations[8] k_a,k_b

$Z \to \infty$; $U=[(1+k_a)(\cos\alpha)(1-s^2)^{1/2}+(b/a)(1+k_c)s\cos\theta\ \sin\alpha]/(1-e^2s^2)^{1/2}$ $W=(1+k_c)\sin\alpha\sin\theta$

At the first stage of Method II the conservation equations are differentiated with respect to s the meridional coordinate and the subsidiary equations for $\partial/\partial s$ are obatined. At next stage the finite differencing for the circumferential derivatives in eqns.(11)-(13) and in the reduced subsidiary equattions is introduced;e.g. the system of equations for U, U_s becomes, with notations a=0,b=1 for W>0, or a=1,b=0 for W<0

$$U_{zz} - VU_z = UU_s/h - KW^2 + (W/r)(U(\theta-a\Delta\theta) - U(\theta-b\Delta\theta))/\Delta\theta + P_s/h \qquad (14)$$

$$U_{szz} - VU_{sz} - V_sU_z = U_s(U/h)_s - (KW^2)_s + (W/r)_s(U(\theta-a\Delta\theta)-U(\theta-b\Delta\theta))/\Delta\theta + (W/r)(U_s(\theta-a\Delta\theta)-U_s(\theta-b\Delta\theta))/\Delta\theta + P_sh_s/h^2 + P_{ss}/h + U_{ss_{Z\to\infty}}U/h \qquad (15)$$

The scheme for Method II is illustrated in Fig. 4 together with grid structure in Method I. The results for the region $0<\theta<\pi$ have been obtained at five values of θ assuming $\partial/\partial\theta(\theta<30^\circ)=\partial/\partial\theta(0)$. For most values of s the computational field was bounded by $\theta=0$ and the separation line at which $U_z=0$ simultaneously with $W_z<0$ at $Z=0$. The position of separation line was determined on solving process so that the most leeside mesh line may not appear in reverse flow region. The variables on the most leeside mesh line for which $0<U_z<0.01$ are evaluated by Method I, since the grid line at more leeside region is required if we use equation (14),(15) for the condition W<0.

Fig.5 shows the angle of separation for the sphere with the spinning axis consistent with the oncoming flow direction(axially symmetric flow). The agreement with the analysis by Hoskin[10] is reasonably good. The results for the ellipsoid of b/a= 1/4 for $b\omega/V_\infty=0, 0.3, 0.6$ are shown in Fig.6 and the zero spin solution is compared with that by Wang[8]. The system based on the omitting of $U_{ss}, U_{s\theta}, U_{\theta\theta}$ is not easy to be converged for the region $|s|>0.8$. Due to spinning the separtion line shifts, in contrast to the leeward shift established for the rptating cylinder or b<<a,toward upstream region since the centrifugal force induces the meridional flow toward equator.

References

1. E.M. Sparrow, H. Quack, and C.J.Boerner, AIAA J. 8(1970) p1936 .
2. M. Holt, Numerical Methods in Fluid Dynamics, Springer Verlag(1977).
3. N. Nishikawa and H. Takase, ASME Paper 78-HT-20(1978).
4. T.C. Lin and S.G. Rubin, AIAA J. 12(1974) p975.
5. B.R. Sanders and H.A. Dwyer, AIAA J. 14(1976) p576.
6. K.I. Babenko et al. NASA TTF-380(1966); The inviscid data for high M_∞ and α should be referred to AGARDO GRAPH 137 by D.J. Jones(1969).
7. R. Agarwal and J.V. Rakich, AIAA Paper 78-65(1978).
8. K.C. Wang, e.g. JFM Vol.72(1975) p49.
9. R.S. Hirsh and T. Cebeci, AIAA Paper 77-683(1977).
10. N.E. Hoskin, extracted from "Boundary Layer Theory" 6th ed Fig. 11.11 p234.

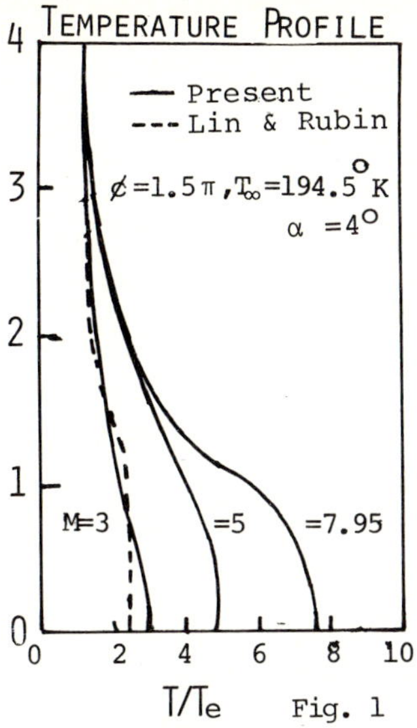

Fig. 1

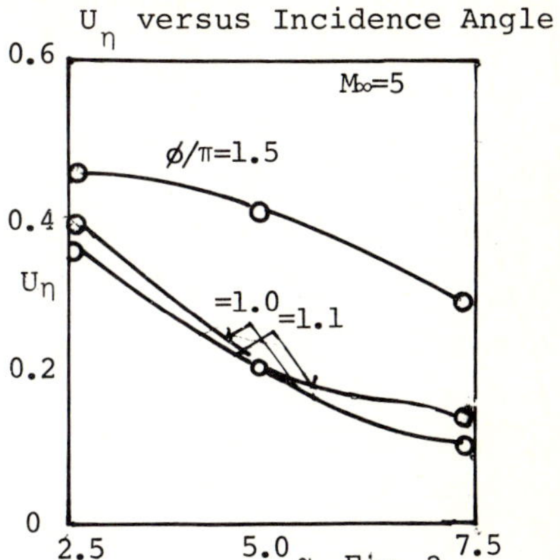

Fig. 2

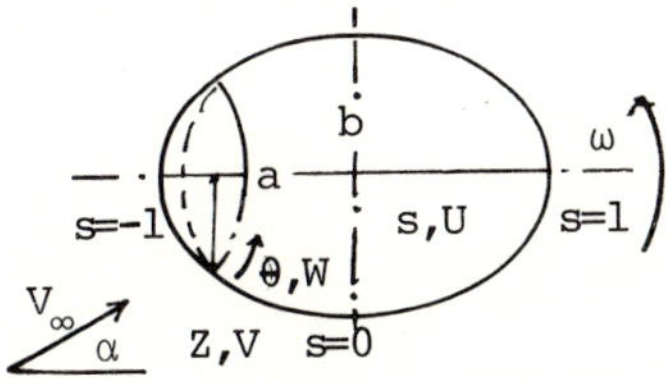

Fig. 3

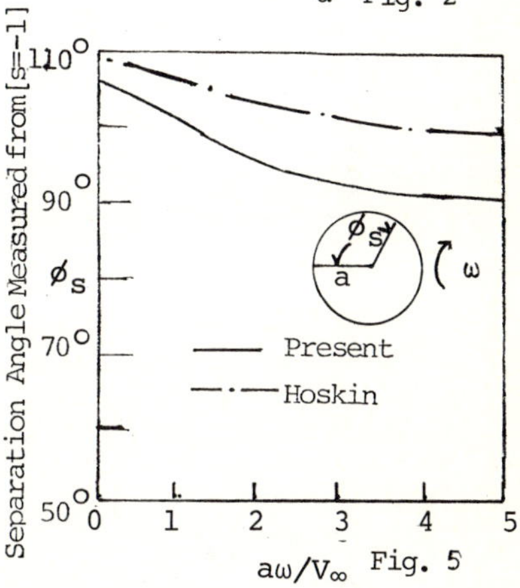

Fig. 5

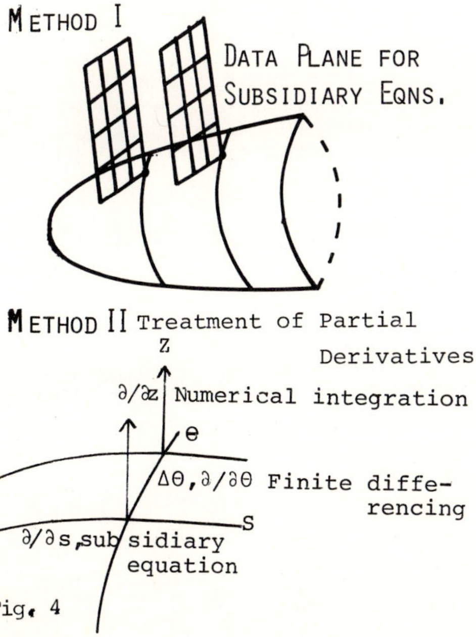

Fig. 4

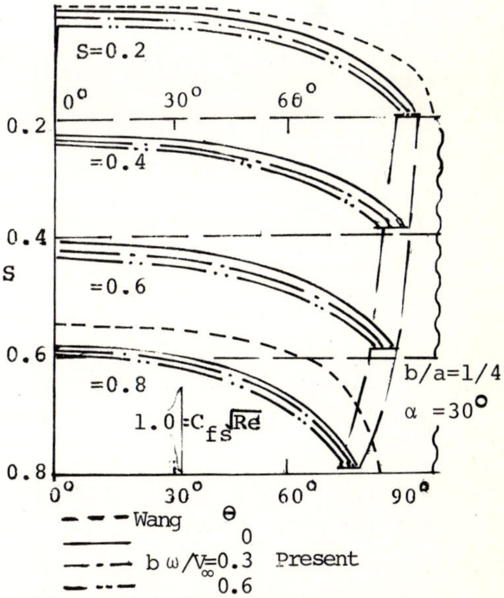

Fig. 6 Meridional Skin Friction

APPLICATIONS OF THE ENERGY ESTIMATES ANALOGOUS TO SAINT-VENANT'S PRINCIPLE TO PROBLEMS OF ELASTICITY AND HYDRODYNAMICS

O.A. Oleinik
Moscow State University
and Institute for Problems of
Mechanics of the Academy of
Sciences of the USSR,
Moscow, USSR

Saint-Venant's principle which is of great importance for the theory of elasticity and constructional engineering was formulated in 1853 [1]. This principle was widely used for the construction of approximate solutions of elasticity problems. In 1965 Saint-Venant's principle was expressed as an a priori energy estimate for solutions of the linear elasticity system and was proved for cylindrical bodies [2]. For elastic bodies of a more general shape such estimates are obtained in [3].

It turns out that energy estimates of this type can be proved for many other partial differential equations and systems and in particular for the system of hydrodynamics. See [4 - 10], [18 - 21].

Energy estimates which express Saint-Venant's principle in the two-dimensional linear theory of elasticity were obtained in papers [II - I2]. These energy estimates do not take into account the variation of the boundary with the increase of the distance from that part of the body where external forces are applied. In papers [I3], [I4], [I8] a new method for the proof of energy estimates analogous to Saint-Venant's principle is given. The estimates obtained there take into account the variation of the boundary. On the basis of these estimates uniqueness theorems for boundary value problems in bounded as well as in unbounded domains in classes of functions with an infinite energy integral are proved, theorems analogous to Phragmen - Lindelöf's theorem for harmonic functions are obtained for solutions of the equations of the two-dimensional linear theory of elasticity and also the behavior of these solutions in a neighbourhood of a nonregular boundary point and at infinity is investigated. Similar questions for solutions of the first boundary value problem and the mixed boundary value problem for the system of linear elasticity are considered in papers [I5 - I9]. Here we mention some results concerning problems of the two-dimensional theory of elasticity and problems of hydrodynamics.

Let Ω be a bounded domain belonging to the half-plane $\{x_1, x_2 : x_1 > 0\}$, $\partial\Omega$ be the boundary of Ω, $S_t = \Omega \cap \{x_1,$

$x_2 : x_1 = t\}$ for $0 < t < T$, $T = \text{const} > 0$. Suppose that S_t consists of a finite number of intervals the ends of which belong to $\partial\Omega$. In Ω consider the boundary value problem

$$\Delta\Delta u = f \ , \ u\big|_{\partial\Omega} = \varphi_1 \ , \ \frac{\partial u}{\partial \nu}\Big|_{\partial\Omega} = \varphi_2 \tag{I}$$

where ν is the exterior normal to $\partial\Omega$. We assume for simplicity that $\partial\Omega$ and the solution $u(x)$ of problem (I) are sufficiently smooth. Let $\Omega_t = \Omega \cap \{x_1, x_2 : x_1 < t\}$. The following theorem can be considered as a version of Saint-Venant's principle.

<u>Theorem I</u> Let $u(x)$ be a solution of problem (I), $f = 0$ in Ω_T, $\varphi_1 = \varphi_2 = 0$ on $\partial\Omega \cap \partial\Omega_T$ Then

$$\int_{\Omega_{t_0}} E(u)\,dx \leq [\Phi(t_0, t_1)]^{-1} \int_{\Omega_{t_1}} E(u)\,dx , \tag{2}$$

$$\max_{\Omega_{t_0}} |u|^2 \leq p(t_0)[\Phi(t_0, t_1)]^{-1} \int_{\Omega_{t_1}} E(u)\,dx \tag{3}$$

where $0 < t_0 < t_1 \leq T$, $E(u) \equiv \sum_{i,j=1}^{2} u^2_{x_i x_j}$, the function $\Phi(x_1, t_1)$ satisfies for $t_0 < x_1 < t_1$ the ordinary differential equation

$$\Phi_{x_1 x_1} - \mu(x_1)\Phi = 0 \tag{4}$$

with the initial conditions

$$\Phi(t_1,t_1)=1 \quad , \Phi_{x_1}(t_1,t_1)=0 \ ;$$

$\mu(t)$ is any continuous function such that

$$0<\mu(t)\le\lambda(t)\equiv \inf_{v\in\mathfrak{M}}\left\{\int_{S_t}E(v)dx_2\left|\int_{S_t}(v_{x_1}^2+v_{x_2}^2-v\,v_{x_1x_1})dx_2\right|^{-1}\right\}, \tag{5}$$

$\mathfrak{M}$ is the class of twice continuously differentiable in a neighbourhood of $\bar{S}_t$ functions satisfying the conditions

$$v=0\ ,\quad v_{x_1}=0\ ,\quad v_{x_2}=0 \qquad \text{on } \bar{S}_t\cap\partial\Omega,$$

$$p(t_0)=\left\{1+\sup_{0\le x_1\le t_0}\left[M(x_1)\right]^{-1}+\sup_{0\le x_1\le t_0}\left[m(x_1)\right]^{-1}\right\},$$

$$0<M(t)\le\inf_{v\in\mathfrak{M}}\left\{\int_{S_t}E(v)dx_2\left[\int_{S_t}v^2dx_2\right]^{-1}\right\},$$

$$0<m(t)\le\inf_{v\in\mathfrak{M}}\left\{\int_{S_t}E(v)dx_2\left[\int_{S_t}\sum_{j=1}^{2}v_{x_j}^2\,dx_2\right]^{-1}\right\}. \tag{6}$$

Let $l(t)$ be the length of the largest interval belonging to S_t. For $\lambda(t)$ the estimate

$$\lambda(t)\geqslant 2\pi^2(l(t))^{-2}$$

is valid. One can take as $\mu(t)$, $M(t)$, $m(t)$ any continuous functions such that

$$0<\mu(t)\le 2\pi^2(l(t))^{-2}$$

$$0 < M(t) \leq (4{,}73)^4 (\ell(t))^{-4},$$

$$0 < m(t) \leq 2\pi^2 (\ell(t))^{-2}.$$

Note that as a function $\varphi(x_1, t_1)$ in estimates (2), (3) one can take any function $\varphi(x_1, t_1)$ which satisfies the conditions

$$|\varphi_{x_1 x_1}(x_1, t_1)| \leq \mu(x_1)\, \varphi(x_1, t_1), \quad \varphi > 0, \; \varphi_{x_1} \leq 0$$

for $t_0 \leq x_1 \leq t_1$ and

$$\varphi(t_1, t_1) = 1, \quad \varphi_{x_1}(t_1, t_1) = 0.$$

This remark is useful for the construction of $\varphi(x_1, t_1)$ in concrete cases. If Ω belongs to the rectangle $\{x_1, x_2 : 0 < x_1 < T, 0 < x_2 < \ell\}$ then

$$[\varphi(t_0, t_1)]^{-1} \leq \frac{1}{2} \exp\{-2\pi (\ell \sqrt{2})^{-1} (t_1 - t_0)\}.$$

Inequality (2) gives in this case the estimate obtained earlier in [I2]. The uniqueness theorems and Phragmen - Lindelöf's type theorems follow immediately from estimate (2) (see [I8]).

Let us consider a domain Ω with a nonregular boundary point at the origin (an angle, an edge, a spike etc.)

<u>Theorem 2</u> Let $u(x)$ be a solution of problem (I) in Ω, $f = 0$ in Ω_T, $\varphi_1 = \varphi_2 = 0$ on $\partial\Omega \cap \partial\Omega_T$, $S_t =$

$=\Omega \cap \{x_1, x_2 : x_1 = t\}$ is not empty for any $t \in (0,T)$. Then

$$\int_{\Omega_T} u^2(x) M(x_1) \Phi(x_1,T,\varepsilon)\,dx \leq \int_{\Omega_T} E(u)\Phi(x_1,T,\varepsilon)\,dx \leq \varepsilon^{-1}\int_{\Omega_T} E(u)\,dx,$$

$$|u(x_1,x_2)|^2 \leq 3\,\varepsilon^{-1}\left[\Phi(x_1,T,\varepsilon)\right]^{-1} M^{-\frac{1}{4}}(x_1)\, m^{-\frac{1}{2}}(x_1)\int_{\Omega_T} E(u)\,dx,\ x\in\Omega_T,$$

where $\varepsilon = \text{const}$, $0 < \varepsilon < 1$, function $\Phi(x_1,T,\varepsilon)$ satisfies the equation

$$\Phi_{x_1x_1} - (1-\varepsilon)\mu(x_1)\Phi = 0$$

and the initial conditions

$$\Phi(T,T,\varepsilon) = 1\ ,\ \Phi_{x_1}(T,T,\varepsilon) = 0\ ;$$

functions $\mu(x_1)$, $M(x_1)$, $m(x_1)$ are defined by relations (5), (6) and it is supposed that $M'(x_1) \leq 0$, $m'(x_1) \leq 0$ for $0 < x_1 \leq T$.

If the domain Ω belongs to the angle $\{x_1, x_2 : |x_2| \leq \frac{1}{2}Ax_1, x_1 > 0, A > 0\}$ then it follows from theorem 2 that

$$|u(x_1,x_2)|^2 \leq C\,|x_1|^{2+s}\int_{\Omega} E(u)\,dx$$

where $s > 0$ and $-s$ is a root of the equation

$$s(s-1) = 2\pi^2(1-\varepsilon)A^{-2},\qquad A = \text{const},$$

the constant C depends on A and ε.

In a similar way one can study the behavior of solutions of problem (I) in an unbounded domain Ω as $x_1 \to \infty$. Let Ω be an unbounded domain and S_t be not empty for $t > T$, $T = \text{const} > 0$. Set

$$\Omega_t = \Omega \cap \{x_1, x_2 : x_1 < t\}, \quad \Omega_t^+ = \Omega \cap \{x_1, x_2 : x_1 > t\}.$$

Theorem 3. Let $u(x)$ be a solution of problem (I) in Ω, $f = 0$ in Ω_T^+, $\varphi_1 = \varphi_2 = 0$ on $\partial\Omega \cap \partial\Omega_T^+$. Then

$$\int_{\Omega_T^+} \mathcal{F}(x_1, T, \varepsilon) E(u)\,dx \leq \varepsilon^{-1} \int_{\Omega_T^+} E(u)\,dx ,$$

$$\int_{\Omega_t^+} E(u)\,dx \leq \left[\mathcal{F}(t, T, 0) \right]^{-1} \int_{\Omega_T^+} E(u)\,dx,$$

$$|u(x_1, x_2)|^2 \leq C(\rho)\Big(1 + \sup_{x_1 < \tau < x_1 + 2\rho} M^{-1}(\tau)\Big) \left[\mathcal{F}(x_1, T, 0)\right]^{-1} \int_{\Omega_T^+} E(u)\,dx,$$

where the constant $C(\rho)$ depends on ρ only, $\varepsilon = \text{const}$, $0 < \varepsilon < 1$, $t > T$. $\mathcal{F}(x_1, T, \varepsilon)$ satisfies for $x_1 > T$ the equation

$$\mathcal{F}_{x_1 x_1} - (1 - \varepsilon)\mu(x_1)\mathcal{F} = 0$$

with initial conditions

$$\mathcal{F}(T, T, \varepsilon) = 1 , \quad \mathcal{F}_{x_1}(T, T, \varepsilon) = 0 ;$$

functions $\mu(x_1)$, $M(x_1)$ are defined by relations (5), (6).

In particular cases one can estimate the growth of $\mathcal{F}(x_1, T, \varepsilon)$ as $x_1 \to \infty$.

Estimates analogous to Saint-Venant's principle are proved by G.A. Yosifian for Stokes' equations (see [20], [21]) . Consider the Stokes system

$$\Delta u + \nabla p = F , \quad div\, u = 0 \quad \left(\nabla = \left(\frac{\partial}{\partial x_1}, \frac{\partial}{\partial x_2}, \frac{\partial}{\partial x_3}\right)\right)$$

in a domain Ω with the boundary conditions

$$u\big|_{\partial\Omega} = 0$$

where $u = (u_1, u_2, u_3)$. Let for symplicity $\Omega = \mathcal{D} \times \{-\infty < x_1 < +\infty\}$, $\mathcal{D}$ is a bounded domain on the plane $\{x_2, x_3\}$ with smooth boundary Set

$$\Omega(\tau, k) = \Omega \cap \{x_1, x_2, x_3 : \tau < x_1 < k\}$$

Suppose that $F \equiv 0$ in $\Omega(R_1, R_1)$. Then for any R and h, $R + h < R_1$, $R > 0$, $h > 0$

$$\int_{\Omega(R,R)} |\nabla u|^2 dx \leq C_1 \exp\{-\alpha h\} \int_{\Omega(R+h, R+h)} |\nabla u|^2 dx , \quad (7)$$

if

$$\int_{\mathcal{D}} u_1(0, x_2, x_3)\, dx_2\, dx_3 = 0 ,$$

where α, C_1 are positive constants depending on $\mathcal{D}$ only.

From estimate (7) Phragmen - Lindelöf's type theorems and uniqueness theorems for the Stokes equations follow immediately. Estimates similar to (7) are proved for a large class of non cylindrical domains Ω . The inequlities of the form (2) , (7) can be used to estimate the error for approximate solutions of the boundary value problems.

REFERENCES

I A.J.C.B. de Saint-Venant, De la torsion des prismes, Mem. présentés par divers savants a l'Acad. des Sciences, t. XIV, Paris (I855), 233-560.

2 R. Toupin, Saint-Venant's principle, Arch. Rat. Mech. and Anal. I8;2 (I965), 83-96.

3 N. Weck, An explicit St. Venant's principle in three-dimensional elasticity, Lecture Notes in Math., vol. 564, Springer, Berlin, (I976), 5I8-526.

4 O.A. Oleinik, G.A. Yosifian, Energy estimates for weak solutions of boundary value problems for second order elliptic equations and their applications, Dokl. Akad. Nauk SSSR, 232, no 6 (I977), I257-I260.

5 O.A. Oleinik, G.A. Yosifian , Boundary value problems for second order elliptic equations in unbounded domains and Saint-Venant's principle. Annali della Scuola Normale Superiore di Pisa, classe di scienze, ser. IV, 4;2, (I977), 269-290.

6 O.A. Oleinik, G A. Yosifian, An analogue of Saint-Venant's principle and the uniqueness of solutions of boundary value problems for parabolic equations in unbounded domains, Uspehi Mat. Nauk, 3I, no 6 (I976), I42-I66.

7 J.K. Knowles, On spatial decay of solutions of the heat equation, Journ. of Appl. Math. and Physics, 22, no 6 (I974), I050-I056.

8 O.A. Oleinik, G.A. Yosifian, On some properties of solutions of

hydromechanics equations in domains with moving boundary, Vestnik Mosk. Univ., mat. mech. 5 (I977), 5-II.

9 O A. Oleinik, On the behavior of solutions of the Cauchy problem and the boundary value problem for parabolic systems of partial differential equations in unbounded domains, Rendiconti di Mat., ser. VI, 8;2,(I975), 545-56I.

I0 O.A. Oleinik, G.A. Yosifian, On removable singularities at the boundary and the uniqueness of solutions of boundary value problems for elliptic and parabolic second order equations. Functional Anal. i Priložen., II, no 3, (I977), 54-67.

II J.K. Knowles, On Saint-Venant's principle in the two-dimensional linear theory of elasticity, Arch. Ration. Mech. Anal., 2I;I (I966), I-22.

I2 J.N. Flavin, On Knowles version of Saint-Venant's principle in two-dimensional elastostatics, Arch. Ration. Mech. Anal., 53;4 (I974), 366-375.

I3 O A. Oleinik, G.A. Yosifian, On Saint-Venant's principle for the plane theory of elasticity, Dokl. Akad. Nauk SSSR, 239, no 3 (I978), 530-533.

I4 O.A. Oleinik, G.A. Yosifian, I.N. Tavhelidze, Estimates of solutions of the biharmonic equation in a neighbourhood of nonregular boundary points and at infinity, Uspehi Mat. Nauk, 33, no 3 (I978),I8I-I82.

I5 O.A. Oleinik, G.A. Yosifian, Saint-Venant's principle for the mixed boundary value problem of the theory of elasticity and its applications, Dokl. Akad. Nauk SSSR, 233, no 5, (I977), 824-827.

I6 O.A. Oleinik, G A. Yosifian, A priori estimates for solutions of the first boundary value problem for the theory of elasticity system of equations and their applications. Uspehi Mat. Nauk, 32, no 5 (I977), I97-I98.

I7 O.A. Oleinik, G.A. Yosifian, On singularities at the boundary points and uniqueness theorems for solutions of the first boundary value problem of elasticity, Comm. in partial differential equations, 2;9, (1977), 937-969.

18 O.A. Oleinik, G.A. Yosifian, Saint-Venant's principle for the plane theory of elasticity and boundary value problems for the biharmonic equation in unbounded domains, Sibirsk. Mat. Ž., 19, no 5 (1978), 1154-1165.

19 I. Kopachek, O.A. Oleinik, On asymptotic properties of solutions of the theory of elasticity system of equations, Uspehi Mat. Nauk, 33, no 5 (1978), 189-190.

20 G.A. Yosifian, On solutions of the hydrodynamics system in unbounded domains, Uspehi Mat. Nauk, 33, no 3 (1978), 134.

21 G.A Yosifian, An analogue of Saint-Venant's principle and uniqueness of solutions of the first boundary value problem for the Stokes system in domains with noncompact boundaries, Dokl. Akad. Nauk SSSR, 242, no 1, (1978), 36-39.

FINITE ELEMENT ANALYSIS OF VISCOUS INCOMPRESSIBLE FLOW AROUND AN OSCILLATING AIRFOIL

Koichi Oshima*, Yuko Oshima**
Yoshio Kuriyama*

* ISAS University of Tokyo
** Dept. Physics, Ochanomizu Univ.

Abstract

The Navier-Stokes equations of viscous incompressible flow around an elliptic airfoil under heaving and pitching motion was numerically analysed using the finite element method. The time relaxation procedure was adopted for several oscillating period, starting with the steady viscous flow condition. It was found that the rear stagnation point moves upstream considerably, though it does exist and no separation takes place.

1. Introduction

In the recent years, the unsteady fluiddynamics has been attracting much renewed attention in the relation to the mechanism of the dynamic stall of oscillating airfoils [1] and of the propulsive locomotions of animals [2]. The linearized potential wing theory for oscillating airfoils [3] has been developed over the years and now most of the fundamental aspects of them are well understood. However, these potential theories are almost useless without assuming the trailing flow condition and this assumption has to be justified based on the viscous flow analysis including the non-linear effects. Recently, several such works have been reported [4].

In this study, the two-dimensional viscous flow pattern around an elliptic airfoil under heaving and pitching motion was numerically analysed. A 20% elliptic airfoil was chosen for the numerical study since no singularity is introduced at the leading or trailing edges and the analytical solutions at low Raynolds number and of the potential flow are known. It is emphasized in this study to cover the wider range of the flow conditions and to present the most comprehensive discussions on the vortex shedding and the trailing edge condition on the oscillating low Raynolds number airfoils.

The Navier-Stokes equations of viscous incompressible flow are reduced into the inhomogeneous non-linear equations of the stream function and the vorticity and solved numerically using the finite element method. In particular, the integral form of the stream function is used to calculate the boundary value of it from the vorticity distribution. The time relaxation procedure is adopted for several oscillating periods, starting with the steady viscous flow condition, until the quasi-periodic flow pattern was obtained.

2. Mathematical Formulation

Let us consider a two-dimensional elliptic airfoil oscillating both of heaving motion V and pitching motion Ω as

$$V = V_a \sin \omega t$$
$$\Omega = \Omega_a \sin(\omega t - \tfrac{\pi}{2}) \tag{1}$$

where V is the velocity of the vertical movement of the mid-chord point of the airfoil, Ω is the angular velocity of the pitching motion. That is, this airfoil has the heving motion with the velocity amplitude V_a and the angular frequency ω, and its pitching motion centered at the mid-chord point has the angular velocity amplitude Ω_a with the same frequency but the phase angle is in advance of a quarter period. Then, using the uniform incoming flow velocity U_0 and the focal distance of the elliptic airfoil section a, the following non-dimensionalized parameters are introduced;

the Raynolds number	$Re = U_0 \cdot 2a / \nu$
the reduced heving amplitude	$\lambda = \Omega_a \cdot a / U_0$
the Rossby number	$Rs = U_0 / \Omega_a \cdot 2a$
the reduced frequency	$K = \omega \cdot 2a / U_0$

The non-dimensionalized heaving and pitching motion, V_0 and σ, are expressed as follows, respectively

$$V_0 = \lambda \sin(Kt/2)$$
$$\sigma = \frac{1}{2R_s} \sin(Kt/2 - \pi/2) \tag{2}$$

The Navier-Stokes equations of incompressible viscous flow are written using the ellipsoidal coordinate system ξ, η fixed to the oscillating airfoil whose oscillation is expressed as Eq.(2)

$$\frac{\partial \omega}{\partial t} + \frac{1}{h^2}\left(\frac{\partial \psi}{\partial \eta}\frac{\partial \omega}{\partial \xi} - \frac{\partial \psi}{\partial \xi}\frac{\partial \omega}{\partial \eta}\right) + 2\frac{\partial \sigma}{\partial t} = \frac{2}{Re}\frac{1}{h^2}\left(\frac{\partial^2 \omega}{\partial \xi^2} + \frac{\partial^2 \omega}{\partial \eta^2}\right)$$
$$\frac{1}{h^2}\left(\frac{\partial^2 \psi}{\partial \xi^2} + \frac{\partial^2 \psi}{\partial \eta^2}\right) = -\omega \tag{3}$$

where ψ is the stream function and ω is the vorticity, h^2 $(= \cosh^2\xi - \cos^2\eta)$ is the scale factor of the coordinate system and the airfoil surface is expressed as $\xi = \xi_0$.

Now the dependent variable ψ and ω are expressed as the sum of the solutions of the potential flow ψ_p and ω_p and the correction terms by the viscous effect $\tilde{\psi}$ and $\tilde{\omega}$, respectively, then the following set of equations result;

$$\omega_p = -2\sigma$$
$$\frac{1}{h^2}\left(\frac{\partial^2 \psi_p}{\partial \xi^2} + \frac{\partial^2 \psi_p}{\partial \eta^2}\right) = -\omega_p \tag{4}$$

$$\frac{\partial \tilde{\omega}}{\partial t} + \frac{1}{h^2}\left(\frac{\partial \psi}{\partial \eta}\frac{\partial \tilde{\omega}}{\partial \xi} - \frac{\partial \psi}{\partial \xi}\frac{\partial \tilde{\omega}}{\partial \eta}\right) = \frac{2}{Re}\frac{1}{h^2}\left(\frac{\partial^2 \tilde{\omega}}{\partial \xi^2} + \frac{\partial^2 \tilde{\omega}}{\partial \eta^2}\right)$$
$$\frac{1}{h^2}\left(\frac{\partial^2 \tilde{\psi}}{\partial \xi^2} + \frac{\partial^2 \tilde{\psi}}{\partial \eta^2}\right) = -\tilde{\omega} \tag{5}$$

The solution of the potential flow equations (4) is known as

$$\psi_p = \exp\xi_0 \cdot \sinh(\xi - \xi_0)\cdot[\sin(\eta + a) + V_0 \cos(\eta + a)]$$
$$+\frac{1}{2}\sigma[\exp(\xi_0 - \xi)\cdot \sinh(\xi - \xi_0)\cdot \cos 2\eta + \frac{1}{2}(\cosh 2\xi - \cosh 2\xi_0)] \tag{6}$$

Then the non-linear equations (5) are solved by the time marching method, in which the non-linear terms are estimated using the previous solution, under the boundary condition of

$$\psi = \frac{\partial \psi}{\partial \xi} = 0 \qquad \text{on the airfoil surface}$$
$$\psi = \psi_p \;, \quad \omega = \omega_p \qquad \text{at infinity} \tag{7}$$

Time marching integration was carried out using implicit formulation starting at the solution of the steady flow with zero incidence angle. At each time step, the equation (5) are formulated using the first order triangular shape function, in which about 400 discretized points are used. The boundary conditions applied for this integration are

$$\psi = 0 \;, \quad \omega = \frac{2\psi(\xi + \Delta\xi, \eta)}{h^2 \Delta\xi^2} \qquad \text{on surface}$$
$$\psi = \psi_p + \tilde{\psi} \;, \quad \frac{\partial \omega}{\partial \xi} = 0 \qquad \text{at outer boundary} \tag{8}$$

where $\tilde{\psi}$ is given as the integral form of Eq.(5) as

$$\tilde{\psi}(\vec{r_0}) = \frac{1}{2\pi}\sum_K \int_S \tilde{\omega} \log\left(\frac{1}{r'}\right) dS + \frac{1}{2\pi}\int_C \frac{\partial \psi_p}{\partial \xi} \log\left(\frac{1}{r'}\right) dC \tag{9}$$

3. Results

Numerical integrations were carried out for the cases of: Re = 40; λ = 0.04, 0.06, 0.1 and 0.2; K = 0.5, 1.0 and 2.0; Rs = 10, 20, 40 and ∞, where the reference length is taken as the chord length instead of 2 a , which is equal to 2.04 a. The computer used is FACOM 230-75 at ISAS University of Tokyo.

An example of the flow pattern obtained is presented in Fig.1, which gives the contours of equi-vorticity lines and instantaneous stream lines with refered to the incoming flow. Since the airfoil is moving, some stream lines cross the airfoil surface. This shows the time sequential flow pattern over the second full period after starting of the oscillation. The virtual angle of attack has its maximum at t/T = 10/8 and 14/8 of the time, at which the rear stagnation point moves considerably forward from the rear trailing edge, and the flow turns around the trailing edge without introducing separation.

Fig.2 shows the leading and trailing stagnation point locations refered to the ellisoidal coordinate axis against the virtual attack angle. The left diagram corresponds to the pure heaving motion and the right is the combined motion of heaving and pitching. The leading stagnation point shows slight hysterisis and agrees fairly well with the potential flow solution. The rear stagnation point locates between those predicted by the irrotational theory and by the Kutta condition. They also show strong hysterisis curves, which are almost symmetrical against the abscissa. Fig.3 is the lift coefficients of the correspondent cases. The hysterisis nature and the decrease of the lift coefficients are clearly recognized.

4. Conclusions

Low Raynolds number flow around an oscillating elliptic airfoil was calculated using the finite element analysis. It is found that the calculated flow pattern around the leading edge shows no much difference from the inviscid solution, but the rear stagnation point moves upstream considerably, though it does exist and no separation takes place for these flow conditions. That is, the Kutta condition is not applicable in these cases. The resulted lift force is greatly reduced from the value predicted by the potential theory, and the shed vortices are not so strong as those theories predict.

References

[1] W.J.McCroskey; J.Fluid Eng. pp 8-38 3 1977

[2] J.Lighthill; "Mathematical Biofluiddynamics" 1977

[3] T.Theodorsen; NACA Rpt 496 1935

[4] U.B.Mehta; AGARD Paper 23 Sept. 1977

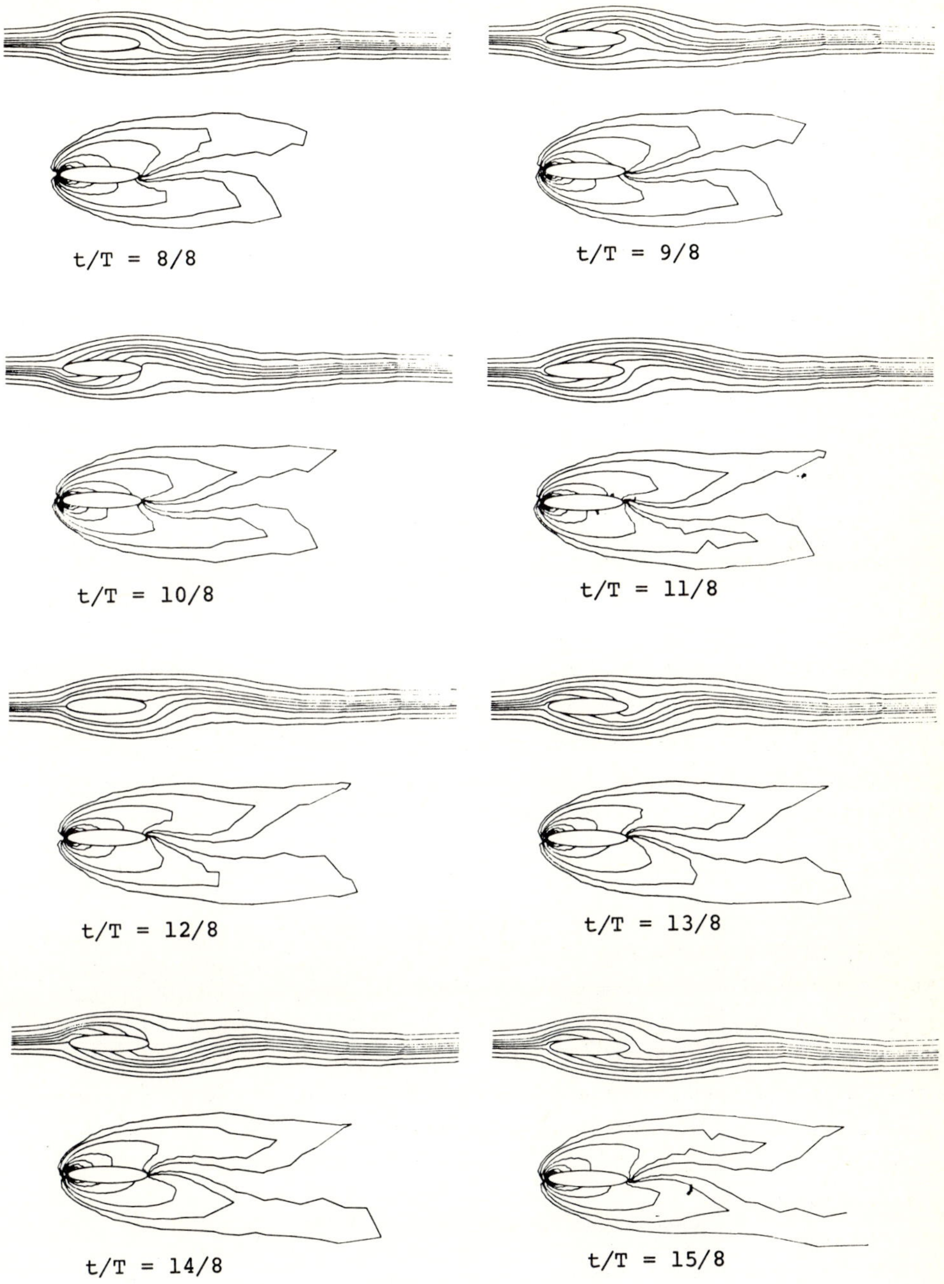

Fig.1 Flow Pattern at K = 2.0 Rs = ∞

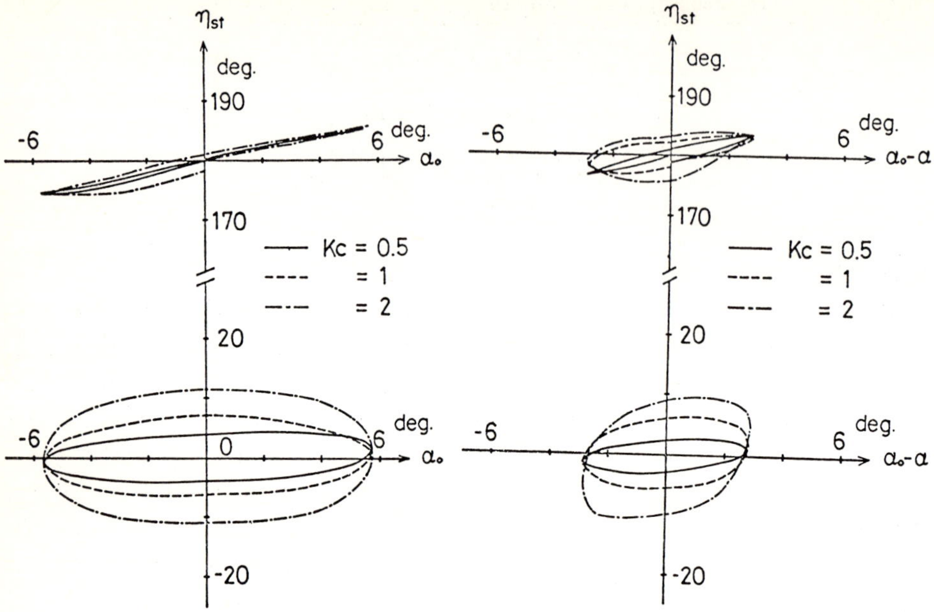

Fig.2 The Leading and Trailing Stagnation Point Locations

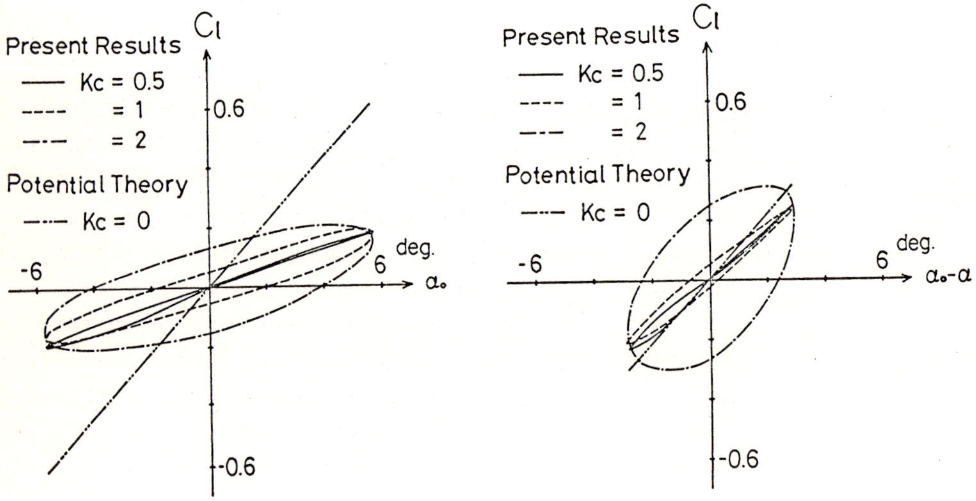

Fig.3 The Hysterisis Curves of the Lift Coefficients

SOME PERMEABLE BOUNDARIES IN MULTIDIMENSIONAL UNSTEADY FLOWS (°)

by M. Pandolfi and L. Zannetti

Istituto di Macchine e Motori per Aeromobili
Politecnico di Torino, Torino, Italy

INTRODUCTION

Unsteady compressible flows are described by hyperbolic partial differential equations. The integration in time of such equations does not only allow for the description of flow transients "per sé", but also for the achievement of the steady configuration of multidimensional compressible rotational flows, according to the "time-dependent" technique. In a basic paper on this matter [1][1], it has been pointed out clearly how important is the numerical treatment of the boundaries and why the algorithm should manage correctly the flow field waves interaction with them.

There are different kinds of boundary:

- fixed or movable
- impermeable or permeable.

A rigid wall represents the simplest case: it is fixed and impermeable. The bow shock ahead of a blunt body, treated as a discontinuity in the time-dependent technique [2], is a movable permeable boundary. A flexible wall with a prescribed pressure distribution, as used in [3] for solving inverse problems, is a movable but impermeable boundary.

In this paper we focuse our attention to fixed and permeable boundaries in the case of subsonic flow. They may simulate entry or exit surfaces about a physical domain. They may also replace physical devices which separate neighboring regions, as for example, in the case of the actuator disc, which simulates the blading in turbomachines.

The numerical procedure at these boundaries is based on the matching of the conditions holding here with the most appropriate arrangement of the flow equations, which is given by compatibility equations. In the following we will show this procedure for the cases of entry, exit and partition boundaries and we will report on the experience we got so far in

(1) Numbers in brackets designate References at the end of paper.

(°) This research has been supported by C.N.R. (contract N. 115.9299 - CT77.00122.07)

different physical problems.
For the sake of simplicity we refer hereafter to 3D unsteady flows described in the cartesian frame (x,y,z) with the corresponding velocity components (u,v,w).
The flow field equations are:

$$
(1)\quad
\begin{aligned}
&P_t+uP_x+vP_y+wP_z+\gamma(u_x+v_y+w_z) = 0 && \text{(continuity)}\\
&u_t+uu_x+vu_y+wu_z+TP_x = 0 && \text{(x-momentum)}\\
&v_t+uv_x+vv_y+wv_z+TP_y = 0 && \text{(y-momentum)}\\
&w_t+uw_x+vw_y+ww_z+TP_z = 0 && \text{(z-momentum)}\\
&S_t+uS_x+vS_y+wS_z = 0 && \text{(energy)}
\end{aligned}
$$

where:

$$P = \ln p; \qquad T = \exp\left(\frac{\gamma - 1}{\gamma}P + \frac{1}{\gamma}S\right)$$

ENTRY BOUNDARY

Suppose that the unsteady flow in a duct has to be investigated (Fig.1). The duct is fed by an upstream reservoir through the entry surface AB. The correct description of such internal unsteady flow should require also the investigation of the unsteady phenomenon, upstream of the entry surface. However, in practical cases, it is almost impossible to describe correctly this upstream unsteadiness; even the use of mappings stretched upstream to a very large distance from the region of interest presents some arbitrariness as regards the upstream actual contours.
It may be then suggested to delimite upstream the domain of investigation with an entry surface AB, where particular conditions have to be imposed. Its use brings in some arbitrariness whatever are the conditions imposed there, but this shortcoming may become as weaker as the entry boundary AB is located more far from the region of interest. Furthermore, in the case the unsteady analysis is used to achieve a steady configuration (time-dependent technique), such a hypothesis presents just the same restrictions as any other classical method.
We suppose the entry surface as a distribution of small nozzles. Each of them is fed with gas at prescribed levels of total pressure at temperature (p_i^o, T_i^o) and the direction of the flow out of them is imposed generally by two angles (α_i, β_i). If the length of such nozzles is assumed short enough, the flow through them is taken as quasi-steady. The following conditions hold across the entry surface (normal to z):

$$
(2)\quad
\begin{aligned}
&u = w\tan\alpha_i; \qquad v = w\tan\beta_i\\
&p\left(1+\frac{\gamma - 1}{2}\frac{q^2}{a^2}\right)^{\gamma/\gamma-1} = p_i^o\\
&T+\frac{\gamma - 1}{2\gamma}q^2 = T_i^o
\end{aligned}
$$

where:

$$q^2 = u^2 + v^2 + w^2 \quad ; \qquad a^2 = \gamma T \text{ (speed of sound)}$$

The entry surface characteristics (p_i^o, T_i^o, α_i, β_i) may be prescribed as non uniform in x,y as well as varying in time in order to simulate unsteady perturbations. The only conditions given by Eq. 1 do not allow for the description of the flow at this boundary (u,v,w,p,T). The further information is provided by the flow field equations and must bring in the nature of the field unsteadiness. By the continuity and z-momentum equations (Eq.1) one gets easily the compatibility equation (in the reference plane z-t) relative to backwards running waves, impinging on the entry surface:

$$(3) \quad (P)' - \frac{\gamma}{a}(w)' = -\gamma(u_x + v_y)$$

where:

$$(\ )' = (\ )_t + u(\ )_x + v(\ )_y + (w-a)(\ )_z$$

At this point we follow the suggestions given in [4,5]. The boundary conditions (Eq.2) are differentiated with respect to the time and, after some rearrangements, one has:

$$(3') \quad P_t + \gamma M^2 \frac{w_t}{w} = \frac{p^o_{it}}{p_i^o} + \frac{\gamma}{2} M^2 \frac{T^o_{it}}{T_i^o} - [\tan\alpha_i(1+\tan^2\alpha_i)\alpha_{it} + \tan\beta_i(1+\tan^2\beta_i)\beta_{it}] \cdot \gamma \frac{w^2}{a^2} \ ; \qquad (M^2 = q^2/a^2)$$

Once the boundary conditions (p_i^o, T_i^o, α_i, β_i) and their time derivatives are prescribed, one has the value of w_t from Eq.2,3'. The w component is then integrated in time with the same procedure used at the interior points. In particular we have adopted the well experimented two levels predictor-corrector scheme suggested by Mac Cormack: it has been slightly modified to ensure the second order of accuracy with reference to the one side derivatives in z. After the new value of w is computed, the remaining four flow properties follow easily from Eq.2. We point out that other particular physical boundary conditions may be easily simulated by writing Eq.2 in more suitable forms.

Similar or slightly different procedures have been suggested in [6,7].

EXIT BOUNDARY

Suppose now to consider an exit subsonic boundary (CD in Fig.1). The problem may be discussed similarly as in the entry case. Even here the assumption to delimit downstream the region of computation presents some arbitrariness from a physical point of view, but it is questionable to expect that something better could be done.

The procedure, we suggest, is based on a physical model, as at the entry. In fact we imagine the exit permeable surface as a distribution of small nozzles fed by the internal flow (with conditions u,v,w,T). These discharge the gas in capacities, with prescribed static pressure (p_e), through a given area constriction (η). Both p_e and η may vary in x,y as well as in time, depending on the requirements of the problem. Since the length of such nozzles is considered as infinitesimal, the following quasi-steady conditions hold through the exit surface:

$$
\begin{aligned}
&pw/T = \eta p_e w_e/T_e && \text{(continuity)}\\
(4)\quad &T + \frac{\gamma - 1}{2\gamma} q^2 = T_e + \frac{\gamma - 1}{2\gamma} w_e^2 && \text{(energy)}\\
&p/p_e = [T/T_e]^{\gamma/\gamma-1} \quad \text{or} \quad S = S_e && \text{(evolution)}
\end{aligned}
$$

These boundary conditions are differentiated in time and matched with the compatibility equation relative to forwards running waves, impinging on the exit:

$$(5)\quad (P)' + \frac{\gamma}{a} (w)' = -\gamma (u_x + v_y)$$

where:

$$(\ )' = (\ )_t + u(\ )_x + v(\ )_y + (w+a)(\ )_z$$

While the derivatives in time of u,v,S are obtained by the corresponding relationships of Eq.1, the time derivative w_t is obtained from Eq.4 (differentiated in time) and 5. The integration in time, according the interior points scheme, provides the new values of u,v,w,S whereas the new level of p follows easily from Eq. 4.
A different approach for exit subsonic boundary is given in [6]. Here, for 2D flows, one parameter (the static pressure) is prescribed at the exit; the remaining flow parameters are then obtained from compatibility equations on forwards running characteristics.

PARTITION BOUNDARY

There are cases where it is required to separate two neighboring regions, with unsteady flow, by a surface of discontinuity which may exibit particular features. Typical example is the actuator disc which replaces the actual blading in turbomachines. The actuator disc deflects the gas stream and provides (in the case of rotor) the corresponding jump of energy because of the mechanical work exchanged between gas and blading.
The description of the flow upstream ("1") and downstream ("2") of the discontinuity requires the prediction of ten unknowns (u,v,w,p,T on the two sides). Being the partition surface infinitesimally thin, these quasi-steady relationships hold:

$$p_2 w_2 / T_2 = p_1 w_1 / T_1$$

$$(6) \qquad T_2 + \frac{\gamma - 1}{2\gamma} q_2^2 = T_1 + \frac{\gamma - 1}{2\gamma} q_1^2 + \Delta T^\circ$$

$$p_2 \left(1 + \frac{\gamma - 1}{2} (q_2/a_2)^2\right)^{\gamma/\gamma - 1} = p_1 \left(1 + \frac{\gamma - 1}{2} (q_1/a_1)^2\right)^{\gamma/\gamma - 1} + \Delta p^\circ$$

Here T^o and p^o denote the variations of total pressure and temperature through the discontinuity. In the case of a rotor these values are evaluated on the base of experiments on the corresponding cascade, in terms of losses and flow deflection, depending on the local relative incidence. The downstream components u_2, v_2 are depending on the flow upstream ("1") and on the geometry of the replaced blading: one has then two further equations. These five relationships represents the boundary conditions. They are differentiated in time and provide five differential equations with the time derivatives on the ten unknowns.

The further five conditions are given by the flow field equations. Namely the x and y momentum and energy equations in the upstream region provide u_t, v_t and S_t. Finally the compatibility equations (Eq.5 upstream and Eq.3 downstream) give the last two needed conditions.

At this stage the new flow properties upstream are obtained by integration in time of the corresponding time derivatives. The new values of the downstream properties follow from the boundary conditions.

NUMERICAL EXAMPLES

a) Ringleb flow [8]

In order to check the accuracy of the algorithm at entry and exit surfaces, we report first about a 2D subsonic flow whose exact configuration is know analytically: the channel flow between two given streamlines of the "Ringleb flow". The shape of the channel and the isobar pattern as achieved through a time-dependent computation is shown in Fig.2a. The computation starts with gas at rest and at a pressure level equal to the theoretical total pressure. The entry nozzles (angle α_i) and the exit constriction (η) are taken from the theoretical configuration. The downstream pressure (p_e) is gradually brought down to the theoretical distribution at the exit. Expansion waves are generated at the exit, travel upstream, interphere with the entry and begin to build up the entry flow. After some paths of waves up and down along the channel, a steady configuration is achieved. The error pattern (with respect to the theoretical static pressure) is reported in Fig.2b. It may be seen that the permeable surface algorithms do not affect the accuracy more than the interior points or rigid wall algorithms and that the errors are kept at very low levels (8 x 10 grid).

b) Transonic swirling flow [9]

Transonic swirling flow in a convergent-divergent nozzle may be computed through a time-dependent technique. The swirling is obtained at the entry by a prescribed distribution of inlet angles (β_i) in the peripherical direction. The transient is generated as in the previous example, except that now the initial exit pressure is kept very low in order to generate here supersonic flow: so the exit points are computed as interior points, with one side derivatives along the nozzle axis. The comparison between Fig.3a,b shows the different pattern of the isobars without and with the swirling.

c) 3D rotational flow in curved ducts [10]

In this case the secondary flow, developed in a curved channel because of entry transversal vorticity, is investigated. The flow is fully 3D and the transient we have computed is the usual. The entry surface simulates a grid with variable drag coefficient in order to generate the entry transversal vorticity. The secondary flow is represented by a vortex in a cross section of the channel and is related to the streamwise generated vorticity (ξ). Fig.4a shows the vortex pattern while Fig.4b refers to the generated streamwise vorticity. Checks on the accuracy with a linearized analytical solution are reported in [10].

d) Flow in axial compressors

We have developed and tested procedures for investigating 2D unsteady flows (axisymmetric flow in low hub-to-tip ratio bladings or non-axisymmetric flow in high hub-to-tip bladings) and we are working presently to the 3D case. The bladings are replaced here by actuator or semiactuator discs. In particular we have investigated the response of a given blading to peripherical distorsions in order to get the prediction of the attenuation or amplification of peripherical perturbations induced at the entry and exit. Steady distorted flow configurations are achieved as the asymptotic result of a transient. However at low mass flow rate in the compressor, no stable solution is obtained but flow instabilities, supported on physical arguments, grow up at the disc and begin to rotate just as rotating stalls. Fig.5 reports the isobars pattern in front of a rotor as it evolves in time: organized waves rotate peripherically in a two cells configuration. More details will be reported in [11].

REFERENCES

[1] Moretti, G.: Importance of Boundary Conditions in the Numerical Treatment of Hyperbolic Equations.
The Physics of Fluid, Vol. 12, 1969.

[2] Moretti, G.; Abbett, M.: A Time-Dependent Computational Method for Blunt Body Flows.
AIAA Journal, Vol. 4, No. 12, 1966.

[3] Zannetti, L.: A Time-Dependent Method to solve the Inverse Problem in Internal Flows. To be published.

[4] Kentzer, C.: Discretization of Boundary Conditions on Moving Discontinuities. Second International Conference on Numerical Methods in Fluid Dynamics, September 1970.

[5] Moretti, G.; Pandolfi, M.: Entropy Layers. Computers and Fluids, Vol. 1, 1973.

[6] Serra, R.: Determination of Internal Gas Flows by a Transient Numerical Technique. AIAA Journal, Vol. 10, No. 5, 1972.

[7] Edelman, R. et al.: Some Aspects of Viscous Chemically Reacting Moderate Altitude Rocket Exaust Plumes. General Applied Science Labs., Westbury, N.Y., March 1970.

[8] Pandolfi, M.; Zannetti, L.: Some Tests on Finite Difference Algorithms for Computing Boundaries in Hyperbolic Flows. GAMM Workshop, University of Stüttgart, February 1977.

[9] Pandolfi, M.: Transonic Swirling Flow in Axisymmetric Nozzles. Meccanica, Vol. 11, No. 3, 1976.

[10] Bussi, G.; Pandolfi, M.: Compressible Inviscid Rotational Flow in Curved Ducts. 2nd GAMM Conference on Numerical Methods in Fluid Dynamics, 1977.

[11] Pandolfi, M.; Colasurdo, G.: Numerical Investigations on the Generation and Development of Rotating Stalls. To be published.

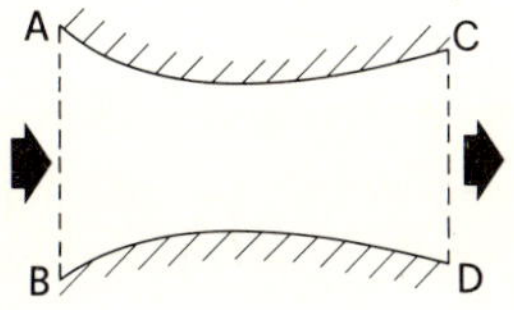

FIG. 1

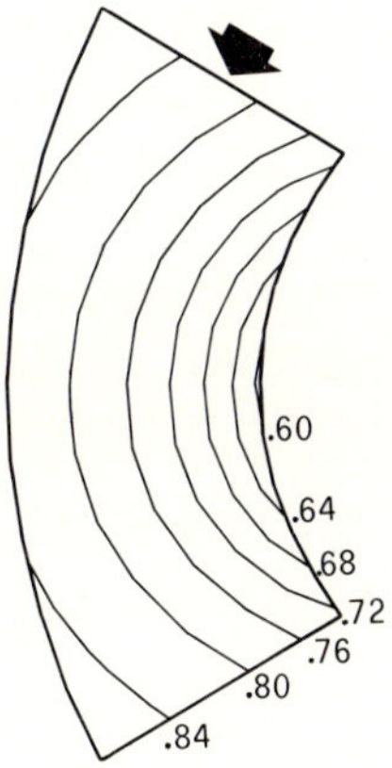

FIG. 2a

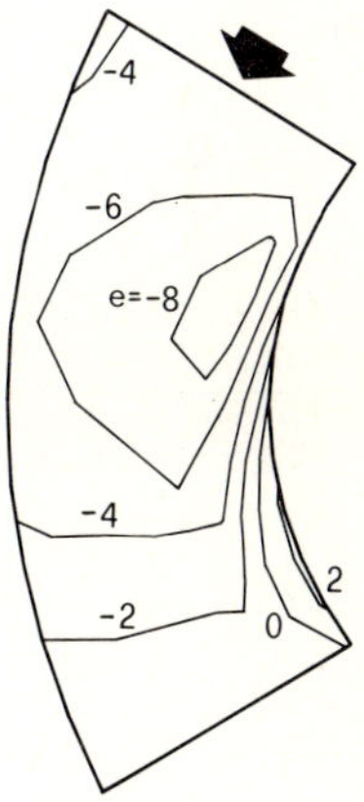

FIG. 2b

$$\left(e = \frac{p_{theor.} - p_{comp.}}{p_{theor.}} \cdot 1000\right)$$

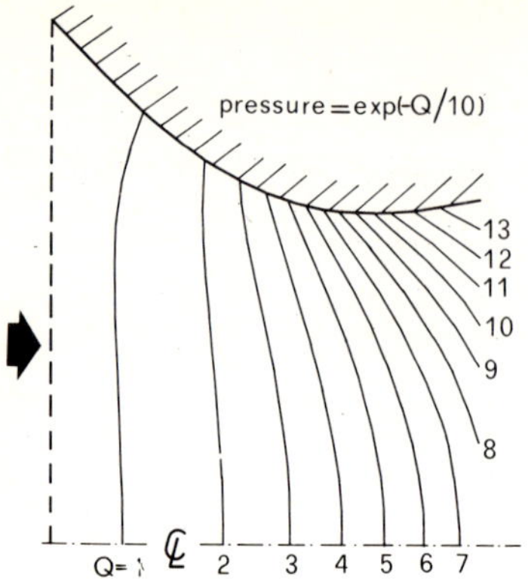

FIG. 3a (no swirling)

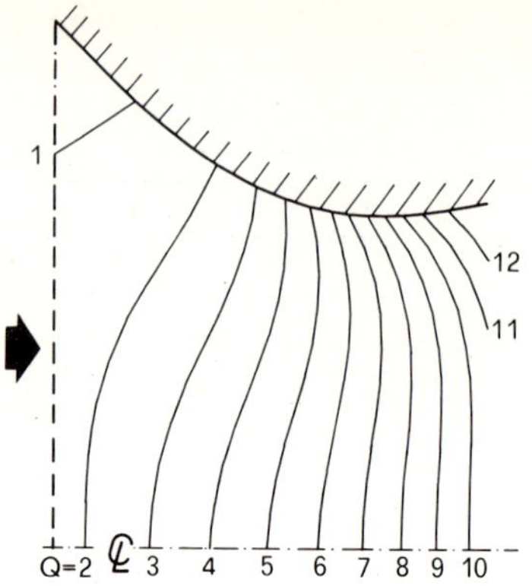

FIG. 3b (swirling flow)
(sonic Line at Q = 6.38)

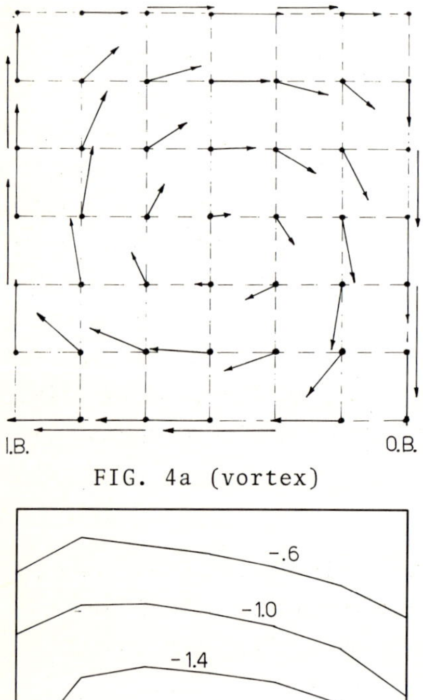

FIG. 4a (vortex)

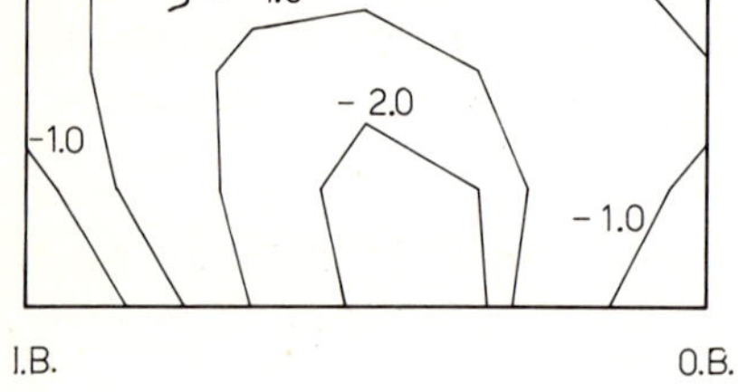

FIG. 4b (streamwise vorticity ξ)

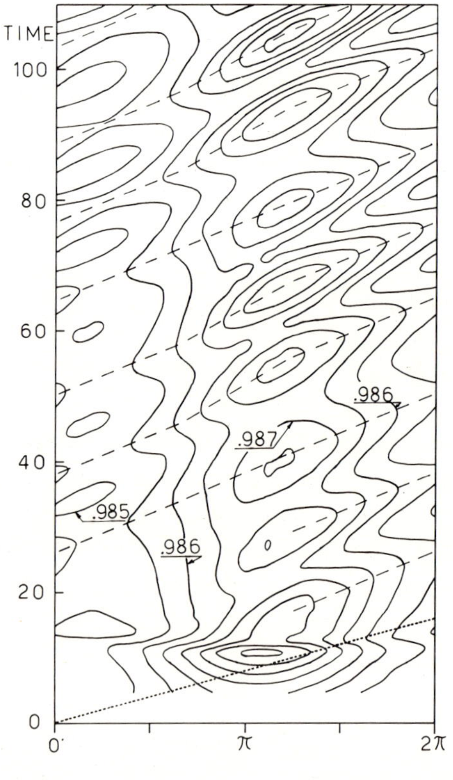

FIG. 5 (the dotted line refers to the path of the rotor)

SUPERSONIC FLOW OF VISCOUS GAS AROUND BODIES AND THE FLOW IN THE WAKE (METHODS OF A NUMERICAL SOLUTION AND COMPUTING RESULTS)

V.M.Paskonov
Research Computer Centre of Moscow
University
USSR

The present work considers a laminar flow with axial symmetry around blunt bodies. We have used the Navier-Stokes model to describe the considered flow. A computed method of an integration of the complete compressible system of Navier-Stokes equations in the whole disturbed flow field is suggested. We are considering here computing results of the laminar flow of a viscous perfect gas around the cone the cylinder with a spherical blunt nose for a wide range of Mach and Reynolds numbers. An accurate investigation of the stream near the stern of the body and the flow in the near and far wake has been carried out. For a high temperature gas flow the method for a solution of the system of Navier-Stokes equations for the multicomponent mixture of the reacting gas with calculations of finite velocities of chemical processes and a diffusion of individual components is suggested.

The study of the flow around limited bodies moving at supersonic speed and the flow in the wake is a complicated and actual problem of the computational aerodynamiecs. The qualitative description of the structure of the flow field may be given on the basis of theoretical and experimental references. Figure 1 represents a scheme of the axial symmetry flow around a blunt cone moving at a supersonic speed with the zero attack angle and moderate Reynolds numbers.

The steady - state solution of the complete unsteady system of Navier-Stokes equations for a viscous perfect heat conduction gas is

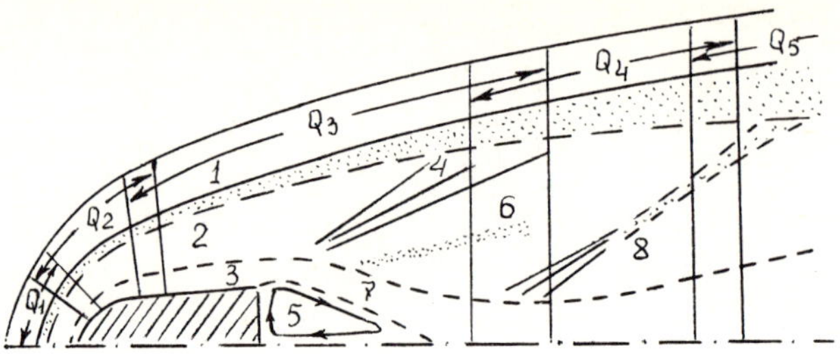

Fig. 1. The general scheme of the flow (1 - the detached bow shock wave, 2 - the inviscous flow, 3 - the boundary layer, 4 - rarefaction waves, 5 - the flow circulation, 6 - the stern wave, 7 - the free boundary layer, 8 - the tail shock wave).

computed by stability method in a total region of a body influence on a stream. No apriori suppositions about a flow field structure and no suppositions about simplifications of the system of equations are used in calculations /1-3/.

On the surface of a streamlined body we use no-slip conditions for velocity and constant temperature. In the considered Reynolds numbers range the flow is supposed laminar and symmetric relatively to the axis $r=o$. We have used two ways for an assignment of the external boundary of an integration region. For moderately small Reynolds numbers ($Re < 10^3$), when it is necessary to take into consideration the detached bow shock wave structure, we choose an external boundary in the process of calculation as a line $r=R(x)$ which is placed in the natural stream at relatively small distance from the shock wave zone. On this line we put a boundary condition which corresponds to the natural stream.

For greater Reynolds numbers ($Re > 10^3$) the influence of a thin shock wave structure on the downstream may be neglected. Therefore we take shock wave as the surface of the discontinuity $r=R(t, x)$, on which we use the unsteady Rankin - Hugoniot conditions with the same approximate correlation, which takes into consideration a linear change of gasdynamics parameters on the downstream near the shock wave.

The reliability of qualitative and quantitative results of computing the flow in a very stretchable region with a complicated character of a solution (shock waves, boundary layers, separations of a flow, the wake) is required to be very accurate. This leads to calculations on detailed mesh systems and it produces great demands to the memory and quick operation of a computer.

We suggest a computer method which allows to avoid the above mentioned difficulties. The method is based on the division of the whole field into a number of overlapping small regions $Q_i (i=1,2...)$ and performing successive calculations in every region beginning with Q_1. On the right outflow boundary in every region Q_i we use the liner extrapolation of unknown functions as a boundary condition. Therefore in region Q_1 all boundary conditions are fixed. In region $Q_1 (i=2,3,...)$ on the upstream left boundary we have used solution from the previous region Q_{i-1} as boundary conditions. The value of the overlapping of regions was determined in the process of calculations dependending on the influence of the downstream region on the upstream region.

Computing regions corresponded to the following stream regions: Q_1 - near the blunt nose, Q_2 - near the side surface, Q_3 - near the stern of the body and flow in the near wake, Q_4 - flow in the wake where the speed near the symmetric axis is subsonic, Q_5 - flow in the far wake.

Computing solution of the Navier - Stokes equations in every region Q_i was performed by means of the difference method. We used some implicit and explicit difference schemes. In the region of the far wake Q_5 we used a transition to the simplified Navier - Stokes equations only in cases where their solution coinsided with the complete Navier - Stokes equations with the necessary accuracy.

As an example we give some results of the computing of the flow around limited bodies of the type "sphere - cylinder" and the blunt nose cone (half angle = 10°) with a vertical bottom cut.

General pictures of the flow around the cylinder (the length $L = 3.6 \cdot r_o$, r_o - the radius of the spherical nose) are given in figure 2 for M=3 Re=200 and in figure 3 for M=20 and Re=300. These cases were computed in the approximation of "the thick shock wave".

Flow structures in the wake of the blunt cone (the length $L = 1.95 \cdot H$, H - the radius of the bottom cut) are given in figure 4 for $M = 15$, $Re = 10^4$, in figure 5 for $M = 20$, $Re = 10^4$, in figure 6 for $M = 20$, $Re = 5 \cdot 10^5$ and in figure 7 for $M = 20$, $Re = 10^6$. These cases were computed in the approximation of "the thin shock wave".

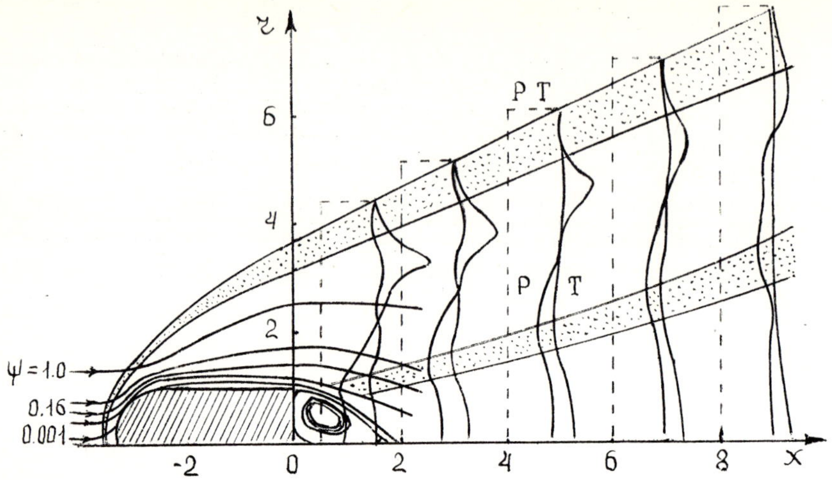

Fig. 2.

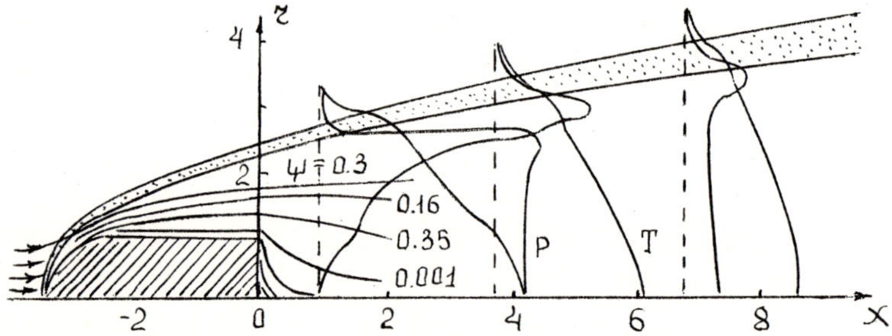

Fig. 3.

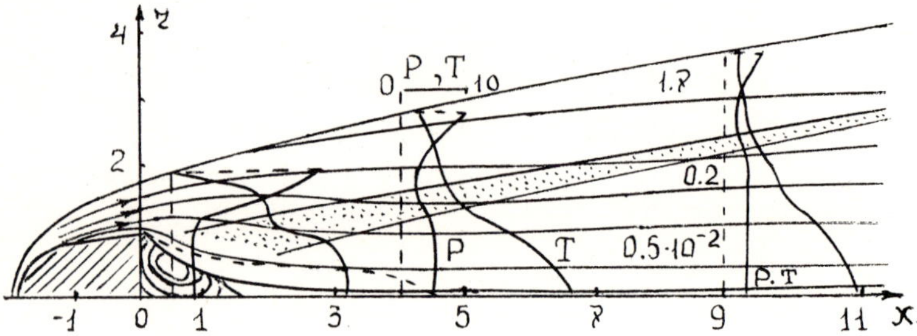

Fig. 4.

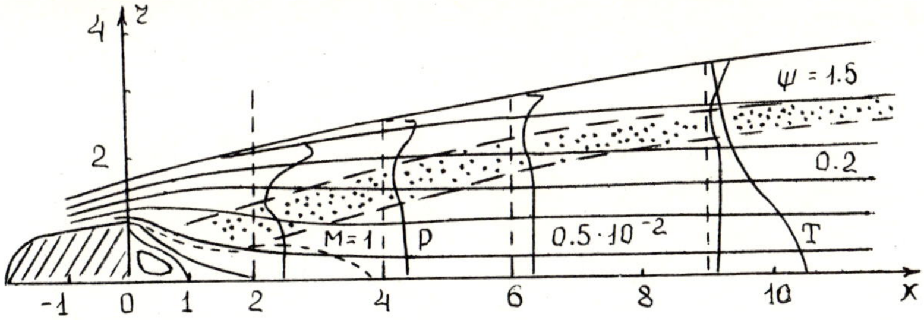

Fig. 5.

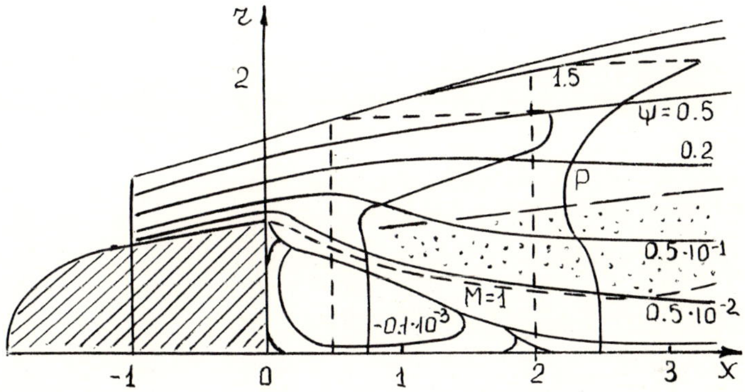

Fig. 6.

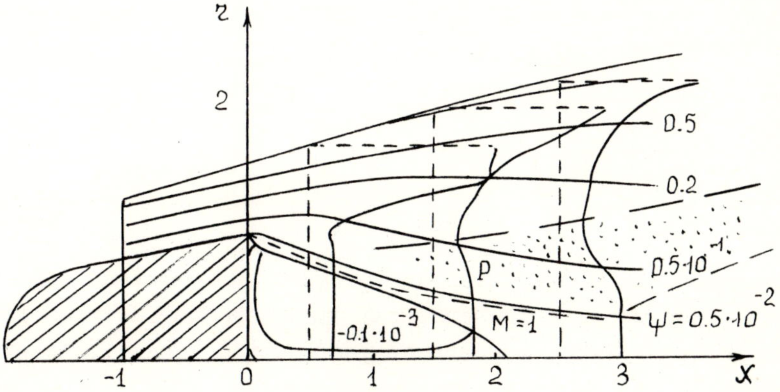

Fig. 7.

When the blunt nose bodies move at supersonic speed the gas warms up to high temperatures in the shock wave and the compressed layer and the gas changes its physical - chemical properties. Therefore we performed computing investigations of the viscous flow of a reacting gas near the spherical nose. To describe such a flow we used a system of equations which may be obtained from Boltzmann equations by means of Enskog-Chapman method and corresponds to the Navier-Stokes approximation. Calculations were performed for seven component gas mixture (model of the air): N_2, O_2, NO_2, NO^+, O, N, e. Seven reactions of the dissociation, recombination, ionization were taken into consideration. Following boundary condition were used. No - slip conditions, constant temperature and conditions of the neutral chemical surface for chemical components were given. As with the perfect gas, approximations of "thick" and "thin" shock waves were considered. As an example we give some results of the computing of the reacting gas near the sphere ($0 \leq \theta \leq 35^{\circ}$) in the atmosphere of the Earth at the altitude H = 72 km for the external stream parameters: $u_{\infty} = 5.8$ km/sec, M = 20, Re = 10^4, for the radius of sphere $r_o = 0.35$ m, the surface temperature $T_w = 9900^{\circ}K$, $1800^{\circ}K$ and the relaxation parameter S = 140 ($S = \rho_{\infty} k_o r_o / \mu_{\infty} u_{\infty}$, k_o depending on the external stream temperature).

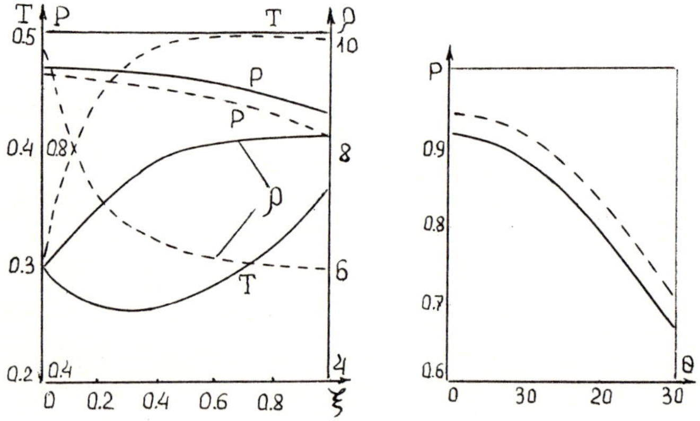

Fig. 8.

Figure 8 gives the comparision of the gasdynamic parameters conduct in the shock wave along the critical line for the reacting gas (continuous lines) and for perfect gas (dotted lines); the pressure distribution along the wall is given as well.

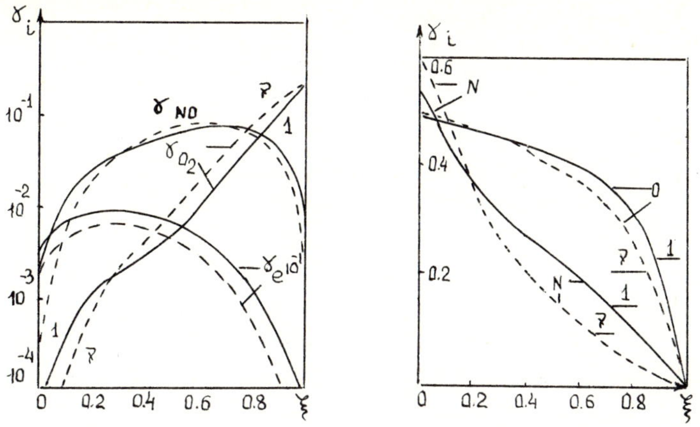

Fig. 9.

Figure 9 presents the chemical components conduct along the critical line (continuous lines) and along the ray $\Theta = 30^{\circ}$ (dotted lines). On the whole the calculations showed that non-equilibrium chemical reactions have a strong influence on the shock layer flow near by the shock wave whereas near by the wall surface transfer processes play the principal role.

Reference

1. Emeljanova Z.M., Ilushenkova A.A., Kokoshinskaya N.S., Pavlov B.M., Paskonov V.M., Petrova L.I., Polyanskiy W.A.
Computing investigation of the real gas flow around limited bodies and the flow in the wake.
"Vychislitel'nye Metody i Programmirovanie", N 23, Moscow Univ., 1974, pp. 3-41.
2. Kokoshinskaya N.S., Pavlov B.M., Paskonov V.M.
Computing of the Supersonic laminar flow around blunt body and the flow in the wake."Aeromechanika i gasovaya dinamika", Moscow, 1976, pp. 160-171.
3. Kokoshinskaya N.S. About the flow in the near laminar wake past the blunt nose cone."Chislennye Metody v Aerodynamike" N 2, Moscow Univ., 1977, pp. 52-68.
4. Petrova L.I. Computing of the reacting gas flow near by the spherical surface on the basis of the Navier - Stokes model. "Chislennye Metody v Aerodynamike", Moscow Univ., N 1, 1976, pp. 3-41.

NUMERICAL MODELS OF TURBULENT CONVECTION BASED ON THE UNSTEADY NAVIER-STOKES EQUATIONS

V.I. Polezhayev, V.L.Gryaznov, A.G. Daikovsky, A.I. Fedoseyev
The USSR Academy of Sciences
Institute for Mechanics Problems

Based on unsteady (nonstationary) Navier-Stokes equations in two dimensions, the present note makes a study of a possibility of numeric modelling of turbulent convection regimes. Preliminary averaging leading to the known Reynolds equations is not applied.

1. The dimensionless Navier-Stokes equations for a nonisothermal fluid in the gravitational field (the Boussinesq approximation) in the variables "vortex,stream function, temperature" [1-3] .

(1) $$\omega_t + u\omega_x + v\omega_y = \omega_{xx} + \omega_{yy} - G_L\theta_y , \quad G_L = g\beta L^3 \Delta T \nu^{-2}$$

(2) $$\psi_{xx} + \psi_{yy} = \omega$$

(3) $$\theta_t + u\theta_x + v\theta_y = P^{-1}(\theta_{xx} + \theta_{yy}) , \quad P = \nu/a$$

where G_L is the Grashof number and P is the Prandtl number, are the initial items.

A problem is set to study the regimes under which the experimentally observed flow is turbulent. In this case the system (1) - (3) has no steady solution, therefore instant values of velocity and temperature and (with subsequent data processing) the average and pulsed characteristics are found. The numeric modelling method developed to realize this approach **comprises** steps as follows:

1) Mesh method computation of the instant values of the unknown fields, $\psi(x,y,t)$, $\theta(x,y,t)$. Implicit difference scheme and the alternating direction method are applied. The essence of the finite -difference scheme for a numeric solution of equations (1) - (3) is in making use of an effective approximation of the boundary conditions for vortex [1-3] ; additional improvements are presented in [4] .The scheme is built on a ununiform mesh with decreasing steps in the near wall area with large velocity gradients. Special attempts are made to optimize the iteration solution for equation (2). This is realized by the alternating direction method with an optimal set of iteration parameters [2,3] being chosen. Paper [4] applies the method of separation of variables with the use of fast Fourier transform.

2) Based on a preliminary analysis of the results, definitions of the time t_1 ,beginning with which the average values are practically independent of the initial data, and statistical data processing to find the average characteristics in the "averaging" interval T

$t_1 \leq t \leq t_1+T$
$$\langle u(x,y)\rangle = \frac{1}{T}\int_{t_1}^{t_1+T} u(x,y,t)\,dt$$

and also pulsational characteristics

$$u' = u - \langle u\rangle\ ,\ \sqrt{\langle u'^2\rangle}\ ,\ \langle u'v'\rangle\ ,\ \langle u'\theta'\rangle$$

and so on.

3) Application of the computer graphics to analyse evolution of large-scale and fine structure of flow fields and temperature.

The values of the temperature fields and of velocity, as well as visual pictures of flow are compared to the experimental data.

A special set of programs designed to realize the technique contains programs of a numerical solution of the system (1) - (3), as well as programs of graphic output and statistical processing.

2. The results of specific computations are illustrated by an example of flow evolution in a closed vertical slot heated from one side, the Rayleigh number relates to the slot height as $R_H = G_H \cdot P = 5.25\cdot 10^{11}$,the Prandtl number P =15, the ratio of the slot sides is H/L =11.2. With these parameters the flow according to the experimental data [5,6] but it is comparatively slow and low frequency (the Reynolds number , $Re = U_{max} \cdot L/\nu$, is of the order of 10^3 , the frequency band is 0-10 Hz). The absence of external perturbations which are of an random nature is of a great importance.

The zero values of velocity ($\Psi = \omega = 0$) and a linear profile of temperature ($\theta(x,y) = y/L$) are given as the initial ones. The boundary conditions are as follows :

$$\Psi = \frac{\partial\Psi}{\partial n} = 0\ ,\quad \theta(x,0) = 0\ ,\ \theta(x,L) = 1\ ,\ \frac{\partial\theta(0,y)}{\partial x} = \frac{\partial\theta(H,y)}{\partial x} = 0$$

These and the values of dimensionless parameters R_H , P , H/L correspond to one of the experiments [5,6] (the working fluid is spirit, the layer width, L , is 60 mm, ΔT =32°C) The computaions have been performed for the 93x39 and 129x39 meshes [2-4] with strong step decresing near solid side surfaces (the minimal distance between the nodes in the near wall area is $0.005\cdot L$).

3. Fig. 1 presents the evolution of large-scale structure of convection in the slot (instant isolines of the stream function). Single-vortex ascend-descend motion (Fig. 1a) occurring in the initial moments grad-

ually transforms into multi-vortex one (Fig. 1b) corresponding to laminar motion in the layer with $R_H \sim 10^7-10^8$ [3,7]. Further convective flow in a kernel splits up and near wall turbulent boundary layer is formed (Fig. 1c). At the turbulent regime, at $t > t_1 = 4.4 \cdot 10^{-2}$ flow in the near wall area consists of a viscous sublayer and a number of small vortexes rising along one of the walls and descending along another (Fig. 2). Vortex motion is accompanied by velocity and temperature pulsations. Fig. 3 shows the cross section $x/L = 5.6$ of instant values of temperature in three points at various distances from the side wall. Evolution of large-scale and fine structures of the near wall turbulent area is illustrated in a more detail in a film shot from the display.

Figs. 4-7 give comparison with the experimental data for the averaged characteristics (in this case x - are the experimental values, —— are the computational results). Fig. 4 presents the vertical distribution of the average temperature along the slot axis ($y = L/2$). The dependence of the local Nusselt number , Nu_x, vercus on that of Rayleigh R_x satisfactorily agrees with the experimental dependence, $Nu_x = 0.108 \cdot R_x^{1/3}$ [5,6] . The profiles of the vertical velocity and its dispersion $\langle u'^2 \rangle^{1/2}$ are presented in Figs. 5,6 for the cross section $x = 5.6 \cdot L$.

The analysis of the results of the present work and their comparison with the experimental data on the basic characteristics of the flow and temperature enable to arrive at a conclusion that the basic features of the generation mechanism of the near wall turbulence in the vertical slot in the range of the Rayleigh numbers under consideration may be satisfactorily described in terms of two-dimension unsteady equations of Navier-Stokes.

REFERENCES

I. Полежаев В.И., Грязнов В.Л. Метод расчета граничных условий для уравнений Навье-Стокса в переменных "вихрь,функция тока". Доклады АН СССР, I974,т.2I9,№2.

2. Грязнов В.Л., Полежаев В.И. Численное моделирование турбулентного режима конвекции в вертикальном слое. Известия АН СССР, МЖГ,I977,№5.

3. Грязнов В.Л., Полежаев В.И. Численное решение нестационарных уравнений Навье-Стокса для турбулентного режима естественной конвекции. Препринт ИПМ АН СССР, I977, № 8I.

4. Дайковский А.Г.,Полежаев В.И., Федосеев А.И. Численное моделирование переходного и турбулентного режимов конвекции на основе нестационарных уравнений Навье-Стокса. Препринт ИПМ АН СССР ,I978, № I0I.

5. Кутателадзе С.С.,Ивакин В.П., Кирдяшкин А.Г.,Кекалов А.И. Тепловая гравитационная конвекция в вертикальном слое при турбулентном режиме течения. В сб."Тепломассообмен 5", т.I,ч.2,Минск,I976.

6. Kutateladze S.S., Kirdyashkin A.G., Ivakin V.P. Turbulent natural convection on a vertical plate and in a vertical layer. Int. J. Heat and Mass Transfer, 1972,v.15.

7. Eckert E.R.,Carlson W.O. Natural convection in an air layer enclosed between two vertical plates with different temperatures. Int. J.Heat and Mass Transfer, 1960,v.2,no. 1/2.

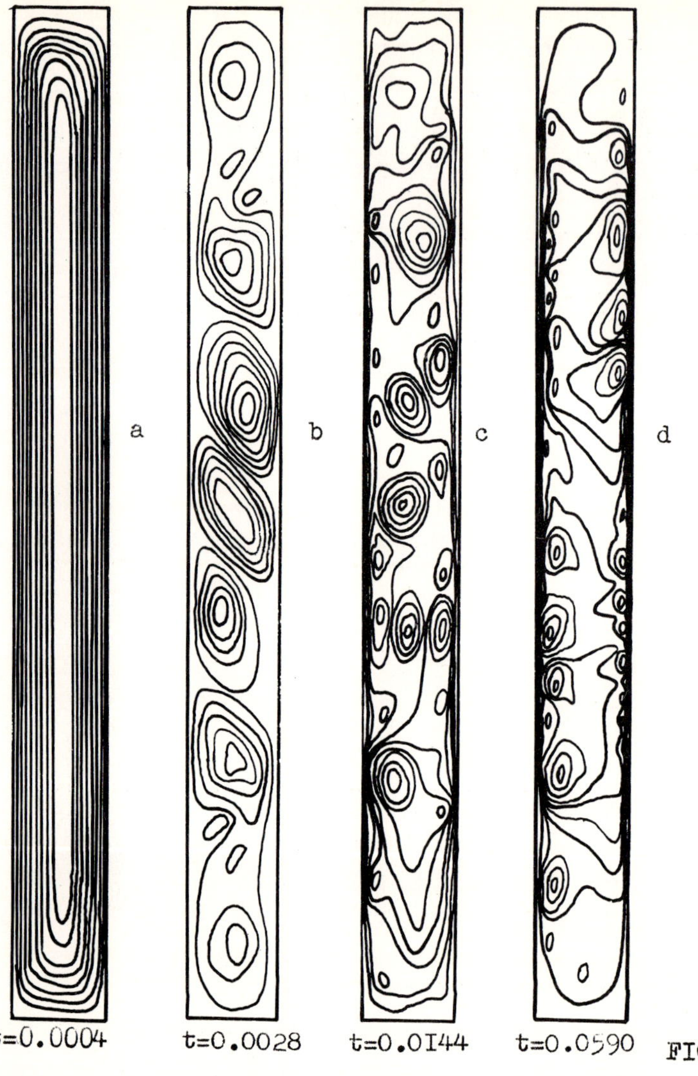

FIG.I

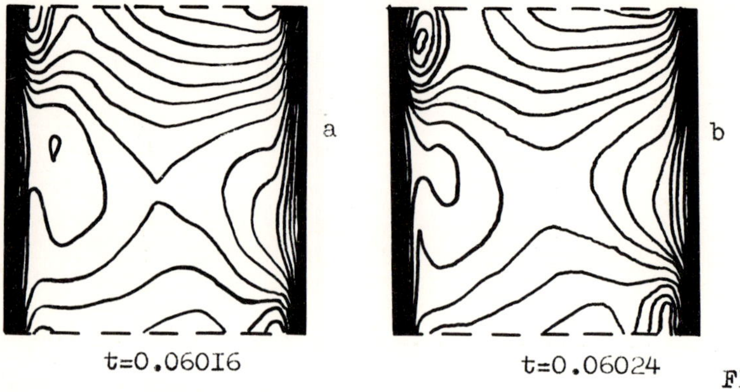

FIG.2

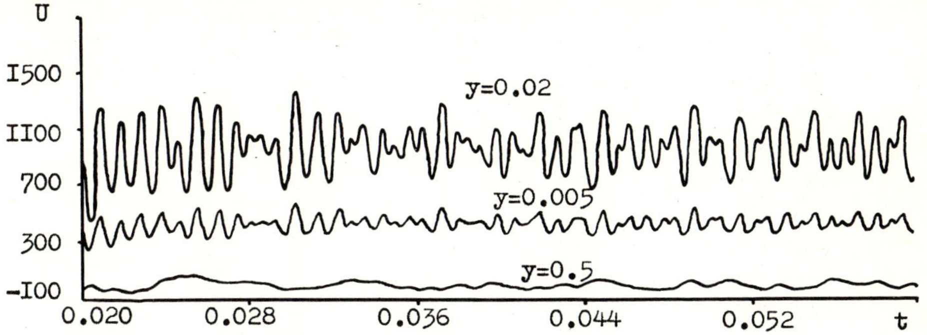

FIG.3

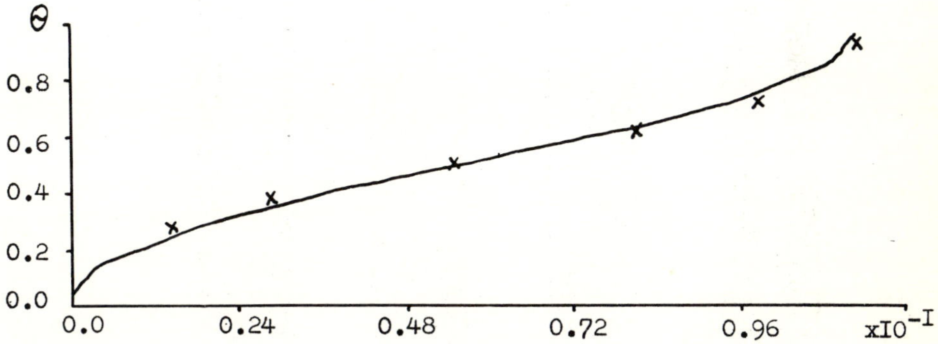

FIG.4

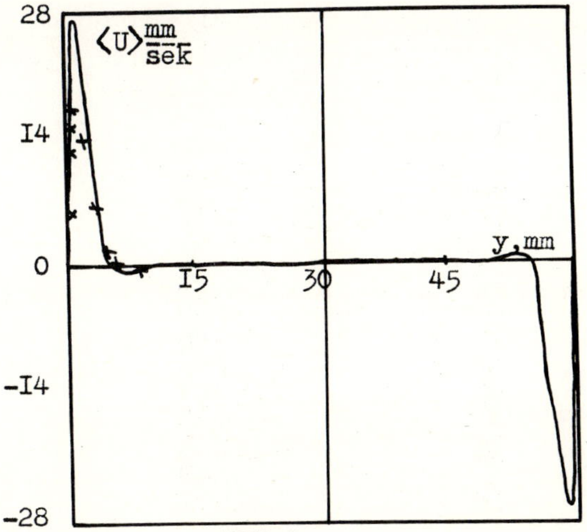

FIG.5

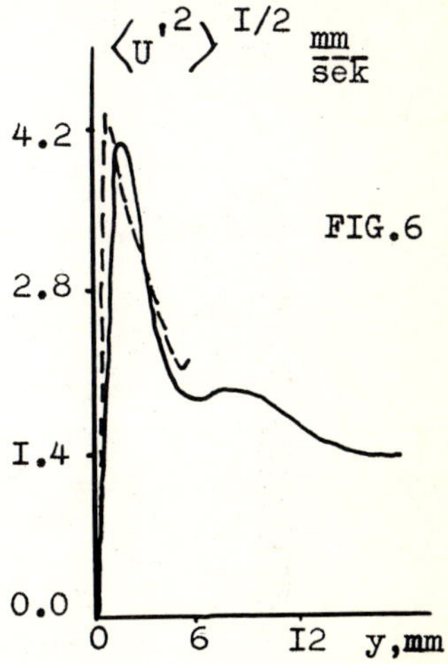

FIG.6

SOLUTION BY NEWTON'S METHOD TO THE STEADY TRANSONIC EULER EQUATIONS

Arthur Rizzi
FFA The Aeronautical Research Institute
of Sweden, S-161 11 BROMMA, Sweden

Introduction

Inviscid flow, of course, possesses none of the many and varied phenomena associated with viscosity. Nevertheless specific details of the inviscid flow such as pressure loadings on bodies and shock patterns throughout the flowfield are still very useful for design applications, particularly if they can be computed easily and cheaply. The most accurate mathematical model for this type of flow is the inviscid equations of gasdynamics (Euler equations). Over the years many numerical methods have been devised to solve these equations for mixed flow, containing both subsonic and supersonic regions in the flowfield. Most of these methods are time dependent as well as time accurate even if the flow is steady, and they therefore usually require a large number of time steps at considerable computer cost to obtain the desired solution. Two recent studies have aimed at improving this situation either by adding extra terms, representing a dashpot mechanism, to the time dependent system[1] or by raising the the order of the system from one to two[2]. Both approaches, however, rely on an analogy with time, either real or artificial, as their iterative variable. But I believe that the introduction of time derivatives into the formulation may give rise to transient phenomena not present in the steady-flow system and which can be "stiff" to resolve. A purely numerical iteration procedure that avoids the explicit use of time, therefore, warrants closer attention for the solution of the steady-flow equations. Based on its success in the solution of the potential equation[3], my idea is to apply the Newton-like iteration procedure to the steady Euler equations. I demonstrate the feasibility of this approach by a transonic converging-diverging nozzle calculation.

Newton Iteration Procedure

Any two-dimensional region enclosed by curved boundaries may be mapped into a rectangle by means of an arbitrary coordinate transformation $\xi = \xi(x, y)$ and $\eta = \eta(x, y)$. Referenced to such a system the steady inviscid axisymmetric equations of gasdynamics can be written[4]

$$\frac{\partial}{\partial \xi} F(q) + \frac{\partial}{\partial \eta} G(q) + H(q) = 0 \tag{1}$$

where $q = [\rho, \rho u, \rho v]$ has the elements of mass, and Cartesian components of momentum, and F, G, and H are vector functions of q

$$F = \frac{1}{D}\left[qU + p \begin{pmatrix} 0 \\ \xi_x \\ \xi_y \end{pmatrix} \right] \qquad G = \frac{1}{D}\left[qV + p \begin{pmatrix} 0 \\ \eta_x \\ \eta_y \end{pmatrix} \right] \qquad H = \frac{v}{yD} q$$

while D is the Jacobian $\partial(\xi,\eta)/\partial(x,y)$ and $U = u\xi_x + v\xi_y$ and $V = u\eta_x + v\eta_y$ define the contravariant components of the velocity. An equation of state

$$p = \frac{\gamma-1}{\gamma}\,\rho\,[h_o - \tfrac{1}{2}(u^2 + v^2)]$$

which assumes that the total energy h_o is constant everywhere completes the system. The solution of Eq. (1) is approached in the following manner, based on Newton's method. Let q^n be an approximate solution, found from either a simpler theory or finite difference solution. We wish to find the corrective term Δq which will improve this approximation. This is done by linearizing F, G, and H around the current approximate q^n, neglecting terms of order two or higher in Δq

$$\begin{aligned} F^{n+1} &= F^n + A^n\Delta q + \ldots \\ G^{n+1} &= G^n + B^n\Delta q + \ldots \qquad (2) \\ H^{n+1} &= H^n + C^n\Delta q + \ldots \end{aligned}$$

where A, B, and C are the Jacobian matrices

$$A = \frac{\partial F}{\partial q} = \frac{1}{D}\begin{bmatrix} 0 & \xi_x & \xi_y \\ -Uu+\xi_x[p/\rho + \frac{\gamma-1}{\gamma}(u^2+v^2)] & U + u\xi_x/\gamma & -\frac{\gamma-1}{\gamma}v\xi_x + u\xi_y \\ -Uv+\xi_y[p/\rho + \frac{\gamma-1}{\gamma}(u^2+v^2)] & v\xi_x - \frac{\gamma-1}{\gamma}u\xi_y & U + v\xi_y/\gamma \end{bmatrix}$$

$$B = \frac{\partial G}{\partial q} = \frac{1}{D}\begin{bmatrix} 0 & \eta_x & \xi_y \\ -Vu+\eta_x[p/\rho + \frac{\gamma-1}{\gamma}(u^2+v^2)] & V + u\eta_x/\gamma & -\frac{\gamma-1}{\gamma}v\eta_x + u\eta_y \\ -Vv+\eta_y[p/\rho + \frac{\gamma-1}{\gamma}(u^2+v^2)] & v\xi_x - \frac{\gamma-1}{\gamma}u\eta_y & V + v\eta_y/\gamma \end{bmatrix}$$

$$C = \frac{\partial H}{\partial q} = \frac{1}{yD}\begin{bmatrix} 0 & 0 & 1 \\ -uv & v & u \\ -v^2 & 0 & 2u \end{bmatrix}$$

Under this linearization Eq. (1) becomes

$$\frac{\partial}{\partial \xi}(A^n \Delta q) + \frac{\partial}{\partial \eta}(B^n \Delta q) + C^n \Delta q = -\left(\frac{\partial F^n}{\partial \xi} + \frac{\partial G^n}{\partial \eta} + H^n\right) \tag{3}$$

a linear boundary-value problem of mixed type which, when solved, yields the unknown Δq. In the derivation of this formula we assumed that q^{n+1} was the exact solution to Eq. (1). When the expansion (2) is broken off after the linear term, however, the Δq defined by Eq. (3) no longer represents the exact correction, although it can be expected to lead to a better approximation to the exact solution. A new estimate $q^{n+1} = q^n + \Delta q$ is then found and the whole process is repeated until the corrections are sufficiently small. Observe that as the Δq grows smaller the left-hand side of Eq. (3) goes to zero and the solution to Eq. (1) is approached. If q^n is not too far from the exact solution, the rate of convergence is quadratic.

Finite Difference Approximation to Equation (3)

To complete the description of the problem boundary conditions on Δq must be added to Eq. (3). Consider subsonic-to-supersonic flow through the 45°-15° conical nozzle displayed in Fig. 1(a) as an illustration of this method.

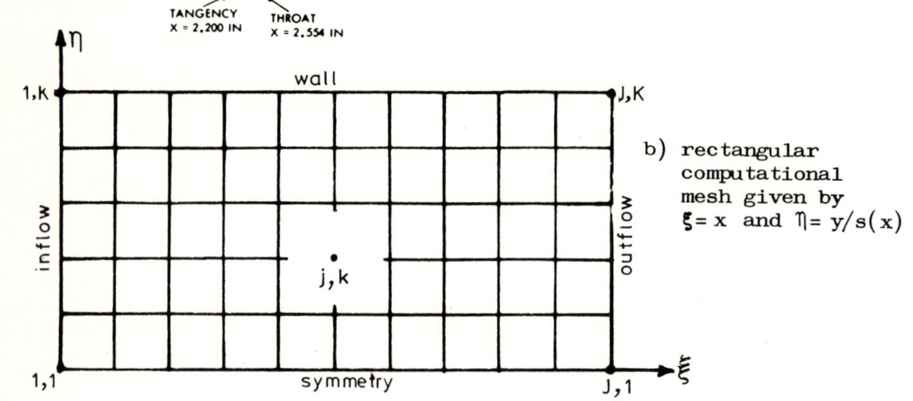

Fig. 1. The geometry of the 45°-15° conical nozzle.

No complete theory for the proper boundary conditions to this problem exists, although a reasonable set[1,3] is: 1) specify u and ρ at inflow, 2) enforce symmetry on the axis and set $v = 0$, and 3) compel the flow to be parallel at the wall $-s'u + v = 0$. Nothing is specified at the supersonic outflow. If q^{n+1} satisfies these boundary conditions then the appropriate set of conditions on Δq follows.

I will now solve Eq. (3) together with the above boundary conditions by finite differences. Introduce an equispaced rectangular mesh on the rectangular region $\xi_{min} \le \xi \le \xi_{max}$ and $0 \le \eta \le 1$ (Fig. 1(b)) so that the mesh point j,k corresponds to the coordinate ξ_j, η_k. The continuous problem for Δq is replaced by the corresponding discrete problem for $\Delta q_{j,k}$. I approximate all derivatives with the same difference schemes that Blomster and Sköllermo[3] found to be stable for their potential flow problem, namely all derivatives at interior points are approximated by central differences, regardless of whether the flow is subsonic or supersonic. The conditions at the boundaries, however, are not sufficient to determine a unique solution to the discrete problem. Additional equations, therefore, are obtained from approximations to Eq. (3) which are centered halfway between the extreme meshlines (see Ref. 3 for details). Since C is indefinite on the axis, it is replaced there by its limiting form as y goes to zero.

Solution to the Linear System

When differenced Eq. (3) together with the boundary conditions result in a linear system of equations $\Lambda X = Y$ which must be solved to yield the solution vector

$$X = [\Delta q_{11}, \Delta q_{12}, \ldots \Delta q_{1K}, \Delta q_{21}, \Delta q_{22}, \ldots \Delta q_{2K}, \ldots \Delta q_{J1}, \Delta q_{J2}, \ldots \Delta q_{JK}]$$

grouped according to the K unknowns along each vertical mesh line. Central differencing gives rise to the coefficient matrix Λ that has the form shown in Fig. 2. It is a block-block tridiagonal matrix that is sparse. The superblocks on the main diagonal contain three sub-block diagonals of 3×3 matrices while the diagonals just above and below the main have, for the interior approximations, nonzero sub-blocks on their main diagonals only. The boundary approximations, however, introduce nonzero elements just off these diagonals, indicated in Fig. 2 by a tilde above the symbols. The total bandwidth of matrix Λ is thus $6K + 11$.

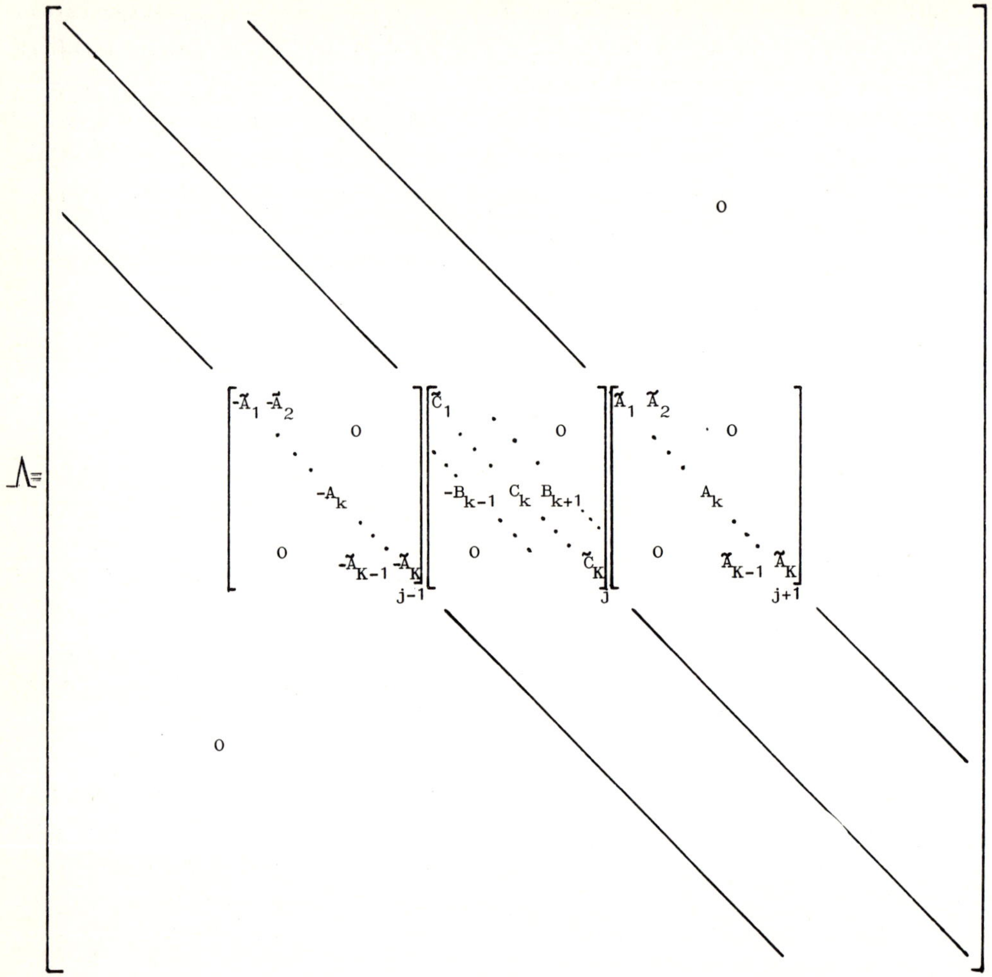

Fig. 2. The block block tridiagonal coefficient matrix Λ.

Normally, if it is possible to do so, a system of equations like the present one is solved by an iterative method. Unfortunately Λ does not possess any of the properties which assure us that the common iterative techniques will work; it is neither positive definite or symmetric, nor even diagonally dominant. I therefore choose to solve it directly by a Gaussian elimination routine that uses row-pivoting.

The method used for the Gaussian elimination is based on algorithms given by Martin and Wilkinson[6] for the factorization of band matrices that do not use any more storage than what is needed in order for the band portion of the matrix, together with some extra diagonals necessary for pivoting, to reside in central memory. However, as the mesh is refined storage requirements still grow rapidly. So I have followed the idea of Blomster and Sköllermo[3] and keep only a small part of the band in memory at any one time (see Fig. 3), and use external storage for the remaining part. In this way, if during the course of the calculation, the coefficient matrix Λ changes so slowly that it can be considered constant, and only the right-hand side varies, then the system can be solved quickly simply by reading the stored information back into the machine and performing the final part of the elimination again.

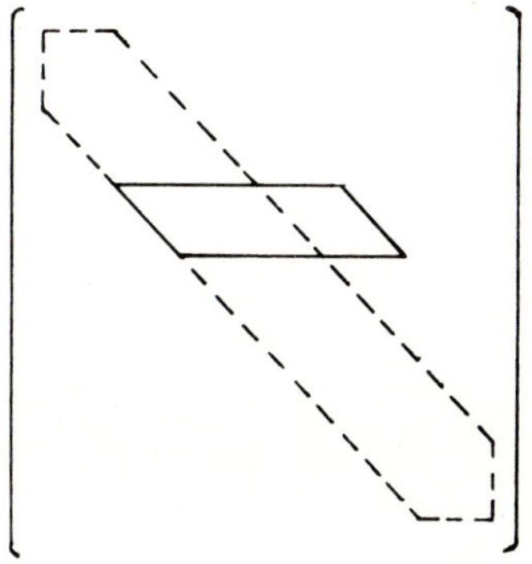

Fig. 3. The portion of the Λ matrix held in main storage during the elimination.

The major part of the computing time is spent on these calculations, and there are ways to make them more efficient. For example, the super-block diagonals of the system contain many zeroes. The maximum number of nonzero elements in any row is only 15, regardless of the mesh size. Wahlund[7], in a similar approach for the blunt-body problem, points out that an important factor for the time needed to solve the system is the rate of fill-in, i.e. the number of zero elements that become nonzero during the elimination procedure. Based on rather detailed analysis of submatrices of the coefficient matrix at several stages during the elimination process, theories[8] exist for pivoting strategies so that the fill-in is held to nearly a minimum. Even by simply recognizing all the zeroes within the band, Wahlund estimates that one can reduce the computational time to one fifth of what it would be without considering the zeroes at all. His figures show that approximately 45 % of the elements in the rows and columns are nonzero at the elimination, compared with 15 % and 20 % nonzero elements in those parts of the rows and columns, respectively, in the original matrix. Furthermore, the proportion of nonzero elements remains constant or even decreases a little when the number of unknowns grows larger.

Computed Results

Transonic flow in a 45°-15° conical nozzle has been calculated by this procedure, and contours of the Mach number are compared in Fig. 4 with other computed and measured results. Within the resolution of Fig. 4, my solution is in complete agreement with that of Blomster and Sköllermo. With a 21 x 8 mesh my results were obtained in 3 iterations, after which all the correction terms were less than 10^{-3}. Without exploiting any of the zero elements the calculation consumed less than one minute of CDC 6600 computer time.

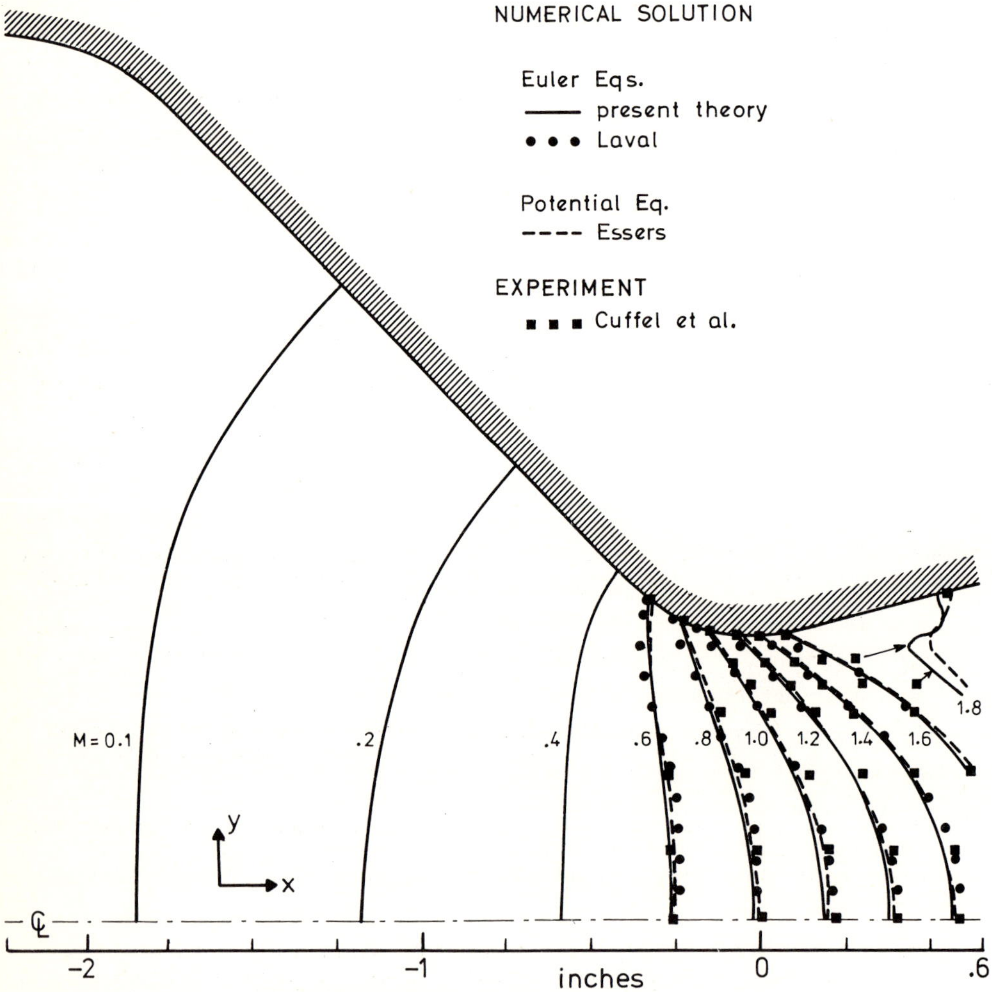

Fig. 4. Mach number contours of the subsonic-to-supersonic flow through the nozzle compared with other computational and experimental results. The grid used was 21 x 8.

Concluding Remarks

Use of Newton's method for the solution of nonlinear problems is of course an old concept. It has been employed before to solve the nonlinear system of difference equations that results from differential governing equations. However, to this author's knowledge, Blomster and Sköllermo are the first to apply Newton's method directly to a first-order system of nonlinear differential equations in order to obtain a linear differential problem which then is approximated by finite difference expressions. Of course for the procedure described here to become truly effective, a fast and efficient algorithm must be found to solve the system of equations. One way around the costly elimination process is that recently proposed by Martin[9] wherein he formulates an iterative procedure which is different from the Newton-type just presented but one which allows use of the fast Cauchy-Riemann solver for the solution to the resulting system. Given the great practical use for these solutions, further efforts at solving the steady equations quickly and inexpensively are fully justified.

References

1. Essers, J-A., "Time-Dependent Methods for Mixed and Hybrid Steady Flows", Computational Fluid Dynamics Lecture Series, VKI, 1978.

2. Chattot, J-J., "Relaxation Approach to the Steady Euler Equations in Transonic Flow", Proc. 3rd AIAA Computational Fluid Dynamics Conf., Albuquerque, 1977.

3. Blomster, J., and Sköllermo, G., "Finite Difference Computation of Steady Transonic Nozzle Flow", Computer Science Dept. Rep.No. 66, Uppsala Univ., 1977.

4. Viviand, H., "Formes Conservatives des Équations de la Dynamique des Gaz", La Recherche Aérospatiale, No. 1, Jan-Feb 1974.

5. Cuffel, R., Back, H., and Massier, P., "Transonic Flow-field in a Supersonic Nozzle with Small Throat Radius of Curvature", AIAA J., Vol. 7. July 1969.

6. Martin, R.S. and Wilkinson, J.H., "Solution of symmetric and unsymmetric band equations and the calculation of eigenvectors of band matrices", Handbook for Automatic Computation, Vol II, pp. 70-92, Springer-Verlag, Berlin, 1971.

7. Wahlund, P., "Finite Difference Methods for Computing the Steady Flow about Blunt Bodies at Supersonic Speeds", Computer Science Dept. Rep.No. 73, Uppsala Univ., 1978.

8. Tewarson, R.P., Sparse matrices, Academic Press. New York, 1973.

9. Martin, E.D., "A Split-Recoupled-Semidirect Computational Technique Applied to Transonic Flow Over Lifting Airfoils", AIAA Paper No.78-11, 1978.

A SIMPLIFIED SPLINE SOLUTION PROCEDURE*

S.G. Rubin and P.K. Khosla
Polytechnic Institute of New York
Aerodynamics Laboratories, Route 110
Farmingdale, New York 11735

1. INTRODUCTION

In recent papers(1-3) the present authors have formulated various higher-order collocation techniques based on polynomial interpolation or Hermitian discretization procedures. Three-point formulations leading to fourth-order and sixth-order methods were considered. In order to obtain equal accuracy the higher-order methods require fewer mesh points when compared with second-order accurate finite difference techniques. This can mean a reduction in computer storage and time. In particular, the fourth-order accurate spline 4 method(1) generally requires one-quarter the number of points, in each coordinate direction, compared to finite-difference solutions of comparable accuracy.

In addition to improvements in accuracy, these methods have certain desirable properties. The discrete equations remain block-tridiagonal in character. In view of the Hermitian formulation, where the derivatives are treated as unknowns, the application of derivative boundary conditions is somewhat more straightforward, and with non-uniform grids the deterioration in accuracy is less severe than with conventional finite-difference methods. However, the resulting system of equations is more complex than those of the usual second-order accurate finite-difference procedures. For a single differential equation, finite-differences lead to a scalar tridiagonal system, whereas the higher order techniques are block tridiagonal; specifically, 2x2 blocks are required for the fourth order spline methods and 3x3 blocks for the sixth order Hermite systems. Consequently, for a given mesh the storage requirements become larger. The solution procedure therefore requires more operations and is more time consuming than the scalar algorithm usually applied for the tridiagonal finite-difference equations. However, for equal accuracy the spline methods require fewer points are therefore still more efficient.

In the present study some of the complexities associated with computer storage and the general spline solution procedure are eliminated. This is accomplished by "uncoupling" the higher-order correction from the coupled lower-order ("finite-difference") system of equations. The present method can therefore result in a significant reduction in computer storage. A 2x2 block tridiagonal system is reduced to two scalar tridiagonal systems; 3x3 blocks reduce to a tridiagonal and a 2x2 block tridiagonal system. In particular, <u>the method can be programmed into a subroutine package to be used with existing finite-difference codes in order to render higher-order ac-</u>

---

*This research is sponsored by the Air Force Office of Scientific Research, USAF, under Contract No. F49620-78-C-0020 Project No. 2307/A1 curacy to the final solutions. This simplicity is achieved by treating the higher-order accurate terms as explicit correctors to a modified finite-difference equation. This correction is evaluated in such a manner that the resulting procedure is consistent and unconditionally stable. A related procedure, the method of deferred corrections, has been discussed by several other authors for ordinary differential equations[4,5]. As is shown herein, if this procedure is extended, so as to achieve full convergence for ODE's or for PDE's in a temporal or spatial marching procedure, an instability can result. The present analysis corrects this deficiency. Finally, it is noteworthy that the Khosla-Rubin difference method,[6] which provides an explicit central-difference corrector to an implicit upwind-difference scheme, is in fact a lower-order adaptation of the method presented herein.

It should be noted that by uncoupling the spline or Hermite corrector and thereby reducing the size of the matrix blocks, it is now possible to obtain higher-order solutions and still maintain a large degree of coupling between the dependent variables in a system of equations. In effect, the equations are coupled solely through the implicit or finite-difference like terms. This is particularly important in both boundary layer and Navier-Stokes calculations, as the rate of convergence is increased considerably when the equations are coupled. The rate of convergence does not appear to be as sensitive to the uncoupling of the higher-order spline terms. In this way we have been able to develop coupled solvers for the boundary layer equations and vorticity (ω)-stream function(Ψ) description of the Navier-Stokes equations. In both cases only (2x2) blocks must be inverted in order to obtain fourth-order accuracy. This procedure has been used with ADI, predictor-corrector, direct solver and strongly implicit two-dimensional solution techniques.

The present method has been tested with several problems: (1) the one-dimensional Burgers equation, (2) the potential flow over a circular cylinder; an extension of Buneman's direct solver is applied, (3) laminar and turbulent boundary layers on flat plates[3]; fully coupled equations are considered, (4) the Navier-Stokes equations describing the incompressible flow in a driven cavity[2]; a spline improved predictor-corrector[7] method for the vorticity transport equation and the extension of Buneman's direct solver for the stream function are applied, and (5) the flow in a channel having a rearward facing step[8]; the coupled (Ψ-ω) Navier-Stokes equations are considered and Stone's strongly implicit method is extended to allow for the iterative spline correction. In all of the problems considered, the solution procedure is considerably simplified over that required for the coupled spline solvers; the final results are of course identical with both the iterative spline and coupled spline methods. The details of the corrector method, stability considerations, and typical calculations are presented in the following sections.

2. ITERATIVE SOLUTION PROCEDURE

In this section, we will describe the numerical technique for solving the governing differential system for the spline 4[1,2] and

Hermite 6[2] procedures. The method will be illustrated here only for the linear Burgers equation. The extension to non-linear multi-dimensional systems of equations is direct. It must be emphasized that the present technique is different from that of the "method of deferred corrections" and is unconditionally stable. In the subsequent discussions, fully implicit time discretization is considered; however, the procedure can be modified for higher-order temporal schemes.

The linear Burgers equation, with ν, c constant, is

$$u_t + c\, u_x = \nu u_{xx} \quad . \tag{1}$$

The proposed iteration procedure is given by the numerical discretization formula (2).* We are primarily concerned with obtaining steady-state solutions so that only a first-order accurate implicit temporal discretization is shown. For more accurate transients, or for spatial marching, higher-order integration formulas can be applied.

$$u_j^{n+1} - u_j^n + \frac{c\Delta t}{h} A \left(\frac{u_{j+1}^{n+1} - u_{j-1}^{n+1}}{2}\right) - B \frac{\nu \Delta t}{h^2} \left(u_{j+1}^{n+1} - 2u_j^{n+1} + u_{j-1}^{n+1}\right)$$

$$= \frac{c\Delta t}{h} \epsilon \left[A\left(\frac{u_{j+1}^n - u_{j-1}^n}{2}\right) - h\, m_j^n\right] \tag{2}$$

$$+ \frac{\nu \Delta t}{h^2} \epsilon \left[h^2 M_j^n - B\left(u_{j+1}^n - 2u_j^n + u_{j-1}^n\right)\right] \quad ,$$

where the constant ϵ is either zero or unity; A and B are two arbitrary constants introduced for stability enhancement. This will become clearer in the following section; m_j and M_j are the higher-order spline approximations to u_x and u_{xx}, respectively. Once convergence is achieved at a given time level, the terms containing A and B cancel so that equation (2), with $\epsilon = 1$, provides a higher-order numerical solution of the Burgers equation (1).

It should be noted that the left-hand side of equation (2) contains terms associated solely with a finite-difference like discretization. Only for A=B=1 does the left-hand side correspond to an exact finite-difference representation. The higher-order spline approximations appear only as explicit corrections on the right-hand side. This solution procedure is by no means unique. Various other forms, which utilize upwind differencing for the convective terms, i.e. those multiplied by coefficient A, can also be considered without changing the ultimate higher-order system. One such discretization can be written for $c \geq 0$ as:

*The discussion contained herein is for a uniform mesh. The procedure is directly applicable and has been used for non-uniform grids.

$$u_j^{n+1}-u_j^n + \frac{c\Delta t}{h} A\ (u_j^{n+1}-u_{j-1}^{n+1})\ -B\ \frac{\nu\Delta t}{h^2}\ (u_{j+1}^{n+1}-2u_j^{n+1}+u_{j-1}^{n+1})$$

$$= \frac{c\Delta t}{h}\ \varepsilon\ [A(u_j^n-u_{j-1}^n)-hm_j^n] \qquad (3)$$

$$+ \frac{\nu\Delta t}{h^2}\ \varepsilon\ [h^2 M_j^n\ -B(u_{j+1}^n-2u_j^n + u_{j-1}^n)]$$

This is an attractive procedure as an initial solution can thus be obtained with a lower-order accurate scheme; e.g., $\varepsilon = 0$ in equation (3). This solution can then be corrected in order to achieve the higher-order accuracy by setting $\varepsilon = 1$ in equation (3). In this manner, a reasonably smooth early solution is used in the curve fits that generate m_j and M_j. This may suppress spurious oscillations appearing in m_j and M_j. These oscillations can sometimes arise from arbitrary initial conditions, with a course grid, or the central-difference solution may be quite inaccurate even if the final spline solution on the same grid is quite reasonable.

If m_j and M_j are obtained not with splines, but with second-order accurate central finite-differences, equation (3) then leads to the K-R[6] predictor-corrector method. This procedure is unconditionally stable for A=B=1 and diagonally dominant for all values of cell Reynolds numbers and courant numbers[6].

A number of special cases of interest are itemized below.

(a) <u>Finite-Differences</u>: A=B=1

Equation (2) with $\varepsilon = 0$ leads to an implicit (scalar tridiagonal) central-difference form of equation (1).

If equation (3) is prescribed and

$$m_j = \frac{u_{j+1} - u_{j-1}}{2h} \quad , \quad M_j = \frac{u_{j+1}-2u_j+u_{j-1}}{h^2}$$

are applied in equation (3), with $\varepsilon = 1$ the K-R[6] scheme is recovered.

(b) <u>Method of Deferred Corrections</u>[4,5]

Equation (2) is initially solved, to a steady-state, with A=B=1 and $\varepsilon = 0$. After convergence is achieved, the right-hand correction, representing the higher-order accurate terms, is switched on by setting $\varepsilon = 1$. These terms are now fixed so that the superscript n is omitted. The time integration is repeated until a new steady-state is achieved. The right-hand side is only updated once so that a convergence of the higher-order terms is not achieved. It will be shown that a straightforward extension to allow for marching in time, or in effect iterate to full higher-order convergence, can in fact be conditionally stable. For spatial marching, such as with non-similar boundary layers, a corresponding difficulty will be encountered. The stability of this method is investigated in more detail in a following section.

(c) Spline 4[1,2]

In this case m_j and M_j are given by,

$$\left.\begin{aligned} M_j &= \frac{1}{2}\left[K_j + \frac{u_{j+1}-2u_j+u_{j-1}}{h^2}\right] \quad , \\ m_j &= \frac{u_{j+1}-u_{j-1}}{2h} + \frac{h}{12}\left(K_{j-1}-K_{j+1}\right) \quad , \\ K_{j+1} + 4K_j + K_{j-1} &= \frac{6}{h^2}\left(u_{j+1}-2u_j + u_{j-1}\right) \quad . \end{aligned}\right] \qquad (4)$$

For additional spline relationships, see reference 1 and 2. In addition to the "implicit" tridiagonal matrix inversion required for u_j, an additional "explicit" tridiagonal inversion for K_j is also required in this solution procedure. However, the original coupled spline 4 calculations[1,2] required a 2x2 block-tridiagonal inversion for $(u,K)_j$. This reduction to a pair of scalar tridiagonal inversions reduces computer storage and allows the higher-order improvement to be programmed as a simple correction to a "finite-difference" code. Only 2 arrays must be stored for the matrix inversions. The coupled 2x2 system requires 6 arrays.

(d) Hermite 6[2]

For the sixth order hybrid Hermite 6 scheme, m_j and M_j are given by

$$\left.\begin{aligned} 7\, m_{j+1} + 16m_j + 7m_{j-1} &= 15\frac{(u_{j+1}-u_{j-1})}{h} + h(M_{j+1}-M_{j-1}), \\ -M_{j+1}+8M_j-M_{j-1} &= 24\,\frac{(u_{j+1}-2u_j+u_{j-1})}{h^2} - 9\,\frac{(m_{j+1}-m_{j-1})}{h} \quad . \end{aligned}\right] \qquad (5)$$

Since the equations for m_j and M_j are themselves coupled, the original 3x3 block-tridiagonal Hermite 6 method is now reduced to one scalar "implicit" tridiagonal inversion for u_j, and a 2x2 block inversion for $(m,M)_j$. Once again computer storage is reduced. Only 6 arrays must be stored for the matrix inversions. For the coupled 3x3 system 12 arrays are required.

In all the cases considered here, the solution to a complex higher-order system of equations has been reduced to the solution of a much simpler tridiagonal equation for u_j alone. The evaluation of the appropriate curve fits for m_j or M_j are uncoupled and treated explicitly. This simplifies the over-all system as well as the treatment of boundary conditions.

3. STABILITY

In this section, we will investigate the stability of the present central (2) and upwind (3) procedures for the spline 4 and Hermite 6 schemes.

(a) Spline 4

The amplification factors as evaluated from either equation (2) or (3), in combination with (4), are as follows:

$$\xi_2 = \frac{1-\beta(1-\cos\theta)\left\{\frac{1-\cos\theta}{2+\cos\theta} + 2(1-B)\right\} + i\alpha\sin\theta\left\{\frac{1+2\cos\theta}{2+\cos\theta} + A-2\right\}}{1 + 2B\beta\,(1-\cos\theta) + iA\alpha\,\sin\theta}, \quad (6)$$

for implicit central-differencing, and

$$\xi_3 = \frac{1+i\alpha\sin\theta\left\{\frac{1+2\cos\theta}{2+\cos\theta} + A-2\right\} + (1-\cos\theta)\left[\alpha A-\beta\left\{\frac{1-\cos\theta}{2+\cos\theta} + 2(1-B)\right\}\right]}{1 + i\alpha A\sin\theta + (\alpha A+2B\beta)\,(1-\cos\theta)}, \quad (7)$$

for implicit upwind-differencing, where $\alpha = \frac{c\Delta t}{h}$ and $\beta = \frac{\nu\Delta t}{h^2}$

From (6) and (7) it can be seen that for the pure diffusion equation, characterized by $\alpha = 0$, $B = 1$ and $\epsilon = 1$, the present procedure is unconditionally stable. If, in addition, convection is included in (1) and $A = 1$, this corresponds to an iterative or marching extension of the "method of deferred corrections"; $|\xi_2|$ and $|\xi_3|$ can however be greater than unity for certain values of α, β and θ. Therefore, in general, this method is only conditionally stable. An upper limit on the values of Δt and h can be obtained from the stability requirement that $|\xi_2|<1$ and $|\xi_3|<1$. However, for $B\geq 1$ and $A\geq 2$, both the schemes (2) and (3) are unconditionally stable. It would appear that the implicit "finite-difference" left-hand side has to be reinforced with increased convection in order to assure stability.

As there are an infinity of stable procedures, the optimal choices for A and B can perhaps also enhance the rate of convergence of the present iterative procedures. This is a topic for further study.

(b) Hermite 6

In this case, the amplification factor as derived from equations (2) and (5) can be written as:

$$\xi = \frac{1+2\beta B(1-\cos\theta)Y + i\alpha AX\sin\theta}{1+2\beta B(1-\cos\theta) + i\alpha A\sin\theta}$$

where $X = 1-9(4+\cos\theta)/A\ell$, $Y = 1-3(11\cos\theta+19)/2B\ell$

and $\ell = 2\cos^2\theta+20\cos\theta+23$.

For $-1\leq\cos\theta\leq1$, the values of $|X|$ and $|Y|$ are always less than unity when $A\geq3$ and $B\geq2$. This insures that $|\xi|<1$. In a similar fashion the stable values of A and B can be determined for the upwind system (3) and (5). For Hermite 6, both diffusion and convection must be reinforced in order to obtain a stable solution. It would appear that as the over-all accuracy of the method is increased, this reinforcement process must also be increased.

4. APPLICATIONS

The iterative spline procedure has been applied to a variety of problems. These are described below. In each case, the present technique reproduces the results of the earlier coupled higher order calculations.

(a) Burgers Equation

Coupled Spline 4 solutions have been presented in reference 1. As noted previously, the present procedure reproduces identical results. A systematic study has shown that comparable CPU times are required for convergence of the original coupled Spline 4 and new iterative Spline 4 procedures. In the course of these tests of the iterative method, it was concluded that the higher-order correction term can be updated every few time-steps, for problems where only the steady state is required. This can lead to some reduction in computer time. The stability characteristics, however, appear to remain unchanged. This study was not optimized.

For $A=B=\varepsilon=1$, we have an iterative extension of the method of deferred corrections. As predicted by the stability analysis, this approach is only marginally stable; instabilities were in fact observed for some of the Burgers equation calculations.

(b) Flow in a Driven Cavity (Navier-Stokes)

Spline 4 solutions for $R_e=100$, 1000 have been previously presented in reference 2. A vorticity-stream function formulation has been considered. The predictor-corrector method[7] was applied for the vorticity equation. Buneman's direct solver was used for the stream function; this is easily extended to the present iterative technique*. The results are once again identical with our coupled Spline 4 solutions. The present Spline 4 method has also been applied to the energy equation in order to obtain the temperature field (thermal cavity).

*The higher-order corrections act as a source term for the solution of a poisson equation, which is solved with a conventional finite-difference discretization

The side wall temperatures are prescribed, and adiabatic conditions are imposed at the upper moving and lower stationary walls. Stone's strongly implicit method[8] has been adapted to the iterative spline technique. The correction again acts as a simple source like effect.

(c) Laminar and Turbulent Boundary Layer

Higher order solutions for laminar and turbulent flow over a flat plate, with and without mass injection, have been presented in reference 3. In that study, the Hermite 6 results were obtained by the present iterative procedure. Spline 4 results have been obtained with both methods. Significantly, large gradients due to injection or turbulent flow conditions are simulated accurately with the iterative procedure. In addition, for non-similar boundary layers, the marching procedure is stable for the A,B values discussed earlier.

(d) Flow in a Channel With a Rearward Facing Step (N-S)

Iterative Spline 4 solutions, for Reynolds number up to 1000, have been obtained[8]. A strongly coupled vorticity-stream function formulation has been developed. This allows for arbitrary time steps, e.g., $\Delta t \approx 10^6$. In effect, a steady-state calculation was performed. The complete procedure, as well as a discussion of non-linear instability or aliasing associated with large cell Reynolds numbers and higher-order approximations, will be discussed in greater detail in reference 8. Once again, by uncoupling the spline correction it becomes a simple matter to modify the coupled strongly implicit finite-difference algorithm.

5. SUMMARY

From the previous discussion it is clear that the present iterative procedure for achieving higher-order accuracy has many advantages.

1. At each time level, a modified finite-difference problem is in effect being solved. Thus the present procedure does not involve any additional coding complexities. Therefore, it is relatively simple to upgrade or "spline" an existing finite-difference program. This has been carried out in the numerous cases described herein.

2. The storage requirements of the present procedure are considerably reduced. For example, with the Spline 4 and Hermite 6 methods, the original coupled procedures for a single differential equation require, for the block inversion algorithm, the storage of 6 and 12 arrays, respectively. In the present case, this is reduced to 2 and 6 arrays respectively. For systems of coupled equations, such as boundary layer or Navier-Stokes problems, the reduction in storage requirements is even more dramatic.

3. The present iterative spline technique has been applied successfully with one-dimensional implicit, and two-dimensional ADI, predictor-corrector, SOR, strongly implicit and direct solver algorithms.

REFERENCES

1. Rubin, S.G. and Khosla, P.K.: AIAA J., 14, 1, pp. 851-858 (1976).
2. Rubin, S.G. & Khosla, P.K.: J. Comp. Phys., 24, pp. 217-244 (1977).
3. Rubin, S.G. & Khosla, P.K.: Comp. & Fluids 5, pp. 241-259 (1977).
4. Fox, L: The Numerical Solution of 2-point Boundary Value Problems in Ordinary Differential Equations. Oxford Univ. Press, Fairlawn, N.J. (1957).
5. Keller, H.B.: Annual Review of Fluid Mechanics, 10, pp. 417-433 (1978).
6. Khosla, P.K. and Rubin, S.G.: Computers and Fluids, 2, pp. 207-209 (1974).
7. Rubin, S.G.: Proc. of the 3rd International Conference on Numerical Methods in Fluid Mech, pp. 146-153, Springer-Verlag, New York (1972).
8. Rubin, S.G. and Khosla, P.K.: In preparation (1978).

A TEST CASE FOR CHECKING COMPUTATIONAL METHODS FOR GAS FLOWS WITH DISCONTINUITIES

V.V. Rusanov
Institute of Applied Mathematics Ac.Sci.
Moscow, USSR

I. Introduction

For the computation of discontinuities in flow fields which are not fixed beforehand but arise as part of the solution, a large number of methods has been developed. For testing the quality of such methods usually very simple exact solutions are used, e.g. a shock of constant strength running into a gas at rest. Such a test is not rigorous enough because in real problems the functions usually have non-zero derivatives on both sides of the discontinuity. With usual methods it is also difficult to judge the accuracy of numerical solutions near boundaries.

In this paper a test procedure is proposed which is based on the well-known model of homentropic flow with adiabatic exponent $K = 3$. In this case, the gasdynamic equations in the continuous regions can be uncoupled and hence the exact solution can be computed rather simply.

2. The equations in the continuous region

Let us consider the system formed by the continuity- and momentum-equation for unsteady one-dimensional gas flow:

$$\rho_t + (\rho u)_z = 0 \qquad (1)$$
$$(\rho u)_t + (\rho u^2 + p)_z = 0$$

where ρ , u and p denote velocity, density and pressure of the gas and an independent variable as index denotes partial differentiation after that variable. Further we put $p = A\rho^k$ and A = const, thereby assuming constant entropy throughout the whole region (homentropic flow); this allows us to drop the energy equation. In non-conservative form the system is written as

$$A w_t + B w_z = 0$$

with $w = \begin{Bmatrix} \rho \\ u \end{Bmatrix}$, $\quad A = \begin{Bmatrix} 1 & 0 \\ u & \rho \end{Bmatrix}$, $\quad B = \begin{Bmatrix} u & \rho \\ u^2 + c^2 & 2\rho u \end{Bmatrix}$ (2)

c is the speed of sound, $c^2 = dp/d\rho = \kappa A \rho^{\kappa-1}$

For the characteristic velocities λ we have the equation

$$Det\,(B - \lambda A) = 0 \tag{3}$$

yielding $\lambda_\pm = u \pm c$

Hence the sytem (I) has the characteristic form:

$$c[\rho_t + (u \pm c)\rho_z] \pm \rho[u_t + (u \pm c)u_z] = 0$$

or, with $c = (\kappa A)^{1/2} \rho^{(\kappa-1)/2}$ and introducing the Riemann variables $r = u + 2c/(\kappa - 1)$, $\quad s = u - 2c/(\kappa - 1)$:

$$r_t + \{[(\kappa+1)r + (3-\kappa)s]/4\} r_z = 0$$
$$s_t + \{[(3-\kappa)r + (\kappa+1)s]/4\} s_z = 0 \tag{4}$$

Specializing for $\kappa = 3$, $A = 1/3$ we have $\rho = c$ and the uncoupled system:

$$r_t + r r_z = 0 \qquad s_t + s s_z = 0 \tag{5}$$

In the region where solution is continuous the system in the conservative form

$$\rho_t + (\rho u)_z = 0$$
$$(\rho u)_t + (\rho u^2 + \rho^3/3)_z = 0 \tag{I}$$

is equivalent to the system

$$r_t + r r_z = 0 \qquad s_t + s s_z = 0$$
$$r = u + \rho \qquad s = u - \rho$$
$$u = (r + s)/2 \qquad \rho = c = (r - s)/2 \tag{II}$$

Because of the numerical equality $\rho = c$ we shall use the notation ρ only.

Remarks:

a) Because ρ is a positive quantity from the physical definition, the condition $r > s$ must hold everywhere.

b) The values r and s are constant on the characteristics which, because of $\lambda_+ = r$ and $\lambda_- = s$, are straight lines. So, if at any point (z_0, t_0) we have $r = r_0$, $s = s_0$, then

$$r = r_0 \quad \text{along} \quad z - z_0 - r_0(t - t_0) = 0$$

and

$$s = s_0 \quad \text{along} \quad z - z_0 - s_0(t - t_0) = 0 \qquad (6)$$

This property greatly simplifies the computation of the exact solution of system (2).

3. Shock waves

Because of the assumption A = const "shocks" in our model are not real gasdynamic shocks but simpler, isentropic discontinuities in the variables u , ρ respectively, in r , s . Nevertheless the most essential features of gasdynamic shocks are retained, providing a realistic environment for the testing of computational methods.

The relations between the values of u , ρ before and after the shock are obtained from system (I) having conservative form. It is known that system (I) has the infinite number of conservative forms and the relations on the shock wave depend on the conservative form. For example system (II) may be written in the form:

$$r_t + (r^2/2)_z = 0 , \qquad s_t + (s^2/2)_z = 0 \qquad (II')$$

System (I) and (II') are equivalent for the continuous solutions but existence of the shocks destroys the equivalency. Thus the relations on the shock for system (II') are uncoupled as the system itself but for system (I) the relations on the shock are coupled. Therefore the discontinuous solutions of (I) could not be reduced to the two independent functions r and s . In following we consider system (I) only which corresponds more to the real gas dynamics equations.

Let D be the velocity of the shock and let the indices I and 2 denote the states before and behind the shock, respectively. The conditions across the shock for system (I), equivalent to the Rankine-Hugoniot relations in the anisentropic case, are

$$\begin{aligned} \rho_2 u_2 - \rho_1 u_1 &= D(\rho_2 - \rho_1) \\ \rho_2 u_2^2 + \rho_2^3/3 - \rho_1 u_1^2 - \rho_1^3/3 &= D(\rho_2 u_2 - \rho_1 u_1) \end{aligned} \qquad (7)$$

Putting $v_i = u_i - D, \quad i = 1,2$

we get

$$\rho_2 v_2 = \rho_1 v_1$$
$$\rho_2 v_2^2 + \rho_2^3/3 = \rho_1 v_1^2 + \rho_1^3/3 \tag{8}$$

The values $v_i > 0$ if region I is on the left of region 2 and $v_i < 0$ in the opposite case.

Introducing a shock-strength $x = \rho_2/\rho_1 = v_1/v_2$ and Mach number $M_i = |v_i|/c_i = |v_i|/\rho_i$ we derive from (8)

$$M_1^2 = (v_1/\rho_1)^2 = x^2(x^2+x+1)/3 \equiv \psi(x)$$

If u_1 , ρ_1 and x are known, system (7) can be solved giving

$$\begin{aligned} \rho_2 &= x\rho_1 & D &= u_1 \pm \rho_1 \psi(x) \\ v_2 &= v_1/x & u_2 &= u_1 \pm \rho_1 \varphi(x) \\ M_1 &= \psi(x) & M_2 &= M_1/x^2 \end{aligned} \tag{9}$$

where $\varphi(x) = (x-1)\sqrt{(x^2+x+1)/(3x)}$

There are two cases to be distinguished:

I) $M_1 < 1$ - the onflow is subsonic with respect to the shock.

Then $x = \rho_2/\rho_1 < 1, \quad |v_1| = |u_1 - D| < c_1 , \quad M_2 > 1$

This case corresponds to a rarefaction shock $(x = \rho_2/\rho_1 < 1)$ it is unstable and can not be realized.

2) $M_1 > 1$ - the onflow to the shock is supersonic, then

$$x = \rho_2/\rho_1 > 1, \quad |v_1| = |u_1 - D| > c_1 , \quad M_2 < 1 \tag{IO}$$

According to the double valued signs in formulae (9) and (IO) there are two possible kinds of shocks named type "s" and type "r", respectively:

a) The shock of type "s"

$$v_1 > \rho_1 > 0$$
$$D < u_1 - \rho_1 = s_1$$
$$D = u_1 - \rho_1 \psi(x)$$

b) The shock of type "r"

$$v_1 < -\rho_1 < 0$$
$$D > u_1 + \rho_1 = r_1$$
$$D = u_1 + \rho_1 \psi(x)$$

a) $$v_2 = \rho_1 \psi(x)/x$$
$$u_2 = u_1 - \rho_1 \varphi(x) < u_1$$
$$r_2 = u_1 + \rho_1 (x - \varphi(x))$$
$$s_2 = u_1 - \rho_1 (x + \varphi(x))$$

b) $$v_2 = -\rho_1 \psi(x)/x$$
$$u_2 = u_1 + \rho_1 \varphi(x) > u_1$$
$$r_2 = u_1 + \rho_1 (x + \varphi(x))$$
$$s_2 = u_1 - \rho_1 (x - \varphi(x)) \qquad (II)$$

The following inequalities hold:

for $1 < x \leq x_0$:

$$s_2 < D < s_1 \leq r_2 < r_1 \qquad\qquad s_1 < s_2 \leq r_1 < D < r_2$$

for $x_0 \leq x < +\infty$

$$s_2 < D < r_2 \leq s_1 < r_1 \qquad\qquad s_1 < r_1 \leq s_2 < D < r_2$$

Here x_0 is the shock strength in the limiting case for which

$$s_1 = r_2 \qquad\qquad r_1 = s_2$$

In both subcases $1 + x_0 - \varphi(x_0) = 1$ with the solution
$$x_0 = 1 + \sqrt{3} + \sqrt{3 + 2\sqrt{3}} = 5.274510\ldots$$

Let us note that the type of the shock ("r" or "s") is not connected with the direction of its motion. This direction is determined by the sign of D only (if D > 0 then the shock is moving to the right and vice versa).

4. The computation of the shock wave trajectory.

Let the trajectory of the shock be $z = Z(t)$ and the shock have the type "r". Let us consider some point $P(z_P, t_P)$ on the trajectory and find the derivative $dZ/dt = D$ from a given set of values r_1 , s_1 , r_2 at the point P . All these invariants may be determined at the point P if r , s are known for $t < t_P$

From equations (IIb) we have

$$u_1 + [x + \varphi(x)]\rho_1 = r_2 \qquad (12)$$

Inserting in (12) the above expressions for u_1 , $\varphi(x)$ yields

$$\omega(x) \equiv (x-1)\left(1 + \sqrt{(x^2 + x + 1)/(3x)}\right) = (r_2 - r_1)/\rho_1 \qquad (13)$$

which is to be solved for x .
The function $\omega(x)$ is monotonic and positive for $x > 0$.

Once x is determined, all quantities in state 2 can be computed using the formulae:

$$\begin{aligned} s_2 &= u_1 - [x - \varphi(x)]\rho_1 \\ D &= u_1 + \rho_1 \varphi(x) \end{aligned} \tag{I4}$$

By integrating $D = dZ/dt$ we find $Z(t)$ along the shock and at each point the value of s_2 . It essential that the characteristics of system (I) to be straight lines which simplifies considerably the determination of z_1 , s_1 , z_2 at P .

The starting point z^* , t^* of a shock of type "r" is determined from the condition that at that point $\partial z/\partial z = \infty$.

5. Centered rarefaction waves

Centered rarefaction waves are so-called "Simple Waves", they depend only on one coordinate $\xi = (z - z_0)/(t - t_0)$ i.e. the point (z_0 , t_0) being the center. Such waves constitute first order discontinuities, i.e. the first derivatives of the variables are discontinuous across the limiting characteristics.
Introducing $z = z(\xi)$, $s = s(\xi)$ into eq. (5) we get

$$(\xi - z) z_\xi = 0 \qquad\qquad (\xi - s) s_\xi = 0 \tag{I5}$$

There are again two cases:

a) Rarefaction wave of type "s"

$$s = \xi , \qquad z = Const$$

$$s_1 < s_2 , \qquad \xi_1 < \xi_2$$

b) Rarefaction wave of type "r"

$$z = \xi , \qquad s = Const$$

$$z_1 < z_2 , \qquad \xi_1 < \xi_2$$

Introducing a wave strength x varying across the wave according to

$$\rho = x\rho_1$$

we have

$$\begin{aligned} u &= (z_1 + \xi)/2 = u_1 - \rho_1(x - 1) & \qquad u &= (\xi + s_1)/2 = u_1 + \rho_1(x - 1) \\ \rho &= (z_1 - \xi)/2 = x\rho_1 & \rho &= (\xi - s_1)/2 = x\rho_1 \\ u_2 &= (z + \xi_2)/2 > u_1 & u_2 &= (\xi_2 + s_1)/2 < u_1 \\ \rho_2 &= (z_1 - \xi_2)/2 < \rho_1 & \rho_2 &= (\xi_1 - s_1)/2 < \rho_1 \end{aligned} \tag{I6}$$

6. The breakdown of initial discontinuity

Let us consider the initial value problem for system (I) with discontinuous initial data:

At $z < z_0$: $\quad u = u_L, \quad \rho = \rho_L$

At $z > z_0$: $\quad u = u_R, \quad \rho = \rho_R$

From the theory it is known that such a discontinuity breaks down by sending one wave to the left (of type "r") and one to the right (of type "s"). The waves may be the shock waves S or rarefaction waves R. The four types of the breakdown are possible:

$$(S_L, S_R), \quad (S_L, R_R), \quad (R_L, S_R), \quad (R_L, R_R)$$

As a result of a breakdown three regions of constant values are formed - the lefthand (u_L, ρ_L) the central (u_0, ρ_0) and the righthand (u_R, ρ_R). Denoting the overall-strength of the waves by x for the lefthand and by y for the righthand

$$x = \rho_0/\rho_L \qquad\qquad y = \rho_0/\rho_R$$

we have

$$x\rho_L = \rho_0 \qquad\qquad y\rho_R = \rho_0 \tag{I7}$$

$$u_L - \rho_L \Phi(x) = u_0 \qquad\qquad u_R + \rho_R \Phi(y) = u_0$$

where

$$\Phi(\sigma) = \begin{cases} \varphi(\sigma) \equiv (\sigma - 1)\sqrt{(\sigma^2 + \sigma + 1)/(3\sigma)} & \sigma \geq 1 \\ \sigma - 1 & \sigma \leq 1 \end{cases}$$

From (I7) x, y, u_0, ρ_0 can be calculated if u_L, ρ_L, u_R, ρ_R are given.
Further we have for

$$x > 1: \; D_L = u_L - \rho_L \varphi(x) \qquad\qquad y > 1: D_R = u_R + \rho_R \varphi(y)$$

$$x < 1: \; \xi_{0L} = u_0 - \rho_0 \qquad\qquad y < 1: \xi_{0R} = u_0 + \rho_0 \tag{I8}$$

$$\xi_L = u_L - \rho_L \qquad\qquad \xi_R = u_R - \rho_R$$

By a suitable choice of the initial data all four types of discontinuity breakdowns are possible.

As we have assumed homentropic flow in our model, there are no contact discontinuities. This greatly simplifies the computation of the exact solution, but on the other hand does not allow to test the ability of a given computational method to handle contact discontinuities. Such a checking, however, can be done by using the linear equations.

8. The construction of the tests

Let for system (I) the initial values $r = r^0(z)$, $s = s^0(z)$ be given on the interval $[Z_1 , Z_2]$ and the boundary conditions (eventually time-dependent) at the boundaries $z = Z_1$ and $z = Z_2$ As system (I) is non-linear, the number of boundary conditions depends on the values of r and s . By an appropriate choice of the functions $r^0(z)$ and $s^0(z)$ and the boundary conditions one can handle, in a region $Z_1 \leq z \leq Z_2$ and $t^0 \leq t \leq T$, the generation of shocks and rarefaction waves as well as their reflection and interaction. The values of $r(z,t)$ and $s(z,t)$ can be calculated all over the region by means of the formulae given above. It is reasonable to find the positions of all shock waves in the region first and then apply eq. (6) for computing r , s and u , ρ .

The information needed for and delivered by any test program consists in several functions of the variable z given analytically or by tables (initial values and boundary conditions as input, the trajectories of shock, the values r , s on both sides of the shocks etc. as output information). The concrete form of these functions depends on the test chosen.

As an example, we describe in the Appendix the test for some special problem.

Appendix

The test for checking the generation and reflection of a single shock.

IA. Initial values and boundary conditions

We choose as initial values (at $t^0 = 0$):

$$r^0(z) = a_0 + a_1(z + a_2)/\sqrt{1 + (z + a_2)^2} \quad \text{with} \quad a_0 > 0,\ a_1 < 0,\ a_2 > 0$$

$$s^0(z) = -r^0(-z)$$

and as boundary conditions:

for $z = Z_1 = -A$: $\quad r = r^{\circ}(-A - rt)$, $\quad A > a_2 > 0$

for $z = Z_1 = 0$: $\quad u = (r + s)/2 = 0$

The first condition simulates the extension to $-\infty$ at the lefthand boundary, the second one represents a solid wall at $z = 0$ (see Fig. IA).

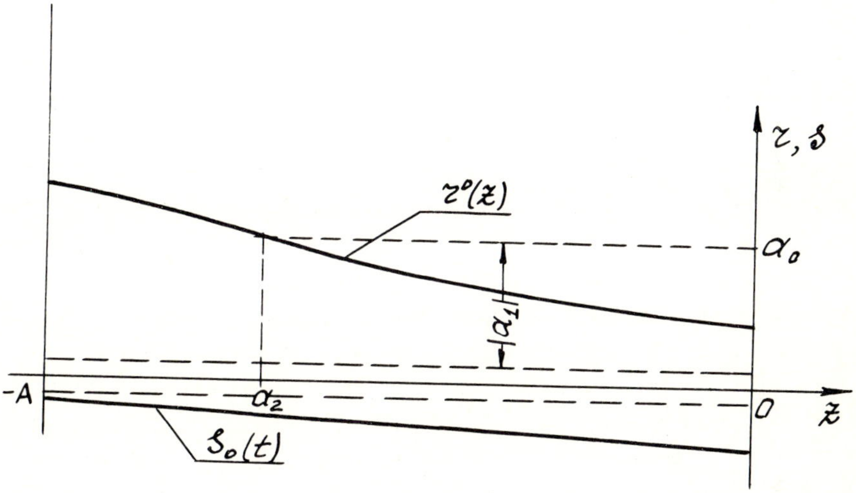

Fig. IA.

2A. The character of the solution

As the time t increases the gradient $\partial r/\partial z$ increases too and at some $t = t_1$ a shock wave of type "r" is generated somewhere (see 4A.). At $t = t_2$ this shock is reflected from the wall $z = 0$ and goes to the left as a shock wave of type "s". At $t = t_3 = T$ it reaches the lefthand boundary $z = -A$.

3A. The computation of the flow in the continuous region

Up to the time t_1 the values of (r, s) are computed as follows. The value of r is computed from the equation:

$$r(z,t) = a_0 + a_1 \xi / \sqrt{1+\xi^2}, \qquad \xi(z,t) = z - rt + a_2 \qquad (1A)$$

The value of s is determined from

$$s(z,t) = -r^{o}(-z+st)$$

where $r^{o}(z)$ is extended on the whole z -axis.

After the shock wave is generated the values of r, s are computed by formulae (6). The characteristics $r = Const$ and $s = Const$ may go to the point given both from the line $t = 0$ (or the boundaries $z = -A$, $z = 0$) and from the shock wave the situation of which has to be computed beforehand.

4A. The computation of the trajectory of the shock wave

The starting point (z_1, t_1) of the shock is determined from the condition $\partial r/\partial z = \infty$:

$$\frac{\partial r}{\partial z} = \frac{\partial}{\partial \xi}(a_0 + a_1 \xi/\sqrt{1+\xi}) \cdot \frac{\partial \xi}{\partial z} =$$

$$= a_1 (1 - t \frac{\partial r}{\partial z})/(1+\xi^2)^{3/2}$$

$$\frac{\partial r}{\partial z} = a_1/(a_1 t + (1+\xi^2)^{3/2})$$

giving $t^* |_{r_z = \infty} = -(1+\xi^2)^{3/2}/a_1$

The smallest value t_1 for t^* arises when $\xi = 0$, application of eqs. (1A) and (6) gives

$$t_1 = -1/a_1 , \qquad r_1 = a_0 , \qquad z_1 = -(a_0/a_1 + a_2)$$

If $z_1 < 0$ the shock wave arises before the reflection at the wall, otherwise it is generated after the reflection of the compression wave.

The computation of the shock wave trajectory has to be done simultaneously with computation of the values (r, s) on both sides of the shock as has been described above. At the initial point (z_1, t_1) $D = r_1 = r_2$ if the shock arises before the reflection $(z_1 < 0)$.

TO THE NUMERICAL SIMULATION AND PROPAGATION OF TSUNAMI ACCORDING TO THE SHALLOW WATER EQUATIONS.

Yu.I.Shokin, An.G.Marchuk, L.B.Chubarov

Institute for Pure & Applied Mechanics of the Siberian Branch of the USSR Academy of Sciencies.

Abstract

The present paper gives the numerical simulation oftsunami generation and propagation in the ocean with real bathymetry. For the describing of this processes nonlinear shallow water equations are used, which are solved with the help of finite-difference method. Here is considered the problem of tsunami generation by bottom displacements, moving with the any given velocity. The computed displacement of water surface and velocity field in the generation area for a typical seismic source (Alexeev, Gusiakov 1973) is used to specify the initial conditions for computing the propagation of tsunami near the Kuril Islands. The wave field of tsunami and mareograms on the shore line are found. The comparison of the results with those for linear model of the tsunami propagation is carried out.

In the general problem of tsunami investigation the problem of numerical simulation of generation, propagation and running over the shore of tsunamis now became quite actual. The physics characteristics of tsunami waves permits us to use the shallow water equations as the mathematical model (J.J.Stoker 1957). The lower results are obtained from the numerical simulation of the processes of generation and propagation of tsunami according to the shallow-water equations with the consideration of inertive, gravitative and convective members.

In the problem of tsunami generation the sistem of nonlinear shallow-water equations which takes into account the bottom deformations is used(Hwang , Divoky 1970). This sistem of the equations is solved by means of explicit finite-difference scheme with artificial viscousity. Here are considered the following problems: tsunami generation as the result of lifting or lowing of the bottom area and the bottom deformations of finite width, moving along the bottom with different speeds. Generation and the first stage of propagation are computed in the rectangle area with the flat or sloping bottom.

The coefficients of artificial viscousity are chosen in order to improve the dispersive and dissipative properties of our difference scheme.

On the fig. 1 one can see the surface after two time steps (3sec) since immidiate (fast) one meter lifting of the rectangular bottom area. For the economy of computing time we took the twice lesser area of computations with the axis of symmetry on its left-back boundary (fig.1). On the axis of symmetry the condition of nonleaking and on the other boundaries the "coming through" conditions are simulated. On the fig. 2 one can see the disturbed surface in the same problem, but after 45 time steps (250 sec). On the fig.3 there is the results tsunami generation by the 0.9 meter lowing of the same bottom area as earlier. Here we see the moment, when the magnitude of the surface disturbance in the center of the source became to be equal to 0.95 m., i.e. greater than the modulus of the initial bottom displacement. It is interesting that there wasn't such effect in the problem of tsunami generation by bottom lifting. On the fig.4 there are the results of computing of tsunami generation problem by the bottom deformations (25 km wide), moving along the bottom at a speed $W = 2C$, where $C = \sqrt{gD}$ -the tsunami propagation velocity in the ocean of the depth D . The slanting waves, which propagate under the angles $\theta = \pm \arccos \frac{C}{W}$ to the direction of moving. This results absolutely suggestes with those, obtained earlier (Novikova, Ostrovsky 1978), using the method of Green's functions. On the fig.5 one can see the results of the analogous problem, but here the velocity of the bottom deformation is equal to $6C$. It is easy to see, that the main directions of tsunami propagation are perpendicular to the moving direction. On the fig.6 one can see the wave picture (the ocean surface) in the problem of tsunami generation by the moving bottom deformation, which changes sign in perpendicular direction, and moves with the velocity $W \gg C$ (the most characteristic speed of Reley waves). The fact that in one direction propagates only one wave and in the opposite - two is interesting.

The computatins also was carried out for the case of sloping bottom. All the given results shows the effectivity of the finite-dif ference method for solving the tsunami generation problems.

While solving the tsunami propagation problem the finite-difference analog of the shallow-water equations is taken as an explicit space-straggered scheme with centered differences. The initial water elevation is computed over the epicentral area for the model source of a dip-slip fault type with the length of 50 km, vertical extension 20 km, at the depth 20 km and the rupture velocity is equal to 3 km/sec. The epicentral area was situated at the south-east corner of the region on the slope of the Kuril deep water trench, where the most of Kuril earthquakes are located.

Fig.7 and 8 shows us the wave picture of interaction of the tsunami wave with the island sistem. On the first of them (fig.7) one can see the tsunami wave, developed into rectilinear front propagated to the Shicotan Island after 16 min since earthquake. The front of the water surface after the tsunami-waves have passed the Little Kuril Islands (at time 42 min) is given on fig.8. Here we can see both reflected and trapped waves. Fig.9 shows positions of the tsunami wave fronts at various moments of time, with the time step equal to 2 min 40 sec; here the epicentral area is shaded. In this case tsunami waves reach the Little Kuril Islands in 10 min since the earthquake, 24 min later waves approach the nothern part of the Kunashir Island and begin to round it. The wave speed in the strait between the Kunashir Island and the Hokkaido is small because the depth in this region is less than 20 m. For this reason waves come up to the whole west coast of the Kunashir from the north. The maximal amplitude (2.8 m) arises at nearest-to-the-source point 1, located at the east coast of the Shicotan Island (fig.10). At the begining of the mareogram we can see the growth of the amplitude up to 1 m, then it decreases to -1.8 m; the subsequent wave train has essentially smaller amplitudes. The first wave has a 12 min period, the posterior periods are shorter.

The additional aim of this study was to compare linear and nonlinear shallow-water models for the tsunami propagation in the ocean with real bathymetry.A similar comparison was made earlier (Chubarov, Shokin 1975) for the model bathymetry. Models of relief used in computations are shown on the insert figure in the upper right of fig. 11. The first model includes the slope from the depth of 4330 m up to 330 m with the angle of tilt equal to 2°30'; the second one includes the same slope up to the depth of 1000 m, the plane segment at this depth with length equal to one wave-length and the following slope up to the depth 330 m with the angle 18'. The wave forms, computed at the points A (the beginning of the slope) are shown on fig.11. As follows from the comparison of the computed wave forms (fig.11) the differences in the behavior of linear and nonlinear waves consists in some phase delay of the linear wave and its amplitude growth. Mainly this difference observed in description of the reflected waves and depends on the presence of the shelf zone in underwater relief. The quantity difference in results increases with the growth of the shelf portion in the relief.

The comparison of two models according to the real bathymetry gives us the same quality difference as it was observed in model computations, but quantity difference is small. This may by explained by the pecularity of the bathymetry of the considered region, which doesn't include a long shelf zone with gentle slopes.

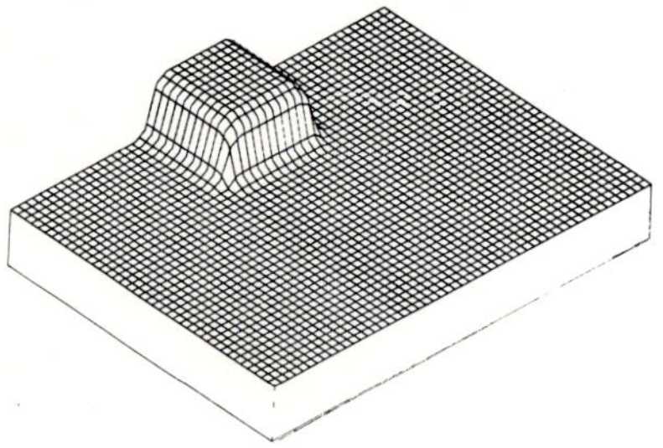

Fig. 1

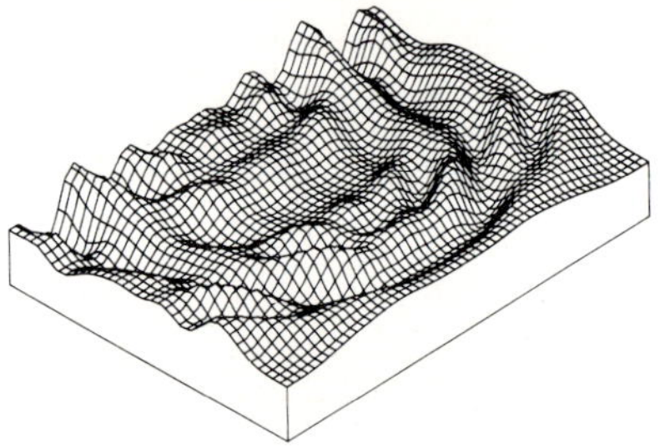

Fig. 2

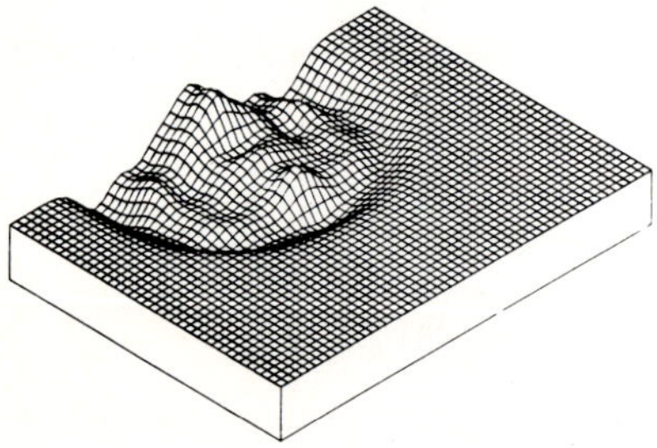

Fig. 3

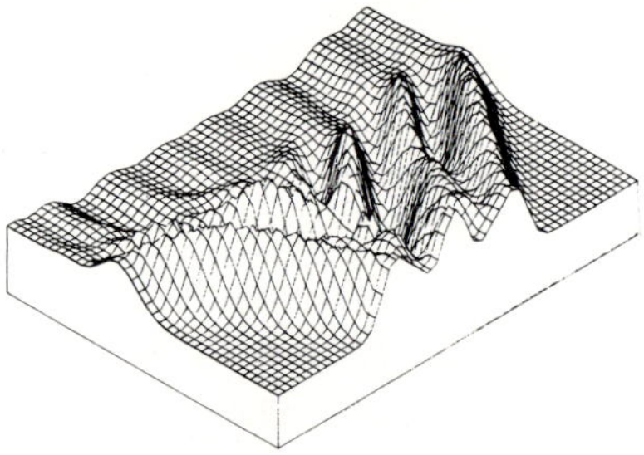

Fig. 4

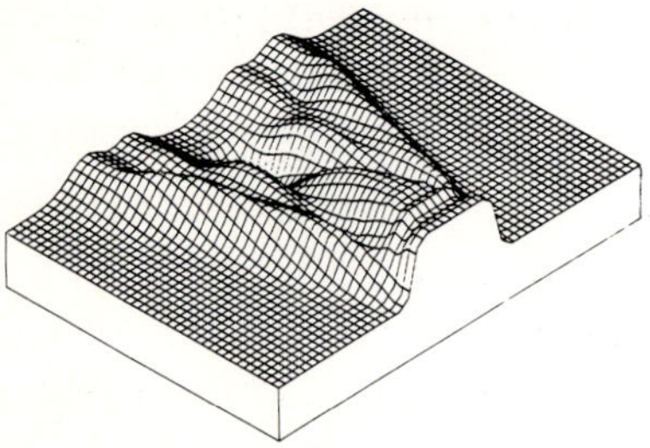

Fig. 5

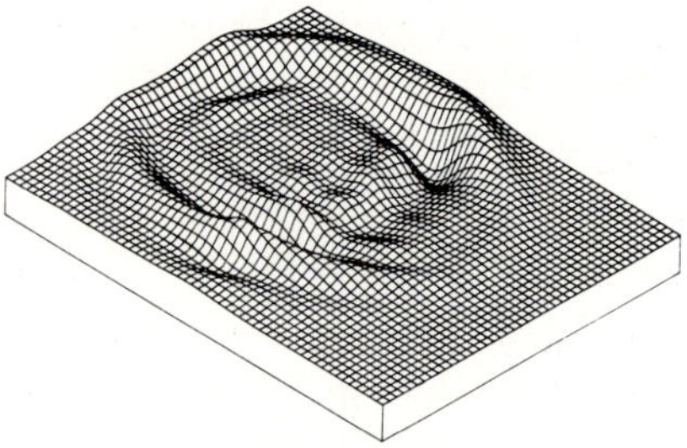

Fig. 6

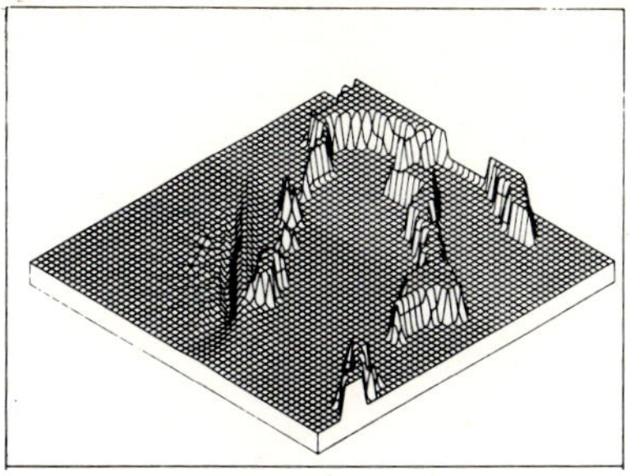

Fig. 7

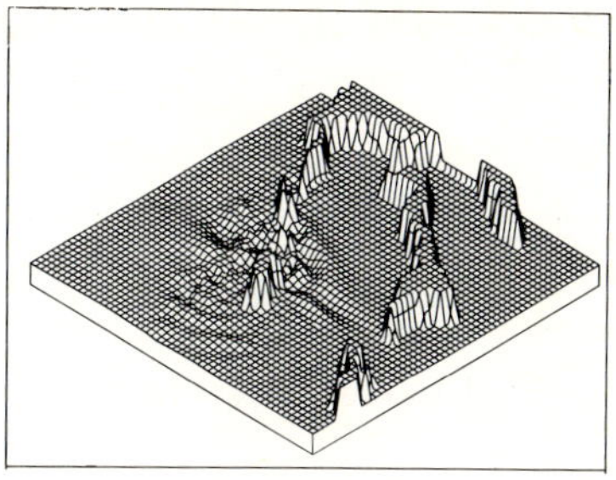

Fig. 8

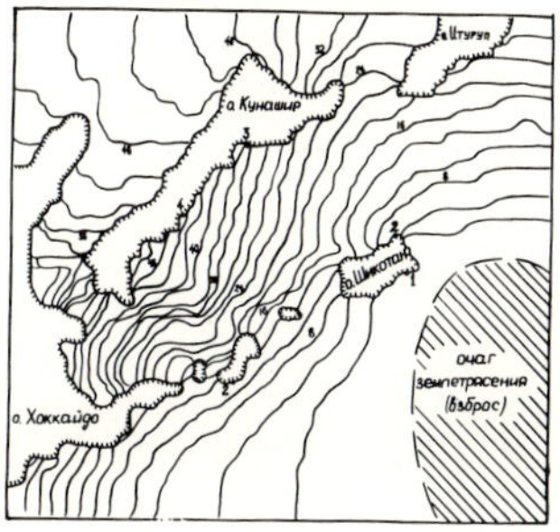

Fig. 9

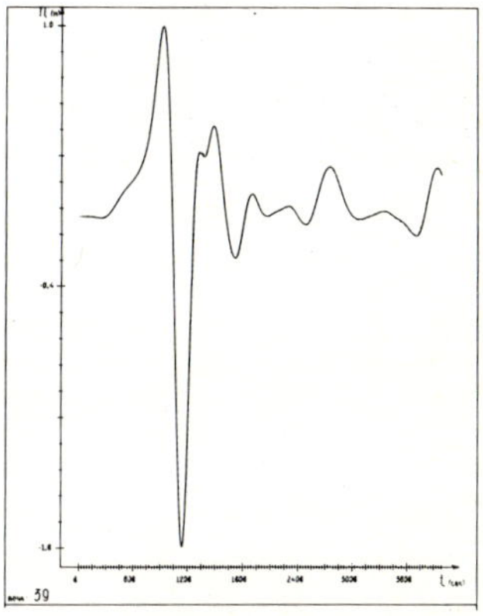

Fig. 10

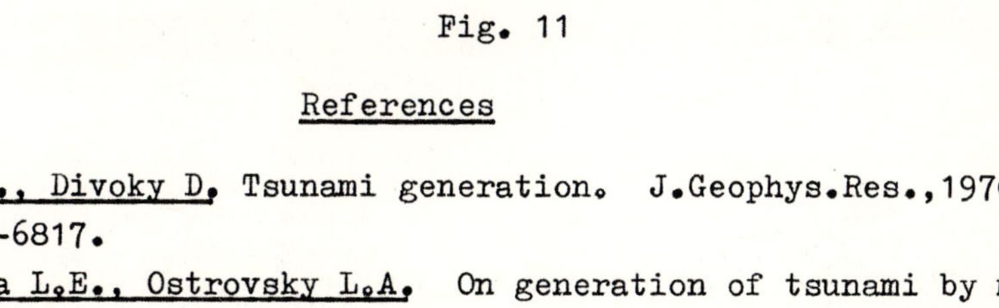

Fig. 11

References

Hwang L., Divoky D. Tsunami generation. J.Geophys.Res.,1970, 75, p. 6802-6817.

Novikova L.E., Ostrovsky L.A. On generation of tsunami by moving deformations of the ocean bottom. Methods for computing of generation and propagation of tsunami. Moscow 1978, p. 88-99.

Alexeev A.S., Gusiakov V.K. Numerical simulation of tsunami and seismic surface waves generation by a submarine earthquake. Trudy VI Konferentsii po difraktsii, vol.2, Moscow-Erevan, 1973.

Stoker J.J. Water waves., 1957.

Chubarov L.B., Shokin Yu.I. Propagation of long waves in shallow water (numerical experiment). Materialy V nauchnoi Konferentsii po matematike i mechanike, vol.1, Tomsk, 1975

A HYBRID RANDOM CHOICE METHOD WITH APPLICATION TO INTERNAL COMBUSTION ENGINES

Gary A. Sod*

Courant Institute of Mathematical Sciences
New York University
New York, New York 10012

and

Lawrence Livermore Laboratory
University of California
P.O. Box 808
Livermore, California 94550

Introduction. Recently a numerical procedure was introduced for solving the one-dimensional equations for an inviscid, radially symmetric flow [8]. The idea of the method was to use operator splitting to reduce this system of equations to two systems of equations; the equations of gas dynamics in Cartesian coordinates and a system of simultaneous ordinary differential equations. This method has been improved and generalized to treat the two-dimensional equations of gas dynamics for an inviscid fluid in a cylindrical geometry [10].

When considering the flow in the cylinder of an internal combustion engine viscous effects cannot be ignored. The effect of viscosity is to create a thin boundary layer which later separates into the fluid. A grid free method for approximating the incompressible boundary layer equations for a flat plate [2] and a cylinder [10] is used. The computational elements are pieces of vortex sheets. The desirability of a grid free method can be seen from analysis which implies that if a grid method is used then several grid points must fall within the

---

*Supported in part by the Sloan Foundation and the Energy Research and Development Administration under Contract No. EY-76-C-02-3077 at the Courant Institute and the U. S. Department of Energy by the Lawrence Livermore Laboratory under contract number W-7405-ENG-48.

boundry layer whose thickness is $O\left(R^{-\frac{1}{2}}\right)$, where R is the Reynolds number. In the problem considered below R ranges from 1000 to 10000.

The computational domain is divided into two regions, one the interior and the other the region near the boundary. Different assumptions about compressibility are made, as well as different numerical methods used, in the two regions. Near the boundary the flow will be considered viscous and imcompressible and in the interior the flow will be considered inviscid and compressible. The coupling of the two methods was originally described by Chorin [2] and generalized and (unfortunately) made more complicated on physical grounds by the author [10].

Outline of the Method for Interior. The two-dimensional equations for an inviscid, non-heat conducting flow in cylindrical coordinates with axial symmetry and zero azimuthal velocity can be written in the form

$$\partial_t \underline{U} + \partial_r \underline{F}(\underline{U}) + \partial_z \underline{G}(\underline{U}) = -\frac{1}{r}\underline{W}(\underline{U})$$

where

$$\underline{U} = \begin{pmatrix} \rho \\ \rho u_r \\ \rho u_z \\ e \end{pmatrix}, \quad \underline{F}(\underline{U}) = \begin{pmatrix} \rho u_r \\ \rho u_r^2 + p \\ \rho u_r u_z \\ (e+p)\, u_r \end{pmatrix}, \quad \underline{G}(\underline{U}) = \begin{pmatrix} \rho u_z \\ \rho u_r u_z \\ \rho u_z^2 + p \\ (e+p)\, u_z \end{pmatrix},$$

and

$$\underline{W}(\underline{U}) = \begin{pmatrix} \rho u_r \\ \rho u_r^2 \\ \rho u_r u_z \\ (e+p)\, u_r \end{pmatrix},$$

where ρ is the density, u_r is the velocity in the radial direction, u_z is the velocity in the axial direction, p is the pressure, e is the energy per unit volume, t is time, and r is the radial distance from the axis of symmetry. We may write

$$e = \frac{p}{\gamma - 1} + \tfrac{1}{2}\rho\left(u_r^2 + u_z^2\right)$$

holding for a pohytropic gas, where γ is the ratio of specific heats (a constant greater than 1).

There are several major problems involved in solving the system (1) directly. The first is the singular term near the axis ($r = 0$), that is, there are singular terms proportional to $1/r$. The second problem is that the momentum equations (the second and third component equations of (1)) cannot be put into conservation form. These problems cause major difficulties near the axis.

In the method developed by the author [8] for one dimensional flows and extended in [10] to two-dimensional flows, both problems have been completely removed without having to resort to any ad-hoc methods.

The first step is to use operator splitting to remove the inhomogeneous term $-\frac{1}{r}\underline{W}(\underline{U})$ from equation (1). The resulting set of equations takes the form

$$\partial_t(\underline{U}) + \partial_r \underline{F}(\underline{U}) + \partial_z \underline{G}(\underline{U}) = 0 \tag{2a}$$

$$d_t \underline{U} = -\frac{1}{r} W(\underline{U}). \tag{2b}$$

The system of equation (2a) represents the two-dimensional equation of gas dynamics in Cartesian coordinates, written in conservation form.

The method used to solve system (2a) is the random choice method introduced by Glimm [4] and developed for hydrodynamics by Chorin [1].

Consider the nonlinear system of hyperbolic equation

$$\partial_t \underline{U} + \partial_x \underline{F}(U) = 0 \tag{3}$$

The random choice method (Glimm's method) is outlined as follows. Divide time into intervals of length Δt and let Δx be the spatial increment. The solution is to be evaluated at time $n\Delta t$ where n is a nonnegative integer at the spatial points $i\Delta x$ where $i = 0, \pm 1, \pm 2, \ldots$ and at time $(n + \frac{1}{2})\Delta t$ at $(i + \frac{1}{2})\Delta x$. This is a two step method. But $\underline{u}_i^n$ approximate $\underline{U}(i\Delta x, n\Delta t)$ and $\underline{u}_{i+\frac{1}{2}}^{n+\frac{1}{2}}$ approximate $\underline{U}((i + \frac{1}{2})\Delta x, (n + \frac{1}{2})\Delta t)$. To find the solution $\underline{u}_{i+\frac{1}{2}}^{n+\frac{1}{2}}$, consider the system (3) along with piecewise constant initial data

$$\underline{U}\,(x, n\Delta t) = \begin{cases} \underline{u}_{i+1}^{n}, & x > (i + \tfrac{1}{2})\,\Delta x, \\ \underline{u}_{i}^{n}, & x < (i + \tfrac{1}{2})\,\Delta x. \end{cases}$$

This defines a sequence of Riemann problems. If $\Delta t < \frac{\Delta x}{(|u|+c)}$, where c is the local sound speed, the waves generated by the individual Riemann problems will not interact. Hence the solution $\underline{v}\,(x,t)$ to the Riemann problem can be combined into a single exact solution. Let ξ_n be an equidistributed random variable which is given by the Lebesque measure on the interval $(-\tfrac{1}{2}, -\tfrac{1}{2})$. Define

$$\underline{u}_{i+\frac{1}{2}}^{n+\frac{1}{2}} = \underline{v}\,((i + \tfrac{1}{2} + \xi_n)\,\Delta x, (n + \tfrac{1}{2})\,\Delta t). \tag{4}$$

At each time step, the solution is approximated by a pieceivise constant function. The solution is the advanced in time exactly and the new values sampled. The method depends on being able to solve the Riemann problem exactly (or very accurately) and inexpensively. The Riemann problem can be efficiently solved by an iterative method due to Godunov [5] and modified by Chorin [1]. As indicated in the review by the author [9] this iterative procedure made the Glimm scheme above 2½ times slower per grid point than most finite difference scheme. A modification in the iterative procedure by the author [10] makes this random method comparable timewise with most finite difference schemes.

The generalization of Glimm's methods to two space dimensions is done by operator splitting. Thus equation (2a) can be written as two sets of one dimensional equations of the form of (3). The two systems of one dimensional equations are solved on a staggered grid in order to ensure consistency. A single two dimensional time step consists of four quarter steps. Each is a sweep in the z or r direction.

Glimm's method will produce perfectly sharp shocks and contact discontinuities. However, for multidimensional problems there are oscillations in a neighborhood of the shocks. This is a dispersive effect due to the operator splitting (used to reduce the number of space dimensions). This is usually not observed in finite difference methods due to the truncation error and the inclusion of artificial viscosity which smooths such oscillations. A numerical study demonstrating this property has been conducted by Colella [3].

To reduce these oscillations an artificial viscosity term is added at each quarter step. The form is a modified version of the artificial viscosity considered by Lapidus [6]. See Colella [3] and Sod [10] and [11] for specialization to Glimm's scheme.

Once the solution of system (2a) $\tilde{\underline{u}}_{i,j}^{n+1}$ is obtained, we have to solve the system of ordinary differential equation (2b). We approximate (2b) by a modified version of the Cauchy-Unler scheme. Randomness, in keeping with the spirit of the Glimm scheme, is introduced which was not in the original method Sod [8]. Let ξ_1 and ξ_2 be the random variable used in the two r sweeps. Then (2b) is approximated by

$$\underline{u}_{i,j}^{n+1} = \underline{u}_{i,j}^{n} - \frac{\Delta t}{(i + \xi_1 + \xi_2)\,\Delta r}\,\underline{W}\,(\tilde{\underline{u}}_{i,j}^{n}). \tag{5}$$

Originally r in (2b) was replaced with $i\Delta r$, here r is replaced with $(i + \xi_1 + \xi_2)\,\Delta r$. This gives solution of (2b) which is in phase with the solution of (2a) in the r-direction.

The system (2b) is only solved at the interior points and the scheme (5) does not require values at $r = 0$. Thus the singularity at the axis is eliminated.

Convergence of the solution of the split system (2a-b) to the solution of equation (1) is proved in [10] for a special case.

Boundary Layer Treatment. Consider the flow past a circular cylinder of constant radius r_0 parallel to its axis ($r = 0$). A boundary layer, initially of zero thickness is formed on the surface of the cylinder. The Prandtl boundary layer equations for a semi-infinite flat plate are a valid approximation. The boundary layer grows parabolically in thickness with increasing z. A stage will be reached when the thickness of the boundary layer can no longer be neglected compared with r_0. So for a long axially symmetric cylinder the boundary layer equations for a flat plate break down. Certain derivatives in r may no longer be neglected. See Schlichting [7].

The boundary layer equations for axially symmetric cylinder (retaining the correct derivatives in r) may be written in the form

$$\partial_t \xi + (\underline{u} \cdot \underline{\nabla}) \xi = \nu \partial_r^2 \xi + \frac{\nu}{r} \partial_r \xi \tag{6a}$$

$$\partial_z (ru_z) + \partial_r (ru_r) = 0, \tag{6b}$$

$$\xi = -\partial_r u_z, \tag{6c}$$

where $\underline{u} = (u_z, u_r)$ with u_z tangent to the boundary and u_r normal to the boundary, ξ is the vorticity, and ν is the viscosity. The boundary conditions are

$$\underline{u} (r_0, z) = 0,$$

$$\underline{u} (\infty, z) = U_\infty (z).$$

We integrate equation (6c) to obtain the tangential component of velocity

$$u_z (r, z) = U_\infty (z) - \int_r^\infty \xi (r, z) \, dr. \tag{7a}$$

With this equation (6b) may be integrated to obtain the normal component of the velocity

$$u_r (r, z) = -\frac{1}{r} \partial_z \int_{r_0}^{r} r \, u_z (r, z) \, dr,$$

which is nonsingular provided $r_0 > 0$. Hence once the vorticity $\xi (r, z)$ is known the velocity field in the boundary layer may be obtained.

The method used is one introduced by Chorin [2] for the boundary layer quations for a flat plate in two-dimentions. Consider a collection of N segments S_i of vortex sheets with intensities ξ_i, i = 1, ..., N. The S_i are straight line segments parallel to the z-axis having length h and center (r_i, z_i). The tangential velocity u_{z_i} of S_i due to the presence of the other segments is obtained by approximating (7a). The normal velocity u_{r_i} of S_i due to the presence of the other segments is obtained by approximating (7b).

Consider the semi-infinite strip $R_i = \{ (r, z) | \, r \geq r_0$ and $z_i - h/2 \leq z \leq z_i + h/2 \}$. The number of interactions among the segments is small since for a segment S_j to influence a given segment S_i it must intersect the strip R_i.

Operator splitting is used to write equations (6a-c) in the form

$$\partial_t \xi + (\underline{u} \cdot \underline{\nabla}) \xi = 0,$$

$$\partial_z (ru_z) + \partial_r (ru_r) = 0, \quad (8)$$

$$\xi = - \partial_r u_z,$$

$$\partial_t \xi = \nu \, \partial_r^2 \xi \, , \quad (9)$$

and

$$\partial_t \xi - \frac{\nu}{r} \partial_r \xi = 0. \quad (10)$$

System (8) is solved deterministically where the sheets S_i are advanced in time according to the tangential (u_{z_i}) and normal (u_{r_i}) components of velocity. Equation (9) is responsible for diffusing the sheets in the r-direction. This equation solved by a random walk. Finally, the sheets are moved along the characteristics of equation (10).

The method of vorticity creation and the coupling of the exterior and interior method will not be discussed here. For details see Chorin [2] and Sod [10].

Application to Motored Engine Chamber. The method is applied to a motored engine, i.e. one that is driven externally. In this talk results will be exhibited at the bottom of the intake stroke, at bottom dead center (the position where the piston is furthest from the head of the cylinder). The engine is assumed to be axially symmetric, with a single intake-exhaust valve located on the chamber axis. The piston motion and valve timing represent an engine timing of 2500 RPM. The bore to stroke ratio is 1 and the intake port is maintained at constant pressure and density. The Reynolds number for the boundary layer calculation is 10^4. When the piston is at bottom dead center the cylinder and intake-exhaust port lie in a 20 x 56 fixed (Eulerian) grid.

Generalization. A random moving grid method in the spirit of Glimm's method has been implemented which in the case of a motored engine chamber gives much better resolution when the piston approaches top dead center (the position where the piston is closest to the cylinder head).

The method of solution is being generalized to a variable geometry [12]. In the application to a motored engine chamber this gives rise to a curved piston head, curved valve face, and beveled edge of valve and port. The main advantage of the method is that by retaining other terms in the boundary layer equations all of the boundary conditions (including piston and valve motion) from both interior and exterior methods can be built into the boundary layer calculation. This will eliminate the problem caused by imposing the boundary condition on the interior method when the boundary is oblique to the mesh. One other feature is that only a minor amount of extra work is required to couple the interior and exterior methods.

Conclusions. Until such time as useful experimental measurements of the flow in a cylinder are made, an assessment of the accuracy of the results cannot be made. However, the results do seem quite plausible.

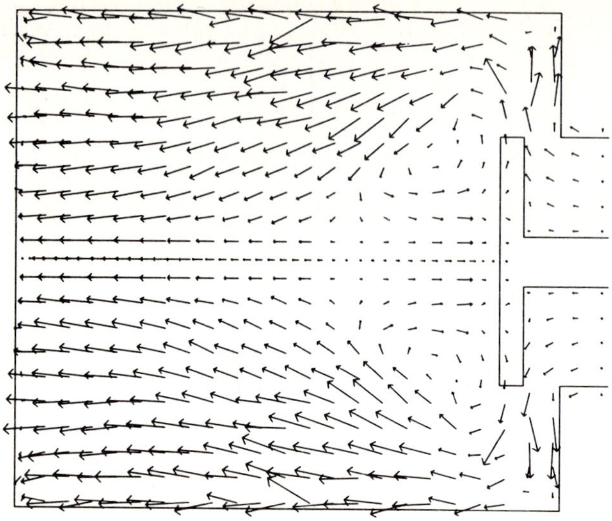

References

[1] A. J. Chorin, J. Comp. Phys., 22, 517 (1976).

[2] ________, "Vortex Sheet Approximation of Boundary Layers, J. Comp. Phys., to appear.

[3] P. Colella, Ph. D. Thesis (1978).

[4] J. Glimm, Comm. Pure Appl. Math., 18, 697 (1965).

[5] S. K. Godunov, Math. Sbornik, 47, 271 (1959).

[6] A. Lapidus, J. Comp. Phys., 2, 154 (1967).

[7] H. Schlichting, Boundary Layer Theory, 6th Ed., McGraw-Hill (1968).

[8] G. A. Sod, J. Fluid Mech., 83, 785 (1977).

[9] ________, J. Comp. Phys., 27, 1 (1978).

[10] ________, "A Numerical Method for Axisymmetric Flows with Application to Internal Combustion Engines," J. Comp. Phys., to appear.

[11] ________ , Numerical Analysis with Application to Fluid Dynamics, Courant Institute Lecture Notes, to appear.

[12] ________ , "A Numerical Method for Axisymmetrics Flow in Variable Geometry with Application to Internal Combustion Engines," to appear.

THE SOLUTION OF THE NAVIER-STOKES EQUATION VIA SUCCESSIVE APPROXIMATIONS

Andrzej Styczek

Instytut Techniki Cieplnej, Politechnika Warszawska

Warszawa, Poland

Abstract. The boundary problem for the Navier-Stokes equation is considered. The solution to the problem is sought-for iteratively. The proof is given - saying that the consecutive approximations of the solenoidal as well as the gradient components of the velocity vector converge. Their limits are determining solution to the original nonlinear system.

The determination of the velocity field $\vec{V}$ and the pressure field p in a steady motion of viscous liquid can be reduced to the boundary problem for the system of N-S equations:

$$-\operatorname{rot}\operatorname{rot}\vec{v}=\lambda[\operatorname{grad}p+(\vec{v}\operatorname{grad})\vec{v}]$$
$$\operatorname{div}\vec{v}=0,\qquad \vec{v}|_{\partial\Omega}=\vec{b},\qquad \oint\vec{n}\vec{b}\,dS=0$$

λ - denotes here the Reynolds number. The pressure p appearing only under the gradient operator involves an additive constant, which can be determined if the value of p is given at an arbitrary point of cosure of the domain of motion.

As it has been proved, the problem formulation as above has at least one solution at any Reynolds number. It has also been proved that for small Reynolds number and limited domain the solution is unique.

However, theorems and their proofs being of fundamental importance - are not constructive: they do not lead to the solution.

The present report is concerned with two or three dimensional flow, and the results obtained will enable us to build a solution for the general case.

Let consider the space $\vec{L}_2$: the norm in $\vec{L}_2$ is a composition of the norms for all components:

$$\|\vec{A}\|^2_{\vec{L}_2} = \sum_{(i)} \|A_i\|^2_{L_2}$$

The following theorem is valid: $\vec{L}_2$ is a simple sum of the subspaces - gradient field and solenoidal field:

$$\vec{L}_2 = \vec{G} \oplus \vec{J}$$

We will take the advantage of this theorem to decompose the nonlinear term of the N - S equations as follows:

$$(\vec{v}\,\mathrm{grad})\vec{v} = \mathrm{grad}\,\Phi(\vec{v},\vec{v}) + \vec{z}(\vec{v},\vec{v})$$

$$\mathrm{grad}\,\Phi \in \vec{G}, \quad \vec{z} \in \vec{J}$$

Naturally div z = 0. Now it is possible to rewrite the N - S equations in the form:

$$-\mathrm{rot\,rot}\,\vec{v} = \lambda\vec{z}(\vec{v},\vec{v}), \qquad \mathrm{grad}[\Phi(\vec{v},\vec{v}) + p] = 0$$

It should be noticed that if the velocity field is known and the decomposition performed then the pressure field is determined. The problem of determining the velocity field may be solved by the succesive approximations in the following way:

$$g\vec{v}_{n+1} + \mathrm{rot\,rot}\,\vec{v}_{n+1} = g\vec{v}_n - \lambda\vec{z}(\vec{v}_n,\vec{v}_n), \quad g > 0,$$

$$\mathrm{div}\,\vec{v}_{n+1} = 0, \qquad \vec{v}_{n+1}|_{\partial\Omega} = \vec{b}$$

The equations which determine the next approximation are linear. They are solvable for any right-hand term. Let the velocity field $\vec{V}_{n+1}$ be presented as a sum:

$$\vec{v}_{n+1} = \vec{j}_{n+1} + \mathrm{grad}\,H_{n+1}$$

H_{n+1} is a harmonic function and $\vec{j}_{n+1}$ is a solenoidal field. Now, we can obtain from the decomposed N - S equations:

$$\mathrm{grad}[H_{n+1} - H_n] = 0$$

Thus, without regarding constants one can write:

$$H_1 = H_2 = H_3 = \dots = H_{n+1} = \dots = H$$

$$\mathrm{div\,grad}\,H = 0, \qquad \frac{dH}{dn}\Big|_{\partial\Omega} = \vec{n}\vec{b}$$

The solenoidal part of the velocity satisfies the equations:

$$g\vec{j}_{n+1} + \mathrm{rot\,rot}\,\vec{j}_{n+1} = g\vec{j}_n - \lambda\vec{z}(\vec{v}_n,\vec{v}_n)$$

$$\vec{j}_{n+1}|_{\partial\Omega} = \vec{b} - \vec{n}(\vec{n}\vec{b})$$

The gradient term of the velocity field doesn't depend on the number n. In order to analyse the convergence, it is sufficient to consider the sequence of the solenoidal terms.

Let us consider the difference:

$$\vec{r}_{n+1} = \vec{j}_{n+1} - \vec{j}_n$$

It is determined by equations:

$$g\vec{r}_{n+1} + \mathrm{rot\,rot}\,\vec{r}_{n+1} = g\vec{r}_n - \lambda\left[\vec{z}(\vec{v}_n,\vec{v}_n) - \vec{z}(\vec{v}_{n-1},\vec{v}_{n-1})\right]$$

$$\vec{r}_{n+1}/_{\partial\Omega} = 0$$

Taking the scalar square of the both sides of the last equation one can obtain:

$$g^2\|\vec{r}_{n+1}\|^2_{\vec{L}_2} + 2g\|\mathrm{rot}\,\vec{r}_{n+1}\|^2_{\vec{L}_2} + \|\mathrm{rot\,rot}\,\vec{r}_{n+1}\|^2_{\vec{L}_2} =$$
$$= g^2\|\vec{r}_n\|^2_{\vec{L}_2} + 2g\lambda\left(\vec{r}_n,\ \vec{z}(\vec{v}_{n-1},\vec{v}_{n-1}) - \vec{z}(\vec{v}_n,\vec{v}_n)\right)$$
$$+ \lambda^2\|\vec{z}(\vec{v}_n,\vec{v}_n) - \vec{z}(\vec{v}_{n-1},\vec{v}_{n-1})\|^2_{\vec{L}_2}$$

Taking into account the decomposition of $\vec{L}_2$ and the fact $\vec{r}_{n+1} \in \vec{J} \perp \vec{G}$ one may write:

$$\left(\vec{r}_n,\ \vec{z}(\vec{v}_{n-1},\vec{v}_{n-1}) - \vec{z}(\vec{v}_n,\vec{v}_n)\right) = \left(\vec{r}_n,\ -(\vec{r}_n\,\mathrm{grad})\vec{v}_{n-1} - (\vec{v}_n\,\mathrm{grad})\vec{r}_n\right)$$

It can be observed, that $\vec{A}$, $\vec{B}$, $\vec{D}$ $\vec{C}^{\infty}_0(\Omega)$

$$(\vec{A}, (\vec{B}\,\mathrm{grad})\vec{D}) = \sum_{(k)}\int_\Omega \frac{\partial}{\partial x_k}\Big(\sum_{(i)} A_i B_k D_i\Big)\,d\Omega$$
$$- \sum_{(i,k)}\int_\Omega D_i A_i \frac{\partial B_k}{\partial x_k}\,d\Omega - \sum_{(i,k)}\int_\Omega D_i B_k \frac{\partial A_i}{\partial x_k}\,d\Omega$$

Moreover, if div $\vec{B} = 0$ as it occurs in the expression $(\vec{r}_n, (\vec{V}_n\mathrm{grad})\vec{r}_n)$ then we see that the transformation may be written as:

$$\left(\vec{r}_n,\ \vec{z}(\vec{v}_{n-1},\vec{v}_{n-1}) - \vec{z}(\vec{v}_n,\vec{v}_n)\right) = -\left(\vec{r}_n,\ (\vec{r}_n\,\mathrm{grad})\vec{v}_{n-1}\right)$$
$$= \left(\vec{v}_{n-1},\ (\vec{r}_n\,\mathrm{grad})\vec{r}_n\right)$$

Using the above result, the Schwartz inequality and the estimation:

$$\|(\vec{A}\,\mathrm{grad})\vec{B}\|^2_{\vec{L}_2} \leq \mathrm{Const}\ \theta^{\frac{n-6}{2}}\left[\theta^2\|\vec{A}\|^2_{\vec{L}_2} + \|\vec{A}\|^2_{\vec{W}_2^{(2)}}\right]\cdot$$
$$\cdot\left[\theta^2\|\vec{B}\|^2_{\vec{L}_2} + \|\vec{B}\|^2_{\vec{W}_2^{(2)}}\right]$$

one can obtain

$$|2g\lambda\left(\vec{r}_n,\ \vec{z}(\vec{v}_{n-1},\vec{v}_{n-1}) - \vec{z}(\vec{v}_n,\vec{v}_n)\right)| \leq$$
$$\leq \frac{2g\lambda\sqrt{\mathrm{const}}}{g^{(6-n)/4}}\left[g^2\|\vec{r}_n\|^2_{\vec{L}_2} + \|\vec{r}_n\|^2_{\vec{W}_2^{(2)}}\right]\cdot\|\vec{v}_{n-1}\|_{\vec{L}_2}$$

Using the same method one can obtain the estimation for the norm of difference:

$$\|\vec{z}(\vec{v}_n,\vec{v}_n)-\vec{z}(\vec{v}_{n-1},\vec{v}_{n-1})\|^2_{\vec{L}_2} \leq \frac{2\cdot const.}{g^{(6-n)/2}}\left[g^2\|\vec{v}_n\|^2_{\vec{L}_2}+\right.$$
$$\left.+\|\vec{v}_n\|^2_{\vec{W}_2^{(2)}}+g^2\|\vec{v}_{n-1}\|^2_{\vec{L}_2}+\|\vec{v}_{n-1}\|^2_{\vec{W}_2^{(2)}}\right]\cdot\left[g^2\|\vec{r}_n\|^2_{\vec{L}_2}+\|\vec{r}_n\|^2_{\vec{W}_2^{(2)}}\right]$$

Let us denote

$$|\vec{A}|^2 = g^2\|\vec{A}\|^2_{\vec{L}_2}+2g\|\mathrm{rot}\,\vec{A}\|^2_{\vec{L}_2}+\|\mathrm{rot}\,\mathrm{rot}\,\vec{A}\|^2_{\vec{L}_2} \geq \frac{1}{const^*}\left[g^2\|\vec{A}\|^2_{\vec{L}_2}\right.$$
$$\left.+\|\vec{A}\|^2_{\vec{W}_2^{(2)}}\right]$$

Now the norm $|r_{n+1}|$ can be estimated in the form:

$$|\vec{r}_{n+1}|^2 \leq g^2\|\vec{r}_n\|^2_{\vec{L}_2}+\frac{2g\lambda\sqrt{const\cdot const^*}}{g^{(6-n)/4}}\|\vec{v}_{n-1}\|_{\vec{L}_2}|\vec{r}_n|^2+$$
$$+\frac{2\lambda^2 const.\, const^*}{g^{(6-n)/2}}\left[|\vec{v}_n|^2+|\vec{v}_{n-1}|^2\right]|\vec{r}_n|^2$$

Using the Friedrichs-Poincare inequality the latter can be rewritten as follows:

$$|\vec{r}_{n+1}|^2 \leq \left[\frac{g^2}{g^2+2g\gamma_1^2+\gamma_2^2}+\frac{2C\lambda g|\vec{v}_{n-1}|}{g^{(6-n)/4}\sqrt{g^2+2g\gamma_1^2+\gamma_2^2}}+\right.$$
$$\left.+2C^2\lambda^2 g^{(n-6)/2}\left(|\vec{v}_n|^2+|\vec{v}_{n-1}|^2\right)\right]\cdot|\vec{r}_n|^2$$

where C=const. > .0, $\gamma_1,\gamma_2 \neq 0$

Choosing the constant number Y such that

$$Y \geq |\vec{v}_n|,\ |\vec{v}_{n-1}|\ \ldots$$

one can obtain the relation:

$$|\vec{r}_{n+1}| \leq \{f(Y)\}|\vec{r}_n|$$

Number Y can be choosen such way, that:

$$\{f(Y)\} = \frac{g^2+\beta(2g\gamma_1^2+\gamma_2^2)}{g^2+2g\gamma_1^2+\gamma_2^2} < 1, \quad 0<\beta<1$$

Solving the last equation one can obtain:

$$Y = \frac{g^{(6-n)/4}}{4C\lambda}\cdot\frac{g}{g^2+2g\gamma_1^2+\gamma_2^2}\left[\sqrt{1+4\beta\frac{2g\gamma_1^2+\gamma_2^2}{g^2}}-1\right]$$

The expression $|\vec{r}_n| \rightarrow |\vec{r}_{n+1}|$ is contracting and consequently:

$$|\vec{v}_p| \leq |\vec{v}_0|\left[1-\sqrt{\frac{g^2+\beta(2g\gamma_1^2+\gamma_2^2)}{g^2+2g\gamma_1^2+\gamma_2^2}}\right]^{-1} \leq Y$$

for each p. The result of the last inequality is an estimation of the norm of the initial element $\vec{v}_0$:

$$|\vec{v}_0| \leq \left(1-\sqrt{\frac{g^2+\beta(2g\gamma_1^2+\gamma_2^2)}{g^2+2g\gamma_1^2+\gamma_2^2}}\right)\cdot\frac{g^{(6-n)/4}}{4\lambda C}\cdot\frac{g}{\sqrt{g^2+2g\gamma_1^2+\gamma_2^2}}\cdot\left[\sqrt{1+4\beta\frac{2g\gamma_1^2+\gamma_2^2}{g^2}}-1\right]$$

This estimation guarantees the convergence of the sequence $\vec{v}_n$ in a Sobolev space. It may be easily shown, that this limit satisfies the nonlinear N - S equations. For $N \to \infty$ there is:

$$\|\operatorname{rot}\operatorname{rot}\vec{v}_\infty + \lambda\vec{z}(\vec{v}_\infty,\vec{v}_\infty)\|^2_{\vec{L}_2} = \|\operatorname{rot}\operatorname{rot}\vec{j}_\infty + \lambda\vec{z}(\vec{v}_\infty,\vec{v}_\infty)\|^2_{\vec{L}_2} = \|g\vec{j}_\infty + \operatorname{rot}\operatorname{rot}\vec{j}_\infty$$

$$+ \lambda\vec{z}(\vec{v}_\infty,\vec{v}_\infty) - g\vec{j}_\infty - g\vec{j}_N - \operatorname{rot}\operatorname{rot}\vec{j}_N - \vec{z}(\vec{v}_{N-1},\vec{v}_{N-1})\cdot\lambda + g\vec{j}_{N-1}\|^2_{\vec{L}_2}$$

$$\leq \left\{ \|\vec{j}_\infty - \vec{j}_N\| + \frac{g}{\sqrt{g^2 + 2g\gamma_1^2 + \gamma_2^2}} \|\vec{j}_\infty - \vec{j}_N\| + \frac{2\gamma\lambda}{g^{(6-n)/4}} \|\vec{j}_\infty - \vec{j}_{N-1}\| \right\}^2 \to 0$$

$$\text{for } \|\vec{j}_\infty - \vec{j}_{N-1}\| \leq \varepsilon_{N_0},\ N > N_0$$

Therefore the condition sufficient for the convergence of the method of succesive approximation of the solenoidal component of the velocity field $\vec{V}$ was obtained. Since the gradient component of $\vec{V}$ is constant, then the sufficient condition for the convergence of the sequence of the succesives approximations for the vector field $\vec{V}$ is only a consequence of the solenoidal component.

References

O.A.Ladyzhenskaya, Matematicheskie voprosy dinamiki vyazkoi neszhimaemoi zhidkosti, "Nauka", Moskva, 1970.

A.Styczek, The solution of Navier-Stokes equations via succesive approximation, Bull.Pol.Acad. de Sc.,ser. de Sc. Techn., Vol.XXV, No.8, 1977.

ON THE CONDENSING OF GRID POINTS IN THE PROCESS OF SOLVING AND HIGH ORDER SCHEMES IN VISCOUS GAS FLOW COMPUTATIONS

A.I. Tolstykh
(USSR)

In order to give the accurate description of the flows with the regions of small characteristic scales (for example, such as boundary layers) it is nesessary to condense the grid points of the mesh. Unfortunately, in many important cases the positions and the dimensions of these regions are unknown beforehand. In order not to use any advance a priori information about the flow field the transformation of one of the fixed coordinates (s , n), for instance n , is introduced /1/ (for the sake of the simplicity only twodimensional case is considered). This transformation depends on velocity components u and v along the s and n axes respectively and has the form

$$\zeta = \zeta(u, v, s, n) \qquad (1)$$

The function ζ must possess the following properties: (a) for the fixed s function $f(s,n) = \zeta(u(s,n), v(s,n), s, n)$ varies rapidly in the same regions as at least one of the functions u and v (the "streching" property); (b) the condition $\partial f/\partial n > 0$ is satisfied. If such a function is constructed, viscous layers with characteristic scales tending to zero with Reynolds number $Re \to \infty$ will be mapped onto the regions in (s , ζ) plane with characteristic scales of the order of O(1). As a result a uniform mesh in the new plane will correspond to that in (s , n) plane with grid points condensing in the high gradient regions. This condensing will be performed automatically while solving the difference equations without knowing a priori the dimensions and the positions of the "stretching" thin layers.

The ways of constructing the function ζ are given in /1/. In general to fulfil the condition (a) the transformation $\zeta = \zeta(z, s, n)$ in which $z = \sqrt{(k(s)u)^2 + (\ell(s)v)^2}$, can be used because with appropriate k and ℓ the function z varies rapidly along the s = const lines intersecting viscous layers. To meet the condition (b) the most general information can be useful. For instance if the variation of z along S = const lines is known to be monotonic within stretching layer (let us assume it to be monotonously increasing, for example) it is sufficent to use the following simple relation:

$$\zeta = a(s)z + b(s)n + c(s)\ , \quad b>0 \qquad (2)$$

where one of the coefficients a or b is chosen to a certain extent arbitrarily (this choice influences the actual thicknes of the viscous layer in the (S , ζ) plane), while the other two coefficients are chosen to satisfy the normalizing condition for the variable ζ . Transformation (2) is appropriate for many interesting problems. Moreover, in some cases after the successful choice of the coordinates (S , n) it is possible to use instead of z in (2) either of the two components of the velocity, for example, u .

It is also possible to envisage the situation when either the flow to be investigated is absolutely unknown or the variable z is nonmonotonic along the S = const = S_0 lines in the viscous layer. In such case it is sufficent for a m -th stage of computations (either for a m-th iteration or for a m-th time level) to determine the intervals of the monotonicity of the function $z_h^{m-1}(n,S_0)$ defined in the grid points during the previous ($m-1$)-th stage. Then it is easy to get the continuous monotonously increasing function $\zeta(n) = \zeta^{m-1}(z,n,S_0)$ which has form (2) for each interval of the monotony of the z^{m-1} . In particular, if one chooses $|a(S_0)|$ = = const = a for all the intervals then transformation (1) can be written in the form

$$\zeta = a \int_0^n \left|\frac{\partial z}{\partial n}\right| dn + b(S_0)n + C(S_0)\ ; \qquad (3)$$

for monotonic z_h^{m-1} it reduces to (2). The choice of the function $\psi(z) = a\int_0^n |\partial z/\partial n|\, dn + C$ is schematically shown in Fig.1.

It is worth noting that after introducing the new variable ζ the Jacobian equation for transformation (1) is added to the basic system of equations; transformation (1) itself can be used for the determination of the physical coordinate n .

2. The construction of the uniform mesh in the computational plane which "adjusts" to the flow field in a physical plane is favorable for further increasing the accuracy of the solutions by means

of the high order schemes. For the model equation

$$\frac{\partial u}{\partial t}+\frac{\partial f(u)}{\partial x}=\frac{\partial}{\partial x}\varepsilon\frac{\partial u}{\partial x}+R(u,x) \qquad (4)$$

the main idea of the implicit conservative third order accurate difference schemes presented in /1/ may be explained in the following way. We introduce operators

$$A_{\pm}^{(x)}=\frac{5}{12}T_{\mp}+\frac{8}{12}E-\frac{1}{12}T_{\pm}\,, \qquad \Delta_{\mp}^{(x)}=\pm E\mp T_{\mp} \qquad (5)$$

defined on the uniform mesh $\omega_h\,(x_n=nh,\quad h=const\,)$. Here T_+ and T_- are the operators of one increment h shift in the positive and negative directions along the x axe, E is the unit operator. For $\varepsilon=0$ the scheme of the order $O(\tau^K+h^3)$ defined on the mesh $\omega_\tau\times\omega_h$ ($\omega_\tau : t_m=m\tau$, τ = const) can be written in the form

$$A_{\pm}^{(x)}\left\langle\frac{\partial u}{\partial t}\right\rangle_n^{m+1}+\frac{\Delta_{\mp}^{(x)}f_n^{m+1}}{h}=A_{\pm}^{(x)}R_n, \qquad f_n^{m+1}=f(u_n^{m+1}) \qquad (6)$$

where $\langle\partial u/\partial t\rangle_n^{m+1}$ is some difference approximation of the order K which uses previous time levels. Scheme (6) will be unconditionally stable in the L_2 norm for $f(u)=au$, a = const (K = 1,2) if for $a>0$ and $a<0$ operators ($A_+^{(x)}$, $\Delta_-^{(x)}$) and ($A_-^{(x)},\Delta_+^{(x)}$) are taken respectively.

In the case $\varepsilon\neq 0$ it is possible to apply the same principle of approximation and to obtain the following scheme

$$A^{(x)}\left\langle\frac{\partial u}{\partial t}\right\rangle_n^{m+1}+\frac{\Delta^{(x)}(f-q)_n^{m+1}}{h}=A^{(x)}R_n, \qquad \frac{\tilde{\Delta}^{(x)}u_n^{m+1}}{h}=\tilde{A}^{(x)}\left(\frac{q}{\varepsilon}\right)_n^{m+1} \qquad (7)$$

after reducing (4) to the set of two equations with respect to variables u and $q=\varepsilon\,\partial u/\partial x$. Here it is assumed that omitted + and - indexes of the $A^{(x)}$ and $\Delta^{(x)}$ operators are chosen as in the scheme (6) and $\tilde{A}^{(x)}$, $\tilde{\Delta}^{(x)}$ are the operators with the indexes different from the chosen ones. The scheme (7) is also unconditionally stable in the "frozen coefficients" approximation.

The matrices of the systems of difference equations (6) and (7) have tridiagonal and block-tridiagonal forms respectively. It can be prooved that for the appropriate boundary conditions these systems are well defined and the "progonka" technique for their solutions is stable.

Schemes (6) and (7) can be transformed into those of the order $O(\tau^{\kappa} + h^{4})$ by their symmetrisation. For this purpose it is sufficient to apply operators $A_o = (A_+^{(x)} + A_-^{(x)})/2$ and $\Delta_o = (\Delta_+^{(x)} + \Delta_-^{(x)})/2$ instead of $A_{\pm}^{(x)}$ and $\Delta_{\mp}^{(x)}$ operators. As in the case of schems /3/ it means using the cubic spline approximations. However, the results of the numerical experiments showed that in many cases nonsymmetric schemes (6) and (7) are preferable because of their better properties of monotony.

In the case of several spatial variables multidimensional analogs of schemes (7) and (9) can be written in many ways as fractional steps schemes /4/; corresponding study is presented in ref.5.

3. Now we illustrate by several examples the applications of the transformations and difference schemes described in the previous sections.

The Burgers equation. In this case $f(u) = u^2/2$, $\varepsilon = \nu$ =const, $F = 0$. After applying transformation (2) with $z = u$ and $n = x - t/2$ one can come to the following boundary value problem: $u \to 1$ ($\xi \to -\infty$) and $u \to 0$ ($\xi \to \infty$). Numerical solutions obtained for $\nu = 10^{-2}$, 10^{-3} and several ratios $b/a = c$ are shown in Fig.2 and Table 1. One can see that shock wave region with the thickness of the order of $O(\nu)$ is mapped onto the region with the characteristic scale of the order of O(1), the actual thikness of this region depending on constant c . The high accuracy of the method is illustrated in Table 1 where functions $n(\xi)$, $u(\xi)$ and $u'_n(\xi)$ are presented together with their exact values $u_{exact}(n)$ and $u'_{exact}(n)$; these data are obtained with numbers of grid points N = 10 and 40.

Couette flow with injection. In this case one can put in (4) $\partial/\partial t \equiv 0$, $R = -P$ = const, $\varepsilon = 1/Re$, $f(u) = v_o u$ where the constants v_o , P and Re are the pressure gradient, velocity of injection and Reynolds number respectively. The boundary conditions are $u(0) = 0$ and $u(1) = 1$. According to the exact solution of this problem the boundary layer with characteristic thickness of the order O(1/Re) exists either on an unmovable wall $y = 0$ ($v_o < 0$) or on a moving one $y = 1$ ($v_o > 0$). The velocity profiles can be either monotonic or have sharp extremums depending on the values of P . Supposing that the position and the thickness of the boundary layer are unknown one can apply the general transformation (3) with $z = u$. The results of the computations for two typical cases of monotonic profile (P = 0.5) and "backward" flow (P = 1.5) are shown in Fig.3 for Re = 10^2, 10^3 and $\mathcal{V}_o$ = -1. Several values of the computed functions and their derivatives are presented with corresponding exact values in Table 2 for the region adjacent to the wall. The concentration of the grid points performed automatically allows to obtain sufficently high accuracy with moderate amount of grid points.

Viscous gas flows. The pressure distributions along a flat plate of finite length are presented in Fig.5 as an example of the solutions of the Navier-Stokes equations. These solutions are obtained for Mach number M_∞ = 10 and various values of the Reynolds number $Re_{\infty L}$, specific heats ratios γ and wall temperature T_w (T_o is the stagnation temperature). Several typical stream lines in the vicinity of the base are shown in Fig.5 (the two dimensional near wake problem, M_∞ = 3, T_w/T_o = 0.4, Re_H = 600 where Re_H is the Reynolds number based on the base height H and temperature T_o). The fragments of the meshes shown in Fig.4 and 5 illustrate the condensing of the grid points in high gradients regions (in case of Fig.4 automatic condensing along the normal to a flat plate is accompanied by the fixed condensing along the surface). Transformation (2) was applied in both cases. The computations of the flow near a blunt body where the schock wave and the boundary layer are simultaneously "stretched" were also carried out /1 /.

REFERENCES

1. Tolstykh A.I. - On the method of numerical solution of the Navier-Stokes equations for the viscous gas in wide domain of the Reynolds number. Doklady Acad. Nauk SSSR, v.210, N 1, 1973 (In Russian).

2. Tolstykh A.I. - On the condensing of grid points in the process of solving and usage the high order schemes in the numerical investigations of the viscous gas flows. Zh.Vichisl.Mat.i Mat.Fiz., v.18, N 1, 1978 (in Russian).

3. Rubin S.G.,Graves R.A. - Viscous flow solution with a cubic spline approximation. - Internat. J. Comput. Fluids., v.3, N 1, 1975.

4. Yanenko N.N. - The method of fractional steps for the solution of multidimensional problems of the mathematical physics. - Nauka, Novosibirsk, 1967 (in Russian).

5. Tolstykh A.I. - On the implicit high order difference schemes for the multidimensional problems. - Zh.Vichisl.Mat.i Mat.Fiz., v.16, N 5, 1976. (in Russian).

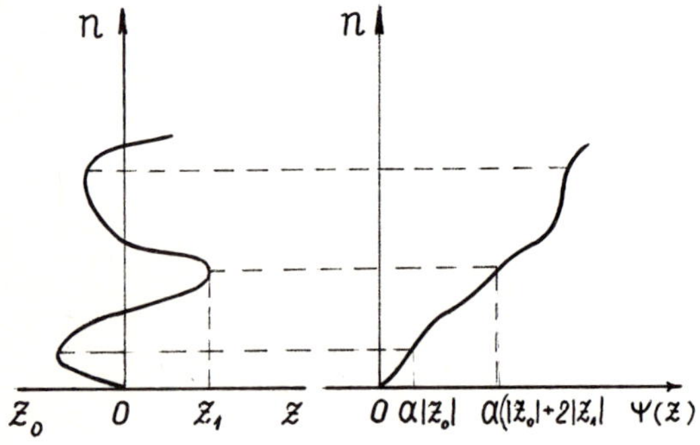

Fig.1

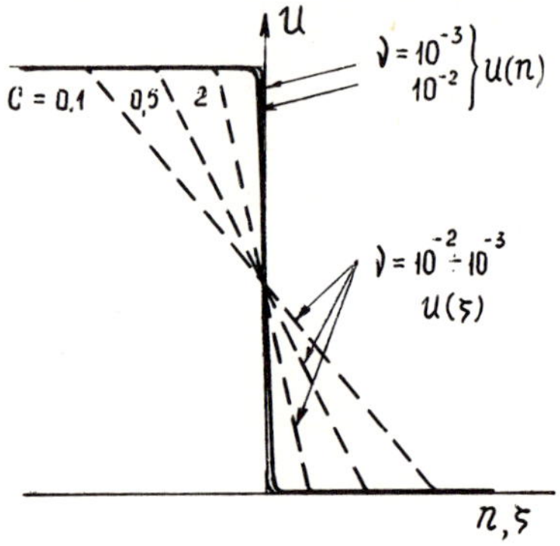

Fig.2

Table 1

$\Delta\zeta = 0.1$, $\nu = 10^{-3}$				$\Delta\zeta = 0.025$, $\nu = 10^{-3}$				
$\lvert\xi\rvert$	$n(\zeta)$	$u(\zeta)$	$u'_x(\zeta)$	$n(\zeta)$	$u(\zeta)$	$u_{exact}(n)$	$u'_n(\zeta)$	$u'_{exact}(n)$
0	0	.50000	125.09	0	.50000	.50000	124.99	125.00
0.2	.00096	.61992	117.90	.000979	.61992	.62004	117.81	117.79
0.4	.00208	.73980	96.337	.00209	.73979	.74001	96.249	96.196
0.7	.00443	.91956	37.071	.00487	.91951	.91961	37.003	36.962
0.9	.39785	1.00021	.92862	.39995	1	1	.92870	0

Table 2

P=1.5, $\varepsilon = 10^{-3}$, $\Delta\zeta = 0.025$					
ζ	$y(\zeta)$	$u(\zeta)$	$u_{exact}(y)$	$u'_y(\zeta)$	$u'_{exact}(y)$
0	0	0	0	-498.49	-498.50
.025	.00021	-.09854	-.09820	-399.62	-399.96
.075	.00088	-.29518	-.29144	-201.99	-205.73
.200	.08987	-.36520	-.36518	1.94439	1.500
.500	.43109	.14664	.14664	1.49934	1.500

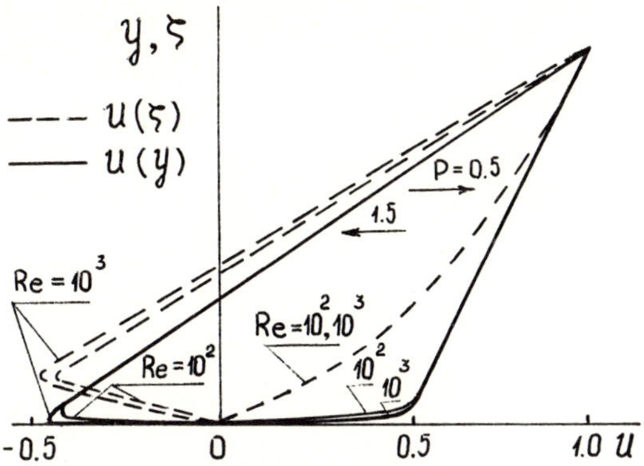

Fig.3

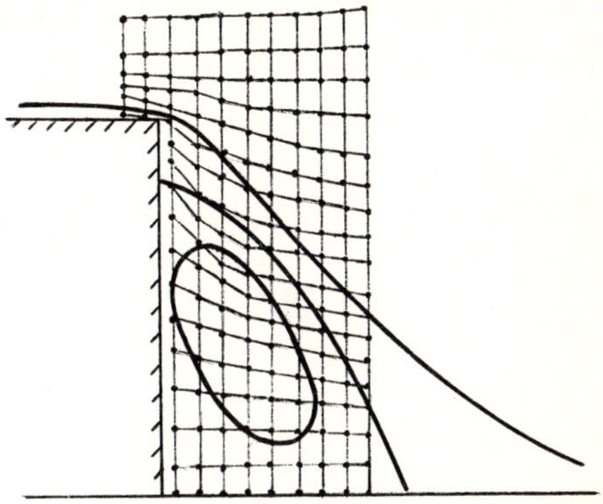

Fig.4

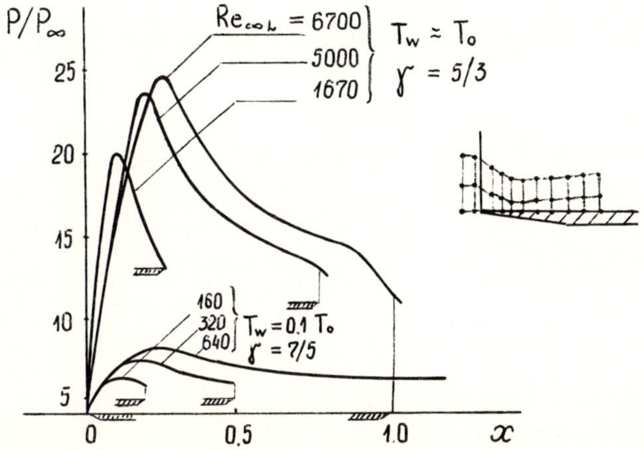

Fig.5

INVESTIGATION OF UNSTEADY SUPERSONIC FLOWS ABOUT BLUNT BODIES

L.I. Turchak

Computing Centre
Academy of Sciences of the USSR, Moscow

Unsteady axisymmetric flows about blunt bodies are studied. The parameters of an oncoming flow and the body shape are given as functions of time. The initial data are chosen to obtain unsteady flows with an arbitrary degree of unsteadiness.

A grid-characteristic method is used for the solution of the task [I]. The corresponding difference equations are given in [2].

The parameters of the gas are determined in a region bounded by a shock wave, a body surface, the axis of symmetry of the body and a ray passing through the cross-section and completely situated in a supersonic flow. Initial relations are presented in [3].

I. The motion of a body with a variable velocity

A flow about a body is considered here, and its velocity is given by the relations:

- harmonic oscillations

$$V = I - A \cos \frac{2\pi t}{T},$$

- retardation of the body

$$V = I - (I - V_I) \sin^2 \frac{\pi t}{2T} \quad (0 < t < T)$$

$$V = V_I \quad (t > T).$$

In the case of longitudinal oscillations the calculations were done for a sphere and a cylinder with various values of Mach number ($M = 2 - I0$), an amplitude ($A = 0.I - 0.3$) and an oscillation period ($T = 2 - 20$).

Fig. I gives the magnitudes of the shock departure ε and pressure p at the stagnation point for a sphere in a flow with $M = I0$, $A = 0.3$ and $T = 6, 4, 2$ (curves I, 2, 3) during one oscillation period. The pressure peaks observed are due to internal shocks originating in a shock layer when the acceleration of the body is large.

With $T > I0$ the distribution of pressure and other parameters of the gas is close to a harmonic one, the solution of the problem is realizable in the linear formulation [4, 5].

A flow about an oscillating cylinder is also considered. The principal distinction is that nonlinear effects become much more apparent.

The calculations of flows for the case of a decelerated body were carried out for Mach numbers $M = 20 \to 4$, $20 \to 2$, $20 \to 1$, $10 \to 2$, $10 \to 1$ at various deceleration times ($T = 1 - 100$). The position of the shock at one of the variants at different times is represented in fig. 2a. With a rapid body deceleration the shock first departs from it, then it occupies a position corresponding to the established value of the body velocity V_1.

To some extent one can judge the flow pattern according to the pressure distribution on the axis of symmetry at the points of the shock and the body. With the increase of the deceleration time ($T > 10$) the particular effects are smoothed out, the distribution of flow parameters is of a monotonous character. Their magnitudes approach those of a corresponding quasistationary flow.

We should single out the body deceleration up to a sonic velocity ($M_1 = 1$) from the calculations done for other initial data. Here we succeeded in following the shock intensity while it is receding from the body. The calculations reveal that with a rapid deceleration the shock intensity considerable excudes the corresponding steady magnitudes.

2. Flows about bodies of a varying shape.

Let a front surface of a spherically blunt body on a trajectory $0 < \theta < \theta_*$ (θ is a pole angle measured from the axis of symmetry) be deformed by a given law. The equation for the body surface has the form

$$r = 2\cos\theta - f(\theta, t).$$

Harmonic oscillations (vibration) of the body surface were considered, i.e.

$$f(\theta,t) = A\left(1 - \frac{\theta^2}{\theta_*^2}\right)\cos\omega t.$$

Fig. 3 presents the pressure distribution at a critical point (solid lines) and a drag coefficient C_x (dashed lines) with $M = 2$ for various oscillation parameters. The calculations show that with the decrease of an oscillation period or with the increase of an amplitude, everage values and pressure maxima grow. Herein the thickness of a shock layer diminishes.

The flow around a body with a rapid deformation of its front surface is also considered. The body deformation equation is given as mentioned above, where

$$f(\theta,t) = A(I - \frac{\theta^2}{\theta_*^2}) \sin^2\frac{\pi t}{2T} \ (0 < t < T).$$

Here A is a deformation magnitude on the axis of symmetry, T is a characteristic time of an unsteady process. With t T the problem of the flow around the resulting concave body is solved prior to the establishment of a steady flow.

Fig. 4 shows the distribution C_x with various initial parameters defining the extent of flow unsteadyness. With a rapid variation of the body shape a considerable increase of the resistance coefficient occurs. With the increase of T a flow pattern approaches a quasistationary one, flow parameters change practically monotonically.

3. Deceleration of bodies entering the atmosphere.

It was shown previously that under some conditions of body variation in shape the drag can rise significantly. Some results are produced here of the calculation of flows about bodies of varying shape at their entry into the earth atmosphere. Account is taken of the variation of density with altitude. The calculations were done primarily for bodies of small densityone of the possible models of the deceleration of meteors in the atmosphere 6 . The initial radius of a ball R = I km, a body mass $m=I0^7$ ton, its velocity V_0= 20 km/sec, the altitude at which deformation starts H_0=I0-50 km. The velocity and the altitude of the body are determined from the equations of motion of its mass center.

The variation of the body shape was of two types - the change in curvature of a front surface and the expansion of the middle cross-section.

In the first case we managed to obtain the following values of the body velocity near the surface of the Earth for some examples

H_0	50	20	20	I0
T	5	I0	5	5
V_I	3.9	4.7	4.6	7.3

For comparison we can show that, with H_0= 50 km in the case of a non-deformed body, V_I= 6.4 km/sec.

In the variants in question, the shock is situated near the body. For it to reced considerably, a rapid deceleration is needed and we failed to obtain it here.

At the entry of an expanding body its front surface is represented as a rotating ellipsoid, whose longitudinal semiaxis does not vary ($a = I$) and the transverse semiaxis increases according to the law : $b = \sqrt{I + Bt^2}$.

Fig. 5 shows the trajectory of motion of an expanding body and the departume of the shock with $H_0 = 50$ km and $V_0 = 20$ km/sec. Curves 0, I, 2 correspond to the values $B = 0, 0.00I, 0.0I$. Herein the velocity magnitudes at the last point calculated for these variants are of the order of the sound velocity and the intensity of the shock remains large even as it recedes from the body. Therefore it may be damaged when landing on the Earth.

References

I. Магомедов К.М., Холодов А.С. Ж-л вычисл.матем.и матем. физ., I969, т.9, № 2.
2. Численное исследование современных задач газовой динамики. Под ред. О.М.Белоцерковского. М., "Наука", I974.
3. Белоцерковский С.М., Турчак Л.И., Холодов А.С. Изв.АН СССР, МЖГ, I975, № 5.
4. Белоцерковский С.М. Изв.АН СССР, ОТН, I956, № 7.
5. Грудницкий В.Г. Труды ВВИА им.Жуковского, I969, в.I253.
6. Петров Г.И., Стулов В.П. Космические исследования, I975, № 4.

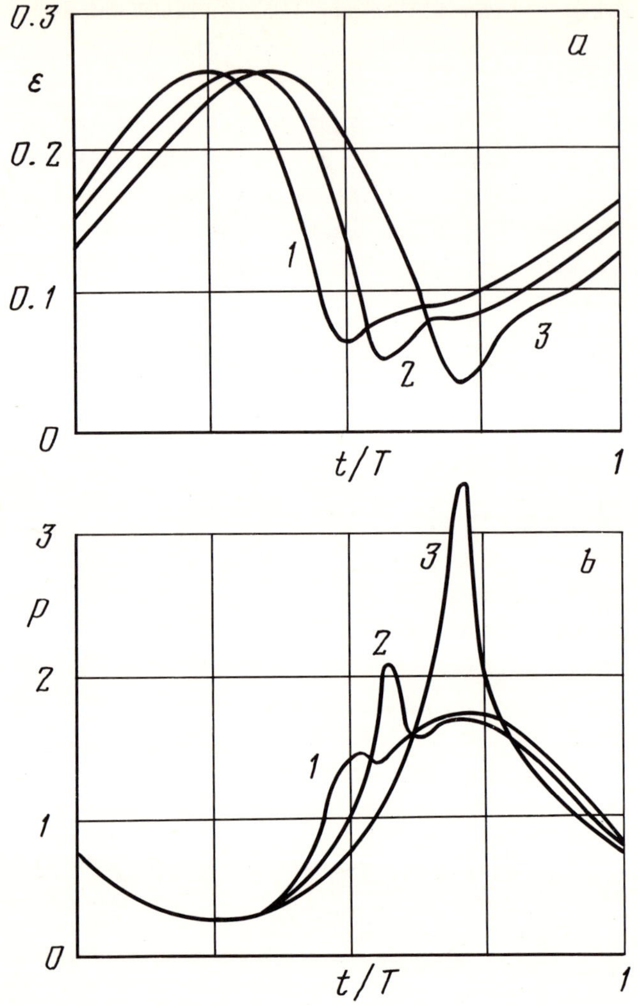

Fig. 1.

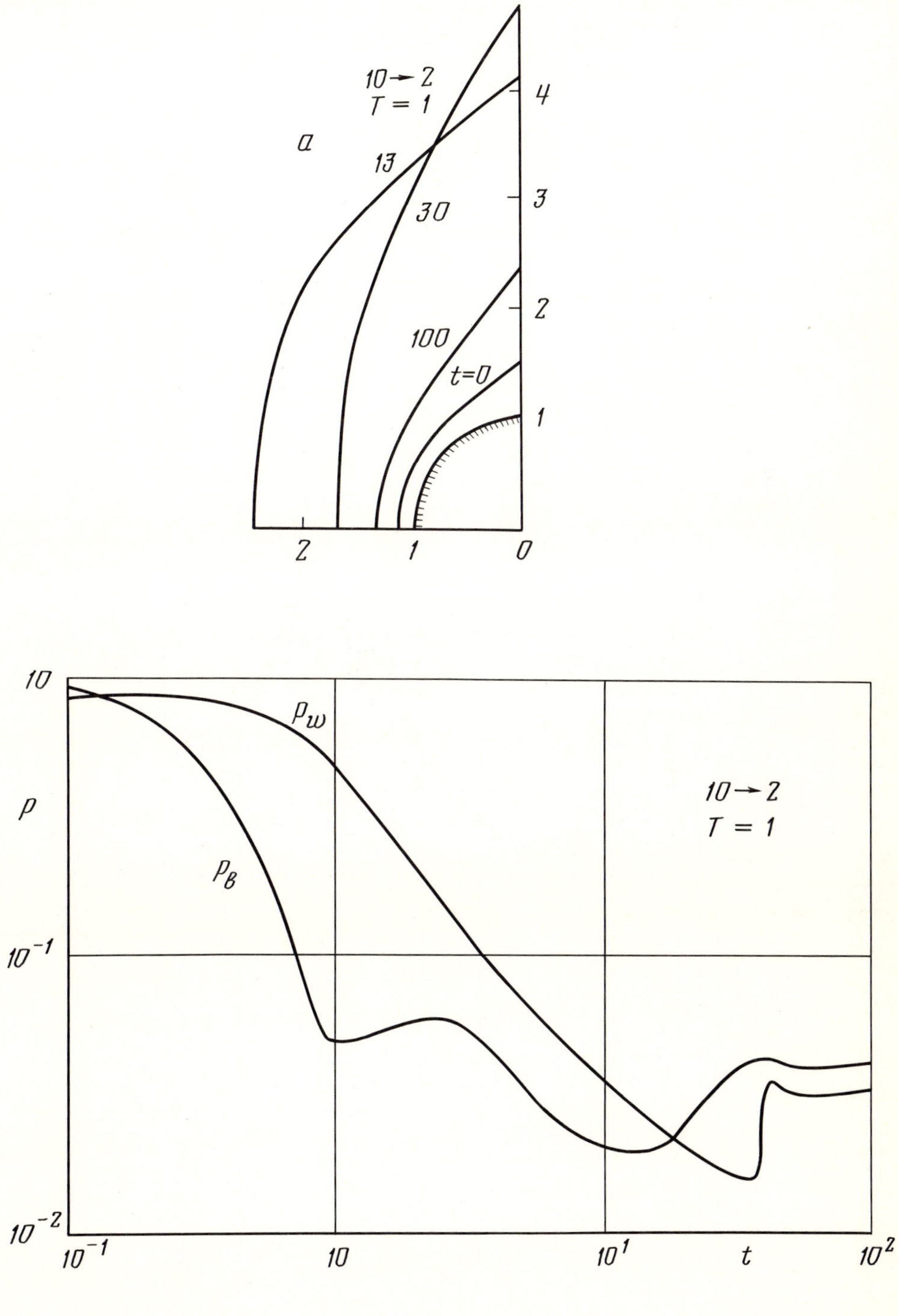
10 → 2
T = 1
a
13
30
100
t=0
4
3
2
1
2
1
0
10
p
p_w
p_B
10 → 2
T = 1
10^{-1}
10^{-2}
10^{-1}
10
10^1
t
10^2

Fig. 2.

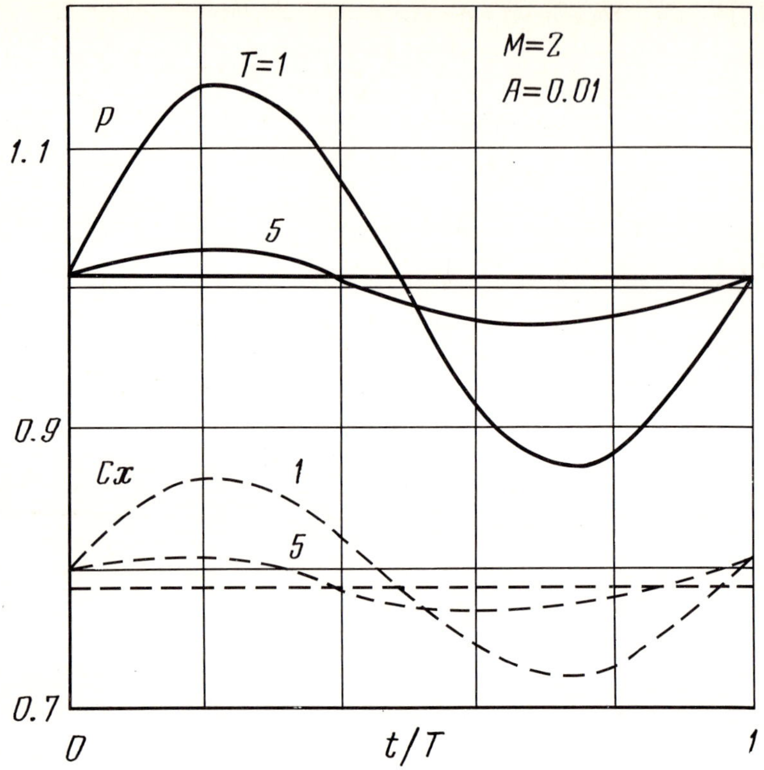

Fig. 3.

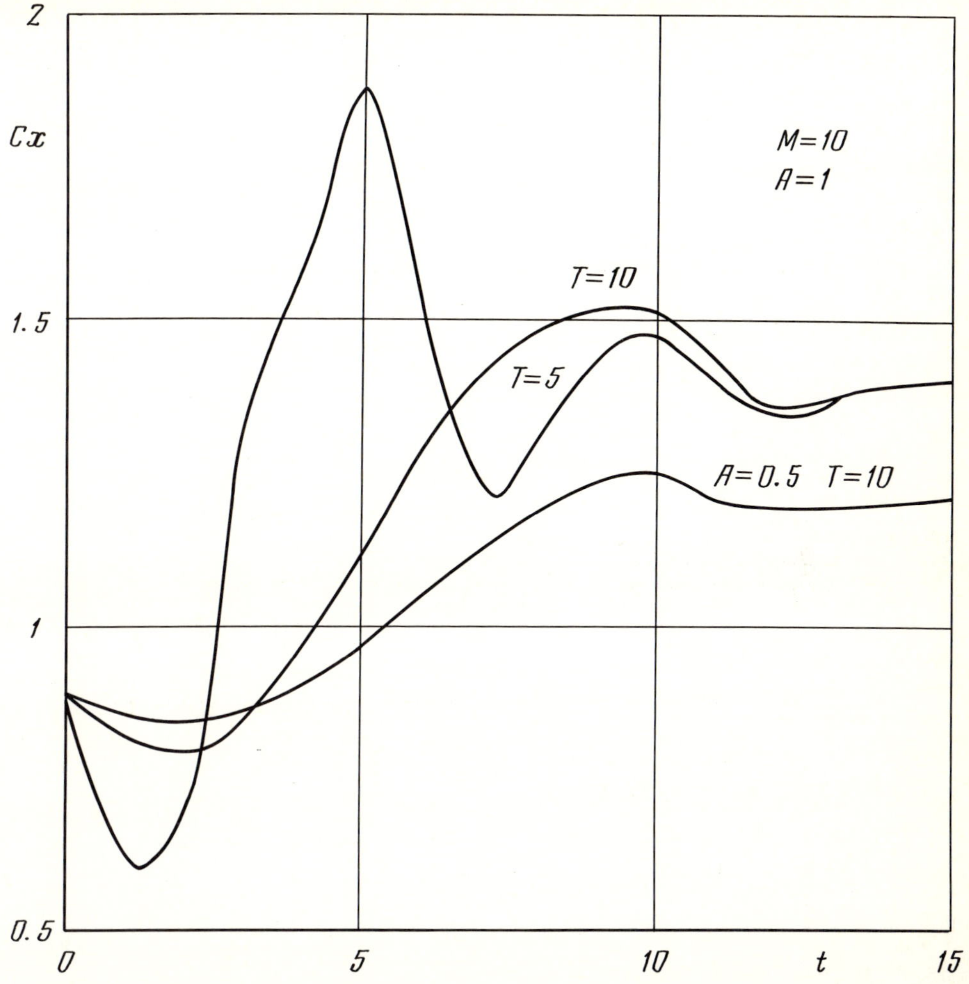

Fig. 4.

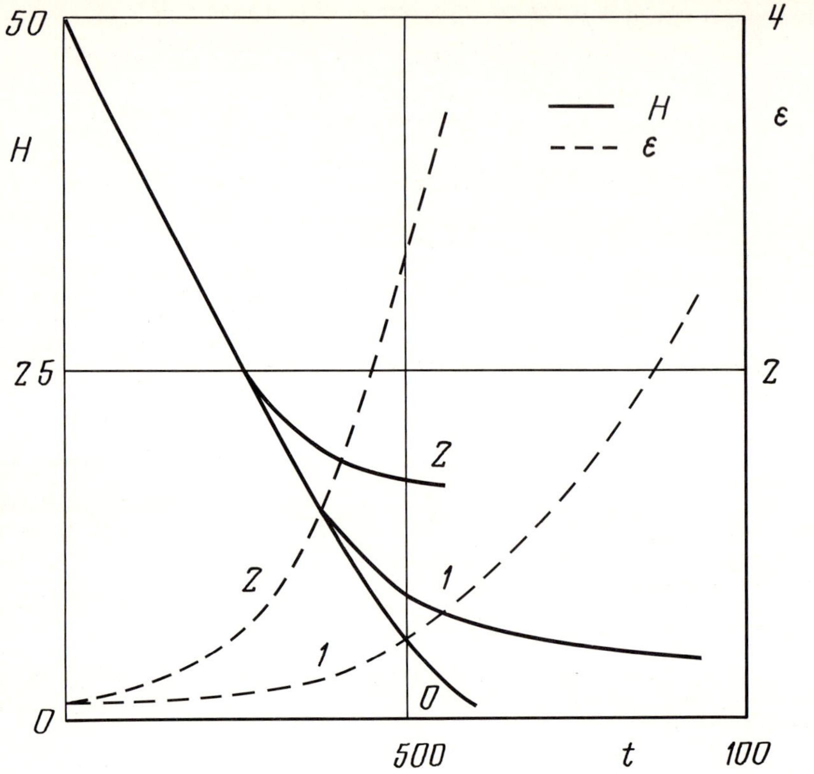

Fig. 5.

CALCULATION OF THE FLOW FIELD IN SUPERSONIC INLETS USING A BICHARACTERISTICS METHOD WITH SHOCK WAVE FITTING[†]

Joseph Vadyak and Joe D. Hoffman
Purdue University, West Lafayette, Indiana 47907, USA

A. R. Bishop
NASA Lewis Research Center, Cleveland, Ohio 44135, USA

INTRODUCTION

The purpose of this investigation was to develop a method for calculating the flow field in an axisymmetric supersonic mixed-compression inlet, such as illustrated in Figure 1, at angle of attack. Compression takes place both in the external flow about the forebody and in the internal flow inside the annulus. A bow shock wave is generated at the forebody tip. An internal shock wave emanates from the cowl lip, and makes a number of reflections before terminating in subsonic flow downstream of the geometric throat.

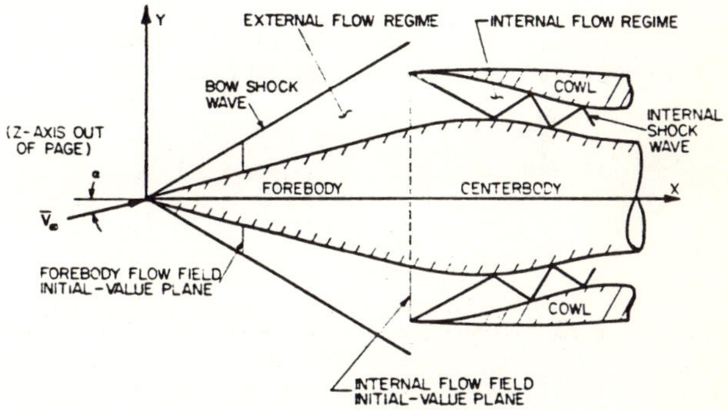

Fig. 1. Supersonic mixed-compression inlet.

The ultimate objective of this work is to develop a combined flow field analysis in which the central core of the flow field is treated as a weakly viscous region, the flow adjacent to the wall is treated as a higher-order boundary layer region, and the two regions are matched near the sonic surface. A weakly viscous flow region is a region in which the influence of molecular transport is of secondary importance as compared to the convective terms and the pressure gradient. Consequently, the transport terms may be treated as forcing functions, and the convective terms and the pressure gradient comprise the principal parts of the equations of motion. For supersonic flow, those equations are hyperbolic.

This paper presents the results of the weakly viscous flow analysis. The flow field, including molecular transport, is computed using the bicharacteristic method devised by Butler (1960). The bow shock wave and the internal shock wave are computed using a three-dimensional shock wave fitting procedure. The boundary layer analysis and the matching of the two flow regions is yet to be accomplished.

GAS DYNAMIC MODEL

The governing equations for the weakly viscous flow region are presented by Vadyak (1977). They are the continuity, momentum, and energy equations,

$$\rho u_x + \rho v_y + \rho w_z + u\rho_x + v\rho_y + w\rho_z = 0 \qquad (1)$$

$$\rho u u_x + \rho v u_y + \rho w u_z + P_x = F_x \qquad (2)$$

$$\rho u v_x + \rho v v_y + \rho w v_z + P_y = F_y \qquad (3)$$

$$\rho u w_x + \rho v w_y + \rho w w_z + P_z = F_z \qquad (4)$$

[†]This work was supported by NASA Lewis Research Center through Grants No. NGR-15-005-162 and No. NGR-15-005-191.

$$uP_x + vP_y + wP_z - a^2(u\rho_x + v\rho_y + w\rho_z) = F_e \tag{5}$$

where ρ is density, u, v and w are cartesian velocity components, P is pressure, F_x, F_y, and F_z are the conventional transport terms in the Navier-Stokes equations, a is the speed of sound, and F_e contains the thermal conduction and viscous dissipation terms. Equations of state are specified by $T = T(P,\rho)$ and $a = a(P,\rho)$, where T is temperature. The viscosity μ and the thermal conductivity κ are specified as functions of temperature only.

CHARACTERISTIC EQUATIONS

A characteristic surface is a surface on which the governing equations may be combined linearly to form an interior operator called the compatibility equation. For steady three-dimensional supersonic flow, two families of characteristic surfaces exist: stream surfaces and wave surfaces. The envelope of all stream surfaces at a point is the streamline, represented by

$$dx_i = u_i \, dt \quad (i = 1,2,3) \tag{6}$$

where t is the time of travel of a fluid particle along the streamline. The compatibility equations valid on stream surfaces are

$$\frac{d_V P}{dt} - a^2 \frac{d_V \rho}{dt} = F_e \tag{7}$$

$$\rho u \frac{d_V u}{dt} + \rho v \frac{d_V v}{dt} + \rho w \frac{d_V w}{dt} + \frac{d_V P}{dt} = u_i F_i \tag{8}$$

where $d_V(\ )/dt$ represents the directional derivative along a streamline. The envelope of all wave surfaces is the Mach conoid, and the lines of tangency between the wave surfaces and the Mach conoid are called bicharacteristics. Bicharacteristics are represented by

$$dx_i = (u_i - an_i)dt \quad (i = 1,2,3) \tag{9}$$

where n_i is the unit normal vector to the wave surface, and t is the time of travel of a fluid particle along the streamline that is the axis of the Mach conoid. The compatibility equation valid along bicharacteristics is

$$\begin{aligned} \rho a n_x \frac{d_\ell u}{dt} + \rho a n_y \frac{d_\ell v}{dt} + \rho a n_z \frac{d_\ell w}{dt} - \frac{d_\ell P}{dt} &= a n_i F_i - F_e \\ &- \rho a^2[(n_x^2 - 1)u_x + (n_y^2 - 1)v_y + (n_z^2 - 1)w_z \\ &+ n_x n_y(u_y + v_x) + n_x n_z(u_z + w_x) + n_y n_z(v_z + w_y)] \end{aligned} \tag{10}$$

where $d_\ell(\ )/dt$ represents the directional derivative along a bicharacteristic. The terms in brackets [ ] in equation (10) represent differentiation in the wave surface in a direction normal to the bicharacteristic direction. Those terms are called cross-derivatives.

At any point in the flow field there exist an infinite number of stream surfaces and wave surfaces. The number of independent compatibility relations cannot exceed the number of independent equations of motion. Rusanov (1963) showed that equations (7) and (8) applied along a stream surface and equation (10) applied along three different wave surfaces form an independent set.

Butler (1960) developed a parametric form for representing the bicharacteristics and the compatibility relation applicable along them. For the bicharacteristics,

$$dx_i = (u_i + c\alpha_i \cos\theta + c\beta_i \sin\theta)dt \qquad (i = 1,2,3) \tag{11}$$

where α_i and β_i are unit vectors that form an orthonormal set with u_i/q, θ is a parametric angle denoting a particular bicharacteristic and has the range $0 \leq \theta \leq 2\pi$, and c is the velocity of divergence of the Mach conoid normal to the streamline

$$c^2 = q^2a^2/(q^2 - a^2) \tag{12}$$

where q is the velocity magnitude. The corresponding parametric form of the compatibility equation is

$$\frac{d_\ell P}{dt} + \rho c(\alpha_i \cos\theta + \beta_i \sin\theta)\frac{d_\ell u_i}{dt} = (c^2/a^2)(F_e - an_iF_i)$$

$$- \rho c^2(\alpha_i \sin\theta - \beta_i \cos\theta)(\alpha_j \sin\theta - \beta_j \cos\theta)\frac{\partial u_i}{\partial x_j} \tag{13}$$

The parametric form of the unit normal n_i is

$$n_i = (a/c)(cu_i/q^2 - \alpha_i \cos\theta - \beta_i \sin\theta) \qquad (i = 1,2,3) \tag{14}$$

Butler also developed a noncharacteristic relation for use in the numerical integration procedure. It is given by

$$\frac{d_V P}{dt} = (c^2/a^2)F_e - (c^2/q^2)u_iF_i - \rho c^2(\alpha_i\alpha_j + \beta_i\beta_j)\frac{\partial u_i}{\partial x_j} \tag{15}$$

NUMERICAL INTEGRATION PROCEDURE

An inverse marching scheme is employed in which the solution is obtained on space-like planes of constant x and on space curves defined by the intersections of the internal shock wave with the solid boundaries. The distance between solution planes is determined by the Courant-Friedrichs-Lewy (CFL) stability criterion. In the vicinity of a shock wave intersection with a solid boundary, the step size is chosen so that the entire intersection falls between two adjacent solution planes. The bow shock wave, the centerbody and the cowl, and planes of flow symmetry, constitute the boundaries of the computational flow regime. The initial data are specified on a plane of constant x. The solution is propagated along the streamlines through the initial-data points. The external flow about the forebody is computed first. Streamlines are inserted as the captured mass flow increases to retain a well dispersed mesh. Internal streamlines are deleted as required to maintain the desired number of computational points. The internal flow field can be computed with or without fitting the internal shock wave. Both the centerbody and the cowl are axisymmetric. The contours may be specified by tabular data or cubic spline fits. The molecular transport terms may be included in the external flow about the forebody and in internal flow without shock wave fitting, but not in internal flow with shock wave fitting.

Four major types of unit processes are employed: interior point, solid boundary point, field-shock wave point, and solid body-shock wave point. A detailed discussion of the unit processes is given by Vadyak (1977). The computational network for an interior point is illustrated in Figure 2. Points (1) to (4) are the intersection points of four rearward running bicharacteristics with the initial-value plane, point (5) is a point on the initial-value plane, and point (6) is the intersection point of the forward running streamline with the solution plane. The flow properties at points (1) to (5) are determined from least squares fit quadratic bivariate interpolating polynomials. The second-order Euler predictor-corrector method is employed to integrate the characteristic and compatibility equations. For the first-order predictor, all of the coefficients and nonhomogeneous terms are calculated at points (1) to (5). For the second-order corrector, all of the coefficients and nonhomogeneous terms are calculated as averages between points (1) to (5) and point (6). However, the cross-derivatives cannot be calculated at point (6) without using information at neighboring points in the solution plane, which violates the domain of dependence of point (6). Butler (1960) developed a procedure that explicitly eliminates the cross-derivatives at point (6) from the finite difference equations. Equation (13) is written for the four values of $\theta = 0, \pi/2, \pi$, and $3\pi/2$, and three linear combinations of those four equations and equation (15) are formed that eliminate the cross-derivatives at point (6).

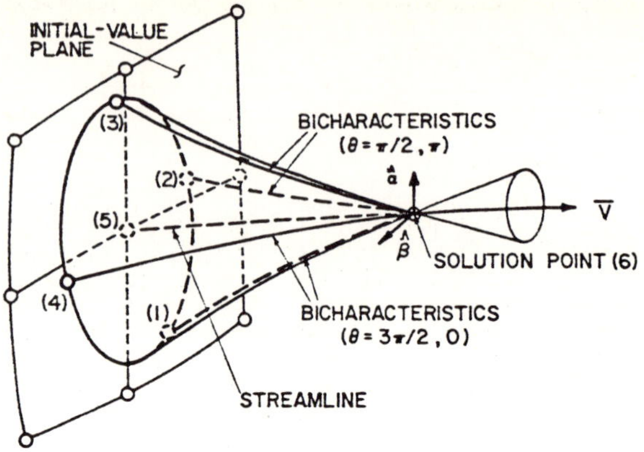

Fig. 2. Interior point finite difference grid.

The computational network for a bow shock wave point is illustrated in Figure 3. Point (2) is the solution point. The flow properties at point (2) on the upstream side of the shock wave are known. The intersections of the shock wave with the initial-value plane and the solution plane define space curves (A) and (B). Point (3) is a predetermined interior solution point which defines the meridional plane, having the angle θ with respect to the y-axis, in which the bow shock wave solution point lies. Point (1) is the intersection point of a bicharacteristic with the initial-value plane, and point (4) is the intersection point of space curve (A) with the meridional plane. A local cartesian coordinate system is employed to describe the shock wave surface. As illustrated in Figure 4, the orientation of the shock wave surface is specified by the unit vector $\hat{n}_s$ which is normal to the shock wave surface and the two unit vectors $\hat{\ell}$ and $\hat{t}$ which are tangent to the surface. $\hat{t}$ lies in the meridional plane and subtends the angle ϕ with the x'-axis. $\hat{\ell}$ lies in the transverse plane and subtends the angle α with the z'-axis.

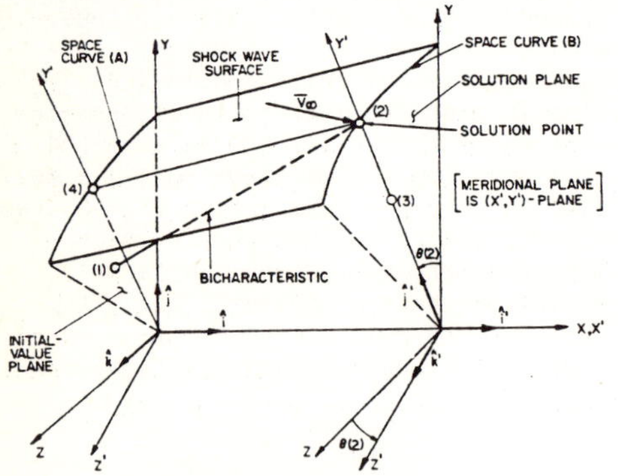

Fig. 3. Bow shock wave point finite difference grid.

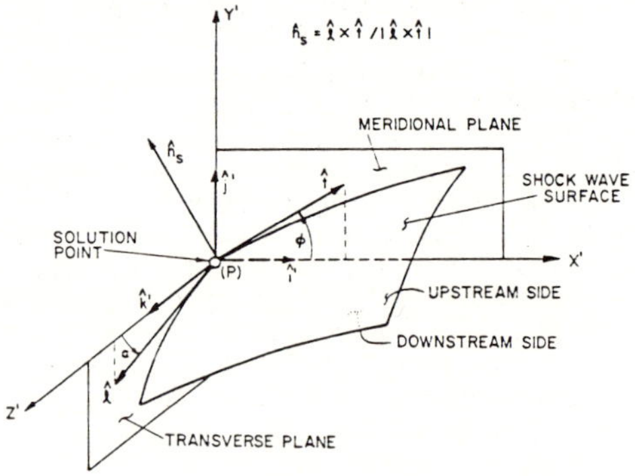

Fig. 4. Shock wave surface unit vectors.

To achieve second-order accuracy in the shock wave point unit process, a predictor-corrector method with global iteration is employed. The interior points and solid boundary points are determined first. Then, the predictor is applied for all of the shock wave points. Finally, the corrector is applied for all of the shock wave points. The tangential unit vectors are given by

$$\hat{\ell} = \sin\alpha\,\hat{j}' + \cos\alpha\,\hat{k}' \quad \text{and} \quad \hat{t} = \cos\phi\,\hat{i}' + \sin\phi\,\hat{j}' \tag{16}$$

and the shock wave normal unit vector $\hat{n}_s$ is given by

$$\hat{n}_s = (\hat{\ell} \times \hat{t})/|\hat{\ell} \times \hat{t}| \tag{17}$$

The value of α is determined from

$$\alpha = tan^{-1}(\frac{1}{r}\frac{dr}{d\theta}) \tag{18}$$

where $r(\theta)$ is obtained by quadratic univariate interpolation in θ along space curves (A) and (B). For the predictor, $\alpha(2)$ is set equal to $\alpha(4)$, and for the global corrector, $\alpha(2)$ is determined along space curve (B). The global iteration is terminated when α converges at each shock wave solution point. The value of ϕ is determined by quadratic univariate interpolation in θ along space curves (A) and (B). For the initial pass through the predictor, $\phi(2)$ is set equal to $\phi(4)$. For the initial pass through the global corrector, $\phi(2)$ from the previous iteration is used. Then, $\phi(2)$ is determined iteratively by the secant method as discussed below. The radial position of point (2) is obtained from

$$r(2) = r(4) + [x(2) - x(4)]\ tan\left\{\frac{1}{2}\ [\phi(2) + \phi(4)]\right\} \tag{19}$$

The local Hugoniot relations are applied at point (2) to obtain the downstream flow properties u(2), v(2), w(2), P(2), and $\rho(2)$. Point (1) is then located iteratively, and the corresponding flow properties determined by interpolation. The parametric unit vector β_i is maintained in the meridional plane of the solution point, and α_i is chosen to be orthogonal to β_i and u_i. The wave surface compatibility relation, equation (13), for $\theta = \pi/2$ is applied between points (1) and (2) to determine the pressure P*(2). The difference between P(2) and P*(2) is driven to within a specified tolerance of zero using the secant method with iteration on the shock wave angle $\phi(2)$.

A solid body-shock wave point unit process is used to determine the flow properties downstream of the shock wave at a point where the shock wave intersects a solid boundary. This unit process is used to determine the solution for the points on the downstream side of the cowl lip shock wave and on the centerbody or cowl on the downstream side of an internal reflected shock wave. For the internal shock wave reflection, the flow properties downstream of the incident shock wave, which constitute the upstream flow properties for the reflected shock wave, are computed by a modified field-shock wave point unit process.

The unit process employed to compute the solution at a shock wave point in the internal flow field is similar to the bow shock wave point unit process, except that the flow properties upstream of the shock wave are determined by a modified interior point unit process. Moreover, modifications to the internal flow shock wave point unit process must be made when the solution point lies on or close to a solid boundary. In some situations the interior point and solid boundary point unit processes must be applied in a modified form. One such instance occurs when the Mach cone, with apex at the interior solution point, intersects not only the initial-value plane but also the internal shock wave and/or a solid boundary. The bicharacteristic intersection points with the shock wave and/or the solid boundary must be determined. The corresponding flow properties are determined by quadratic trivariate interpolating polynomials.

RESULTS

Selected results are presented by Vadyak (1977). Some comparisons with experimental data are made. Results are compared with the numerical procedure developed by Presley (1975). Transport processes are neglected in all of these results.

Internal flow calculations were performed for the Mach 3.5 mixed-compression inlet described by Syberg and Hickcox (1973). This inlet has a conical forebody. Consequently, the initial data at the cowl lip axial station were obtained by Jones' method (1968). Extensive boundary layer removal is employed in this inlet.

Since the present analysis does not compute the boundary layer nor account for boundary layer removal, good agreement between computed and experimental results cannot be expected in regions of high viscous interaction.

Figure 5 presents the pressure ratio (static/free-stream stagnation) versus the distance ratio x/R_C (axial position/cowl lip radius) for the design condition of $M_\infty = 3.5$, zero centerbody translation ($\Delta x/R_C = 0$), and zero angle of attack ($\alpha = 0$ deg), for three cases: (1) the 2-D method of characteristics program of Anderson, et al. (1976) with 50 radial points across the annulus, (2) the present method with 11 radial points, and (3) the present method with 21 radial points. For the case with 11 radial stations (9 streamline points and an upstream and downstream shock wave point), good overall agreement is observed. A slight smearing downstream of the second intersection of the shock wave with the centerbody, and a shifting of the locations of the shock wave-solid boundary intersections are present. For the case with 21 radial stations, the agreement is excellent. The only differences are in the regions enclosed by the circles. The pressure distribution behind the second shock wave-centerbody reflection is correctly predicted, and the locations of the shock wave-solid boundary intersections agree very well. Comparisons of numerical and experimental data are shown in Figure 6. Generally speaking, good agreement is observed. The best agreement is obtained away from the regions of boundary layer removal.

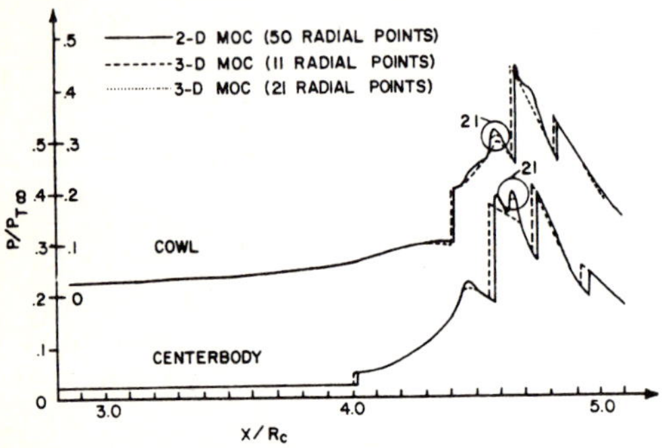

Fig. 5. Comparison of two- and three-dimensional results for design point.

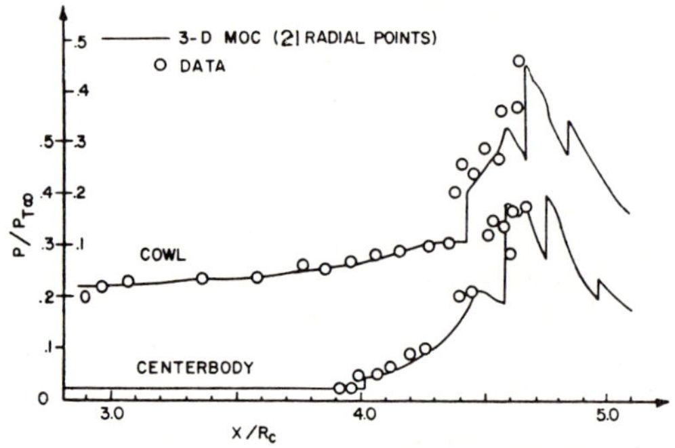

Fig. 6. Comparison of computed and experimental data for design point.

At a given free-stream Mach number, the centerbody must be translated forward as the angle of attack is increased to maintain supersonic flow through the geometric throat. Numerical results were obtained for two off-design conditions: (1) $M_\infty = 2.5$ with a centerbody translation $\Delta x/R_C = 0.855$, and (2) $M_\infty = 3.3$ with $\Delta x/R_C = 0.356$, both for $\alpha = 0$, 3.0, and 5.0 deg. For all of the computations, 21 circumferential stations were employed in the computed sector (half-plane) with 11 radial stations (9 streamline points and an upstream and downstream shock wave point).

Figure 7 presents the results for off-design case (1) for $\alpha = 3.0$ deg. Generally speaking, good overall agreement between theory and experiment is obtained except in regions of high viscous interaction and boundary layer bleed. The maximum deviation in mass flow rate at any solution plane from the initial mass flow rate is 0.44 percent. Figure 8 presents the computed pressure distributions on the centerbody for $\alpha = 0$, 3.0, and 5.0 deg.

Figure 9 presents the results for off-design case (2) for $\alpha = 3.0$ deg. Fairly good overall agreement between the analysis and experiment is obtained until regions of high viscous interaction and boundary layer removal are reached. The maximum deviation of the mass flow rate at any solution plane is 0.67 percent.

The present method was compared to the shock capturing method developed by Presley (1975). That method employs the second-order accurate MacCormack method (1969). Figure 10 compares the pressure distributions given by the method of characteristics scheme to those given by the shock capturing technique for the case of $M_\infty = 3.3$, $\alpha = 3.0$ deg, and $\Delta x/R_C = 0.356$. The shock capturing algorithm employed 11 circumferential stations and 21 radial stations. For the most part, good agreement between the two analyses is obtained. In the method of characteristics solution, however, the shock wave-solid boundary intersections are more sharply defined. This effect is to be expected, since in the shock-capturing technique shock waves are not discretely fitted but rather are smeared over a number of mesh points.

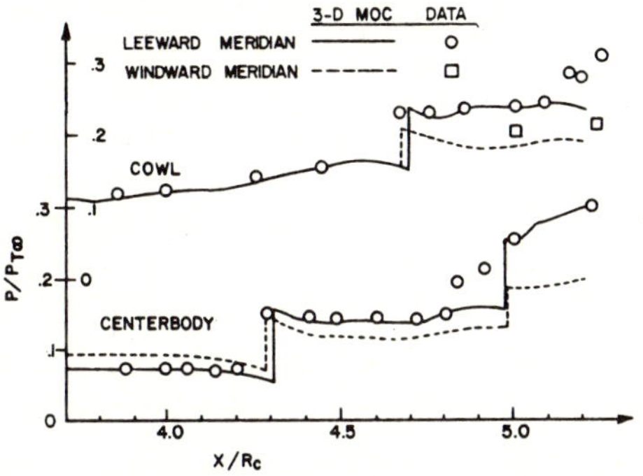

Fig. 7. Comparison of computed and experimental data for off-design case (1).

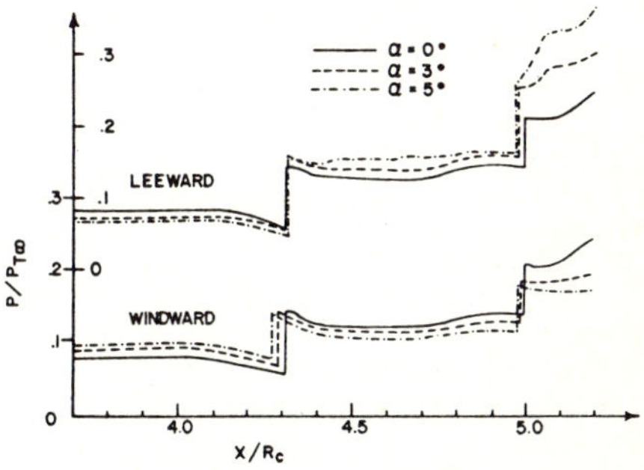

Fig. 8. Computed pressure distributions for several angles of attack for off-design case (1).

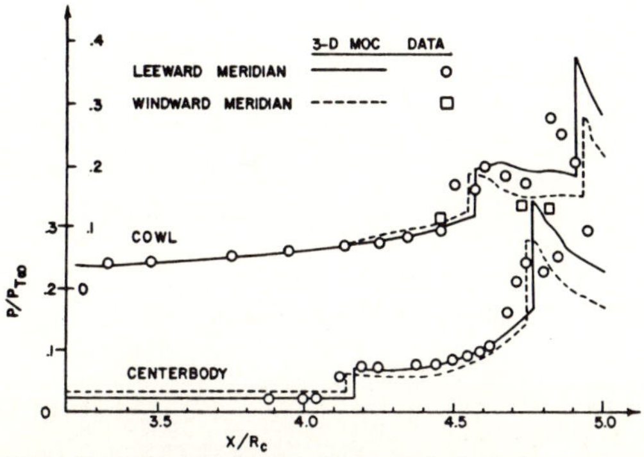

Fig. 9. Comparison of computed and experimental data for off-design case (2).

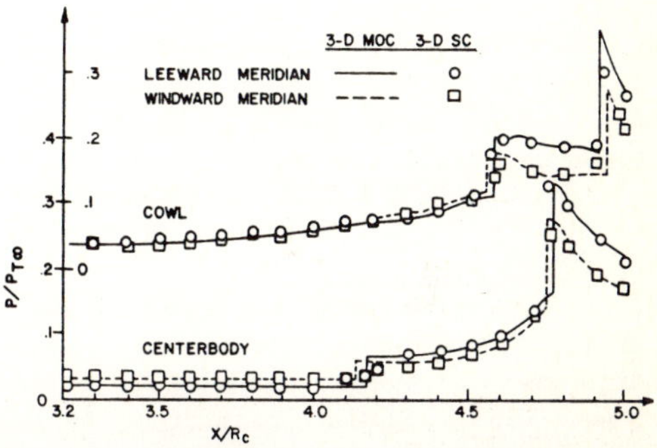

Fig. 10. Comparison of method of characteristics with shock capturing technique for off-design case (2).

REFERENCES

Anderson, B. H., Tassa, Y., and Reshotko, E., AIAA Paper 76-426 (1976).
Butler, D. S., Proc. Roy. Soc. (London), A, 255, 232-252 (1960).
Jones, D. J., National Research Council of Canada Report LR-507 (1968).
MacCormack, AIAA Paper 69-354 (1969).
Presley, L. L., AIAA Paper 75-1214 (1975).
Rusanov, V. V., Zh. Vych. Mat. i Mat. Fiz., 3, 508-527 (1963).
Syberg, J. and Hickcox, T. E., NASA CR-2187 (1973).
Vadyak, J., Ph.D. Thesis, Purdue University (1977).

BISPECTRAL MEASUREMENTS IN TURBULENCE COMPUTATIONS

C. W. Van Atta
University of California, San Diego
La Jolla, California U. S. A.

INTRODUCTION

Attempts to study the dynamics of spectral energy transfer and wavenumber interactions in homogeneous turbulence have led to the development of new numerical techniques in both experimental research and theoretical computations. The experiments are subject to basic constraints which severely limit the kinds of information which can be obtained. Turbulence simulations based on the three-dimensional Navier Stokes equations are not subject to some of these basic constraints. In simulations, one is not limited to one-dimensional wavenumber space and fluctuating components of velocity and velocity gradients in all three directions are not difficult to determine simultaneously. Many of the quantities of physical interest cannot in fact be measured experimentally. There is a need for close cooperation between experimentalists and those performing turbulence simulations. The experimentalist needs to be able to compare those quantities which he can measure with the results of carefully done detailed computations, before going on to much more difficult measurements of other quantities. However, up till the present, none of the simulations have yielded results which can be compared with the kind of experimental results to be discussed here. The purpose of this paper is to discuss the experimental information presently being obtained and to present some of the more physically interesting results, in the hope of stimulating a dialogue between those working on the numerical methods and experimentalists.

THE EXPERIMENTAL APPROACH

The only previous results for spectral energy transfer in homogeneous turbulence are those for the net rate of transfer of energy to a given one- or three-dimensional wavenumber from all other wavenumbers (see e.g., Van Atta and Chen, 1968). Similar results can be obtained from three-dimensional numerical simulations. However, they give no information about the physics of the problem of interest here, which is the detailed balancing of the triadic wavenumber interactions in turbulence, which occur between triplets of wavenumbers. To determine the strength of such interactions, one must measure third-order spectra, which are commonly called bispectra.

Two different experimental approaches have been outlined recently in order to attack this problem. The first is described by Lii et al. (1976) and the second was developed by Wilson (1974).

In the first approach, the rate of transfer of kinetic energy from one one-dimensional wavenumber k to another one ℓ is in principle determined from the sum of measurements of six different bispectra. Only two of these have been measured to date, and the remaining ones are more difficult to measure. The general behavior to be expected of the latter, and even their algebraic signs, are unknown. Unfortunately, all terms in the sum must be measured before one can draw any conclusions about the magnitude and sign of the net measured energy transfer. Clearly, if results of numerical simulations were available for the bispectra, and if these results were consistent with the few available measurements, then one could confidently draw useful conclusions regarding the net transfer between two arbitrary wavenumbers. This would help to clarify the physical processes involved, and might furnish a means of critical appraisal of various approximations used in analytical closures adopted for use in turbulence computations.

Progress to date using this approach has been summarized by Helland et al. (1978). No useful physical interpretations regarding spectral transfer have as yet arisen from it, and we shall not consider it further here.

Here we consider in some detail the results of an alternative approach developed by Wilson (1974). This formulation has the advantage of yielding physically interpretable results from a much more restricted class of measurements. In fact, some restricted conclusions about the magnitude and direction of the spectral transfer between two different wavenumbers may be drawn from measurements of single bispectra of the type needed for the first approach.

Following Wilson, with the velocity field (u is the downstream velocity component) represented by its Fourier coefficients u_k,

$$u(x) = \Sigma u_k = \Sigma (C_k/k) \sin(kx + \theta_k)$$

the time rate of change of the averaged kinetic energy per unit volume in the ℓth Fourier component of the x velocity component is

$$\frac{\partial}{\partial t}\left(\frac{\overline{u_\ell^2}}{2}\right) = \overline{-u_\ell \, \Sigma u_m \frac{\partial \Sigma u_k}{\partial x}} + \overline{\nu u_\ell \nabla^2 \Sigma u_k} \; \ldots \qquad (1)$$

Here, we are interested in the inertial transfer term (first term on the right hand side) which describes the net rate of arrival of energy at wavenumber ℓ due to inertial transfer by interactions between downstream velocity components. The last term describes the rate of loss of downstream energy at wavenumber ℓ due to viscosity.

Each term in the double sum of the inertial transfer term can be written as the sum of three contributions, each one of which arises from interactions between three different wavenumbers. These contributions, called types 1, 2, and 3 by Wilson, are,

$$1. \quad -\frac{C_\ell C_m C_k}{4\ell m} \cos(\theta_\ell - \theta_m + \theta_k) \qquad \text{if} \quad \ell - m + k = 0$$

$$2. \quad -\frac{C_\ell C_m C_k}{4\ell m} \cos(\theta_\ell - \theta_m - \theta_k) \qquad \text{if} \quad \ell - m - k = 0 \qquad (2)$$

$$3. \quad +\frac{C_\ell C_m C_k}{4\ell m} \cos(\theta_\ell + \theta_m - \theta_k) \qquad \text{if} \quad \ell + m - k = 0$$

In terms of A_k and B_k, the cosine and sine coefficients obtained from a finite Fourier transform of a record of $\frac{\partial u}{\partial x} \sim \frac{1}{U}\frac{\partial u}{\partial t}$, excluding the factor $\pm 1/\ell m$ the above expressions for all three types are just equal to the real part of the bispectrum of $\partial u/\partial x$, where the bispectrum is given, for type 2, by

$$B(k,m) = (A_k + iB_k)(A_\ell - iB_\ell)(A_m + iB_m) \qquad \ell = k + m$$

The other two interactions produce the same numerical values of the bispectrum for given values of k, ℓ, and m. Thus the bispectrum need be computed only once, and the differences in the three interactions arise from the factor $1/\ell m$

which is constrained by the different relations between the three interacting wavenumber for each of the interactions.

The division into three types of interactions has no particular physical significance, but the constraints on ℓ, m and k can be interpreted in terms of the relative scale sizes involved in each interaction. The division does furnish a convenient ordering for obtaining various sums relating to transfer between differing scales or transfer to a particular wavenumber.

PRESENT EXPERIMENTS AND DATA ANALYSIS

The data discussed in the present analysis were obtained in three experiments, two in the atmospheric boundary layer and one in laboratory grid turbulence. Descriptions of the experiments and details of the numerical methods used to obtain the bispectra may be found in Wilson (1974), Lii et al. (1976) and Helland et al. (1978). Wilson's data were obtained over water, while the UCSD atmospheric data were obtained over land. Features common to both of these sets of data may reveal universal characteristics of turbulence, only slightly influenced by boundary conditions. The comparison with the grid data then furnishes further information on universal features of bispectral transfer in turbulent flows as well as a comparison with a flow with a much lower value for the turbulence Reynolds number (of order 10 as compared with 10^3 for the atmosphere).

Both sets of atmospheric data and the grid data covered 3-4 minutes of real time, and a total of roughly 10^6 sampled data points were used for the bispectral analysis. Some important parameters relating to the experimental conditions, data sampling, and numerical analysis are given in Table 1. To reduce the variance, bispectral estimates over shorter blocks of 2048 or 4096 samples were averaged, after estimates were averaged over square neighborhoods within each block. For Wilson's data the grouping scheme was chosen to give estimates of the means of the triple products at approximately equally spaced values of the logarithm of the wavenumber, while in the UCSD data estimates at equally spaced values of the wavenumber were obtained.

Evaluation of type 2 interactions was much more straightforward and convenient using equally spaced data than with logarithmically spaced data. All of the basic bispectral data used here are available as tabulated values in the references previously cited. The contributions of the three types of interactions were computed by dividing the elements of each bispectral array by the proper values of ℓm

TABLE 1
Parameters of the Experiments

	U (m/s)	z (height) (m)	Sampling Rate (Hz)	Kolmogorov Microscale (cm)	Resolved Bispectral Frequency Interval
UCSD Atmospheric boundary layer	3.76	5.66	4172	0.08	25 Hz or 15 Hz
UCSD Grid turbulence	7.7	----	4172	0.05	25 Hz
UBC Atmospheric boundary layer	7.8	1.5-2.0	6000	0.07	logarithmic

in each case. For type 1, the horizontal and vertical axes of the array are chosen as ℓ and k, respectively. In this case, $m = \ell + k$ increases proceeding up a column in the array. For type 2, the corresponding axes are chosen as m and k, and the values for a given ℓ are obtained along a diagonal line in the array with $\ell = k + m$. For type 3, the horizontal and vertical axes are chosen as ℓ and m, respectively. In this case, $k = \ell + m$ increases proceeding up a column in the array. One now has three arrays describing the three types of interactions, and the next step in the analysis depends upon the kind of information desired. For example, for a fixed value of ℓ one can examine (as Wilson did) the rate of transfer of energy to wavenumber ℓ from each of the three types of interactions individually, or the net rate of transfer to wavenumber ℓ due to the sum of transfers for the three types of interaction. Or, one can examine the transfer to wavenumber ℓ as a function of the value of another contributing wavenumber. Wilson did this for each of the interactions using k or m for the second wavenumber, but did not attempt to determine the net transfer as function of a common second wavenumber, as will be done here.

EXPERIMENTAL RESULTS

If one is interested only in the total net transfer to wavenumber ℓ, then for a fixed value of ℓ one merely sums over all possible values of k and m. The summing proceeds vertically in the arrays for type 1 and 3, and along a diagonal for type 2. The results of this operation for both the UCSD atmospheric high frequency "global" data and grid turbulence data are shown in Fig. 1. The individual contributions from each type of interaction as well as their sum are shown. The data are presented in Figs. 1 and 2 as functions of frequencies ℓ, m and k in units of Hz (cps). Corresponding wavenumbers are equal to $2\pi/U$ times the corresponding frequency in Hz under the "frozen turbulence" hypothesis. The results for the two flows are quite similar in appearance, and both exhibit the same main features as found by Wilson for atmospheric turbulence over water. The contributions for type 1 and 2 interactions are always positive, and for type 3 are always negative. For the atmospheric data, for frequencies greater than 20 Hz, the type 1 and type 3 interactions decrease fairly rapidly with increasing frequency, while those of type 2 decrease much more slowly. The low frequency UCSD atmospheric analysis (from 15 to 150 Hz) yielded similar results, with indication of a maximum value for type 2 at a frequency of 30 Hz, similar to the maximum found by Wilson at a slightly higher frequency. The sum is negative for low frequencies, indicating a net loss of energy at low wavenumbers. The sum becomes positive for frequencies greater than about 45 Hz, exhibits a broad positive maximum up to 150 Hz, and then decreases slowly toward zero as the frequency increases. The center of the broad maximum occurs near the frequency (80 Hz) at which the dissipation spectrum reaches its maximum value. The present atmospheric data exhibit considerably smoother variation with frequency than do Wilson's data, especially for type 2 and type 3. This may be due to the different methods of two-dimensional frequency averaging used to reduce the variance for the two experiments. The grid turbulence data exhibit all the same features noted for the atmospheric data, with corresponding features being shifted to higher frequencies for the grid data as would be expected applying the usual Kolmogorov scaling. If f_k is the Kolmogorov frequency corresponding to the Kolmogorov length (see Table 1), the crossover point from negative to positive net energy transfer occurs at f/f_k equal to 0.06 for the atmospheric boundary layer and 0.08 for the grid turbulence. The plateau region of positive energy transfer would also scale with Kolmogorov variables, since it is found to scale with the dissipation peak. Both of these features are similar to the behavior found for the net spectral energy transfer term T(k) for grid turbulence by Van Atta and Chen (1968) for the more familiar net spectral energy transfer term T(k), which has a crossover point of $f/f_k = 0.2$ and for $T(k) \geq 0$ scales with Kolmogorov variables. The decrease in the dimensionless crossover point with increasing Reynolds number observed here was also found in T(k) for laboratory flows by Helland et al. (1977). The present net spectral transfer term due to interactions of the downstream velocity component thus bears many similarities to T(k), which contains contributions from the cross stream components as well. Both terms show a net loss of energy at low wavenumbers and a net gain at high wavenumbers where most of the dissipation occurs. However, this sum of the three types of interactions does not tell us anything about the relative strengths of interactions between different wavenumbers. Such information can be obtained for fixed ℓ if no summing is done over k or m. For fixed ℓ, one can plot the values in the columns of the arrays for type 1 and type 3 interactions versus k and m, respectively, and the values on diagonals of the type 2 array versus k or m. One thus determines the relative contributions due to each type of interaction between wavenumber ℓ and wavenumber k or m. It is found that in each case the individual contributions in the array are almost always of the same sign as the net transfer for that type of interaction, and the largest contributions are always of the same sign. For type 1, at low and intermediate frequencies, contributions from both larger $(k, m > \ell)$ and smaller $(k, m < \ell)$ wavenumbers are important, while for the highest wavenumbers the main contribution comes from smaller wavenumbers. However, for a large intermediate range the contributions have a maximum near ℓ, and are large and of the same sign for both larger and smaller wavenumbers. This implies large transfers of energy from smaller to larger scales and is not consistent with the usual picture of an energy cascade down the spectrum. A similar picture is found for type 3, although relatively less energy is transferred from higher to lower wavenumbers. By the manner in which they are defined, the type 2 interactions can transfer energy to wavenumber ℓ only from wavenumbers k and m which are less than ℓ, because $\ell = k + m$ for this interaction. At this point one should remember that the individual interactions of the three types do not have physical significance, and that arguments about energy transfer based on any one type of interaction alone may not have any meaning. Since only the sum of the three types of interactions is meaningful when considering the net spectral transfer to wavenumber ℓ, it seems likely that only this sum will be physically interpretable when examining transfer between two arbitrary wavenumbers. To investigate this hypothesis, the net energy transfer between wavenumbers ℓ and k was determined by summing the terms having the same (or as nearly as possible the same) values of ℓ and k for the three types of interactions. Each sum thus consisted of at most three terms, and a minimum of one for some of the computations when $k = \ell$. Type 1 always contributes for each value of k. Type 2 contributes beginning with the second value of k because of the constraint $\ell = m + k$, but does not contribute beyond $k = \ell - m$. Because of the constraint $k = m + \ell$, type 3 contributes only for $k > \ell$. The results for several values of ℓ are shown in Fig. 2. The value of the sum of the contributions of the three types of interactions, $S(k;\ell)$, has been multiplied by k so that equal areas on the semilogarithmic plot correspond to equal contributions to the energy transfer.

The sum $S(k;\ell)$ is the net rate of energy transfer to wavenumber ℓ from wavenumber k, due to inertial interactions of triplets of wavenumbers. The sum is taken over all possible values of the third wavenumber, m, subject to the constraints given in equation (2). According to equation (1) the wavenumbers ℓ and m are associated with the velocity, while wavenumber k corresponds to the velocity derivative. A positive value of S denotes energy transfer from wavenumber k to wavenumber ℓ, while a negative value of S denotes a transfer of energy from wavenumber ℓ to wavenumber k. The shapes of the curves for all values of ℓ are remarkably similar. S is always positive for $k < \ell$, is nearly zero for $k = \ell$, and is always negative for $k > \ell$. Thus, the net energy transfer due to the interactions is always from lower to higher wavenumbers. The net transfer for $k = \ell$ must be essentially zero if the present interpretation of S is physically meaningful. The rapid change from positive to negative values of S in the immediate neighborhood of $k = \ell$ is, of course, associated with the fact that for $k \leq \ell$ the type 3 interaction does not contribute to S, while for $k > \ell$ the type 2 interaction does not contribute and the type 3 interaction is much larger than the type 1. The positive and negative contributions are sharply peaked at wavenumbers which differ from ℓ by at most a factor of two. For the atmospheric data, the maximum transfer rate to wavenumber ℓ occurs at $k/\ell \sim 0.7$-0.8 for wavenumbers in the inertial subrange and near the dissipation maximum in the energy spectrum, the ratio increasing somewhat for larger values of ℓ. The maximum rate of transfer from wavenumber ℓ occurs at $k/\ell \sim 1.3$-1.4 for the same range of ℓ, with the ratio decreasing somewhat for larger ℓ.

The area between the positive part of the kS curve and the zero transfer level represents all energy being transferred to ℓ, while the area between the negative part of the kS curve and the zero transfer axis gives the total energy being transferred from wavenumber ℓ to larger wavenumbers. Nearly all of the transfer to ℓ comes from a range of wavenumbers which are at most a factor of three to four smaller than ℓ, while nearly all the energy transferred from ℓ goes to wavenumbers which are at most a factor of three larger than ℓ. The total transfer to wavenumber ℓ is usually somewhat larger than the transfer from ℓ to larger wavenumbers. This difference may be due to the energy which is lost by viscous dissipation directly at wavenumber ℓ, with less energy being available for the inertial cascade to higher wavenumbers than that which arrived from lower wavenumbers.

The grid turbulence data are similar to the atmospheric data in most respects. The maximum positive values of kS occur for $k/\ell \sim 0.6$-0.9, generally increasing with ℓ, while the maximum negative values occur for $k/\ell \sim 1.1$-1.4 and the ratio decreases with ℓ. Part of the systematic variation is probably artificial, due to the small number of points obtained in the neighborhood of the extreme values (especially for low frequencies).

SUMMARY

Bispectral measurements have been used to experimentally study spectral energy transfer due to wavenumber interactions in turbulent flows. Very similar results are obtained for both atmospheric boundary layer and grid turbulence data. The very limited results indicate that most of the transfer of energy is fairly localized in wavenumber space, involving wavenumbers differing at most by a factor of three or four. The data are consistent with the idea of a continuous cascade of energy to higher wavenumbers developed originally by A. N. Kolmogorov (1941). Further progress may depend upon comparison with bispectral computations carried out for numerical simulations of turbulence. Such simulations could yield valuable information about interactions which are difficult or impossible to study experimentally. The results, when compared with available experiments, would also provide a fundamental test of any direct simulation or turbulence model.

REFERENCES

Helland, K.N., Van Atta, C.W. & G.R. Stegen (1977) J. Fluid Mech. 79, 337-359.

Helland, K.N., Lii, K.S. & M. Rosenblatt (1977) Proc. Symp. on Applications of Statistics, P.R. Krishaniah, ed.

Kolmogorov, A.N. (1941) Dokl. Akad. Nauk SSSR 30, 301.

Lii, K.S., Rosenblatt, M. & C.W. Van Atta, J. Fluid Mech. 77, 45-62.

Van Atta, C.W. & W. Y. Chen (1969) J. Fluid Mech. 38, 743-763.

Wilson, J.R. (1974) Ph.D. Thesis, University of British Columbia.

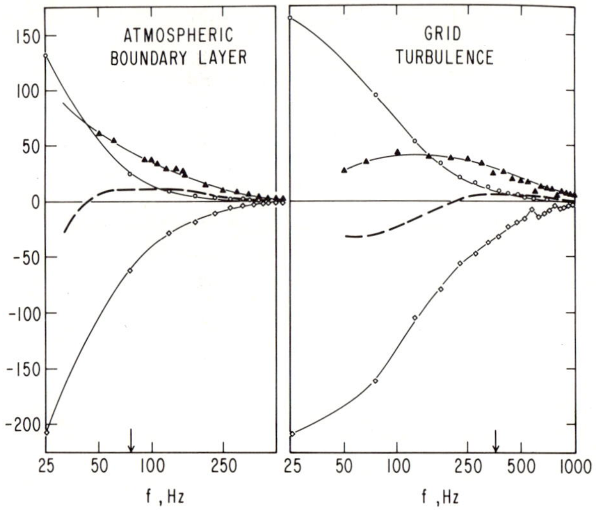

Fig. 1. Rate of energy transfer to frequency f due to three types of interactions and their sum. Vertical scale gives relative contributions (arbitrary scale). Arrows indicate the maxima in the dissipation spectra. ○, Type 1. ▲, Type 2. ◇, Type 3. ---, Sum.

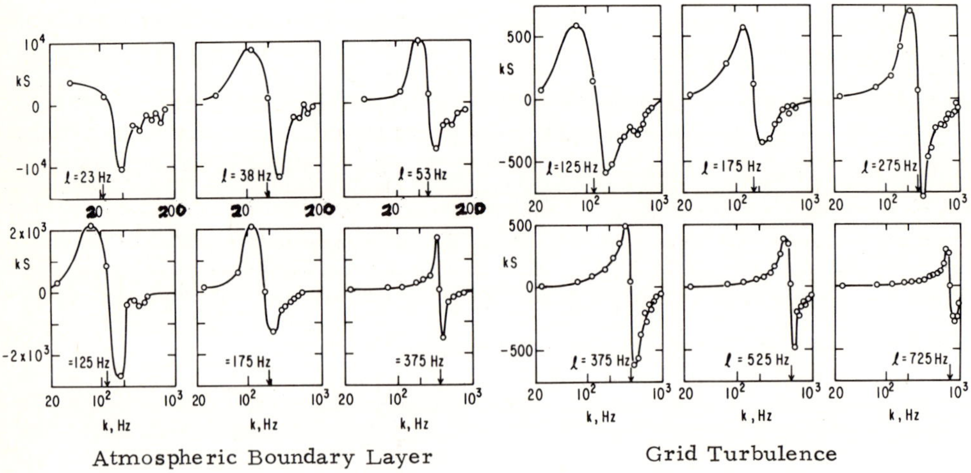

Fig. 2. Rate of energy transfer to frequency ℓ from frequency k. Arrows indicate values of ℓ. Vertical scale gives relative contributions (arbitrary scales).

ON THE APPLICATION OF ORTHOGONAL POLYNOMIALS IN THE THEORY OF ELASTICITY

T.S. Vashakmadze

Institute of Applied Mathematics
Tbilisi State University
Tbilisi/USSR

Summary

The first section deals with the questions connected with the definition of a stress-deformed state of isotropic elastic media for linear problems of elasticity in the domain $\Omega=\Omega_1 \cup \Omega_2$, when $\Omega_1 = D(x,y) \times (\bar{h}(x,y), \overset{+}{h}(x,y))$, $\text{mes}\,\Omega_2$ is small.

The Schwarz method is used. In the domain Ω_2 the problem is solved by the shooting method. In the domain Ω_1, depending on h, different orthogonal systems are chosen and, on account of the projection method, the initial problem is approximated by a finite system of multipointed equations defined in $\bar{D}$. The algorithm is proposed by which the finding of the solution of multipointed equations is reduced to a successive solution of a series of differential problems with three unknown functions. In the case when h is finite the multipointed system is equivalent to Vekua's system.

In the second section the techniques proposed are used for studying N.N. Yanenko's linear model, corresponding to the Navier-Stokes equation.

The paper presents the calculation methods for a class of multidimensional linear problems of the mathematical physics, which belong to the scientific trend predetermined by the late academician I.Vekua.

1. The methods for constructing a calculation scheme for a class of multidimensional problems will be illustrated by an example of the basic linear static problems of three-dimensional elasticity of a homogeneous isotropic body in the case of connected domains $\Omega=\Omega_1 \cup \Omega_2$ when $\Omega_1 = D(x,y)\times(\bar{h}(x,y), \overset{+}{h}(x,y))$ is a cylindrical domain, $\Omega_2 \subseteq \Omega_3 \subset \Omega$ $\text{mes}\,\Omega_3 = O(\varepsilon^\alpha)$, $\alpha > 0$, ε is a small parameter. The results stated below are a direct corollary of the author's papers [1-5].

It may be assumed without loss of generality that

$$\Omega_3 = \begin{cases} \varnothing, & \text{when either } \overset{\pm}{h} = \pm\infty \text{ or } \bar{h}=0,\ \overset{+}{h}=\infty, \\ O(\varepsilon^\alpha), & \text{when } \Omega \text{ is a finite domain}. \end{cases}$$

Following the Schwarz method, the initial problem may be divided into two sequences of differential problems:

a) Find a state of stress and strain of the elastic body in the domain Ω_1; denote the solution by u_n ($n = 0,1,2,\ldots$).

b) Find a state of stress and strain of the elastic body in the domain Ω_3; denote the solution by v_n ($n = 0,1,2,\ldots$).

To find v_n we may resort to the methods of the theory of potential and approximate solutions of integral equations worked out by prof. V.D. Kupradze and his followers in the case of a sufficiently smooth boundary $\partial\Omega_3$ and to T.A. Cruse's algorithms in the case when Ω_3 is a technical domain.

However, we think it more expedient to use the shooting method modified for three-dimensional problems in the case of small domains [3]. Replacing the derivatives with respect to the direction of the small parameter by difference relations, we may obtain the approximate solution $\bar{v}_n$ by an explicit scheme as a solution of the Cauchy problem. Using the boundary conditions on $\partial\Omega_2$, we may define the solution more precisely. Since the Cauchy problem is non-correct, for a strongly elliptic system we choose only a fixed number of pitches, say, on the basis of a priori estimates.

After that the problem is reduced to finding the solution u_n in the cylindrical domain Ω_1.

Depending on h, we choose different coordinate systems formed on the basis of orthogonal systems and, using the Galerkin-Petrov method, the initial problem is approximated in Ω_1 by a system of multi-pointed operator equations defined in the domain D. Denote the facial surface of the body by $S^{\pm}=\partial\Omega_1 \cap_{z=\pm h}$ and the lateral surface by $S=\partial\Omega_1\setminus S^{\pm}$.

Assume that on $S^{\pm}$ we are given anyone of the five classical boundary conditions.

For a finite h we may construct a monoparametric class of operator equations which is equivalent to Vekua's system of differential equations in the case of thin prismatic shells.

The following statements are true:

The finding of approximate solutions $\bar{u}_n$ is equivalent to the inversion of a series of twodimensional problems:

a) in the case when the conditions on $S^{\pm}$ are of the Dirichlet type, the components of the approximate solution $\bar{u}_{n,N}$, where N denotes the number of approximation, are found as a solution of a five-pointed operator equation (see, for example, (2.8), p. 34, of ref. [4])

b) in the case when on $S^{\pm}$ the external stress vector is given, in the domain Ω_1 the approximate solution is reduced to finding a solution of system (2.1) [4].

The solution of system (2.8) [4] is obtained by means of the operator factorization, while that of system (2.1) [4] by means of an explicit scheme.

c) in the case of contact boundary conditions one part of operator equations has a structure corresponding to the Dirichlet condition, while another part coincides with a system chosen according to the case b).

d) the case when on $S^{\pm}$ a linear combination of displacement and external stress vectors is given coincides with the case b).

A full analogy is observed if instead of Legendre polynomials we use, as it is done for finite h, the orthogonal Laguerre polynomials when $\bar{h}=0$, $\overset{+}{h}=+\infty$ and the Hermite-Chebyshev polynomials when $\overset{\pm}{h}=\pm\infty$.

In these cases when there are no stresses on $S^{\pm}$ the multipointed operator equations have the form

$$\begin{aligned} &l_2 \overset{m}{u} + \mu\gamma^2 \overset{m}{u}_+ - \gamma(\lambda+\mu)\,\mathrm{grad}\,\overset{m}{u}_3 + \sum_{k=m+1}^{\infty}\Big(2\mu\gamma \overset{k}{u} - \mu(k-m)\overset{k+1}{u}_+ - \\ &-(\lambda+\mu)\,\mathrm{grad}\,\overset{k}{u}_3\Big) = -\overset{m}{F}_+, \\ &\mu\Delta\overset{m}{u}_3 + \gamma^2(\lambda+2\mu)\overset{m}{u}_3 - \gamma(\lambda+\mu)\,\mathrm{div}\,\overset{m}{u}_+ + \sum_{k=m+1}^{\infty}\Big(2(\lambda+2\mu)\gamma\overset{k}{u}_3 - \\ &-(\lambda+2\mu)(k-m)\overset{k}{u}_3 - (\lambda+\mu)\,\mathrm{div}\,\overset{k}{u}_+ = -\overset{m}{F}_3\,; \\ &\qquad\qquad (0\le\gamma\le 1) \end{aligned} \tag{1a}$$

$$\begin{aligned} &l_2\overset{m}{u} + \lambda\,\mathrm{grad}\,\overset{m+1}{u}_3 - \mu(m+1)\,\mathrm{grad}\,\overset{m}{u}_3 - (m+1)\mu\overset{m+1}{u}_+ = -\overset{m}{F}_+, \\ &\mu\Delta\overset{m}{u}_3 + \mu\,\mathrm{div}\,\overset{m+1}{u} - \lambda\,\mathrm{div}\,\overset{m}{u}_+ - (m+1)(\lambda+2\mu)\overset{m+1}{u}_3 = -\overset{m}{F}_3 \\ &\qquad\qquad (m=1,2,\dots) \end{aligned} \tag{2a}$$

Here l_2 is the operator of plane elasticity of an isotropic elastic body, $(u_+=(u_1,u_2),u_3)$ are the components of the unknown vector, F is the known vector.

If the boundary conditions on S have, for example, the form

$$\left(\frac{\partial}{\partial z}A_0 + A_1\right)U(x,y,z) = -\Phi(x,y,z),$$

where A_i are the matrix linear operators not depending on z, they are reduced as follows:

$$A_1\overset{m}{u} - \gamma^2 A_0\sum_{k=m}^{\infty}\overset{k}{u} = -\overset{m}{\Phi}, \quad (x,y)\in S \tag{1b}$$

in the case of a semi-infinite cylindrical body and as

$$A_1 \overset{m}{u} - A_0 \overset{m-1}{u} = -\overset{m}{\varphi}, \quad (x,y) \in S \tag{2B}$$

when $\overset{\pm}{h} = \pm\infty$.

The investigation of the systems (1a) and (2a) is reduced for finite h to the following differential problems:

$$\begin{aligned} & l_2 u + \mu\gamma^2 u_+ - (\lambda+\mu)\gamma \operatorname{grad} u_3 = -F_+, \\ & \mu\Delta u_3 + (\lambda+2\mu)\gamma^2 u_3 - (\lambda+\mu)\operatorname{div} u_+ = -F_3, \quad (x,y)\in D \end{aligned} \tag{3a}$$

$$A_1 u - \gamma^2 A_0 u = -\varphi, \qquad (x,y) \in S \tag{3b}$$

and

$$l_2 u_+ - \alpha \operatorname{grad} u_3 = -F_+, \; \mu\Delta u_3 - \rho \operatorname{div} u_+ = -F_3, \; (x,y) \in D \tag{4a}$$

$$A_1 u = -\varphi \qquad (x,y) \in S \tag{4b}$$

The investigation and finding of an approximate solution of systems (3) and (4) may be conducted by using, for example, the scheme of refs. [3-5].

II. The techniques of Sect. I are applicable to the study of some problems of hydrodynamics.

Let, for example, be given the linearized iterated scheme (1.28), (1.29) [6] (pp. 644-645), which, for brevity, will be rewritten as

$$\begin{aligned} & \alpha \frac{\partial w^{(n+\frac{1}{2})}}{\partial t} + \frac{\partial}{\partial x}\beta_1(w^{(n)}) = \gamma^{(n)} \frac{\partial^2 w^{(n+\frac{1}{2})}}{\partial x^2}, \\ & \alpha \frac{\partial w^{(n+1)}}{\partial t} + \frac{\partial}{\partial y}\beta_2(w^{(n+\frac{1}{2})}) = \gamma^{(n)} \frac{\partial^2 w^{(n+1)}}{\partial y^2}, \\ & (n = 0, 1, 2, \dots) \end{aligned} \tag{5}$$

where the following notations are introduced:

$w = (u, v, p)^T$, $\alpha = (0{,}5;\ 0{,}5;\ 0{,}5\,\varepsilon)$, $\beta_1 = (u^2+p, uv, u)^T$,

$\beta_2(w) = (uv, v^2+p, v)$, $\gamma = (\mu(1+f_1), \mu(1+f_2), f_3)^T$.

Consider the Problem B [6] (p. 645) when a viscous incompressible liquid is flowing in the rectangular chamber $[0 \le x \le 1, \overset{-}{h} \le y \le \overset{+}{h}]$ with holes along the straight lines $x = 0,1$.

For the finite h system (5) takes the form

$$\alpha\left(\frac{\partial}{\partial t} - \gamma^{(n)}\frac{\partial^2}{\partial x^2}\right)\left(a_{i-1}\overset{i-2}{w}{}^{(n+\frac{1}{2})} - c_i \overset{i}{w}{}^{(n+\frac{1}{2})} + a_{i+1}\overset{i+2}{w}{}^{(n+\frac{1}{2})}\right) = g_i^{(n)} \tag{6a}$$

$$\begin{aligned} & \alpha \frac{\partial}{\partial t}\left(a_{i-1}\overset{i-2}{w}{}^{(n+1)} - (c_i + \gamma^{(n)})\overset{i}{w}{}^{(n+1)} + a_{i+1}\overset{i+2}{w}{}^{(n+1)}\right) = g_i^{(n+\frac{1}{2})} \\ & \left(i = 1, 2, \dots, N, \quad a_i = \frac{2}{2i+1}, \quad c_i = a_{i-1} + a_{i+1}\right) \end{aligned} \tag{6b}$$

The boundary conditions on the straight lines $X = 0,1$ are reduced similarly if it is assumed that the quantities occurring in (5) and in the boundary conditions are summable and the condition of the type

$$\frac{\partial p}{\partial x} = \mu \frac{\partial^2 u}{\partial x^2} - \frac{1}{2} \frac{\partial u^2}{\partial x}$$

is to linearized by analogy with scheme (5).

The system of equations (6) together with the boundary conditions is solvable exactly by the factorized scheme [1].

In the case of the channel of infinite length the initial system is replaced by a multipointed system which is constructed by means of equations (1) or (2) depending on h.

References

1. Vashakmadze T.S. The three-pointed operator equation of the shell theory (in Russian). Sem.Inst. Appl. Math., Paper Reviews, No. 8, 1973, pp. 23-28.

2. Vashakmadze T.S. Some numerical methods of solution of boundary-value problems for shells and plates (in Russian). Works of the 1st All-Union School on the Theory and Numerical Methods of Calculation of Shells and Plates. Tbilisi, "Metsniereba", pp. 291-298.

3. Vashakmadze T.S. To the calculation of elastic shells of variable thickness (in Russian). Proc. Xth All-Union Conf. Shell and Plate Theory, vol. 1, Tbilisi, "Metsniereba" , 1975, pp. 23-30.

4. Vashakmadze T.S. On the multipointed operator equations of the theory of shell and the theory of elasticity (in Russian), The yearbook of higher schools in engineering mechanics, Sofia, pp.33-38.

5. Vashakmadze T.S. The investigation of the operator of Vekua of the elastic shell theory (in Russian), Complex analysis and its applications, "Nauka", Moscow, 1978, pp.102-107.

6. Yanenko N.N., Okuntsov V.V., Zakharov Yu.N., Shokin Yu.I., Panichkin A.V. On two iteration scemes for the solution of a stationary system of Navier-Stokes equations (in Russian), Complex analysis and its applications, "Nauka", Moscow, 1978, pp.638-652.

NUMERICAL MODELING OF ATMOSPHERIC POLLUTION*

George F. Widhopf and Leslie Glatt
The Aerospace Corporation
Los Angeles, California 90009

INTRODUCTION

Throughout this decade, there has been concern regarding the effect of various activities of man on the earth's protective ozone shield. The importance of investigating the effect that these pollutants may have on our environment is apparent and, thus, a number of international programs have been formed to investigate these potential atmospheric pollution problems. Since it is very difficult to perform experiments in the atmosphere which can detect the small, time-dependent change in ozone concentration which may result from the introduction of various pollutants, mathematical models of the atmosphere must be relied upon to make state-of-the-art estimates of the present and the potential future effect on the upper atmosphere.

Accurate numerical methods must therefore be utilized along with realistic physical models of the important mechanisms controlling both the hydrodynamic and chemical structure of the atmosphere. Since the primary concern is a determination of changes in the distribution of ozone, which is itself chemically dependent on the formation, destruction, and transport of many chemical species, the number of atmospheric chemical species whose distribution needs to be determined in a model is large, being on the order of 20-30. This complexity is further compounded because of the large range of chemical reaction time scales (τ_c) encountered throughout the computational domain. For the present computational domain of interest, which extends from the surface to the stratopause (approximately 50 km) and from north to south pole, these reaction times can range from fractions of a second to months. Thus, due to the large chemical system involved, together with long physical simulation times of interest (approximately five to ten years for an aircraft pollution problem and up to 30 to 50 years for a chloroflurometha‌ne pollution problem), the accuracy, stability and efficiency of the solution algorithm are very important. Straightforward explicit schemes are very inefficient for such a system, and other more efficient algorithms must be used.

This paper describes a detailed time-dependent, two-dimensional photochemical model of the atmosphere which has been developed for the U.S. Department of Transportation's Federal Aviation Administration High Altitude Pollution Program. The model is phenomenological in nature in which the hydrodynamic variables (mean atmospheric density, temperature, turbulent diffusion coefficients and mean meridional winds) are either specified from observations or obtained indirectly from observations as a function of time during the year. These variables are used to solve the system of species conservation equations for the meridional distribution of trace species throughout the year. The purpose of this model is to study the chemical structure of the natural atmosphere and also to estimate the potential effect of aircraft emissions on ozone. Transport by mean meridional winds and large-scale turbulent eddies, which control the distribution of many long-lived trace species in the atmosphere, is included in the present model. Thus, this type of model can be used to investigate the significant latitudinal as well as vertical distributions of species in the natural atmosphere together with the important latitudinal variations in the pollutant effects of aircraft emissions. A detailed description of the basic model is included in Widhopf, et al. [1977].

*This work was supported in part by the High Altitude Pollution Program of the Federal Aviation Administration, U.S. Department of Transportation.

CHEMICAL MODEL

The chemical system considered in this investigation includes the following species: $O(^1D)$, $O(^3P)$, O_2, O_3, N, NO, NO_2, NO_3, N_2, N_2O, N_2O_5, H, OH, H_2O, H_2O_2, HNO_3, CO and CH_4. Smog type reactions initiated by the oxidation of methane by OH, which have been shown to be potentially important in the lower regions of the atmosphere, in particular for the evaluation of aircraft emissions effects through the work of Hidalgo and Crutzen [1977] and Widhopf, et al. [1977], are also included. These reactions involve the species CH_3, CHO, CH_2O, CH_3O, CH_3O_2, and CH_3O_2H. The specific reaction system and the associated reaction rate coefficients used in this investigation are tabulated in Tables I and II. Table II lists additional reactions and updated reaction rates that are recommended by the NASA-CFM study [1977]. Calculations have been performed using both of the chemical systems.

Computation of the absorption of solar radiation is an integral step in determining the chemical structure of the atmosphere since many of the important reactions in the atmosphere are photochemical processes. The diurnally averaged local photolysis rates J_i are calculated at every location in the atmosphere at each time step by a technique developed by Kramer and Widhopf [1978], using the solar flux data compiled by Ackerman [1971]. The time variation of the solar zenith angle with latitude and solar declination is included in the determination of the photolysis rates J_i. The absorption cross sections utilized to compute J_i for the various species are outlined in Widhopf [1975].

The effect of chlorine in the atmosphere was not included in this investigation. From NASA studies [1977], the inclusion of chlorine should attenuate the effect of NO_x destruction of ozone due to the formation of $C\ell ONO_2$. These effects are currently being investigated.

NUMERICAL SCHEME

As stated in the Introduction, an accurate and efficient numerical scheme is required since the problem involves large numbers of species, whose chemical reaction time scales vary over orders of magnitude, and the numerical simulations must be carried on for many years. In this model, an accurate and efficient time-implicit finite difference scheme has been employed to solve for the short-lived species ($\tau_c < 2$ days) [$O(^1D)$, $O(^3P)$, O_3, N, NO, NO_2, NO_3, N_2O_5, H, OH, HO_2 H_2O_2 and $C_xH_yO_z$]. Advective and diffusive terms that are important in determining the time-dependent distributions of the species are treated using a leap-frog and a Dufort-Frankel finite difference scheme, respectively. The longer-lived ($\tau_c \geq 2$ days) species (N_2O, HNO_3, H_2O, CO and CH_4) are handled in an explicit manner.

The time-implicit method makes use of a second order accurate method developed by Widhopf and Victoria [1973]. In this method, the chemical production/loss term $\mathring{\omega}_i$, at a specific mesh point and at the new time level n+1, is approximated by the expansion

$$\mathring{\omega}_i^{n+1}(Y_i, \rho, T) = \omega_i^n + \sum_{i=1}^{N}\left(\frac{\partial \mathring{\omega}_i}{\partial Y_i}\right)^n \left(Y_i^{n+1} - Y_i^n\right) + \left(\frac{\partial \mathring{\omega}_i}{\partial \rho}\right)^n \left(\rho^{n+1} - \rho^n\right)$$

$$+\left(\frac{\partial \mathring{\omega}_i}{\partial T}\right)^n \left(T^{n+1} - T^n\right)$$

where the index i denotes the species i, Y_i the corresponding mass fraction, T the temperature, ρ the density, n the current time level of the computation, and N the number of species considered. All partial derivatives of $\mathring{\omega}_i$ are analytically computed and evaluated at the current time level n. In addition, $\mathring{\omega}_i^n$ is approximated by

TABLE I. CHEMICAL REACTIONS AND RATE COEFFICIENTS

	REACTION	RATE COEFFICIENT[a]
1.	$O(^3P) + O_3 \rightarrow 2O_2$	$1.9(10)^{-11}$ exp[-2300/T]
2.	$O_2 + h\nu \rightarrow 2O(^3P)$	J_2
3.	$O_3 + h\nu \rightarrow O(^3P) + O_2$	J_3
4.	$NO_2 + h\nu \rightarrow O(^3P) + NO$	J_4
5.	$O(^3P) + O_2 + M \rightarrow O_3 + M$	$1.07(10)^{-34}$ exp[510/T]
6.	$O(^3P) + NO_2 \rightarrow O_2 + NO$	$9.1(10)^{-12}$
7.	$O_3 + NO \rightarrow O_2 + NO_2$	$9(10)^{-13}$ exp[-1200/T]
8.	$O_3 + NO_2 \rightarrow O_2 + NO_3$ $\vert NO_3 + h\nu \rightarrow 2/3[NO_2 + O(^3P)] + 1/3[NO + O_2]\vert$	$1.23(10)^{-13}$ exp[-2470/T]
9.	$O_3 + OH \rightarrow O_2 + HO_2$	$1.6(10)^{-12}$ exp[-1000/T]
10.	$NO + HO_2 \rightarrow OH + NO_2$	$2.3(10)^{-13}$
11.	$O(^3P) + H_2O \rightarrow OH + OH$	0.0
12.	$OH + NO_2 + M \rightarrow HNO_3 + M$	$\dfrac{2.76(10)^{-13}\ \exp[880/T]}{1.166(10)^{18}\ \exp[220/T] + [M]}$
13.	$HNO_3 + h\nu \rightarrow OH + NO_2$	J_{13}
14.	$HO_2 + O_3 \rightarrow OH + O_2 + O_2$	$1(10)^{-13}$ exp[-1250/T]
15.	$HO_2 + O(^3P) \rightarrow OH + O_2$	$3(10)^{-11}$
16.	$OH + HO_2 \rightarrow H_2O + O_2$	$2(10)^{-11}$
17.	$OH + HNO_3 \rightarrow H_2O + NO_3$ $\vert NO_3 + h\nu \rightarrow 2/3[NO_2 + O(^3P)] + 1/3[NO + O_2]\vert$	$8.9(10)^{-14}$
18.	$H_2O_2 + h\nu \rightarrow OH + OH$	J_{18}
19.	$H_2O_2 + OH \rightarrow H_2O + HO_2$	$1.7(10)^{-11}$ exp[-910/T]
20.	$HO_2 + HO_2 \rightarrow H_2O_2 + O_2$	$1.7(10)^{-11}$ exp[-500/T]
21.	$O_3 + h\nu \rightarrow O_2 + O(^1D)$	J_{21}
22.	$O(^1D) + M \rightarrow M + O(^3P)$	$2.2(10)^{-11}$ exp[92/T]
23.	$N_2O + h\nu \rightarrow N_2 + O(^1D)$	J_{23}
24.	$N_2O + O(^1D) \rightarrow N_2 + O_2$	$5.7(10)^{-11}$
25.	$N_2O + O(^1D) \rightarrow NO + NO$	$5.7(10)^{-11}$
26.	$NO + h\nu \rightarrow N + O(^3P)$	J_{26}
27.	$N + O_2 \rightarrow NO + O(^3P)$	$1.02(10)^{-14}T$ exp[-3130/T]
28.	$N + NO \rightarrow N_2 + O(^3P)$	$2.7(10)^{-11}$
29.	$N + NO_2 \rightarrow NO + NO$	0.0
30.	$N_2 + O(^1D) + M \rightarrow N_2O + M$	$2.8(10)^{-36}$
31.	$NO_2 + N \rightarrow N_2O + O(^3P)$	$1.4(10)^{-12}$
32.	$O(^1D) + H_2O \rightarrow OH + OH$	$2.32(10)^{-10}$
33.	$O(^1D) + CH_4 \rightarrow OH + CH_3$	$1.38(10)^{-10}$
34.	$OH + O(^3P) \rightarrow O_2 + H$	$4.2(10)^{-11}$
35.	$H + O_2 + M \rightarrow HO_2 + M$	$2.08(10)^{-32}$ exp[290/T]
36.	$H + O_3 \rightarrow OH + O_2$	$1.23(10)^{-10}$ exp[-562/T]
37.	$NO + O(^3P) + M \rightarrow NO_2 + M$	$3.96(10)^{-33}$ exp[940/T]
38.	$OH + OH \rightarrow H_2O + O(^3P)$	$1(10)^{-11}$ exp[-550/T]
39.	$N + O_3 \rightarrow NO + O_2$	$5.7(10)^{-13}$
40.	$HO_2 + h\nu \rightarrow OH + O(^3P)$	J_{40}
41.	$OH + CH_4 \rightarrow H_2O + CH_3$	$2.36(10)^{-12}$ exp[-1710/T]
42.	$2OH + M \rightarrow H_2O_2 + M$	$2.5(10)^{-33}$ exp[2500/T]
43.	$H_2O_2 + O(^3P) \rightarrow OH + HO_2$	$2.75(10)^{-12}$ exp[-2125/T]
44.	$CO + OH \rightarrow H + CO_2$	$\log_{10}K = -12.95 + 3.94(10)^{-4}T$
45.	$CH_2O + h\nu \rightarrow H_2 + CO$	J_{45}
46.	$CHO + O_2 \rightarrow HO_2 + CO$	$5(10)^{-12}$
47.	$CH_3 + O_2 + M \rightarrow CH_3O_2 + M$	$2.6(10)^{-31}$
48.	$CH_3O_2 + NO \rightarrow CH_3O + NO_2$	$1.5(10)^{-12}$ exp[-500/T]
49.	$CH_3O_2 + HO_2 \rightarrow CH_3O_2H + O_2$	$3.0(10)^{-11}$ exp[-500/T]
50.	$CH_3O_2H + h\nu \rightarrow CH_3O + OH$	J_{50}
51.	$CH_3O + O_2 \rightarrow CH_2O + HO_2$	$1.6(10)^{-13}$ exp[-3300/T]
52.	$CH_2O + h\nu \rightarrow H + CHO$	J_{52}
53.	$CH_2O + OH \rightarrow H_2O + CHO$	$1.4(10)^{-11}$

[a] Units in sec^{-1}, $cm^3 sec^{-1}$ and $cm^6 sec^{-1}$ for unimolecular, bimolecular and trimolecular reactions.

TABLE II. ADDITIONAL REACTIONS AND NEW NASA (1977) RATES

REACTION	RATE COEFFICIENT[a]	REACTION	RATE COEFFICIENT[a]
7. $O_3 + NO \rightarrow O_2 + NO_2$	$2.1\ 10^{-12}\ exp[-1450/T]$	53. $CH_2O + OH \rightarrow H_2O + CHO$	$3\ 10^{-11}\ exp[-250/T]$
8. $O_3 + NO_2 \rightarrow O_2 + NO_3$	$1.2\ 10^{-13}\ exp[-2450/T]$	54. $NO_3 + NO_2 + M \rightarrow N_2O_5 + M$	$1.9\ 10^{-30}(3+5C)/\{(1+5C)(3+C)\}$
9. $O_3 + OH \rightarrow O_2 + HO_2$	$1.5\ 10^{-12}\ exp[-1000/T]$		$C = 5\ 10^{-19}[M]$
10. $NO + HO_2 \rightarrow OH + NO_2$	$8\ 10^{-12}$	55. $N_2O_5 + M \rightarrow NO_3 + NO_2 + M$	$D(4+6F)/\{(1+6F)(4+F)\}$
12. $OH + NO_2 + M \rightarrow HNO_3 + M$	$\log_{10}(k) = -AT/(B+T) - 0.5 \log_{10}(T/280)$		$D = 2.2\ 10^{-5}\ exp[-9700/T]$
14. $HO_2 + O_3 \rightarrow OH + O_2 + O_2$	$7.3\ 10^{-14}\ exp[-1275/T]$		$E = 5.7\ 10^{14}\ exp[-10600/T]$
15. $HO_2 + O \rightarrow OH + O_2$	$3.5\ 10^{-11}$		$F = B[M]/E$
16. $OH + HO_2 \rightarrow H_2O + O_2$	$3\ 10^{-11}$	56. $NO_3 + h\nu \rightarrow 2/3(NO_2 + O) + 1/3(NO + O_2)$	
17. $OH + HNO_3 \rightarrow H_2O + NO_3$	$8\ 10^{-14}$	57. $N_2O_5 + h\nu \rightarrow NO_2 + NO_3$	
19. $H_2O_2 + OH \rightarrow H_2O + HO_2$	$1\ 10^{-11}\ exp[-750/T]$	58. $NO_3 + NO \rightarrow 2NO_2$	$8.7\ 10^{-12}$
20. $HO_2 + HO_2 \rightarrow H_2O_2 + O_2$	$2.5\ (10)^{-12}$	59. $NO_2 + O + M \rightarrow NO_3 + M$	$1\ 10^{-31}$
24. $N_2O + O(^1D) \rightarrow N_2 + O_2$	$5.5\ 10^{-11}$	60. $CO_2 + h\nu \rightarrow O + CO$	
25. $N_2O + O(^1D) \rightarrow NO + NO$	$5.5\ 10^{-11}$	61. $H_2 + O(^1D) \rightarrow OH + H$	$9.9\ 10^{-11}$
27. $N + O_2 \rightarrow NO + O$	$5.5\ (10)^{-12}\ exp[-3220/T]$	62. $O(^1D) + CH_4 \rightarrow H_2 + CH_2O$	$1.4\ 10^{-11}$
28. $N + NO \rightarrow N_2 + O$	$8.2\ 10^{-11}\ exp[-410/T]$	63. $O + CH_4 \rightarrow CH_3 + OH$	$3.5\ 10^{-11}\ exp[-4450/T]$
30. $N_2 + O(^1D) + M \rightarrow N_2O + M$	$3.5\ 10^{-37}$	64. $CH_2O + O \rightarrow OH + CHO$	$2\ 10^{-11}\ exp[-1450/T]$
31. $NO_2 + N \rightarrow N_2O + O$	$2\ 10^{-11}\ exp[-800/T]$		
34. $OH + O \rightarrow O_2 + H$	$4.2\ 10^{-11}$		
37. $NO + O + M \rightarrow NO_2 + M$	$1.55\ 10^{-32}\ exp[584/T]$		
39. $N + O_3 \rightarrow NO + O_2$	$5\ 10^{-12}\ exp[-650/T]$	$A = A_1 + A_2z + A_3z^2 + A_4z^3$	
42. $2OH + M \rightarrow H_2O_2 + M$	$1.25\ 10^{-32}\ exp[900/T]$	$B = B_1 + B_2z + B_3z^2$	
44. $CO + OH \rightarrow H + CO_2$	$1.4\ 10^{-13}$	$A_1 = 31.62273$	$B_1 = -327.372$
46. $CHO + O_2 \rightarrow HO_2 + CO$	$6\ 10^{-12}$	$A_2 = -0.258304$	$B_2 = 44.5586 \quad z = \log_{10}\{0.94[N_2]\}$
48. $CH_3O_2 + NO \rightarrow CH_3O + NO_2$	$1.5\ 10^{-12}$	$A_3 = -0.0889287$	$B_3 = -1.38092$
49. $CH_3O_2 + HO_2 \rightarrow CH_3O_2H + O_2$	$2.5\ 10^{-12}$	$A_4 = 2.520173(10)^{-3}$	

[a]Units in sec^{-1}, $cm^3 sec^{-1}$ and $cm^6 sec^{-1}$ for unimolecular, bimolecular and trimolecular reactions.

the following:

$$\mathring{\omega}_i^n = \frac{\mathring{\omega}_i^{n+1} + \mathring{\omega}_i^{n-1}}{2\Delta\tau} .$$

The use of these relations in the governing species conservation equations results in a linear set of coupled equations for Y_i^{n+1}. (For this problem, the time variations of ρ and T are specified.) These equations are coupled only in time and not in space, and thus the technique results in a solution of a set of N_s equations at each mesh point. The stability and accuracy of the scheme is discussed by Widhopf and Victoria [1973].

This time-implicit algorithm overcomes the "stiff" nature of the governing equations which results from the wide range of chemical time scales of the problem. For the current numerical system, the allowable time step is determined by the convective time-step limitation, which yields a maximum time step of a few days. In order to simplify the calculation and reduce the N_s matrix size, only those species whose shortest chemical time scales are less than two days throughout the computational domain need to be solved using the time-implicit algorithm. All other long-lived species are solved in a straightforward explicit manner. This combination of numerical algorithms has proven to be computationally stable and accurate. The simulation of one complete yearly cycle requires approximately 15 minutes on a CDC 7600 and includes all radiative flux calculations. This time is comparable with other techniques which utilize a "family" approach in solving the governing equations.

In the family approach, species that chemically interact with each other over relatively short time scales are grouped into a so-called "family". In contrast with the individual species, the chemical change in the family concentration occurs over a relatively longer time scale. Since the equation for the behavior of the family concentration is more stable than the equations for its members due to the difference in chemical time scales, accurate solutions of the family concentration can be obtained using larger time steps. The required concentration of the individual members of the family are first approximately determined using a suitable stable numerical technique and are then adjusted to correspond to the more accurate family concentration. Due to the relatively short time scales for these individual species, the effects of transport are neglected in the computation of their concentration. This adjustment is necessary since the numerical technique used in solving for the individual species generally does not conserve atoms in a closed chemical system. The choice of the individual members of the family as well as the adjustment procedures are not unique. These required correction procedures must be carefully performed in order to insure the accuracy of the calculation. Other long-lived species not included in families are computed individually. Careful implementation of this technique has been shown to be computationally efficient and accurate in other models.

As discussed, in the present method the concentration of each species is determined by accurately solving a conservation equation governing that species' time history. Thus, potential problems that can be encountered in the family approach, such as the correct "family" grouping and the correction procedures needed to insure accuracy, are not encountered in the present method. Also, it is not necessary to inherently assume that some individual species are not influenced by transport. As stated previously, model computational times are nearly equivalent for both techniques.

TRANSPORT

The model advective and turbulent diffusive transport have been developed using observational data and tested and refined by calculating the time-dependent transport of various radioactive tracers (C-14, W-185, Zr-95) which have resulted from past atmospheric nuclear weapons tests. These model simulations are in good agreement with available time-dependent data of the meridional distributions of these tracers and the decay of their total atmospheric burden. Further discussion of the transport and associated tracer simulations is included in Widhopf, et al. [1977].

MODEL RESULTS

Using this model, the distribution of the species in the natural atmosphere was calculated over the entire yearly cycle. In approximately four years of numerical simulation, the results converged to a periodic solution, within one-tenth of one percent in the total ozone column throughout succeeding years. A comparison of the ozone column variation with time during the year from the south to the north pole, as computed using the chemical system in Table I, is shown in Figure 1. The corresponding observed distribution is depicted in Figure 2. As can be seen, the calculated results are in good agreement with data. The model results for the meridional distribution of ozone concentration as well as the distribution of other chemical species in the natural atmosphere are also in good agreement with data. It should be mentioned that these results were calculated using observed levels of the distribution of water vapor and, thus, water vapor was not actively modeled. In addition, the species NO_3 and N_2O_5 were not considered in this calculation. These species can be important sinks for NO and NO_2 in regions of prolonged periods of darkness (e.g., at high latitudes).

The effect of aircraft NO_x emissions on the ozone layer was studied using as a basis the aforementioned natural atmosphere calculations. Estimated 1990 aircraft NO_x emissions (upper bound) from a projected fleet of approximately 140 Concorde-Tupolev supersonic type aircraft plus approximately 10,000 present and advanced type subsonic aircraft were supplied by the Federal Aviation Administration and are given in Table III. The results of the effect of these emissions on the ozone column are shown in Figures 3 and 4. For this case, the ozone perturbation is positive at all latitudes throughout the year. This results from ozone production below approximately 15 km through the methane-smog cycle as a consequence of the low altitude subsonic NO_x emissions. The emissions from the higher flying supersonic aircraft which are only about 20 percent of the total NO_x emissions result in a correspondingly small ozone reduction above approximately 15 km. However, the small increase in ozone below 15 km is slightly larger than the small ozone decrease above 15 km, resulting in the positive perturbation shown. The significant meridional variation is depicted in Figure 5. This result, together with the results of Hidalgo and Crutzen [1978], demonstrates the potential importance of the effect of subsonic NO_x injections on the ozone layer when "smog" reactions are included. More work must be performed to study these smog reactions and associated rates before a more definite statement can be made regarding these results.

For this chemical system, the aircraft NO_x emissions have a small effect on the ozone column (≤ 1.5 percent). This reduced effect, and indeed a sign change, is a consequence of reduced projections of supersonic NO_x emissions from earlier estimates, the inclusion of subsonic aircraft emissions, and improved measurements of the chemical rate constants. The largest effect is seen to occur at latitudes where the largest percentage of the population lives. This latitudinal distribution is important for any environmental determination of the associated effect on the world population. Further discussion of these results is included in Widhopf, et al. [1977].

Although aircraft emit water vapor along with NO_x, the effect of water vapor on ozone has not been studied as extensively as the effect of NO_x emissions because of their different ozone destruction rates as determined by the rates contained in Table I. In this regard, it should be pointed out that the hydroxyl

TABLE III. 1990 WORLDWIDE AIRCRAFT NO_x EMISSIONS, HIGH ESTIMATES, kg/yr*

	ALTITUDES - km											
Latitude	6-8	8-9	9-10	10-11	11-12	12-13	13-14	14-15	15-16	16-17	17-18	18-19	Total
N 60+	3.35E6	3.03E6	1.43E7	1.31E7	1.46E7	1.31E6	9.99E5	4.06E5	1.72E6	2.59E6	2.10E6	1.43E6	5.894E7
50-60	2.15E7	2.59E7	9.44E7	1.06E8	9.09E7	8.26E6	3.72E6	2.12E6	6.57E6	1.03E7	8.06E6	3.71E6	3.814E8
40-50	7.60E7	8.70E7	1.79E8	2.79E8	1.62E8	2.48E7	4.59E6	2.09E6	4.17E6	6.96E6	5.46E6	2.36E6	8.334E8
30-40	7.74E7	9.20E7	1.67E8	3.09E8	1.72E8	2.97E7	3.11E6	1.74E6	1.30E6	2.73E6	2.07E6	8.63E5	8.589E8
20-30	2.61E7	2.83E7	6.74E7	1.02E8	6.92E7	8.73E6	1.55E6	1.20E6	8.06E5	1.90E6	1.71E6	4.67E5	3.094E8
10-20	1.11E7	1.18E7	2.65E7	4.28E7	3.99E7	3.67E6	4.74E5	1.54E5	3.24E5	5.38E5	4.22E5	1.71E5	1.379E8
0-10	4.80E6	5.14E6	1.50E7	1.82E7	1.36E7	1.26E6	1.73E5	0	2.91E5	4.08E5	3.44E5	1.63E5	5.938E7
10-0	3.31E6	3.77E6	1.22E7	1.38E7	1.09E7	8.65E5	1.38E5	0	3.01E5	4.22E5	3.56E5	1.65E5	4.623E7
20-10	2.74E6	3.21E6	1.14E7	1.52E7	1.15E7	1.11E6	3.15E5	1.32E5	1.10E5	2.19E5	1.58E5	7.52E4	4.617E7
30-20	3.67E6	4.01E6	9.47E6	1.37E7	8.66E6	9.31E5	5.10E4	0	9.85E4	1.38E5	1.16E5	6.63E4	4.091E7
40-30	4.01E6	4.63E6	6.62E6	1.18E7	6.14E6	1.21E6	8.64E4	5.16E4	1.56E4	4.74E4	2.84E4	6.22E3	3.464E7
50-40	2.36E5	3.05E5	3.19E5	8.28E5	4.46E5	9.29E4	1.5 E1	0	0	0	0	0	2.227E6
60-50	4.77E4	3.79E4	2.99E4	2.52E4	1.04E4	1.45E3	0.97	0	0	0	0	0	1.526E5
S 60+	0	0	0	0	0	0	0	0	0	0	0	0	0
Total	2.343E8	2.691E8	6.036E8	9.255E8	5.999E8	8.197E7	1.521E7	7.894E6	1.571E7	2.625E7	2.082E7	9.477E6	2.810E9

*Oliver, R.C., et al. [1977]

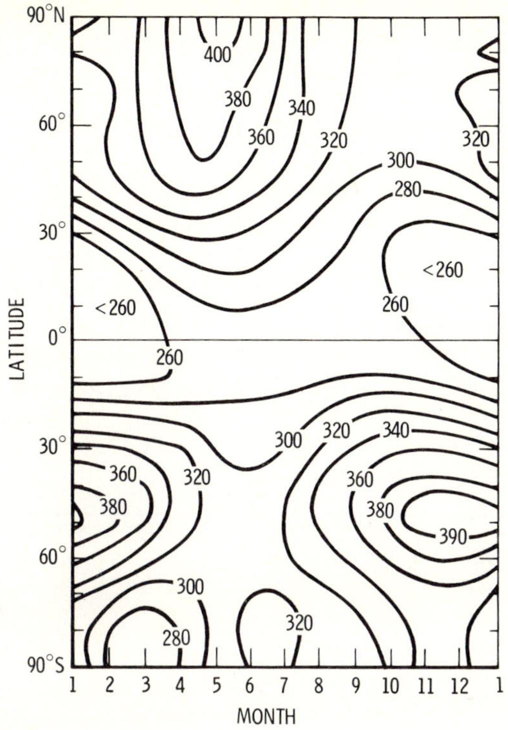

Fig. 1. Calculated Monthly Variation of the Total Ozone Column as a Function of Latitude (10^{-3} cm at STP)

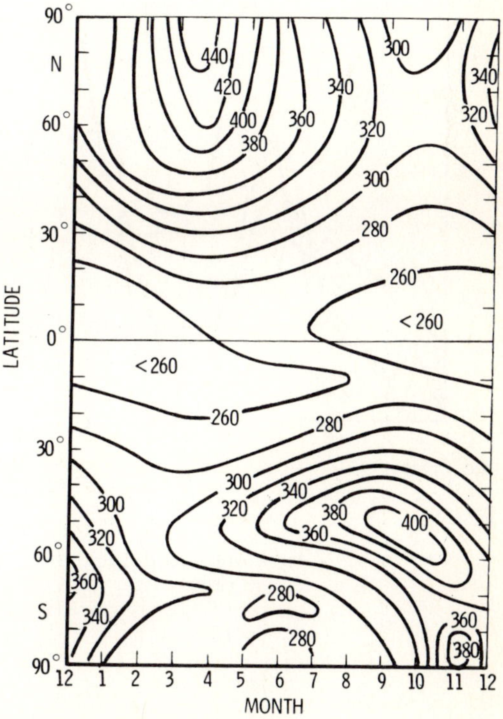

Fig. 2. Observed Monthly Variation of the Total Ozone Column as a Function of Latitude (10^{-3} cm at STP)

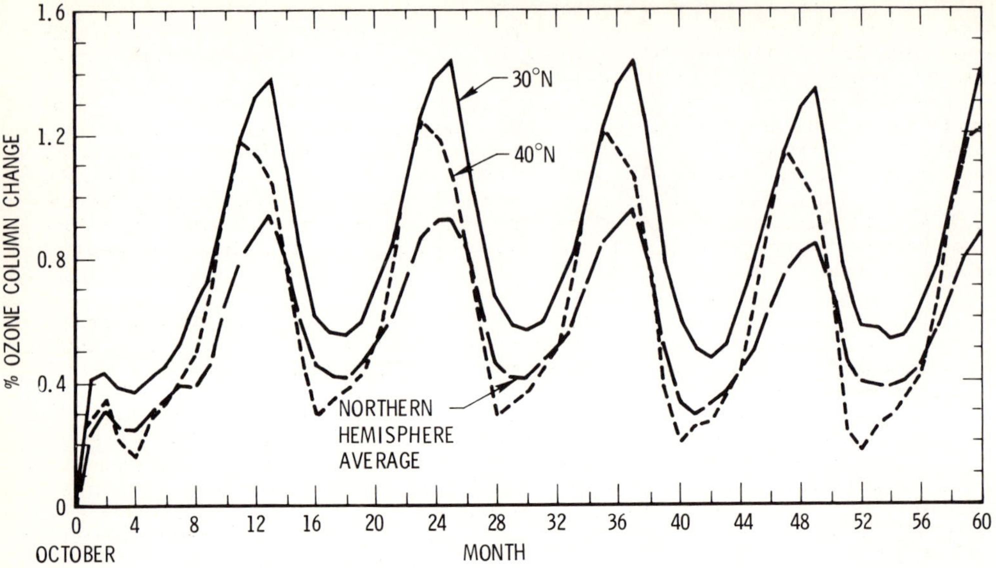

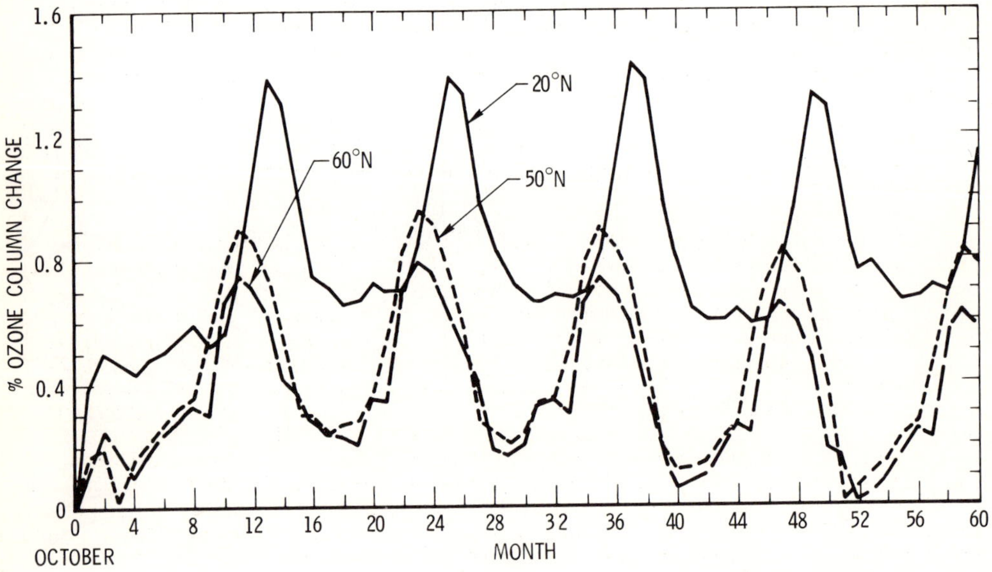

Figs. 3, 4. Calculated Temporal Ozone Change Resulting from Aircraft NO_x Emissions from a Combined Subsonic and Supersonic Fleet Projected to be Operational in 1990

reaction rates are very difficult to measure directly and thus have in the past been obtained by indirect means. However, in the past year new direct measurements of the reaction of NO with HO_2, by Howarth and Evenson [1977], has increased the rate of this reaction by a factor of approximately 35 over that listed in Table I. This has led to an increased importance of hydroxyl chemistry and, thus, it is now much more important to actively model both the chemistry and transport of water vapor in the atmosphere. The inclusion of water vapor modeling is necessary not only because water vapor is emitted by aircraft, but also because it is the source of hydroxyls in the natural atmosphere. Since it also has a major sink in the troposphere, through rainout, the proper modeling of this sink will aid in modeling other washout sinks which are as important for the proper description of other species such as HNO_3.

In this regard, a water vapor model has been included in the atmospheric model. This model is a result of further development of a one-dimensional and two-dimensional steady state water vapor modeling effort described by Glatt and Widhopf [1978]. Here, the average rainout rate as a function of latitude [Junge, 1963] has been used to describe the average rainout in the troposphere. The rainout is treated as a first order removal mechanism proportional to the local water vapor concentration in the troposphere and removed at the measured average precipitation rate. The time-dependent surface boundary condition is a relative humidity specification using the work of Manabe and Wetherald [1967]. Typical calculated water vapor profiles during October are compared with available tropospheric data [Oort and Rasmussen, 1971], and are shown in Figure 6. As can be seen, the exponential decrease of the water vapor distribution with increased altitude in the troposphere is calculated correctly, as is the average stratospheric concentration [Harries (1976)]. (It should also be pointed out that there are not many measurements of water vapor in the stratosphere and, thus, the concentration is not known to within a factor of approximately two, as indicated by the data bars.) In addition, it should be emphasized that the water vapor distribution in the stratosphere is strongly influenced by transport near the tropopause and, thus, is very sensitive to this transport specification. The calculated precipitation rates are also in relative agreement with data, as shown in Figure 7, another indication of the adequacy of the water vapor model. Thus, the agreement between the model results and available data lends credence to the overall model transport specification in the troposphere and the model exchange rate between the troposphere and the stratosphere.

With the availability of an active water vapor model, the natural atmosphere was recalculated including the increased rate for $NO + HO_2$ as measured by Howarth and Evenson [1977]. Included in this calculation are other updated reaction rates and additional species (NO_3 and N_2O_5) as indicated in Table II. Here, the modeling has been changed from a strict diurnal average to an average nighttime and average daytime modeling which is needed to correctly model the nighttime production of N_2O_5. The corresponding calculated results for the variation of the ozone column as a function of latitude and time of year are shown in Figure 8. The contours are similar to those shown in Figures 1 and 2; however, the ozone level is seen to be approximately 20 to 30 percent higher than either that measured or calculated with the old rate. This difference is primarily due to the increase in the $NO + HO_2$ reaction rate which results in, among other things, the increased conversion of NO into NO_2 and the increased production of the NO_x sink, HNO_3. Since NO_2 is much less effective in catalytically reducing ozone than NO, and more NO_x is stored in HNO_3, this reaction rate change has resulted in an overall increase in ozone in the calculated natural atmosphere.

It should be pointed out that in this calculation the vertical transport above 20 km has been reduced from that used in the previously described natural atmospheric calculation. This change was determined not to affect any of the previous inert tracer results, since for these cases the major portion of the tracer was always below 25 km. This transport modification reduced the ozone overprediction to the present level of 20 to 30 percent. Further reduction does not seem to be possible if the model results for other species distributions at high altitudes are to

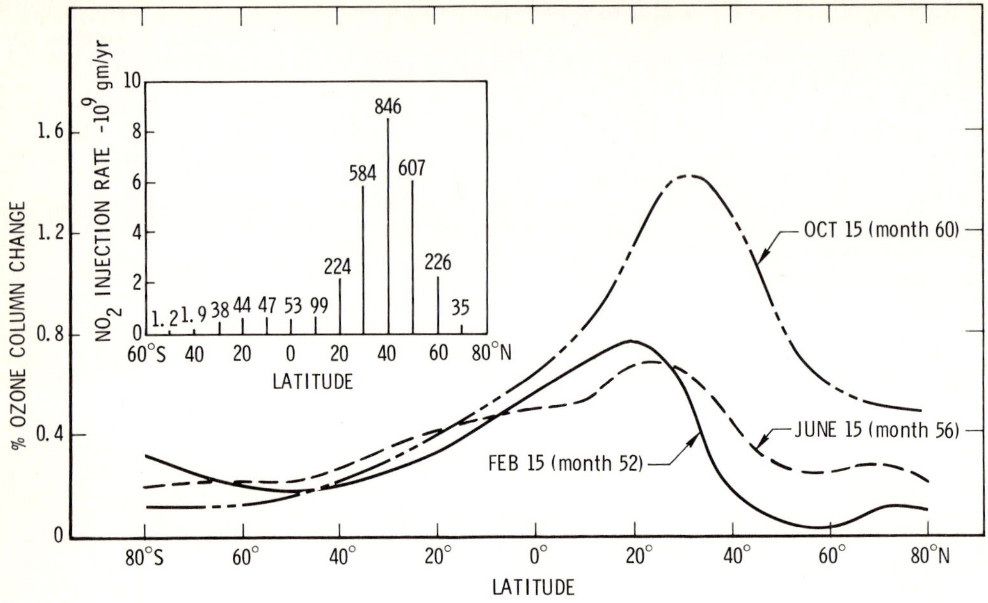

Fig. 5. Latitudinal Variation of Ozone Column Change

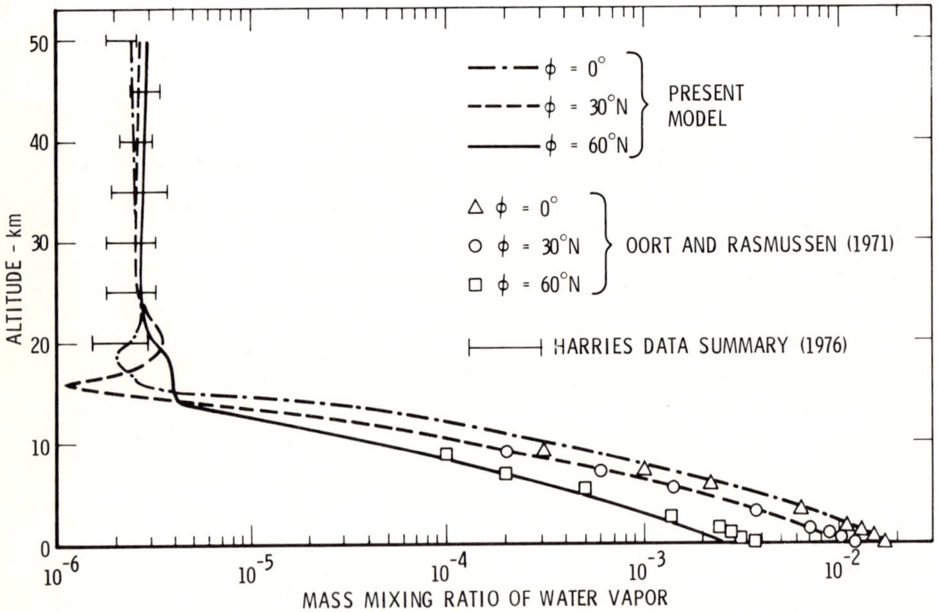

Fig. 6. Water Vapor Profile in Natural Atmosphere in October

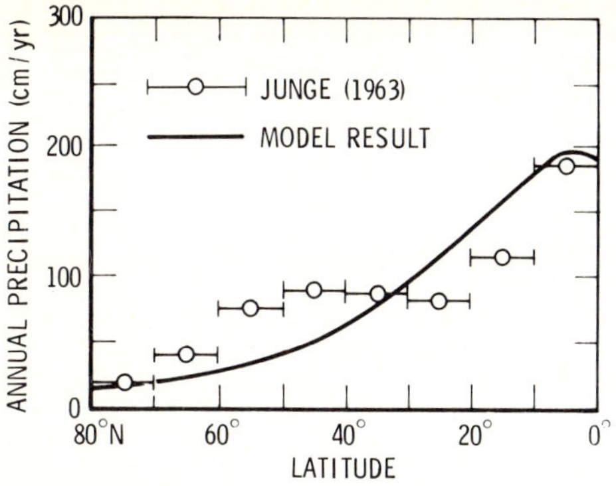

Fig. 7. Comparison of Measured and Calculated Average Annual Precipitation Rate

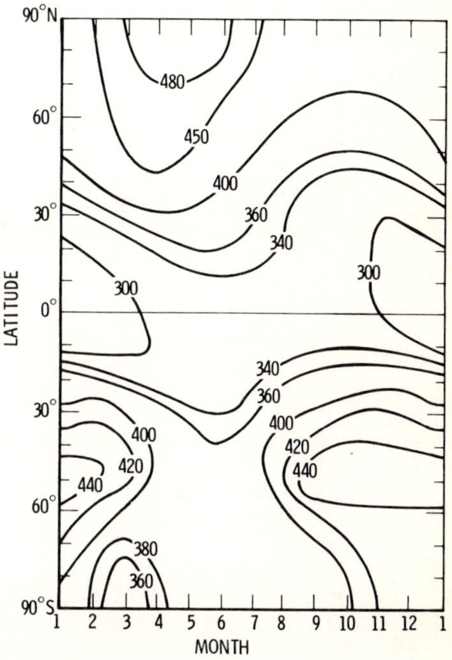

Fig. 8. Calculated Monthly Variation of the Total Ozone Column as a Function of Latitude Using Chemical System as Modified by Table II (10^{-3} cm at STP)

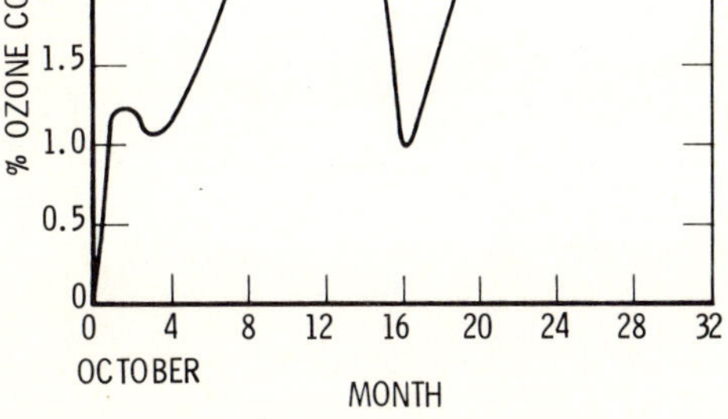

Fig. 9. Calculated Temporal Ozone Change Resulting from NO_x and HO_x Emissions from a Combined Subsonic and Supersonic Fleet of Aircraft (Table II Chemistry)

agree with available measurements (specifically N_2O and CH_4). Thus, the results indicate at this time that another chemical rate could also be inaccurate or an additional important chemical mechanism (perhaps HO_2NO_2) is not presently included in the model. These questions are actively being pursued at the present time, and candidate reactions are suggested later in the text.

In order to estimate the effect of combined NO_x and HO_x aircraft emissions, a calculation was performed to inject NO_x at the rate specified in Table III with the following modification: the rate of injection above 15 km was tripled. This effectively triples the number of supersonic aircraft considered. The HO_x was simultaneously injected at a rate 75 times the NO_x rate, which corresponds to the ratio of the HO_x to NO_x emission indices. The results of this calculation are shown in Figure 9. For this case the effect of the combined NO_x and HO_x emissions is to increase the local zone column above the level calculated in the natural atmosphere. The maximum increase is approximately 3.5 percent, somewhat higher than the 1.5 percent obtained in the previous calculations where HO_x was not injected. The variation with time of year is quite similar to the previous calculation, in which the minimum effect occurs during late winter-early spring and the maximum effect occurs during late summer-early autumn. This calculation was only carried out for approximately two years since, as in the previous NO_x case, the changes are almost periodic from year to year after one-half a year.

In order to explain the increase in total ozone column obtained for this calculation, the chemical mechanisms for the production/depletion of ozone must be investigated. An analysis of the results indicates different chemical mechanisms are important in the stratosphere and troposphere. In the stratosphere the level of ozone is increased in the injection case due to the decreased effectiveness of the catalytic destruction cycle. Injection of NO_2 depletes ozone through the cycle

$$NO_2 + O(^3P) \longrightarrow NO + O_2 \qquad R6$$

$$NO + O_3 \longrightarrow NO_2 + O_2 \qquad R7$$

However, the injection of H_2O forms OH and HO_2 through the cycle

$$O(^1D) + H_2O \longrightarrow OH + OH \qquad R32$$

$$O_3 + OH \longrightarrow O_2 + HO_2 \qquad R9$$

This increased concentration of HO_2 reduces NO through the reaction

$$HO_2 + NO \longrightarrow OH + NO_2 \qquad R10$$

Although O_3 is depleted through R9, the depletion of O_3 through R7 has been reduced considerably due to the reduction in NO through R10. In addition, NO_2 is reduced through the reaction

$$OH + NO_2 + M \longrightarrow HNO_3 + M \qquad R12$$

and this in turn reduces the NO produced by the photodissociation of NO_2, i.e.

$$NO_2 + h\nu \rightarrow NO + O(^3P) \qquad R4$$

The net result is a decrease in the effect of the catalytic reduction of ozone through reactions R6 and R7.

In the troposphere, the effect of the combined injection of NO_2 and H_2O is a production of ozone. This ozone increase is due to the increase in the concentration of $O(^3P)$. As pointed out by Widhopf, et al. [1977], in the strict NO_2 injection case the increase in $O(^3P)$ was initiated by the oxidation of methane by OH which then cycled through the smog chain to produce HO_2. The HO_2 reacted with NO to produce NO_2 which then photodissociated to produce $O(^3P)$ as follows:

$$HO_2 + NO \rightarrow OH + NO_2 \qquad R10$$

$$NO_2 + h\nu \rightarrow NO + O(^3P) \qquad R4$$

The additional injection of H_2O increases the effect of the smog cycle on the production of $O(^3P)$ since H_2O is converted to OH by the reaction

$$O(^1D) + H_2O \rightarrow OH + OH \qquad R32$$

The concentration of $O(^3P)$ is also increased through the reaction

$$OH + OH \rightarrow H_2O + O(^3P) \qquad R38$$

The increase of $O(^3P)$ from both NO_2 and H_2O injection increases the ozone through the reaction

$$O_2 + O(^3P) + M \rightarrow O_3 + M \qquad R5$$

Thus, the combined effect of the injection of NO_x and HO_x for these calculations is to increase ozone in both the troposphere and stratosphere. It should be pointed out that these results are based on the new reaction rate measurement for R10 which has been increased by a factor of 35 over previous recommended value (Table I). Since the mechanisms in both stratosphere and troposphere are strongly dependent on the reaction $HO_2 + NO \rightarrow OH + NO_2$, and have resulted in a substantial increase in the ozone columns in the natural atmosphere above observed levels, one should not draw any immediate definite conclusions in regard to the magnitude of the production of ozone for this NO_x and HO_x pollution case. The increased ozone levels in the natural atmosphere are particularly disturbing, and it is possible that other reaction rates involving the hydroxyls are still inaccurate and could influence the present results. This is an important area of future investigation, and possible candidates are hydroxyl reactions involving production or depletion of $O(^3P)$ and those depleting ozone; for example

$$O_3 + OH \rightarrow O_2 + HO_2 \qquad R9$$

$$HO_2 + O_3 \rightarrow OH + O_2 + O_2 \qquad R14$$

$$HO_2 + O(^3P) \rightarrow OH + O_2 \qquad R15$$

$$OH + O(^3P) \rightarrow O_2 + H \qquad R34$$

$$OH + OH \rightarrow H_2O + O(^3P) \qquad R38$$

$$HO_2 + h\nu \rightarrow OH + O(^3P) \qquad R40$$

These results emphasize the immediate need for much better measurements of these hydroxyl reaction rates. They also demonstrate the usefulness of numerical models in the development of a more focused understanding of atmospheric chemistry and the distribution of these important trace species in the stratosphere and troposphere, together with providing current estimates of the magnitude of potential pollution problems.

REFERENCES

Ackerman, M., "Ultraviolet Solar Radiation Related to Mesospheric Processes," Mesospheric Model and Related Experiments, Springer-Verlag, Berlin, pp. 149-159, 1971.

Glatt, L. and Widhopf, G.F., "Aircraft HO_x and NO_x Emission Effects on Stratospheric Ozone and Temperature (NASA report in preparation), 1978.

Harries, J.E., "The Distribution of Water Vapor in the Stratosphere," Rev. of Geophysics and Space Sciences, 14, pp. 565-575, 1976.

Hidalgo, H. and Crutzen, P.J., "The Tropospheric and Stratospheric Composition Perturbed by NO_x Emissions of High Altitude Aircraft," J. Geophysics Res., 82, pp. 5833-5866, 1977.

Howarth, C.J. and Evenson, K.M., "Kinetics of the Reaction of HO_2 Radicals with NO," Geophys. Res. Letters (in preparation), 1977.

Junge, C.E., Air Chemistry and Radioactivity, Academic Press, New York and London, p. 10, 1963.

Kramer, R. and Widhopf, G.F., "On the Evaluation of Diurnally Averaged Photodissociation Rates in Atmospheric Photochemical Models," J. Atmos. Sci. (in preparation), 1978.

London, J., "The Distribution of Total Ozone in the Northern Hemisphere," Beitrage zur Physik der Atmosphäre, 26, pp. 254-263, 1963.

Manabe, S. and Wetherald, R.T., "Thermal Equilibrium of the Atmosphere with a Given Distribution of Relative Humidity," J. Atmos. Sci., 24, pp. 241-259, 1967.

"Chlorofluoromethane and the Stratosphere," R.D. Hudson, ed., NASA Reference Publication 1010, NASA Goddard Space Flight Center, Greenbelt, Maryland.

Oliver, R.C., Bauer, E., Hidalgo, H., Garner, K.A., and Wasylklwskyj, W., "Aircraft Emissions: Potential Effects on Ozone and Climate, a Review and Progress Report,", Department of Transportation, Washington, D.C., p. 404, 1977.

Oort, A.H. and Rasmussen, E.M., "Atmospheric Circulation Statistics," NOAA Prof. Paper No. 5, Geophysical Fluid Dynamics Laboratory, Princeton, New Jersey, 1971.

Sticksel, P.R., "The Annual Variation of Total Ozone in the Southern Hemisphere," Mon. Wea. Rev., 98, pp. 787-788, 1970.

Widhopf, G.F. and Victoria, K.J., "On the Solution of the Unsteady Navier-Stokes Equations Including Multicomponent Finite Rate Chemistry," Computers and Fluids, 1, pp. 159-184, 1973.

Widhopf, G.F., "A Two-Dimensional Photochemical Model of the Stratosphere Including Initial Results of Inert Tracer Studies," Proc. of the Fourth Conf. on CIAP, T.M. Hard and A.J. Broderick, eds., Department of Transportation, DOT-TSC-OST-75-38, February 1975, pp. 316-331, 1975.

Widhopf, G.F., Glatt, L., and Kramer, R.F., "Results of a Phenomenological Time-Dependent Two-Dimensional Photochemical Model of the Atmosphere," AIAA J., 15, pp. 1322-1330, 1977.

INTEGRAL-REPRESENTATION APPROACH FOR TIME-DEPENDENT VISCOUS FLOWS*

J. C. Wu and Y. M. Rizk
Georgia Institute of Technology
Atlanta, Georgia 30332, U.S.A.

INTRODUCTION

This paper describes a novel approach for the numerical simulation of time-dependent incompressible viscous flows. The distinguishing feature of this approach is the formulation of the entirety of the time-dependent flow problem as integral representations prior to the process of discretization. The integral representations are completely equivalent to the more familiar differential Navier-Stokes and continuity equations. Numerical procedures based on the use of the integral representations, however, offers several distinct advantages.

It is convenient to utilize the concept of vorticity, defined as the curl of the velocity vector, and separate the overall incompressible viscous flow problem into its kinetic and kinematic aspects [1]. The kinematic aspect of the problem, describing the instantaneous relationship between the velocity and the vorticity fields, is familiarly expressed as an elliptic differential equation for the velocity vector. The kinetic aspect, describing the transport of vorticity in the fluid, is familiarly expressed as a differential vorticity transport equation. This transport equation is elliptic in the case of steady flows and is parabolic in its time-space relation in the case of time-dependent flows. An analogy between the kinetic and kinematic aspects therefore exists for steady flows, but not for time-dependent flows.

It has been shown [1] that the elliptic differential equation for the kinematic aspect of the viscous flow problem may be recast into an integral representation for the velocity vector. Using this integral representation, it becomes possible to compute the velocity values in the flowfield explicitly, point by point. This unique ability is absent in other methods. It permits the confinement of the solution field to only the vortical region of the flowfield. By eliminating the potential region of the flowfield, which occupies the major part of the total flowfield in external flows at high Reynolds numbers, the computing need for such flows is drastically reduced [2].

Integral representation for the kinematic aspect offers several other major advantages in computation [2,3,4]. In earlier efforts exploring these advantages, the kinetic aspect of the problem was kept in the differential form. Certain limitations experienced by conventional numerical methods using totally differential formulations were found to persist. Relatively recently, the present authors initiated a program of research with the goal of establishing numerical approaches, based totally on integral representations, that may transcend these limitations. It was found that, for steady flows, the kinetic aspect of the problem is readily recast into an integral representation for the vorticity vector by virtue of its analogy to the kinematic aspect [5]. Furthermore, numerical procedures previously developed for the computation of velocity values (kinematics) are easily adopted for the computation of vorticity values. The development of a numerical approach for steady flows based totally on integral representations has therefore progressed rapidly [6]. For time-dependent flows, the kinetic aspect of the problem is dissimilar to the kinematic aspect. An integral representation for the vorticity vector [7] formally rather different from its steady flow counterpart was derived. This new integral representation necessitated the development of entirely new numerical procedures. The work reported here represents the first successful application of the totally integral-representation approach to the time-dependent viscous flow problem.

* This research is supported by the U.S. Army Research Office under Grant Number DAAG29-75-G-0147; and by the Office of Naval Research N00014-75-C-0249.

MATHMATICAL FORMULATION AND PHYSICAL INTERPRETATION

In this paper, time-dependent incompressible viscous flows involving two spatial dimensions past non-rotating finite solids are considered. Integral representations for such flows are expressible as:

$$\vec{v}\,(\vec{r},t) = -\frac{1}{2\pi}\int_R \frac{(\vec{\omega}_o)_{t_o=t} \times (\vec{r}_o-\vec{r})}{|\vec{r}_o-\vec{r}|^2} dR_o + \vec{v}_\infty \tag{1}$$

and

$$\vec{\omega}\,(\vec{r},t) = \iint_R (F\omega_o)_{t_o=o}\, dR_o + \int_o^t dt_o \iint_R \omega_o\, (\vec{v}_o \cdot \vec{\nabla}_o F)\, dR_o$$
$$+ \nu\int_o^t dt_o \oint_B (F\vec{\nabla}_o \omega - \omega_o \vec{\nabla}_o F) \cdot \vec{n}_o\, dB_o \tag{2}$$

where $\vec{v}$ is the velocity vector, $\vec{\omega}$ is the vorticity vector defined as the curl of the velocity vector, $\vec{r}$ is the position vector, t is the time, R is the region occupied by the fluid where the vorticity is non-zero, B is the solid boundary, $\vec{n}$ is the unit normal vector on B directed away from R, ν is the kinematic viscosity, $\vec{v}_\infty$ is the freestream velocity, the subscript "o" designates that a variable, its differentiations, or its integrations is performed in the $\vec{r}_o$ space, and F is the fundamental solution of a parabolic differential equation given by

$$F = \frac{1}{4\pi\nu(t-t_o)} \exp\left\{-\frac{|\vec{r}_o - \vec{r}|^2}{4\nu(t-t_o)}\right\} \tag{3}$$

Equation (1) describes the kinematic aspect of the problem and is completely equivalent to the differential continuity and vorticity definition equations specialized to two-dimensional flows past non-rotating solids with "no-slip" velocity condition on B. Integral representations for more general circumstances, e.g. three-dimensional flows, flows past rotating solids, flows with other than no-slip boundary conditions, as well as derivations of these integral representations are available in [1,2,7]. The specialized integral representation, Eq. (1), is readily interpreted as a statement of the Biot-Savart law for distributed vortex lines.

Equation (2) describes the kinetic aspect of the problem and is equivalent to the familiar differential vorticity transport equation. This equation is derived in [7]. The fundamental solution F gives the vorticity distribution in an infinite unlimited stationary fluid at the time level t as a result of diffusion of a vortex line (concentrated vorticity) of unit strength located at r_o at a preceeding time level t_o. The first integral in Eq. (2) therefore represents the effect of an initial (t_o=o) vorticity distribution. In actual flows, a stationary fluid cannot co-exist with a non-zero vorticity field. Thus, the vorticity distribution changes as a result of convective process which continually effects the subsequent redistribution of vorticity in the fluid. The effects of this process is represented by the second integral in Eq. (2). The third integral in Eq. (2) represents the effects of continual generation (and depletion) of vorticity on the solid boundary B. This process of vorticity generation is a result of the no-slip condition as discussed in [1]. Because the velocity is zero on B, the generated vorticity leaves B only through diffusion, as is shown by Eq. (2).

The significance of the above observations in analytical studies of the viscous flow problem has not yet been fully explored. The advantages of utilizing Eqs. (1) and (2) in numerical solutions are more evident to the present authors and are discussed in the next section of this paper.

NUMERICAL PROCEDURE

The overall numerical procedure to advance the solution by one time interval, say Δt, consists of a kinetic step and a kinematic step. In the kinetic step, the initial time level (t=o) in Eq. (2) is taken to be the time level $t-\Delta t$. Known values of velocity and vorticity at this time level are used and Eq. (2) is solved to obtain a new set of vorticity values at the time level t. In the kinematic step, the newly computed vorticity values are used and Eq. (1) is solved to obtain new velocity values corresponding to the new vorticity values. The kinetic and kinematic steps complete a computation loop. Numerical solution of the problem progresses with repeated executions of this loop, simulates the physical processes of vorticity diffusion, convection, and generation, and offers time-dependent solutions as well as steady-state or periodic (vortex shedding) solutions in the limit of large time, as the case may be.

It is obvious from Eq. (1) that the velocity values can be computed explicitly, point by point. It is also obvious from Eq. (2) that in computing new vorticity values, there is no need to know the velocity values in the non-vortical region of the flow. The entire solution field can be therefore confined to the vortical region. Many options are available for the numerical quadrature of the integrals in Eqs. (1) and (2). The computational advantage offered by carefully developed procedures exploiting the ability to confine the solution field to the vortical region is drastic. Other advantages derived from a better physical insight of the problem, made possible by the integral representations, are described briefly below.

In recent years, the critical importance of employing "boundary-fitted" [8] grid systems in numerical solutions of the Navier-Stokes equation has become well known. Extensive efforts have been devoted by many researchers to develop coordinate transformation and finite-element methods for this purpose. With the entire problem formulated as integral representations, well developed techniques of conventional finite-element methods [7] can be adopted for the construction of boundary-fitted grids. The solution field, confined to the vortical region by the use of integral representations, is divided into cells and the boundary B is divided into segments. In finite-element terminology, the cells and the segments are referred to as elements. The integrals in Eqs. (1) and (2) with respect to spatial variables are replaced by sums of element integrals, each of which is over an individual element. Suitable interpolation functions are introduced to represent flow variables in each element,and the element integrals are intergrated analytically. In this manner, algebraic equations [5] expressing velocity values explicitly in terms of the nodal values of vorticity are obtained from Eq. (1). For the integration with respect to time in Eq. (2), flow variables can be represented by suitable "time interpolation-functions" in the "time element" Δt. For simplicity, the flow variables are taken to be linear functions of time within each time element. Algebraic equations obtained from Eq. (2) by carrying out the integrations with respect to time analytically is arranged in the form:

$$\{\omega\} = [A]\ \{u\omega\} + [B]\ \{v\omega\} + [C]\ \{\omega_B\} + [D]\ \{h_B\} + \{f\} \tag{4}$$

It should be emphasized, before discussing Eq. (4), that the process of discretization just described permits the use of elements of any desired shape and size. Furthermore, different interpolation functions of any desired type and order of accuracy can be used for different elements. In cases where analytical integrations are not obtained, the element integrals can be evaluated numerically and the results stored efficiently. The above described process therefore offers an inherent flexibility not only in accomodating complex boundary geometries but also in selecting proper numerical representations for the flow variables. In addition, since the various kinetic processes are represented by separate integrals in Eq. (2), different techniques of numerical quadrature can be utilized to compute the effects of different processes. Furthermore, a specific technique ideally suited for one flow region may not be well suited for another region. With the integral representations, it is simple to develop hybrid methods using different techniques to treat different regions, and there is no need to use the same numerical techniques throughout the entire solution field. These features of the integral-representation approach have not yet been fully utilized. The results of the present study, however, have conclusively demonstrated that this approach is well suited for separated flows at high Reynolds numbers. In such flows, there usually exist regions of boundary layer, recirculation, and wake,each with its own characteristics and in each the processes of convection and diffusion maybe proceeding at vastly different rates.

In Eq. (4), $\{\omega\}$, $\{u\omega\}$, $\{v\omega\}$, and $\{f\}$ are Nxl matrices, N being the number of nodes in the solution field. $\{\omega_B\}$ and $\{h_B\}$ are M x 1 matrices, M being the number of nodes located on the boundary B. u and v are Cartesian velocity components. ω_B and h_B are respectively the value and the normal gradient of vorticity on B. The matrix $\{f\}$ is dependent only on flow variables at the initial time level $t-\Delta t$. All other one-column matrices are to be evaluated at the new time level t. The coefficient matrices $[A]$, $[B]$, $[C]$, and $[D]$ are sparsely populated and depend only on the node locations and the time inteval Δt. An iterative technique has been developed in the present study for solving Eq. (4) at each new time level. In principle, values of $\{u\omega\}$, $\{v\omega\}$, $\{\omega_B\}$, and $\{h_B\}$, calculated for any given iteration, can be placed into Eq. (4) to compute new values of $\{\omega\}$ at nodes not on B for the subsequent iteration. New values of $\{u\omega\}$, $\{v\omega\}$, $\{\omega_B\}$, and $\{h_B\}$ can then be computed to complete the iteration loop.

In actual computation, Eq.(4) is first re-arranged in a more convenient form. Values of u and v at non-boundary nodes are computed from values at the initial time level $t-\Delta t$ using Eq. (1) and, for simplicity, are not iterated. $\{u\omega\}$ and $\{v\omega\}$ on the non-boundary points are therefore easily obtained by scalar multiplications. Boundary values of $u\omega$ and $v\omega$ are zero because of the no-slip condition. It has been shown that ω_B values can be expressed as linear functions of ω values on non-boundary nodes through a kinematic restriction [3] which is generally valid for viscous flows. With $\{\omega_B\}$, $\{u\omega\}$, and $\{v\omega\}$ computed, the only unknowns in Eq. (4), when applied on the boundary nodes, are $\{h_B\}$. New values of $\{h_B\}$ can therefore, be expressed in terms of non-boundary values of $\{\omega\}$ Expressing $\{h_B\}$ and $\{\omega_B\}$ in terms of $\{\omega\}$ and rearranging the results through a matrix inversion procedure, one obtains from Eq. (4),

$$\{\omega\} = [F]\ \{u\omega\} + [G]\ \{v\omega\} + \{g\} \qquad (5)$$

where $\{g\}$ depends only on flow variables at the initial time level, the coefficient matrices $[F]$ and $[G]$ depend only on the locations of nodes and the time inteval Δt. Equation (5) is convenient for the iterative solution of $\{\omega\}$. New values of $\{\omega\}$ calculated for each iteration is used to obtain new values of $\{u\omega\}$ and $\{v\omega\}$ which is then placed into Eq. (5) to compute $\{\omega\}$ values for the subsequent iteration. This process continues until $\{\omega\}$ converges to prescribed limits. The coefficient matrices $[F]$ and $[G]$ are computed only once and stored for use in all time loops using the same time inteval Δt. The computation procedure is very efficient.

In Eq. (4), the terms containing $\{\omega_B\}$ and $\{h_B\}$ represent the diffusion of newly generated vorticity from the solid boundary B. With any reasonably small time step Δt, the effective distance through which this vorticity diffuses is small. In the present study, the solution field is divided into an outer region far from B, where the terms $[C]\ \{\omega_B\}$ and $[D]\ \{h_B\}$ in Eq. (4) are neglected because the pertinent elements in $[C]$ and $[D]$ are negligibly small, and an inner region adjacent to B, where these terms are important. This division not only simplifies the computation of $\{\omega\}$ in the outer region but also reduces the computation of the coefficient matrices $[F]$ and $[G]$ in Eq. (5).

RESULTS AND DISCUSSIONS

The ability of the integral-representation approach to treat attached and separated flows has been tested by solving the problems of flows past a finite flat plate aligned with the freestream and past a circular cylinder. In each problem, the motion of the fluid results from an impulsively started motion of the solid. The time-dependent solution is carried to a sufficiently large time level so that asymptotic steady-state results are obtained.

A steady-state mid-plate tangential velocity profile obtained for the flat plate problem is shown in Figure 1. As expected, the present result agrees well with the well known Blasius' profile for a semi-infinite plate at a station half the finite plate length from its leading edge. The present results shows a slight velocity overshoot ($u > U$), which is expected because of the favorable pressure gradient along the plate caused by the displacement thickness. The Reynolds number for the flow based on the plate length L and the freestream velocity U is 1000. Triangular elements with linear interpolation functions are used. By using an "expanding" grid system, i.e., increasing the size of the element with the distance from the plate, the total number of nodes is kept to 403, with 42 nodes in the inner region.

Computed streamlines and equi-vorticity contours are shown in Figure 2, in the upper and lower halves of the figure respectively, for a flow past a circular cylinder at a Reynolds number of 40 based on the freestream velocity and the cylinder diameter. The flow patterns shown are for a asymptotically steady-state time level. Elements formed by "expanding" polar coordinate lines are used, with a total of 800 nodes in the solution field, 80 nodes in the inner region. The computed pressure coefficient, C_P, i.e. the pressure difference divided by the freestream kinetic energy, on the cylinder surface is shown in Figure 3. The polar angle Θ is measured from the forward stagnation point on the cylinder. Very good agreement exists between computed results and experimental data [9].

It is well known that the shearing stress at solid boundaries in a viscous flow is directly proportional to the vorticity ω_B on the boundary, the constant of proportionality being the negative of the absolute viscosity. It is easy to show, by considering the familiar Navier-Stokes equation as applied at the solid boundary, that the tangential pressure gradient is similarly proportional to the normal vorticity gradient h_B on the solid boundary. These relations have been utilized in the present work as well as in conventional numerical methods to compute forces acting on solid bodies in viscous flows. In conventional methods, ω_B and h_B are computed using one-sided differences formulas by extrapolating known values on non-boundary points. The results of these computations are sensitive to the grid spacing near the wall and to the order of accuracy of the difference formulas. In the present work, highly accurate methods for computing ω_B and h_B are established by solving integral equations [3] . In Figure 4, values of h_B computed using different methods are shown. The difference between various results are significant, particularly between results obtained using difference formulas of first and second orders of accuracy.

For the circular cylinder problem treated, about 60 seconds of CDC-6600 CPU time is required, on the average, to advance the numerical solution by one unit dimensionless time, i.e., the physical time inteval during which the cylinder moves relative to the freestream by one diameter. This efficiency in computation as well as the accuracy of solution for a separated flow problem is particularly encouraging since the present study represents only an initial stage of development of the integral-representation approach. The considerable insight into the interplay between the physical and numerical aspects of the viscous flow problem, brought into focus by the use of integral representations, is expected to continue contributing to future development of the present numerical approach as well as other approaches.

REFERENCES

1. Wu, J.C. and Thompson, J.F., "Numerical Solution of Time-Dependent Incompressible Navier-Stokes Equations Using an Integro-Differential Formulation", Journal of Computers and Fluids, Vol. 1, No. 2, pp. 197-215, 1973.

2. Wu, J.C., Sankar, N.L., and Sampath, S., "A Numerical Study of Unsteady Viscous Flows Around Airfoils", AGARD Conference Proceedings No. 227, pp. 24-1-18, 1978.

3. Wu, J.C., "Numerical Boundary Conditions for Viscous Flow Problems", AIAA Journal, Vol. 14, No. 8, pp. 1042-1049, 1976.

4. Wu. J.C., Spring, A.H., and Sankar, N.L., "A Flowfield Segmentation Method for the Numerical Solution of Viscous Flow Problems", Lecture Notes in Physics, Vol. 35, pp. 452-457, Springer-Verlag, 1974.

5. Wu, J.C., and Wahbah, M.M., "Numerical Solution of Viscous Flow Equations Using Integral Representations", Lecture Notes in Physics, Vol. 59, pp. 448-453, Springer-Verlag, 1976.

6. Wahbah, M.M., "Computation of Internal Flows with Arbitrary Boundaries Using the Integral Representations Methods", Georgia Institute of Technology, School of Aerospace Engineering Report, 1978.

7. Wu, J.C., "Finite Element Solution of Flow Problems Using Integral Representations", Proc. of Second International Symposium on Finite Element Methods in Engineering, International Centre for Computer Aided Design, Conference Series No. 2/76, pp. 205-216, 1976.

8. Thompson, J.F., Thames, F.C., and Mastin, C.W., "Automatic Numerical Generation of Body-Fitted Curvilinear Coordinate System for Field Containing any Number of Arbitrary Two-Dimensional Bodies", Journal of Computational Physics, Vol. 15, No. 3, 1974.

9. Grove, A.S., Shair, F.H., Petersen, E.E., and Acrivos, A., "An Experimental Investigation of the Steady Separated Flow Past a Circular Cylinder", Journal of Fluid Mechanics, Vol. 33, Part, pp. 60-80, 1964.

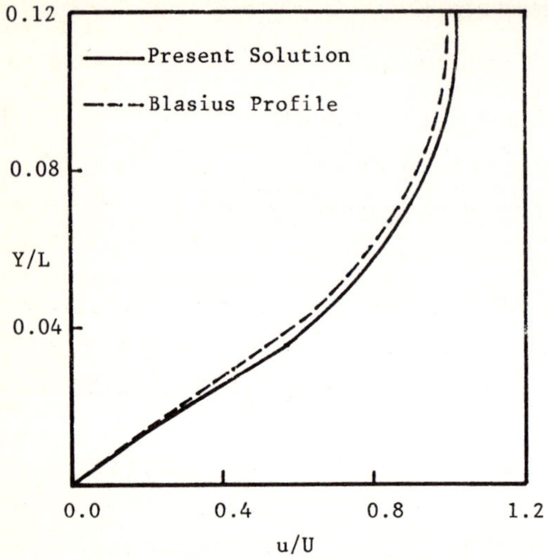

Figure 1. Midplate Velocity Profile

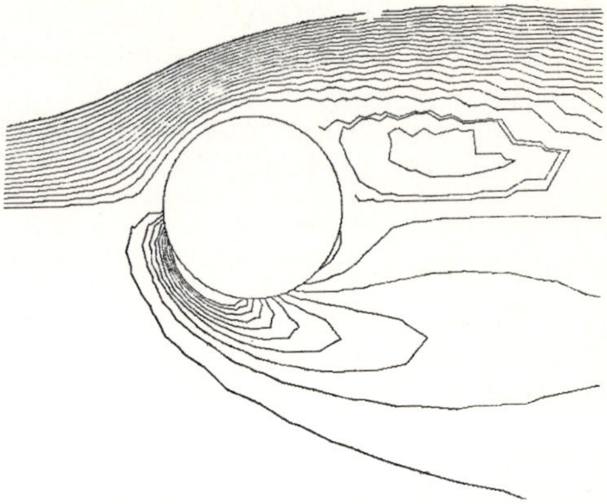

Figure 2. Flow Pattern Around Circular Cylinder

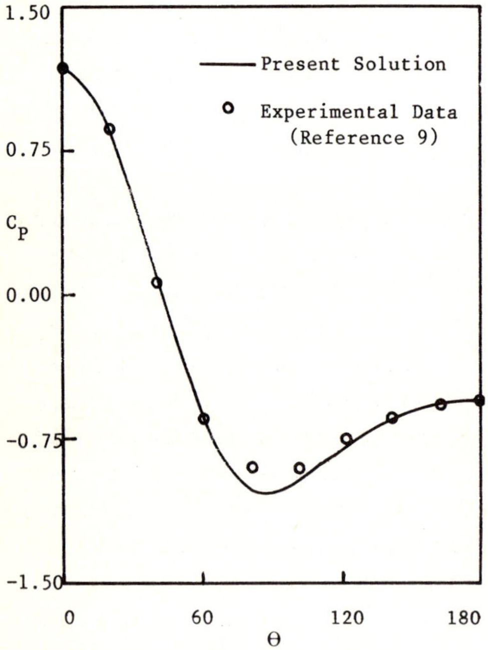

Figure 3. Pressure Distribution on Circular Cylinder

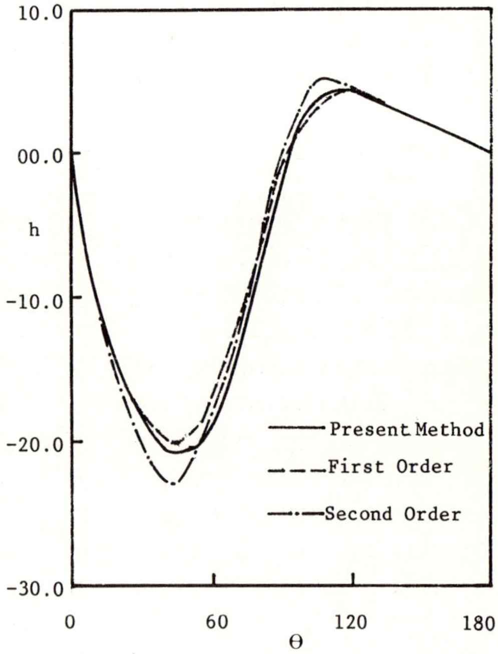

Figure 4. Normal Gradient of Vorticity on Circular Cylinder

ON SOME METHODS FOR THE NUMERICAL SIMULATION OF FLOWS WITH COMPLEX STRUCTURE

N.N.Yanenko, V.M.Kovenya, V.D.Lisejkin,
V.M.Fomin, E.V.Vorozhtsov
Institute of Theoretical & Applied Mechanics
USSR Academy of Sciences, Novosibirsk 630090

This paper represents a survey of the investigations carried out at the Institute of Theoretical & Applied Mechanics of the Siberian Division of the USSR Academy of Sciences. The questions of construction of efficient difference schemes for the numerical integration of a compressible viscous heat-conducting gas based on splitting in terms of the physical processes and the space variables in conjuction with adapting to the flow moving difference grids are considered. The questions of the interpretation of the shock-capturing computational results of the gas-dynamical problems solution are also considered, in particular, the problem of the shock wave localization in the numerical simulation of flows of an inviscid, non-heat-conducting compressible gas.

The methods presented are especially efficient in computations of flows with c o m p l e x s t r u c t u r e where the singular zones and elements are present (e.g. shock waves and transitions, contact stripes, boundary layers, double and triple shock configurations etc.).

1. Until present the g e o m e t r i c approach to the construction of moving grids for the solution of gas dynamical problems was the basic one. In this approach the grid was determined by the domain under consideration rather than by the solution itself. The approach to be proposed is d y n a m i c, that is the grid is constructed according to the solution of the boundary-value problem adapting to it in the process of a computation. Necessary equations governing the mesh are obtained as Euler equations on the basis of certain variational principles and form a coupled equation system simultaneously with the equations of the flow. More generally, one can speak not about a grid, but about its continuous image being a continuous m a p p i n g of the Eulerian coordinates x^1, x^2, x^3, t space into the space of moving coordinates q^1, q^2, q^3, t . This mapping determines the information medium movement simultaneously with the material medium movement. The functionals underlaying the variational principle may be of quite a different nature. It may by functionals re-

lated to a certain measure of the grid deformation as well as to the proximity to the Lagrangian mesh. The presence of comparatively narrow transition zones with large gradients requires a mesh refinement in these zones. But the most general approach seems to be related to the use of the error functional when the mapping function should minimize in some norm the approximation error. This determines automatically the deformation as well as the mesh refinement.

At first let us formulate a variational principle accounting for the "non-lagrangianness", deformation and refinement measure of the mesh [1]. Consider for simplicity the two-dimensional case. Let us denote by $x^1(q^1, q^2, t)$, $x^2(q^1, q^2, t)$ the transformation of the new coordinates q^1, q^2, t into the Cartesian ones and by $q^1(x^1, x^2, t)$, $q^2(x^1, x^2, t)$ let us denote the transformation of the Cartesian coordinates into the new ones. We regard the Jacobian $\Delta = \det \frac{\partial x^i}{\partial q^j}$ as a measure of the mesh refinement and the quantity $h_1 = \left(\frac{\partial q^1}{\partial x^1} - \frac{\partial q^2}{\partial x^2}\right)^2 + \left(\frac{\partial q^1}{\partial x^2} + \frac{\partial q^2}{\partial x^1}\right)^2$ characterizing the deviation from the conformal mapping as a measure of the mesh deformation. We regard the quantity $h_2 = \sum_{i=1}^{2} (\tilde{u}^i - \tilde{\vartheta}^i)^2$ as a measure of the "non-lagrangianness" where $\tilde{u}^i$, $\tilde{\vartheta}^i$ are the velocity components of the medium and of the mesh, respectively, in the local basis $e^i = \frac{\partial x}{\partial q^i}$. Let us define the function φ by the condition

$$\varphi = \varepsilon_1 h_1 + \varepsilon_2 h_2 + \varepsilon_3 h \cdot \Delta^{\alpha} \tag{I.I}$$

where ε_i, α are parameters, h is a function depending on the gradients of hydrodynamic quantities. The variational principle for the mesh construction is formulated as a minimization problem for a functional

$$\Phi(q^i) = \int_X \varphi \, dx^1 dx^2 dt \, . \tag{I.2}$$

The integration in (I.2) is carried out over the domain X, since in this case the hydrodynamic quantities are not varied. When minimizing (I.2) the first item in (I.I) tends to make the grid conformal, the second one - Lagrangian, the third one - refined. The extremal surface satisfies the Euler equation

$$\frac{\partial}{\partial t}\varepsilon_2 \Delta(\tilde{u}^i - \tilde{\vartheta}^i) + \frac{\partial}{\partial q^j}\Big[\varepsilon_2 \Delta(\tilde{u}^i - \tilde{\vartheta}^i)(\tilde{u}^j - \tilde{\vartheta}^j) + \\ + \varepsilon_1 \Delta g^{ij} - \delta^{ij}\varepsilon_2 h_2 \Delta - \delta^{ij}\varepsilon_3 \alpha h \Delta^{\alpha+1}\Big] = 0 \tag{I.3}$$

where

$$g^{ij} = \sum_{\alpha} \frac{\partial q^i}{\partial x^{\alpha}} \cdot \frac{\partial q^j}{\partial x^{\alpha}} \, , \qquad \delta^{ij} = \begin{cases} 0 \, , & i \neq j \\ 1 \, , & i = j \end{cases} .$$

When $\varepsilon_2 = \varepsilon_3 = 0$, the equation (I.3) coincides with the one for a mesh defined by Winslow in [2] . Adding to (I.3) the gas dynamical equations written in tensor form [3]

$$\partial \Delta \tilde{\sigma}^{i0}/\partial t + \partial \Delta \tilde{\sigma}^{ij}/\partial q^j + \Delta \Gamma^i_{\alpha\beta} \tilde{\sigma}^{\alpha\beta} = 0 \qquad \text{(I.4)}$$

one obtaines a closed equation system with respect to the variable $F = \{ \rho, \tilde{u}^i, e, \tilde{\vartheta}^i \}$.

For some problems of gas dynamics, e.g. computation of the flow around convex bodies, it is sufficient to refine the mesh in one coordinate direction. In this case one has to find such one-dimensional transformation $x = x(q, t)$ that the gradients $\partial f_i / \partial q$ (f_i are the hydrodynamic functions) be not great with respect to the new coordinate q . We define the transformation as the minimization of functionals

$$\varphi_1(q,t) = \int_X \left[\sum a_i \left| \frac{\partial f_i}{\partial q} \right|^{\alpha_i} + b \ln \Delta \right] dx \, ,$$

$$\varphi_2(q,t) = \int_X \left[\sum a_i \left| \frac{\partial f_i}{\partial q} \right|^{\alpha_i} + b \Delta^{\beta} + c \Delta^{-\eta} \right] dx \, . \qquad \text{(I.5)}$$

where α_i, b , c , η , β are the real parameters, a_i are the functions of hydrodynamic quantities gradients. The mapping $x(q,t)$ satisfies the Euler equation

$$\frac{\partial}{\partial q} \left[\left(\sum a_i \alpha_i \left| \frac{\partial f_i}{\partial q} \right|^{\alpha_i} + b \right) \frac{\partial x}{\partial q} \right] = 0 \, , \qquad \text{a)}$$

$$\frac{\partial}{\partial q} \left[\sum a_i \alpha_i \left| \frac{\partial f_i}{\partial q} \right|^{\alpha_i} \frac{\partial x}{\partial q} + b \cdot \beta \left(\frac{\partial x}{\partial q} \right)^{\beta+1} - c \cdot \eta \left(\frac{\partial x}{\partial q} \right)^{-\eta+1} \right] = 0 . \text{b)} \qquad \text{(I.6)}$$

In the variational principles (I.2), (I.5) the functional is chosen in such a way that the requirements imposed a priori on the mesh are satisfied. In papers [4-5] the variational principle for the mesh construction is formulated by the condition of the error functional minimization.

Let $\mathcal{L}(u) = 0$ be a system of differential equations with respect to the vector function u and let $\Lambda(u_i)$ be an approximation to $\mathcal{L}$ on a moving grid. Denote by ε the leading term of the error in u and by R the leading term in the approximation error for $\mathcal{L}$. As it is shown in [5] , ε and R are bound by the equation $\mathcal{L}_1(\varepsilon) = R$ where $\mathcal{L}_1$ is the variation equation for the system $\mathcal{L}(u) = 0$. Since the approximation error R depends on the difference grid, ε does it also. Therefore it is worthwhile to find such a mesh for which ε is minimum in some norm. If the norm is taken in the form $\|\varepsilon\|^{\alpha} = \int_X A(\varepsilon)\,dx$, the problem of such a mesh construction reduces to the minimization problem for a functional $\int_X A(\varepsilon)dx$ with a connection $\mathcal{L}_1(\varepsilon) = R$. Moreover, one can find a mesh from the condition of the norm minimization of R , that is to define the functional by the relationship $\varphi(q) = \int_X A(R)dx$.

Another approach to the mesh construction minimizing the error is also of interest. From the equation $\mathcal{L}_1(\varepsilon) = R$ at $R = 0$ one gets $\varepsilon = 0$. It is obvious that the mesh for which $R = 0$ would be optimum. Therefore the optimum mapping is found if the equation $R = 0$ is soluble with respect to the mesh points coordinates x_i. But the equation $R = 0$ is insoluble for arbitrary scheme Λ consistent with $\mathcal{L}$. In this case the problem of the optimum mesh construction is solved simultaneously with the problem of the construction of such a scheme Λ for which the equation $R = 0$ is soluble with respect to the mesh points coordinates x_i. In this case the optimum mesh increases by unity the order of approximation. For example, the optimum mesh for the first order scheme

$$(u_{i+1} - u_i)/(x_{i+1} - x_i) = f(u_i, x_i)$$

consistent with the ordinary differential equation $du/dx = f(u,x)$ is given by the equation

$$x_{i+1} - x_i = \Delta q \cdot (|f_{i+1}| + \Delta q)^{-1/2}$$

where Δq is the mesh step.

2. For the numerical solution of an equation system of the viscous heat-conducting compressible gas an implicit difference scheme based on splitting in terms of the physical processes and the space variables is proposed. This scheme possesses the property of the absolute stability and conservation (in the stationary case) and is operated by scalar sweeps which renders it economical. When solving multidimensional equations the splitting in terms of the space variables involved is made (cf., e.g. [6]) and in each spatial direction the splitting in terms of physical processes is used. At first let us describe the schemes in Eulerian Cartesian coordinates (x^1 , x^2, x^3, t).

A) One-dimensional case. Splitting in terms of physical processes. Let

$$\frac{\partial \vec{F}}{\partial t} + \frac{\partial \vec{w}}{\partial x} = 0 \qquad (2.I)$$

be a system of the Navier-Stokes equations (or the equations of gas dynamics when $\mu \equiv \lambda \equiv 0$) where

$$\vec{F}=\begin{pmatrix}\rho\\ \rho u\\ E\end{pmatrix},\quad E=\rho(e+u^2/2),\quad \vec{w}=\begin{pmatrix}\rho u\\ \rho u^2+p-\frac{4}{3}\mu\frac{\partial u}{\partial x}\\ u(E+p)-\lambda\frac{\partial T}{\partial x}-\frac{4}{3}\mu u\frac{\partial u}{\partial x}\end{pmatrix}.$$

Let us represent the equation (2.I) in the form of a splitting in terms of physical processes

$$\frac{\partial \vec{f}}{\partial t}+\sum_{j=1}^{2}\Omega_j\vec{f}=\vec{\mathcal{F}}. \tag{2.2}$$

Differential matrix operator Ω_1 takes into account the convective and viscous terms and Ω_2 - the terms with the pressure in the momentum equation and the terms of the form $\operatorname{div}\vec{u}$ in the continuity and energy equations:

$$\vec{f}=\begin{pmatrix}\rho\\ u\\ e\end{pmatrix},\quad \vec{\mathcal{F}}=\begin{pmatrix}0\\ 0\\ \frac{4}{3}\frac{\mu}{\rho}\left(\frac{\partial u}{\partial x}\right)^2\end{pmatrix},\quad \Omega_2=\begin{pmatrix}0 & \rho\frac{\partial}{\partial x} & 0\\ a^2\frac{\partial}{\partial x} & 0 & b^2\frac{\partial}{\partial x}\\ 0 & c^2\frac{\partial}{\partial x} & 0\end{pmatrix},\quad q=\frac{1}{\rho}\begin{pmatrix}0&0&0\\ 0&\frac{4}{3}&0\\ 0&0&\frac{\gamma}{P_r}\end{pmatrix},$$

$$\Omega_1=u\frac{\partial}{\partial x}I-q\frac{\partial}{\partial x}\mu\frac{\partial}{\partial x},\quad a^2=\frac{1}{\rho}\frac{\partial p}{\partial \rho},\quad b^2=\frac{1}{\rho}\frac{\partial p}{\partial e},\quad c^2=p/\rho.$$

Here I is the unit matrix. Introducing the matrix $A=(\partial\vec{F}/\partial f)^{-1}$ let us represent the equation system (2.I) in the form

$$\partial\vec{f}/\partial t+A\,\partial\vec{w}/\partial x=0. \tag{2.3}$$

The equations (2.2) - (2.3) are equivalent in their differential form, but the equation system (2.3) is more preferable for the numerical realization because in the stationary case the divergence form is used (the matrix A is nonsingular).

For the numerical solution of the equation system (2.3) let us consider the difference scheme of splitting in terms of the physical processes [7]

$$(E+\tau\alpha_1\Omega^{\ell}_{1h})\vec{\xi}^{\,n+1/2}=-\tau\left[A^n_h\left(\frac{\partial\vec{w}}{\partial x}\right)^n_h\right]^{\kappa},$$

$$(E+\tau\alpha_2\Omega^{\ell}_{2h})\vec{\xi}^{\,n+1}=\vec{\xi}^{\,n+1/2},$$

$$\vec{f}^{\,n+1}=\vec{f}^{\,n}+\vec{\xi}^{\,n+1} \tag{2.4}$$

where Ω^{ℓ}_{jh}, $\left[\left(\frac{\partial\vec{w}}{\partial x}\right)^n_h\right]^{\kappa}$ are the difference approximations of corresponding differential operators Ω_j, $\partial\vec{w}/\partial x$ ($j=1,2$) with the order ℓ, κ ($\ell,\kappa=1,2$). The weight multipliers α_j are chosen from the stability condition of a difference scheme. Eliminating auxiliary fractional steps one obtains difference scheme in integral steps of the universal algorithm type [7] :

$$C^{\ell}\frac{\vec{f}_h^{n+1}-\vec{f}_h^{n}}{\tau}=-\left[A_h^n\left(\frac{\partial \vec{w}}{\partial x}\right)_h^n\right]^{\kappa},\quad C^{\ell}=\prod_{j=1}^{2}(E+\tau\alpha_j\Omega_{jh}^{\ell}). \qquad (2.5)$$

The difference scheme (2.5) or the equivalent to it scheme (2.4) approximate the equation (2.3) with order $O(\tau+h^{\ell})$ and in the stationary case with an order κ .

The stabilizing operator C^{ℓ} is called consistent if $\ell=\kappa$ and inconsistent otherwise. The difference scheme (2.4) is operated by scalar tridiagonal (ℓ = I) or pentadiagonal (ℓ = 2) sweeps at each fractional level. It is convenient to choose the inconsistent operator C^1 as a stabilizing operator in the process of obtaining the stationary solution by the relaxation method. In this case the difference scheme (2.4) is operated by tridiagonal scalar sweeps, has the error $O(h^{\kappa})$ in the stationary limit and is conservative: the matrix A is nonsingular, and the stationary equations $\left[\left(\frac{\partial \vec{w}}{\partial x}\right)_h^n\right]^{\kappa}=0$ are written in the form of difference conservation laws. It can be shown by the spectral method that the difference scheme (2.5) for the system of linearized equations is unconditionally stable at $\alpha_j \geq 1/2$ in the case of the consistent operator C^{ℓ} (ℓ = 1,2) at $\alpha_j \geq 1$ in the case of the nonconsistent operator C^1 (κ = 1,2). At the first fractional step $n+1/2$ all the difference equations are resolved by scalar sweeps independently of each other (the matrix Ω_{1h}^{ℓ} has nonzero elements only on the principal diagonal). At the second fractional step one obtains a (2 ℓ + 1) - diagonal difference equation with respect to ξ_u^{n+1} performing the elimination of ξ_{ρ}^{n+1} and ξ_e^{n+1} from the momentum equation. This equation can be solved by a scalar sweep, afterwards the quantities ξ_{ρ}^{n+1} and ξ_e^{n+1} are determined using the known ξ_u^{n+1}. At last the new values of functions are found from the last equation (2.4).

B) Multidimensional case. The splitting in terms of physical processes and space variables. The system of the multidimensional equations of the viscous compressible gas

$$\frac{\partial \vec{F}}{\partial t}+\sum_{j=1}^{N}\frac{\partial \vec{w}_j}{\partial x^j}=\vec{G}$$

where N is the dimension of equations (N = 1,2,3), $\vec{w}_j$ are mass, momentum and total energy fluxes in the direction x^j, may be represented in the form of a splitting in terms of space variables and physical processes

$$\frac{\partial \vec{f}}{\partial t}+\sum_{j=1}^{2N}\Omega_j\vec{f}=\vec{L}$$

or in the equivalent form

$$\frac{\partial \vec{f}}{\partial t} = -A\left(\sum_{j=1}^{N} \frac{\partial \vec{w}_j}{\partial x^j} - \vec{G}\right). \qquad (2.6)$$

Here $A = (\partial \vec{F} / \partial \vec{f})^{-1}$, $\vec{F} = \vec{F}(\rho, \rho u_1, \dots, \rho u_N, E)$, $\vec{f} = \vec{f}(\rho, u_1, \dots, u_N, e)$. The differential matrix operators Ω_S ($S = 1, \dots, N$) take into account the convection and viscous terms (without mixed derivatives) in the direction of x_j, and Ω_{S+N} are the terms with pressure and the terms of the form $\operatorname{div} \vec{u}$ in each direction. The vectors $\vec{L}$, $\vec{G}$ contain the mixed derivatives.

For the numerical solution of the system of equations (2.6) the difference scheme of the universal algorithm type is constructed analogously to the one-dimensional case

$$C^{\ell} \frac{\vec{f}_h^{n+1} - \vec{f}_h^{n}}{\tau} = -A\left[\left(\sum_{j=1}^{N} \frac{\partial \vec{w}_j}{\partial x^j} - \vec{G}\right)_h^{n}\right]^{\kappa}, \quad C^{\ell} = \prod_{j=1}^{2N} (E + \tau \alpha_j \Omega_{jh}^{\ell}). \qquad (2.7)$$

This scheme is absolutely stable, fully consistent and is conservative (in the stationary case). The scheme (2.7) is operated by scalar tri- or pentadiagonal sweeps.

In the numerical computations of the flows of complex structure the coordinate transformation $q^i = q^i(x^j)$ mapping the initial domain onto the unit square ($j, i = 1, \dots, N$) is used. The scheme (2.7) is written analogously for the equations of gas dynamics in new coordinates (I.4). The properties of the absolute stability are conserved. The coordinate transformation is constructed on the basis of the variational principle (I.5) to get small steps of the difference grid in the corresponding direction in the domain of large gradients.

Numerical computations of the supersonic flow of a viscous heat-conducting gas around axisymmetric bodies were carried out using the above difference scheme (2.7) and the method of moving grids. The computations showed that the difference scheme (2.7) is absolutely stable also for the nonlinear equations. In the Figures 1,2 the distribution of the velocity components on the symmetry axis and in the cross-section $S = \pi/2$ is given for various Reynolds numbers. A coordinate system fitted to the body surface was chosen for computations.

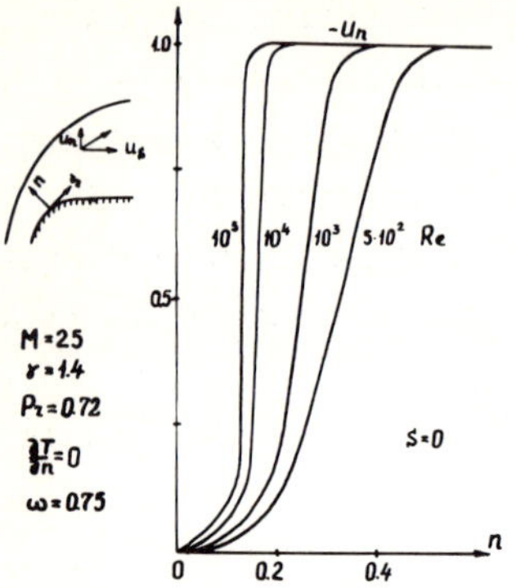

Fig.1

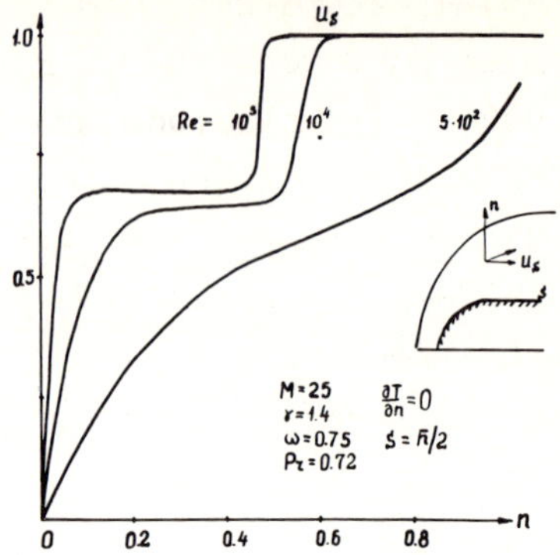

Fig.2

The application of a moving grid having a small number of nodes (≈ 20) enables one to obtain the solution until $Re \approx 10^4$ being in a good agreemnt with computations of previous authors. In the Fig.3 the computing mesh obtained in the relaxation process is shown

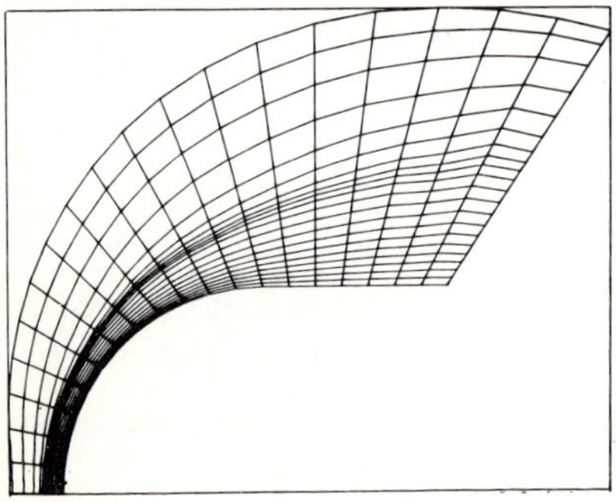

Fig.3

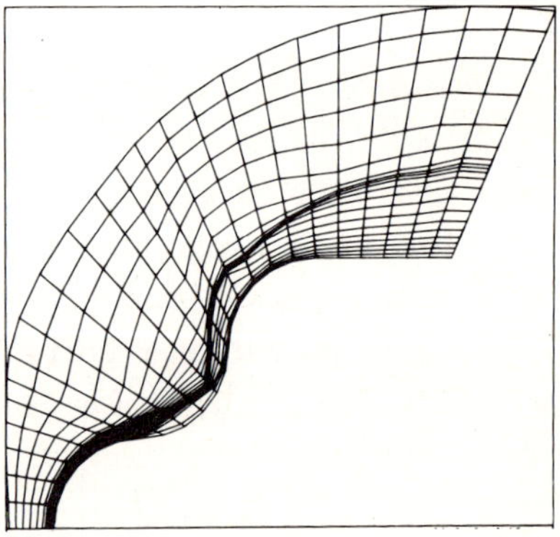

Fig.4

for a variant $M = 25$, $Re = 3 \cdot 10^3$, $\gamma = 1.2$, $\omega = 0.75$. The difference grid for a non-convex body is given in Fig.4 ($M = 15$, $\gamma = 1.2$, $\omega = 0.5$, $Re = 3 \cdot 10^3$, $T_W = 15\,T_\infty$). It is observed a strong condensation of coordinate lines in the domains corresponding to the boundary layer and the shock wave. In the non-convex domain a recurrent flow region is formed.

The computations carried out enable one to draw a conclusion on the efficiency of the difference scheme proposed and the moving grids method.

3. As it is known [8-9], the amount of information obtained in the numerical solution of multidimensional problems exceeds by several orders of magnitude the one being of real interest for a research worker. In this connection the questions of the development and the foundation of algorithms for the localization of singularities using the results of the gas-dynamical shock-capturing computations arise, in particular, the questions of the differential analysers of shock waves, i.e. the algorithms for the determination of the shock wave location within the computational domain.

It should be noted that the existing algorithms for the localization of shock waves when using the homogeneous difference schemes have not been founded well so far. First of all one can relate to these algorithms the ones based on the isolines condensation, on the maximum of the artificial viscosity or the gradients of some functions sought for. In papers [10-13] whose review is made below the first theory is described apparently which justifies some methods for locating shock waves by the shock-smoothing computation results of the one-dimensional gas-dynamic problems.

Before stating the main notions of this theory let us consider a comparatively simple example. Let $u_\mu(x,t)$ be the exact solution of the equation

$$\frac{\partial u_\mu}{\partial t} + u_\mu \frac{\partial u_\mu}{\partial x} = \mu \frac{\partial^2 u_\mu}{\partial x^2}, \quad \mu = \mathrm{Const} > 0 \tag{3.I}$$

with the initial condition

$$u_\mu(x,0) = \begin{cases} u_1, & x < x_0 \\ u_2, & x > x_0 \end{cases} \tag{3.2}$$

where u_1, u_2, x_0 are constants and $u_1 > u_2$. Let us indicate the following properties of the solution of the problem (3.I), (3.2): for arbitrary fixed $t > 0$ there exists such a point (x_c, t) at which the solution graphs of the problem (3.I), (3.2) obtained for the different values of the constant $\mu > 0$ intersect. Furthermore,

$$\max_x \left| \frac{\partial u_\mu}{\partial x} \right| = \left| \frac{\partial u_\mu}{\partial x} \right|_{x = x_c}.$$

And, finally, the locus of points (x_c, t) forms in the (x, t) plane a line which is the discontinuity trajectory in the solution $u(x,t)$ of a problem

$$\frac{\partial u}{\partial t} + u \frac{\partial u}{\partial x} = 0,$$

$$u(x,0) = \begin{cases} u_1, & x < x_0 \\ u_2, & x > x_0 \end{cases}, \quad u_1 > u_2. \tag{3.3}$$

Thus there exists a point at which the quantity $u_\mu(x, t_0)$ does not depend on μ in the profiles of $u_\mu(x_c, t_0)$ obtained for different μ and this point coincides with the exact position of the discontinuity in the solution of a problem (3.3). This point was called the centre of a "smeared" discontinuity. The above properties of the solution $u_\mu(x, t)$ are fulfilled exactly at the initial data function $u_\mu(x, 0)$ having a form of a step and seem to be fulfilled approximately for the arbitrary profile with a shock wave.

Let us consider a finite difference shock-smoothing scheme consistent with the one-dimensional equations of gas dynamics. Generally the numerical solution obtained by such a scheme depends on grid parameters h, τ where h is the step of a space grid and τ is the time step. Put in accordance to the first or the second order difference scheme the differential approximation (first or second, respectively) [14-16] . We shall consider the solutions of the progressive wave class of the equations of the differential approximation type, i.e. the solutions depending only upon the variable $x' = x - Dt$ where x is the space coordinate, t is time, D is the stationary shock wave speed. As it is known, the solution of the progressive wave type is determined with accuracy up to the translation along the x' axis. Let $\vec{w}(x', x'_0, h, \tau) = \{ w_1(x', x'_0, h, \tau), \ldots, w_n(x', x'_0, h, \tau) \}$ be a vector function of the solution of the progressive wave class of the equations of the differential approximation type having the form

$$F_i = \left(\vec{w}, \frac{\partial \vec{w}}{\partial x'}, h, \tau\right) = 0, \quad i = 1, \ldots, n \tag{3.4}$$

at the boundary conditions

$$w(x', x'_0, h, \tau) = \begin{cases} \vec{W}_1, & x' \to -\infty \\ \vec{W}_2, & x' \to +\infty \end{cases}, \tag{3.5}$$

where x'_0 is an arbitrary constant having the same dimension as the variable x' and $\vec{W}_1$, $\vec{W}_2$ are constant vectors satisfying the Hugoniot conditions, the subscripts "1" and "2" refer to the states behind and before the shock front, respectively. The solution $\vec{w}(x', x'_0, h, \tau)$ we call a smeared shock wave.

Definition 1. Let us call a smeared shock wave center the point in the solution $\vec{w}(x', x'_0, h, \tau)$ of a progressive wave class of the equations (3.4) of the differential approximation type at the boundary conditions (3.5) at which the values of the components of the vector $\vec{w}$ do not depend on h and τ for each fixed value x_0'.

We say further that the smeared shock wave center is unique if there exists the only value of vector $\vec{w}$ components which does not depend on the values of constant quantities h, τ, x_0'.

Unlike the smeared shock wave we understand the finite difference shock wave as a solution of the finite difference equations approximating the shock wave and calculated at the discrete grid points in the (x, t) plane. Let us introduce a notion of the finite difference shock wave center analogously to the notion of a smeared shock wave center as a point which is located in the finite difference shock wave zone, belongs to the solutions of the finite-difference equations at the same t obtained by using two different values of constant steps h, τ : $h = h_1$, $\tau = \tau_1$, $h = h_2$, $\tau = \tau_2$, where $|h_1 - h_2| + |\tau_1 - \tau_2| \neq 0$, and determined at the (x, t) points being not nodes with the help of a linear interpolation.

Definition 2. By the differential analyser of a shock wave we mean the algorithm which enables one to find the finite difference shock wave center coordinates in a computational cell by shock-smoothing calculation results.

In the papers [10-12] requirements to the artificial viscosity introduced additively into the pressure were formulated at which there exists the unique smeared shock wave centre. By the use of the results obtained some forms of the artificial viscosoty used in gas-dynamical computations were analyzed [10-12] and the limitations on q were indicated providing that the abscissa of the point of $\max q$ is at the distance from the discontinuity not exceeding h. In Fig.5 the density and velocity graphs are presented in the vicinity of a shock wave obtained when using the two-step Lax-Wendroff scheme with the quadratic artificial viscosity q introduced additively into the pressure. Here $h_1 = 1/40$, $h_2 = 2h_1$, $h_3 = 0.5h_1$, the Courant number $K \approx 0.25$ and the shock wave has already propagated the distance of $\approx 23 h_1$ [12].

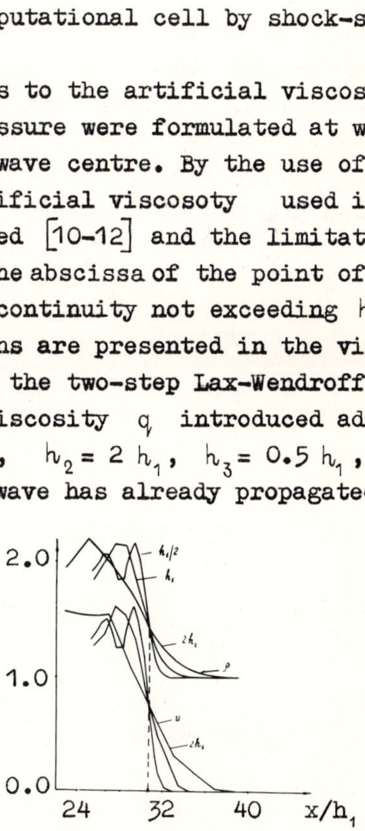

Fig.5

Of course, the finite difference shock wave centre determines the location of a shock front with an error; but the results of numerous calculations show that the accuracy of localization of the shock wave on the basis of the centre notion is higher than the accuracy of the scheme. For example, in the computation given in Fig.5 the above error obtained when determining the discontinuity location did not exceed $0.1\, h_1$.

In the presence of the approximation viscosity in a shock-smoothing scheme it proves to be advisable to use the differential approximations of a difference scheme for the shock capturing. In the paper [10] a multiparametric family of conservative difference schemes approximating the system

$$\frac{\partial u}{\partial t} + \frac{\partial \varphi(u)}{\partial x} = 0$$

was introduced whose first differential approximation has the form

$$\frac{\partial \vartheta}{\partial t} + \frac{\partial \varphi(\vartheta)}{\partial x} = \frac{\partial}{\partial x}\left[C(\vartheta, h, \tau) \frac{\partial \vartheta}{\partial x}\right]$$

where C is the matrix of approximation viscosity coefficients. It is shown that under corresponding conditions on the matrix C there exists the unique smeared shock wave centre and the property

$$\max_x \|C\| = \|C\|_{x=x_c}$$

holds. In the papers [10-12] the question of the existence and uniqueness of the wave centre was investigated when using a series of known finite difference schemes consistent with the system of one-dimensional gas-dynamical equations in the Eulerian variables. In Fig.6 an example of the calculation of a stationary shock wave motion is presented obtained by the modified Harlow's particle-in-cell method [11] when τ / h = const. The intersection point abscissa of the solution graphs for different τ and h coincides with the exact shock front position within the error of $\approx 0.2\, h_1$.

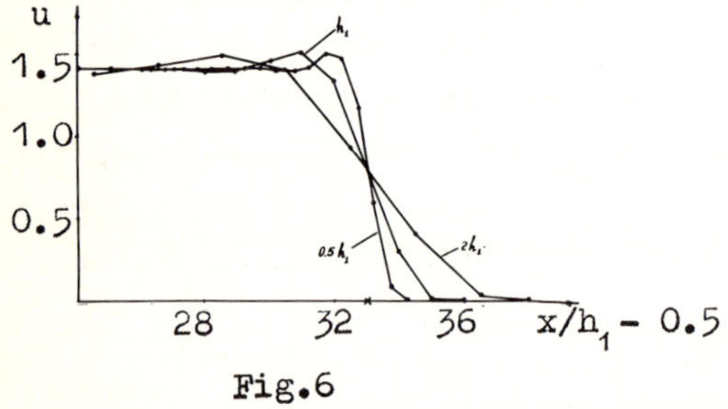

Fig.6

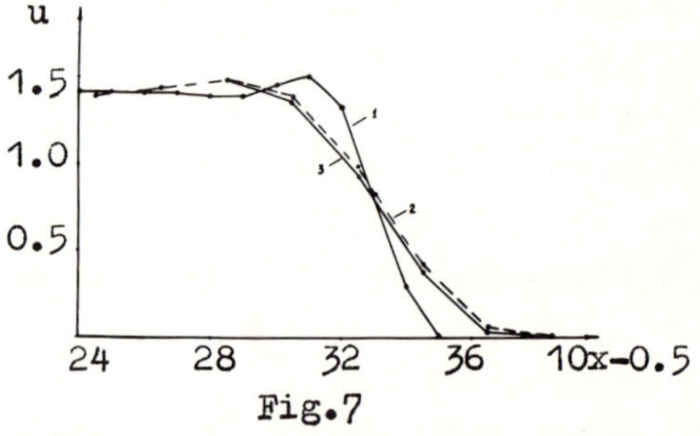

Fig.7

In Fig.7 the calculation example is given for the same problem obtained by Harlow's method when $\tau / h \neq$ const. This example shows the dependence of the finite difference shock wave centre existence on the constancy of the relation τ / h when using the particle-in-cell method.

The theory elaborated has been successfully used for the solution of multidimensional problems of gas dynamics with large deformations of the computational domains and boundaries. Examples of the application of this theory are given for high velocity impact problems in [18].

References

1. N.N.Yanenko, N.T.Danaev, V.D.Lisejkin. On the variational method for the grids construction. Čisl. Metody Mekh. Splošnoi Sredy, Novosibirsk, 1977, 8, No.4, 157-163.
2. A.M.Winslow. Numerical solution of the quasilinear Poisson equation in a nonuniform triangle mesh. J.Comput.Phys., 1966, 1, No.2, 149-172.
3. N.T.Danaev, V.D.Lisejkin, N.N.Yanenko. On the method of moving coordinates in gas dynamics. In: Problems of Mathematical Physics and Computational Mathematics. Moscow, "Nauka", 1977, 107-115.
4. A.N.Tikhonov, A.D.Gorbunov. Error estimations of the Runge-Kutta methods and the choice of optimum meshes. Ž.Vyčisl.Mat.i Mat.Fiz., 1964, 4, No.2, 232-242.
5. V.D.Lisejkin, N.N.Yanenko. On the choice of optimum difference grids. Čisl.Metody Mekh. Splošnoj Sredy, Novosibirsk, 1977, 8, No.7, 100-104.
6. N.N.Yanenko. The method of fractional steps. Springer-Verlag, Berlin and New York, 1971.
7. N.N.Yanenko and V.M.Kovenya. A difference scheme for solving the multidimensional equations of gas dynamics. Soviet Math. Dokl., 1977, 18, No.1, 260-264.
8. J.Neumann. Theory of selfreproducing automates. University of Illinois Press, 1966.
9. V.V.Rusanov. Processing and analysis of computation results for multidimensional problems of aerohydrodynamics. Lect.Notes Phys., 1973, 18, 154-162.
10. N.N.Yanenko, E.V.Vorožcov and V.M.Fomin. Differential analyzers of shock waves. Soviet Math.Dokl., 1976, 17, No.2, 358-362.
11. E.V.Vorožcov, V.M.Fomin, N.N.Yanenko. Differential analysers of shock waves. Applications of the theory. Čisl.Metody Mekh. Splošnoj Sredy, Novosibirsk, 1976, 7, No.6, 8-23.
12. V.M.Fomin, E.V.Vorozcov and N.N.Yanenko. Differential analyzers of shock waves: theory. Computers & Fluids, 1976, 4,No.3/4,171-183.
13. E.V.Vorožcov. Numerical tests of differential analyzers of shock waves. Čisl. Metody Mekh. Splošnoj Sredy, Novosibirsk, 1977, 8, No.2, 12-27.

14. B.L.Roždestvenskii and N.N.Yanenko. Quasilinear systems and their applications to the dynamics of gases. Moscow, "Nauka", 1968.

15. Yu.I.Shokin, N.N.Yanenko. On the relation between correctness of the first differential approximations and the stability of difference schemes for hyperbolic systems of equations. "Matem. Zametki", 1968, 1, No.5, 493-502.

16. N.N.Yânenko, Yu.I.Shokin. On the group classification of difference schemes for a system of one-dimensional equations of gas dynamics. In: Some Problems of Mathematics and Mechanics. Leningrad, "Nauka", 1970, 277-283.

17. J.Neumânn, R.Richtmyer. A method for the numerical calculation of hydrodynamic shocks. J.Appl.Phys., 1950, 21, No.3, 232-237.

18. N.N.Yanenko, E.A.Kroshko, V.D.Lisejkin, V.M.Fomin, V.P.Shapeev, and Yu.A.Shitov. Methods for the construction of moving grids for problems of fluid dynamics with big deformations. Lect. Notes Phys., 1976, 59, 454-459.

DESIGN CRITERIA AND GENERATION OF OPTIMUM FINITE ELEMENT MESHES*

S. M. Yen and K. D. Lee
Coordinated Science Laboratory
University of Illinois
Urbana, Illinois 61801, USA

INTRODUCTION

The finite element method for numerical solution of fluid dynamic problems has several distinct advantages over other numerical schemes. Implementation of complex boundary geometries or refinement of elements in regions of particular interest is easier to accomplish due to the freedom in choosing mesh structure, element shape and size, and order of approximation. Complex boundary geometry and boundary conditions can be treated rather easily or more efficiently. No interpolation or extrapolation would be necessary and the computation can proceed directly in the physical space. Varying physical properties such as the nonhomogeneous situation and moving boundary geometry can be treated with less difficulty.

The application of the finite element method is comparatively more complex and needs more bookkeeping work and therefore more storage. These disadvantages in the finite element method may be offset by its advantages when it is used to solve some specific problems for which the finite difference method may encounter serious difficulties. For example, for the problems with complex geometry, the finite difference method may require the use of a numerical transformation which involves complex implementation, and cause additional sources of errors.

The objective of our study is to develop an efficient finite element method for fluid flow problems by using an optimum grid system generated by a numerical transformation. We have made a parametric study of the effect of mesh systems on the accuracy of the numerical solutions and, on the basis of this study, have formulated a set of criteria to select the shape and the structure of elements in a mesh system to minimize the discretization errors. We have developed two methods that meet these criteria to generate numerically optimum mesh systems for finite element computations. We have applied the methods developed to several fluid flow problems.

SECTOR METHOD

We shall describe briefly below the sector method we propose to implement the finite element method. Let us consider the following matrix equation for each finite element:

$$[K^e]\{\phi_i\} = \{0\} \tag{1}$$

where $[K^e]$ = the element stiffness matrix and ϕ_i = flow variable (say, potential function) at node i. The system stiffness matrix can be constructed by

$$[K] = [\sum_{e=1}^{N_e} k_{ij}^e] \tag{2}$$

where N_e is the total number of elements in the computational domain. The system matrix equation to be solved becomes

$$[K]\{\phi_i\} = \{R_i\} \tag{3}$$

in which the right hand side is obtained from the boundary conditions.

When the finite element method is applied to larger and more complex problems, it becomes increasingly more difficult to construct the matrix equation and to solve it. We have developed an iterative scheme called the sector method (Lee, 1976) that is more efficient to implement. As shown in Fig. 1, a sector is defined by a combination of elements surrounding a node or nodes. Its boundaries denote the finite

*This work was supported by the National Science Foundation under Grant ENG 75-15050.

cut-off zone of influence of the interior node or nodes. The flow variable at an interior node is, thus, affected only by those at other nodes in that sector. The sector stiffness matrix is constructed by the element stiffness matrices of member elements in that sector.

$$[K^s] = [k^s_{ij}] = [\sum_{e=1}^{N_e} k^e_{ij}] \qquad (4)$$

where N_e is the number of elements in that sector. The solution procedure is to construct the sector stiffness matrix for each sector and to iterate by sweeping all the sectors.

The sector method provides a way to avoid the tedious data handling in constructing the system stiffness matrix and facilitates the treatment of boundary conditions. Since it is used with an iterative scheme, the values at non-Dirichlet boundary points are updated separately from the field computation after each iterative step.

The accuracy of the finite element method can be increased by using smaller elements and/or higher-order polynomials. The degree of freedom in discretization is limited by considerations of economy and computer capacity. For a given degree of freedom, it is essential to choose a proper mesh system and a finite element algorithm to obtain the maximum possible accuracy.

OPTIMUM MESH SYSTEM

We have made a detailed study of the accuracy of finite element calculations by using the sector method to conduct numerical experiments on the potential flow problem around an elliptic cylinder (Lee, 1976). The parameters considered are the mesh structure, the shape and the size of elements, the element type and the interpolation functions, the boundary conditions, the degree of freedom, the range of finite cut-off of domain, the accommodation of boundary geometry, and the eccentricity of the elliptic cylinder. On the basis of this study, we have formulated the following criteria for designing an optimum mesh system:

1. The mesh structure should be regular and of simple pattern.
2. The element shape should also be as regular and equilateral as possible.
3. The distribution of size of elements should be consistent with that of flow variables to insure maximum uniformity in the error distribution.
4. Higher-order interpolation functions are necessary if derivatives of the field variable are important in the problem.
5. For non-Dirichlet boundary conditions, smaller boundary elements or higher-order interpolation functions should be used to compensate for the reduced order of accuracy on that boundary.
6. The mesh system should accommodate the boundary geometry as accurately as possible.
7. Since the range of finite cut-off of the domain is closely related to the choice of degree of freedom, an infinite domain should be cut-of in relation to the required computational accuracy and economy.
8. The computational mesh should facilitate efficient numbering of grid system and data handling.

NUMERICAL GENERATION OF OPTIMUM MESH SYSTEM

We shall present here two methods each of which generates numerically a mesh system that meets very well the guidelines outlined above. The numerical transformation of a mesh system has been considered by Winslow (1962) and Chu (1971) for internal geometries and extended by Thompson (1974) to external flow problems. While these methods do not produce mesh systems that meet all the requirements specified here, they did provide information that was useful to us in the development of our method.

In our methods, we propose to map an orthogonal system of grid lines on the field of computation. A set of equipotential lines T = constant is placed between the inner and outer boundaries which themselves are equipotential lines. The other set of lines P = constant is orthogonal to the set T = constant. Such a grid system, if it exists, does accommodate requirements of the boundary surfaces and yields a regular and uniform distribution of the shape and structure of the elements.

The set of orthogonal grid lines is to be generated by solving numerically two Laplace equations. Let us consider external flow problems with a simple-connected flow geometry as shown in Fig. 2. The inner boundary Ω_I is specified by the function $f_I(X,Y) = 0$, and the outer boundary Ω_0 by $f_0(X,Y) = 0$. At Ω_I and Ω_0 the values of T are set to be equal to T_I and T_0 respectively and the grid lines P = constant are set perpendicular to the boundaries. At the intersection boundaries Ω_R and Ω_L, the continuity conditions are applied. This formulation leads to the Laplace equations for T and P and the appropriate boundary conditions.

Method A

The mesh system is generated by solving the following set of equations and boundary conditions

$$(X_P^2 + Y_P^2)X_{TT} + (X_T^2 + Y_T^2)X_{PP} = 0.$$
$$(X_P^2 + Y_P^2)Y_{TT} + (X_T^2 + Y_T^2)Y_{PP} = 0. \qquad (5)$$

$$\text{At } \Omega_I': \quad f_I(X,Y) = 0, \; Y_T\cos\theta_I - X_T\sin\theta_I = 0.$$
$$\text{At } \Omega_0': \quad f_0(X,Y) = 0, \; Y_T\cos\theta_0 - X_T\sin\theta_0 = 0. \qquad (6)$$
$$\text{At } \Omega_R' \text{ and } \Omega_I': \quad X_R = X_L, \; Y_R = Y_L.$$

where θ_I and θ_0 are the direction cosines at the inner and outer boundaries respectively. Each term in Equations (5) is of the same degree of differentiation with respect to T or P; therefore, the distribution of P can be chosen arbitrarily to meet the computational requirement of a particular problem. For example the second conditions in Equations (6) can be replaced by a Dirichlet condition which makes the boundary conditions sufficient:

$$\text{At } \Omega_0': \quad X = X_0, \; Y = Y_0. \qquad (7)$$

The fact that P can be arbitrarily chosen does sometimes result in slow convergence or distortion of the grid system. In such cases, the resulting mesh system is no longer orthogonal but still optimal. Hence, such a distortion will not have a significant effect on the accuracy of computation since the finite element method solves for the field variables directly in the physical plane.

Figure 3 shows the optimum mesh system for an irregularly shaped cylinder generated by Method A.

Method B

This method is a modification of Method A. Instead of solving the set of nonlinear partial differential equations in Equations (5), the following two Laplace equations are solved for the numerical transformation:

$$X_{TT} + X_{PP} = 0$$
$$Y_{TT} + Y_{PP} = 0 \qquad (8)$$

with the same boundary conditions as in Method A. Compared to Method A, this transformation implies that Equations (8) is a special case of Equations (5) when

$$X_T^2 + Y_T^2 = X_P^2 + Y_P^2 \qquad (9)$$

This type of transformation actually occurs in the elliptic-cylindrical coordinates. Method B may be looked upon as an attempt to find an orthogonal transformation of elliptic-cylindrical type for a domain of arbitrary geometry.

This modified method can be used for any arbitrary body geometry with convex curvature, and the resulting mesh system is very similar to that of Method A. For the bodies with concave corners, Method B does not work. However, this method also gives an optimum mesh system for bodies with sharp edges as in airfoils. An optimum mesh system for an airfoil by this modified method is shown in Fig. 4.

Instead of generating a mesh system of quadrilateral elements, triangular elements of an equilateral type can be used as shown in Fig. 5. An optimum mesh system can also be produced in the internal flow problem by both methods. For example, for the internal flow problem around an ellipse between solid walls, Method B generates an optimum mesh system as shown in Fig. 6.

Figure 7 shows the initial mesh system used in solving the problem of a symmetrical hydrofoil.

APPLICATION OF THE OPTIMUM MESH SYSTEM

We have applied our finite element method using the optimum mesh system to several problems. We shall show here two examples each of which requires special treatment of the boundary conditions and the computational domain near the boundary. We have solved several free surface potential flow problems, including that of a symmetrical hydrofoil as shown in Fig. 7. In these problems, we restructure the elements near the free surface in order to accommodate the free surface geometry, as shown in Fig. 8. We use the finite element method to make the field calculation of the velocity potential and the finite difference method to update the height and the velocity potential at the free surface. The schematic of the problem of a symmetrical hydrofoil is shown in Fig. 9. The evolution of the wave height obtained from the solution of this problem is shown in Fig. 10.

We have also been studying the implementation of our method to the inviscid and viscous flows around an airfoil. In these problems, we need to consider the accurate treatment of the Kutta condition and the appropriate discretization of the field of computation near the trailing edge. The node distributions and the sector shapes near the trailing edge for the two mesh systems (with circular outer boundary) are shown in Fig. 11. Because of the difference in sector shape in that region, the calculation of the sector stiffness matrix requires special handling. However, we believe the Neumann and Kutta conditions can be treated more accurately than the finite difference method.

REFERENCES

Chu, W. H. (1971), J. Comp. Phys., Vol. 8, pp. 392-408.

Lee, K. D. (1976), Report T-30, Coord. Science. Lab., Univ. of Illinois.

Thompson, J. F., Thames, F. D., and Martin, C. W. (1974), J. Comp. Phys., Vol. 15, pp. 299-319.

Winslow, A. M. (1963), UCRL Report 7312, Lawrence Rad. Lab., Univ. of Cal.

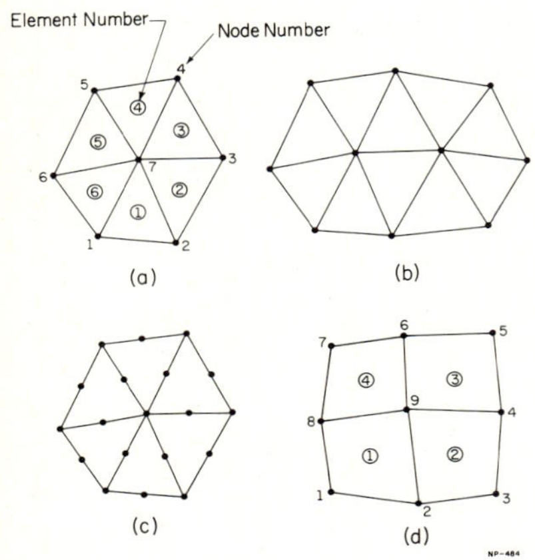

Fig. 1. Finite element sectors: (a) Six triangular elements, one interior node, (b) Ten triangular elements, two interior nodes, (c) Six triangular elements, seven interior nodes, (d) Four quadrilateral elements, one interior node.

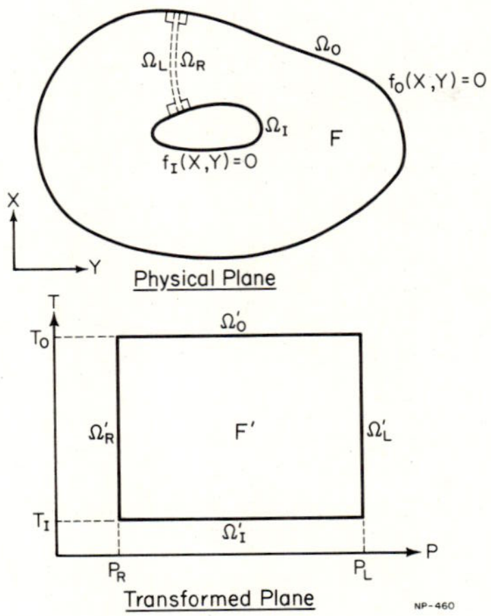

Fig. 2. Schematic representation of flow geometry in the physical plane (X,Y) and the transformed plane (T,P).

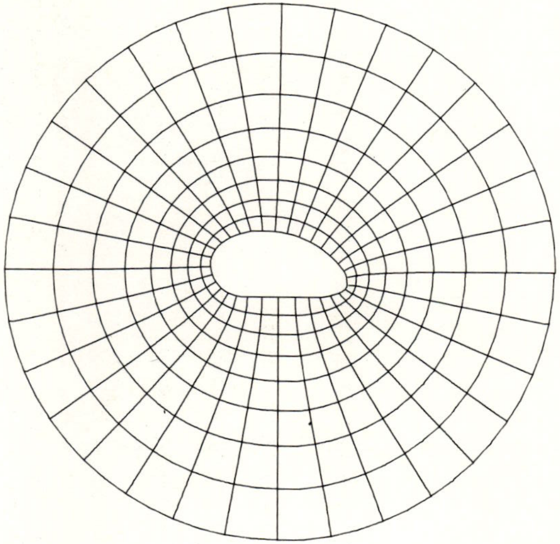

Fig. 3. Optimum mesh system - irregularly shaped cylinder.

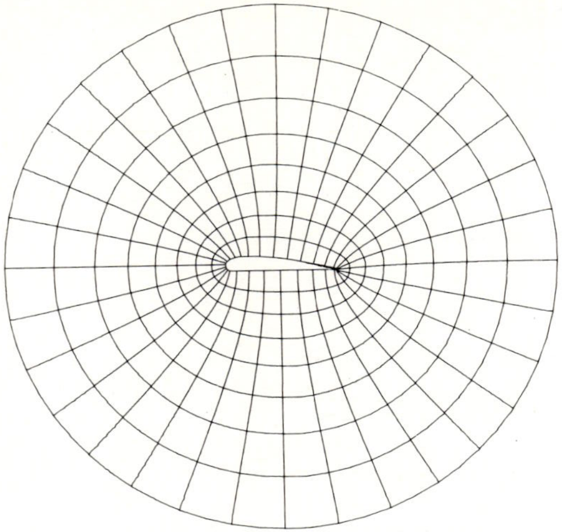

Fig. 4. Optimum mesh system - airfoil.

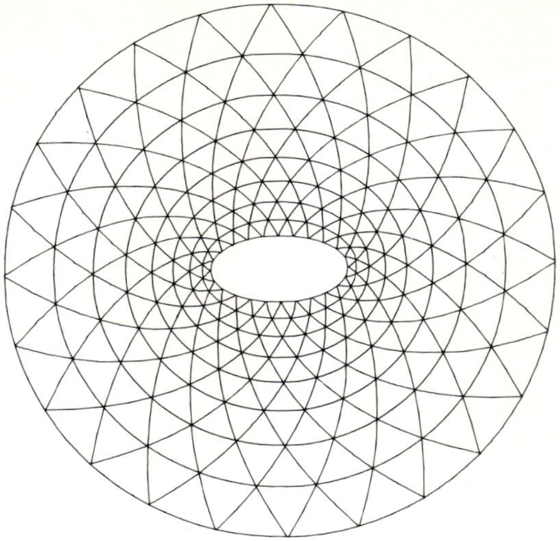

Fig. 5. Mesh with equilateral triangular elements.

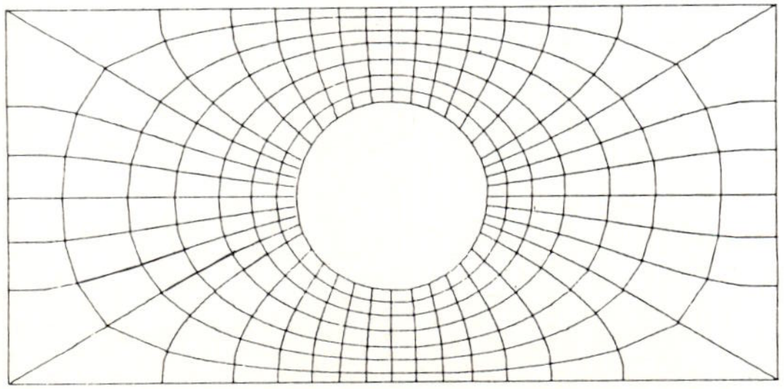

Fig. 6. Optimum mesh system - internal flow.

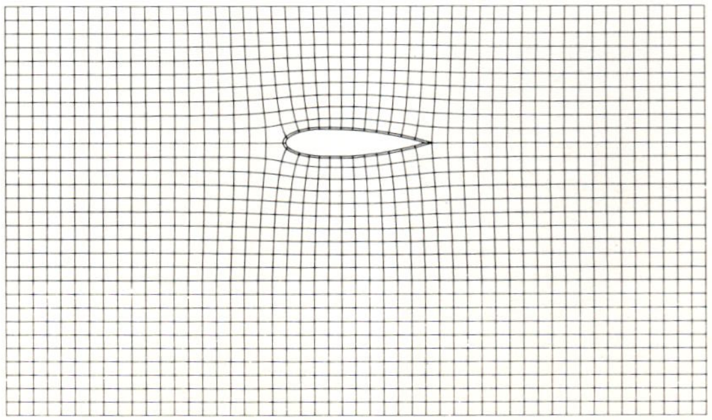

Fig. 7. Initial mesh system - a symmetrical hydrofoil.

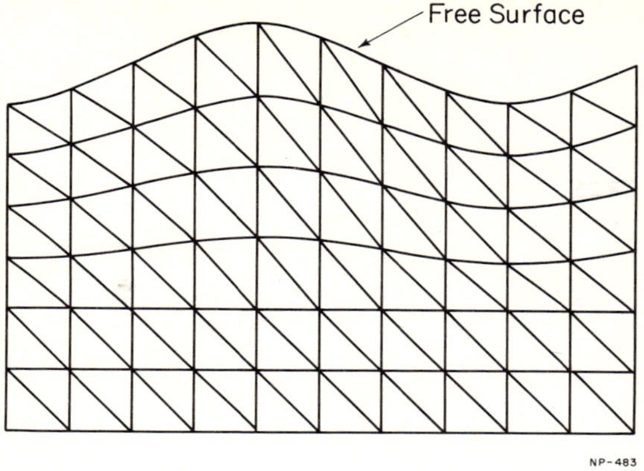

Fig. 8. Restructuring elements near free surface.

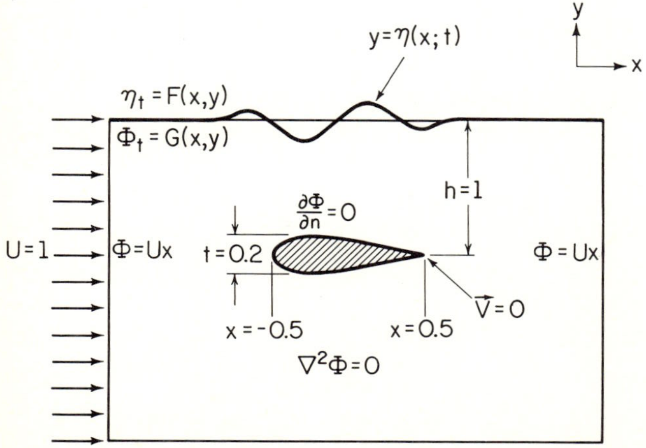

Fig. 9. Schematic of the hydrofoil problem.

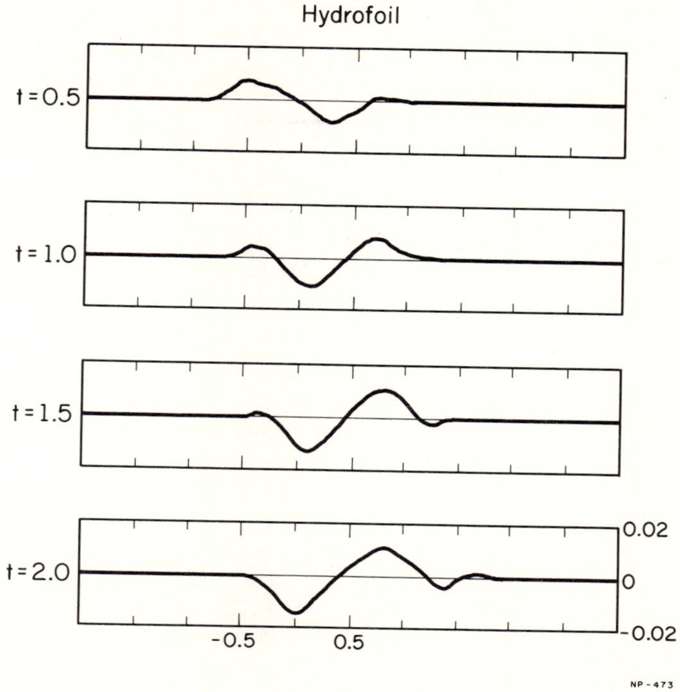

Fig. 10. Development of wave heights.

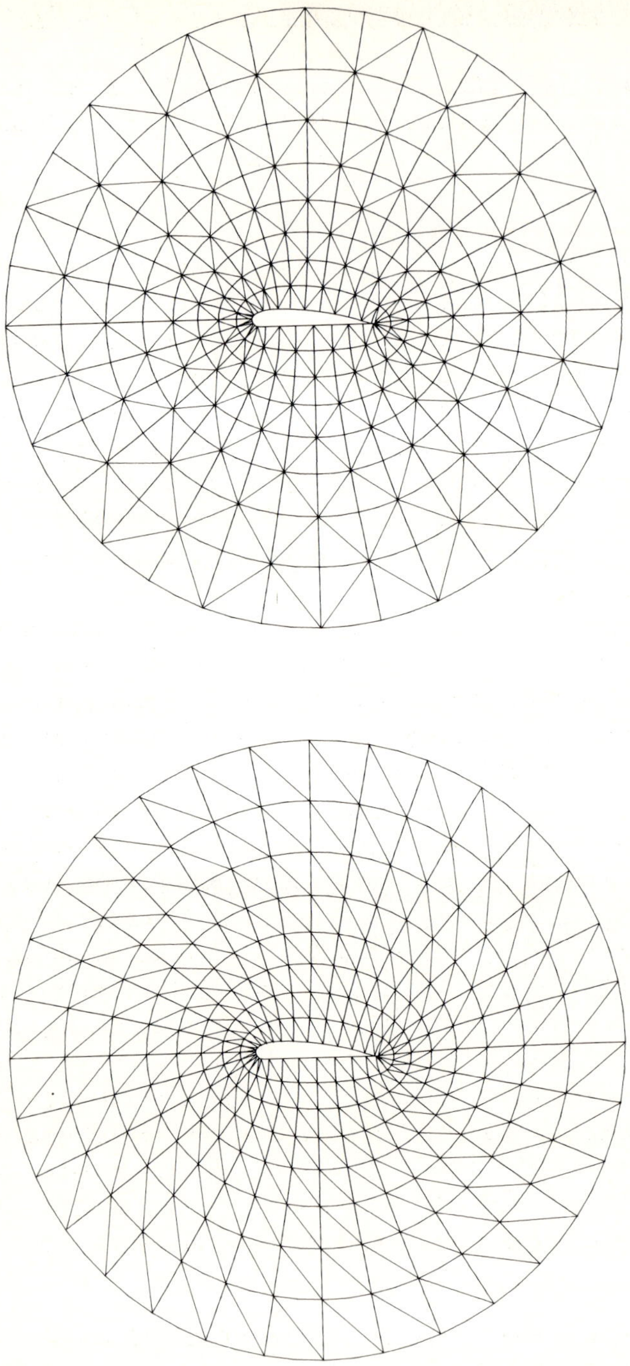

Fig. 11. Two mesh systems for the airfoil problem.

SELF-ADAPTED ALGORITHMS IN PROBLEMS OF GAS DYNAMICS

A.V. Zabrodin, G.P. Prokopov, V.A. Cherkashin
Institute of Applied Mathematics Academy of Sciences
Moscow, USSR

A considerable progress in computer technology not being underestimated though, it should be noted that, as one's experience shows, the complexity of continuum mechanics problems generated by reality tends to grow more rapidly as compared with the computer capabilities. The complex problems (e.g. the calculation of a flow around flight vehicles, etc.) can be solved only with effective computational algorithms being available, but even then the computations are carried out at the limit of computer capabilities. Thus, the search for new approaches to constructing effective algorithms is an urgent problem.

The structure of a solution to non-stationary gasdynamic problems is of a specific nature that shows itself in existing the smooth regions separated by particular lines or surfaces present in initial data or arising during the process evolution. If one does not take this into consideration and uses, e.g., the homogeneous difference schemes of run-through computations while constructing the computational algorithm, then an approximation of discontinuous functions with unknown discontinuity locations has to be made. By rather simple reasoning it may be shown that in order to approximate with an accuracy to ε , a bounded discontinuous function in the norm of space L_1 (or its difference analog) through its values in points (with the discontinuity locations being unknown), the grid of the $C_0 \varepsilon^{-(1+d)}$ nodes should be provided, where d is the space dimension, the unity accounts for the time variable t , C_0 is a certain constant depending on the size of considered region for specifying the function. Hence, if the difference schemes of run--through computations with the fixed number of operations per point are used to find the solution, then one should perform the work

$$A = C_1 \varepsilon^{-(1+d)}$$

Note that increasing the order of approximation of the difference scheme used for finding the discontinuous solution does not improve the exponent in this estimate. The similar remark may also be made about implicit difference schemes that do not require limiting the time step because of stability loss. Thus, the amount of work is defined rather by the way of describing the difference functions, under which the discontinuities are not isolated, than by the essence of a problem. On the other hand, the discontinuities being isolated and taken into account, the considerable decrease of the work amount is possible. It can be done by means of isolating the smooth regions and using the schemes with an approximation order corresponding to the smoothness of the desired solution [I] . For example, to obtain, in the above norm, the deviation of the order of ε on the solution with no discontinuities according to the first-order scheme, one needs the grid with the node number of the order of $\varepsilon^{-1/2(1+\alpha)}$. However, while the non-stationary gasdynamic problems being solved numerically, isolating and tracking the discontinuities is extremely difficult due to their interaction. To isolate and take into considerations all the discontinuities is, in fact, an impracticable problem. So one can speak only of isolating the main discontinuities, as for the other (secondary) ones, they can be calculated by "smearing" with the use of run-through computational techniques. For the first-order schemes it is expedient to "smear" the discontinuities with intensities of the order of $\varepsilon^{1/2}$. By applying the above arguments, concerning the work amount estimation, to the discontinuous functions with the discontinuity values of the order of $\varepsilon^{1/2}$ and assuming that the evaluation of errors of different kinds is carried out under the superposition principle, one may hope to obtain the accuracy ε on the grid with the node number $\varepsilon^{-1/2(1+\alpha)}$ for the first-order scheme.

Another note should be made in connection with the fact that in practice the calculation of discontinuity interactions, conserving the isolated fronts, is not always possible. Due to this, some of the discontinuities have to be "smoothed" and then isolated again. If one bases on conserving the node number of the grid of the order of $\varepsilon^{-1/2(1+\alpha)}$ to obtain the accuracy ε , then the regions, where the calculation of discontinuous solutions is allowed without isolating the discontinuities, should be chosen of the order of $\varepsilon^{1/2}$. Thus, in the regions small in space and in the intervals small in time the reconstruction of the discontinuity system may be performed to simplify the calculations essentially.

The development of algorithms to carry out the calculation with isolated fronts, the location of which is not known beforehand and which can arise during this calculation, is quite a problem. One of the effective methods for numerical integration of quasi-linear, hyporbolic-type, partial differential equations is discussed in monograph [2] along with some examples of applying this method to a wide range of non-stationary gasdynamic problems and some other sections of continuum mechanics.

The essence of the above method is that the main discontinuities isolated during the calculation are described as coordinate lines. With these, using special algorithms, a moving, curvilinear difference grid is constructed to describe the desired solution. The choice of unknown functions for describing the motion is made to match the local reference frame in order that their components be slowly variable along the coordinate lines. The difference grid may be connected with the behavior of desired solution, and it is constructed depending on the point-to-point solution change, either being dense or thinning out in accordance with the change value. Hence, while calculating a specific problem one may take into account and reflect its peculiarities, carrying out the calculations under the conditions most fit for describing the problem. If a priori information on the solution is not sufficient one may first perform a crude calculation and, on analyzing it, then take into account the problem peculiarities in subsequent calculations. The accuracy for calculating the main factors may be increased at the expense of making the secondary factors more crude. We call the algorithms thus constructed self-adapted.

Consider two problems as examples.

The first problem is on interaction of a shock wave with an axisymmetric finite cylinder. The cylinder is placed in a circular tube so that its axis of symmetry coincide with that of the tube (Fig.I).

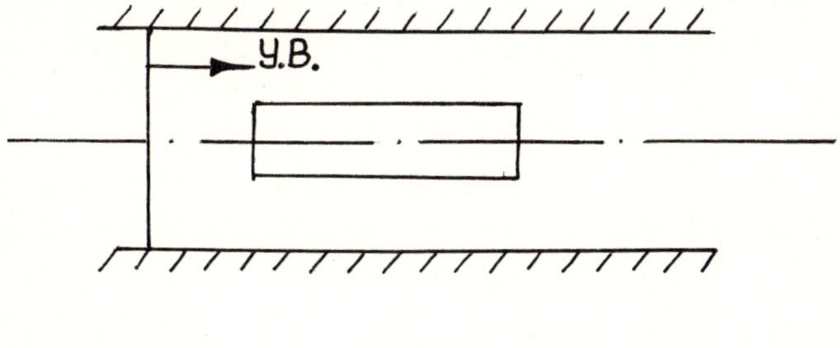

Fig. I

The shock front is perpendicular to the axis of symmetry.

After the shock has reached the left cylinder base the falling shock will generate a complicated configuration of waves including the bow shock reflected by body and the shock propagating along the cylinder side. After the latter has reached the right cylinder base the diffraction will start. As a result of diffraction the pattern becomes more complex with additional discontinuities arising. Fig.2 shows the line pattern for constant Mach numbers when the diffracted shock DD' approaches the symmetry axis. In Fig. 2 one can see the main peculiarities: the bow shock with the front partially isolated by the line AA' by this time, the expansion wave coming from the left cylindrical base, the impedent wave near the cylinder side surface, the expansion wave and separation surface CC coming from the base edge. In this calculation we isolate the diffracted shock front DD' and the part of stagnation wave, arising in such cases between the expansion and shock waves [3] , [4] .

As the diffracted shock converges on the symmetry axis at some distance from the base the reflected shock forms with the suspended tail shock wave developing in time

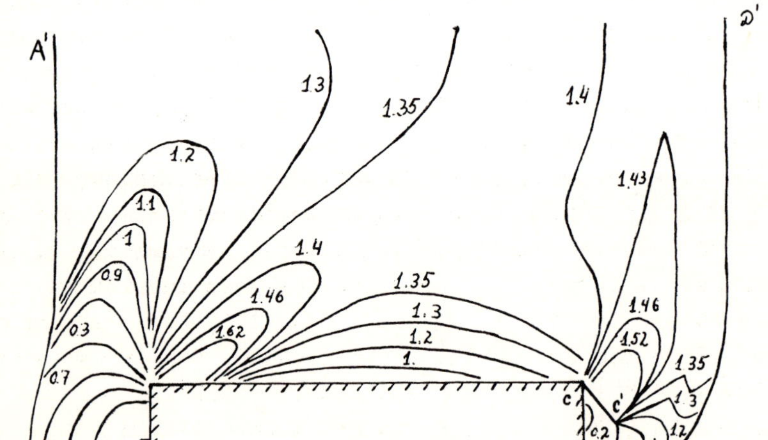

Fig. 2

Now we go over to the second problem (Fig. 3). The problem is on the supersonic flow around an axisymmetric body. The latter represents a spherically blunted cylinder. From the blunt edge a spike is set forth with a cone head capattached [5] .

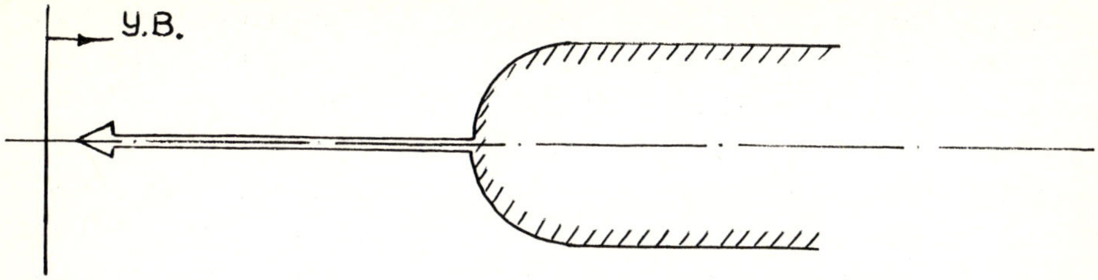

Fig. 3

The flow under investigation is very complex and needs an adequate mathematical model for its investigation. Assume that the basic events are described in the ideal gas approximation. But to describe the vortex separated flows seems to be impossible basing on the corresponding set of stationary gasdynamic equations. Therefore, we consider a non-stationary process and study its evolution in time.

The problem is formulated as follows. A shock wave, with the given parameters behind the shock, flows about the given body at zero incidence (see Fig. 3).

Let us not discuss the incident-shock-front-body interaction. We only note that, behind the cone head cap, the conditions develop near the spike surface such that the reflected wave, in interaction between the incident front and spherical blunt nose, reconstructs the flow while moving upstream and leaves the stalling region behind. Within this region the circulation flow forms. Let the region, separated from the outside free stream flow by a surface coming from the cone cap, be called an attached cone. Due to the attached cone in front of the body, a suspended shock wave, less intensive than the strong bow shock, is formed. The calculations were carried out for two flow Mach numbers: M_∞ = I.8 and M_∞= I.3. For the Mach number M_∞ = I.8 we had eventually a quasi-steady flow that is presented in Fig. 4 with the constant Mach number lines.

One can see here the attached bow shock wave AA' in front of the cone cap, and the suspended bow shock CC' in front of the spherical blunt nose. The generatrix of the attached cone is denoted by BCD. Within the attached cone the flow is vortex, and the pressure slightly exceeds that of the free-stream flow. By using the pressure distribution obtained the drag coefficient was computed according to the relation

$$C_D = \frac{D}{qA} - \frac{p_\infty}{q}$$

where $D = \int_S p\, n_x\, dS$ is the body drag, A is the cross section area, $q = \frac{1}{2}\rho_\infty V_\infty^2$ is the dynamic pressure, p_∞ is the static pressure in the free-stream flow. For the spike body $C_D = 0.367$ and for the body without spike $C_D = 0.8I$. Hence, the spike decreases the drag coefficient by more than a factor of two.

In case $M_\infty = I.8$ the calculations gave the quasi-stationary solution, while for $M_\infty = I.3$ the solution behavior varies in time. One can observe that the qualitative pattern of flow that took place at $M_\infty = I.8$ is sometimes violated at $M_\infty = I.3$. In the latter case the attached cone surface eventually becomes swollen, and at a certain time the spherical blunt nose gets into the attached cone with the body finding itself in the "shadow" of the cone cap. As a result the outflow of matter from the separation region takes place. After this, the attached cone is gradually decreases and the qualitative pattern is restored. The pattern of flow reconstruction appeared twice during the computational period. When the body was caught by the shadow of the cone cap the body drag decreased and was approximately $C_D = 0.I8$. When the pattern was restored the drag increased and was about 0.43. The drag of the body without spike was $C_D = 0.75$.

Thus, in solving this problem we obtained, within the frame of the model of ideally compressible fluid, the flow about given body with the separation region formed.

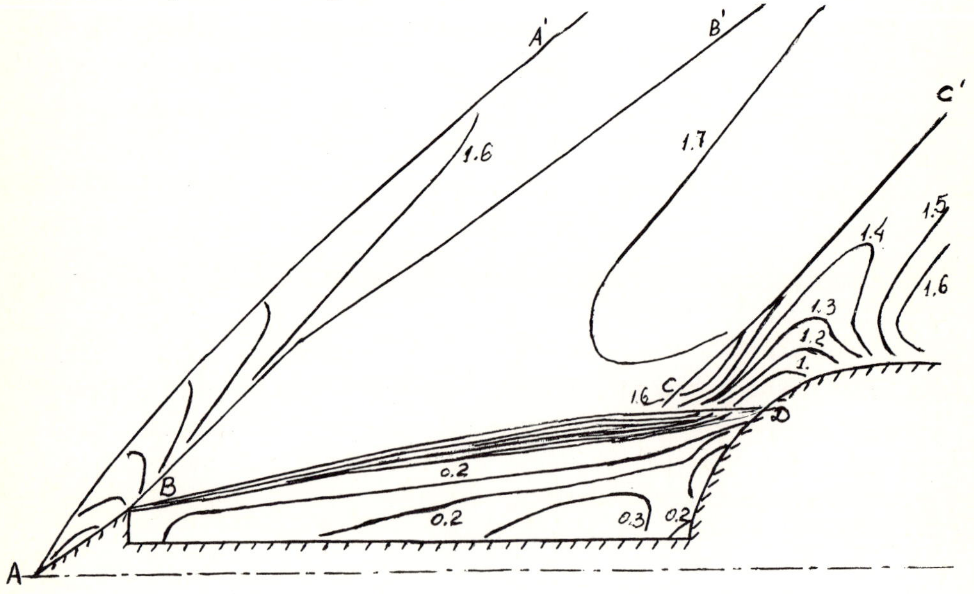

Fig. 4

Acknowledgement

Authors are deeply grateful to Mstislav V. Keldysh for his help in stating the problem, useful discussions and his invariable attention to our work.

R e f e r e n c e s

I. Babenko, I.I., On an approach to the evaluation of computational algorithm quality, Preprint IPM, N 7, I974.

2. Godunov, S.K., Zabrodin, A.V., Ivanov, M.YA., Kraiko, A.N., Prokopov, G.P., Numerical solution of multi-dimensional gasdynamic problems, M., "Nauka", I976.

3. Zabrodin, A.V., Prokopov, G.P., Cherkashin, V.A., On shock diffraction on the base of stream-line body, IV Vsesouznii S-ezd po Teoreticheskoji i Prikladnoji Mekhanike, Kiev, I976.

4. Bazhenova, T.V., Gvozdjova, L.G., Komarov, V.S., Sukhov, B.G., Study of diffraction of strong shock waves on convex angles, Izv.Akad.Nauk SSSR, MJG, I973, N 4.

5. Paul K. Chang. Separation of Flow. V.I, 2

APPLICATION OF LAGRANGIAN INVARIANTS TO THE CALCULATION OF THREE-DIMENSIONAL ROTATIONAL FLOWS OF A PERFECT FLUID

by Radyadour Kh. ZEYTOUNIAN

59650 Lille I University (France)
Scientific Advisor to Office National d'Etudes et de Recherches Aérospatiales (ONERA)
92320 Châtillon (France)

SUMMARY

The governing equations can be put in conservative form $DI_\nu/Dt = 0$, where $\nu = 1$ to 5, and $D/Dt = \partial/\partial t + v_\alpha \, \partial/\partial x_\alpha$ ($\alpha = 1, 2, 3$) is the material derivative in the fluid characterized by the velocity field $v_\alpha(t,x_\alpha)$; $I_\nu(t,x_\alpha)$ are the scalar fields related to the velocity field $\mathbf{v}_\alpha$, to the specific entropy $S(t,x_\alpha)$ and to the density $\rho(t,x_\alpha)$ of the flow considered.

If at time zero $t = 0$ the initial values v^o_α, S^o and ρ^o are given, then the values $I_\nu(0,x_\alpha)$ are known. It is shown in the paper that we can then determine the values of v_α, S and ρ at any time $t > 0$, using the conservation equations $DI_\nu/Dt = 0$.

INTRODUCTION

Lagrangian invariants can be used to characterize the flow of a perfect fluid. It is obvious that these invariants should be very useful for determining the nature of the flow to which they apply. On the one hand, these invariants should actually allow the determination of the nature of the flow but, on the other hand, they may also be used to estimate the error made during an approximate calculation.

Let us note that for a compressible, three-dimensional flow of perfect fluid, the specific entropy S is a Lagrangian invariant, as $DS/Dt = 0$. In incompressible (or more generally barotropic) three-dimensional flow, the circulation Γ , along a closed curve L within the flow, is also a Lagrangian invariant, as it satisfies the relation $D\Gamma/Dt = 0$. Always in barotropic three-dimensional flow, the vortex vector $\underline{\omega}$ satisfies the Helmholtz equation :

$$\frac{D}{Dt}\left(\frac{1}{\rho}\,\underline{\omega}\right) = \left(\frac{1}{\rho}\,\underline{\omega}.\underline{\nabla}\right)\underline{v} \quad , \qquad (1)$$

$\underline{v}$ being the velocity vector, $\underline{\nabla}$ the nabla operator and ρ the density.

In the case of plane flow, when $\underline{\omega}$ has a single component, perpendicular to the flow plane, this component is maintained along the trajectories of the plane flow - the vortex, once created, is simply convected by the fluid; this property has been judiciously used by Chorin (1973) for the calculation of plane rotational flows. Let us also note that the fact that Γ is a Lagrangian invariant and has been used, in particular, by Belotserkowskii (1977); as Rehbach (1977) correctly noted, the fact that the $\underline{\omega}$ vector varies - according to (1) - during its displacement as a function of the local distorsions of the velocity field, preserves the circulation of the vortex line of which $\underline{\omega}$ is a part.

In barotropic three-dimensional flow we can, instead of equation (1), consider following Lagrangian invariant, obtained by Ertel and Rossby (1949) :

$$\frac{D}{Dt}\{\frac{1}{\rho}\, \underline{\omega}.(\underline{v}-\underline{\nabla}\, H_b)\} = 0 \ , \qquad (2)$$

where

$$H_b = \int_0^t (\frac{v_i^2}{2} + U - \int \frac{dp}{p})\, dt$$

is the mechanical (barotropic) action defined for each element of fluid trajectory, once we assume that the external forces are derived, from a potential U ; p being the pressure.

Up till now, it does not seem that, advantage had been taken of (2) in calculations of barotropic three-dimensional flows - an interesting application of (2) could consist in verifying to what degree it is actually satisfied during the approximate numerical calculation of barotropic (or incompressible) three-dimensional flows.

One more remark about integration of equation (1) : this equation is in fact equivalent to the so-called Cauchy equations - see section 17 of Serrin (1959) :

$$\omega_i = \frac{\rho}{\rho^o}\, \omega_j^o \frac{\partial x_i}{\partial x_j^o} \ . \qquad (3)$$

Lastly, it is interesting to note that the introduction of "generalized" velocity fields :

$$\underline{w} = \underline{v} - \beta\, \underline{\nabla}\, S \qquad (4)$$

where $\beta = \int_0^t T\, dt \ (\Longleftrightarrow \frac{D\beta}{Dt} = T)$,

T being the temperature, allows one to reformulate in "barotropic" form various results concerning compressible barocline flows (when $\underline{\nabla}\, p \wedge \underline{\nabla}\, \rho \neq 0$ and $D\Gamma/Dt \neq 0$). In particular we may write, for the generalized vortex :

$$\underline{\Omega} = \underline{\nabla} \wedge \underline{w} \equiv \underline{\omega} + \underline{\nabla}\, S \wedge \underline{\nabla}\, \beta$$

the classical Helmholtz equation (1) :

$$D\Omega/Dt = (\underline{\Omega}.\underline{\nabla})\, \underline{v} - (\underline{\nabla}.\underline{v})\, \underline{\Omega} \ . \qquad (5)$$

We can also see that the vortex $\underline{\omega}$ in a barocline medium of ideal fluid, initially (t = 0) irrotational, is generated by the vector $\underline{\nabla}\, \beta \wedge \underline{\nabla} S$ *).

This passage from the baroclinic case the classical barotropic case by means of equation (4) should permit the application of numerical calculation methods, valid in barotropic, to the baroclinic case - we are thinking, in particular, of flows with vortex sheets.

LAGRANGIAN INVARIANTS

Let :

$$H = \int_0^t (\frac{v_i^2}{2} + U - h)\, dt \ , \qquad (6)$$

---

*) For more detail, one may consult chapter III of our Notes - Zeytounian (1974) - as well as our article (in Russian) Zeytounian (1977).

be the mechanical (baroclinic) action defined for each element of fluid trajectory, the function h being the specific enthalpy, with

$$\underline{\nabla}\, h = T\, \underline{\nabla}\, S + 1/\rho\, \underline{\nabla}\, p \ .$$

Let us introduce the velocity.

$$\underline{u} = \underline{v} - \underline{\nabla}\, H \ . \tag{7}$$

It then appears that the equations governing perfect fluid flow :

$$\left.\begin{aligned} &D\underline{v}/Dt = \underline{\nabla}\, U - 1/\rho\, \underline{\nabla}\, p \ ; \\ &\frac{D \operatorname{Log} \rho}{Dt} + \underline{\nabla}.\underline{v} = 0 \ ; \\ &DS/Dt = 0 \ , \end{aligned}\right\} \tag{8}$$

may be associated with the following system of Lagrangian invariants :

$$\frac{D}{Dt}\left\{\frac{1}{\rho}\,(\underline{\nabla} \wedge \underline{u}).\ \underline{\nabla} S\right\} \equiv \frac{DI_1}{Dt} \ , \tag{9}$$

$$\frac{D}{Dt}\left\{\frac{1}{\rho}\,\underline{u}.(\underline{\nabla} S \wedge \underline{\nabla}\, I_1)\right\} \equiv \frac{DI_2}{Dt} \ , \tag{10}$$

$$\frac{D}{Dt}\left\{\frac{1}{\rho}\,\underline{u}.(\underline{\nabla} S \wedge \underline{\nabla}\, I_2)\right\} \equiv \frac{DI_3}{Dt} \ , \tag{11}$$

$$\frac{D}{Dt}\left\{\frac{1}{\rho}\,\underline{\nabla} S.(\underline{\nabla}\, I_1 \wedge \underline{\nabla}\, I_2)\right\} \equiv \frac{DI_4}{Dt} \ , \tag{12}$$

$$\frac{DS}{Dt} \equiv \frac{DI_5}{Dt} \ . \tag{13}$$

To this system of Lagrangian invariants (9 through 13) we must add the definition of H, according to (6) :

$$\frac{DH}{Dt} = \frac{v_i^2}{2} - h + U \ . \tag{14}$$

The Lagrangian invariants have been obtained by Hollmann (1964); one may also consult the article by Gavrilin and Zaslavskii (1970) on this point as well as chapter II of our Notes (1974).

We should remark that the existence of Lagrangian invariants (9 through 13) is connected to the fact that the surfaces of constant entropy are not, in a general way, vortex surfaces (meaning that we exclude the so-called "oligotropic" flows). The system (9 through 14) is entirely equivalent to the system of equations (8). We can naturally construct an infinity of Lagrangian invariants which satisfy, in particular, the functional relation :

$$\underline{u}/\rho.(\underline{\nabla} S \wedge \underline{\nabla}\, I_n) = I_{n+1} \ , \tag{15}$$

(n = 1, 2, 3 ...);
however, among all the Lagrangian invariants (15) there are only two that are linearly independent.

CALCULATION SCHEME OF v_i, ρ AND S USING LAGRANGIAN INVARIANTS

By hypothesis, we assume that at initial time t = 0 the five fields v_i ρ and S are given, i.e. that we know the initial values v_i^o, ρ^o and S^o. According to (6), we also have $H^{o} \equiv 0$ and hence $u_i^o \equiv v_i^o$ according to the definition (7) of vector $\underline{u}$. Thus the five scalar fields I_ν^o are known and we shall be able, in principle, to compute them at time Δt from the conservation equations (9 through 13); for this, it suffices to apply a finite difference scheme, for instance, the method of fractional steps, or a more elaborate method such as that proposed by Yanenko and Shokin (1974). In any case, we may thus assume that the $I_\nu^{\Delta t}$ are known.

Now we must decide how to calculate at time Δt the values of $v_i^{\Delta t}, \rho^{\Delta t}$ and $S^{\Delta t}$ fields.

First, we have :

$$S^{\Delta t} \equiv I_5^{\Delta t} \; . \tag{16}$$

From (12) we can determine :

$$\rho^{\Delta t} = \frac{(\underline{\nabla} S)^{\Delta t} . [(\underline{\nabla} I_1)^{\Delta t} \wedge (\underline{\nabla}\, I_2)^{\Delta t}]}{I_4^{\Delta t}} \; . \tag{17}$$

Lastly, from the three Lagrangian invariants (9 through 11) :

$$\left.\begin{aligned} I_1^{\Delta t} &= 1/\rho^{\Delta t} [\underline{\nabla} \wedge \underline{u})^{\Delta t} . (\underline{\nabla} S)^{\Delta t}] \; ; \\ I_2^{\Delta t} &= 1/\rho^{\Delta t} \underline{u}^{\Delta t} . [(\underline{\nabla} S)^{\Delta t} \wedge (\underline{\nabla}\, I_1)^{\Delta t}] \; ; \\ I_3^{\Delta t} &= 1/\rho^{\Delta t} \; \underline{u}^{\Delta t} . [(\underline{\nabla} S)^{\Delta t} \wedge (\underline{\nabla}\, I_2)^{\Delta t}] \; , \end{aligned}\right\} \tag{18}$$

we must deduce $\underline{u}^{\Delta t}$, i.e. more precisely, the three components $u_i^{\Delta t}$.

From the second and third equations (18) we can easily express the components $u_1^{\Delta t}$ and $u_2^{\Delta t}$ as a function of $u_3^{\Delta t}$:

$$\left.\begin{aligned} u_1^{\Delta t} &= A_1^{\Delta t} + B_1^{\Delta t}\, u_3^{\Delta t} \; ; \\ u_2^{\Delta t} &= A_2^{\Delta t} + B_2^{\Delta t}\, u_3^{\Delta t} \; , \end{aligned}\right\} \tag{19}$$

the four coefficients $A_1^{\Delta t}$, $B_1^{\Delta t}$, $A_2^{\Delta t}$ and $B_2^{\Delta t}$ being known functions of $I_1^{\Delta t}$, $I_2^{\Delta t}$, $I_3^{\Delta t}$, $\rho^{\Delta t}$ and $S^{\Delta t}$. If we now combine (19) with the first equation (18), we obtain a partial derivative differential equation of first order for the component $u_3^{\Delta t}$; this last equation can be put in the following form :

$$\mathcal{A}^{\Delta t} \frac{\partial u_3^{\Delta t}}{\partial x_1} + \mathcal{B}^{\Delta t} \frac{\partial u_3^{\Delta t}}{\partial x_2} + \mathcal{C}^{\Delta t} \frac{\partial u_3^{\Delta t}}{\partial x_3} = \mathcal{D}^{\Delta t} \; , \tag{20}$$

where, once more, the coefficients $\mathcal{A}^{\Delta t}$, $\mathcal{B}^{\Delta t}$, $\mathcal{C}^{\Delta t}$ and $\mathcal{D}^{\Delta t}$, are known functions of $I_1^{\Delta t}$, $\rho^{\Delta t}$, $S^{\Delta t}$ and the partial derivatives $A_1^{\Delta t}$, $A_2^{\Delta t}$, $B_1^{\Delta t}$ and $B_2^{\Delta t}$, relative to x_i, i = 1, 2, 3.

The problem is then to integrate equation (20); to this end, we shall have to assume that $u_3^{\Delta t}$ takes a given value at each point of a given surface S_o. But from the viewpoint of the kinematics of our flow, it is not $u_3^{\Delta t}$, but

$v_3^{\Delta t} = u_3^{\Delta t} + (\partial H/\partial x_3)^{\Delta t}$ that is given on S_o. We circumvent this difficulty by estimating the function H by an approximation of its integral between 0 and Δt :

$$H^{\Delta t} = (\frac{v_i^2}{2} + U - h)^o \ \Delta t \ . \qquad (21)$$

Let us return to the integration of equation (20). We first note, in accordance with Lainé (1946, p. 231) that the left hand side of (20), can always be reduced to $(A^{\Delta t})^2 + (B^{\Delta t})^2 + (C^{\Delta t})^2 = 1$, which expresses the derivative of the function $u_3^{\Delta t}$ in the direction of the given vectorial field of components $(A^{\Delta t}, B^{\Delta t}, C^{\Delta t})$ by an eventual modification of $D^{\Delta t}$. We are thus led to consider the integral lines of the system :

$$\frac{dx_1}{A^{\Delta t}} = \frac{dx_2}{B^{\Delta t}} = \frac{dx_3}{C^{\Delta t}} \ . \qquad (22)$$

Let us consider some of these lines, $\ell_1, \ell_2, \ell_3, \ldots$ crossing the given surface S_o at points $O_1, O_2, O_3, \ldots$ We shall now determine the function $u_3^{\Delta t}$ on each of them, noting that it is reduced on this surface to a function of one variable, e.g. the curvilinear abscissa σ measured from the surface S_o, and that this function satisfies the ordinary differential equation :

$$du_3^{\Delta t}/d\sigma = D^{\Delta t} \ (x_i(\sigma)), \qquad (23)$$

in accordance with equation (20) and the contraint $(A^{\Delta t})^2 + (B^{\Delta t})^2 + (C^{\Delta t})^2 = 1$.

We then take the integral of this equation which, for $\sigma = 0$, is reduced to the value given on S_o, at the foot of the chosen ℓ_j .

In conclusion, $\underline{u}^{\Delta t}$ and $H^{\Delta t}$ being known, we can return to the velocity field $\underline{v}^{\Delta t}$ so that, at time Δt, we are brought back to the original problem.

CONCLUDING REMARKS

The computational scheme outlined above for the calculation of v_i, ρ and S from the Lagrangian invariants (9 through 13), account being taken of the definition of H by (14) shows that we can answer positively the question initially raised. Of course a coherent calculation algorithm remains to be developed which actually accounts for the configuration of the domain in which the flow takes place and, more particularly, the conditions to be imposed at the boundaries (material or not) of this domain. It is clear that this development poses a rather significant numerical problem. The actual success of the present theory will of course depend up on the tractibility of this problem.

Acknowledgments

The author expresses his thanks to ONERA for its financial support which permitted presentation of this paper at the Tbilisi meeting.

REFERENCES

BELOTSERKOVSKII S.M. (1977). Annual Rev. Fluid. Mech., No. 9, Stanford Univ. Press, pp. 469-494.

CHORIN A. (1973). J. Fluid Mech., Vol. 57, part 4, pp. 785-796.

ERTEL H. and ROSSBY C.G. (1949). Sitz - Berlin, Dtsh. Akad. Wiss. KI. f. Maths. u. allg. Naturw., Nr. 1, pp. 1-11.

GAVRILIN B.L. and ZASLAVSKII M.M. (1970). Soviet Physics Doklady, 15, 5, pp. 436-438.

HOLLMANN G. (1964). Arch. Météor. Géophys., série A, 14, pp. 1-13.

LAINE E. (1946). Précis d'Analyse Mathématique, tome II, Paris, Vuibert.

REHBACH C. (1977). La Recherche Aérospatiale, n° 1977-5, pp. 289-298.

SERRIN J. (1959). Mathematical principles of classical fluid mechanics, Handbuch der Physik, Band VIII/I, Springer-Verlag, pp. 125-263.

YANENKO N.N. and SHOKIN YU.I. (1974). Proceedings of Symposium on Continuous mechanics and related problems of analysis (Tbilisi, 23-29, IX, 1971) Vol. II, pp. 292-306.

ZEYTOUNIAN R. Kh. (1974). Notes sur les écoulements rotationnels de fluides parfaits. Lecture Notes in Physics, Vol. 27, Springer-Verlag.

(1977). Méthodes numériques en mécanique du continu, tome 8, N° 5, pp. 71-101.

SIXTH INTERNATIONAL CONFERENCE
ON NUMERICAL METHODS IN FLUID DYNAMICS

Tbilisi, U.S.S.R., 20-25 June, 1978

LIST OF THE PARTICIPANTS
(all academic titles are omitted)

ABRAMOV A.A.	Computing Center, Acad.of Sci.of the U.S.S.R., 40 Vavilova str., Moscow B-333 U. S. S. R.
AKAMATSU T.	Department of Mechanical Engineering, Kyoto University, Kyoto 606, JAPAN
ALEKSIN V.A.	Institute of the Problems of Mechanics, Acad.of Sci.of the U.S.S.R., IOI Avenue Vernadskii, Moscow II7526, U. S. S. R.
ANDREJEV G.N.	Institute of Mechanics, Moscow State University, I Avenue Michurin, Moscow B-234, U. S. S. R.
ASTAFJEVA N.M.	Institute of Space Research, Acad. of Sci. of the U.S.S.R., 88 Profsojuznaja str., Moscow B-485, U. S. S. R.
BABAKOV A.V.	Computing Center, Acad. of Sci. of the U.S.S.R., 40 Vavilova str., Moscow B-333, U. S. S. R.
BABENKO K.I.	Institute of Applied Mathematics, Acad. of Sci. of the U.S.S.R., Miusskaja pl., 4, Moscow A-47, U. S. S. R.
BABIKOV P.E.	Central Aerohydrodynamical Institute, I40I60 Zhukovskii-3, U. S. S. R.
BALAJAN G.S.	Interkosmos, Leninskii prosp. I4, Moscow B-7I, U. S. S. R.

BALLHAUS W.F.	M.S. 202-I, NASA Ames Research Center, Moffet Field, California 94035, U. S. A.
BASKIN V.E.	Central Aerohydrodynamical Institute, I40I60 Zhukovskii-3, U. S. S. R.
BAZZHIN A.P.	Central Aerohydrodynamical Institute, I40I60 Zhukovskii-3, U. S. S. R.
BEGALISHVILI N.A.	Hydrometorological Institute ZAKNIGMI, I50 Av.Plekhanov, 3800I2, Tbilisi, U. S. S. R.
BEGIASHVILI G.A.	Institute of Cybernetics, ul.S.Euli 5, Tbilisi 380086, U. S. S. R.
BELJAEVA T.V.	Irkutsk State University, I Karl Marx Avenue, Irkutsk 664003, U. S. S. R.
BELOTSERKOVSKII O.M.	Computing Center, Acad.of Sci. of the U.S.S.R. 40 Vavilova str., Moscow B-333, U. S. S. R.
BELOTSERKOVSKII S.M.	Central Aerohydrodynamical Institute, Moscow, U. S. S. R.
BELOTSERKOVSKII S.O.	Moscow Physico-Technical Institute, Dolgoprudnyi I4I700, U. S. S. R.
BEREZIN Y.A.	Institute of Theoretical and Applied Mechanics, Siberian Branch of the Acad.of Sci.of the U.S.S.R., Novosibirsk 90, U. S. S. R.
BLAGOSKLONOV V.I.	Moscow Physico-Technical Institute, Dolgoprudnyi I4I700, U. S. S. R.
BLINOVSKII M.S.	Central Aerohydrodynamical Institute, I40I60 Zhukovskii-3, U. S. S. R.
BONTOUX P.	Institut de Mecanique des Fluides de Marseille, I3003 Marseille, FRANCE
BRATOS M.	Institute of Fundamental Technical Research, Polish Acad.of Sci., Swietokrzyska 2I, 00-049 Warszawa, POLAND

BRUSHLINSKII K.V.	Institute of Applied Mathematics, Acad.of Sci.the U.S.S.R., Miusskaja pl.4, Moscow A-47, U. S. S. R.
BRYKINA I.G.	Institute of Mechanics, Moscow State University, I Avenue Michurin, Moscow B-234, U. S. S. R.
BUDUNOV N.F.	Irkutsk State University, I Karl Marx Avenue, Irkutsk 664003, U. S. S. R.
BUNEMAN O.	Institute for Plasma Research, Stanford University, Via Crespi, Stanford, California 94305, U. S. A.
BYRKIN A.P.	Moscow Pedagogical Institute, I M.Pirogovskaja str., Moscow, U. S. S. R.
CABANNES H.	Universite de Paris YI, Mecanique Theorique, Tour 66-9, Quai Saint-Bernard, 75 Paris-5^e, FRANCE.
CHATTOT J.J.	ONERA, 92320 Chatillon, FRANCE.
CHEREMISSIN F.G.	Computing Center, Acad.of Sci.of the U.S.S.R., 40 Vavilova,str., Moscow B-333, U. S. S. R.
CHERKASHIN V.A.	Institute of Applied Mathematics, Acad.of Sci.of the U.S.S.R., Miusskaja pl. 4, Moscow A-47, U. S. S. R.
COLLINS R.	Universite de Technologie de Compiegne B.P.233, 60206 Compiegne, FRANCE.
COOPER M.	Office of Naval Research, Arlington, Virginia 22217, U. S. A.
COOPER R.D.	Office of Naval Research, 800 North Quincy Street, Arlington, Virginia 22217, U. S. A.
CUVELIER C.	Delft University of Technology, Dept. of Mathematics, Delft, THE NETHERLANDS.
DAVITAJA A.A.	Hydrometeorological Institute ZAKNIGMI, I50 Av. Plekhanov, 380012 Tbilisi, U. S. S. R.

DAVYDOV Y.M.	Computing Center, Acad.of Sci.of the U.S.S.R. 40 Vavilova str., Moscow B-333, U. S. S. R.
DIJKSTRA D.	Twente University of Technology, Dept. of Appl. Mathematics, Enschede, THE NETHERLANDS
DORODNICYN A.A.	Computing Center, Acad.of Sci.of the U.S.S.R. 40 Vavilova str.,Moscow B-333, U. S. S. R.
DZHISHKARIANI A.V.	Institute of Mathematics, Georgian Acad.of Sci., Z.Rukhadze str. I, Tbilisi 93, U. S. S. R.
EBANOIDZE T.A.	Computing Center, Georgian Acad.of Sci., Akurskaja str.8, Tbilisi I5, U. S. S. R.
EGELI G.	Magyar Tudoma nyaos Akademia, Muszaki Mechanika i Tanszeki Munkakozossneg XI Muegytem rkp 3, I52I Budapest, HUNGARY
EISEMAN P.R.	Scientific Research Associates, Inc. P.O. Box 498, Glastonbury, Connecticut 06033, U. S. A.
FAVORSKII A.P.	Institute of Applied Mathematics, Acad.of Sci.of the U.S.S.R., Miusskaja pl.4, Moscow A-47, U. S. S. R.
FILIMONOV I.M.	Moscow Physico-Technical Institute, Dolgoprudnyi I4I700, U. S. S. R.
FILIPPOV B.V.	Leningrad State University, University Embankment 7/9, Leningrad B-I64, U. S. S. R.
FILIPKOWSKI S.	Institute of Fundamental Technical Research, Polish Acad.of Sci., Swietokrzyska 2I, 00-049 Warszawa, POLAND.
FOERSTER K.	Institut für Raumfahrtantriebe, Universität Stuttgart, D-7000 Stuttgart 80,Pfattenwaldring 2I, WEST GERMANY

FONAREV A.S. Central Aerohydrodynamical Institute, I40I60 Zhukovskii-3, U. S. S. R.

FRUEHAUF H.H. Institut für Raumfahrtantriebe, Universität Stuttgart, D-7000 Stuttgart 80, Pfaffenwaldring 2I, WEST GERMANY

FURSENKO A.A. Physico-Technical Institute, Acad.of Sci. of the U.S.S.R., Politekhnicheskaja str. 26, I9402I Leningrad, U. S. S. R.

GAJEWSKI P. Politechnika Warszawska, ul.Nowowiejska 24, 00-665 Warszawa, POLAND.

GAL TCHEN T. Dept.of Physics, University of Toronto, Toronto, Ontario N5S IA7, CANADA.

GEGELIA T.G. Inst.of Appl.Mathematics, Tbilisi State University, 2 Universitetskaja str., Tbilisi 43, U. S. S. R.

GEGECHKORI Z.G. Tbilisi State University, Tbilisi, U. S. S. R.

GERSHBEIN E.A. Institute of Mechanics, Moscow State University, I Avenue Michurin, Moscow B-234, U. S. S. R.

GILINSKII M.M. Institute of Mechanics, Moscow State University, I Avenue Michurin, Moscow B-234, U. S. S. R.

GLAZ H.M. Department of the Navy, Naval Ordnance Laboratory, White Oak, Silver Spring, Md.209I0, U. S. A.

GOGODZE I.K. Inst.of the Control Systems, Georgian Acad.ofSci., 32 Pekinskaja str., Tbilisi 42, U. S. S. R.

GOLOVACHEV Y.P. Physico-Technical Institute, Acad.of Sci. of the U.S.S.R., 26 Politekhnicheskaja str., Leningrad I9402I, U. S. S. R.

GOLOVICHEV V.I. — Inst.of Theor.and Appl.Mechanics, Siberian Branch, Acad.of Sci.of the U.S.S.R., Novosibirsk 90, U. S. S. R.

GOLOVIZNIN V.M. — Kurchatov Institute of Atomic Energy, Acad.of Sci.of the U.S.S.R., Moscow, U. S. S. R.

GOLOVKIN M.A. — Central Aerohydrodynamical Institute, I40I60 Zhukovskii-3, U. S. S. R.

GORDEZIANI D.G. — Inst.of Appl.Mathematics, Tbilisi State University, 2 Universitetskaja str., Tbilisi 43, U. S. S. R.

GORELIK G.E. — Inst.of Heat and Mass Transfer, Belorussian Acad.of Sci., I5 Podlesnaja, Minsk 220728, U. S. S. R.

GORGIDZE A.Y. — Georgian Polytechnic Institute, Tbilisi, U. S. S. R.

GRABITZ G. — Institut für Strömungsforschung MPI, D-3400 Göttingen, Böttingerstrasse 6-8, WEST GERMANY

GRANBERG I.G. — Inst.of Atmospheric Physics, Acad.of Sci. of the U.S.S.R., 3 Pyzhevskii per., Moscow I09017, U. S. S. R.

GRIDNEV N.P. — Inst.of Theor.and Appl.Mechanics, Siberian Branch Acad.of Sci.of the U.S.S.R. Novosibirsk 90, U. S. S. R.

GRIGORENKO Y.M. — Inst.of Mechanics, Ukrainian Acad.of Sci., 3 Nesterova str., Kiev 57, U. S. S. R.

GRIGORJEV Y.P. — Cheliabinsk State University, Cheliabinsk, U. S. S. R.

GRJAZNOV V.L. — Institute of the Problems of Mechanics, Acad.of Sci.of the U.S.S.R., IOI Avenue Vernadskii, Moscow II7526, U. S. S. R.

GRODZOVSKII G.L. Moscow Physico-Technical Institute, Dolgoprudnyi I41700, U. S. S. R.

GROMOV V.G. Institute of Mechanics, Moscow State University, I Avenue Michurin, Moscow B-234, U. S. S. R.

GUIU-ROUX J. Universite Paris-Sud, Bat.425, 9I400 Orsay, FRANCE

GUSHCHIN V.A. Computing Cemter, Acad.of Sci.of the U.S.S.R., 40 Vavilova str., Moscow B-333, U. S. S. R.

HIRSCH CH. Dept.of Fluid Mechanics, Vrije Universiteit Brussel, Bruxelles, BELGIUM

HIRSH R.S. Douglas Aircraft Company, 3855 Lakewood Boulevard, Long Beach, California 90846, U. S. A.

HOLT M. University of California, Berkeley, Mechanical Engineering, Berkeley, California 94720, U. S. A.

HUSSAINI M.Y. NASA Ames Research Center, SST 202-I, Moffett Field, California 94035, U. S. A.

INOUYE M. NASA Ames Research Center, P.O.Box 532, Los Gatos, California 95030, U. S. A.

ISTOMIN V.I. Institute of Chemical Technology, Moscow II5409, U. S. S. R.

IVERINMAZASHVILI Z.S. Tbilisi State University, Tbilisi, U. S. S. R.

IZVOLSKII V.A. Inst.of Appl.Mathematics, Acad.of Sci. of the U.S.S.R., Miusskaja pl.4, Moscow A-47, U. S. S. R.

JAKOVLEV V.Y. Central Institute of Scientific Information and Technico-Economical Research, Moscow M-I48, U. S. S. R.

JAMI A. — Ecole Nationale Superieure de Techniques Avancees, 32 Boulevard Victor, 75015 Paris, FRANCE

JAVORSKAIA I.M. — Inst.of Space Research, Acad.of Sci.of the U.S.S.R., 88 Profsojuznaja str., Moscow B-485, U. S. S. R.

KABZE Z. — Polytechnical Institute, ul.Wyspianskiego 27, 50-370 Wroclaw, POLAND

KALIS H.E. — Computing Center, Latvia State University, 29 Rainis Boulevard, Riga, U. S. S. R.

KARPOV K.A. — Computing Center, Acad.of Sci.of the U.S.S.R., 40 Vavilova str., Moscow B-333, U. S. S. R.

KARTSIVADZE I.N. — Tbilisi State University, Tbilisi, U. S. S. R.

KHARAZISHVILI G.I. — Inst.of the Control Systems, Georgian Acad.of Sci., 32 Pekinskaja str., Tbilisi 42, U. S. S. R.

KHATIASHVILI G.M. — Computing Center, Georgian Acad.of Sci., 8 Akurskaja str., Tbilisi 15, U. S. S. R.

KHOLODOV A.S. — Moscow Physico-Technical Institute, Dolgoprudnyi 141700, U. S. S. R.

KHVOLES A.R. — Inst.of Appl.Mathematics, Tbilisi State University, 2 Universitetskaja str., Tbilisi 43, U. S. S. R.

KIKNADZE L.G. — Computing Center, Georgian Acad.of Sci., 8 Akurskaja str., Tbilisi 15, U. S. S. R.

KIKNADZE L.S. — Inst.of Appl.Mathematics, Tbilisi State University, 2 Universitetskaja str., Tbilisi 43, U. S. S. R.

KLOPFER G.H.	NASA Ames Research Center, Moffett Field, California 94035, U. S. A.
KOBAKHIDZE T.N.	Hydrometeorological Institute ZAKNIGMI, I50 Av.Plekhanov, 380012 Tbilisi, U. S. S. R.
KOGARKO B.S.	Institute of Chemical Physics, Acad.of Sci.of the U.S.S.R., Vorobjevskoje shosse 2b, Moscow B-234, U. S. S. R.
KOLESNIKOV A.F.	Inst.of the Problems of Mechanics, Acad. of Sci.of the U.S.S.R., IOI Av.Vernadskii, Moscow II7526, U. S. S. R.
KOLESNIKOV Y.B.	Institute of Physics, Latvian Acad.of Sci., Salaspils-I, Latvia, U. S. S. R.
KOMAROVSKII L.V.	University of Tomsk, Tomsk 50, U. S. S. R.
KORDADZE R.A.	Inst.of Appl.Mathematics, Tbilisi State University, 2 Universitetskaja str., Tbilisi 43, U. S. S. R.
KOROBEINIKOV V.P.	Steklov Institute of Mathematics, Acad. of Sci.of the U.S.S.R., 42 Vavilova str., Moscow B-333, U. S. S. R.
KOVENJA V.M.	Inst.of Theor.and Appl.Mechanics, Siberian Branch, Acad.of Sci.of the U.S.S.R., Novosibirsk 90, U. S. S. R.
KOZEL K.	Strojni fakulta ČVUT, Suchbatarova 4, I60 00 Praha 6, CHEKHOSLOVAKIA
KRIVTSOV V.M.	Computing Center, Acad.of Sci.of the U.S.S.R., 40 Vavilova str., Moscow B-333, U. S. S. R.
KULAKOV A.A.	Academy of Sciences of the U.S.S.R., Leninskii Prosp.I4, Moscow B-7I, U. S. S. R.

KUPRADZE V.D. Academy of Sciences of Georgia, Mathematics and Physics, I Z.Rukhadze str., Tbilisi 93, U. S. S. R.

KVESELAVA D.A. Computing Center, Georgian Acad.of Sci., 8 Akurskaja str., Tbilisi I5, U. S. S. R.

LAINE S. Helsinki University of Technology, Dept. of Mech.Eng., SF-02I50 Espoo I5, FINLAND

LANGLOIS W.E. IBM Research Laboratory, San Jose, California 95I24, U. S. A.

LAPTAURI N.A. Institute of Physics, Georgian Acad.of Sci., 6 Guramishvili str., Tbilisi 77, U. S. S. R.

LARIN O.B. Institute of Mechanics, Moscow State Universizy, I Av.Michurin, Moscow B-234, U. S. S. R.

LATYSHEV A.F. Inst.of Theor.and Appl.Mechanics, Siberian Branch, Acad.of Sci.of the U.S.S.R., Novosibirsk 90, U. S. S. R.

LAU R.L. Office of Naval Research Branch Office, IO30 East Green Street, Pasadena, California 9II06, U. S. A.

LAVRENTJEV M.A. Academy of Sciences of the U.S.S.R., Leninskii Prosp.I4, Moscow B-7I, U. S. S. R.

LEBEDEV M.G. Research Computing Center, Moscow State University, Moscow B-234, U. S. S. R.

LENOIR M. Ecole Nationale Superieure de Techniques Avancees, 32 Boulevard Victor, 75015 Paris, FRANCE

LEONTJEVA N.V. Physico-Technical Institute, Acad.of Sci. of the U.S.S.R., 26 Politekhnicheskaja str., Leningrad I9402I, U. S. S. R.

LERAT A.	Ecole Nationale Superieure d'Arts et Metiers, 75013 Paris, FRANCE
LEVCHENKO V.Y.	Inst.of Theor.and Appl.Mechanics, Siberian Branch, Acad.of Sci.of the U.S.S.R., Novosibirsk 90, U. S. S. R.
LEVIN V.A.	Institute of Mechanics, Moscow State University, I Av.Michurin, Moscow B-234, U. S. S. R.
LEVY R.	Scientific Research Associates, Inc. P.O.Box 498, Glastonbury, Connecticut 06033, U. S. A.
LIFANOV I.K.	Central Aerohydrodynamical Institute, Moscow, U. S. S. R.
LISEIKIN V.D.	Inst.of Theor.and Appl.Mechanics, Siberian Branch, Acad.of Sci.of the U.S.S.R., Novosibirsk 90, U. S. S. R.
LISOVETS V.S.	Cheliabinsk State University, Cheliabinsk, U. S. S. R.
LYAKHOV V.N.	Inst.of High Temperatures, Acad.of Sci. of the U.S.S.R., Korovinskoje shosse, Moscow I274I2, U. S. S. R.
MACCARIO J.	Universite de Technologie de Compiegne, Departement de Genie Biologique, BP 233, 60206 Compiegne Cedex, FRANCE
MAKHVILADZE G.M.	Inst.of the Problems of Mechanics, Acad. of Sci.of the U.S.S.R., IOI Av.Vernadskii, Moscow II7526, U. S. S. R.
MAKSIMOV V.P.	Inst.of Theor.and Appl.Mechanics, Siberian Branch, Acad.of Sci.of the U.S.S.R., Novosibirsk 90, U. S. S. R.
MAMALADZE Y.G.	Inst.of Physics, Georgian Acad.of Sci., 6 Guramishvili str., Tbilisi 77, U. S. S. R.

MANDZHAVIDZE G.F. Inst.of Appl.Mathematics, Tbilisi State University, 2 Universitetskaja str., Tbilisi 43, U. S. S. R.

MARKOV A.A. Inst.of the Problems of Mechanics Acad. of Sci.of the U.S.S.R., IOI Av.Vernadskii, Moscow II7526, U. S. S. R.

MARKOV V.V. Steklov Institute of Mathematics, Acad. of Sci.of the U.S.S.R., 42 Vavilova str., Moscow B-333, U. S. S. R.

MAROTI L. Magyar Tudoma nyaos Akademia, Muszaki Mechanika i Tanszeki Munkakozossneg XI Muegytem rkp 3, I52I Budapest, HUNGARY

MARTUZANE B.Y. Computing Center, Latvian State University, 29 Rainis Boulevard, Riga, U. S. S. R.

MARTUZANE E.N. Computing Center, Latvian State University, 29 Rainis Boulevard, Riga, U. S. S. R.

MARTYNENKO O.G. Inst.of Heat and Mass Transfer, Belorussian Acad.of Sci., I5 Podlesnaja, Minsk 220728, U. S. S. R.

MASLOV L.A. Central Aerohydrodynamical Institute, I40I60 Zhukovskii-3, U. S. S. R.

MCCRACKEN M. Indiana University, Dept.of Mathematics, Swain Hall-East, Bloomington, Indiana 4740I, U. S. A.

MCRAE D.S. Air Force Flight Dynamics Laboratory, NASA Ames Research Center, Moffett Field, Calif. 94035, U. S. A.

MELADZE G.V. Tbilisi State University, Tbilisi, U. S. S. R.

MELLER N.A. Computing Center, Acad.of Sci.of the U.S.S.R., 40 Vavilova str., Moscow B_333 U. S. S. R.

MELNIK W.L.	University of Maryland, Aerospace Eng. Dept., College Park, Maryland 20742, U. S. A.
MIKHAILOV G.S.	Institute of Mechanics, Gorky State University, Gorky, U. S. S. R.
MIKHAILOV Y.J.	Central Aerohydrodynamical Institute, I40I60 Zhukovskii-3, U. S. S. R.
MINAILOS A.N.	Central Aerohydrodynamical Institute, I40I60 Zhukovskii-3, U. S. S. R.
MINOSTSEV V.B.	Institute of Mechanics, Moscow State University, I Av.Michurin, Moscow B-234, U. S. S. R.
MIRSKII V.N.	Institute of Mechanics, Moscow State University, I Av.Michurin, Moscow B-234, U. S. S. R.
MOISEEVA I.N.	Institute of Mechanics, Moscow State University, I Av.Michurin, Moscow B-234, U. S. S. R.
NAZHESTKINA E.I.	Inst.of Appl.Mathematics, Acad.of Sci. of the U.S.S.R., Miusskaja pl.4, Moscow A-47, U. S. S. R.
NEILAND V.Y.	Central Aerohydrodynamical Institute, I40I60 Zhukovskii-3, U. S. S. R.
NEUBERGER A.W.	Inst.für Angew.Gasdynamik der DFVLR, Linder Höhe, D-5000 Köln 90, WEST GERMANY
NISHIKAWA N.	Faculty of Engineering, Chiba University, Yayoi I-33 Chiba, JAPAN
NIELSEN H.B.	Inst.of Numer.Analysis, Techn.University of Denmark, Bld.30I, DK 2800, Lyngby, DENMARK
NOGOTOV E.F.	Inst.of Heat and Mass Transfer, Belorussian Acad.of Sci., I5 Podlesnaja,

Minsk 220728, U. S. S. R.

NOVITSKII A.G. Inst.of Oceanology, Acad.of Sci. of the U.S.S.R., 23 Krasikova str., Moscow B-2I8, U. S. S. R.

OBOLASHVILI E.I. Inst.of Mathematics, Georgian Acad.of Sci., I Z.Rukhadze str., Tbilisi 93, U. S. S. R.

ODISHARIA M.G. Inst.of Mathematics, Georgian Acad.of Sci., I Z.Rukhadze str., Tbilisi 93, U.S.S.R.

OLEINIK O.A. Moscow State University, Lenin Hills, Moscow B-234, U. S. S. R.

OSHIMA K. Institute of Space and Aeronautical Science, Komaba, Meguro-ku, Tokyo, JAPAN

OSIPOV V.Z. Inst.of mathematics, Georgian Acad.of Sci., I Z.Rukhadze str., Tbilisi 93, U. S. S. R.

PÄRNPUU A.A. Computing Center, Acad.of Sci.of the U.S.S.R., 40 Vavilova str., Moscow B-333, U. S. S. R.

PARULAVA B.V. Computing Center, Georgian Acad.of Sci., 8 Akurskaja str., Tbilisi I5, U.S.S.R.

PAVERMAN V.S. Inst.of Physics, Georgian Acad.of Sci., 6 Guramishvili str., Tbilisi 77, U.S.S.R.

PAVLOV B.M. Research Computing Center, Moscow State University, Moscow B-234, U. S. S. R.

PASKONOV V.M. Research Computing Center, Moscow State University, Moscow B-234, U. S. S. R.

PINCHUKOV V.I. Inst.of Theor.and Appl.Mechanics, Siberian Brauch, Acad.of Sci.of the U.S.S.R., Novosibirsk 90, U. S. S. R.

PIROGOV V.B.	Moscow Physico-Technical Institute, Dolgoprudnyi I4I700, U. S. S. R.
PIRUMOV U.G.	Moscow Aviation Engineering Institute, 4 Volokolamskoje shosse, Moscow, U.S.S.R.
PIVA R.	Institute of Aerodynamics, University of Rome, Rome, ITALY
POLASEK J.	Strojni fakulta ČVUT, Suchbatarova 4, I60 00 Praha 6, CHEKHOSLOVAKIA
POLEVIKOV V.K.	Inst.of Heat and Mass Transfer, Belorussian Acad.of Sci., I5 Podlesnaja, Minsk 220728, U. S. S. R.
POLEZHAEV V.I.	Inst.of the Problems of Mechanics, Acad. of Sci.of the U.S.S.R., IOI Av.Vernadskii, Moscow II7526, U.S.S.R.
PROKOPOV G.P.	Inst.of Appl.Mathematics, Acad.of Sci.of the U.S.S.R., Miusskaja pl.4, Moscow A-47, U.S.S.R.
PRZEKWAS A.	Polytechnical Institute, ul.Wyspianskiego 27, 50-370 Wroclaw, POLAND
PYZIK J.	Politechnical Institute, ul.Nowowiejska 24, Warszawa 00-665, POLAND
RADEV S.P.	Inst.of Math.and Mech.with Comput.Center, Bulgarian Acad.of Sci., P.O.Box 373 I000 Sofia, BULGARIA
RADVOGIN Y.B.	Inst.of Appl.Mathematics, Acad.of Sci. of the U.S.S.R., Miusskaja pl.4, Moscow A-47, U. S. S. R.
REICHENBACH R.E.	European Research Office, 223, Old Marylebone Road, London NWI 5TH, ENGLAND
RESHETIN A.G.	Moscow Physico-Technical Institute, I4I700 Dolgoprudnyi, U. S. S. R.

REYNOLDS W.C. Stanford University, Mech.Eng.Dept., Stanford, Calif. 94305, U. S. A.

RIZZI A. The Aeronautical Research Institute of Sweden, B.O.Box II02I, S-I6I II Bromma II, SWEDEN

ROESNER K. Inst.für Strömungslehre und Strömungsmaschinen, Universität Karlsruhe, Kaiserstrasse I2 7500 Karlsruhe I, WEST GERMANY

ROSLJAKOV G.S. Research Computing Center, Moscow State University, Moscow B-234, U. S. S. R.

ROSTOMASHVILI Z.I. Institute of Physics, Georgian Acad.of Sci., 6 Guramishvili str., Tbilisi 77, U. S. S. R.

ROZHDESTVENSKII B.L. Inst.of Appl.Mathematics, Acad.of Sci. of the U.S.S.R., Miusskaja pl.4, Moscow A-47, U. S. S. R.

RUSANOV V.V. Inst.of Appl.Mathematics, Acad.of Sci. of the U.S.S.A., Miusskaja pl.4, Moscow A_47, U. S. S. R.

SABADOCZ L. Magyar Tudoma nyaos Akademia, Muszaki Mechanika i Tanszeki Munkakozossneg XI Muegytem rkp 3, I52I Budapest, HUNGARY

SAGDEEV R.Z. Inst.of Space Research, Acad.of Sci.of the U.S.S.R., 88 Profsojuznaja str., Moscow B-485, U. S. S. R.

SAMARSKII A.A. Inst.of Appl.Mathematics, Acad.of Sci. of the U.S.S.R., Miusskaja pl.4, Moscow A-47, U. S. S. R.

SANIKIDZE DJEM.G. Computing Center, Georgian Acad.of Sci., 8 Akurskaja str., Tbilisi I5, U.S.S.R.

SANIKIDZE DJUM.G. Inst.of Cybernetics, Georgian Acad.of Sci., 5 S.Euli str., Tbilisi 86, U.S.S.R.

SARMIN E.N.	Computing Center, Acad.of Sci.of the U.S.S.R., 40 Vavilova str., Moscow B-333, U. S. S. R.
SATOR F.G.	Inst. d'Aerodynamigue, EPF-Lausanne, En Vernay, CH-I024 Ecublens, SWITZERLAND
SAVELJEV V.V.	Inst.of Appl.Mathematics, Acad.of Sci. of the U.S.S.R., Miusskaja pl.4, Moscow A-47, U. S. S. R.
SAVINOV K.G.	Inst.of Mechanics, Moscow State University, I Av.Michurin, Moscow B-234, U. S. S. R.
SCHIPPERS H.	Mathematical Center, 2e Boerhaavestraat 49, Amsterdam, THE NETHERLANDS
SELEZOV I.T.	Inst.of Hydromechanics, Ukrainian Acad. of Sci., 8/4 Zheliabova str., Kiev 57, U. S. S. R.
SEVERINOV L.I.	Moscow Physico-Technical Institute, I4I700 Dolgoprudnyi, U. S. S. R.
SHAPAKIDZE L.D.	Inst.of Mathematics, Georgian Acad.of Sci., I Z.Rukhadze str., Tbilisi 93, U.S.S.R.
SHARAJEV D.Y.	Ural Scientific Center, Acad,of Sci.of the U.S.S.R., Institute of Mathematics and Mechanics, ul.Kowalewskoi I6, Sverdlovsk GSP-384, U. S. S. R.
SHARIKADZE D.V.	Tbilisi State University, Tbilisi, U. S. S. R.
SHASHKIN A.P.	Inst.of Theor.and Appl.Mechanics, Siberian Branch, Acad.of Sci.of the U.S.S.R., Novosibirsk 90, U. S. S. R.
SHCHENNIKOV V.V.	Moscow Pedagogical Institute, I M. Pirogovskaja str., Moscow, U. S. S. R.
SHESTAKOV A.I.	L-402, Lawrence Livermore Laboratory, University of California, P.O.Box 808 Livermore, California 94550, U. S. A.

SHIDLOVSKAJA L.V. Inst.of Atmospheric Physics, Acad.of Sci.of the U.S.S.R., 3 Pyzhevskii per., Moscow I09017, U. S. S. R.

SHIDLOVSKY V.P. Computing Center, Acad.of Sci.of the U.S.S.R., 40 Vavilova str., Moscow B-333, U. S. S. R.

SHIPILIN A.V. Computing Center, Acad.of Sci.of the U.S.S.R., 40 Vavilova str., Moscow B-333, U. S. S. R.

SHMIDT A.A. Physico-Technical Institute, Acad.of Sci.of the U.S.S.R., 26 Politekhnicheskaja str., Leningrad I9402I, U.S.S.R.

SHMYGLEVSKII Y.D. Computing Center, Acad.of Sci.of the U.S.S.R., 40 Vavilova str., Moscow B-333, U. S. S. R.

SHOKIN Y.I. Inst.of Theor.and Appl.Mechanics, Siberian Branch, Acad.of Sci.of the U.S.S.R., Novosibirsk 90, U. S. S. R.

SHURSHALOV L.V. Computing Center, Acad.of Sci.of the U.S.S.R., 40 Vavilova str., Moscow B-333, U. S. S. R.

SIMUNI L.M. Oceanology Institute (Leningrad Branch) Vasiljevskii Ostrov, 30 Ist line, Leningrad I99053, U. S. S. R.

SKRIPACH B.K. Central Aerohydrodynamical Institute, I40I60 Zhukovskii-3, U. S. S. R.

SLEPTSOV A.G. Inst.of Theor.and Appl.Mechanics, Siberian Branch, Acad.of Sci.of the U.S.S.R., Novosibirsk 90, U.S.S.R.

SLOMCZINSKA I. Inst.of Fundamental Technical Research, Polish Acad.of Sci., Swietokrzyska 2I, 00-049 Warszawa, POLAND

SOBOLSKII P.E. Voronezh State University,Voronezh, U. S. S. R.

STOJANOV T.V.	Inst.of Math.and Mech.with Comput.Center, Bulgarian Acad.of Sci., P.O.Box 373, 1000 Sofia, BULGARIA
STYCZEK A.	Politechnika Warszawska, ul.Nowowiejska 24, 00-665 Warszawa, POLAND
SULIKASHVILI R.S.	Inst.of Mathematics, Georgian Acad.of Sci., I Z.Rukhadze str., Tbilisi 93, U. S. S. R.
SUSLOV O.N.	Institute of Mechanics, Moscow State University, I Av.Michurin, Moscow B-234, U. S. S. R.
TAGANOV G.I.	Central Aerohydrodynamical Institute, 140160 Zhukovskii-3, U. S. S. R.
TARKASHVILI C.T.	Inst.of Physics, Georgian Acad.of Sci., 6 Guramishvili str., Tbilisi 77, U.S.S.R.
TIKHOMIROV B.P.	Inst.of Appl.Mathematics, Acad.of Sci. of the U.S.S.R., Miusskaja pl.4, Moscow A-47, U. S. S. R.
TIKHONOV N.N.	Cheliabinsk State University, Cheliabinsk, U. S. S. R.
TISHKIN V.F.	Inst.of Appl.Mathematics, Acad.of Sci.of the U.S.S.R., Miusskaja pl.4, Moscow A-47, U. S. S. R.
TKACHENKO B.K.	Moscow Physico-Technical Institute, 141700 Dolgoprudnyi, U. S. S. R.
TKHELIDZE M.G.	Tbilisi State University, Tbilisi, U.S.S.R.
TOLSTYKH A.I.	Computing Center, Acad.of Sci.of the U.S.S.R., 40 Vavilova str., Moscow B-333, U. S. S. R.
TSKHADAJA F.G.	Inst.of Appl.Mathematics, Tbilisi State University, 2 Universitetskaja str., Tbilisi 43, U. S. S. R.

TSULADZE M.G.	Computing Center, Georgian Acad.of Sci., 8 Akurskaja str., Tbilisi I5, U.S.S.R.
TURCHAK L.I.	Computing Center, Acad.of Sci.of the U.S.S.R., 40 Vavilova str., Moscow B-333, U. S. S. R.
VAN ATTA C.W.	University of California, San Diego, Dept.of Appl.Mach.and Engng Sciences, Mail Code B-0I0, La Jolla, California 92093, U. S. A.
VAN DE VOOREN A.I.	University of Groningen, Mathematical Institute, P.O.Box 800, Groningen, THE NETHERLANDS
VASHAKMADZE T.S.	Inst.of Appl.Mathematics, Tbilisi State University, 2 Universitetskaja str., Tbilisi 43, U. S. S. R.
VASILJEV L.E.	Central Aerohydrodynamical Institute, I40I60 Zhukovskii-3, U. S. S. R.
VOROZHTSOV E.V.	Inst.of Theor.and Appl.Mechanics, Siberian Branch, Acad.ofSci.of the U.S.S.R., Novosibirsk 90, U. S. S. R.
VOSKRESENSKII G.P.	Inst.of Appl.Mathematics, Acad.ofSci.of the U.S.S.R., Miusskaja pl.4, Moscow A-47, U. S. S. R.
WESSELING P.	Technological Institute Delft, Julianalaan I32 Delft, THE NETHERLANDS
WIDHOPF G.F.	The Aerospace Corporation, Los Angeles, California 90009, U. S. A.
WIMMER M.	Universität Karlsruhe, Kaiserstrasse I2 7500 Karlsruhe I, WEST GERMANY
WU J.C.	School of Aerospace Engineering, Georgia Institute of Technology, Atlanta, Georgia 30332, U. S. A.
YANENKO N.N.	Inst.of Theor.and Appl.Machanics, Siberian Branch, Acad.of Sci.of the U.S.S.R., Novosibirsk 90, U. S. S. R.

YEN S.M.	Coordinated Science Laboratory, University of Illinois, Urbana, Ill. 61801, U. S. A.
ZABRODIN A.V.	Inst.of Appl.Mathematics, Acad.of Sci. of the U.S.S.R., Miusskaja pl.4, Moscow A-47, U. S. S. R.
ZANDBERGEN P.J.	Technical University Twente, P.O.Box 217, Enschede, THE NETHERLANDS
ZANNETTI L.	Istituto di Macchine e Motori per Aeromobili, Politechnico di Torino, Torino 10129, ITALY
ZAPRJANOV Z.	Inst.of Math.and Mach.with Computing Center, Bulgarian Acad.of Sci., P.O.Box 373, 1000 Sofia, BULGARIA
ZAWISTOWSKA E.	Inst.of Fundamental Technical Research, Polish Acad.of Sci., Swietokrzyska 21, 00-049 Warszawa, POLAND
ZEYTOUNIAN R.	Universite de Lille I, B.P.36 Villeneuve d'Ascq 59650, FRANCE
ZHILIN Y.L.	Moscow Aviation Engineering Institute, 4 Volokolamskoje shosse, Moscow, U.S.S.R.
ZHURAVLEVA G.S.	Irkutsk State University, I Karl Marx Street, Irkutsk 664003, U. S. S. R.
ZUBER I.	State Research Institute of Mechanical Engineering, 250 97 Praha 9-Bechovice, CHEKHOSLOVAKIA